Lecture Notes in Artificial Intelligence 1244

Subseries of Lecture Notes in Computer Science
Edited by J. G. Carbonell and J. Siekmann

Lecture Notes in Computer Science

Edited by G. Goos, J. Hartmanis and J. van Leeuwen

Springer
Berlin
Heidelberg
New York
Barcelona
Budapest
Hong Kong
London
Milan
Paris
Santa Clara
Singapore
Tokyo

Dov M. Gabbay   Rudolf Kruse
Andreas Nonnengart
Hans Jürgen Ohlbach (Eds.)

Qualitative and Quantitative Practical Reasoning

First International Joint Conference
on Qualitative and Quantitative Practical
Reasoning, ECSQARU-FAPR'97
Bad Honnef, Germany, June 9-12, 1997
Proceedings

 Springer

Series Editors
Jaime G. Carbonell, Carnegie Mellon University, Pittsburgh, PA, USA
Jörg Siekmann, University of Saarland, Saarbrücken, Germany

Volume Editors

Dov M. Gabbay
Hans Jürgen Ohlbach
Imperial College of Science, Technology and Medicine, Dept. of Computing
180 Queen's Gate, London SW7 2AZ, U.K.
E-mail: (dg/h.ohlbach)@doc.ic.ac.uk

Rudolf Kruse
Otto-von-Guericke-Universität, Fakultät für Informatik
Universitätsplatz 2, D-39106 Magdeburg, Germany
E-mail: kruse@iik.cs.uni-magdeburg.de

Andreas Nonnengart
Max-Planck-Institut für Informatik
Im Stadtwald, D-66123 Saarbrücken, Germany
E-mail: nonnenga@mpi-sb.mpg.de

Cataloging-in-Publication Data applied for

Die Deutsche Bibliothek - CIP-Einheitsaufnahme

Qualitative and quantitative practical reasoning : proceedings /
First International Joint Conference on Qualitative and Quantitative
Practical Reasoning, ECSQARU FAPR '97, Bad Honnef, Germany,
June 9 - 12, 1997. Dov Gabbay ... (ed.). - Berlin ; Heidelberg ; New
York ; Barcelona ; Budapest ; Hong Kong ; London ; Milan ; Paris ;
Santa Clara ; Singapore ; Tokyo : Springer, 1997
  (Lecture notes in computer science ; Vol. 1244 : Lecture notes in
  artificial intelligence)
  ISBN 3-540-63095-3

CR Subject Classification (1991): I.2, F.4.1

ISBN 3-540-63095-3 Springer-Verlag Berlin Heidelberg New York

© Springer-Verlag Berlin Heidelberg 1997
Printed in Germany

Typesetting: Camera ready by author
SPIN 10550340   06/3142 – 5 4 3 2 1 0   Printed on acid-free paper

Preface

It has become apparent in the last decades that human practical reasoning demands more than traditional deductive logic can offer. From both a philosophical and an engineering perspective the analysis and mechanisation of human practical reasoning requires a subtle understanding of pragmatics, dialectics, and linguistics, even psychology. Philosophers, software engineers, and AI researchers have similar ambitions in this respect; they all try to deepen our understanding of human reasoning and argumentation.

Various aspects of human practical reasoning have resulted in the development of non-monotonic logics, default reasoning, modal logics, belief function theory, Bayesian networks, fuzzy logic, possibility theory, and user modelling approaches, to name a few. These are new and active areas of research with many practical applications and many interesting, as yet unsolved, theoretical problems.

This volume contains the accepted and the invited papers for the ECSQARU/-FAPR '97, the first international joint conference on quantitative and qualitative practical reasoning.

The three predecessors of ECSQARU '97 were sponsored and organized by the consortium of DRUMS (Defeasible Reasoning and Uncertainty Management Systems, ESPRIT III BRA 6156). The goal of this project, which involved 21 European universities and research organizations, was to develop techniques in the fields of belief change, non-monotonic deduction, inconsistency in reasoning, abduction, efficient inference algorithms, and dynamic reasoning with partial models. FAPR '96 was sponsored by MEDLAR, the European project on practical reasoning, which involved 15 major European groups in mechanised deduction in qualitative practical reasoning.

The purpose of the ECSQARU/FAPR is to introduce these communities to each other, compare the current state of research, and make research available to all researchers involved.

This year particular attention was directed to special tutorials and invited sessions which were organised by leading researchers from the "quantitative" and the "qualitative communities" respectively.

We are indebted to the program committee for their effort and thought in organizing the program, to the invited speakers, and to the presenters of the tutorials. Moreover, we gratefully acknowledge the contribution of the many referees who were involved in the reviewing process. Special thanks go to Christine Harms who ensured that the event ran smoothly.

March 1997

Dov M. Gabbay
Rudolf Kruse
Andreas Nonnengart
Hans Jürgen Ohlbach

Program Committee

Contents

Multisensor Data Fusion
in Situation Assessment Processes

Alain Appriou

ONERA
BP 72, 92322 Châtillon Cedex
France

Abstract. To identify or localize a target, multisensor analysis has to be able to recognize one situation out of a set of possibilities. To do so, it uses measurements of more or less doubtful origin and prior knowledge that is understood to be often poorly defined, and whose validity is moreover difficult to evaluate under real observation conditions. The present synthesis proposes a generic modeling of this type of information, in the form of mass sets of the theory of evidence, with closer attention being paid to the most common case where the data originates from statistical processes. On the one hand robust target classification procedures can be achieved by applying appropriate decision criteria to these mass sets, on the other hand they can be integrated rigorously into a target tracking process, to reflect the origin of the localization measurements better.

1 Problem Formulation

Ordinarily, when analyzing a situation, the available sensors have to be used under unfavorable conditions, inducing uncertainties at different levels :

- measurements that are imprecise, erroneous, incomplete, or ill-suited to the problem,

- ambiguous observations (*e.g.* a position or velocity measurement not necessarily related to the object in question),

- knowledge (generated by learning, models, and so forth) that is, in theoretical terms, incomplete, poorly defined, and especially more or less representative of reality, in particular in light of the varying context.

Moreover, the disparity of the data delivered by the various sensors, which is intended to remedy the individual insufficiencies of each, requires a detailed evaluation of each of them, based on any exogenous information that might characterize their pertinence to the problem at hand and the context investigated, while such information is itself often very subjective and imprecise.

Theories of uncertainty offer an attractive federative framework in this context. But they run up against a certain number of difficulties in practice : interpretation and modeling of the available information in appropriate theoretical frameworks, choice of an association architecture and combination rules, decision principles to be adopted, constraints concerning the speed and volume of the necessary computations.

To provide solutions to these questions, we will first consider a generic problem in which we attempt to characterize the likelihood of I hypotheses H_i theoretically listed in an exhaustive and exclusive set E. These hypotheses may typically concern the presence of entities, target or navigation landmark identities, vector or target localization, or the status of a system or of a situation.

Such a likelihood function may then be integrated either into :

- a choice strategy, to declare the most likely hypothesis (target identification, intelligence, and so on),

- a filtering process (such as target tracking or navigation updating),

- a decision aid process for implementing means of analysis, electronic warfare, or intervention.

The likelihood functions we want have to be developed from the data provided by J sensors S_j. Each of them is assumed to be associated with processes that extract a measurement or a set of measurements s_j, pertinent to the targeted discrimination function, from the raw signals or images it generates.

In the framework of the generic problem we will be considering first, we assume that each measurement s_j can be used to generate I criteria C_{ij}, on the basis of any *a priori* knowledge, having values in [0, 1] capable of characterizing the likelihood of each hypothesis H_i. A quality factor q_{ij} with values in [0, 1] is also associated with each likelihood C_{ij}. Its purpose is to express the aptitude of the criterion C_{ij} to discriminate the hypothesis H_i under the given observation conditions, on the basis of a dedicated learning process or exogenous knowledge. This factor includes mainly the confidence that can be accorded to the validity of the *a priori* knowledge used for generating C_{ij}. As concerns, for example, the representativity of a learning process in a varying context, it will typically depend on the quality, volume, and exactness of the available preliminary data, and on any marring of the corresponding measurements.

The developments presented are conducted in the theory of evidence framework [4], which happens to be the broadest and best-suited to the interpretation of the data considered, and also the most federative in terms of synergy with the related processes (especially as concerns the stochastic filtering). The generic problem considered thus leads to an axiomatic search for the mass sets on the set of hypotheses H_i capable of ensuring the synthesis of the set of (C_{ij}, q_{ij}).

A similar approach is also followed in the most commonly encountered concrete cases in which we have statistical learning $p(s_j/H_i)$ of each measurement s_j under the different hypotheses H_i, using axioms specific to this latter situation. Nevertheless, the corresponding models remain coherent with those found for the generic problem.

The models developed in the two cases are then used for an application to target classification, thanks to a suitable decisional procedure.

Lastly, the same models are used for a tracking application involving multiple, varied, moving targets in a dense environment, on the basis of observations output by a set of disparate and possibly delocalized sensors. The hypotheses considered are then the joint identity and localization hypotheses. Thanks to a set approach to the localization problem, the discernment frameworks can be conveniently managed to allow all the available data to be merged and, at the same time, generate an exact set of information that can be injected into a Bayesian filter of usual form. In addition to a richer and more appropriate exploitation of the data, the concept proposed integrates the classification function into the tracking function, which cannot be done formally with the usual probabilistic approaches, and intrinsically performs the matching of disparate multisensor data.

2 Generic Model

The very general problem of discrimination introduced in section 1 will be considered here in the practical case of interest when the criteria C_{ij} are generated by separate information channels, for which reason they are differentiated according to their pertinence by the factors q_{ij}. We further assume that we are in the most frequently encountered context where the criteria C_{ij} taken separately are always at least of refutation value, in the sense that, when zero, this guarantees that the associated hypothesis H_i is not verified.

This leads to a formal construction of the problem on the basis of two axioms :

Axiom 2.1 : Each of the I*J pairs [C_{ij}, q_{ij}] constitutes a distinct source of information having the focal elements H_i, $\neg H_i$, and E, in which the frame of discernment E represents the set of the I hypotheses.

Axiom 2.2 : When $C_{ij} = 0$ is valid ($q_{ij} = 1$), we can assert that H_i is not verified.

Axiom 2.1 requires that I*J mass sets $m_{ij}(.)$ be generated from the I*J respective pairs [C_{ij}, q_{ij}]. For each, the mass of focal elements H_i, $\neg H_i$, and E is at first defined by the value of the corresponding criterion C_{ij}, which can be interpreted in terms of credibility or plausibility of H_i. Axiom 2.2 then limits the number of allowable models to two. Including the confidence factor q_{ij} for C_{ij} by discounting at the rate ($1-q_{ij}$) provides the corresponding mass sets $m_{ij}(.)$ [1] :

Model 1 :

$$m_{ij}(H_i) = 0 \tag{2.1}$$
$$m_{ij}(\neg H_i) = q_{ij}*(1-C_{ij}) \tag{2.2}$$
$$m_{ij}(E) = 1-q_{ij}*(1-C_{ij}) \tag{2.3}$$

Model 2 :

$$m_{ij}(H_i) = q_{ij}*C_{ij} \tag{2.4}$$
$$m_{ij}(\neg H_i) = q_{ij}*(1-C_{ij}) \tag{2.5}$$
$$m_{ij}(E) = 1-q_{ij} \tag{2.6}$$

A mass set $m(.)$ synthesizing all the evaluations is then obtained by computing the orthogonal sum of the different mass sets $m_{ij}(.)$ in the framework of each model :

$$m(.) = \bigoplus_{i,j} m_{ij}(.) \tag{2.7}$$

It should be noted that Model 1 is consonant, and therefore lends itself (but it alone) to interpretation in the framework of possibility theory.

The practical determination of the C_{ij} and q_{ij} terms is still, in all cases, a problem specific to the type of application at hand. In most cases in practice, suitable information is provided thanks to stochastic knowledge. The following section covers this type of situation.

3 Model With Statistical Learning

The problem dealt with now assumes that each of the measurements s_j has first been subjected to a learning of the *a priori* probability distributions $p(s_j/H_i)$, under the various hypotheses Hi. Most systems do in fact allow a certain number of preliminary measurements in different real or simulated situations, from which histograms can be generated to get a numerical or analytical model of the distributions $p(s_j/H_i)$. The I*J values of probability density $p(s_j/H_i)$ associated respectively with the J local measurements s_j constitute the inputs for the processes discussed hereafter.

If we consider the most common case, where the measurements s_j can be assumed to be statistically independent, since the sensors are generally chosen for the complementary nature of the data they generate, the likelihood of each hypothesis H_i can be established immediately by the Bayesian approach, which typically calls for an evaluation of the *a posteriori* probability $P(H_i/s_1,...,s_J)$ of each hypothesis H_i using :

$$P(H_i/s_1,...,s_J) = \{[\prod_j p(s_j/H_i)]*P(H_i)\} / \sum_k \{[\prod_j p(s_j/H_k)]*P(H_k)\} \qquad (3.1)$$

in which $P(H_i)$ designates its *a priori* probability.

However, this kind of approach quickly runs into difficulty when the real observation conditions differ from the available learning conditions, or when the measurement bank is not sufficient for a suitable learning process. The lack of control that can be seen at this level in most applications does in effect lead us to use distribution models that turn out to be more or less representative of the data actually encountered. In addition, it is often difficult to find a set of *a priori* probabilities $P(H_i)$ capable of reflecting the real situation with fidelity.

3.1 Specification of the Problem and Solutions

What we want to do here is to find a modeling based solely on the knowledge of $p(s_j/H_i)$ and capable of integrating any information concerning the reliability of the various distributions, whether this come from a more or less partial knowledge of the observations conditions or from a qualification of a data bank.

According to the generic approach introduced in section 1, any available qualitative information is assumed to be synthesized in the form of $I*J$ coefficients $q_{ij} \in [0,1]$, each being representative of a degree of confidence in the knowledge of each of the $I*J$ distributions $p(s_j/H_i)$.

Dealing with this problem in the terms of evidence theory requires finding, for each source S_j, a model of its I *a priori* probabilities $p(s_j/H_i)$ and their I respective confidence factors q_{ij} in the form of a mass set $m_j(.)$. Since the sources S_j are distinct, a global evaluation $m(.)$ can then be obtained by computation of the orthogonal sum of the $m_j(.)$. The appropriate frame of discernment is of course the set of the I *a priori* listed hypotheses H_i .

To do this, we conduct an exhaustive and exact search of all the models that might satisfy three fundamental axioms in the context considered. These three axioms are chosen beforehand on the basis of their legitimacy in most of the applications concerned. They are :

Axiom 3.1 : Consistency with the Bayesian approach in the case where the learned distributions $p(s_j/H_i)$ are perfectly representative of the densities actually encountered ($q_{ij}=1$, $\forall i,j$) and where the *a priori* probabilities $P(H_i)$ are known.

Axiom 3.2 : Separability of the evaluation of the hypotheses H_i ; that is, each probability must be considered as a distinct source of information generating a particular mass set $m_{ij}(.)$, mainly capable of integrating the confidence factor q_{ij} specific to it. We thus require that each mass set $m_j(.)$ be the orthogonal sum of the I mass sets $m_{ij}(.)$ considered for $i \in [1,I]$. Also, considering the way the $p(s_j/H_i)$ probabilities are generated, the focal elements of the mass set $m_{ij}(.)$ can be only H_i, $\neg H_i$, or E, where the frame of discernment E is the set of hypotheses H_i.

Axiom 3.3 : Consistency with the probabilistic association of the sources ; for independent sources Sj and densities $p(s_j/H_i)$ perfectly representative of reality, the modeling procedures retained must lead to the same result if we compute the orthogonal sum of the $m_j(.)$ modeled from the $p(s_j/H_i)$ or if we model directly the joint probabilities $p(s_1,...,s_J/H_i)$ given by :

$$p(s_1,...,s_J/H_i) = \prod_j p(s_j/H_i) \tag{3.2}$$

The search for models satisfying these three axioms is conducted by progressively restricting the set of possible models, taking the axioms into account in the order stated. After suitable developments [1,3], this leads to only two models, that meet the decomposition :

$$m_j(.) = \bigoplus_i m_{ij}(.) \tag{3.3}$$

Model 1 is particularized by :

$$m_{ij}(H_i) = 0 \tag{3.4}$$
$$m_{ij}(\neg H_i) = q_{ij}*\{1-R_j*p(s_j/H_i)\} \tag{3.5}$$
$$m_{ij}(E) = 1-q_{ij}+q_{ij}*R_j*p(s_j/H_i) \tag{3.6}$$

and *Model 2* by :

$$m_{ij}(H_i) = q_{ij}*R_j*p(s_j/H_i)/\{1+R_j*p(s_j/H_i)\} \tag{3.7}$$
$$m_{ij}(\neg H_i) = q_{ij}/\{1+R_j*p(s_j/H_i)\} \tag{3.8}$$
$$m_{ij}(E) = 1-q_{ij} \tag{3.9}$$

In both cases, the normalization factor R_j is theoretically constrained by :

$$R_j \in [0, (\max_{s_j,i}\{p(s_j/H_i)\})^{-1}] \tag{3.10}$$

Nevertheless, the specificity of the function used to generate model 2 allows R_j to be simply a positive number for this model in practice.

3.2 Tie-in With the Generic Problem and Comments

The problem of discriminating on the basis of a statistical learning process, as described in section 3, is in fact a special case of the general problem discussed in section 2. Both models provided by (2.1) to (2.3) and (2.4) to (2.6) in section 2 are in fact strictly equivalent to the two models found here in (3.4) to (3.6) and in (3.7) to (3.10), if we adopt the following respective definitions for the C_{ij} :

for model 1 : $C_{ij} = R_j * p(s_j/H_i)$ (3.11)

for model 2 : $C_{ij} = R_j * p(s_j/H_i)/[1 + R_j * p(s_j/H_i)]$ (3.12)

in which R_j is still, of course, the normalization gain constrained by (3.10).

This outcome is in fact legitimate if we note that Axiom 2.1 is expressed directly by Axiom 3.2, and that the solutions required by Axioms 3.1 and 3.3 automatically verify Axiom 2.2. Axioms 3.1 and 3.3 simply make it possible to specify the inclusion of the particular information $p(s_j/H_i)$ in the expression for the criterion, C_{ij} .

Furthermore, whether for the generic problem or for the problem with statistical learning, the models proposed always have the faculty of handling incomplete data sets. The absence of C_{ij} or $p(s_j/H_i)$ data, characterized by $q_{ij}=0$, leads to a trivial corresponding mass set ($m_{ij}(E)=1$], which remains perfectly defined. Moreover, the elementary models $m_{ij}(.)$ are simply defined on $\{H_i, \neg H_i\}$ with no requirement concerning the content of $\neg H_i$, so that the association of sources defined in different frames of reference is automatically performed.

Lastly, when the data s_j are discrete values (local identity declarations, for example), the generalized Bayes theorem defined by P. SMETS in the framework of evidence theory can be applied, for the case of statistical learning, to the cartesian product of the set of data and set of hypotheses. It then strictly yields Model 1 developed here.

4 Target Classification

The target classification function consists in recognizing the type of target observed, or even identifying a target, on the basis of the different discriminating features s_j delivered by the sensors S_j used. So the question is to designate the most likely hypothesis H_i* in light of the information generated. Such a decision, which is immediate when a probability can be associated *a posteriori* with each hypothesis, becomes quite delicate when the evaluations are presented in terms of the mass sets of evidence theory. The whole difficulty revolves around the non-exclusivity of the evaluations, which raises the practical problem of interpretation and relative inclusion of the masses attached to those focal elements of cardinal 2 or greater, in the designation of a unique singleton.

This problem, which is general to evidence theory and unavoidable in the present context, has been addressed thanks to different global approaches when no other *a priori* basis for discriminating among the H_i is available [1,3]. All of them converge to the same decisional procedure, which consists in retaining the most likely hypothesis H_i^* such that :

$$Pl(H_i^*) = \max_{i \in [1,I]} \{ Pl(H_i) \} \tag{4.1}$$

4.1 Solutions Involved

This criterion can be applied to the two models provided for the generic problem, leading to the two respective solutions :

Solution 1 : $\max_i \{ \prod_j [1-q_{ij}*(1-C_{ij})] \}$ (4.2)

Solution 2 : $\max_i \{ \prod_j [1-q_{ij}*(1-C_{ij})]/[1-q_{ij}*C_{ij}] \}$ (4.3)

It should be noted that Solution 1 also meets a maximum credibility criterion. The simplicity of the calculations and ease of use of these solutions is also worth noting.

Correlatively, criterion (4.1), applied to the two models obtained for dealing with statistical learning, generate the following two solutions, respectively :

Solution 1 : $\max_i \{ \prod_j [1-q_{ij}+q_{ij}*R_j*p(s_j/H_i)] \}$ (4.4)

Solution 2 : $\max_i \{ \prod_j ([1-q_{ij}+R_j*p(s_j/H_i)]/[1+(1-q_{ij})*R_j*p(s_j/H_i)]) \}$ (4.5)

in which R_j is still constrained by (3.10).

Let us note that in this case, when all the q_{ij} are 1, *i.e.* when the distributions $p(s_j/H_i)$ are perfectly representative of reality, the two approaches do in fact reduce to a maximum likelihood procedure.

Implementation of Solution 1 using neuro-fuzzy techniques further brings out an automatic learning of the factors q_{ij} for complex situations [5].

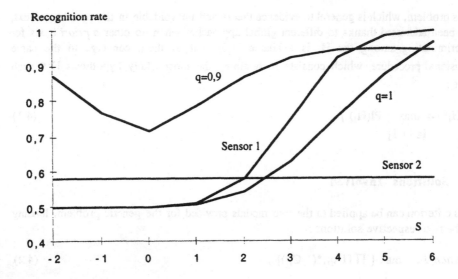

Fig. 1. Classification with unreliable learning

4.2 Illustration

Figure 1 shows the mean probability of good recognition provided by the simulation of 2 sensors for a problem of discrimination between 2 hypotheses H_1 and H_2. In this very simple example both sensors are similar, as regards either their *a priori* good discrimination capability, or the unreliability of their learning concerning hypothesis H_2, in relation with an anticipated possible evolution of the context.

More precisely, available learnings are given by normal distributions :

$P(s_1/H_1) = P(s_2/H_1) = N(0,1)$, with $q_{11} = q_{12} = 1$
$P(s_1/H_2) = P(s_2/H_2) = N(6,1)$, with $q_{21} = q_{22} = q$

while measurements actually simulated fellow :

$P(s_1/H_1) = P(s_2/H_1) = N(0,1)$
$P(s_1/H_2) = N(S,1)$, $P(s_2/H_2) = N(2,1)$

So in this test sensor 2 has effectively a wrong knowledge about H_2, and the reliability of sensor 1 varies in function of the signal S due to H_2. This is in accordance with the choice of factors q_{ij} that expresses a situation where a severe error concerning H_2 may occur simultaneously on both sensors. In this context our attention has to focus on the values of S much lower than 6, *e.g.* typically S<4. Then the curves of figure 1 emphasize the robustness of our approach (q=0,9), as regards either the probabilistic approach, that is a particular case of our method (q=1), or each sensor alone, which the probabilistic approach does not achieve.

Moreover, the aptitude of the q_{ij} factors to integrate linguistic or subjective information, considering the low sensitivity of the results to the choice of a given value for these coefficients, must be pointed out [3].

5 Target Tracking

The problem dealt with here is that of tracking a moving target of any possible nature, in a dense environment, using observations delivered by a set of disparate and possibly delocalized sensors. One of the main purposes is to overcome the problem of spurious sources present in the vicinity of the target. These sources may be due to intelligent countermeasures, artifacts, or vehicles that are untracked, for operational or technical reasons. A situation of major practical interest appears when the tracking is initialized on objects that are very close together or even at the same point, such as when a fighter plane enters an airspace hidden behind or close to an airliner. Simultaneous tracking of multiple targets may also be suitably handled with the proposed approach [1], but will not be considered here.

Unlike classical methods, the concept proposed performs a filtering directly on the discriminatory features available in the different resolution cells of each sensor, rather than on plots provided by a detection procedure.

Although it constitutes no particular limitation on the concept proposed, the discussion here presumes that the target tracked is the only one of its particular identity in the space being processed, and that a given resolution cell contains at most one target of any given identity.

5.1 General Principle

The technique used for the filtering aspects is inspired directly from the Probabilistic Data Association Filter (PDAF) family of methods developed by Y. BAR SHALOM from the ordinary KALMAN filter, to handle multiple detections [7]. These methods differ essentially from the KALMAN filter by the estimate updating phase, in which they proceed in two steps :

- First the statistical gating selects the detected plots located in a given vicinity of the predicted position. The vicinity is determined so as to contain the target with an *a priori* probability greater than a given threshold.

- Then the estimate and its covariance are updated on the basis of an innovation determined by linear combination of the innovations individually due to each plot retained as potential successor of the processed track. The weighting coefficients are the *a priori* probabilities for each of these plots to actually be due to the target, considering the detection and false alarm probabilities of the detector used, the predicted position and its covariance, and the statistical gating threshold.

In a first approximation, the method proposed here can be interpreted as a PDAF whose detection would operate at minimum threshold, with Detection Probability = False Alarm Probability = 1. At the level of the statistical gating, then, this is equivalent to retaining and processing one plot per resolution cell located within the vicinity defined around the predicted position.

The "*a priori*" probability that weights the innovation due to each of these plots in updating the estimate is, on the other hand, modified to reflect the likelihood of the identity present in the corresponding cell — information generated from the recognition of identity features extracted from the signal isolated by the spatial resolution of the sensors.

The special development of the weightings then necessary for the innovation requires two indispensable ideas to be defined :

- The sensors are said to be "aligned" if they break the validation gate down into the same resolution cells. For convenience here, the sensors are assumed to be classed in groups of sensors that are aligned among themselves, while two sensors of two different groups are necessarily unaligned. Each sensor will thus be denoted S_j^l, where l designates to which of the L groups of aligned sensors the sensor in question belongs, and j is its sequence number within the group of J sensors.

- If, for a group l of aligned sensors, x^{ln} designates the nth of N resolution cells of nonzero intersection with the validation gate, then the sensors in question "resolve" the gate if the gate entirely includes each x^{ln}.

5.2 Procedure description

The extraction of features in each resolution cell x^{ln} by each sensor S_j^l is assumed to provide information of the type considered by the generic model (section 2), or more specifically by the model with statistical learning (section 3), such that these models are applicable. We therefore have I*J mass sets $m_{ij}(.)$ per resolution cell x^{ln}, with each of them being defined either by (2.1) to (2.3), or (2.4) to (2.6), or (3.4) to (3.6), or (3.7) to (3.9). So they are, from now on, denoted $m_{ij}^{ln}(.)$, by reference to the resolution cell x^{ln} to which it relates, and their respective frames of discernment are denoted $E_i^{ln}=\{H_i^{ln}, \neg H_i^{ln}\}$. It will be noted that the use of the models established above is advantageous in light of their suitability to the problems generally encountered, but that this is not indispensable : the discussion here starts with any given mass sets $m_{ij}^{ln}(.)$, which can be obtained by any other means.

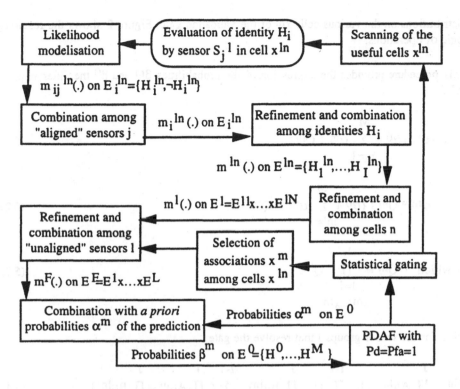

Fig. 2. Combination processing

The procedure therefore consists in combining the various sources $m_{ij}{}^{ln}(.)$, each being specific to a sensor $S_j{}^l$, a resolution cell x^{ln}, and a particular evaluated identity H_i. The combination is performed in such a way as to provide the likelihood of each possible distribution of identity hypotheses (including target absence) on the M resolution cells x^m of the validation gate. The x^m cells are the intersections of the x^{ln} cells of the various groups l of sensors, so that the combination processes applied offer the best spatial resolution at the end of the process.

Thanks to a special property of evidence theory, the combination of the resulting likelihoods with the *a priori* localization probabilities (α^0, α^m) of the tracked target delivered by the filter prediction, directly generate the *a posteriori* probabilities (β^0, β^m) of the target in question. The probabilities α^m and β^m are relative to the presence of the target in the cell x^m (H^m), while the probabilities α^0 and β^0 concern its absence in the gate (H^0). The probabilities β^m and β^0 are used to weight the innovation due to each of the cells x^m in the estimate update, as was introduced above.

Considering the nature of the problem, the required combinations must be performed by orthogonal sum of all of the sources, to obtain their conjunction. This must be done in the finest common frame of discernment, which is the set E^F of the possible identity distributions on the various cells x^m of the validation gate. Figure 2 shows the resulting logic of operations.

This procedure provides the expression of the probabilities β^m and β^0 the filter requires [6] :

$$\beta^0 = \alpha^0 / \{\alpha^0 + \sum_{m=1}^{M} \alpha^m * Q^m\} \tag{5.1}$$

$$\beta^m = \alpha^m * Q^m / \{\alpha^0 + \sum_{m'=1}^{M} \alpha^{m'} * Q^{m'}\} \tag{5.2}$$

$$\text{in which : } Q^m = \prod_{\substack{l=1 \\ x^m \subset x^{ln}}}^{L} Q^{ln} \tag{5.3}$$

with, for the sensor groups l that resolve the gate :

$$Q^{ln} = \prod_{j=1}^{J} A_{Ij}^{ln} / \{1 - \prod_{i=1}^{I-1} (1 - \prod_{j=1}^{J} B_{ij}^{ln}) + \sum_{i=1}^{I-1} (\prod_{j=1}^{J} A_{ij}^{ln} - \prod_{j=1}^{J} B_{ij}^{ln})\} \tag{5.4}$$

and, for the sensor groups l that do not resolve the gate :

$$Q^{ln} = \prod_{j=1}^{J} A_{Ij}^{ln} / \{1 - \prod_{i=1}^{I} (1 - \prod_{j=1}^{J} B_{ij}^{ln}) + \sum_{i=1}^{I} (\prod_{j=1}^{J} A_{ij}^{ln} - \prod_{j=1}^{J} B_{ij}^{ln})\} \tag{5.5}$$

In both cases, H_I designates the identity of the tracked target, and the coefficients A_{ij}^{ln} and B_{ij}^{ln} represent, respectively, the expressions :

$$A_{ij}^{ln} = \{m_{ij}^{ln}(H_i^{ln}) + m_{ij}^{ln}(E_i^{ln})\} / \{m_{ij}^{ln}(\neg H_i^{ln}) + m_{ij}^{ln}(E_i^{ln})\} \tag{5.6}$$

$$B_{ij}^{ln} = m_{ij}^{ln}(E_i^{ln}) / \{m_{ij}^{ln}(\neg H_i^{ln}) + m_{ij}^{ln}(E_i^{ln})\} \tag{5.7}$$

The resulting filter will hereafter be designated the Multiple Signal Filter (MSF).

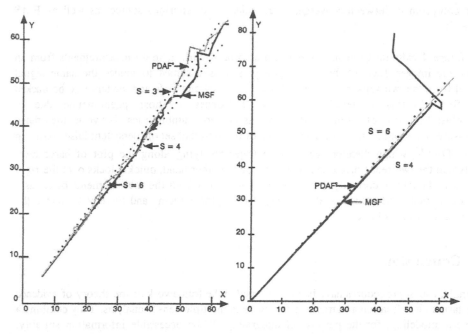

Fig. 3. 1 2D radar   **Fig. 4.** 1 pulse radar + 1 optronic imager

5.3 Illustration

Two very simple simulations are used to illustrate some of the potential advantages of MSF [1,6], with reference to the most suitable classical method. The latter consists in an usual PDAF, associated with a classification before tracking that aims at declaring at first the identity present in each resolution cell, thanks to a Maximum Likelihood criterion. It will be noted PDAF[C].

The only attribute considered in each resolution cell is the observed signal level, with this level being characterized by its *a priori* probability density under the various possible identity hypotheses : $N(0,1)$ for no target, and $N(S,1)$ when one is present, with S being able to take different values depending on the target considered.

The trajectories simulated are straight and level, at constant speed, approaching the sensors colocalized in (0,0) head-on. The dynamic model used in the filter is the same as the one that generates the trajectories. The only error introduced at the level of the filter concerns the track position and velocity initializations. The real trajectories are in dotted lines and the estimated ones in solid lines.

In figure 3 three targets (S=3, S=4, S=6) are observed by a 2D radar (azimuth and range), and the one we are trying to track is target S=4, which is, therefore, hemmed in between two targets : one weaker in signal power and the other stronger. Under these particularly difficult conditions, the MSF converges much faster, and on the right target, while the PDAF[C] can only lock onto the more powerful one, hampered by the unavoidable limitations of his detection phase. This emphasizes the inability of the PDAF[C] to meet the compromise between convergence and rejection of spurious sources as well as FMS does it.

In figure 4 radar range measurements are associated with azimuth measurements from an optronic imager. Each of the two targets present is assumed to induce the same signal level from the two sensors, which is respectively S=4 and S=6, and the target to be tracked is S=4. In this context plot processing induces a ghost phenomenon due to misclassifications of target S=6, *e.g.* association ambiguities between the range measurements and azimuth measurements, of the kind that set off persistent false alarms in the PDAF^C, which therefore generates a trajectory lying along the plot of barycenters between the two real trajectories. The MSF on the other hand, quickly locks onto the right target and tracks it correctly, thanks to its ability to match the measurements better as a function of the identities of the targets that originate them, and thereby to reject the incorrect identities better.

6 Conclusion

Simple, robust procedures have been developed in the framework of the theory of evidence for target classification and tracking functions in multiple-sensor analysis. They combine a generic modeling, for the purpose of integrating all the accessible information suitably, with original classification and tracking concepts starting with arbitrary discriminating mass sets, in order to offer a centralized and global approach to the problems handled. A formal integration of the data matching, combining, classification, and tracking functions is possible with the proposed approach, with all the advantages this implies according to function performances and robustness.

References

1. A. Appriou : Uncertain data aggregation in classification and tracking processes. In "Aggregation of evidence under fuzziness", Studies in Fuzziness, Physica Verlag, 1997.
2. A. Appriou : Classification par fusion de données incertaines multi-senseurs. In "Multisensor multitarget data fusion, tracking and identification techniques for guidance and control applications", AGARDOGRAPH 337, october 1996.
3. A. Appriou : Probabilités et incertitude en fusion de données multi-senseurs. Revue Scientifique et Technique de la Défense, n°11, 1991-1, pp 27-40.
4. G. Shafer : A mathematical theory of evidence. Princeton University Press, Princeton, New Jersey, 1976.
5. M.C. Perron-Gitton : Apport d'une approche neuro-floue dans un contexte de fusion de données basé sur la théorie de l'évidence. IPMU' 94, Paris, 4-8 juillet 1994.
6. A. Appriou : Multiple signal tracking processes. Aerospace Science and Technology, n° 2, February 1997.
7. Y. Bar Shalom, T.E. Fortmann : Tracking and data association. Academic Press, New York, 1988.

Dependency Mining in Relational Databases

Siegfried Bell[1]

Daimler–Benz AG
Research and Technology F3S/E
email: siegfried.bell@dbag.ulm.daimlerbenz.com
89013 Ulm Germany

Abstract. Semantic query optimisation promises to free users from the need of understanding the intricacies of databases when making an efficient query. The aim of semantic query optimisation is to use knowledge for reformulating a query into one that may require less answering time than the original query. Most approaches have the disadvantage of presuming this knowledge to be given by an expert or stated in the data dictionary as integrity constraints. This drawback can be overcome by using discovered knowledge.

Discovering data about data in databases, i.e. metadata, entails a new point of view, because only states of databases are considered. A consequence of this new view is that data dependencies as metadata and their relationships have to be extended by an expanded axiomatisation in order to minimise the database access in the discovery process.

In this paper, the expanded implication problem is discussed in order to decide entailment of functional dependencies. Results are an axiomatisation of functional dependencies, and the corresponding inference relation. The approach also discuss general properties of data mining approaches in relational databases.

1 Introduction

In general, the process of *dependency mining* in databases may described as follows: Assume that S is a language of a particular class of dependencies. $test(r,s)$ is a function which evaluates to *true* if $s \in S$ and s holds in r, otherwise to *false*. The task of discovering a complete set of dependencies which holds in r is defined by: Find a set $V \subseteq S$ with $s \in V$ if and only if $test(r,s)$ evaluates to *true*. This definition can be transformed into a simple enumeration algorithm provided that S is finite for a given database:

$$V := \{\}$$
$$\textbf{for each } s \in S \textbf{ do}$$
$$\qquad \text{if } test(r,s) \text{ then } V := V \cup \{s\}$$

The algorithm has the disadvantage that a lot of redundant dependencies are tested. Therefore, the algorithm is improved by a consequence relation $Cn(V)$

---

[1] This work was carried out at Dortmund University, Informatik VIII.

which computes all logical consequences of a set of dependencies, i.e. $s \in Cn(V)$ if s is a consequence of V.

$$V := \{\}$$
for each $s \in S$ **do**
  if $s \notin Cn(V)$
    then if $test(r, s)$ **then** $V := V \cup \{s\}$

There are still redundant tests in the algorithm. For example, if functional dependencies are discovered in r: $A \to B \in V$ and $test(r, C \to B)$ evaluates to *false*. Thus, the test $test(r, C \to A)$ is redundant, because $C \to A$ cannot be valid in r. This shows, that the test is avoided by the inference $\neg s \notin Cn(V)$ provided that S is finite.

$$V := \{\}$$
for each $s \in S$ **do**
  if $s \notin Cn(V)$ and $\neg s \notin Cn(V)$
    then if $test(r, s)$ **then** $V := V \cup \{s\}$
                else $V := V \cup \{\neg s\}$

This approach is presented in the second section, where we discuss an expanded axiomatisation of functional dependencies based on negative functional dependencies, i.e. a syntactical characterisation of Cn. In the third section we present an expanded inference relation, and show that the inference process is bounded by $O(n^2)$, i.e. a test function of the implication of functional dependencies. This approach can be improved by integrating the consequence relation in the enumeration process in order to avoid enumerating redundant dependencies and computing the consequences. Thus, we propose an improved discovery process as follows:

$$V := \{\}$$
for each $s \in S$ **and** $s \notin Cn(V)$ **and** $\neg s \notin Cn(V)$ **do**
  if $test(r, s)$
    then $V := V \cup \{s\}$
    else $V := V \cup \{\neg s\}$

In section four we introduce the concept of foreign subsets of attributes to integrate the consequence relation into the enumeration process. We present an algorithm of discovering functional dependencies which enumerates only non–redundant functional dependencies. Finally, we use the result of discovering functional dependencies to discover key dependencies. In the last section, we discuss related work, and present some results on an empirical evaluation of our proposed data mining process.

But, keep in mind that the discovered dependencies are no longer integrity constraints, because the validity depends on the state of the database. In contrast, constraints are valid in all allowed states of a database.

2 Axiomatisation of Functional Dependencies

In this section we discuss functional dependencies and their axiomatisation. We assume familiarity with the definitions of relational database theory (for an overview see for example [Kanellakis, 1990]) and the basic properties of the classical consequence relation Cn. The capital letters $A, B, C, \ldots$ denote attributes, and X, Y, Z denote attribute sets. We do not distinguish between an attribute A and an attribute set $\{A\}$. Remember that every attribute is associated with a set of values, called its *domain*. Functional dependencies are defined as usual. By simplicity, we neglect nulls in the database.

The consequences of a set of dependencies are defined as follows: F is a set of functional dependencies, then $X \to Y$ is a consequence of F or $X \to Y \in Cn(F)$: whenever a relation satisfies F, then it satisfies $X \to Y$. According to [Paredaens et al., 1989] a sound and complete axiomatisation of functional dependencies is given as follows:

Definition 1 (Axiomatisation). X, Y and Z are sets of attributes. An axiomatisation of functional dependencies is given by:

$F1 : (Reflexivity)$ If $X \subseteq Y$ then $Y \to X$

$F2 : (Augmentation)$ If $W \subseteq V$ then $\frac{X \to Y}{XV \to YW}$

$F3 : (Transitivity)$ $\frac{X \to Y, Y \to Z}{X \to Z}$

Some well known rules likes union and complementation are logically implied by this system. The closure of attributes X regarding a set of functional dependencies F is defined as: $\overline{X}_F = \{Y | X \to Y \in Cn(F)\}$.

Janas introduced functional *independencies* in [Janas, 1988] to mirror functional dependencies. But they are meant for a different purpose: Independencies are not semantical constraints on the data, rather a support for the database designer in the task of identifying functional dependencies.

In [Paredaens et al., 1989] *afunctional dependencies*[1] are introduced. They are a kind of semantic constraints and much stronger than Janas functional independencies. In order to improve the discovery process, we introduce *negative functional dependencies*, which are equivalent to Janas functional independencies:

Definition 2 (Negative Functional Dependency). $X \not\to Y$ denotes a negative functional dependency. A relation r satisfies $X \not\to Y$ ($r \models X \not\to Y$), if there exist tuples t_1, t_2 of r with $t_1[X] = t_2[X]$ and $t_1[Y] \neq t_2[Y]$.

The consequences of positive and negative functional dependencies are defined as follows: F is a set of positive and F' a set of negative functional dependencies. $Cn(F \cup F') := \{\sigma |$ for each relation r if $r \models F \cup F'$, then $r \models \sigma\}$, where σ is a positive or negative functional dependency. $F \cup F'$ is called inconsistent, if there is no relation r with $r \models F \cup F'$.

---

[1] The definition of the AD $X \not\to Y$ requires that for each tuple t there exists a tuple t' so that $t[X] = t'[X]$ and $t[Y] \neq t'[Y]$.

An important property of the relationship between positive and negative functional dependencies is that the inference of positive functional dependencies is not affected by the presence of negative functional dependencies. This is stated by lemma 4. Since an extension of a result in [Beeri et al., 1984] makes the following proofs easier, we first discuss this extension. Beeri et al. investigate the relationship between functional dependencies and Armstrong relations. They show that for each set of functional dependencies F and a functional dependency σ, which is not a consequence of F, a relation r with two tuples exists and F holds in r, but σ does not hold in r.

Lemma 3. *F and F' are sets of positive and negative functional dependencies respectively and $|F'| = k$.*
   If F and F' are consistent, then a relation r exists with $2 \cdot k$ tuples and F and F' are valid in r.

Proof: According to the result of Beeri et al. we can assume k tables with 2 tuples and two different values. It holds, that F is valid in each of the tables. Also at least one element of F' is valid in the tables. Without loss of generality we assume, that the values in each table are pairwise different. If the k tables are unioned to a big table r, then F holds in r. Also F' holds in r, because for each element of F': two tuples exist in order to make F' valid. □

Lemma 4. *F is a set of positive functional dependencies, F' a set of negative functional dependencies and $F \cup F'$ is consistent.*
$X \to Y \in Cn(F \cup F')$ if and only if $X \to Y \in Cn(F)$.

Proof: (if) is trivial by monotonicity of Cn.
(only-if) Assume that $X \to Y \notin Cn(F)$ and $X \to Y \in Cn(F \cup F')$. Then there must be a relation r with $r \models F$ and $r \not\models X \to Y$ according to Beeri et al. This means, there are tuples t_1 and t_2 in r with $t_1[X] = t_2[X]$ and $t_1[Y] \neq t_2[Y]$. We can add for each element $V \not\to W \in F'$ two tuples which satisfy $V \not\to W$ according to lemma 3. Any values are assigned to the remaining attributes without affecting F, because $F \cup F'$ is consistent. Remember that our domains are countably infinite, which ensures that we can use new values if needed. The expanded relation satisfies F and $F \cup F'$ by construction, but not $X \to Y$, which is a contradiction. □

The next important observation is that negative functional dependencies do not interact in the process of inference, because there exist at least two tuples for each negative functional dependencies, but we cannot identify them. For example, we cannot conclude the negative functional dependency $X \not\to Z$, if the given dependencies are: $X \not\to Y$ and $Y \not\to Z$.

Lemma 5. *$F' = \{S_1 \not\to T_1, \ldots, S_n \not\to T_n\}$ is a set of negative functional dependencies.*
   If $X \not\to Y \in Cn(F')$, then there exists a negative functional dependency $S_i \not\to T_i \in F'$ with $X \subseteq S_i$ and $T_i \cap Y \neq \{\}$.

Proof: (by contradiction) Assume that $X \not\to Y \in Cn(F')$ and for each $S_i \not\to T_i \in F'$ it holds that $T_i \cap Y = \{\}$ or $X \not\subseteq S_i$. We construct two relations and show that a relation exists with: if $r \models F'$ and $S_i \not\to T_i \notin F'$, then $r \not\models X \not\to Y$. This implies that $X \not\to Y \notin Cn(F')$, which is a contradiction to the assumption.

The first relation r_1 with $2n$ tuples is constructed by assigning pairwise different values to each attribute of each tuple. An exception is only S_i with $S_i \not\to T_i \in F'$ for $1 \leq i \leq n$: $t_i[S_i] = t_{i+n}[S_i]$. It follows that F' holds in r_1. If $X \not\subseteq S_i$ for $i \in 1, \ldots, n$, then $r_1 \not\models X \not\to Y$, because $t_j[X] = t_k[X]$ for all j and k with $j \neq k$. But this is a contradiction, therefore $X \subseteq S_i$.

The second relation r_2 with $2n$ tuples is constructed as follows: each attribute is assigned the same value, except for each $S_i \not\to T_i \in F'$, $t_i[S_i] = t_{i+n}[S_i]$ and $t_i[T_i] \neq t_{i+n}[T_i]$. Then F' holds in r_2. Assume that $T_i \cap Y = \{\}$ for every $1 \leq i \leq n$. If $t_i[X] = t_{i+n}[X]$, then $t_i[Y] = t_{i+n}[Y]$. Therefore $X \not\to Y$ holds in r_2. $\qquad\square$

The proof shows, that for a $S_i \not\to T_i \in F'$ with $X \subseteq S_i$ and $T_i \cap Y = Z$, that $S_i \not\to Z$ holds in each relation r, if F' and $X \to Y$ holds in r.

Lemma 6. $F' = \{S_1 \not\to T_1, \ldots, S_n \not\to T_n\}$ *is a set of negative functional dependencies.*
If $X \not\to Y \in Cn(F')$ with $X \subseteq S_i$ and $T_i \cap Y = Z$ for i with $1 \leq i \leq n$, then $S_i \not\to Z \in Cn(F')$.

Proof: We construct a relation r_3 as r_2 with $t_i[S_i] = t_{i+n}[S_i] = 0$, $t_i[T_i \setminus Z] = 1$, $t_{i+n}[T_i \setminus Z] = 1$, $t_i[Z] = t_{i+n}[Z] = 0$ and $t_i[Y \setminus Z] = t_{i+n}[Y \setminus Z] = 0$. Therefore, it holds that $r_3 \models S_i \not\to T_i$, $r_3 \models F'$, $r_3 \not\models S_i \not\to Z$, but $r_3 \not\models X \not\to Y$. This is a contradiction to the assumption. Therefore $S \not\to Z$ holds. $\qquad\square$

An axiomatisation of functional dependencies in general has already been given by Janas [Janas, 1988], which establishes an inference relation $\vdash_{Janas}$.

Definition 7 (Janas Axiomatisation). An Axiomatisation is given by Janas:

1. $\dfrac{X \not\to Y}{X \not\to YZ}$

2. $\dfrac{XZ \not\to YZ}{XZ \not\to Y}$

3. $\dfrac{X \to Y, X \not\to Z}{Y \not\to Z}$

We show by a counterexample that this inference relation is not complete, i.e., there exists some $X \not\to Y$ with $X \not\to Y \in Cn(F \cup F')$ and $F \cup F' \not\vdash_{Janas} X \not\to Y$.

Lemma 8. *The following inference rule is correct:*

$$\frac{X \to Y, Z \not\to Y}{Z \not\to X}$$

Proof: trivial by assuming that the conclusion is not satisfied and Armstrong's Axioms. $\qquad\square$

Lemma 9. $\{X \to Y, Z \not\to Y\} \not\vdash_{Janas} Z \not\to X$.

Proof: Assume that X, Y and Z are disjoint. Then the first and the second rule cannot be applied to infer $Z \not\to X$. Thus, the third rule can be applied only. But Z is not in the closure of Y and F, i.e. $\{X \to Y\} \not\vdash_{Janas} Y \to Z$. Thus, $Z \not\to X$ cannot be inferred. $\qquad\square$

Corollary 10. *The axiomatisation by Janas is not complete.*

Instead, we propose the following axiomatisation:

Definition 11 (Axiomatisation). An inference relation $\vdash_{fi}$ is given by an axiomatisation of the positive functional dependencies and the following inference rules:

$$F4: \quad \frac{V \not\to YU, U \subseteq V}{V \not\to Y}$$

$$F5: \quad \frac{X \to Y, X \not\to Z}{Y \not\to Z}$$

$$F6: \quad \frac{Y \to Z, X \not\to Z}{X \not\to Y}$$

For example, the negative functional dependency $X \not\to YZ$, which is a consequence of Janas's first inference rule, can be inferred by $\vdash_{fi}$ as follows: we infer $YZ \to Y$ by Armstrong's Axiom and use *F6* to infer $X \not\to YZ$ from $X \not\to Y$ and $YZ \to Y$. The inference rule *F4* reflects lemma 5, because negative functional dependencies can only be inferred from a set of negative functional dependencies by this rule.

Theorem 12 (Correctness). *The inference rules of definition 11 are correct.*

Proof: (Correctness) By Lemma 4 it is sufficient to show the soundness of *F4*, *F5* and *F6* w.r.t. negative functional dependencies:

- (F4) We have to show that $V \not\to Y \in Cn(\{V \not\to YU\})$. This means that each relation that satisfies $V \not\to YU$ must satisfy $V \not\to Y$. By definition there are tuples t_1, t_2 with $t_1[V] = t_2[V]$ and $t_1[YU] \neq t_2[YU]$. Since $U \subseteq V$, it follows that $t_1[U] = t_2[U]$, thus $t_1[Y] \neq t_2[Y]$ and $t_1[V] = t_2[V]$ and therefore $V \not\to Y$.
- (F5) $\{X \to Y, X \not\to Z\}$ means there are tuples t_1, t_2 with $t_1[X] = t_2[X]$, $t_1[Z] \neq t_2[Z]$ and for all tuples, particularly for t_1, t_2 $t_1[X] = t_2[X]$ and $t_1[Y] = t_2[Y]$. Then $t_1[Y] = t_2[Y]$ and $t_1[Z] \neq t_2[Z]$ and therefore $Y \not\to Z$.
- (F6) see Lemma 8 $\qquad\square$

The axiomatisation and the lemma 6 implies following corollary:

Corollary 13. *F is a set of positive functional dependencies, F' a set of negative functional dependencies.*
If $X \not\to Y \in Cn(F \cup F')$ then there exists a $S \not\to T \in F'$ and $X \not\to Y \in Cn(F \cup \{S \not\to T\})$.

In order to prove completeness of the axiomatisation, we introduce a new concept:

Definition 14 (Base of Positive Functional Dependencies). The base of a set X of attributes according a set F of positive functional dependencies is defined as follows:

$$\underline{X_F} = \{Y \mid Y \to X \in Cn(F)\}$$

The base of a set of attributes is a set of sets, whereas the set of all attributes is an element. An example clarifies the definition.

Example 15. $F = \{AB \to C, CD \to EG\}$ is a set of functional dependencies. The base of EG with respect to F is: $\underline{EG_F} \supseteq \{ABD, CD, EG\}$.

The completeness of the axiomatisation is proven as follows:

Theorem 16 (Completeness). *The axiomatisation of definition 11 is complete.*

Proof: According to lemma 4 it is sufficient to show completeness with respect to negative functional dependencies. F is a set of positive functional dependencies and F' a set of negative functional dependencies, and both are consistent. We show: if $X \not\to Y \in Cn(F \cup F')$, then $F \cup F' \vdash_{fi} X \not\to Y$. According to lemma 5 it follows that $X \not\to Y \in Cn(F \cup \{R \not\to S\})$ for a $R \not\to S \in F'$. Without loss of generality, we assume that $R \cap S = \{\}$. Otherwise we apply $F4$.

- $X \nsubseteq \overline{R_F}$. We construct a relation r with two tuples and assign each attribute of $\overline{R_F}$ the same value 0. We assign a 0 for the first row to the remaining attribute, and a 1 to the second row. It holds that $R \not\to S$ is valid in r and each dependency of F holds in r, but $X \not\to Y$ holds not in r. We show each of the conditions:

r	$U - \overline{R_F}$	$\overline{R_F}$
	$0 \ldots$	$0 \ldots$
	$1 \ldots$	$0 \ldots$

1. $R \not\to S$ holds in r: trivial
2. F holds in r: We show by inferring a contradiction by assuming that $P \to Q \in F$ and $P \to Q$ holds not in r. It follows that $P \subseteq \overline{R_F}$ and $Q \nsubseteq \overline{R_F}$. It follows that $R \to P$ does not hold in r and $R \not\to Q$ does not hold in r. By applying $F5$ it follows that $P \not\to Q$ does not hold in r. Since this is a contradiction, it follows that F must hold in r.
3. $X \not\to Y$ holds not in r: $X \nsubseteq \overline{R_F}$

Therefore, it follows that $X \subseteq \overline{R_F}$.

- In this part, we use the notion of a base of positive functional dependencies.

1. $Y \in \underline{S_F}$, which implies that $S_i \in \underline{S_F}$ and $Y = S_i$ and $Y \to S \in Cn(F \cup \{R \not\to S\})$. By the inference rule $F6$ it follows that $F \cup \{R \not\to S\} \vdash_f R \not\to Y$. By assumption $X \subseteq \overline{R_F}$ and $R \to X \in Cn(F \cup \{R \not\to S\})$ it is possible by $F5$ to infer $X \not\to Y$, this implies $F \cup \{R \not\to S\} \vdash_f X \not\to Y$, which we have to show.

2. $Y \notin \underline{S_F}$. We construct a relation r with F holds in r, $R \not\to S$ holds in r and $X \not\to Y$ holds not in r, and $Y' := Y - \overline{R_F}$.

r	$U - (\overline{R_F} \cup \overline{Y'_F})$	$\overline{R'_F}$	$\overline{Y'_F}$
	$0 \dots$	$0 \dots$	$0 \dots$
	$1 \dots$	$0 \dots$	$0 \dots$

(a) F holds in r: We infer a contradiction to the assumption that for a $P \to Q \in F$ it holds that: $P \to Q$ does not hold in r. This is only possible, if $P \subseteq overlineR_F \cup \overline{Y_F}$ and $Q \not\subseteq overlineR_F \cup \overline{Y_F}$. Q is a set of attributes $Q_1, \dots, Q_k$, then for each Q_j with $1 \le j \le k$ it is in the closure of R or Y'. This is a contradiction to the assumption.

(b) $R \not\to S$ holds in r: trivial

(c) $X \not\to Y$ holds not in r: trivial

This implies that a relation r exists with $F \cup \{R \not\to S\}$ holds in r and $X \not\to Y$ holds not in r, which is a contradiction to the assumption to $X \not\to Y$ follows from $F \cup \{R \not\to S\}$.

Corollary 17 (Completeness and Correctness). *F and F' are sets of positive and negative functional dependencies, respectively and $F \cup F'$ is consistent. $X \not\to Y \in Cn(F \cup F')$ if and only if $F \cup F' \vdash_{fi} X \not\to Y$.*

3 Inference of Functional Dependencies

Entailment of functional dependencies is discussed by studying if $F \vdash X \to Y$ holds where F is a set of functional dependencies, and $n = |F|$. This can be decided in linear time with appropriate data structures, c.f. [Kanellakis, 1990]. We construct an algorithm for testing negative functional dependencies based on lemma 6:

function $F \cup F' \vdash_{fi} X \not\to Y$;
begin
  for each $V \not\to W \in F'$ **do**
    for each $Z \subseteq W \cap V$ **do**
      if $F \vdash V \to X$ **and** $F \vdash Y \to W \backslash Z$
        then return Yes;
  return No;
end;

It is obvious that testing negative functional dependencies takes $O(m \cdot n^2)$ time where $n = max(F, F')$ and m is the number of attributes. If we demand that for each $S_i \not\to T_i \in F'$ and $S_i \cap T_i = \{\}$, then testing takes $O(n^2)$ time. Correctness and completeness follow immediately from the previous section.

4 Enumeration of Functional Dependencies

In this section we show the integration of the above presented inference relation in the enumeration process by using the inference rules to prune the set of attribute on the left hand side.

The discovery of functional dependencies is guided by the concept of a *search space*, which consists of a set of attribute. The search space is defined as follows: X is called a search space, if we want to discover all dependencies whose left–hand–side are a subset of X. The search space can be suitable restricted to enumerate only non redundant dependencies by the notion of *foreign subsets*:

Definition 18 (Foreign Subset). A *quotient* of a set X according a set $Y = \{e_1, \ldots, e_n\}$ with $Y \subset X$ is a set of sets $FS(X, Y) = \{M_1, \ldots, M_n\}$ with: for each element e_i with $1 \leq i \leq n$ of Y exists a set $M_i := X \backslash \{e_i\}$.
A quotient of several sets $FS^*(\{M_1, \ldots, M_n\}, Y)$ with respect to a set Y is defined as follows: $FS^*(\{M_1, \ldots, M_n\}, Y) = FS(M_1, Y) \cup \ldots \cup FS(M_n, Y)$, and if $Z_1, Z_2 \in FS^*(\{M_1, \ldots, M_n\}, Y)$ and $Z_1 \subset Z_2$, then remove $Z1$.
A sequence of quotients of sets is defined as follows:
$$FS^+(U, S) := FS^*(\ldots FS^*(FS^*(U, S_1), S_2), \ldots, S_n)$$

Foreign subsets divide a search space in sets. The benefit of this definition is shown by the following lemma for functional dependencies:

Lemma 19 (Restriction of the Search Space). *r is a relation, $Y \to A$ and $X \to A$ are different valid functional dependencies in r with $Y \not\subset X$, and for each M_i with $i \in \{1, \ldots, n\}$ it holds, that no element is a subset of another, and do not contain A.*
If $FS^*(\{M_1, \ldots, M_n\}, Y) = \{M_1', \ldots, M_m'\}$ and $Y \subseteq M_j$ and $X \subseteq M_k$ with $j, k \in \{1, \ldots, n\}$, then it holds that $X \subseteq M_i'$ with $1 \leq i \leq m$

Proof: (if) Assume $M_i' \to A$ is a valid functional dependency. There are three cases, how M_i' has been generated.
(Case 1) $M_i' = M_j$ for $j \in \{1, \ldots, n\}$: This is trivial, because $X \subseteq M_k$.
(Case 2) $M_i' = M_j \backslash e_1$, e_1 is an attribute of M_j and $e_1 \in Y$. If it is the case that $e_1 \notin X$, then $X \subseteq M_i'$.
(Case 3) $M_i' = M_j \backslash e_1$ and $e_1 \in X$. But this can not be the case, because we must have chosen the wrong M_i', otherwise it is for all $e \in Y$, that $e \in X$. This is a contradiction to the assumption that $X \neq Y$ and both determine A. $\qquad \Box$

The notion of *foreign subsets* is used in the enumeration of functional dependencies in the process of data mining, c.f. the recursive function *find_fds* in figure 1. $\overline{X_\tau}$ is the closure of X with respect to the set of functional dependencies τ. *neg-closure(Rhs, τ)* is the set of negative functional dependencies, which is computed by the inference relation of section 3. Each search is pruned, if no non redundant left hand side of a functional dependency is contained in it. A bottom–up search is done in each possible search space until a key is discovered. After the search space is again pruned by the new dependency, the next dependency is search until all search spaces are empty.

```
procedure find_fds(M, Rhs, τ, r, Result)
begin
    if M ≠ {} then
        Mᵢ ∈ M and Mᵢ ≠ {}
        M := M\Mᵢ
        if test_fds(Mᵢ → Rhs, r) then
            X ∈ neg-closure(Rhs, τ) and X ⊆ Mᵢ
            X' := X̄ᵣ
            while not test_fds(X' → Rhs, r) do
                X' := X' ∪ {A} with A ∈ Mᵢ and A ∉ X'
            τ := τ ∪ {X' → Rhs}
            find_fds(FS(M, X'), τ, r, Result)
            else  Result := {}
        else  Result := {}
end
```

Fig. 1. Algorithm *find_fds*

The process of data mining in general is sketched in figure 3. For each relation and each attribute the recursive function *find_fds* is called. The attribute is used as the right hand side of the functional dependency. The set M is initialised with the primary keys. If no keys are defined, then the set of all attributes is used. The already discovered dependencies are added to the result and in the next step used. The main properties of the algorithm are:

```
function test_fds(X → A, r) : bool
    begin
    if  1 = SELECT MAX (COUNT(DISTINCT A))
            FROM r
            GROUP BY X
        then return yes
        else  return no
    end
```

Fig. 2. Algorithm *test_fds*

Lemma 20 (Termination). *The algorithm* discover_fd *stops.*

Proof: The algorithm terminates, because the search spaces are reduced at every call of this recursive procedure. □

For stating correctness of the algorithm, we present the test function *test_fds* with respect to a relational database interface SQL/2. The test counts the distinct values in each group X. If two or more values exists, then the functional dependency $X \rightarrow A$ is not valid.

Completeness follows immediately from lemma 19.

Corollary 21 (Correctness and Completeness). *The discovery process of valid functional dependencies by the algorithm* discover_fd *is correct and complete.*

The discovered functional dependencies are in canonical normal form, which is easy to see. The canonical form is introduced by Maier in [Maier, 1980].

```
procedure discover_fd(τ, Result)
begin
      for  relation r do
        for  each attribute Rhs of r do
            M' is a set of the attributes of r without Rhs
            S is a set of the defined keys
            M := FS⁺(M', S)
            find_fds(M, Rhs, τ, r, Result)
            for  each Sᵢ ∈ S do
                Result:= Result ∪{Sᵢ → Rhs}
      end
```

Fig. 3. Algorithm *discover_fd*

Definition 22 (Canonical Normal Form). F is a set of functional dependencies.

F is canonical, if for each $X \rightarrow Y \in F$ it holds that:

1. Y is a single attribute and
2. there is no X' with $X' \subset X$ and $X' \rightarrow Y \in Cn(F)$

The number of functional dependencies can exponentially grow according the number of attributes in a relation, which is a consequence of the results on Armstrong relations by Beeri et al. [Beeri et al., 1984]. This establishes an upper bound of the algorithm.

In order to obtain a more precise description about the complexity of the proposed algorithm, it make sense to consider a slightly different complexity measure. By using the result as additional input of a turing machine, it is possible to give more precise boundaries.

Theorem 23. *Assume that the input of a turing machine consist of attributes and all valid functional dependencies according a given database.*

The discovery of functional dependencies is bounded by n with n is the number of attributes.

Proof: Obviously after n steps a dependency is discovered. If the dependencies consist of less than n attributes, then less steps are needed according the length. □

If the discovered dependencies are used for semantic query optimisation, then changes of the database have to be reflected in the dependencies, which is called dependency maintenance. We refer the reader for a discussion of maintaining functional dependencies to [Bell, 1995].

These results may easily adapted for key dependencies, because key dependencies are a special case of functional dependencies. We omit the translated terminology of key dependencies here.

5   Evaluation, Related Work, and Conclusions

We implemented a system in order to discover functional dependencies and connected it to a ORACLE DBMS. We used a real world database for testing the system. A result was that more dependencies were valid in the database than expected, or defined. A second result was that the time performance was mainly bounded by the number of attributes, and not on the number of tuples of the database as expected. This is quite reasonable, because the number of attribute determines the number of possible functional dependencies. The details and the use of discovered functional dependencies and keys are discussed in [Bell, 1996] with respect to semantic query optimisation.

5.1   Related Work

Gottlob and Libkin [Gottlob and Libkin, 1990] have shown that the MAX-set, introduced by Mannila and Räihä [Mannila and Räihä, 1986], can be written and interpreted as functional independencies. But in both works a closed world is assumed, which makes the concept of negative functional dependencies superfluous, because the negative functional dependencies are only an alternative way of representing positive functional dependencies. In contrast to this, we are not forced to know all functional dependencies. Our system still works if we have only proper subsets of functional dependencies in order to prevent the worst cases of exponentially many functional dependencies in a relation. This makes their approaches to ours absolutely incomparable.

Comparable to our approach in order to discover functional dependencies, there are similar's by Mannila and Räihä [Mannila and Räihä, 1991], Schlimmer

[Schlimmer, 1993], Savnik and Flach [Savnik and Flach, 1993], and Dehaspe et al. [Dehaspe et al., 1994]. Mannila and Räihä have investigated the problem of inferring functional dependencies from example relations in order to determine database schemes. But they do not use a complete inference relation regarding independencies. Savnik and Flach have investigated a special data structure for the functional dependencies. Briefly, they start with a bottom–up analysis of the tuples and construct a *negative cover*, which is a set of negative functional dependencies. In the next step they use a top–down search approach. They check the validity of a dependency by searching for negative functional dependencies in the negative cover. Also, the negative cover is not complete regarding a classical consequence relation. Schlimmer uses a top–down approach too, but in conjunction with a hash–function in order to avoid redundant computations. However, he does not use a complete inference relation even regarding functional dependencies. Also do Dehaspe et al. because their inferences are based on Θ–subsumption. In addition, the verification is based on theorem proving which is not suitable for real world databases.

In general, these authors do not use a relational database like OracleV7 or any other commercial DBMS. In such case, we argue that the proposed algorithm and approaches have to redesign according the set oriented interface of a relational database system. For example, the concept of the negative cover has only advantages if the tuples can be accessed directly, i.e. the tuples are stored in the main memory as Prolog–facts. Savnik and Flach have introduced it because the complexity for testing contradiction of the functional dependencies is reduced.

In contrast to these approaches our purpose is different because we maintain the discovered functional dependencies in order to use them all the time by semantic query optimisation. In addition, we argue that by using a relational database system, the higher complexity of the complete inference relation is justified by the size of a real world database. Thus, our approach has two advantages: it does not presumes the closed world assumption, which does not make any sense in a knowledge discovery or machine learning environment, cf. [Bell and Weber, 1993]. The second advantage is, that this approach guarantees minimal database access by completeness of the axiomatisations.

5.2 Conclusions

We have presented a more detailed view of the implication or membership problem $\sigma \in Cn(F)$. We use the concept and the axiomatisation of negative dependencies in our system twofold: First, negative dependencies help to minimize the number of accesses to the database w.r.t. the discovery of dependencies, because the alternative to a complete inference would be a more or less exhaustive test of functional dependencies on the database. Usually, real world databases are very large, the number of tuples is much larger than the number of attributes. Thus, the main costs of database management systems are caused by reading from secondary memory. Therefore, a single saved database query makes up for the costs of inferring positive and negative functional dependencies.

We improved this approach by integrating the inference process in the enumeration of the dependencies. The advantages are, that only non redundant dependencies are considered as possible valid dependencies. In general, we discussed a logical view on data mining in relational databases by a dependency mining approach. We have pointed out, that the interactions between the discovered dependencies help to improve the discovery process, because real-world database are huge and the search space of the target data have to be pruned in a suitable way.

References

[Beeri et al., 1984] Beeri, C., Dowd, M., Fagin, R., and Statman, R. (1984). On the structure of armstrong relations for functional dependencies. *Journal of the ACM*, 31(1):30–46.

[Bell, 1995] Bell, S. (1995). Discovery and maintenance of functional dependencies by independencies. In Fayyad, U., editor, *First International Conference on Knowledge Discovery in Databases*. AAAI, AAAI-Press.

[Bell, 1996] Bell, S. (1996). Deciding distinctness of query result by discovered constraints. In *Practical Application of Constraint Technology*. Practical Application Company.

[Bell and Weber, 1993] Bell, S. and Weber, S. (1993). A three-valued logic for inductive logic programming. Technical Report 4, Dortmund University, Computer Science VIII, 44221 Dortmund Germany.

[Dehaspe et al., 1994] Dehaspe, L., Laer, W. V., and Raedt, L. D. (1994). Applications of a logical discovery engine. In *ILP*.

[Gottlob and Libkin, 1990] Gottlob, G. and Libkin, L. (1990). Investigations on armstrong relations, dependency inference, and excluded functional dependencies. *Acta Cybernetica*, 9(4).

[Janas, 1988] Janas, J. M. (1988). Covers for functional independencies. In *Conference of Database Theory*. Springer, Lecture Notes in Computer Science 338.

[Kanellakis, 1990] Kanellakis, P. (1990). *Formal Models and Semantics, Handbook of Theoretical Computer Science*, chapter Elements of Relational Database Theory, 12, pages 1074 – 1156. Elsevier.

[Maier, 1980] Maier, D. (1980). Minimum covers in the relational database model. *Journal of the ACM*, 27(4):664 – 674.

[Mannila and Räihä, 1986] Mannila, H. and Räihä, K.-J. (1986). Design by example: An application of armstrong relations. *Journal of Computer and System Science*, 33.

[Mannila and Räihä, 1991] Mannila, H. and Räihä, K.-J. (1991). *The design of relational databases*. Addison-Wesley.

[Paredaens et al., 1989] Paredaens, J., de Bra, P., Gyssens, M., and van Gucht, D. (1989). *The Structure of the Relational Database Model*. Springer Verlag Berlin Heidelberg.

[Savnik and Flach, 1993] Savnik, I. and Flach, P. (1993). Bottum-up indution of functional dependencies from relations. In Piatetsky-Shapiro, G., editor, *KDD-93: Workshop on Knowledge Discovery in Databases*. AAAI.

[Schlimmer, 1993] Schlimmer, J. (1993). Using learned dependencies to automatically construct sufficient and sensible editing views. In Piatetsky-Shapiro, G., editor, *KDD-93: Workshop on Knowledge Discovery in Databases*. AAAI.

Syntactic Combination of Uncertain Information: A Possibilistic Approach [1]

Salem BENFERHAT – Didier DUBOIS – Henri PRADE

Institut de Recherche en Informatique de Toulouse (I.R.I.T.)
Université Paul Sabatier – 118 route de Narbonne
31062 Toulouse 4 – France
Email: {benferha, dubois, prade}@irit.fr

Abstract: This paper proposes syntactic combination rules for merging uncertain propositional knowledge bases provided by different sources of information, in the framework of possibilistic logic. These rules are the counterparts of combination rules which can be applied to the possibility distributions (defined on the set of possible worlds), which represent the semantics of each propositional knowledge base. Combination modes taking into account the levels of conflict, the relative reliability of the sources, or having reinforcement effects are considered.

1 - Introduction

In many situations, relevant information is provided by different sources. Information is also often pervaded with uncertainty. This uncertainty may directly reflect the reliability of the source, or may be attached to the information provided by the source itself. Taking advantage of the different sources of information usually requires to perform some combination operation on the pieces of information, and leads to a data fusion problem. The way this problem is tackled depends on the way the information is represented, which in turn is contingent on the nature of the information. On the one hand, pieces of information pertaining to numerical parameters are usually represented by distribution functions (in the sense of some uncertainty theory). These distributions are directly combined by means of operations which are in agreement with the uncertainty theory used, and which yield a new distribution on the set of the possible values of the considered parameter. On the other hand, information may be also naturally expressed in logical terms, especially in case of (symbolic) information pertaining to properties, which may be, however, pervaded with uncertainty. In this case, some uncertainty weights are attached to the logical formulas. Although similar issues are raised in the two frameworks, like the handling of conflicting information, the two lines of research in numerical data fusion (e.g., Abidi and Gonzalez, 1992; Flamm and Luisi, 1992) and in symbolic information combination (e.g., Baral et al., 1992; Cholvy, 1992; Dubois et al., 1992; Benferhat et al., 1993) have been investigated independently, and are unequally developed (the second trend being much more recent and the proposals still preliminary).

Possibility theory (Zadeh, 1978; Dubois and Prade, 1988a) offers an uncertainty modelling framework for dealing with possibility distributions representing fuzzy

---

[1] This is an abbreviated version, where proofs are also omitted, of a paper entitled "From semantic to syntactic approaches to information combination in possibilistic logic" to appear in: Aggregation of Evidence under Fuzziness (B. Bouchon-Meunier, ed.), Physica Verlag.

information. This information may pertain to the value of numerical parameters, or to variables ranging in discrete and finite domains. Moreover, possibilistic logic formulas (Dubois et al., 1994) can be handled in a purely syntactic way (as classical logic formulas associated with a weight) in complete agreement with a semantic interpretation in terms of a possibility distribution over the set of possible worlds, which represents the set of possibilistic logic formulas under consideration. The combination of possibility distributions representing ill-known data has been studied by two of the authors in the last past years (Dubois and Prade, 1988b, 1992), and different combination modes have been proposed depending on the nature of the conflict between the sources and their reliability.

In this paper, we apply these combination modes to the possibility distributions associated with the sets of uncertain propositions provided by each source of information and we look for the syntactic counterparts of these combination on the sets of possibilistic formulas. Section 2 gives the necessary background on possibilistic logic. Section 3 presents the semantical combination modes on possibility distributions, while Section 4 gives their syntactic counterparts applied to the possibilistic propositional logic bases corresponding to each sources.

2 - Possibilistic Logic

2.1 - Possibility Distribution and Possibilistic Entailment

In this paper, we only consider a finite propositional language denoted by $\mathcal{L}$. We denote by $\models$ the classical consequence relation, Greek letters α, β,... represent formulas. Let Ω be the finite set of interpretations of the propositional logic language $\mathcal{L}$.

A possibility distribution is a mapping π from Ω to the interval $[0,1]$. π is said to be normal if $\exists \omega \in \Omega$, such that $\pi(\omega)=1$. By convention, π represents some background knowledge about the possible states of the real world; $\pi(\omega)=0$ means that the state ω is impossible, and $\pi(\omega)=1$ means that nothing prevents ω from being the real world. When $\pi(\omega)>\pi(\omega')$, ω is a preferred candidate to ω' for being the real state of the world.

A possibility distribution π induces two mappings grading respectively the possibility and the certainty of a formula ϕ:

- the possibility degree $\Pi(\phi) = \max\{\pi(\omega) \mid \omega \models \phi\}$ which evaluates to what extent ϕ is consistent with the available knowledge expressed by π (Zadeh, 1978). It satisfies the characteristic property:
$$\forall \phi \ \forall \psi \ \Pi(\phi \vee \psi) = \max(\Pi(\phi), \Pi(\psi));$$
- the necessity (or certainty) degree $N(\phi) = \inf\{1 - \pi(\omega) \mid \omega \models \neg\phi\}$ which evaluates to what extent ϕ is entailed by the available knowledge. We have:
$$\forall \phi \ \forall \psi \ N(\phi \wedge \psi) = \min(N(\phi), N(\psi)).$$

The duality between necessity and possibility is expressed by $N(\phi) = 1 - \Pi(\neg\phi)$. Given a possibility distribution π, a notion of preferential entailment $\models_\pi$ can be defined.

Definition 1: An interpretation ω is a π-preferential model of a formula ϕ w.r.t. π, which is denoted by $\omega \models_\pi \phi$, iff:

i) $\omega \models \phi$, ii) $\pi(\omega) > 0$ and iii) $\not\exists \omega'$, $\omega' \models \phi$ and $\pi(\omega') > \pi(\omega)$.

A π-preferential model of ϕ is a normal state of affairs where ϕ is true.

Definition 2: A formula ψ is said to be a *conditional conclusion* of a fact ϕ, with background knowledge π, which is denoted by $\phi \models_\pi \psi$, iff
  i) $\Pi(\phi) > 0$, and
  ii) each π-preferential model of ϕ satisfies ψ, i.e., $\forall \omega$, $\omega \models_\pi \phi \Rightarrow \omega \models \psi$.

2.2 - From Possibilistic Knowledge Bases to Possibility Distributions

In particular, a possibilistic knowledge base made of one formula $\{(\phi \ a)\}$ is represented by the possibility distribution:

$$\forall \omega \in \Omega, \ \pi_{\{(\phi \ a)\}}(\omega) = 1 \qquad \text{if } \omega \models \phi$$
$$= 1 - a \qquad \text{otherwise.}$$

Thus, π_Σ can be viewed as the result of the conjunctive combination of the $\pi_{\{(\phi_i \ a_i)\}}$'s using the min operator, that is, a fuzzy intersection. The possibility distribution π_Σ is not necessarily normal, and $\text{Inc}(\Sigma) = 1 - \max_{\omega \in \Omega} \pi_\Sigma(\omega)$ is called the degree of inconsistency of the knowledge base Σ.

Lastly, note that several syntactically different possibilistic knowledge bases may have the same possibility distribution as a semantics counterpart. In such a case, it can be shown that these knowledge bases are equivalent in the following sense: their a-cuts, which are classical knowledge bases, are logically equivalent in the usual sense, where the a-cut of a possibilistic knowledge base Σ is the set of classical formulas whose level of certainty is greater than or equal to a.

2.3 - Possibilistic Inference

The possibilistic logic inference can be performed at the syntactical level by means of a weighted version of the resolution principle:

$$\frac{(\phi \vee \psi \ a)}{(\neg \phi \vee \delta \ b)}$$
$$(\psi \vee \delta \ \min(a,b))$$

It has been shown that $\text{Inc}(\Sigma)$ corresponds to the greatest lower bound that can be obtained for the empty clause by the repeated use of the above resolution rule and using a refutation strategy. Proving $(\psi \ a)$ from a possibilistic knowledge base Σ comes down to deriving the contradiction $(\perp \ a)$ from $\Sigma \cup \{(\neg \psi \ 1)\}$ with a weight $a > \text{Inc}(\Sigma)$. It will be denoted by $\Sigma \vdash (\psi \ a)$. This inference method is as efficient as classical logic refutation by resolution, and can be implemented in the form of an A*-like algorithm (Dubois et al., 1987). This inference method is sound and complete with respect to the possibilistic semantics of the knowledge base. Namely (Dubois et al., 1994):

$\phi \models_{\pi_\Sigma} \psi$ if and only if $\Sigma' \vdash (\psi \; a)$ with $\Sigma' = \Sigma \cup \{(\phi \; 1)\}$, and $a > \mathrm{Inc}(\Sigma \cup \{(\phi \; 1)\})$.

Clearly possibilistic reasoning copes with partial inconsistency. It yields non-trivial conclusions by using a consistent sub-part of Σ, consistent with ϕ, which corresponds to formulas belonging to the layers having sufficiently high levels of certainty. Moreover, possibilistic logic offers a syntactic inference whose complexity is similar to classical logic.

3 - Merging Possibility Distributions

In (Dubois and Prade, 1992), several propositions have been made to address the problem of combining n uncertain pieces of information represented by n possibility distributions $\pi_{i=1,n}$ (encoding the knowledge of n experts or sources of information about some parameters of interest), into a new possibility distribution. All the proposed combination modes are defined at the semantical level, and their syntactic counterpart will be investigated in Section 4. We assume that the experts who provide the π_i's use the same universe of discourse to describe their information, and also use the same scale to evaluate the levels of uncertainty.

Let us now review the different combination modes on possibility distributions.

3.1 - Idempotent Conjunctive and Disjunctive Modes

The basic combination modes in the possibilistic setting are the conjunction (i.e., the minimum) and the disjunction (i.e., the maximum) of possibility distributions. Namely define:

$$\forall \omega, \; \pi_{cm}(\omega) = \min_{i=1,n} \pi_i(\omega), \qquad \qquad \text{(CM)}$$

$$\forall \omega, \; \pi_{dm}(\omega) = \max_{i=1,n} \pi_i(\omega). \qquad \qquad \text{(DM)}$$

If the information provided by a source k is less (resp. more) specific than the information given by all the others then $\pi_{dm} = \pi_k$ (resp. $\pi_{cm} = \pi_k$). The conjunctive aggregation makes sense if all the sources are regarded as equally and fully reliable since all values that are considered as impossible by one source but possible by the others are rejected, while the disjunctive aggregation corresponds to a weaker reliability hypothesis, namely, in the group of sources there is at least one reliable source for sure, but we do not know which one. If $\forall \omega, \; \pi_{cm}(\omega)$ is significantly smaller than 1 the conjunctive mode of combination is debatable since in that case at least one of the sources or experts is likely to be wrong, and a disjunctive combination might be more advisable. It is why disjunctive combinations of possibility distributions should be regarded as *cautious* operations, while conjunctive combinations are more *adventurous*. Besides, if two sources provide the same information $\pi_1 = \pi_2$, the result of the conjunctive, or of the disjunctive combination is still the same distribution; indeed min (resp. max) is the only idempotent conjunction (resp. disjunction) connective.

3.2 - Other t-Norm and t-Conorm-Based Combination Modes

The previous combination modes based on maximum and minimum operators have no reinforcement effect. Namely, if expert 1 assigns possibility $\pi_1(\omega)<1$ to interpretation ω, and expert 2 assigns possibility $\pi_2(\omega)<1$ to this interpretation then overall, in the conjunctive mode, $\pi(\omega)=\pi_1(\omega)$ if $\pi_1(\omega)<\pi_2(\omega)$, regardless of the value of $\pi_2(\omega)$. However since both experts consider ω as rather impossible, and if these opinions are independent, it may sound reasonable to consider ω as less possible than what each of the experts claims. More generally, if a pool of independent experts is divided into two unequal groups that disagree, we may want to favor the opinion of the biggest group. This type of combination cannot be modelled by the minimum operation, nor by any idempotent operation. What is needed is a reinforcement effect. A reinforcement effect can be obtained using a triangular norm operation other than min in case of conjunctive, thus adventurous, combination, and a triangular conorm operation other than max for disjunctive, thus cautious, combination.

Definition 3: A triangular norm (for short t-norm) tn is a two place real-valued function whose domain is the unit square $[0,1]\times[0,1]$ and which satisfies the following conditions:
1. $\text{tn}(0,0) = 0$, $\text{tn}(a,1) = \text{tn}(1,a) = a$ (boundary conditions);
2. $\text{tn}(a,b) \leq \text{tn}(c,d)$ whenever $a\leq c$ and $b\leq d$ (monotonicity);
3. $\text{tn}(a,b) = \text{tn}(b,a)$ (symmetry);
4. $\text{tn}(a, \text{tn}(b,c)) = \text{tn}(\text{tn}(a,b), c)$ (associativity).
A triangular conorm (for short t-conorm) ct is a two place real-valued function whose domain is the unit square $[0,1] \times [0,1]$ and which satisfies the conditions 2-4 given in the previous definition plus the following boundary conditions:
5. $\text{ct}(1,1)=1$, $\text{ct}(a,0)=\text{ct}(0,a)=a$.
Any t-conorm ct can be generated from a t-norm through the duality transformation:
$$\text{ct}(a,b) = 1 - \text{tn}(1-a, 1-b)$$
and conversely. The basic t-norms are the minimum operator, the product operator and the t-norm $\max(0, a+b-1)$ sometimes called "Lukasiewicz t-norm" (since it is directly related to Lukasiewicz many-valued implication). The duality relation respectively yields the following t-conorms: the maximum operator, the "probabilistic sum" $a+b-ab$, and the "bounded sum" $\min(1, a+b)$.

In particular, several authors (e.g., Dubois and Prade, 1992; Boldrin and Sossai, 1995) have proposed to use Lukasiewicz operator for combining possibility distributions defined by (for two possibility distributions π_1 and π_2):

$$\forall\omega, \pi_{LM}(\omega) = \max(0, \pi_1(\omega) + \pi_2(\omega) - 1). \tag{LM}$$

Like the product-based combination, the (LM) combination mode is not idempotent; moreover the reinforcement effect with Lukasiewicz t-norm is more drastic since the combination of low degrees (i.e., $\pi_1(\omega) + \pi_2(\omega) < 1$) may result into a null degree of possibility which expresses complete impossibility.

We shall denote by π_{tn} and π_{ct} the possibility distributions resulting from the combination using a t-norm operator tn and a t-conorm operator ct respectively.

3.3 - Normalisation Rules

Disjunctive combination preserves normalisation (for any triangular conorm). When conjunctive combination mode $\mathcal{C}$ provides subnormal results ($\nexists\omega$, $\pi_{\mathcal{C}}(\omega)=1$), we may think of renormalizing $\pi_{\mathcal{C}}$. Indeed, let $h(\pi_{\mathcal{C}})=\max_{\omega}\{\pi_{\mathcal{C}}(\omega)\}$. It estimates to what extent there exists at least one interpretation which is possible according to each source. Thus, $h(\pi_{\mathcal{C}})$ is a degree of consistency of the pieces of information provided by the different sources. When $h(\pi_{\mathcal{C}})<1$ there is a partial inconsistency between the sources. There are several ways to renormalize $\pi_{\mathcal{C}}$ into $\pi_{N-\mathcal{C}}$. The minimal requirements for $\pi_{N-\mathcal{C}}$ are:

1) $\exists\omega$, $\pi_{N-\mathcal{C}}(\omega) = 1$,
2) if $\pi_{\mathcal{C}}$ is normal then $\pi_{N-\mathcal{C}} = \pi_{\mathcal{C}}$,
3) $\forall\omega,\omega'$, $\pi_{\mathcal{C}}(\omega)<\pi_{\mathcal{C}}(\omega')$ iff $\pi_{N-\mathcal{C}}(\omega)<\pi_{N-\mathcal{C}}(\omega')$.

Note that the third condition entails that only interpretations having possibility degrees equal to $h(\pi_{\mathcal{C}})$ can receive value 1 in the normalisation process. In the following we consider three noticeable renormalizing procedures defined by the three following equations where it is assumed that $h(\pi_{\mathcal{C}})>0$.[2]

- (N-1) $\pi_{N1-\mathcal{C}}(\omega) = \dfrac{\pi_{\mathcal{C}}(\omega)}{h(\pi_{\mathcal{C}})}$.

- (N-2) $\pi_{N2-\mathcal{C}}(\omega) = 1$ if $\pi_{\mathcal{C}}(\omega) = h(\pi_{\mathcal{C}})$
         $= \pi_{\mathcal{C}}(\omega)$ otherwise.

- (N-3) $\pi_{N3-\mathcal{C}}(\omega) = \pi_{\mathcal{C}}(\omega) + (1 - h(\pi_{\mathcal{C}}))$.

All the normalization rules get rid of inconsistency since $h(\pi_{Ni-\mathcal{C}})=1$ for $i=1,3$, and thus the inconsistency degree of any possibilistic knowledge base associated to $\pi_{Ni-\mathcal{C}}$ for $i=1,3$ is zero. The normalisation based on the equation (N-1) is the most usual one and is anologuous to the one used in probability theory. As (N-1), normalisation (N-2) is related to a conditioning in possibility theory. Indeed, the conditioning of a possibility distribution π by a formula ϕ held for true is either defined by (Dubois and Prade, 1988a):

$- \pi(\omega|\phi) = 1$ if $\pi(\omega) = \max\{\pi(\omega) \mid \omega\models\phi\} = \Pi(\phi)$
$= \pi(\omega)$ if $\omega\models\phi$ and $\pi(\omega)<\Pi(\phi)$
$= 0$ if $\omega\not\models\phi$;

in a purely ordinal setting (where the grades are only encoding a linear ordering), or using [0,1] as a ratio scale, by

$- \pi(\omega|\phi) = \dfrac{\pi(\omega)}{\Pi(\phi)}$ if $\omega\models\phi$.
$= 0$ if $\omega\not\models\phi$.

---

[2] When $h(\pi_{\mathcal{C}})=0$ there is a total disagreement between the sources, and conjunctive combination makes no sense in this case.

Clearly, these two types of conditioning are particular cases of (N-2) and (N-1) respectively (viewing the set of models of ϕ as a binary possibility distribution). The third normalisation rule (N-3), first introduced by Yager (1987) in the Dempster-Shafer framework, transforms the amount of conflict $1-h(\pi_{\mathcal{C}})$ into a level of total ignorance. Indeed, any interpretation ω in Ω gets a possibility degree at least equal to $1-h(\pi_{\mathcal{C}})$.

In general, an associative combination rule does not remain associative under normalization $(N-i)_{i=1,3}$. However, in the case of rule (N-1), associativity holds if $\pi_{\mathcal{C}}$ is defined by means of the product t-norm (Dubois and Prade, 1988b). When associativity does not hold, we have first to combine all the possibility distributions before normalizing the result.

3.4 - Conflict-Respectful Combination Rule

As explained above the renormalization erases the conflict between the sources expressed by the subnormalization. Even if it is good that the result of the combination focuses on the values on which all the sources partially agree (in the sense that none of them gave to these values a possibility equal to 0), it would be better that the result also takes into account the conflict in some way. A natural idea for taking into account a partial conflict is to discount the result given by (N-i) for i=1,3, by a weight corresponding to the lack of normalization, i.e., $1-h(\pi_{\mathcal{C}})$, thus partial inconsistency is transformed into partial uncertainty. Namely $1-h(\pi_{\mathcal{C}})$ is viewed as the degree of possibility that both sources are wrong since when $h(\pi_{\mathcal{C}})=1$ nothing suggests that a source is wrong. This discounting leads to the following modified conjunctive combination rule:

$$\forall \ \omega, \ \pi_{CoR-N_i-\mathcal{C}}(\omega) = \max(\pi_{N_i-\mathcal{C}}(\omega), 1-h(\pi_{\mathcal{C}})) \qquad \text{(CoRM)}$$

The amount of conflict $1-h(\pi_{\mathcal{C}})$ induces a uniform level of possibility for all values outside the ones emerging in the subnormalized intersection of π_i's, i.e., the result of the combination $(N-i-\mathcal{C})$ is not fully certain. Note that this rule leaves $\pi_{N-3-\mathcal{C}}$ unchanged since N-3 already incorporates a discounting effect.

3.5 - Prioritized Aggregation of Expert Opinions

When combining information conjunctively or disjunctively, we may give priority to some sources over others, and thus discount the information provided by a less prioritary source if this information is not in agreement with the information provided by a more prioritary source. In the following, we only consider min and max-based combinations. Then the following combination rule has been proposed (for the case of two possibility distributions, π_1 and π_2, where π_1 is more reliable than π_2) (Dubois and Prade, 1988c; Yager, 1991):

$$\forall \omega, \ \pi_{CPM1>2}(\omega) = \min(\pi_1(\omega), \max(\pi_2(\omega), 1-h(\pi_{cm})). \qquad \text{(CPM)}$$

Note that in case of total conflict, i.e., $h(\pi_{cm})=0$, only the information which has priority, represented by π_1, is retained ($\pi_{CPM1>2} = \pi_1$). More generally, if $1-h(\pi_{cm})\geq$

0.5 then for any ω, $\pi_{CPM1>2}(\omega)=\min(\pi_1(\omega),1-h(\pi_{cm}))$.[3] Moreover, if there is no conflict then the usual conjunctive combination is performed: $\pi_{CPM1>2}=\pi_{cm}$. The CPM rule is easily interpreted as the conjunctive combination of the information supplied by source 1 and the information supplied by source 2, the latter being discounted by a certainty coefficient $h(\pi_{cm})$, such that the degree of possibility that source 2 is wrong is $1-h(\pi_{cm})$.

The disjunctive counterpart of (CPM) has also been proposed by Dubois and Prade (1988d) and Yager (1991)

$$\forall\omega, \pi_{DPM1>2}(\omega) = \max(\pi_1(\omega), \min(\pi_2(\omega),h(\pi_{cm}))). \qquad \text{(DPM)}$$

The effect of this rule is to "truncate" the information supplied by the less prioritary source (by performing $\min(\pi_2,h(\pi_{cm}))$ in the above formula), while it is disjunctively combined with information given by source 1. Again if the two sources disagree ($h(\pi_{cm})=0$) then $\pi_{DPM1>2} = \pi_1$; if $h(\pi_{cm})=1$ then $\pi_{DPM1>2} = \pi_{dm}$.

3.6 - Grading the Reliability of the Sources

We assume that a relative level of reliability λ_j is attached to each source s_j. These reliability levels can be used for weighting conjunctive or disjunctive combinations. A normalisation condition is supposed to hold for the λ_j's, namely $\max_j\lambda_j=1$, i.e., the most reliable source(s) is graded by 1. We only consider min and max-based combination again. The weighted disjunctive combination mode is defined by (e.g., Dubois et al., 1992):

$$\forall\omega, \pi_{DRM}(\omega) = \max_{j=1,n} \min(\pi_j(\omega),\lambda_j) \qquad \text{(DRM)}$$

i.e., π_{DRM} is obtained as a weighted union of the possibility distributions associated with each source. If all the λ_j's are equal to 1 then the disjunction mode (DM) applied to the π_j's is recovered. If $\lambda_j=0$ the information provided by the source s_j is not taken into account. For intermediary λ_j, only the possible worlds whose possibility degree is low (which are rather impossible) are taken into account.

The conjunctive counterpart of (DRM) is defined by (e.g., Dubois et al., 1992):

$$\forall\omega, \pi_{CRM}(\omega) = \min_{j=1,n} \max(\pi_j(\omega), 1-\lambda_j) \qquad \text{(CRM)}$$

For $\lambda_j=1$, $\forall j$, the min combination mode is recovered. Note that (CRM) and (DRM) are similar to (CPM) and (DPM). One may think of combining them in order to take into account both the conflict of the sources and the reliability degrees attached to the sources. Assuming that $\lambda_1=1>\lambda_2$ (we give the priority to the most reliable source), for the case of two possibility distributions this leads to:

$$\pi_{CPRM1>2} = \min(\pi_1, \max(\pi_2, 1-\lambda_2, 1-h(\pi_{cm}))); \qquad \text{(CPRM)}$$

---

[3] Since $\pi_{CPM1>2}(\omega)$ can be rewritten as $\max(\min(\pi_1(\omega),\pi_2(\omega)), \min(\pi_1(\omega), 1-h(\pi_{cm}))$ and $h(\pi_{cm})\leq0.5 \Rightarrow 1-h(\pi_{cm})\geq0.5\geq \min(\pi_1,\pi_2)$.

$$\pi_{DPRM1>2} = \max(\pi_1, \min(\pi_2, \lambda_2, h(\pi_{cm}))). \qquad\qquad \text{(DPRM)}$$

These types of prioritized combination rules have been proposed by Yager (1991) and studied in detail by Kelman (1996).

4 - Syntactical Combination Modes

In Section 3, several propositions have been made to aggregate possibility distributions. In this section, we are interested in the combination of n possibilistic knowledge bases $\Sigma_{i=1,n}$ provided by n sources. Each possibilistic knowledge base Σ_i is associated with a possibility distribution π_i which is its semantical counterpart (i.e., $\forall \phi, \psi, \exists a > Inc(\Sigma_i \cup \{(\phi\ 1)\})$, $\Sigma_i \cup \{(\phi\ 1)\} \vdash (\psi\ a)$ iff $\phi \models_{\pi_i} \psi$). So we need to identify syntactical combination modes C on the Σ_i's which are the counterpart of the combination rules $\complement$ on the π_i's reviewed in the previous section. More formally, given a semantic combination rule $\complement$ we look for a syntactic combination C such that:

$$\complement(\pi_{\Sigma_1}, ..., \pi_{\Sigma_n}) = \pi_{C(\Sigma_1, ... \Sigma_n)}.$$

or equivalently,

$$\phi \models_{\pi_{\complement}} \psi \quad \text{iff} \quad \Sigma_C \cup \{(\phi\ 1)\} \vdash (\psi\ a) \text{ with } a > Inc(\Sigma_i \cup \{(\phi\ 1)\}),$$

where $\pi_{\complement} = \complement(\pi_{\Sigma_1}, ..., \pi_{\Sigma_n})$ and $\Sigma_C = C(\Sigma_1, ... \Sigma_n)$.

4.1 - t-Norm and t-Conorm-Based Combination Modes

In the framework of substructural logics, Boldrin and Sossai (1995) have proposed an extension of possibilistic logic where a second "and" connective (other than the min-based conjunction), defined from Lukasiewicz t-norm, is introduced for combining information from distinct independent sources. At the syntactic level this new conjunction applied to two one-formula knowledge bases $\Sigma_1 = \{(\phi\ a)\}$ and $\Sigma_2 = \{(\psi\ b)\}$ results in three possibilistic formulas: $\Sigma_{LM} = \{(\neg\phi\vee\psi\ b), (\phi\vee\neg\psi\ a), (\phi\vee\psi\ \min(1,a+b))\}$ which expresses an adventurous combination mode. Note that this knowledge is equivalent to the following: $\Sigma_{LM} \equiv \{(\psi\ b), (\phi\ a), (\phi\vee\psi\ \min(1,a+b))\} = \Sigma_1 \cup \Sigma_2 \cup \{(\phi\vee\psi\ \min(1,a+b))\}$, the equivalence being understood as the equality of the associated possibility distributions. This can be easily checked. Indeed, there are two different cases:

- If $\omega \models \phi$ or $\omega \models \psi$ then this trivially implies that $\omega \models \phi\vee\psi$. Hence:

$$\pi_{\Sigma_{LM}}(\omega) = 1 = \max(0, \pi_1(\omega) + \pi_2(\omega) - 1) \quad \text{if } \omega \models \phi\wedge\psi \quad (\text{since } \pi_1(\omega) = \pi_2(\omega) = 1)$$

$$= 1 - b = \max(0, \pi_1(\omega) + \pi_2(\omega) - 1) \quad \text{if } \omega \models \phi\wedge\neg\psi \quad (\text{since } \pi_1(\omega) = 1)$$

$$= 1 - a = \max(0, \pi_1(\omega) + \pi_2(\omega) - 1) \quad \text{if } \omega \models \neg\phi\wedge\psi \quad (\text{since } \pi_2(\omega) = 1)$$

- If $\omega \not\models \phi$ and $\omega \not\models \psi$ then this implies that $\omega \not\models \phi\vee\psi$. Hence:

$$\pi_{\Sigma_{LM}}(\omega) = 1 - \max(a, b, \min(1, a+b))$$

$$= 1 - \min(1, a+b)$$

$$= \max(0, 1-a-b)$$

$$= \max(0,1-(1-\pi_1(\omega)) - (1-\pi_2(\omega))) \quad \text{(since } \pi_1(\omega)=1-a \text{ and } \pi_2(\omega)=1-b)$$
$$= \max(0,\pi_1(\omega) + \pi_2(\omega)-1).$$

The previous remark can be generalized to the case of general possibilistic knowledge bases and to any t-norms and t-conorms. Let $\Sigma_1=\{(\phi_i\ a_i) \mid i\in I\}$ and $\Sigma_2=\{(\psi_j\ b_j) \mid j\in J\}$. Namely, we have the following result.

Proposition 1: Let π_{tn} and π_{ct} be the result of the combination based on the t-norm tn and the t-conorm ct. Then, π_{tn} and π_{ct} are respectively associated with the following knowledge bases:

$\Sigma_{tn} = \Sigma_1\cup\Sigma_2\cup\{(\phi_i\vee\psi_j\ ct(a_i,b_j)) \mid (\phi_i\ a_i)\in\Sigma_1 \text{ and } (\psi_j\ b_j)\in\Sigma_2\}$ (adventurous
 where ct is the t-conorm dual to the t-norm tn. combination)
$\Sigma_{ct} = \{(\phi_i\vee\psi_j\ tn(a_i,b_j)) \mid (\phi_i\ a_i)\in\Sigma_1 \text{ and } (\psi_j\ b_j)\in\Sigma_2\}$ (cautious
 where tn is the t-norm dual to the t-conorm ct. combination)

The syntactic counterpart of idempotent conjunctive and disjunctive modes is thus respectively the two following possibilistic knowledge base (letting tn=min and ct=max),

$$\Sigma_{min} = \Sigma_1\cup\Sigma_2\cup\{(\phi_i\vee\psi_j\ \max(a_i,b_j)) \mid (\phi_i\ a_i)\in\Sigma_1 \text{ and } (\psi_j\ b_j)\in\Sigma_2\}$$
$$= \Sigma_1 \cup \Sigma_2$$
$$\Sigma_{max} = \{(\phi_i\vee\psi_j\ \min(a_i,b_j)) \mid (\phi_i\ a_i)\in\Sigma_1 \text{ and } (\psi_j\ b_j)\in\Sigma_2\}.$$

Of course, Σ_{min} may be inconsistent and the handling of inconsistency is simply achieved by using the possibilistic entailment; see Section 2. Note that Σ_{max} is always consistent (provided that Σ_1 or Σ_2 is consistent), and that a formula ψ is considered as a plausible consequence of Σ_{max} if ψ is inferred both from Σ_1 and Σ_2. In the fuzzy case, the fuzzy set of possibilistic consequences of Σ_{max} can also be viewed as the fuzzy intersection of the fuzzy set of possibilistic consequences of Σ_1 and the fuzzy set of those of Σ_2.

4.2 - Normalization

This section discusses the effect of the normalization at the syntactical level. Let π_C be a sub-normalized possibility distribution obtained by combining π_1 and π_2 using a conjunction operator C. Let Σ_C be the possibilistic knowledge base associated with π_C and built using the result of the previous section. Let $h(\pi_C)=\max_\omega\{\pi_C(\omega)\}$. Then we have:

Proposition 2: 1. (N-1) $\pi_{N1-C}(\omega)=\dfrac{\pi_C(\omega)}{h(\pi_C)}$ is associated with:

$$\Sigma_{N1-C} = \left\{\left(\phi_i\ 1 - \frac{1-a_i}{h(\pi_C)}\right) \mid (\phi_i\ a_i) \in \Sigma_C \text{ and } a_i > 1-h(\pi_C)\right\}.$$

2. (N-2) $\pi_{N2-C}(\omega) = 1$ if $\pi_C(\omega) = h(\pi_C)$
$$= \pi_C(\omega) \quad \text{otherwise}$$

is associated with: $\Sigma_{N2-C}=\{(\phi_i\ a_i) \mid (\phi_i\ a_i)\in\Sigma_C \text{ and } a_i>1-h(\pi_C)\}$.

3. (N-3) $\pi_{N3\_\mathcal{C}}(\omega) = \pi_{\mathcal{C}}(\omega) + (1-h(\pi_{\mathcal{C}}))$ is associated with:
$$\Sigma_{N3\_\mathcal{C}} = \{(\phi_i \; a_i - (1-h(\pi_{\mathcal{C}}))) \mid (\phi_i \; a_i) \in \Sigma_{\mathcal{C}} \text{ and } a_i > 1-h(\pi_{\mathcal{C}})\}.$$

Note that $\Sigma_{N1\_\mathcal{C}}$ can be defined from $\Sigma_{N3\_\mathcal{C}}$ as follows:

$$\Sigma_{N1\_\mathcal{C}} = \left\{\left(\phi_i \; \frac{b_i}{h(\pi_{\mathcal{C}})}\right) \mid (\phi_i \; b_i) \in \Sigma_{N3\_\mathcal{C}}\right\}.$$

All the normalization procedures maintain all the formulas of $\Sigma_{\mathcal{C}}$ whose certainty degrees are higher than the inconsistency degree of $\Sigma_{\mathcal{C}}$. The normalization based on (N-2) just forgets the presence of conflicts and does not modify the certainty degrees of the pieces of information encoded by $\pi_{\mathcal{C}}$. The two other normalization procedures modify the certainty degrees of the formulas retained in $\Sigma_{\mathcal{C}}$. Let us observe that all the normalization modes diminishes the certainty levels of formulas in the wide sense. This diminution is more important with (N3) than with (N1).

4.3 - Conflict-Respectful, Prioritized and Reliability-Based Combination Modes

Let us first define two parametered functions which operate on a possibilistic knowledge base Σ:

- *Truncate(Σ,a)*, which consists in forcing the inconsistency level of Σ to be at least equal to a (which inhibits the formulas in Σ having certainty degrees strictly less than a), and
- *Discount(Σ,a)* which basically consists in decreasing to level a the certainty degree of the formulas of Σ whose certainty is higher.

More formally, we have:

Truncate $(\Sigma,a)=\{(\phi \; b) \mid (\phi \; b) \in \Sigma \text{ and } b \geq a\} \cup \{(\bot \; a)\}$
Discount $(\Sigma,a)=\{(\phi \; a) \mid (\phi \; b) \in \Sigma \text{ and } b \geq a\} \cup \{(\phi \; b) \mid (\phi \; b) \in \Sigma \text{ and } b < a\}$.

It is easy to check that each possibilistic conclusion of Truncate(Σ,a) or of Discount(Σ,a) is also a possibilistic conclusion of Σ. The converse is false. The following proposition lays bare the semantical counterparts of these two functions.

Proposition 3: Let Σ be a possibilistic knowledge base, and π its associated possibility distribution. Then:
- Truncate (Σ,a) is associated with $\pi'=\min(\pi,1-a)$,[4] and
- Discount (Σ,a) is associated with $\pi'=\max(\pi,1-a)$.

---

[4] Indeed, $h(\pi') \leq 1-a$ and thus the level of inconsistency is at least equal to a. Moreover, using equation (1) of Section 2.2., it is clear that the formulas $(\phi_i \; b_i)$ with $b_i < a$ have no influence in π' since $1-a < 1-b_i$.

The corresponding syntactic counterparts of conflict-respectful, prioritized and reliability-based combination modes are immediate using the discussion of the previous sub-sections, we get the corresponding possibilistic knowledge bases:

$$\Sigma_{CoR-Ni-C} = \text{Discount}(\Sigma_{Ni-C}, h(\pi_C))$$
$$C_{CPM1>2}(\Sigma_1, \Sigma_2) = \Sigma_1 \cup \text{Discount}(\Sigma_2, h(\pi_{cm})),$$
$$C_{DPM1>2}(\Sigma_1, \Sigma_2) = C_{dm}(\Sigma_1, \text{Truncate}(\Sigma_2, 1-h(\pi_{cm}))),$$
$$C_{DRM}(\Sigma_i)_{i=1,n} = C_{dm} (\text{Truncate}(\Sigma_i, 1-\lambda_i))_{i=1,n},$$
$$C_{CRM}(\Sigma_i)_{i=1,n} = \bigcup_i \text{Discount}(\Sigma_i, \lambda_i).$$

where C_{dm} denotes the combination operator yielding the possibilistic knowledge base Σ_{dm}, and $C_X(\Sigma_i)_{i=1,n}$ is the syntactic combination operator whose results is associated with π_X for X=CPM1>2, PM1>2, DRM, CRM.

Note that the addition of reliability degrees and priorities aims either to decrease the certainty degrees of highest formulas, or simply to remove formulas which are not sufficiently entrenched.

5 - Concluding Discussions

This paper has proposed a preliminary investigation of syntactic combination modes which can be applied to layered propositional logic knowledge bases, and which are the counterparts of possibilistic combination rules applied to the possibility distributions encoding the semantics of the layered bases. This can be viewed as a logical analog of the fusion of belief networks in agreement with the combination laws of probability distributions, see (Matzkevick and Abramson, 1992).

This work belongs to a larger research trend which also encompasses belief revision where syntactic encoding of possibilistic revision methods have been devised (e.g., Dubois and Prade, 1996; Williams, 1996). However, in belief revision the input information does not play a symmetrical role with respect to the knowledge we start with, while in this paper all the sources play symmetrical roles. Moreover the syntactic combination methods are exact counterparts of semantic ones. As a consequence they are drastic since they use truncation effects in the presence of inconsistency. All formulas in layers below the inconsistency level are lost. This is called the drowning effect (Benferhat et al., 1993). One might thing of taking into account the syntax of the knowledge bases when combining, and especially when restoring consistency, in order to delete formulas only when necessary.

Lastly, in this paper the weights attached to formulas were understood as uncertainty levels and the combination laws which have been considered are compatible with this view. They might be also interpreted in terms of priority levels, thus expressing preferences rather than uncertainty. In such a case, we would have to look for counterparts of fuzzy set aggregation operations which are meaningful in multiple criteria evaluation when preference profiles can be viewed as fuzzy sets.

Acknowledgements: The authors wish to thank Serafin Moral for helpful comments on an early draft of this paper. This work is partially supported by the European working group FUSION.

References

Abidi M.A., Gonzalez R.C. (eds) (1992) Data Fusion in Robotics and Machine Intelligence. Academic Press, New York.

Baral C., Kraus S., Minker J., Subrahmanian (1992) Combining knowledge bases consisting in first order theories. Computational Intelligence, 8(1), 45-71.

Benferhat S., Cayrol C., Dubois D., Lang J., Prade H. (1993) Inconsistency management and prioritized syntax-based entailment. Proc. of the 13th Inter. Joint Conf. on Artificial Intelligence (IJCAI'93), Chambéry, France, Aug. 28-Sept. 3, 640-645.

Boldrin L., Sossai C. (1995) An algebraic semantics for possibilistic logic. Proc of the 11th Conf. Uncertainty in Artifucial Intelligence (P. Besnard, S. Hank, eds.),27-35.

Cholvy F. (1992) A logical approach to multi-sources reasoning. In: Applied Logic Conference: Logic at Work, Amsterdam.

Dubois D., Lang J., Prade H. (1987) Theorem proving under uncertainty — A possibility theory-based approach. Proc. of the 10th Inter. Joint Conf. on Artificial Intelligence, Milano, Italy, August, 984-986.

Dubois D., Lang J., Prade H. (1992) Dealing with multi-source information in possibilistic logic. Proc. of the 10th Europ. Conf. on Artificial Intelligence (ECAI'92) Vienna, 38-42.

Dubois D., Lang J., Prade H. (1994) Possibilistic logic. In: Handbook of Logic in Artificial Intelligence and Logic Programming — Vol 3: Nonmonotonic Reasoning and Uncertain Reasoning (Dov M. Gabbay et al., eds.), Oxford Univ. Press, 439-513.

Dubois D., Prade H. (1988a) Possibility Theory — An Approach to Computerized Processing of Uncertainty. Plenum Press, New York.

Dubois D., Prade H. (1988b) Representation and combination of uncertainty with belief functions and possibility. Computational Intelligence, 4, 244-264.

Dubois D., Prade H. (1988c) Default reasoning and possibility theory. Artificial Intelligence, 35, 243-257.

Dubois D., Prade H. (1988d) On the combination of uncertain or imprecise pieces of information in rule-based systems. A discussion in a framework of possibility theory. Int. Journal of Approximate Reasoning, 2, 65-87.

Dubois D., Prade H. (1990) Aggregation of possibility measures. In: Multiperson Decision Making Using Fuzzy Sets and Possibility Theory (J. Kacprzyk, M. Fedrizzi, eds.), Kluwer Academic Publ., 55-63.

Dubois D., Prade H. (1992) Combination of fuzzy information in the framework of possibility theory. In: Data Fusion in Robotics and Machine Intelligence (M.A. Abidi, R.C. Gonzalez, eds.) Academic Press, New York, 481-505.

Dubois D., Prade H. (1996) Belief revision with uncertain inputs in the possibilistic setting. Proc. of the 12th Conf. on Uncertainty in Artificial Intelligence (E. Horvitz et al., eds.), Portland, 236-243

Flamm J., Luisi T. (Eds.) (1992) Reliability Data and Analysis. Kluwer Academic Publ.

Kelman A. (1996) Modèles flous pour l'agrégation de données et l'aide à la décision. Thèse de Doctorat, Université Paris 6, France.

Matzkevich I., Abramson B. (1992) The topological fusion of Bayes nets. Proc. of the 8th Conf. on Uncertainty in Artificial Intelligence (D. Dubois et al., eds.), pp. 191-198.

Williams M.A. (1996) Towards a practical approach to belief revision: Reason-based change. Proc. of the 5th Conf. on Knowledge Representation and Reasoning Principles (KR'96).

Yager R.R. (1987) On the Dempster-Shafer framework and new combination rules. Information Sciences, 41, 93-138.

Yager R. R. (1991) Non-monotonic set-theoritic operators. Fuzzy Sets and Systems 42, 173-190.

Zadeh L.A. (1978) Fuzzy sets as a basis for a theory of possibility. Fuzzy Sets and Systems, 1, 3-28.

A Coherence-Based Approach to Default Reasoning

Salem BENFERHAT - Laurent GARCIA

IRIT, CNRS-Université Paul Sabatier, 118, route de Narbonne
31062 TOULOUSE Cedex - France
email: {benferhat, garcia}@irit.fr

Abstract. In the last 15 years, several default reasoning systems have been proposed to deal with rules having exceptions. Each of these systems has been shown to be either cautious (where some intuitive conclusions do not follow from the default base), or adventurous (some debatable conclusions are inferred). However, the cautiousness and the adventurous aspect of these systems are often due to the incomplete way of describing our knowledge, and that plausible conclusions depend on the meaning (semantics) assigned to propositional symbols. This paper mainly contains two parts. The first part discusses, with simple default bases (where the used symbols have no a priori meaning), which assumptions are assumed when a given conclusion is considered as intuitive. The second part investigates a local approach to deal with default rules of the form "generally, if α then β" having possibly some exceptions. The idea is that when a conflict appears (due to observing exceptional situations), we first localize the sets of pieces of information which are responsible for conflicts. Next, using a new definition of specificity, we attach priorities to default rules inside each conflict. Lastly, three proposals are made to solve conflicts and restore the consistency of the knowledge base. A comparative study with some existing systems is given.

1. Introduction

One of the most important problems encountered in knowledge based systems is the handling of exceptions in generic knowledge. A rule having exceptions (also called default rule or conditional assertion) is a piece of information of the form "generally, if α is believed then β is also believed", where α and β are assumed here to be propositional logical formulas. In the presence of a fact, one may jump to conclusions which are just plausible and can be revised in the light of a new fact.

Several proposals have been done to deal with default information. Gabbay (1985), Lehmann and his colleagues (Kraus et al., 1990; Lehmann and Magidor, 1992) and Gärdenfors and Makinson (1994) have tried to provide basic set of properties (or postulates) that a default reasoning system should satisfy. However, inference machinery emerged from these proposals are either very cautious (like System P of Kraus et al. (1990)), or adventurous (like the rational closure (Lehmann and Magidor, 1992)). This means that these postulates are not enough to define a default reasoning system which correctly provides intuitive conclusions. In our opinion, what is missing is to characterize which assumptions are assumed when a given postulate is applied or when a given conclusion is derived. For instance, even if it is agreed that the contraposition rule (i.e., "normally, from α infer β" implies that "normally, from $\neg\beta$ infer $\neg\alpha$") is not a desired postulate, we believe that there are particular cases where this rule can be applied. The first part of this paper (Section 2) tries to explain, with sample default bases, what "intuitive" conclusion means, and what are the implicit assumptions done to derive plausible conclusions.

The second part of the paper (Section 3) investigates a local and coherent-based approach to deal with default information. This approach is based on restoring the consistency of the default base due to observing exceptional situations. Contrary to some existing systems, like System Z of Pearl (1990), we neither proceed to a global handling of conflicts, nor to a global ranking of the defaults rules. Independent conflicts (sets of pieces of information which are responsible for the inconsistency of the database) are handled separately, and hence inferring unwanted conclusions is avoided. Moreover, priorities, based on a new definition of specificity, are attached to default rules inside each conflict. We propose three ways to solve conflicts. One of these ways (the interesting one) is compared with some existing default reasoning systems.

2. What "Intuitive" Conclusions Mean?

The cautiousness and the adventurous aspect of default reasoning systems are often due to the incomplete way of describing the knowledge, and that plausible conclusions depend on the meaning assigned to propositional symbols. Two different meaning assignments to propositional symbols in the same set of rules can lead to different sets of intuitive conclusions. Indeed, when we give a precise meaning to the symbols in a default base, we are leaning to infer conclusions which are based on general information about the real world that are not explicitly mentioned in the default base. We will illustrate this situation in the example of Figure 1. Clearly, one of the important problems in the common sense reasoning remains how to correctly write our knowledge.

In this section, we try to explain with simple default bases (where the used symbols have no a priori meaning) which assumptions are assumed when a given conclusion is considered as intuitive. This section is not intended to give a detailed discussion of the meaning of default rules and ways to deal with them but just to point out, with some examples, what implicit assumptions should be made to do inferences in the presence of incomplete information.

In the following, $\mathcal{L}$ denotes a finite propositional language. Formulas of $\mathcal{L}$ are denoted by Greek letters α, β, δ,... T represents tautology(ies), and $\perp$ denotes any inconsistent formula. We write a default rule "generally, if α then β" as $\alpha \rightarrow \beta$. The implication "$\rightarrow$" is a *non-classical* arrow, and it should not be confused with material implication ($\neg \alpha \lor \beta$). Default rules considered here are not general default rules in the sense of Reiter (1980), but more a sort of normal default rules or conditional assertions in the sense of Kraus & al. (1990). A *default base* is a set $\Delta = \{\alpha_i \rightarrow \beta_i, i=1,...,n\}$ of default rules.

Before starting the analyzing of some default bases, we consider the following principle as fundamental:
Auto-deductivity principle: For any default $\alpha \rightarrow \beta$, if we observe α and only α then β must be considered as intuitive.

Example 1: Let us consider a simple default base containing only one default rule $\Delta = \{x \rightarrow y\}$. Given an observed fact $\neg y$, we are interested to know if x (resp. $\neg$x) follows or not.
• If we accept *x* then this means that the observation of $\neg$y is due to the existence of an exceptional x. For us this conclusion is non-intuitive and disagrees with the idea of default reasoning where it is preferred to believe in normal situations rather than in the

exceptional ones. Moreover, we can check that any possibility distribution where N(y|x)>0 and N(x|¬y)>0 implies $\prod(y)>\prod(\neg y)$ which means that we have a prior default information "generally, y is accepted" (see (Dubois & al., 1994) for a complete overview of possibility theory). This is of course non-intuitive. As far as we know, there is no existing system which infers x.

• Now if we assume that ¬x is plausible then this means that the contraposition rule (from α→β deduce ¬β→¬α) is accepted. However this conclusion is intuitive, since ¬y does not contradict our knowledge base. Moreover accepting ¬x means that the rule x→y has a very small number of exceptions. Here, we do not have any prior default information that is implied from all possibility distributions where N(y|x)>0 and N(¬x|¬y)>0. Some systems like System Z (1990), Geffner's conditional entailment (1992) infer such a conclusion.

• If *neither x nor ¬x* is inferred, then this means that we do not decide between the two previous cases. This is a very cautious attitude because we take into consideration the two possibilities which make sense since the two following default bases Δ_1={x→y, ¬y→x} and Δ_2={x→y, ¬y→¬x} are consistent. In a probability theory, it is both possible to construct a probability distribution where P(y|x)>0.5 and P(x|¬y)>0.5 and another where P(y|x)>0.5 and P(¬x|¬y)>0.5. Some systems, like Reiter's default logic (1980) or System P of Kraus et al. (1990), prefer to choose this attitude.

In the approach that we will develop in Section 3, we consider, from the rule x→y and a fact ¬y, that the conclusion ¬x is plausible, and more generally we will assume the following consistency principle:

Consistency principle: When an observed fact does not contradict the default base, then plausible conclusions are exactly the same as the ones of classical logic; hence, all the properties of the classical logic are considered valid.

The following examples consider the case when adding a fact contradicts the default base.

Example 1 (continued): We add a second default rule (y→z) to the base of Example 1 and obtain Δ = {x→y, y→z}.

$$x \longrightarrow y \longrightarrow z$$

Figure 1

Given a fact x∧¬z, we are interested to know if y can be inferred or not? Clearly, adding this fact leads to an inconsistent knowledge base which means that at least one of these two rules should not be applied. Let us consider two cases:

• From x∧¬z, we prefer inferring ¬y. This intuitively means that the rule x→y meets more exceptions than the rule y→z (namely, we prefer to give priority to y→z rather than to x→y). Indeed, if y→z accepts a very small number of exceptions then we are leaning to apply the "contraposition rule". Intuitive situation of this case can be obtained by interpreting x by "students", y by "young" and z by "have a special rate for travelling". The fact "Tom is a student and has no special rates for travelling" leads intuitively to prefer concluding "Tom is not young". The reason is that the rule "young people have special rates for travelling" accepts a very small number of exceptions, while it is not very surprising to find some students which are not young. Note that using the possibilistic logic machinery, we can check that from (¬x∨y, a),

$(\neg y \vee z, b), (x \wedge \neg z, 1)$ with $0 < a < b \le 1$ the conclusion $\neg y$ follows. As far as we know, there is no system which infers $\neg y$ from $x \wedge \neg z$.

• If from $x \wedge \neg z$, we prefer to infer y then applying the contraposition rule to $y \rightarrow z$ in this case can be debatable. In Figure 1, let us interpret x by "Danish people", y by "tall people" and z by "play basket-ball". In the presence of Bjarne Riis (who is Danish and does not play basket-ball), one would like to infer that he is a tall person. The reason is that it is less surprizing to find a tall person which does not play basket-ball rather than to find a small Danish person. Similarly to the precedent case, in possibilistic logic, from $(\neg x \vee y, a), (\neg y \vee z, b), (x \wedge \neg z, 1)$ with $0 < b < a \le 1$ we can check that the conclusion y follows. A probabilistic interpretation where y is inferred from $x \wedge \neg z$ can be the following. Let denote by $|\alpha|$ the number of complete situations (e.g. classical interpretations) where α holds. Let us assume that the two default rules are such that $|x \wedge y|/|x| = |y \wedge z|/|y| = a$ with $a > 0.5$. When there is no information about the proportion of α being β, we assume that $|\alpha \beta| = |\alpha \neg \beta|$. Then, using these *strict* assumptions, we conclude that $|xy \neg z| = a.(1-a) > |x \neg y \neg z| = (1-a)/2$, since $a > 0.5$, and hence, the conclusion y is preferred. An example of default reasoning system where y is inferred from $x \wedge \neg z$ is Reiter's default logic.

Of course there is also a cautious attitude where neither y nor $\neg y$ is inferred. Most of the existing systems (e.g. System P (Kraus & al., 1990) , System Z (Pearl, 1990), Boutilier's approach (1992), ε-belief functions (Benferhat et al., 1995)) adopt this attitude.

The previous example shows that the meaning assignment to propositional symbols can alter our decisions. It also illustrates the need of providing explicit information to get a desired conclusion. This can be done by adding some ordering between default rules. In (Benferhat & al., 1994) a possibility theory-based encoding of pieces of information of the form "in the context α, δ has no influence on β" has been suggested as a way to express explicit information. In the previous example, adding the explicit independence information "in the context x, z (resp. $\neg z$) has no influence on y" leads to infer y in the presence of $x \wedge \neg z$ in the possibility theory framework, while without this independence information neither y nor $\neg y$ is inferred in this framework.

The other reason why y should be preferred to $\neg y$ in Figure 1 is to use the specificity principle. Indeed the rule $x \rightarrow y$ is preferred to the rule $y \rightarrow z$. Specificity (Touretzky, 1984 ; Simari and Loui, 1992) is at the core of defeasible reasoning, and it is modelled by some preference relation, generally extracted from the syntax of the knowledge base, which guarantees that results issued from sub-classes override those obtained from super-classes. The following well-known triangle example illustrates the specificity principle:

Example 1 (continued): Now, let us expand the example of Figure 1 by adding the rule $x \rightarrow \neg z$. So, we have $\Delta = \{x \rightarrow y, y \rightarrow z, x \rightarrow \neg z\}$.

Figure 2

This is the famous penguin example when x is interpreted by "penguin", y by "birds" and z by "fly". Given an observed fact x then it is well agreed that y and ¬z are plausible conclusions of Δ. The main argument used to justify these conclusions is that the class x is more specific than the class y. Figure 2 in some way is similar to Figure 1 when we have a fact x∧¬z. The difference is that in Figure 1, ¬z is observed, while ¬z is deduced in Figure 2. However we have no problems to conclude y in Figure 2 while this conclusion in general does not follow in Figure 1! Let's notice that Reiter's default logic doesn't allow to infer neither z nor ¬z but the principle of specificity is inherent to most of the inference relations devoted to default reasoning.

Example 1 (continued): Irrelevance
Now let us again consider the example of Figure 2. We assume that the language contains four propositional symbols x, y, z and v (where v does not appear in the default base). We are interested to know if from a fact *x and v* we get ¬z or not? Some systems, like System P of Kraus et al. (1990), do not allow to infer neither z nor ¬z. This cautious attitude is justified by the fact that we have no information to decide if "x and v" is an exceptional x with respect to the property "¬z" or not. This problem is known as the irrelevance problem. In the following, we consider the next principle as valid:
Irrelevance Principle: Let δ be a propositional formula composed of propositional symbols which do not appear in the default base; if some conclusion ψ is a plausible consequence of a given fact φ w.r.t. Δ, then ψ is also a plausible consequence of a given fact φ∧δ w.r.t. Δ.
Reiter's default logic, System Z and its extensions have no problem to satisfy the irrelevance problem.

Let us study now the importance of specificity beside the length of the path of deduction. For this, let us consider the example given in the following Figure 3:
Δ={a1→a2, a2→a3, ..., an-1→an, a1→¬an}.

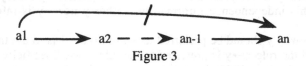

$$a1 \longrightarrow a2 \dashrightarrow an\text{-}1 \longrightarrow an$$

Figure 3

We want to know what must be deduced from a1. We have yet discussed, in Figure 2, the case where n=3. Let now consider the case where n=4 and let us give two interpretations to propositional symbols.
A meaning that can be applied to this case is again the penguin example where a1 is interpreted by "penguin", a2 by "bird", a3 by "has feathers" and a4 by "fly". Here, we want to deduce that a penguin has feathers. This solution follows the idea of Touretzky (1984) or Moinard (1987), where they justify this conclusion by applying the specificity principle. Another meaning is to consider that a3 is interpreted by "moving in the sky" (the other variables have the same meaning). Here, we want to deduce that a penguin does not move in the sky (since, intuitively, it is very surprising to find a non-flying object moving in the sky). Of course, there is a cautious alternative where neither an-1 nor ¬an-1 is inferred. That's the choice of System Z or Geffner's conditional entailment.

In both situations, we are using the knowledge of the real world to decide what conclusions we want to infer (here, we implicitly use the properties of penguins). This example can be seen as an extension of the example of Figure 1 but here the facts a2 and a4 are not given but deduced. So, it should be natural to keep on deducing the same results.

We must notice that some systems like Pearl's System Z (1990) or Geffner's system (1992) do not deduce an-1 (nor a3, ..., an-2) from this base. The reason is that their definition of specificity is based on the notion of contradiction between rules i.e. a rule is specific if, when it is verified, there is an inconsistency due to the presence of other rules. But these definitions do not take into account that specificity can also exist between classes when no exception explicitly holds. If we take the base $\Delta=\{A\rightarrow B, B\rightarrow C\}$, the two rules are not conflicting and are in the same level even if $A\rightarrow B$ is more specific than $B\rightarrow C$. Here, the notion we advocate is rather to consider that a class is more specific when it concerns a subset of properties w.r.t. a certain set with more properties (the rules are more precise if they are applied to the subset). This idea does not take into account the notion of exceptionality and, in fact, refines it since, when a subclass is exceptional, it concerns a subset of another set (which is the general class).

3. A Local Approach to Deal with Default Information

3.1. Why a Local Approach?

We have seen in the previous section that the inference of debatable conclusions can be due to some reasonable assumptions done by default reasoning systems in the absence of complete information. However some systems, like System Z of Pearl(1990) (or equivalently the rational closure of Lehmann and Magidor (1992)), infers debatable conclusions that are due to the global handling of inconsistency appearing when we learn some new fact. To illustrate this situation, let us consider the following example:

Example 2 (ambiguity): Let $\Delta = \{q\rightarrow p, r\rightarrow\neg p\}$. This is the famous Nixon diamond example: "Republicans normally are not pacifists" and "Quakers normally are pacifists". Clearly, from q we prefer to infer p, and from r we prefer to infer ¬p and this is justified by the auto-deductivity principle. But however, if we are interested to know if Nixon, who is a Republican and a Quaker, is a pacifist or not, then we prefer to say nothing. This is intuitively satisfying. Now let us add to this example three further rules (not related to pacifism), which give more information about Quakers: "Quakers are Americans", "Americans like base-ball" and "Quakers do not like base-ball". So, let $\Delta = \{q\rightarrow p, r\rightarrow\neg p\} \cup \{q\rightarrow a, q\rightarrow\neg b, a\rightarrow b\}$. This is illustrated by the following Figure 4:

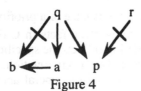

Figure 4

Given a fact q∧r, we have two conflicts, one A = {r∧q, q→p, r→¬p} is related to the pacifism and the other B={q, q→a, q→¬b, a→b} is related to playing base-ball. These two conflicts are independent and, hence, from q∧r we prefer to infer a,¬b but neither p nor ¬p is intuitively considered as a plausible conclusion. Some systems like System Z, prefer to infer p rather than ¬p. The reason is that the two conflicts are not handled independently but globally, and hence, in System Z, for the conflict B the class q is a specific class and is given a higher priority over all the general classes especially over the class r, therefore the debatable conclusion p is inferred!

There is another reason where handling default information locally is recommended. It concerns the case of default bases containing cycles. Let's take the example of the base Δ={A→B, B→C, C→A}. It is illustrated by the following Figure 5.

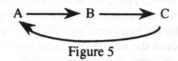

Figure 5

A global treatment of rules does not allow to make a difference between the rules and put the three rules in the same level. This is intuitively satisfying. Considering the fact A∧¬C, a global treatment does not allow to infer B. With a local treatment, from the fact A∧¬C, only the two rules A→B and B→C are involved in the inconsistency and we do not take into account the rule C→A (namely, C→A is not activated). Priority is given to the rule A→B upon B→C using the specificity principle. Then, B can be inferred.

3.2. Handling Stratified Conflicts
In the example of Figure 4, we have seen that, when it is possible, conflicts must be handled independently and that defaults must be ranked locally inside each conflict. A same default can be involved in different conflicts. This section gives a formal definition of conflicts and ways to solve them.

We represent a fact by a set of its prime implicates $\mathcal{F}$={p_1, p_2, ..., p_n}. We denote by Σ_Δ = {¬α_i∨β_i / α_i→β_i ∈ Δ} the material counterpart of Δ obtained by replacing each default α_i→β_i in Δ by its corresponding material implication ¬α_i∨β_i. $\mathcal{F}$ and Σ_Δ are assumed to be consistent. Only, $\mathcal{F}$ ∪ Σ_Δ can be inconsistent.

Definition 1: A subbase C of Σ_Δ∪$\mathcal{F}$ is called a conflict if it satisfies the two following properties: ① C ⊢ ⊥ (C is classically inconsistent) and ② ∀ φ∈C, C -{φ} ⊬⊥ (C is minimal w.r.t. the set inclusion relation).

For each conflict C, we assume the existence of a priority relation <$_C$ between rules inside this conflict. Formulas representing facts in C (i.e., $\mathcal{F}$∩C) get always the highest rank. Intuitively, this ordering reflects the specificity relation and we will see in Section 3.3. one way to compute this priority relation. In fact this ordering is only used to determine which defaults contain a general antecedent class, in other way which defaults are susceptible to be given up. Hence a conflict C can be simply

viewed as a couple $(\underline{C}, \overline{C})$ where:

$$\underline{C} = \{\varphi \mid \varphi \in C \text{ and } \nexists \ \psi \in C, \varphi <_C \psi\}, \text{ and}$$

$$\overline{C} = C - \underline{C}.$$

Intuitively, $\underline{C}$ contains defaults which can be non-pertinent for the fact and $\overline{C}$ contains defaults which are surely pertinent. Of course $\underline{C}$ must be strictly included in Σ_Δ. We

will denote by C=$\{C_i = (\underline{C}_i, \overline{C}_i) \mid C_i$ is a conflict in Δ and $\mathcal{F}\}$ the set of all the conflicts of $\Sigma_\Delta \cup \mathcal{F}$.

When $\Sigma_\Delta \cup \mathcal{F}$ is consistent, we simply apply classical logic to deduce plausible conclusions from $\Sigma_\Delta \cup \mathcal{F}$. However, an interesting case is when C is not empty. Hence we have to solve all the conflicts inside C. Solving a conflict C in C means removing at least one formula in C, and more precisely in $\underline{C}$. Of course, we can have several possibilities, and the following subsections present three of them.

3.2.1. A Cautious Consequence Relation
The first way to do is to remove all the defaults which are in the last layer of each conflict in C. Then we get the following non-monotonic consequence relation:

Definition 2: Let Del = $\{\neg\alpha_i \vee \beta_i \ / \ \neg\alpha_i \vee \beta_i \in \Sigma_\Delta$ and $\exists C \in C$ s.t. $\neg\alpha_i \vee \beta_i \in \underline{C}\}$ be the set of deleted formulas; let E = Σ_Δ- Del be the remaining set. A formula ψ is said to be a *cautious conclusion* of Σ_Δ and $\mathcal{F}$ if and only if E $\cup$ $\mathcal{F} \vdash \psi$.

The main drawback of this method is that it deletes too many rules. The inference is then too cautious as we can see with the following example:

Example 3: Let Δ=$\{rr \rightarrow w , rr \rightarrow ll , rr \rightarrow \neg f , w \rightarrow f , ll \rightarrow r , a \rightarrow \neg f \vee \neg r \}$,

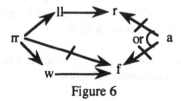

Figure 6

where rr, w, ll, f, a, r respectively mean "roadrunners", "have wings", "have long legs", "fly", "animals" and "run".
We are interested to know the properties of Beep-Beep which is a roadrunner animal (a∧rr). Clearly, there is no problem to infer that Beep-Beep has wings (w) and long legs (ll). Concerning the properties of flying and running we have two conflicts stratified in the following way:

$$A=\{\underline{A}=\{w \rightarrow f\}, \overline{A}=\{rr \rightarrow w, rr \rightarrow \neg f\}\} \text{ and}$$

$B=\{\underline{B}=\{w{\to}f, ll{\to}r, a{\to}{\neg}fv{\neg}r\}, \overline{B}=\{rr{\to}w, rr{\to}ll\}\}.$

The stratification is supposed to be given. A and B are the only conflicts in the knowledge base. Intuitively, the conflict A is easily solved since the $\underline{A}$ contains exactly one formula and hence w→f has to be removed. Hence we prefer inferring that Beep-Beep does not fly. The conflict B concerns the property of running where we have a reason to believe in r composed of {a, rr, rr→ll, ll→r } and a reason to believe in ¬r composed of B={a∧rr, rr→w, w→f, a→¬fv¬r}}. However, the last reason contains a formula w→f which is defeated, namely has to be removed using the conflict A. Hence the last reason is contestable and therefore we prefer to infer that Beep-Beep runs.

Unfortunately, the cautious inference relation does not allow us to infer that Beep-Beep runs. Indeed, using the previous definition, the sets $\underline{A}$ and $\underline{B}$ are deleted from Σ_Δ then we get the remaining base E = {rr→w, rr→ll, rr→¬f}. Hence, we cannot infer r from E and a∧rr.

3.2.2. Determining the First Solvable Conflicts

In the previous example, we have seen that solving a conflict (here the conflict A) can lead to solve other conflicts (the conflict B). Hence, it is very important to decide which conflicts must be first solved. The aim of this sub-section is to define a ranking between conflicts which indicates their solving influence:

Definition 3: A conflict A has a *positive influence* on a conflict B (or solving A solves B), denoted by A $\leq_I$ B, iff ① $\underline{A} = \underline{B}$ or, ② $\underline{A} \cap \underline{B} \neq \emptyset$ but $\underline{B} \nsubseteq \underline{A}$.

Note that, when $\underline{A} \subset \underline{B}$, removing any rule from $\underline{A}$ necessarily leads to solve B. However, if $\underline{A} \cap \underline{B} \neq \emptyset$ then removing a rule from $\underline{A}$ can possibly solve the conflict B (if the removed rule belongs both to A and to B). Note that $\underline{A} \subset \underline{B}$ implies $\underline{A} \cap \underline{B} \neq \emptyset$. To define the first conflicts to solve, we represent the relation $\leq_I$ by a graph where the nodes are the set of conflicts and an edge is drawn from A to B iff A has a positive influence on B.

Definition 4: A conflict A is said to be in the *set of first solvable conflicts*, denoted by min(C), if and only if there is no conflict B such that ① there is a path from B to A, but ② there is no path from A to B.

The two following sub-sections present two approaches which use min(C) to define less cautious non-monotonic consequence relations.

3.2.3. A Less Cautious Consequence Relation

The idea in this approach is the following: for each conflict C in min(C) we remove from Σ_Δ all the formulas which are in $\underline{C}$, and from C all the solved conflicts (conflicts containing at least one formula of $\underline{C}$). We repeat again the previous step until solving all the conflicts. This approach is described by the following algorithm:

a. Let Del = ∅, C be the set of conflicts in $\mathscr{F} \cup \Sigma_\Delta$.
b. Repeat until C = ∅
      For each C in min(C),
      b.1. Del = Del ∪ <u>C</u>,
      b.2. C = C - {C' / C'∈C and C'∩<u>C</u> ≠ ∅}
c. Return E = Σ_Δ - Del.

Let E be the sub-set of Σ_Δ obtained in step c. Then a formula ψ is said to be a *RFS-consequence* (RFS for Removing the First Solvable conflicts) of $\mathscr{F} \cup \Sigma_\Delta$ iff ψ is a classical consequence of $\mathscr{F} \cup$ E.

Clearly, any cautious conclusion of $\mathscr{F} \cup \Sigma_\Delta$ is also a RFS-consequence of $\mathscr{F} \cup \Sigma_\Delta$. The converse is false. Unfortunately the RFS-consequence can lead to adventurous conclusion as it is illustrated by the following example:

Example 4:
Let Δ = {p→b, p→¬f, b→f, mw→f}

Figure 7

where p, b, f, mw respectively means "penguins", "birds", "fly", "have metal wings". Given a fact mw∧p, we are in the situation of ambiguity and we have no reason to believe in f or in ¬f. We have two stratified conflicts:

    A={<u>A</u>={p→¬f, mw→f}, Ā=∅} and

    B={<u>B</u>={b→f}, B̄={p→b, p→¬f}}.

Clearly, solving A can lead to solve B (if the rule p→¬f is removed from A). Then it is interesting to try to solve A then B. Unfortunately, the RFS-consequence removes all the formulas of <u>A</u>={p→¬f, mw→f} from Σ_Δ, this leads to infer "f" which is of course non-intuitive.

Note that this approach of solving conflicts is inspired from the ideas of Williams (1996) where her proposal is done in the context of belief revision with a partial epistemic entrenchment.

3.2.4. A Universal Consequence Relation

In the two previous ways of restoring the consistency of $\mathscr{F} \cup \Sigma_\Delta$ only one sub-set, denoted by E, of Σ_Δ is generated. This sub-set in general is not a maximal consistent sub-set of Σ_Δ (namely, there exists a formula ϕ in Σ_Δ and ϕ is not in E while $E \cup \{\phi\} \cup \mathcal{F}$ is classically consistent).

Approaches based on generating one extension are either cautious or adventurous. The cautious entailment and the RFS-entailment described in the previous sub-sections are examples of such approaches. The cautious or the adventurous aspects are due to the fact that when we solve a conflict C, all the formulas of $\underline{C}$ are removed.

In this last approach, we only remove one formula to solve a given conflict. Of course, we consider all the different possibilities and this leads to compute several possible extensions rather than to generate one extension.

The following short algorithm gives one way to construct one extension:

a. Let Del = $\varnothing$, C be the set of conflicts in $\mathcal{F} \cup \Sigma_\Delta$.
b. Repeat until $C = \varnothing$
   b.1. Let C a conflict in min(C),
   b.2. Let ϕ a formula in $\underline{C}$,
      b.2.1 $C = C - \{C' / C' \in C \text{ and } \phi \in C'\}$
      b.2.2. Del = Del $\cup \{\phi\}$
c. Return E = Σ_Δ - Del.

We denote by $\mathcal{E}(\mathcal{F} \cup \Sigma_\Delta)$ the set of all extensions E_i obtained using the previous algorithm (using all the possible cases in b.1. and all the possible cases in b.2.).

Definition 5: A formula ψ is said to be a universal consequence of $\mathcal{F} \cup \Sigma_\Delta$ iff for each extension E_i in $\mathcal{E}(\mathcal{F} \cup \Sigma_\Delta)$ we have $E_i \cup \mathcal{F} \vdash \psi$.

Clearly, each cautious conclusion of $\mathcal{F} \cup \Sigma_\Delta$ is also a universal conclusion of $\mathcal{F} \cup \Sigma_\Delta$ but the converse is false. Indeed, if we consider Example 3, it is enough to notice that r is a universal consequence of a∧rr but it is not a cautious conclusion.

Moreover, RFS-consequence relation and the universal consequence relation are incomparable. Indeed, it is enough to consider Example 4 to see that the conclusion f is a RFS-consequence of mw∧p while it is not a universal consequence. If we consider the base {x→y, x→z, ¬y∨¬z, y→t, z→t}, we can notice that, assuming ¬b∨¬c is not in the general layer of a conflict, t is a universal consequence of x but not a RFS-consequence.

3.3. Specificity-Based Default Ranking

In the description of the three different ways of dealing with inconsistency caused by the addition of a fact $\mathcal{F}$ to the default base Δ, we have assumed that each conflict is stratified into two sub-sets: $\underline{C}$ which contains rules with the general antecedent classes, and $\overline{C}$ which contains rules with the specific antecedent classes.

This section gives how to compute $\underline{C}$ for a given conflict C. We will denote by $C_{\mathcal{F}}$ the factual part of C (namely $C_{\mathcal{F}} = C \cap \mathcal{F}$), and by C_Σ the knowledge part of C (namely $C_\Sigma = C \cap \Sigma$). We propose the following definition of specificity:

Definition 6: A formula $\neg\phi\vee\psi$ of C_Σ is said to be in $\underline{C}$ iff there is no rule $\neg\phi'\vee\psi'$ in C_Σ with $\phi\not\equiv\phi'$ such that $\{\phi, \psi\} \cup C_\Sigma \vdash \phi'$.

In the previous definition, it is easy to check that default rules considered in $\underline{C}$ are the most general ones. Indeed a rule is in $\underline{C}$ if activating this rule (by considering true its antecedent) does not activate any other rules. We can check that the stratification of the conflicts given in the Examples 2, 3 and 4 can be obtained using the previous definition.

This definition of specificity extends the one of Touretzky (1984). It is clear that, if we reduce the rules we use to the form of the ones used by Touretzky, the notion of specificity is the same since this notion uses paths of deduction. The difference is that, with our notion, we can deal with formulas. Although it is extended, our notion remains easy to compute.

Our notion of specificity is more refined than the one used by Pearl in his System Z (1990). If we refer to the end of the discussion of Section 2, we must notice that our notion of specificity allows to infer all the results which are not exceptional. Let remind that it corresponds to the base $\Delta=\{A\rightarrow B, B\rightarrow C\}$. Considering $A\wedge\neg C$, our notion allows to consider that $A\rightarrow B$ has priority on $B\rightarrow C$ (then B can be inferred).

The main drawback of our notion is that it can't really deal with rules which form a cycle. But we can notice that it is impossible to have a cycle in a conflict. Suppose it is possible to have a cycle. This means that we can delete at least one rule and obtain the same results. So, when a conflict has a cycle, it is not minimal. This contradicts the definition of a conflict. Let's take the example of Figure 5: $\Delta=\{A\rightarrow B, B\rightarrow C, C\rightarrow A\}$. Considering $A\wedge\neg C$, we do not take into account the rule $C\rightarrow A$ and no cycle is in the conflict. Our definition is then useful in this view.

To sum up, we can say that our notion of specificity is adapted to the treatment of rules having exceptions when dealing with default is made in a local way.

3.4. Comparative Study

This section compares the behaviour of the universal consequence relation (where the stratification of conflicts is based on Definition 6) with some existing default reasoning systems. We will not discuss the behaviour of the cautious entailment or of the RFS-entailment because the first one is too cautious (see Example 3) while the later can lead to non-intuitive and adventurous conclusions (see Example 4).

Let us start this comparative study with System Z. This system proceeds in the following way: first it partitions the set of default rules Δ in $(\Delta_1,...,\Delta_n)$ (this partition is based on a definition of specificity called *tolerance* where Δ_1 contains the most specific rules and Δ_n the most general rules), and next for a given fact ϕ System Z selects one sub-set E of Δ such that: $\{\phi\}\cup\Delta_1\cup...\cup\Delta_i$ is consistent but $\{\phi\}\cup\Delta_1\cup...\cup\Delta_{i+1}$ is inconsistent. An inference in system Z is simply defined by: ψ follows from Δ and a fact ϕ if ψ is classical consequence of $E\cup\{\phi\}$ (for more details see (Pearl, 1990)). Concerning the set of inferred conclusions, the two approaches are incomparable. Indeed, for example, in Figure 4 System Z infers the non-intuitive conclusion p from $(q\wedge r)$ while in our approach neither p nor $\neg p$ is inferred (which is satisfying). Moreover, System Z can be cautious and more precisely suffers from a so-called "blocking property inheritance", illustrated by the following figure:

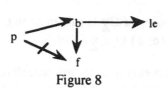

Figure 8

where b→le means that generally birds have legs. System Z does not allow to infer that "penguins have legs" while with our approach we get this intuitive conclusion. Our approach removes only concerned formula and no more.

Several solutions have been proposed to deal with the blocking property inheritance. Boutilier (1992) uses Brewka's preferred subtheories (1989) in System Z to define a new nonmonotonic inference relation. The idea is to start with the default base $(\Delta_1,...,\Delta_n)$ stratified by System Z and construct a preferred subtheory E from Δ by adding to the set of observed facts $\mathcal{F}$ as many formulas of the set Δ_1 as possible (with respect to consistency criterion) then as many as possible formulas of the set Δ_2, and so on. Lastly, Boutilier defines the set of plausible conclusions as a set of assertions which hold in all preferred subtheories of Δ. This approach partially remedies to the blocking inheritance problem, but however it leads to increase the set of plausible conclusions provided by System Z. Hence, using preferred sub-theories do not block the inference of adventurous conclusions produced by System Z. There are also other solutions to the "blocking property inheritance", like the lexicographical approach, based on selecting some preferred sub-theories which contains a highest number of formulas with lower rank (see (Benferhat et al., 1993) and (Lehmann, 1993) for more details)). This approach not only generates undesirable conclusions, like in example of Figure 4, but can be very syntax dependent. For example, duplicating the same formula can change the set of plausible conclusions. For instance, if we take a variant of Nixon diamond, namely $\Delta=\{q\rightarrow p, r\rightarrow \neg p, e\rightarrow p\}$, applying the lexicographical approach to the given fact $e\wedge q\wedge r$ leads to infer p, while in our approach neither p nor $\neg p$ are inferred. Moreover, if we consider the example of Figure 1, then neither Boutilier's approach nor the lexicographical approach allow us to infer the conclusion y from the fact "$x\wedge \neg z$", while in our approach we get it.

Besides, the notion of extensions used in our approach to define the universal consequence relation is not the same as the one used in Reiter's default logic since the direction of the arrow in more restrictive for Reiter. For instance from $\Delta=\{x\rightarrow y\}$ (in Reiter's notation we write x:y/y) and $\mathcal{F}=\{\neg y\}$, using algorithm given in Section 3.2.4., we get one extension $E=\{\neg y, \neg x\vee y\}$ while in Reiter's default logic we obtain $E=\mathcal{F}$. Moreover, Default logic does not use specificity criteria to prefer one extension over the other (see example of Figure 3). Delgrande and Schaub (1994), following ideas of Reiter and Criscuolo (1981), have suggested to transform some normal defaults, in the default base, to semi-normal defaults. This will lead to eliminate unwanted extension. However, their approach suffers from some limits, due to the use of default logic, such as generating a non-maximal extension (e.g., from $\Delta=\{x\rightarrow y\}$ and $\mathcal{F}=\{\neg y\}$ we do not get $\neg x$ even if $\{\neg y, \neg x\vee y\}$ is consistent).

Geffner's conditional entailment (1992) and our system are not comparable. If we consider bases according to Figure 3, with Geffner's system it is not possible to deduce a3, ..., an-1 while our system can. On the other hand, if we take the example of the base $\Delta=\{A\wedge Y\rightarrow U, U\rightarrow A, U\rightarrow\neg W, A\rightarrow W\}$, Geffner deduce $\neg W$ from $A\wedge Y$ while it is not possible by our method (the minimality of the conflict leads to not consider the rule $U\rightarrow A$ and, then, to not between $U\rightarrow\neg W$ and $A\rightarrow W$). The wanted result depends on the meaning given to the symbols. Geffner provides an interpretation of propositional symbols where $\neg w$ follows. However, we can also give another meaning where the intuition is the opposite. This problem should be studied more precisely and discussed more deeply since the intuition is not obvious here.

Finally, the universal consequence satisfies the auto-deductivity, the consistency and the irrelevance principles discussed in Section 2. Moreover we get the desired conclusions in examples discussed in Section 2.

4. Conclusion

This paper has tried to point out that intuitive conclusions depend on general information that we have about the real world and which are not explicitly mentioned in the default base. A future work will explore this direction by providing general principles which indicate the implicit assumptions done to considering that some conclusion is plausible or not.

The proposed notion of local and coherence-based approach is appealing for several reasons. First, it extends the classical logic when the observing fact does not contradict the default. Next, it is modular in the sense that the step of computing the specificity ordering of the defaults is independent of the step of solving conflicts. Hence, if one prefers another definition of specificity, then it is not very hard to change our method. Thirdly, our approach to deal with conflicts is local and hence independent conflicts are solved separately. This is not easy to do with approaches based on a global handling of inconsistency, like System Z, where some formulas are removed while they are outside the conflicts. Lastly, the universal consequence relation appears to be particularly attractive. It is not too cautious and avoids to infer unwanted conclusions. A future work will be a deep comparison with other existing systems, especially with the one of Dung's notion of argument (1993).

Acknowledgements

We would like to thank Didier DUBOIS and Henri PRADE for their useful comments.

References

S. Benferhat, C. Cayrol, D. Dubois, J. Lang & H. Prade (1993). Inconsistency management and prioritized syntax-based entailment. Proc. of the 13th Intern. Joint Conf. on A. I. (IJCAI'93). 640-645.

S. Benferhat, D. Dubois & H. Prade (1994). Expressing Independence in a Possibilistic Framework and its Application to Default Reasoning. Proc. of the 11th European Conf. on A. I. (ECAI'94). 150-154.

S. Benferhat, A. Saffiotti & P. Smets (1995) Belief functions and default reasoning. Proc. of the 11th conf. on Uncertainty in Artificial Intelligence (UAI'95). 19-26.

C. Boutilier (1992). What is a Default priority?. Proc. of the 9th Canadian Conf. on Artificial Intelligence (AI'92). 140-147.

G. Brewka (1989). Preferred subtheories: an extended logical framework for default reasoning. Proc. of the 11th Intern. Joint Conf. on A. I. (IJCAI'89). 1043-1048.

J. P. Delgrande & T. H. Schaub (1994). A general approach to specificity in default reasoning. Proc. of the 4th Intern. Conf. on Princ. of Knowledge Representation and Reasoning (KR'94). 146-157.

D. Dubois, J. Lang & H. Prade (1994). Possibilistic logic. In: Handbook of Logic in A. I. and Logic Programming, vol. 3. Oxford University Press. 439-513.

P.M. Dung (1993). On the acceptability of arguments and its fundamental role in nonmonotonic reasoning and logic programming. Proc. of the 13th Intern. Joint Conf. on A. I. (IJCAI'93). 852-857.

D. Gabbay (1985). Theoretical foundations for non-monotonic reasoning in expert systems. In: Logics and models of Concurrent Systems. Springer Verlag. Berlin. 439-457.

P. Gärdenfors & D. Makinson (1994). Nonmonotonic inference based on expectations. In: Artificial Intelligence, 65. 197-245.

H. Geffner (1992). Default reasoning: causal and conditional theories. MIT Press.

S. Kraus, D. Lehmann & M. Magidor (1990). Nonmonotonic reasoning, preferential models and cumulative logics. In: Artificial Intelligence, 44. 167-207.

D. Lehmann (1993). Another perspective on default reasoning. Technical report. Hebrew University, Jerusalem.

D. Lehmann & M. Magidor (1992). What does a conditional knowledge base entail? Artificial Intelligence, 55. 1-60.

Y. Moinard (1987). Donner la préférence au défaut le plus spécifique. Actes du 6ème congrès AFCET-RFIA. 1123-1132.

J. Pearl (1990). System Z: A natural ordering of defaults with tractable applications to default reasoning. Proc. of the 3rd Conf. on Theoretical Aspects of Reasoning about Knowledge (TARK'90). 121-135.

R. Reiter (1980). A logic for default reasoning. In: Artificial Intelligence, 13. 81-132.

R. Reiter & Criscuolo (1981). On interacting defaults. Proc. of the 7th Intern. Joint Conf. on A. I. (IJCAI'81). 270-276.

G. R. Simari & R. P. Loui (1992). A mathematical treatment of defeasible reasoning and its implementation. Artificial Intelligence, 53. 125-157.

D. S. Touretzky (1984). Implicit ordering of defaults in inheritance systems. Proc. of the 1984 National Conf. on A. I. (AAAI'84). 322-325.

M. A. Williams (1996). Towards a Practical Approach to Belief Revision: Reason-Based Change. Proc. of the 5th Intern. Conf. on Princ. of Knowledge Representation and Reasoning (KR'96).

A Syntactical Approach to Data Fusion

Paolo Bison, Gaetano Chemello,
Claudio Sossai and Gaetano Trainito

Ladseb-CNR
Institute of Systems Science and Biomedical Engineering
of the National Research Council
Corso Stati Uniti 4, I-35127 Padova, Italy

Abstract. An extended version of the Logic of Possibility is proposed as the formal basis for a data-fusion technique. The basic concepts underlying the approach are summarized and discussed. The method has been applied to a real-world problem of noisy sensor data fusion: the position estimation of an autonomous mobile robot navigating in an approximately and partially known office environment. Several test runs have evidenced the adequacy of the approach in interpreting and disambiguating the information coming from two independent perceptual sources, in combination with abstract common-sense knowledge.

1 Introduction

We present a syntactical method for data fusion, based on a valid and complete sequent calculus [BS96] for Possibilistic Logic [DLP94]. We apply this method to a concrete robotic problem: localization in an unknown environment. Two questions naturally arise:

- why possibility theory?
- why logic?

As we shall see through the examples in the experimental section, we choose to express real-world information by a pair (E, α), where E is an event and $\alpha \in [0, 1]$ a degree.

We must now find an operational semantics. If, for example, α is calculated as an event frequency, (the ratio of the occurrences of E over the occurrences of all events), then the intended semantics is that of probability, in the frequentist sense. In our experimental activity though, we have often met degrees calculated in a completely different way, in essence, through learning and/or interpolation techniques.

The basic idea is that learning techniques have been found adequate to estimate degrees of similarity or dissimilarity between prototypical models of objects and observed objects.

If we think of the possible configurations assumed by the observations of an object as possible models of the object, then the degree of similarity with the prototype determines a fuzzy set expressing an order of preference (objects more similar to the prototype are preferred) on the observations of the object.

Note that the same object, if observed from different positions or with different sensors, could give different degrees of similarity with the prototype. So, if we assume that the degree of similarity is a truth value, we must decide which of the two is the "right" value, or we must define a way of combining them. Now, if we assume that the degree does not determine a truth value, but an informational state, the two different values refer simply to two different informational states, corresponding to different points of observations or to different sources (sensors). This means that the intended semantics does not refer to the domain of *vagueness*, but to that of *uncertainty* [HHV95], and justifies our choice of local semantics (*forcing*) versus a many-valued logical apparatus (*fuzziness*).

But these are exactly the concepts underlying the semantics of possibilistic logic [DLP91]. The degree of similarity has nothing in common with the frequency with which the object is observed. In fact, objects with degree of similarity close to 1 are often the most difficult to observe.

It is worthwhile to notice that with this technique no a priori hypothesis is required for the initial possibilistic distributions.

This method refers only to the degrees of the observations of the elementary events. As we shall see in the examples, the estimation of the whole system state is generally the result of the combination of the degrees of the elementary events. In our experiment, the estimated robot location in the environment is obtained by combining the degrees of the observations of several events, like the various sensor measurements of different features, with pieces of abstract knowledge, i.e. not affected by uncertainty.

The process of information fusion is a key factor in the achievement of practically useful results. The main idea is that information fusion provides an approximation of the (implicit) distribution which determines the degree of preference (fuzzy set) on all the possible states (models) of the system.

In essence, the object of this work is the study of a formal method for combining the observations in order to estimate the current state from the observations and the abstract knowledge available.

2 The Role of Logic

Once we have chosen a way of coding real-world information, the following problem arises:

Problem 1. Given a set of observations and a body of abstract knowledge, how do we calculate the degree of an event E, not directly observable?

Possibility and necessity measures are non-compositional, therefore even if we know the degrees of the elementary events that constitute E, we cannot, in general, calculate the degree of E as a combination of the degrees of the elementary components.

To solve Problem 1, we can use two important properties of possibility distributions:

- the pointwise order relation over the set of possibility distribution, where $\pi_1 \leq \pi_2$ means that π_1's information content is larger than π_2's;

- the principle of minimum specificity: given a set of distributions, there is always one and only one distribution with minimum information content.

Combining these properties, we can calculate the least informative distribution that satisfies the observations and then use this to calculate the degree of E. We will see that this is exactly what sequent calculus does. At this point, as everything can be done semantically, one could argue about the use of a proof system or logic altogether. The answer is twofold:

- the normal form theorem (Theorem 7) shows that the algorithm to disassemble formulae and to find the least informative one, within the given constraints, is far from trivial, especially if we allow T-norm combination;

- the above statement "find the least informative..." does not imply that a calculation method is available; giving a constructive definition is equivalent to providing a proof system, as the proof of the validity and completeness theorem (Theorem 6) shows.

But let us see why logic gives us something more. If, in addition to the set of distributions, we have a logical language with which we can represent the events, then an important property can be described.

As we will see, the observations can be represented from inside the language: (E, r) is expressed by the formula $(1 - r) \oplus E$. At this point, a duality appears between information (distributions) and logic (formulae). The dual of each distribution π is the set of formulae true in that distribution, i.e. the formulae *forced* by π [Be85, MM92, TZ71], and the dual of each sentence is the set of distributions that make the sentence true. In this framework, it is easy to show that the dual of a formula is exactly the truth value of that formula. This allows us to define truth-functional semantics for measure-based logics, that is, logics for uncertainty.

Let us indicate the least informative distribution of the dual of a formula A with π_A. The validity and completeness theorem $(A \vdash E \text{ iff } \pi_A \leq \pi_E)$ says that from A we can prove E if and only if the information content of A (i.e. π_A) is greater than the information content of E. Hence sequent calculus performs the following operation: the proof $A \vdash x \oplus E$ constrains x to a value such that the distribution determined by $x \oplus E$ is less informative than the distribution determined by A. Hence the necessity of E computed with π_A satisfies $N_{\pi_A}(E) \geq 1 - x$. This shows how we can use the syntax of the logic to describe a formal algorithm to solve Problem 1. Doing all this without logic would require reinventing logic.

3 Local Possibilistic Logic (LPL)

3.1 The Semantics

We assume the following language, where α for any $\alpha \in [0,1]$ are constants[1]; $\mathcal{L}_0$ is the set of atomic propositions.

formula ::= atomic_proposition | α| formula & formula | formula $\otimes$ formula |
formula $\oplus$ formula | formula $\rightarrow$ formula

Negation is defined as $\neg A =_{\text{def}} A \rightarrow 0$. We take $\mathcal{L}$ to be the set of formulae; it is convenient to define $\mathcal{L}_1$ as the set of formulae with no occurrences of α constants for any $\alpha \in (0,1)$ — notice that 0 and 1 are in $\mathcal{L}_1$. We use uppercase Latin letters (A, B, C,...) for formulae, while reserving L, M, N for $\mathcal{L}_1$-formulae, and uppercase Greek letters (Γ, Δ, ...) for multisets of formulae; the lowercase Greek letters α and β always represent real numbers in $[0,1]$.

The language is the same as in [Pa79], where our & corresponds to $\wedge$ and our $\otimes$ to $\oplus$. Our choice of the connectives differs from Pavelka's, because we want to stress the proximity of our logic to substructural logics in the style of [Gi87].

Let $\mathcal{W}$ denote the set of classical propositional models for the language $\mathcal{L}_1$; each element $w \in \mathcal{W}$ is a valuation for atomic propositions. Let $\mathcal{P} = \{\pi : \mathcal{W} \longrightarrow [0,1]\}$ denote the set of functions from $\mathcal{W}$ to the real interval $[0,1]$, which we interpret as unnormalized possibility distributions. The set $\mathcal{P}$ is equipped with the usual order $\leq$ ($\pi_1 \leq \pi_2$ iff for any w it holds that $\pi_1(w) \leq \pi_2(w)$); the lattice operations $\vee$ and $\wedge$ on possibility functions are defined with respect to the order $\leq$; $\langle \mathcal{P}, \vee, \wedge \rangle$ is a complete lattice, which we endow with a monoidal operator $\times$ chosen among continuous T-norms.

Let us denote by $\overline{\alpha}$, $\overline{\beta}$, $\overline{\gamma}$, etc. the functions constantly valued α, β, γ etc.

We define the operation $\rightarrow$ as the residuated of $\times$: $\pi_1 \rightarrow \pi_2 = \bigvee\{\pi : \pi \times \pi_1 \leq \pi_2\}$. The negation is defined consequently as $\neg \pi = \pi \rightarrow \overline{0}$

Definition 2. For any continuous T-norm $\times$ we define a function $\|\cdot\|_\times : \mathcal{L} \longrightarrow \mathcal{P}$ as follows (from now on we omit the subscript $\times$):

$$\|p\| =_{\text{def}} \lambda w. \begin{cases} 1 & \text{if } w \models p \\ 0 & \text{otherwise} \end{cases}$$

$$\|\alpha\| =_{\text{def}} \overline{\alpha}$$

$$\|A \,\&\, B\| =_{\text{def}} \|A\| \wedge \|B\|$$

$$\|A \oplus B\| =_{\text{def}} \|A\| \vee \|B\|$$

$$\|A \otimes B\| =_{\text{def}} \|A\| \times \|B\|$$

$$\|A \rightarrow B\| =_{\text{def}} \|A\| \rightarrow \|B\|$$

---

[1] when the context is clear, we will use the same symbols for numbers and constants representing the numbers from inside the language

Under the above definition, it holds that:

$$\|\neg A\| = \neg\|A\|$$

$$\|L\| = \lambda w. \begin{cases} 1 & \text{if } w \models L \\ 0 & \text{otherwise} \end{cases} \quad \text{for any } \mathcal{L}_1\text{-formula } L$$

$$\|\alpha \oplus L\| = \lambda w. \begin{cases} 1 & \text{if } w \models L \\ \alpha & \text{otherwise} \end{cases} \quad \text{for any } \mathcal{L}_1\text{-formula } L$$

This third equivalence suggests a natural way of representing standard possibilistic logic inside our logic: consider that the least informative possibilistic function π satisfying the condition $N_\pi(X) \geq \alpha$ (where N_π is the necessity measure associated to π) is:

$$\pi = \lambda w. \begin{cases} 1 & \text{if } w \in X \\ 1 - \alpha & \text{otherwise} \end{cases}$$

So the token of information $N_\pi(X) \geq \alpha$ can be represented here by $\beta \oplus L$, where $\beta = 1 - \alpha$ and $X = \{w \mid w \models L\}$. Given a formula L and a distribution π, we will call *necessity* of L computed with π any number α such that $N_\pi(L) \geq \alpha$. This can be represented from inside the logic using the formula $A = (1 - \alpha) \oplus L$. We will call the value $(1 - \alpha)$ the *degree* of L in the formula A. The same token of information is represented by the couple (L, α) in [DLP94]. We now state the following:

Fact 3. The set $\mathcal{B} = \{\pi \in \mathcal{P} \mid \text{ for all } w \in \mathcal{W}, \ \pi(w) \in \{0, 1\}\}$ is a Boolean algebra contained in the structure $\mathcal{P}$.

Note that $\mathcal{L}_1$-formulae are mapped in the Boolean algebra, so it is natural to expect that they provide an exact copy of first-order logic embedded in the language $\mathcal{L}$. We introduce the notions of *local validity* (Def. 4) and of *semantic entailment* (Def. 5) as follows:

Definition 4. For any $\pi \in \mathcal{P}$ and for any closed formula A, we say that π *forces* A (written: $\pi \Vdash A$) iff $\pi \leq \|A\|$.

Definition 5. For any closed formulae A and B, we define: $A \models B$ iff, for any $\pi \in \mathcal{P}$, $\pi \Vdash A$ implies $\pi \Vdash B$

Since in this context the notion of *forcing* is linked to the order on possibility distributions, we also have (this is the form we will most frequently use):

$$A \models B \text{ iff } \|A\| \leq \|B\|$$

The notion of local validity defined by the forcing relation corresponds to the notion of validity in [DLP94].

3.2 The Proof System of LPL

We present the calculus in a sequent form, since we are dealing with a connective ($\otimes$, corresponding to a T-norm) which in general is not idempotent. The calculus will therefore lack the structural rule of contraction; in fact a weak form of contraction is allowed, which only can be applied to the sublanguage $\mathcal{L}_1$. Note that the calculus is relative to the particular choice of the T-norm, and validity and completeness theorem is parametric in the T-norm.

1. *Structural rules:*

$$\text{id)} \quad A \vdash A \qquad\qquad \text{cut)} \quad \frac{\Gamma \vdash B \quad \Delta, B \vdash C}{\Delta, \Gamma \vdash C}$$

$$\text{ex L)} \quad \frac{\Gamma, B, A, \Delta \vdash C}{\Gamma, A, B, \Delta \vdash C} \qquad \text{we L)} \quad \frac{\Gamma \vdash B}{\Gamma, A \vdash B}$$

$$\mathcal{L}_1 \text{ con)} \quad \frac{\Gamma, A, L \vdash B \quad A \vdash L}{\Gamma, A \vdash B} \qquad \text{for any } \mathcal{L}_1 \text{ formula } L$$

2. *Logical rules:*

$$\&) \quad \frac{\Gamma, A \vdash C}{\Gamma, A \& B \vdash C} \quad \frac{\Gamma, B \vdash C}{\Gamma, A \& B \vdash C} \qquad\qquad \frac{\Gamma \vdash A \quad \Gamma \vdash B}{\Gamma \vdash A \& B}$$

$$\otimes) \quad \frac{\Gamma, A, B \vdash C}{\Gamma, A \otimes B \vdash C} \qquad\qquad \frac{\Gamma \vdash A \quad \Delta \vdash B}{\Gamma, \Delta \vdash A \otimes B}$$

$$\oplus) \quad \frac{\Gamma, A \vdash C \quad \Gamma, B \vdash C}{\Gamma, B \oplus A \vdash C} \qquad\qquad \frac{\Gamma \vdash A}{\Gamma \vdash A \oplus B} \quad \frac{\Gamma \vdash B}{\Gamma \vdash A \oplus B}$$

$$\to) \quad \frac{\Gamma \vdash A \quad \Delta, B \vdash C}{\Gamma, \Delta, A \to B \vdash C} \qquad\qquad \frac{\Gamma, A \vdash B}{\Gamma \vdash A \to B}$$

$$1) \quad \frac{\Gamma \vdash A}{\Gamma, 1 \vdash A} \qquad\qquad\qquad \Gamma \vdash 1$$

$$0) \quad \Gamma, 0 \vdash A \qquad \neg\neg) \quad \neg\neg L \vdash L \quad \text{if } L \in \mathcal{L}_1$$

3. *Distributivities and Numerical rules:*
   $\otimes$-& distr) $(A \otimes C) \,\&\, (B \otimes C) \vdash (A \,\&\, B) \otimes C$
   $\oplus$-& distr) $(A \oplus C) \,\&\, (B \oplus C) \vdash (A \,\&\, B) \oplus C$

   $\qquad\qquad$ S') $\beta \vdash \alpha$ $\qquad\qquad\qquad\qquad$ for any $\beta \leq \alpha$
   $\qquad$ $\otimes$ def) $\alpha \otimes \beta \dashv\vdash \gamma$ $\qquad\qquad$ where $\gamma = \alpha \times \beta$
   $\qquad\qquad$ $\neg$ def) $\neg\alpha \dashv\vdash \gamma$ $\qquad\qquad$ where $\gamma = \alpha \to 0$

In [BS96] we give a semantics and proof system also for the predicative part of the language.

The validity and completeness theorem has the following form:

Theorem 6. *The proof system of LPL is valid and complete with respect to the given semantics, i.e., for any closed multiset Γ and any formula B, the sequent $\Gamma \vdash B$ is proved iff $\bigotimes \Gamma \models B$.*

4 A Syntax for Data Fusion

To see how the proof system of LPL can be used for applicative purposes, we first introduce a "canonical form" theorem, which provides a normal form to the formulae of the language of LPL:

Theorem 7. *Any formula A is provably equivalent in the calculus of LPL to an &-formula, i.e. a formula* $A' = \&_{i \in I}(\alpha_i \oplus L_i)$ *where* L_i *are* $\mathcal{L}_1$*-formulae.*

The normal form looks particularly attractive if we remember that the formula $\alpha \oplus L$ corresponds to the semantical constraint $N(L) \geq 1 - \alpha$, which uniquely determines a possibility distribution. Remembering that the necessity measure is compositional (only) with respect to the & connective, i.e. $N(A\&B) = min\{N(A), N(B)\}$, the normal form theorem gives us a method to compute the distribution implicitly defined by a sentence of the language.

The proof of the normal form theorem is completely constructive (see [BS96]), hence it gives an effective algorithm, defined by induction on the length of the formula, to compute the implicit possibilistic distribution. Let D be a formula and n its length (remember that $A \dashv\vdash B$ means that A and B are logically equivalent):

1. n=1
   - $D = \alpha \dashv\vdash \alpha \oplus 0$ (we remind the reader that $0 \in \mathcal{L}_1$).
   - $D = P \dashv\vdash 0 \oplus P$.
2. n+1
   Let $A \dashv\vdash \&_{i \in I}(\alpha_i \oplus A_i)$ and $B \dashv\vdash \&_{j \in J}(\beta_j \oplus A_j)$, and $H = I \cup J$ (where I and J are supposed to be disjoint sets). For any $K \subseteq H$, let

$$\alpha_K = \begin{cases} \bigwedge_{i \in I \setminus K} \alpha_i & \text{if } I \setminus K \neq \emptyset \\ 1 & \text{otherwise} \end{cases}$$

$$\beta_K = \begin{cases} \bigwedge_{j \in J \setminus K} \beta_j & \text{if } J \setminus K \neq \emptyset \\ 1 & \text{otherwise} \end{cases}$$

$$C_K = \neg \left(\&_{k \in K} A_k \,\&\, \&_{k \in H \setminus K} \neg A_k \right)$$

then
   - $D = A \oplus B \dashv\vdash \&_{K \subseteq H}(\alpha_K \oplus \beta_K) \oplus C_K$
   - $D = A \to B \dashv\vdash \&_{K \subseteq H}(\alpha_K \to \beta_K) \oplus C_K$
   - $D = A \otimes B \dashv\vdash \&_{i \in I, j \in J}(\alpha_i \oplus A_i) \,\&\, (\beta_j \oplus A_j) \,\&\, ((\alpha_i \otimes \beta_j) \oplus (A_i \oplus A_j))$
   - $D = \forall x A(x) \dashv\vdash \&_{i \in I} \alpha_i \oplus \forall x A_i(x)$

If $\Gamma = \{A_1, ..., A_n\}$ is a context of a sequent, then it is equivalent to the formula $A_1 \otimes ... \otimes A_n$, hence to every context of a sequent it is associated a unique possibility distribution that we will call π_Γ.

Now we can describe how this machinery can be used to represent and fuse, from inside the logical system, real-world, i.e. affected by uncertainty, and abstract, i.e. symbolic, information:

- code real-world information using a suitable set of graded formulae $\{A_1, ..., A_n\}$, i.e. where numerical constant appears.
- code symbolic knowledge using the classical subsystem, i.e. formulae where no numerical constants appear,

– fuse them using the rules for data fusion as presented in [DP94], obtaining a multiset of formulae Γ.

If the above operations are adequate, then π_Γ is an approximation of the "real" and unknown distribution $\bar{\pi}$, i.e. $\pi_\Gamma \geq \bar{\pi}$.

Now the problem is: how can we use this information? The sequent calculus gives an answer to this question, as it follows from the following facts:

– if there is a proof of $\Gamma \vdash \alpha \oplus A$ then, from the validity of the proof system, we have that $\pi_\Gamma \Vdash \alpha \oplus A$ and hence $N_{\pi_\Gamma}(A) \geq 1 - \alpha$;
– every proof tree of $\Gamma \vdash \alpha \oplus A$ terminates with two types of leaves:
  1. purely logical, i.e. of the form $L \vdash L$
  2. purely numerical, i.e. of the form $\beta \vdash \alpha$
  The purely numerical leaves give the constraints which α must satisfy: $\beta \vdash \alpha$ is an axiom if and only if $\beta \leq \alpha$. Hence we can use the sequent calculus to determine the necessity value of a formula A: start searching a proof of $x \oplus A$ with x undetermined, and if a proof is found, it gives the constraints sufficient to determine the value of x.
– if there is α such that $\pi_\Gamma \Vdash \alpha \oplus A$ then, due to the completeness property, there is a proof such that its constraints determine β with the property $\beta \geq \alpha$;
– note that Γ can contain abstract or purely symbolic knowledge, i.e. formulae where no numerical constants appear.

5 A Localization Experiment

We have developed an experimental testbed in which a team of two robots [BT96], a larger navigating "master" robot and a smaller "slave" robot, mutually assist each other towards a number of cooperative tasks: localization, map learning and updating, and navigation to pre-defined operating points. The slave robot carries a beacon (a point-like lamp), and the master a beacon-tracker (a video camera mounted on a pan-tilt unit). Typically, the slave follows the master and acts as a local portable landmark, by waiting at strategic points in the navigation environment. We assume a certain amount of a priori knowledge, roughly corresponding to the blueprint of the building, but no details are given about furniture and other non-structural elements.

In this experiment we use two types of sensors: a sonar belt and the beacon tracker. The first one consists of twelve sonars arranged in a ring and it is used to detect front or back occlusions and to compute, through a line-fitting algorithm, the presence of walls on the robot's sides. Wall detection is done using the three sonars on each lateral side. The last 10 significant readings are collected (readings from positions too close to each other are discarded) and passed, separately for each sonar, to a least-squares line-fitting algorithm, which returns 3 lines, each with a dispersion value Δ. The best-fitting line is chosen and its Δ is normalized to $[0, 1]$:

$$\delta_w = min\left(\frac{\Delta}{\Delta_{max}}, 1\right)$$

where Δ_{max} is a threshold value for dispersion, beyond which the points certainly do not represent a wall.

The second sensor tracks the beacon light continuously while the robot moves. Various techniques can be adopted to exploit this facility for localization. A very basic method is to compute the area, i.e. the number of pixels, of the lamp spot in the image. This value can give us a rough measure of the distance of the robot with respect to the light beacon, since the area decreases with the distance.

The scene used in the experiment is an indoor environment, as shown in Fig. 1a, consisting of two connected rooms, r_1 and r_2. Six "relevant places", denoted in the figure as $p_1, \ldots, p_6$, have been defined. The star marks the position of the slave robot holding the beacon.

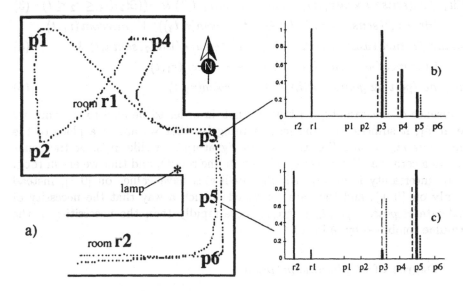

Fig. 1. The rooms with the relevant places (a), and the necessity values computed at p_3 (b) and p_5 (c)

Each place, with the exception of p_6, is labeled with a sensory signature, the light spot size, collected off line by manually positioning the robot at each place. At place p_6 the beacon cannot be seen.

No exact metrics are available, all we know about the environment is the set of places with their signatures.

The dotted line in Fig. 1a shows a path followed by the robot during a run, starting from r_2, going through p_5, p_3, p_1, p_2, p_4, p_3, p_5, p_6, and back to r_2. During the motion, the robot's heading, the twelve sonar readings and the area of the light spot were collected and saved in a file at a sampling interval of 10 cm approximately. These saved values are used by the localization system to compute, at each point, the necessity values for the set of places.

6 Sensor Fusion

The preliminary task of our system is to use real sensor data to obtain degrees of reliability for atomic predicates. In the following we will show some of the sequents used by the robot for self-localization with respect to a place in the environment, say p_1, and the ways to combine those sequents, using the sensor data, to obtain the necessity for the robot to be at place p_1 at time t. This will be denoted by predicate $at\_place(p_1, t)$.

$$at\_place\_type(pt_a, t) \ \& \ in\_room(r_1, t) \vdash at\_place(p_1, t) \tag{1}$$

$$(sonar\_feature(corner, north, west, t) \ \& \ at(corner)) \ \oplus \tag{2}$$
$$visual\_signature(s_a, t) \vdash at\_place\_type(pt_a, t)$$

$$(\exists t_1 \leq t)(sense\_room(r_1, t_1) \oplus enter\_room(r_1, t_1)) \ \& \ \neg((\exists t_2)(t_1 \leq t_2 \leq t) \tag{3}$$
$$(\exists r \neq r_1)(sense\_room(r, t_2) \oplus enter\_room(r, t_2))) \vdash in\_room(r_1, t)$$

$$sonar\_feature(wall_1, north, t) \ \& \ sonar\_feature(wall_2, south, t) \ \& \tag{4}$$
$$(distance(wall_1, wall_2) > 3) \vdash sense\_room(r_1, t)$$

$$at\_place(p_5, t) \ \& \ going(north) \vdash enter\_room(r_1, t) \tag{5}$$

In absence of uncertainty, (1) tells us that, in order to be able to say that we are at p_1, it suffices that the sensors recognize that we are at a place of the same type pt_a as p_1's (i.e. the camera signature s_a is like p_1's, or the sonar detects a corner at North-West and close to the robot), and that we are in room r_1. If uncertainty is introduced, the predicates have values on $[0, 1]$, instead of only on $\{0, 1\}$, and they are combined in such a way that the necessity of predicate $at\_place\_type(pt_a, t)$ grows more rapidly than the necessities of the formulae combined by $\oplus$ in (3).

6.1 Degree of $at\_place\_type(pt_a, t)$

Let
$$L = sonar\_feature(corner, north, west, t) \ \& \ at(corner)$$
$$M = visual\_signature(s_a, t)$$

and assume that we have computed the degree γ of L and the degree δ of M and we want to calculate the degree of $A = at\_place\_type(pt_a, t)$. We can assume that the two sources, the sonar and the camera, are independent sources of information; hence we can fuse the two observations in a single formula $(\gamma \oplus L) \otimes (\delta \oplus M)$.

Using the proof system of LPL [BS96], we can prove the following sequent:

$$(L \oplus M) \rightarrow A, (\gamma \oplus L) \otimes (\delta \oplus M) \vdash (\gamma \otimes \delta) \oplus A$$

From this and the following proof:

$$\frac{\dfrac{\vdash (L \oplus M) \rightarrow A \quad (L \oplus M) \rightarrow A, (\gamma \oplus L) \otimes (\delta \oplus M) \vdash (\gamma \otimes \delta) \oplus A}{\vdash (\gamma \oplus L) \otimes (\delta \oplus M) \quad (\gamma \oplus L) \otimes (\delta \oplus M) \vdash (\gamma \otimes \delta) \oplus A}}{\vdash (\gamma \otimes \delta) \oplus A}$$

we have that the degree of A, i.e. the degree of being at place type pt_a in the t-th observation, is $\gamma \otimes \delta$.

Note that the logic allows to perform some basic operations of data fusion:

1. use abstract knowledge, i.e. the classical formula $(L \oplus M) \to A$ in the cut-rule of the first line;
2. combine and use the observations affected by uncertainty. In fact the formula $(\gamma \oplus L) \otimes (\delta \oplus M)$ is the result of:
   - describe from inside the logic the uncertainty contained in the observations: this is done by writing the two formulae: $\gamma \oplus L$, $\delta \oplus M$,
   - fuse the above information using $\otimes$, because we assume the independence of the sources;

The computed necessity of A is higher than the necessities of L and M: this is due to the non-idempotency of the T-norm and consequently of the $\otimes$ connective. This is a crucial point that justifies the use of two different connectives (& and $\otimes$) for the "and".

6.2   Degrees of the Atomic Predicates

The degrees of the other formulae in the right part of the sequents (1)–(5) can be computed with the same technique, starting from the degrees of the atomic predicates, coded in the following axioms:

$$\vdash \beta_1 \oplus sonar\_feature(occlusion, direction, t)$$
$$\vdash \beta_2 \oplus sonar\_feature(wall, direction, t)$$
$$\vdash orthogonal(north, west)$$
$$\vdash orthogonal(west, north)$$
$$\vdash \beta_3 \oplus at(wall)$$
$$\vdash \beta_4 \oplus at(occlusion)$$
$$\vdash \beta_5 \oplus visual\_signature(s_a, t)$$

where:

$$\beta_1 = \begin{cases} 1 \text{ if } occlusion\_distance \leq 1 \text{ meter} \\ 0 \text{ if } occlusion\_distance > 1 \text{ meter} \end{cases}$$

$\beta_2 = \delta_w$, where δ_w is the dispersion computed as described in Sect. 5

$\beta_3 = min\,(wall\_distance/2, 1)$, where $wall\_distance$ is the distance from the robot of the interpolating line

$\beta_4 = min\,(occlusion\_distance/2, 1)$, where $occlusion\_distance$ is the distance from the robot of an obstacle detected by the sonar

$\beta_5 = \delta_c$, where δ_c is the degree of similarity between a prototype image signature of the light in p_1, called s_a, and the one detected by the camera during the experiment.

7 Experimental Results and Future Work

The places have been chosen so that some of them are not distinguishable by the sonar (places p_3 and p_4 in Fig. 1a), others by the camera (p_3 and p_5).

The data collected by moving the robot in the rooms were analyzed by an off-line program which, for each point in the path, computed the necessity of being at the relevant places.

Two types of T-norms were tested, Łukasiewicz ($\alpha \times_L \beta = max(\alpha + \beta - 1, 0)$) and product. In our experiments, Łukasiewicz T-norm performs better than the product, because it gives much more emphasis to the fusion process, due to the following inequality: $\alpha \times_L \beta \leq \alpha \cdot \beta$

The results are encouraging, in the sense that the system is able to identify correctly the places while the robot is navigating along the path: this means that in the relevant places the computed necessity values reach 1.

In particular, they show that through sensor fusion it is possible to choose the correct place among different relevant places which have the same sensor signature.

For example, when the robot is at p_3, the sonar information allows to see a wall on the robot's right side and an occlusion in front of the robot. As already said, this sonar signature is common both to place p_3 and place p_4 and it is not possible to distinguish between these two places using sonar data only; however, if we combine this information with the information coming from the video sensor, the system is able to identify place p_3 correctly, as shown in Fig. 1b, which plots the necessity values for each place when the robot is at p_3: the greater the necessity, the more likely is the robot to be at the corresponding place. For rooms, the solid line shows the necessity of being in the corresponding room: in the example of Fig. 1b, the robot has previously identified room r_1. For the places, the solid line corresponds to the resulting global necessity, the dotted line to the necessity due to the vision sensor, and the dash-dotted line to the necessity computed from the sonars. In this case, places p_3 and p_4 have the same necessity values according to the sonars, but, thanks to the fusion with video data, p_3 has the highest resulting necessity.

In Fig. 1c we show the necessity values at p_5 with the robot heading south. Here we can see that when the robot recognizes p_5, the necessity of being in room r_2 grows to 1, while the necessity of being in r_1 gets much smaller. Therefore, the necessity of being at p_3, which would be 0.7 if only the place type were considered, is reduced to 0.1 (this is shown by the short continuous line partially covering a dotted line at p_3 in Fig. 1c).

Only at one point, out of 400, the system was misled by a spurious sonar reading and produced a false positive, i.e. a necessity value of 1 when the robot was clearly far from any relevant place.

The robustness can be increased by using the camera in a less basic way and by adding further sensor information, such as the odometry, or logical information, e.g. topological constraints coming from the relative positions of the relevant places.

8 Acknowledgments

We are indebted to Philippe Smets and Alessandro Saffiotti of Iridia, Didier Dubois, Henry Prade, Jérôme Lang and Salem Benferhat of IRIT-CNRS. This work was partially supported by a CNR-CNRS joint project (CNR Code 132.3.1) and the Fusion Project of the European Community.

References

[Be85]   Bell J.L.: Boolean-valued models and independence proofs in set theory. Oxford University Press, Oxford (1985)

[BT96]   Bison P., Trainito G.: A Robot Duo for Cooperative Autonomous Navigation. In H. Asama, T. Fukuda, T. Arai and I. Endo (Eds.). Distributed Autonomous Robotic Systems 2, Springer Verlag, Saitama, Japan (1996)

[BS96]   Boldrin L., Sossai C.: Local Possibilistic Logic. Journal of Applied Non-Classical Logic (to appear)
          (http://www.ladseb.pd.cnr.it/infor/papers/jancl_BoSo.pdf)

[DLP91]   Dubois D., Lang J., Prade H.: Fuzzy sets in approximate reasoning, Part 2: Logical approaches. Fuzzy Sets and Systems 40 (1991) 203–244

[DLP94]   Dubois D., Lang J., Prade H.: Possibilistic logic. In: D. Gabbay, C. Hogger and J. Robinson (eds.). Handbook of Logic in Artificial Intelligence and Logic Programming, vol. 3, Clarendon Press (1994)

[DP94]   Dubois D. , Prade H.: Possibility theory and data fusion in poorly informed environments. Control Eng. Practice 2:5 (1994) 811–823

[Gi87]   Girard J.Y. : Linear logic. Theoretical computer science 50 (1987) 1–101

[HHV95]   Hajek P., Harmancova D., Verbrugge R.: A Qualitative Fuzzy Possibilistic Logic. International Journal of Approximate Reasoning 12 (1995) 1–19

[MM92]   Mac Lane S., Moerdijk I.: Sheaves in geometry and logic: a first introduction to topos theory. Springer-Verlag, New York (1992)

[Pa79]   Pavelka L.: On fuzzy logic I, II, III. Zeitschr. f. math. Logik und Grundlagen d. Math 25 (1979) 45–52; 119–131; 447–464

[TZ71]   Takeuti G., Zaring W.M.: Introduction to Axiomatic Set Theory. Graduate Texts in Mathematics 1, Springer-Verlag, New York (1971)

Some Experimental Results on Learning Probabilistic and Possibilistic Networks with Different Evaluation Measures

Christian Borgelt and Rudolf Kruse

Department of Information and Communication Systems
Otto-von-Guericke-University of Magdeburg
D-39106 Magdeburg, Germany

e-mail: borgelt@iik.cs.uni-magdeburg.de

Abstract. A large part of recent research on probabilistic and possibilistic inference networks has been devoted to learning them from data. In this paper we discuss two search methods and several evaluation measures usable for this task. We consider a scheme for evaluating induced networks and present experimental results obtained from an application of INES (Induction of NEtwork Structures), a prototype implementation of the described methods and measures.

1 Introduction

Since reasoning in multi-dimensional domains tends to be infeasible in the domains as a whole — and the more so, if uncertainty is involved — decomposition techniques, that reduce the reasoning process to computations in lower-dimensional subspaces, have become very popular. For example, decomposition based on dependence and independence relations between variables has extensively been studied in the field of graphical modeling [16]. Some of the best-known approaches are Bayesian networks [22], Markov networks [19], and the more general valuation-based networks [28]. They all led to the development of efficient implementations, for example HUGIN [1], PULCINELLA [27], PATHFINDER [11] and POSSINFER [7].

A large part of recent research has been devoted to learning such inference networks from data [4, 12, 8]. In this paper we examine two search methods and several evaluation measures usable for this task. We consider how to evaluate an induced network and present some experimental results we obtained from an application of INES (Induction of NEtwork Structures), a prototype implementation of the described search methods and evaluation measures.

2 Probabilistic and Possibilistic Networks

The basic idea underlying probabilistic as well as possibilistic networks is that under certain conditions a multi-dimensional distribution can be decomposed without much loss of information into a set of (overlapping) lower-dimensional distributions. This set of lower-dimensional distributions is usually represented as a hypergraph, in which there is a node for each attribute and a hyperedge for each distribution of the decomposition. To each node and to each hyperedge a projection of the multi-dimensional distribution (a *marginal distribution*) is assigned: to the node a projection to its attribute and to a hypergraph a projection to the set of attributes connected by it. Thus hyperedges represent direct influences the connected attributes have on each other, i.e. how constraints on the value of one attribute affect the probabilities or possibilities of the values of the other attributes in the hyperedge.

Reasoning in such a hypergraph is done by propagating evidence, i.e. observed constraints on the possible values of a subset of all attributes, along the hyperedges. This can be done with *local computations*, usually restricted to a single hyperedge, if certain axioms are fulfilled [28].

Probability theory allows for local computations that are especially simple. Evidence entered into a node is first *extended* to the hyperedge along which it is to be propagated by multiplying the joint probability distribution associated with the hyperedge with the quotients of the posterior and prior probability of the values of the node. Then it can be *projected* to any other node contained in the hyperedge by simply summing out the other attributes (computing the new marginal distribution). A similar scheme can be derived for networks with directed edges to which a conditional probability distribution is assigned.

Possibilistic networks can be based on an interpretation of a degree of possibility that rests on the context model [6, 17]. In this model possibility distributions are interpreted as information-compressed representations of (not necessarily nested) random sets, a degree of possibility as the one-point coverage of a random set [21]. This interpretation allows to construct possibilistic networks in analogy to probabilistic networks. Only the propagation functions have to be replaced, namely the product (for extension) by the minimum and the sum (for projection) by the maximum.

Both types of networks, probabilistic as well as possibilistic, can be induced automatically from data. An algorithm for this task consists always of two parts: an evaluation measure and a search method. The evaluation measure estimates the quality of a given decomposition (a given hypergraph) and the search method determines which decompositions (which hypergraphs) are inspected. Often the search is guided by the value of the evaluation measure, since it is usually the goal to maximize or to minimize its value. In the following two sections we describe two search methods and several evaluation measures.

3 Search Methods

There is a large variety of search methods usable for learning inference networks. In principle any general heuristic search method is applicable, like hill climbing, simulated annealing, genetic algorithms etc. But to keep things simple, since our emphasis is on evaluation measures, we consider only two methods: optimum weight spanning tree construction and greedy parent selection.

The construction of an optimum weight spanning tree was suggested already in [3]. An evaluation measure (in [3]: mutual information) is computed on all possible edges (two-dimensional subspaces) and then the Kruskal algorithm is applied to determine a maximum or minimum weight spanning tree.

Greedy parent selection is used in the K2 algorithm described in [4]. To narrow the search space and to avoid cycles in the resulting hypergraph a topological order of the attributes is defined. A topological order of the nodes of a directed graph satisfies: If there is a directed edge from a node A to a node B, then A precedes B in the order. Fixing a topological order restricts the permissible graph structures, since the parents of an attribute can be selected only from the attributes preceding it in the order. A topological order can either be stated by a domain expert or derived automatically [30].

Parent attributes are selected using a greedy search. At first an evaluation measure (in [4]: the g-function) is calculated for the child attribute alone, or — more precisely — for the hyperedge consisting only of the child attribute. Then in turn each of the parent candidates (the attributes preceding the child in the topological order) is temporarily added to the hyperedge and the evaluation measure is computed. The parent candidate yielding the highest value of the evaluation measure is selected as a first parent and is permanently added to the hyperedge. In the third step all remaining candidates are added temporarily as a second parent and again the evaluation measure is computed for each of the resulting hyperedges. As before, the parent candidate yielding the highest value is permanently added to the hyperedge. The process stops, if either no more parent candidates are available, a given maximal number of parents is reached or none of the parent candidates, if added to the hyperedge, yields a value of the evaluation measure exceeding the best value of the preceding step. The resulting hypergraph contains for each attribute a (directed) hyperedge connecting it to its parents (provided parents where added).

4 Evaluation Measures

In this section we review some evaluation functions that can be used for learning inference networks from data. All of them estimate the quality of single hyperedges and are based on the empirical probability or possibility distributions found in the database: If N is the number of tuples in the database and N_i the number of tuples in which attribute A has value a_i, then $P(a_i) = \frac{N_i}{N}$.

4.1 Measures for Learning Probabilistic Networks

The basic idea of several evaluation measures used for learning probabilistic networks is to compare the joint distribution with the product of the marginal distributions. This seems to be reasonable, since the more these two distributions differ, the more dependent the attributes are on each other. Other approaches include Bayesian estimation and minimization of description length.

The χ^2-Measure The χ^2-measure directly implements the idea to compare the joint distribution and the product of the marginal distributions by computing their squared difference. For two attributes A and B it is defined as

$$\chi^2 = \sum_{i,j} N \frac{(P(a_i)P(b_j) - P(a_i, b_j))^2}{P(a_i)P(b_j)},$$

where N is the number of tuples in the database to learn from.

This version of the χ^2-measure is sufficient, if the optimum weight spanning tree method is used, since then only two-dimensional edges have to be evaluated. But for learning hypergraphs, e.g. with the greedy parent search method, we need an extension to more than two attributes. Such an extension can be obtained in two ways, the first of which is to define for m attributes $A^{(1)}, \ldots, A^{(m)}$

$$\chi_1^2 = \sum_{i_1,\ldots,i_m} N \frac{\left(\prod_{k=1}^m P(a_{i_k}^{(k)}) - P(a_{i_1}^{(1)}, \ldots, a_{i_m}^{(m)})\right)^2}{\prod_{k=1}^m P(a_{i_k}^{(k)})},$$

i.e. to compare the joint probability with the product of the single attribute marginal probabilities. The second extension is especially suited for learning directed hyperedges and consists simply in viewing the (candidate) parent attributes as one pseudo-attribute, i.e. if $A^{(1)}, \ldots, A^{(m-1)}$ are the (candidate) parents of attribute $A^{(m)}$, then

$$\chi_2^2 = \sum_{i_1,\ldots,i_m} N \frac{\left(P(a_{i_1}^{(1)}, \ldots, a_{i_{m-1}}^{(m-1)})P(a_{i_m}^{(m)}) - P(a_{i_1}^{(1)}, \ldots, a_{i_m}^{(m)})\right)^2}{P(a_{i_1}^{(1)}, \ldots, a_{i_{m-1}}^{(m-1)})P(a_{i_m}^{(m)})}.$$

If not stated otherwise, all measures described for two attributes in the following can be extended in these two ways.

Entropy-based Measures In [3] the (two-dimensional) edges of a tree-decomposition of a multi-dimensional distribution are selected with the aid of *mutual information*. Under the name of *information gain* this measure was later used for the induction of decision trees [23, 24], which is closely related to learning inference networks (with directed edges): A hyperedge consisting of an attribute and its parents can be seen as a decision tree with the restriction that all leaves have to lie on the same level and all decisions in the same level of the tree have to be made on the same attribute.

Mutual information implements the idea to compare the joint distribution and the product of the marginal distributions by computing the logarithm of their quotient. For two attributes A and B mutual information is defined as

$$I_{\text{mutual}} = \sum_{i,j} P(a_i, b_j) \log_2 \frac{P(a_i, b_j)}{P(a_i)P(b_j)} = H_A + H_B - H_{AB} = I_{\text{gain}},$$

where H is the Shannon entropy [29]. It can be shown, that mutual information is always greater or equal to zero, and equal to zero, if and only if the joint distribution and the product of the marginal distributions coincide [18]. Hence it can be seen as measuring the difference between the two distributions. In the interpretation as information gain, it measures the information (in bits) gained about the value of one attribute from the knowledge of the value of the other attribute.

When using information gain for decision tree induction, it was discovered that information gain is biased towards many-valued attributes. To adjust for this bias the *information gain ratio* was introduced, which is defined as the information gain divided by the entropy of the split attribute [23, 24]:

$$I_{gr} = \frac{I_{gain}}{H_A} = \frac{I_{gain}}{-\sum_i P(a_i) \log_2 P(a_i)}.$$

Transferred to learning inference networks this means to divide the information gain by the entropy of the parent attributes. (Obviously this is only applicable when directed edges are used. Otherwise there would be no "split attribute" in contrast to the "class attribute.") In the two extensions to more than two attributes either the sum of the entropies of the marginal distributions of the parent attributes (first version) or the entropy of the marginal distribution of the pseudo-attribute formed from all the parent attributes (second version) forms the denominator.

An alternative is the *symmetric information gain ratio* defined in [20], which is the information gain divided by the entropy of the joint distribution:

$$I_{sgr}^{(1)} = \frac{I_{gain}}{H_{AB}} = \frac{I_{gain}}{-\sum_{i,j} P(a_i, b_j) \log_2 P(a_i, b_j)}.$$

Because of its symmetry this measure is also applicable for undirected edges. Another symmetric version that suggests itself is to divide by the entropy sum of the single attribute distributions, i.e.

$$I_{sgr}^{(2)} = \frac{I_{gain}}{H_A + H_B} = \frac{I_{gain}}{-\sum_i P(a_i) \log_2 P(a_i) - \sum_j P(b_j) \log_2 P(b_j)}.$$

It is easy to see that this measure leads to the same edge selections as the previous one. Nevertheless it is useful to consider both measures, since their effects can differ, if weighting is used (see section 5).

The measures discussed above are all based on Shannon entropy, which can be seen as a special case (for $\beta \to 1$) of *generalized entropy* [5]:

$$H^\beta(p_1, \ldots, p_r) = \sum_{i=1}^r p_i \frac{2^{\beta-1}}{2^{\beta-1} - 1} (1 - p_i^{\beta-1})$$

Setting $\beta = 2$ yields the *quadratic entropy*

$$H^2(p_1, \ldots, p_r) = \sum_{i=1}^r 2p_i(1 - p_i) = 2 - 2 \sum_{i=1}^r p_i^2.$$

Using it in a similar way as Shannon entropy leads to the so-called Gini index:

$$\text{Gini} = \frac{1}{2}(H_A^2 - H_{A|B}^2) = \sum_{j=1}^{n_B} P(b_j) \sum_{i=1}^{n_A} P(a_i|b_j)^2 - \sum_{i=1}^{n_A} P(a_i)^2,$$

a well known measure for decision tree induction [2, 31]. Here only the second type of extension to more than two attributes is applicable. A symmetric ratio can be derived, but only for two attributes:

$$\text{Gini}_{\text{sym}} = \frac{H_A^2 - H_{A|B}^2 + H_B^2 - H_{B|A}^2}{H_A^2 + H_B^2}.$$

Bayesian Measures In the K2 algorithm [4] as an evaluation measure the g-function is used, which is defined as

$$g(A, \text{par}_A) = c \cdot \prod_{j=1}^{n_{\text{par}_A}} \frac{(n_A - 1)!}{(N_{.j} + n_A - 1)!} \prod_{i=1}^{n_A} N_{ij}!,$$

where A is an attribute and par_A the set of its parents. n_{par_A} is the number of distinct instantiations (value vectors) of the parent attributes that occur in the database to learn from and n_A the number of values of attribute A. N_{ij} is the number of cases (tuples) in the database in which attribute A has the ith value *and* the parent attributes are instantiated with the jth value vector, $N_{.j}$ the number of cases in which the parent attributes are instantiated with the jth value vector, that is $N_{.j} = \sum_{i=1}^{n_A} N_{ij}$. c is a constant prior probability, which can be set to 1, since usually only the relation between the values of the evaluation measure for different sets of parent attributes matters.

The g-function estimates (for a certain value of c) the probability of finding the joint distribution of the attribute and its parents that is present in the database. That is, assuming that all network structures are equally likely, and that, given a certain structure, all conditional probability distributions compatible with the structure are equally likely, it uses Bayesian reasoning to compute the probability of the network structure given the database from the probability of the database given the network structure.

MDL-based Measures Information gain can also be seen as measuring the reduction in the description length of a dataset, if the values of a set of attributes are encoded together (one symbol per tuple) instead of separately (one symbol per value). The *minimum description length principle* [26] in addition takes into account the information needed to transmit the coding scheme, thus adding a "penalty" for making the model more complex by enlarging a hyperedge. We consider the two types of minimum description length functions stated in [15] for decision tree induction.

Coding based on relative frequencies:

$$L_{\text{gain}}^{(1)} = \qquad \log_2 \frac{(N_{..} + n_A - 1)!}{N_{..}!(n_A - 1)!} + N_{..}H_A \qquad\qquad L_{\text{prior}}^{(1)}$$

$$- \sum_{j=1}^{n_B} \left(\log_2 \frac{(N_{.j} + n_A - 1)!}{N_{.j}!(n_A - 1)!} + N_{.j}H_{A|b_j} \right) \qquad L_{\text{post}}^{(1)}$$

The first term in each line states the costs for transmitting the frequency distribution. Intuitively, this is done by transmitting the page number for a code book listing all possible distributions of N cases on n_A attribute values. The second term in each line describes the costs to actually transmit the value assignments.

Coding based on absolute frequencies:

$$L_{\text{gain}}^{(2)} = \qquad \log_2 \frac{(N_{..} + n_A - 1)!}{N_{..}!(n_A - 1)!} + \log_2 \frac{N_{..}!}{N_1!\cdots N_{n_A}!} \qquad L_{\text{prior}}^{(2)}$$

$$- \sum_{j=1}^{n_B} \left(\log_2 \frac{(N_{.j} + n_A - 1)!}{N_{.j}!(n_A - 1)!} + \log_2 \frac{N_{.j}!}{N_{1j}!\cdots N_{n_Aj}!} \right) \qquad L_{\text{post}}^{(2)}$$

Again the first term in each line describes the costs for transmitting the frequency distribution, the second term the costs for transmitting the value assignments. In this version the value assignments are also coded as a page number for a code book listing all possible assignments of values to cases for a given absolute frequency distribution. This measure is closely connected to the g-function described above. More precisely, it is $\log_2(g) = \log_2(c) + L_{\text{post}}^{(2)}$.

4.2 Measures for Learning Possibilistic Networks

In analogy to the probabilistic setting the idea of some of the measures for learning possibilistic networks is to compare the joint distribution with the minimum (instead of the product) of the marginal distributions. Other approaches are based on nonspecificity measures.

All measures described in this section are extended to more than two attributes in the first possible way and not by combining some of them into a pseudo-attribute. The reason is that in possibilistic networks edges are undirected, since it is difficult to define a conditional possibility distribution.

Comparison-based Measures For probabilistic networks both the χ^2-measure and mutual information compare directly the joint distribution and the product of the marginal distributions; the former by the difference, the latter by the quotient. Hence the idea suggests itself to apply the same scheme to possibilistic networks, replacing the product by the minimum and the sum by the maximum.

We thus obtain for two attributes A and B

$$d_{\chi^2} = \sum_{i,j} \frac{(\min(\pi(a_i), \pi(b_j)) - \pi(a_i, b_j))^2}{\min(\pi(a_i), \pi(b_j))}$$

as the analogon of the χ^2-measure, and

$$d_{\mathrm{mi}} = -\sum_{i,j} \pi(a_i, b_j) \log_2 \frac{\pi(a_i, b_j)}{\min(\pi(a_i), \pi(b_j))}$$

as the analogon of mutual information. Since both measures are always greater or equal to zero, and zero if and only if the two distributions coincide, they can be seen as measuring the difference between the two distributions. Just as for information gain it may be a good idea to divide d_{mi} by the sum of the logarithms of the possibility degrees of the joint distribution to remove (or at least to reduce) a possible bias.

Nonspecificity-based Measures A possibilistic evaluation measure can also be derived from the U-uncertainty measure of *nonspecificity* of a possibility distribution [14], which is defined as

$$\mathrm{nsp}(\pi) = \int_0^{\sup(\pi)} \log_2 |[\pi]_\alpha| d\alpha$$

and can be justified as a generalization of Hartley information [10] to the possibilistic setting [13]. $\mathrm{nsp}(\pi)$ reflects the expected amount of information (measured in bits) that has to be added in order to identify the actual value within the set $[\pi]_\alpha$ of alternatives, assuming a uniform distribution on the set $[0, \sup(\pi)]$ of possibilistic confidence levels α [9].

The role nonspecificity plays in possibility theory is similar to that of Shannon entropy in probability theory. Thus the idea suggests itself to construct an evaluation measure from nonspecificity in the same way as information gain and (symmetric) information gain ratio are constructed from Shannon entropy.

By analogy to information gain we define *specificity gain* as

$$S_{\mathrm{gain}} = \mathrm{nsp}(\pi_A) + \mathrm{nsp}(\pi_B) - \mathrm{nsp}(\pi_{AB}).$$

This measure is equivalent to the one defined in [9]. In addition, just like information gain ratio and symmetric information gain ratio, *specificity gain ratio*

$$S_{\mathrm{gr}} = \frac{S_{\mathrm{gain}}}{\mathrm{nsp}(\pi_A)} = \frac{\mathrm{nsp}(\pi_A) + \mathrm{nsp}(\pi_B) - \mathrm{nsp}(\pi_{AB})}{\mathrm{nsp}(\pi_A)}$$

and *symmetric specificity gain ratio* in either of the two forms

$$S_{\mathrm{sgr}}^{(1)} = \frac{S_{\mathrm{gain}}}{\mathrm{nsp}(\pi_{AB})} = \frac{\mathrm{nsp}(\pi_A) + \mathrm{nsp}(\pi_B) - \mathrm{nsp}(\pi_{AB})}{\mathrm{nsp}(\pi_{AB})}$$

or

$$S_{\mathrm{sgr}}^{(2)} = \frac{S_{\mathrm{gain}}}{\mathrm{nsp}(\pi_A) + \mathrm{nsp}(\pi_B)} = \frac{\mathrm{nsp}(\pi_A) + \mathrm{nsp}(\pi_B) - \mathrm{nsp}(\pi_{AB})}{\mathrm{nsp}(\pi_A) + \mathrm{nsp}(\pi_B)}$$

can be defined.

5 Missing Values

In real world databases often a substantial number of values is missing. This, of course, poses problems for any learning algorithm for probabilistic networks, because it is not completely clear how to handle missing values when evaluating edges. Thus, when constructing a learning algorithm, it is often required that there are no missing values. In [4] and [12] this constraint is stated explicitly. But it is obvious that such an assumption does not lead to much, because it severely restricts the domain of application of the algorithm. Missing values are just too frequent in the real world for such a requirement.

When trying to handle missing values, the idea that comes to mind first is to ignore tuples possessing one or more of them, thus enforcing the above assumption. In some cases, where the number of tuples with missing values is small, this may be sufficient, but often a substantial part of the database has to be discarded in this way. Other approaches include replacing a missing value by a new distinct element *unknown* added to the domain of the corresponding attribute, thus transforming it into a normal value, or imputing the most frequent, an average, or a random value. But such approaches can distort the frequency distribution present in the database and hence may either lead to spurious dependences between attributes or conceal existing dependences.

Therefore in our implementation we refrained from using either of the above approaches, but tried the following scheme: Since the evaluation measures we use are local measures, they require only part of the tuple to be known in order to be computable. Hence, when evaluating a hyperedge, we ignore only those tuples in the database, in which a value is missing in one of the attributes contained in the hyperedge. Other attributes may be known or unknown, we do not care. In this way at least part of the information contained in a tuple with missing values can be used.

If an edge is evaluated using the described scheme, the resulting value of the measure for different edges can refer to different numbers of tuples. Hence for some measures, e.g. the g-function, the χ^2-measure and the reduction of description length measures, it is necessary to normalize their value, such that it refers to single tuples. To achieve this, the measure is simply divided by the number of tuples used to compute it.

In addition one may consider weighting the worth of an edge with the fraction of the tuples it was calculated on. This idea stems from learning decision trees with an information gain measure [24], where based on such a weighting a possible split attribute yielding a smaller gain is sometimes preferred to a split attribute yielding a higher gain, if due to a low probability of knowing the latter attribute the *expected gain* is higher for the first. By similar reasoning such weighting can be justified for learning inference networks. An edge may represent a stronger dependence of the connected attributes, but due to missing values it is less likely that this dependence can be exploited for inferences. In such a situation it may be preferable to use an edge connecting attributes whose dependence is weaker, but can be exploited more often. Nevertheless, in our experiments we used unweighted measures, since weighting is not applicable to all measures and thus would have rendered some of the results incomparable.

For learning possibilistic networks, of course, missing values are no problem at all, since possibility theory was designed especially to handle such kind of uncertainty information. In the random set view, a missing value simply represents a random set containing all values of the underlying domain. Hence neither removing tuples nor additional calculations are necessary.

6 Evaluating Learned Networks

Learning probabilistic and possibilistic networks with local evaluation measures and search methods like greedy parent selection are heuristic methods. Hence it is not guaranteed that an optimal solution will be found. Thus the question arises: Can we find a way to assess a learned network, or at least a way to compare the quality of two networks?

For probabilistic networks a simple evaluation scheme can be derived from the idea underlying the Bayesian measure of the g-function described above. From a given network — dependence structure and (conditional) probabilities — the probability of each tuple in the database can be calculated. Multiplying these probabilities yields the probability of the database given the network structure, provided that the tuples are independent. If we assume all networks to have the same prior probability, the probability of the database given the network can be interpreted as a direct indication of the network quality. Although it is not an absolute measure, since we cannot determine an upper bound for this probability, networks can be compared with it.

The only problem with this method is the treatment of missing values, since for tuples with missing values no definite probability can be calculated. For our prototype program we decided on the following scheme: Every missing value of a tuple is instantiated in turn with each possible value, and for each resulting (completely known) tuple the probability is determined. Then the minimum, average, and maximum of these probabilities are computed. We thus arrive at a minimum, average, and maximum value for the probability of a database, of which the average may be the best to use. It is obvious that this method is applicable only, if the number of missing values per tuple is fairly small, since otherwise the number of tuples to be examined gets too large.

In theory such a global evaluation method can be used directly to learn a network. We only need to add a search method to traverse the space of possible solutions. Each candidate selected by the search method is then evaluated using the global evaluation function. But this is not very practical. The main reason is that evaluating a network in the way described can take fairly long, especially if there are missing values. If they abound, even a single network cannot be evaluated in reasonable time. Since during a search a large number of networks has to be inspected, the learning time can easily exceed reasonable limits. Nevertheless it may be worthwhile to examine such an approach.

We now turn to evaluating possibilistic networks. Unfortunately we cannot compute a degree of possibility for the whole database, but we can use a similar

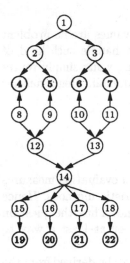

1 – parental error
2 – dam correct?
3 – sire correct?
4 – stated dam ph.gr. 1
5 – stated dam ph.gr. 2
6 – stated sire ph.gr. 1
7 – stated sire ph.gr. 2
8 – true dam ph.gr. 1
9 – true dam ph.gr. 2
10 – true sire ph.gr. 1
11 – true sire ph.gr. 2

12 – offspring ph.gr. 1
13 – offspring ph.gr. 2
14 – offspring genotype
15 – factor 40
16 – factor 41
17 – factor 42
18 – factor 43
19 – lysis 40
20 – lysis 41
21 – lysis 42
22 – lysis 43

The grey nodes correspond to observable attributes. Node 1 can be removed to simplify constructing the clique tree for propagation.

Fig. 1. Domain expert designed network for the Danish Jersey cattle blood type determination example

approach. From the propagation method of possibilistic networks it is obvious that the degree of possibility derivable from the network can only be greater or equal to the (true) degree of possibility derivable from the database. Hence, the better a network approximates the possibility distribution represented by the database, the smaller the sum of the possibility degrees over the tuples in the database should get. For tuples with missing values we use a similar approach as above. For each completely known tuple compatible with the tuple missing some values, the degree of possibility is determined from the network and the minimum, average, and maximum of these degrees is computed. Then these are summed for all tuples in the database. To be in accordance with the ideas underlying possibility theory, the maximum value may be the proper quality measure. If one commits to using the maximum, computation is significantly simplified, since a completely known tuple compatible with a tuple with missing values and having the maximum degree of possibility of all such tuples can easily be determined without inspecting all compatible tuples. Hence, with this restriction, a learning algorithm for possibilistic networks based on this global evaluation method may be a noteworthy alternative.

7 Experimental Results

The experiments described in this section were conducted with a prototype program called INES (Induction of NEtwork Structures). It contains the two search methods (optimum weight spanning tree construction and greedy parent selection) and all evaluation measures described above.

network	cond.	$\log_2(P_{\text{avg}})$
indep.	0	-11632
orig.	24	-7451
db. prob.	24	-5105
I_{gain}	21	-5221
$I_{\text{sgr}}^{(1)}$	21	-5254
Gini_{sym}	21	-5505
χ^2	21	-5221

network	cond.	$\log_2(P_{\text{avg}})$
$\log_2(g)/N$	24	-4620
I_{gain}	32	-4357
I_{gr}	20	-5243
$I_{\text{sgr}}^{(1)}$	24	-4736
Gini	32	-4357
χ^2/N	33	-4373
$L_{\text{gain}}^{(1)}/N$	24	-4704
$L_{\text{gain}}^{(2)}/N$	26	-4620

Table 1. Evaluation of probabilistic networks obtained by optimum weight spanning tree construction (left bottom) and by greedy parent selection (right) on the Danish Jersey cattle data. As a reference point evaluations of a network with independent nodes, of the original expert designed network, and of the expert designed network with probabilities determined from the database (left top) are added.

Although we tested INES on several databases from the UCI machine learning repository (e.g. flags, solar flare, mushroom, vote etc.), we chose to present here the results obtained on the Danish Jersey cattle blood type determination example [25], because it has the advantage that there is a probabilistic network designed by domain experts (see figure 1), which can be used as a baseline for result evaluation. Results on other datasets do not differ significantly.

The Danish Jersey cattle blood type determination example consists of the domain expert designed network, whose structure is shown in figure 1, and a database containing 500 tuples over the twenty-two attributes of the network. Only eight of the attributes, those shaded in the network, are actually observable. Several tuples of the database contain missing values.

As a baseline for comparisons we first evaluated a network without any edges (isolated nodes), the domain expert designed network, and the domain expert designed network with probabilities determined from the database. Their evaluation shows that the database seems to be a little distorted and not really fitting the domain expert designed model, since the evaluation of the original network is considerably worse than that of the network with adjusted probabilities.

We then tested inducing probabilistic networks on this dataset. For each of the symmetric measures described in section 4 (I_{gain}, I_{sgr}, Gini_{sym}, and χ^2), we constructed an optimum weight spanning tree and evaluated it on the database. The results are shown in the bottom left of table 1 (since we did not use weighting, the results for the two symmetric information gain ratios are the same, hence only one is shown). Although they are restricted to 21 edges because of the tree structure, they come fairly close to the evaluation of the adapted original network. Inspecting the learned networks in more detail reveals that their structure is indeed very close to the domain expert designed network.

network	cond.	$\sum \pi_{\min}$	$\sum \pi_{\text{avg}}$	$\sum \pi_{\max}$
indep.	0	139.8	141.1	158.2
db. poss.	24	137.3	137.7	157.2
d_{mi}	33	115.6	118.1	147.2
S_{gain}	35	122.9	123.8	146.1
S_{gr}	27	129.9	131.4	154.1
$S_{\text{sgr}}^{(1)}$	33	123.6	124.7	147.2
d_{χ^2}	35	122.3	123.2	145.2
d_{mi}	21	117.6	119.4	144.3
S_{gain}	21	123.3	124.9	148.8
$S_{\text{sgr}}^{(1)}$	21	121.1	123.8	148.3
d_{χ^2}	21	120.3	122.5	143.5

Table 2. Evaluation of possibilistic networks obtained by optimum weight spanning tree construction (top) and by greedy parent selection (bottom) on the Danish Jersey cattle blood type determination data.

In a third step we induced networks with the greedy parent selection method. We selected a topological order compatible with the domain expert designed network and restricted the number of parents to two. To compute the evaluation measures for attributes with two parents, we combined the parents into one pseudo-attribute, i.e. we extended the measure in the second possible way. Evaluations of the learned networks are shown on the right in table 1. With the exception of the information gain ratio network they all perform better than the original structure. A closer inspection reveals that they do so by exploiting additional dependences present in the database but not in the domain expert designed network.

From this table one may infer that information gain, Gini index and χ^2-measure lead to the best results, since given the networks learned with these measures, the average probability of the database is highest. But taking into account the number of conditions selected, which is highest for these measures, the suspicion arises that the good evaluation results are obtained by "over fitting" the data. This hypothesis was confirmed by an experiment on two artificial datasets generated from the domain expert designed network. On the test dataset the evaluation results of the networks learned with these measures where considerably lower than on the dataset they were learned from. The effect seems to be less pointed for information gain than for Gini index and χ^2-measure. Nevertheless, the bias in favour of many-valued attributes was clearly visible, since with information gain the *offspring genotype* attribute (6 values) was selected as a parent attribute for the *lysis* attributes instead of the *factor* attributes (2 values) as in the domain expert designed network. The results also confirmed that forming some kind of ratio reduces the bias.

Evaluations of learned possibilistic networks are shown in table 2. It is not surprising that the expert designed network (with possibility degrees determined from the database) performs badly, since the possibilistic scheme exploits a different type of dependence. But it is remarkable that allowing larger edges to be learned by using the greedy parent selection method seems not to improve the results over optimum weight spanning trees, although this may be due to the restrictions imposed by the topological order. The strength and weaknesses of the measures seem to be similar to those of the analogous measures for learning probabilistic networks.

8 Conclusions

In this paper we considered two search methods (optimum weight spanning tree construction and greedy parent selection) and a large number of local evaluation measures for learning probabilistic and possibilistic networks. The experimental results, which we obtained with the prototype program INES, show that a problem of some evaluation measures is that they lead to the selection of too large edges (in terms of the number of attributes as well as in terms of the number of attribute values), resulting in some kind of "over fitting" (information gain, Gini index, χ^2-measure). For probabilistic networks the best results seem to be achievable with the symmetric information gain ratio, the g-function and the minimum description length measures. For possibilistic networks d_{mi}, the analogon of mutual information, seems to yield the best results.

Acknowledgments

We are grateful to J. Gebhardt for fruitful discussions and to S.L. Lauritzen and L.K. Rasmussen for making the Danish Jersey cattle blood type determination example available for us.

References

1. S.K. Andersen, K.G. Olesen, F.V. Jensen, and F. Jensen. HUGIN — A shell for building Bayesian belief universes for expert systems. *Proc. 11th Int. J. Conf. on Artificial Intelligence*, 1080–1085, 1989
2. L. Breiman, J.H. Friedman, R.A. Olshen, and C.J. Stone. *Classification and Regression Trees*, Wadsworth International Group, Belmont, CA, 1984
3. C.K. Chow and C.N. Liu. Approximating Discrete Probability Distributions with Dependence Trees. *IEEE Trans. on Information Theory* 14(3):462–467, IEEE 1968
4. G.F. Cooper and E. Herskovits. A Bayesian Method for the Induction of Probabilistic Networks from Data. *Machine Learning* 9:309–347, Kluwer 1992
5. Z. Daróczy. Generalized Information Functions. *Information and Control* 16:36–51, 1970
6. J. Gebhardt and R. Kruse. A Possibilistic Interpretation of Fuzzy Sets in the Context Model. *Proc. IEEE Int. Conf. on Fuzzy Systems*, 1089-1096, San Diego 1992.

7. J. Gebhardt and R. Kruse. POSSINFER — A Software Tool for Possibilistic Inference. In: D. Dubois, H. Prade, and R. Yager, eds. *Fuzzy Set Methods in Information Engineering: A Guided Tour of Applications*, Wiley 1995
8. J. Gebhardt and R. Kruse. Learning Possibilistic Networks from Data. *Proc. 5th Int. Workshop on AI and Statistics*, 233–244, Fort Lauderdale, 1995
9. J. Gebhardt and R. Kruse. Tightest Hypertree Decompositions of Multivariate Possibility Distributions. *Proc. Int. Conf. on Information Processing and Management of Uncertainty in Knowledge-based Systems*, 1996
10. R.V.L. Hartley. Transmission of Information. *The Bell Systems Technical Journal* 7:535–563, 1928
11. D. Heckerman. *Probabilistic Similarity Networks*. MIT Press 1991
12. D. Heckerman, D. Geiger, and D.M. Chickering. Learning Bayesian Networks: The Combination of Knowledge and Statistical Data. *Machine Learning* 20:197–243, Kluwer 1995
13. M. Higashi and G.J. Klir. Measures of Uncertainty and Information based on Possibility Distributions. *Int. Journal of General Systems* 9:43–58, 1982
14. G.J. Klir and M. Mariano. On the Uniqueness of a Possibility Measure of Uncertainty and Information. *Fuzzy Sets and Systems* 24:141–160, 1987
15. I. Kononenko. On Biases in Estimating Multi-Valued Attributes. *Proc. 1st Int. Conf. on Knowledge Discovery and Data Mining*, 1034–1040, Montreal, 1995
16. R. Kruse, E. Schwecke, and J. Heinsohn. *Uncertainty and Vagueness in Knowledge-based Systems: Numerical Methods*. Springer, Berlin 1991
17. R. Kruse, J. Gebhardt, and F. Klawonn. *Foundations of Fuzzy Systems*, John Wiley & Sons, Chichester, England 1994
18. S. Kullback and R.A. Leibler. On Information and Sufficiency. *Ann. Math. Statistics* 22:79–86, 1951
19. S.L. Lauritzen and D.J. Spiegelhalter. Local Computations with Probabilities on Graphical Structures and Their Application to Expert Systems. *Journal of the Royal Statistical Society, Series B*, 2(50):157–224, 1988
20. R.L. de Mantaras. A Distance-based Attribute Selection Measure for Decision Tree Induction. *Machine Learning* 6:81–92, Kluwer 1991
21. H.T. Nguyen. Using Random Sets. *Information Science* 34:265–274, 1984
22. J. Pearl. *Probabilistic Reasoning in Intelligent Systems: Networks of Plausible Inference (2nd edition)*. Morgan Kaufman, New York 1992
23. J.R. Quinlan. Induction of Decision Trees. *Machine Learning* 1:81–106, 1986
24. J.R. Quinlan. *C4.5: Programs for Machine Learning*, Morgan Kaufman, 1993
25. L.K. Rasmussen. *Blood Group Determination of Danish Jersey Cattle in the F-blood Group System*. Dina Research Report no. 8, 1992
26. J. Rissanen. A Universal Prior for Integers and Estimation by Minimum Description Length. *Annals of Statistics* 11:416–431, 1983
27. A. Saffiotti and E. Umkehrer. PULCINELLA: A General Tool for Propagating Uncertainty in Valuation Networks. *Proc. 7th Conf. on Uncertainty in AI*, 323–331, San Mateo 1991
28. G. Shafer and P.P. Shenoy. Local Computations in Hypertrees. Working Paper 201, School of Business, University of Kansas, Lawrence 1988
29. C.E. Shannon. The Mathematical Theory of Communication. *The Bell Systems Technical Journal* 27:379–423, 1948
30. M. Singh and M. Valtorta. An Algorithm for the Construction of Bayesian Network Structures from Data. *Proc. 9th Conf. on Uncertainty in AI*, 259–265, Morgan Kaufman, 1993
31. L. Wehenkel. On Uncertainty Measures Used for Decision Tree Induction. *Proc. IPMU*, 1996

Information Fusion in Logic:
A Brief Overview

Laurence Cholvy[1] and Anthony Hunter[2]

(1) ONERA-CERT
2 avenue Ed Belin
31055 Toulouse, FRANCE
and
(2) Department of Computer Science
University College London
Gower Street
London WC1E 6BT, UK

Abstract. Information fusion is the process of deriving a single consistent knowledgebase from multiple knowledgebases. This process is important in many cognitive tasks such as decision-making, planning, design, and specification, that can involve collecting information from a number of potentially conflicting perspectives or sources, or participants. In this brief overview, we focus on the problem of inconsistencies arising in information fusion. In the following, we consider reasoning with inconsistencies, acting on inconsistencies, and resolving inconsistencies.

1 Introduction

Many tasks that an intelligent agent performs such as decision-making, planning, design, and specification, often involve collecting information from a number of potentially conflicting perspectives or sources, or participants with different views, and forming a single combined view or perspective — a synthesis, or consensus.

Consider requirements engineering. The development of most large and complex systems necessarily involves many people, each with their own perspectives on the system defined by their knowledge, responsibilities, and commitments. Inevitably, the different perspectives of those involved in the process intersect, giving rise to conflicts. From a logics perspective, these conflicts can be viewed as logical contradictions or inconsistencies. While classical logic is a rich and useful formalism in requirements engineering, it does not allow useful reasoning in the presence of inconsistency: the proof rules of classical logic allow any formula of the language to be inferred. Hence, classical logic does not provide a means for continued deduction in the presence of inconsistency.

Traditionally, inconsistency in logic has been viewed as a problem that requires immediate rectification. In [GH91,GH93], it was proposed that inconsistency is not necessarily a problem, as long as we know how to act in the presence of it. Indeed, inconsistencies can be viewed as being useful, since they can help to direct a cognitive activity. Certaintly, premature resolution of inconsistency can result in the loss of valuable information, and constrain an overall problem solution.

Ultimately, a single consistent view is required from a set of multiple views. Information fusion is the process of deriving this single consistent view. Whilst theoretical approaches such as belief revision [AGM85,Gar88,DP97], databases and knowledgebase updating [FKUV86,KM89,Win90,Som94], and combining knowledgebases (for example [DLP92,Mot93,BKMS92,BKMS91]) are relevant, information fusion addresses a wider range of issues raised by practical imperatives in applications such as requirements engineering.

The problem of information fusion appears in many fields, such as gathering beliefs or evidence, developing specifications, and merging regulations. The types of information to be modelled can differ, depending on the application. Frequently, they can be beliefs (describing what things are or what they are supposed to be in the real world) or they can be norms (describing how things should be in an ideal world). And of course, the type of logical formalism used to model the information depends on the type of information. Possibilities include classical logic, belief logics and deontic logics, though, potentially, any logic may be used.

But, whatever the type of the logic, the problem of information fusion raises the crucial problem of inconsistencies. Whilst the aim is to build a consistent set of information, the status of this set of information will differ, depending on the type of information. For example, in gathering beliefs, the aim of the fusion process is to build a consistent set of beliefs — a consistent representation of the real world — whereas in regulation merging, the aim is to build a consistent regulation i.e, a set of rules which consistently specifies an ideal world [Cho97].

In the following, we discuss some of the features of information fusion in logic. In particular, we focus on the problems of inconsistencies arising in information fusion, including the management of inconsistent information, reasoning with inconsistencies, and resolving inconsistencies.

2 Reasoning with inconsistencies

In practical reasoning, it is common to have "too much" information about some situation. In other words, it is common for there to be classically inconsistent information in a practical reasoning problem. The diversity of logics proposed for aspects of practical reasoning indicates the complexity of this form of reasoning. However, central to this is the need to reason with inconsistent information without the logic being trivialised.

Classical logic is trivialised because, by the definition of the logic, any inference follows from inconsistent information *(ex falso quodlibet)* as illustrated by the following example. From the set of formulae α, $\neg\alpha$, $\alpha \to \beta$, δ, reasonable inferences might include α, $\neg\alpha$, $\alpha \to \beta$, and δ by *reflexivity*; β by *modus ponens*; $\alpha \wedge \beta$ by *conjunction introduction*; $\neg\beta \to \neg\alpha$ and so on. In contrast, trivial inferences might include γ and $\gamma \wedge \neg\delta$.

For classical logic, trivialisation renders the reasoning useless, and therefore classical logic is obviously unsatisfactory for handling inconsistent information. A possible solution is to weaken classical logic by dropping some of the inferencing capability, such as for the C_ω paraconsistent logic [dC74], though this kind of weakening of the proof rules means that the connectives in the language do not behave in a classical fashion [Bes91]. For example, disjunctive syllogism does not hold, $((\alpha \lor \beta) \land \neg\beta) \to \alpha$, whereas modus ponens does hold. Variations on this theme include [AB75,Arr77,Bat80,PR84,PRN88,CFM91]. Alternative compromises on classical logic include three-valued logic [Lin87], four-valued logic [Bel77], quasi-classical (QC) logic [BH95], and using a form of conditioanl logic [Roo93]. Another approach is to reason with classically consistent subsets of inconsistent information. This has given rise to a number of logics (for example [MR70,Wag91,BDP93,BCD$^+$93,BDP95,EGH95]) and truth maintenance systems (for example [Doy79,Kle86,MS88]). These options from C_ω through to reasoning with maximally consistent subsets behave in different ways with data. None can be regarded as perfect for handling inconsistency in general. Rather they provide a spectrum of approaches. However, in all of them, the language is based on that of classical logic, and the aim is to preserve features of classical reasoning. For a review of these approaches see [Hun97].

Modal logics have also been developed for reasoning with inconsistent information. For instance, in [FH86], Farinas and Herzig define a modal logic for reasoning about elementary changes of beliefs which is a kind of conditional logic. The accessibility relation associated with the modality aims to capture the relation between a world and the worlds which are obtained after adding a piece of information. This logic has more recently been extended in order to take into account the dependances that may exist between the different pieces of information [FH92,FH94] and the influence these dependence links have in the updating process. Another modal conditional logic defined for belief revision is described in [Gra91]. Modal logic has also been proposed for consistent reasoning with inconsistent information by using only a consistent subset of beliefs [Lin94]. For a review of modal logics in handling inconsistent information see [MvdH97].

Problems of handling default knowledge are closely related to that of handling inconsistent information. Indeed, implicit in default, or non-monotonic, reasoning is the need to avoid trivialization due to conflicting defaults [BH97]. There are a variety of non-monotonic logics for handling default knowledge (for reviews see [Bes89,Bre91,GHR94]) with different strategies for avoiding inconsistency by selecting preferred consistent subsets of formulae. These include preference for more specific information [Poo85], ordered theory presentations [Rya92], preferred subtheories [Bre89], explicit preferences [Pra93], and prioritised syntax-based entailment [BCD$^+$93]. Non-monotonic logics therefore offer means for analysing inconsistent information and preferences over that information.

The ability to reason with inconsistencies is important, since it allows inconsistent information to be explored and analysed. However, there is also a need to act on inconsistencies, and eventually to resolve inconsistencies. We address these two topics in the next two sections.

3 Acting on inconsistencies

Immediate resolution of inconsistency by arbitrarily removing some formulae can result in the loss of valuable information. This can include loss of information that is actually correct and also loss of information that can be useful in managing conflicts. Immediate resolution can also unduly constrain cognitive activities such as problem solving and designing.

Identifying the appropriate inconsistency handling strategy depends on the kinds of inconsistency that can be detected and the degree of inconsistency tolerance that can be supported. Possible kinds of actions include:

Circumventing the inconsistent parts of the information. This can be viewed as ignoring the inconsistency, and using the rest of the information regardlessly. This may be appropriate in order to avoid inconsistent portions of the information and/or to delay resolution of the inconsistency. This includes using the logical techniques discussed in Section 2. Isolating inconsistency is a special case where the minimally inconsistent subset of the information is not used in the reasoning – it is isolated — but not deleted.

Ameliorating inconsistent situations by performing actions that "improve" these situations and increase the possibility of future resolution. This is an attractive approach in situations where complete and immediate resolution is not possible (perhaps because further information is required from elsewhere), but where some steps can be taken to "fix" part or some of the inconsistent information. This approach requires techniques for analysis and reasoning in the presence of inconsistency.

Sequencing of conflicts so that some conflicts are addressed before others. The criteria for sequencing are diverse but may include:

> **Granularity of inconsistency:** Some conflicts are more significant than others. Furthermore, it is possible that solving a less significant conflict before a more significant conflict may unduly constrain the allowed solutions for the more significant conflict. Solving bigger or more important conflicts first means that we need to order the conflicts. For instance, when desiging a building, first solve the conflicts that are about its "function" (deciding what this building is for will determine its height, its surface...), then solve the conflicts about the "inside" of the building (this will determine the number of rooms, the exact places of the walls...), and then solve the conflicts about "decoration" (this will finally determine the style of the curtains and the colour of the walls,...). For this, the notion of topichood of information, such as in [CD89,CD92,Hun96], is potentially useful. This could allow us to say that "inconsistent set X is about topic T", and may be used in determining the significance of the inconsistency.

Temporality of inconsistency: Other temporal constraints can impose an ordering on the sequence of resolution of conflicts. For example, in a building project, delaying the resolution of conflicts about the position of a wall is temporally more sensitive than delaying the resolution of conflicts about the colour of the paint on the wall, since the construction of the wall needs to be completed before the wall is painted. Clearly the colour of the paint can be chosen before the position of the wall.

For the above examples, it can be seen that there is an overlap in granularity and temporality of inconsistencies. However, an example of equally significant inconsistencies. where the first is more temporally sensitive, is choice of paint for the ceiling of a room, and a choice of colour for the walls. And an example where some inconsistencies are far more significant than others, but where ultimately they are of equal temporal sensitivity, is in a book that is about to be published.

Resolving inconsistencies altogether by correcting any mistakes or resolving conflicts. This depends on a clear identification of the inconsistency and assumes that the actions required to fix it are known.

Circumventing, ameliorating and sequencing inconsistency all imply that the resolution of conflicts or inconsistency is delayed. In practice applications usually involve multiple conflicts of diverse kinds and significance. As a result, a combination of circumventing, ameliorating and sequencing of inconsistency is required.

4 Resolving inconsistencies

In order to resolve an inconsistency intelligently, as opposed to arbitrarily, we require appropriate information. Information that is manipulated in information fusion can be partitioned, and is often represented, in different ways. The information includes the extra information required for combination.

Object-level information: The information to be combined.

Combination information: The information used to facilitate combination. This is composed of meta-level information and domain information.

  Meta-level information: Information about information. For example, information about

  − The sources of the object-level information
  − The reliability of sources
  − Preferences about the information

  Domain information: Information on the context or domain of the object-level. This is used to constrain the combination process. Examples of domain information include integrity constraints such as "everybody is either a man or a woman" and "a cube has 6 sides". Domain information can be uncertain, such as for example heuristics. Though using uncertain information significantly increases the difficulty of combining object-level information.

Combination information constitutes extra information that can be used by the combination process in order to combine the object-level information. Neither domain information nor meta-level information needs to be in the same formalism as the object-level information. The only constraint on the formalism is that it can be used by the combination process.

To illustrate these concepts, consider the following example. We have two sources of information S1 and S2 and we wish to combine their object-level information.

S1: The colour of the object is blue.
S2: The colour of the object is green.

We also have the following domain information and meta-level information.

Domain: Green and blue are different colours.
Meta: The domain information is more reliable than source S1.
Meta: Source S1 is more reliable than source S2.

In forming the combined information, we can accept the information from S1, because it is consistent and from the most reliable source. However, we cannot now add the information from S2 since it would cause an inconsistency.

For example, assuming an ordering over development information is reasonable in software engineering. First, different kinds of information have different likelihoods of being incorrect. For example, method rules are unlikely to be incorrect, whereas some tentative specification information is quite possibly incorrect. Second, if a specification method is used interactively, a user can be asked to order pieces of specification according to likelihood of correctness.

As a second example, assuming an order between different sources is reasonable in the case when merging beliefs or evidence provided by those different information sources [Cho94,Cho93]. In particular, it can be useful to order sources according to the topics of the information they provide — in effect adopting context-sensitive ordering over the sources [CD94,Cho95]. Indeed, assuming only one ordering over the different information sources is not very realistic, given that frequently a source can be assumed to be more reliable than a second source, on one topic, but less reliable on another topic.

As a third example, in the domain of regulation merging, assuming an ordering over regulations can be useful to consistently reason with rules [CC95]. This is related to the use of priorities in argumentation for legal reasoning, in particular [Pra93,PS95,PS96,RD96,TdT95].

There are number of ways that this approach can be developed. First, there are further intuitive ways of deriving orderings over formulae and sets of formulae. These include ordering sets of formulae according to their relative degree of contradiction [GH97]. Second, there are a number of analyses of ways of handling ordered formulae and sets of ordered formulae such as dicussed above in Section 2 on logics for inconsistent information.

5 Discussion

In general, information fusion is a difficult problem. Given the difficulty, there is the need for the following:

- Inconsistency management during information fusion, to track inconsistencies and minimize the negative ramification of inconsistency.
- A range of logics for reasoning and analysis of inconsistent information, to allow continued use of inconsistent information, and to facilitate resolution.
- Extra information, called combination information, to enable the resolution of inconsistencies during information fusion.
- Interactive support for information fusion where the support system offers suggestions but the user controls the fusion process.

Despite the difficulties, there are practical reasoning applications where information fusion is likely to be of significant import: Take for example managing inconsistencies in the development of multi-perspective software development [FGH+94,HN97], where inconsistencies can be detected using classical logic, information surrounding each inconsistency can be used to focus continued development, and actions are taken in a context-dependent way in response to inconsistency.

Acknowledgements

This work was partly funded by the CEC through the ESPRIT FUSION project.

References

[AB75]   A Anderson and N Belnap. *Entailment: The Logic of Relevance and Necessity.* Princeton University Press, 1975.

[AGM85]   C Alchourrón, P Gardenfors, and D Makinson. On the logic of theory change: Partial meet functions for contraction and revision. *Journal of Symbolic Logic*, 50:513–530, 1985.

[Arr77]   A Arruda. On the imaginary logic of NA Vasilev. In A Arruda, N Da Costa, and R Chuaqui, editors, *Non-classical logics, model theory and computability.* North Holland, 1977.

[Bat80]   D Batens. Paraconsistent extensional propositional logics. *Logique et Analyse*, 90–91:195–234, 1980.

[BCD+93] S Benferhat, C Cayrol, D Dubois, J Lang, and H Prade. Inconsistency management and prioritized syntax-based entailment. In *Proceedings of the Thirteenth International Joint Conference on Artificial Intelligence.* Morgan Kaufmann, 1993.

[BDP93]   S Benferhat, D Dubois, and H Prade. Argumentative inference in uncertain and inconsistent knowledge bases. In *Proceedings of Uncertainty in Artificial Intelligence*, pages 1449–1445. Morgan Kaufmann, 1993.

[BDP95]   S Benferhat, D Dubois, and H Prade. A logical approach to reasoning under inconsistency in stratified knowledge bases. In *Symbolic and Quantitative Approaches to Reasoning and Uncertainty*, volume 956 of *Lecture Notes in Computer Science*, pages 36–43. Springer, 1995.

[Bel77]   N Belnap. A useful four-valued logic. In G Epstein, editor, *Modern Uses of Multiple-valued Logic*, pages 8–37. Reidel, 1977.

[Bes89]   Ph Besnard. *An Introduction to Default Logic*. Springer, 1989.

[Bes91]   Ph Besnard. Paraconsistent logic approach to knowledge representation. In M de Glas M and D Gabbay D, editors, *Proceedings of the First World Conference on Fundamentals of Artificial Intelligence*, pages 107–114. Angkor, 1991.

[BH95]   Ph Besnard and A Hunter. Quasi-classical logic: Non-trivializable classical reasoning from inconsistent information. In C Froidevaux and J Kohlas, editors, *Symbolic and Quantitative Approaches to Uncertainty*, volume 946 of *Lecture Notes in Computer Science*, pages 44–51, 1995.

[BH97]   Ph Besnard and A Hunter. Introduction to actual and potential contradictions. In *Handbook of Defeasible Reasoning and Uncertainty Management*, volume 3. Kluwer, 1997.

[BKMS91] C. Baral, S. Kraus, J. Minker, and V.S. Subrahmanian. Combining multiple knowledge bases. *IEEE Trans. on Knowledge and Data Engineering*, 3(2), 1991.

[BKMS92] C Baral, S Kraus, J Minker, and V Subrahmanian. Combining knowledge-bases of first-order theories. *Computational Intelligence*, 8:45–71, 1992.

[Bre89]   G Brewka. Preferred subtheories: An extended logical framework for default reasoning. In *Proceedings of the Eleventh International Conference on Artificial Intelligence*, pages 1043–1048, 1989.

[Bre91]   G Brewka. *Common-sense Reasoning*. Cambridge University Press, 1991.

[CC95]   L. Cholvy and F. Cuppens. Solving normative conflicts by merging roles. In *Proceedings of the fifth International Conference on Artificial Intelligence and Law*, Washington, May 1995.

[CD89]   F. Cuppens and R. Demolombe. How to recognize interesting topics to provide cooperative answering. *Information Systems*, 14(2), 1989.

[CD92]   S. Cazalens and R. Demolombe. Intelligent access to data and knowledge bases via users' topics of interest. In *Proceedings of IFIP Conference*, pages 245–251, 1992.

[CD94]   L. Cholvy and R. Demolombe. Reasoning with information sources ordered by topics. In *Proceedings of Artificial Intelligence : Methods, Systems and Applications (AIMSA)*. World Scientific, Sofia, september 1994.

[CFM91]   W Carnielli, L Farinas, and M Marques. Contextual negations and reasoning with contradictions. In *Proceedings of the International Joint Conference on Artificial Intelligence (IJCAI'91)*, 1991.

[Cho93]   L. Cholvy. Proving theorems in a multi-sources environment. In *Proceedings of IJCAI*, pages 66–71, 1993.

[Cho94]   L. Cholvy. A logical approach to multi-sources reasoning. In *Proceedings of the Applied Logic Conference*, number 808 in Lecture notes in Artificial Intelligence. Springer-Verlag, 1994.

[Cho95]   L. Cholvy. Automated reasoning with merged contradictory information whose reliability depends on topics. In *Proceedings of the European Conference on Symbolic and Quantitative Approaches to Reasoning and Uncertainty (ECSQARU)*, Fribourg, July 1995.

[Cho97]   L. Cholvy. Reasoning about merged information. In *Handbook of Defeasible Reasoning and Uncertainty Management*, volume 1. Kluwer, 1997.

[dC74]   N C da Costa. On the theory of inconsistent formal systems. *Notre Dame Journal of Formal Logic*, 15:497–510, 1974.

[DLP92]   D. Dubois, J. Lang, and H. Prade. Dealing with multi-source information in possibilistic logic. In *Proceedings of ECAI*, pages 38–42, 1992.

[Doy79]   J Doyle. A truth maintenance system. *Artificial Intelligence*, 12:231–272, 1979.

[DP97]   D Dubois and H Prade. *Handbook of Defeasible Reasoning and Uncertainty Management*, volume 1. Kluwer, 1997.

[EGH95]   M Elvang-Goransson and A Hunter. Argumentative logics: Reasoning from classically inconsistent information. *Data and Knowledge Engineering Journal*, 16:125–145, 1995.

[FGH⁺94] A Finkelstein, D Gabbay, A Hunter, J Kramer, and B Nuseibeh. Inconsistency handling in multi-perspective specifications. *IEEE Transactions on Software Engineering*, 20(8):569–578, 1994.

[FH86]   L. Farinas and A. Herzig. Reasoning about database updates. In Jack Minker, editor, *Workshop of Foundations of deductive databases and logic programming*, 1986.

[FH92]   L. Farinas and A. Herzig. Revisions, updates and interference. In A. Fuhrmann and Rott H, editors, *Proceedings of the Konstanz colloquium in logic and information (LogIn-92*. DeGruyter Publishers, 1992.

[FH94]   L. Farinas and A. Herzig. Interference logic = conditional logic + frame axiom. *International JOurnal of Intelligent Systems*, 9(1):119–130, 1994.

[FKUV86] R Fagin, G Kuper, J Ullman, and M Vardi. Updating logical databases. *Advances in Computing Research*, 3:1–18, 1986.

[Gar88]   P Gardenfors. *Knowledge in Flux: Modelling the Dynamics of Epistemic States*. MIT Press, 1988.

[GH91]   D Gabbay and A Hunter. Making inconsistency respectable 1: A logical framework for inconsistency in reasoning. In *Fundamentals of Artificial Intelligence*, volume 535 of *Lecture Notes in Computer Science*, pages 19–32. Springer, 1991.

[GH93]   D Gabbay and A Hunter. Making inconsistency respectable 2: Meta-level handling of inconsistent data. In *Symbolic and Qualitative Approaches to Reasoning and Uncertainty (ECSQARU'93)*, volume 747 of *Lecture Notes in Computer Science*, pages 129–136. Springer, 1993.

[GH97]   D Gabbay and A Hunter. Negation and contradiction. In *What is negation?* Kluwer, 1997.

[GHR94]   D Gabbay, C Hogger, and J Robinson. *Handbook of Artificial Intelligence and Logic Programming*, volume 3. Oxford University Press, 1994.

[Gra91]   G. Grahne. A modal analysis of subjonctive queries. In R. demolombe, L. farinas, and T. Imielinski, editors, *Workshop on nonstandard queries and answers*, Toulouse, 1991.

[HN97]   A Hunter and B Nuseibeh. Analysing inconsistent specifications. In *Proceedings of 3rd International Symposium on Requirements Engineering*, pages 78–86. IEEE Computer Society Press, 1997.

[Hun96]   A Hunter. Intelligent text handling using default logic. In *Proceedings of the Eighth IEEE International Conference on Tools with Artificial Intelligence (TAI'96)*, pages 34–40. IEEE Computer Society Press, 1996.

[Hun97]   A Hunter. Paraconsistent logics. In *Handbook of Defeasible Reasoning and Uncertainty Management*. Kluwer, 1997.

[Kle86]   J De Kleer. An assumption-based TMS. *Artificial Intelligence*, 28:127–162, 1986.

[KM89]   H Katsuno and A Medelzon. A unified view of propositional knowledgebase updates. In *Proceedings of the Eleventh International Joint Conference on Artificial Intelligence*, 1989.

[Lin87]   F Lin. Reasoning in the presence of inconsistency. In *Proceedings of the National Conference on Artificial Intelligence (AAAI'87)*, 1987.

[Lin94]   J Lin. A logic for reasoning consistently in the presence of inconsistency. In *Proceedings of the Fifth Conference on Theoretical Aspects of Reasoning about Knowledge*. Morgan Kaufmann, 1994.

[Mot93]   A. Motro. A formal framework for integrating inconsistent answers from multiple information sources. Technical Report ISSE-TR-93-106, George Mason University, 1993.

[MR70]   R Manor and N Rescher. On inferences from inconsistent information. *Theory and Decision*, 1:179–219, 1970.

[MS88]   J Martins and S Shapiro. A model of belief revision. *Artificial Intelligence*, 35:25–79, 1988.

[MvdH97]   J Meyer and W van der Hoek. Modal logics for representing incoherent knowledge. In *Handbook of Defeasible Reasoning and Uncertainty Management, Volume 3*. Kluwer, 1997.

[Poo85]   D Poole. A logical framework for default reasoning. *Artificial Intelligence*, 36:27–47, 1985.

[PR84]   G Priest and R Routley. Introduction: Paraconsistent logics. *Studia Logica*, 43:3–16, 1984.

[Pra93]   H Prakken. An argument framework for default reasoning. In *Annals of mathematics and artificial intelligence*, volume 9, 1993.

[PRN88]   G Priest, R Routley, and J Norman. *Paraconsistent logic*. Philosophia, 1988.

[PS95]   H. Prakken and G. Sartor. On the relation between legal language and legal argument : assumptions, applicability and dynamic priorities. In *Proc. Fifth Conference on Artificial Intelligence and Law*, University of Maryland, May, 1995.

[PS96]   H. Prakken and G. Sartor. A system for defeasible argumentation with defeasible prorities. In *Proc. of FAPR'96*, May, 1996.

[RD96]   L. Royakkers and F. Dignum. Defeasible reasoning with legal rules. In *Proc of DEON'96, Sesimbra*. Springer, 1996.

[Roo93]   N Roos. A logic for reasoning with inconsistent knowledge. *Artificial Intelligence*, 57(1):69–104, 1993.

[Rya92]   M Ryan. Representing defaults as sentences with reduced priority. In *Principles of Knowledge Representation and Reasoning: Proceedings of the Third International Conference*. Morgan Kaufmann, 1992.

[Som94]   Leá Sombé. *Revision and updating in knowledgebases*. Wiley, 1994.

[TdT95]   Y Tan and L Van der Torre. Why defeasible deontic logic needs a multi preference semantics. In Ch. froidevaux and J. Kohlas, editors, *Quantitative and Qualitative Approches to Reasoning and Uncertainty*, number 946 in Lectures notes in Artificial Intelligence. Springer, 1995.

[Wag91]   G Wagner. Ex contradictione nihil sequitur. In *Proceedings of the International Joint Conference on Artificial Intelligence (IJCAI'91)*, 1991.

[Win90]   M Winslett. *Updating logical databases*. Cambridge University Press, 1990.

Focusing vs. Belief Revision: A Fundamental Distinction When Dealing with Generic Knowledge

Didier DUBOIS and Henri PRADE

Institut de Recherche en Informatique de Toulouse (I.R.I.T.)
Université Paul Sabatier – 118 route de Narbonne
31062 Toulouse Cedex 4 – France
Email: {dubois, prade}@irit.fr

Abstract: This paper advocates a basic distinction between two epistemic operations called focusing and revision, which can be defined in any, symbolic or numerical, representation framework which is rich enough for acknowledging the difference between factual evidence and generic knowledge. Revision amounts to modifying the generic knowledge when receiving new pieces of generic knowledge (or the factual evidence when obtaining more factual information), while focusing is just applying the generic knowledge to the reference class of situations which exactly corresponds to all the available evidence gathered on the case under consideration. Various settings are considered, upper and lower probabilities, belief functions, numerical possibility measures, ordinal possibility measures, conditional objects, nonmonotonic consequence relations.

1 - Introduction

Some basic modes of belief change have been laid bare by Levi (1980): an *expansion* corresponds to adding the new piece of information without rejecting previous beliefs, a *contraction* is the converse operation by which some piece of information is lost, and *revision* itself strictly speaking corresponds to accepting a piece of information that partially contradicts previous beliefs and modifying the latter accordingly. Revision and expansion are coherent in the sense that they coincide when the input information is consistent with previous beliefs. Another mode of belief change that has been more recently recognized is called updating (Katsuno and Mendelzon, 1991). The difference between updating and revision is that an updating operation takes into account the fact that the world referred to in the body of knowledge, has evolved, and so the set of beliefs must evolve accordingly; on the contrary, belief revision presupposes that some previous beliefs were wrong, and are corrected by the input information.

A last distinction, which is often absent from the symbolic approaches but which probabilists often endorse, even if not always explicitly, is between revision and what we have called 'focusing' (Dubois and Prade, 1992a). This is the topic of this paper. It relies on the idea that a body of knowledge often consists in two parts: background knowledge and particular pieces of evidence on a case at hand. The input can either alter the background knowledge (revision), or complete the available evidence, which then points to a new reference class (focusing).

To handle the distinction properly, we separate the generic knowledge K pertaining to classes of situations, from the factual evidence E pertaining to a particular case to which the generic knowledge is applied. In the following the term 'knowledge' refers to generic information, while information about a case is referred to as '(factual) evidence'. These two kinds of information may convey some form of partial ignorance. Factual evidence consists of information gathered on a case at hand, or the description of the actual world in a given situation. This information can be more or less precise and more or less reliable. Generic knowledge pertains to a class of situations considered as a whole, but does not refer to a particular case. Sometimes this class of situations is well defined (it is a population in the sense of statistics) and the generic knowledge is of a frequentist nature. Sometimes the relevant class of population is much more vaguely described (as in the famous "birds fly" example), and the generic knowledge describes rules of plausible inference of the form "if all I know is p then plausibly conclude q" (when knowledge is pervaded with exceptions). The levels of confidence expressed by the rules can be numerically modelled via a conditional probability for instance, or be handled in a purely ordinal setting (as in default reasoning). The factual evidence, which might be also somewhat uncertain, would be however assumed to be represented by propositional statements in the following (except if the contrary is explicitly acknowledged). The difference between generic knowledge and factual evidence can be illustrated by a diagnosis problem. The generic knowledge of a clinician consists in his/her knowledge about the links between the diseases and the symptoms and the distribution of the diseases in the population (in practice, the likelihoods and the prior probabilities). The factual evidence consists in the symptoms collected from the patient under consideration.

The next section provides a general discussion of focusing versus revision. See also Dubois and Prade (1994a), Dubois et al. (1996) for detailed discussions on this topic. Section 3 recalls focusing rules in numerical settings while Section 4 summarizes how focusing is handled in non-numerical settings. Section 5 briefly points out further developments and provides concluding remarks.

2 - Focusing vs. Revision — A General Discussion

The distinction between evidential and generic information is crucial for a proper understanding of belief revision processes and commonsense inference. By applying the generic knowledge K to the evidence E, a set C of consequences can be produced. Thus C is made of default conclusions, or of propositions having high probabilities, according to the representation framework used in K. Generally speaking, C is the set of the most plausible conclusions which can be inferred from E and K. Clearly, the consequences in C are open to revision and should be defeasible. Indeed when more evidence is available and E is changed into E', the consequences which can be obtained from E' and K may differ, and even contradict those derivable from E and K. In Gärdenfors(1988)'s view of revision, no distinction is made between evidence of knowledge. Beliefs are supposed to be maintained in a so-called, (propositional) belief set B closed under logical consequences. This belief set is to be revised when an input information is received which contradicts B. The lack of distinction between evidence and knowledge in Gärdenfors' approach put all the information at the same level and limits the expressive power. However, considering revision and nonmonotonic reasoning as the two sides of the same coin (Makinson and Gärdenfors, 1991), reconciles the two views. Indeed if we consider that the belief set B corresponds exactly to the set of plausible consequences C, the epistemic entrenchment which underlies any revision process obeying the postulates proposed by Alchourrón, Gärdenfors and Makinson (see Gärdenfors, 1988) can be turned into a plausibility ordering between interpretations which implicitly gives birth to a conditional knowledge base K (Gärdenfors and Makinson, 1994). Then, revising B = C into C' when a new information is received (i.e., when E is expanded into E') amounts to compute C' as the set of propositions p such that the default rule E' ~> p is a piece of conditional knowledge deducible from K using the rational closure entailment proposed by Lehmann and Magidor (1992). But this corresponds to a particular modelling of the generic knowledge supposed to be stored in K. The distinction between E, K and C leaves open the way the generic knowledge is represented (K may also be modelled in terms of a family of probability measures, or by a Bayesian network, for instance).

An example of probabilistic knowledge base that can be represented by means of a family of probability measures consists in a set of statements of the form "most A_i's are B_i's", "few A_j's are B_j's", ... etc. More generally K = {"Q_i A_i's are B_i's", i = 1,n} where Q_i is an imprecise quantifier of the form $[\alpha,1]$ (for "most") or $[0,\alpha]$ (for "few"), or $[\alpha,\beta]$, expressing an ill-known proportion. Each such statement can be modelled by a constraint of the form $P(B_i \mid A_i) \in Q_i$. The evidence E is supposed to be incomplete but not uncertain here. It is represented by a proposition or an event. Assume a new piece of information arrives. The problem addressed here is what to do in order to account for the new information in the pair (K,E) so as to produce new plausible conclusions. When the input information is evidential it only affects the evidence E. Then, focusing consists in applying the generic knowledge to the context of interest as described by the evidential information (including the new one), in order to answer questions of interest pertaining to this particular situation. In such a process the knowledge should remain completely unchanged. Namely we know that the case at hand is in class A (E = A). Then, considering the above probabilistic example, suppose it is asked whether the case at hand has property B. What is to be computed is $P(B \mid A)$. However since there maybe more than one probability measure restricted by the statements in K, only upper and lower bounds of $P(B \mid A)$ can be computed.

This has to be contrasted with the situation where the input information is generic. It thus should be understood as a new constraint refining the available knowledge. In such a case, a genuine revision takes place since the knowledge has to be modified in order to incorporate the new piece of information as a new constraint (except if it is redundant with respect to what is already known). In the case of revision, the new piece of information is of the form $P(A) = 1$, which means that all cases (that the knowledge base refers to) satisfy property A. Then K becomes K' ∪ {$P(A) = 1$} and new values for the upper and lower bounds for $P(B)$ where B is a property of interest can be then computed. These bounds do not coincide with the ones of $P(B \mid A)$ calculated above in the focusing case, and may even not exist if K implies that $P(A) \neq 1$. Revision of generic knowledge would correspond to the following types of example: a probability distribution over sizes of adults in a given country is available, and some input information comes in that nobody in the population is more than 6ft tall. These remarks make it clear that revising is not focusing. However, note that for both revision and focusing on the basis of uncertainty functions, such as probability measures, the basic tool is the conditioning operation.

A revision process can be iterated, and it progressively modifies the initial knowledge base. Iterated focusing only refines the body of evidence and the same generic knowledge is applied to a new reference class of situations pointed at by the evidence. In some problems, only evidential information is available, and new pieces of possibly uncertain evidence lead to a non-trivial revision process due to uncertainty, as in Shafer (1976). It is also worth pointing out that in case of multiple sources of information, the factual evidence is obtained through a symmetric combination process (for instance by applying Dempster's rule of combination in the belief function framework). A similar process should apply at the generic knowledge level if K is built from several sources.

The distinction between focusing and (generic) revision can only be made in uncertainty frameworks where the difference between generic knowledge and factual evidence can be captured and where generic knowledge is represented and processed as distinct from factual evidence. Such a distinction is not relevant in propositional logic for instance, since every piece of information takes the form of a propositional sentence. In the setting of Bayesian probability the distinction is conceptually meaningful but focusing and revising are both expressed by the same conditioning rule, that is Bayes rule. Hence some disputes among probabilists as to the meaning of conditioning. The most widely found view (e.g., De Finetti, 1974) is that Bayes rule operates a change of reference class, namely going from a prior to a posterior probability is not a revision process. Posterior probabilities are precalculated and the input information just prompts the selection of a particular posterior. With this view, the prior probability together with the likelihoods determine a unique joint probability over a space of interest, construed as a body of generic knowledge (the clinician's experience on a certain disease), and conditioning means integrating factual evidence so as to configurate the generic knowledge properly with respect to the reference class of the object on which this factual evidence bears (test results on the patient). This point of view, which is shared by the rule-based expert system literature, culminates with the advent of Bayesian networks.

The revision view of conditioning is typically advocated by people working in probability kinematics. Philosophers like Jeffrey (1983), Domotor (1985), Williams (1980), and Gärdenfors (1988), understand Bayes rule as a revision process, by which a prior probability P is changed into a new one P' due to the input information A. This latter view is supported by maximal cross-entropy arguments whereby it is established that Bayes rule obeys a minimal change requirement. This school of thought is called probability kinematics: the input information A is understood as the discovery that $P'(A) = 1$ while the prior is $P(A) < 1$, and must be modified. The possibility that the input information is itself probabilistic is commonly envisaged in probability kinematics: the input information enforces the probability of some event to take on a value which differs from the prior value. In this view the input information is at the same level as the prior information: both are (possibly uncertain) factual evidence or pieces of generic knowledge.

3 - Numerical Settings

3.1 - Belief Functions — Upper and Lower Probabilities

While revision and focusing do coincide in probability theory as already said, they no longer coincide in more general settings such as belief functions or upper and lower probabilities. Revision in belief function theory is defined by Dempster rule of conditioning that combines the conjunctive revision mode of the crude partial ignorance model, with Bayes rule normalization underlying a stability of the degrees of uncertainty in relative value. Let Bel be a belief function defined by $Bel(A) = \sum_{\emptyset \neq E_i \subseteq A} m(E_i)$ from a set of masses m_i on focal elements E_i's such that $\sum_i m(E_i) =$

1, $E_i \subseteq \Omega$, and Pl the associated plausibility measure $Pl(A) = 1 - Bel(\bar{A})$. Indeed Dempster rule of conditioning can be described as follows:

- Given the new piece of information A, turn each focal element E_i into $E_i \cap A$, and attach the mass m_i to it, adding the masses m_i and m_j if $E_i \cap A = E_j \cap A$.
- Renormalize the masses allocated to non-empty subsets $E_i \cap A \neq \emptyset$ as with Bayes rule, so as to reallocate the masses m_i such that $E_i \cap A = \emptyset$ proportionally.

It is well-known (Shafer, 1976) that it comes down to compute the expected degree of potential support:

$$Pl(B \mid A) = \frac{Pl(A \cap B)}{Pl(A)}. \tag{1}$$

Interestingly, Dempster's rule of conditioning can also be derived in a pure upper and lower probability context. If the belief function is viewed as characterizing a set of probabilities $\mathbb{P} = \{P \leq Pl\}$, then it can be proved that:

$$Pl(B \mid A) = \frac{Pl(A \cap B)}{Pl(A)} = \sup\left\{\frac{P(A \cap B)}{P(A)}, P \leq Pl, P(A) = Pl(A)\right\}. \tag{2}$$

This result is due to the fact that the constraint $P(A) = Pl(A)$ never forbids $\sup\{P(A \cap B), P \leq Pl, P(A) = Pl(A)\}$ to be equal to $Pl(A \cap B)$, i.e., we can always have $P(A) = Pl(A)$ and $P(A \cap B) = Pl(A \cap B)$ for the same probability measure P, if Pl is a plausibility function. Equation (2) makes it clear the kind of revision at work with Dempster rule in the upper and lower probabilities context: the constraint $P(A) = Pl(A)$ corresponds to the selection of a maximum likelihood probability, which is very usual in statistics. As a generalization of this principle, Moral and De Campos (1991) have suggested that the magnitude of $P(A)$ reflects the possibility of accepting the corresponding conditional probability $P(\cdot \mid A)$ in the updated set of probabilities.

Gilboa and Schmeidler (1992) have given a decision-theoretic interpretation of $Pl(B \mid A)$ in the setting of upper and lower probabilities, i.e., when the set $\mathbb{P}$ does not necessarily represent a belief function. Indeed equation (6) does hold when Π satisfies order-2 subadditivity only, that is $P^*(A \cup B) \leq P^*(A) + P^*(B) - P^*(A \cap B)$. The reason is that the property stating that we can always have $P(A) = P^*(A)$ and $P(B) = P^*(B)$ for a subset B of A for the same probability measure P is characteristic of order-2 subadditivity (see Huber, 1981). However equation (2) does not hold for upper probabilities deriving from any set of probabilities $\mathbb{P}$.

The alternative conditioning rule, already suggested by Dempster (1967), De Campos et al. (1990), Fagin and Halpern (1989) and Jaffray (1992) and here referred to as focusing, is defined in the setting of upper and lower probabilities as:

$$P^*_A(B) = \sup\left\{\frac{P(A \cap B)}{P(A)}, P \leq P^*\right\} = \frac{P^*(A \cap B)}{P^*(A \cap B) + P_*(A \cap \bar{B})} \tag{3}$$

where P^* denotes an upper probability function and P_* the associated lower probability with $P^*(A) = 1 - P_*(\bar{A})$. As already said, the idea of focusing is to compute the probability of B in the state when A is supposed to be true without making any assumption about how the set of probabilities should be revised if A were actually true, especially without considering any probability measure in the set $\{P \leq P^*\}$ as impossible except those such that $P(A) = 0$. It leads to performing a sensitivity analysis on Bayes rule, when the probability function ranges over the set $\{P \leq P^*\}$. Interestingly, if $P^* = Pl$ is a plausibility function then P^*_A is still a plausibility function. Note that while Dempster rule of conditioning (2) can easily be iterated, this is not so with the focusing rule, because the set $\{P(\cdot \mid A), P \leq P^*\}$ is generally a proper subset of the set $\{P, P \leq P^*_A\}$ induced by the upper probability bounds obtained via focusing (see Jaffray, 1992) even if P^* is a plausibility measure. But iterating the focusing operation by "conditioning again" P^*_A on a new event C makes no sense. What must be done is to start from the original function P^* and focus on the new reference class $A \cap C$.

P^*_A is generally much less informative than $P^*(\cdot \mid A)$ (applying (2) with P^* instead of Pl) and even sometimes less informative than P^* itself. This has been noticed by Kyburg (1987) for belief functions where $Pl_A \geq Pl(\cdot \mid A) \geq Bel(\cdot \mid A) \geq Bel_A$ holds. For instance if $A \cap E_i \neq \emptyset$, and $E_i \not\subseteq A \; \forall i$, then $P^*_A(B) = 1$ and $P^*_A(\bar{B}) = 1$, $\forall B \neq A; B \subseteq A$, i.e., we get a total ignorance function on the referential set A. This would be surprising in a learning process where A is a new piece of information and revision should improve our knowledge. The reason is that focusing is not made for revision and achieves no learning. To make it clear, suppose that the set of probabilities $\mathbb{P}$ represents knowledge stored in a database. Then $P^*_A(B)$ is part of the response of a query asking for the probability of being in B for an individual in A (focusing rule). The response should give $P_{*A}(B) = 1 - P^*_A(\bar{B})$ as well. On the other hand $P^*(B \mid A)$ is the result of modifying the database by enforcing $P(A) = 1$ (revision).

Focusing can be justified in terms of belief functions only, namely $Pl_A(B)$, as obtained in (3), can be viewed as the upper limit of a family of belief functions obtained by transferring for all focal elements such that $E_j \cap A \neq \emptyset$, $E_j \cap \bar{A} \neq \emptyset$ only one part of the mass m_j to the set $E_j \cap A$; see (De Campos et al., 1990). Each possible partial mass transfer from each focal element E_j to $E_j \cap A$ determines a possible way of changing the belief function into another one. The focusing process refrains from selecting among these potentially resulting belief functions and rather considers their lower envelope.

Examples: Consider the following small knowledge base

$$P(\text{young} \mid \text{student}) \geq 0.9$$
$$P(\text{single} \mid \text{young}) \geq 0.7$$

The focusing rule deals with the following problem: Tom is a student; what is the probability that he is single. It can be checked that P(single | student) is totally unknown: $P_*(\text{single} \mid \text{student}) = 0$, $P^*(\text{single} \mid \text{student}) = 1$. In particular nothing is known about P(student). The revision rule consists in entering the new piece of knowledge P(student) = 1, that expresses the fact that the knowledge base applies only to a population of students. Then it can be deduced that P(young) = P(young | student) ≥ 0.9 and P(single) = P(single | student) $\geq$ P(single | young) $\cdot$ P(young) = 0.63. However it is a mistake to model "Tom is a student" by P(student) = 1 since the former statement applies to Tom only while P(student) = 1 refers to the whole population under concern.

Another example illustrating the difference between focusing and revision can be given in the setting of belief functions: a die has been thrown by a player but he does not know the outcome yet. The player expects a rather high number (5 or 6), does not believe too much in a medium number (3, 4), and almost rules out the possibility of getting a small number (1, 2). He assigns a priori probability masses m({5,6}) = 0.7, m({3,4}) = 0.2 and m({1,2}) = 0.1 to focal elements. If he asks himself what if the outcome were not a six, then he has to focus on the situations where he would get no 6 and compute $\text{Bel}_A(B)$, $\text{Pl}_A(B)$ for A = {1, 2, 3, 4, 5}, and B ⊆ {1, 2, ..., 5}, where $\text{Bel}(B) = \Sigma_{E \subseteq B} m(E)$. For instance

$$\text{Bel}_A(5) = 0 = \text{Bel}(5); \text{Pl}_A(5) = \frac{0.7}{0.7 + 0.3} = 0.7 = \text{Pl}(5).$$

This is because he cannot rule out the situation where his belief in outcome 5 precisely would be zero. But if a friend tells him that the outcome is not a 6, then by Dempster's conditioning one gets Bel(5 | A) = 0.7. What happens is that, insofar as outcome 5 receives all the mass, 6 is now definitely ruled out. ∎

In Dempster's approach to belief function theory, a finite probability space (Ω, P) is projected to a set U through a multiple-valued mapping Γ thus inducing belief and plausibility functions on U, such that

$$\forall B \subseteq U, \text{Bel}(B) = P(B_*) / P(U^*) \text{ and } \text{Pl}(B) = P(B^*) / P(U^*)$$

where $B_* = \{\omega \mid \Gamma(\omega) \subseteq B\}$, $B^* = \{\omega \mid \Gamma(\omega) \cap B \neq \varnothing\}$ are the lower and upper images of B via Γ. Revision by a constraint A ⊆ U comes down to modifying the multiple-valued mapping $\Gamma: \Omega \to 2^U$ into $\Gamma_A: \Omega \to 2^A$ such that $\Gamma_A(\omega) = \Gamma(\omega) \cap A$, and normalizing the result. Denoting B_{*A} and B^{*A} the lower and upper images of B via Γ_A, yields

$$\text{Pl}(B \mid A) = P(B^{*A}) / P(A^{*A}) = P((B \cap A)^*) / P(A^*)$$

since $B^{*A} = (B \cap A)^* = \{\omega \mid \Gamma_A(\omega) \cap A \cap B \neq \varnothing\}$ as defined above. On the contrary focusing on A ⊆ U comes down to envisaging all the possible inverse images of A through mappings f: $\Omega \to$ U compatible with Γ, letting

$$\Gamma^{-1}(A) = \{f^{-1}(A) \mid f \in \Gamma\},$$

where f ∈ Γ means ∀ω, f(ω) ∈ Γ(ω), and to condition P on all possible such $f^{-1}(A)$ in Ω. Noticing that $\Gamma^{-1}(A) = \{C \mid A_* \subseteq C \subseteq A^*\}$, it can be proved that (Dubois and Prade, 1994a)

$$\text{Pl}_A(B) = \max_{A_* \subseteq C \subseteq A^*} P(B^* \mid C). \tag{4}$$

In other words, focusing on B ⊆ U comes down to conditioning on an ill-known event in Ω, due to the imprecision expressed via the multiple-valued mapping. On the contrary, Dempster's conditioning rule assumes that C = B* in the above theorem. This result may help in the practical computing of the focusing rule since it comes down to maximizing over a finite set (Ω is assumed to be finite).

3.2 - Focusing in Numerical Possibility Theory

Since numerical possibility measures are particular cases of plausibility functions that are max-decomposable, it makes sense to envisage possibility degrees as upper bounds of probabilities and to apply conditioning (3) to possibility and necessity functions Π and N:

$$\forall B, \quad \Pi_A(B) = \frac{\Pi(B \cap A)}{\Pi(B \cap A) + N(\bar{B} \cap A)} \tag{5}$$

and
$$N_A(B) = 1 - \Pi_A(\bar{B}) = \frac{N(B \cap A)}{N(B \cap A) + \Pi(\bar{B} \cap A)}. \tag{6}$$

When $N(A) > 0$, (5) and (6) do correspond to $\Pi_A(B) = \sup\{P(\cdot \mid A) \mid P \in \mathcal{P}(\Pi)\}$ and $N_B(A) = \inf\{P(\cdot \mid A) \mid P \in \mathcal{P}(\Pi)\}$, where $\mathcal{P}(\Pi) = \{P \mid P(B) \leq \Pi(B), \forall \, B \subseteq U\}$. Despite the strong probabilistic flavor of the above conditioning, it has been shown (Dubois and Prade, 1996) that Π_A and N_A are still possibility and necessity measures, i.e., Π_A is max-decomposable. This result was also recently pointed out by Walley (1996) for possibility measures. Moreover, given a possibility measure Π on a set U, the conditional set-function Π_A is associated with the possibility distribution

$$\begin{cases} \pi_A(u) = \max\left(\pi(u), \dfrac{\pi(u)}{\pi(u) + N(A)}\right) & \text{if } u \in A \\ \pi_A(u) = 0 & \text{if } u \notin A \end{cases} \tag{7}$$

with $\Pi_A(B) = \sup_{u \in B} \pi_A(u)$. The following properties are easy to check and expected as characterizing a focusing operation:

- if $N(A) = 0$, $\pi_A(u) = 1$, $\forall \, u \in A$ (total ignorance inside A when A is not an accepted belief, i.e., $N(A) > 0$ does not hold). This means that if nothing is known about A, focusing on A does not bring any information;
- $N_A(B) = N_A(A \cap B) = N_A(\bar{A} \cup B)$ (same property for Π_A). It means that when we focus on A, inquiring on B, $A \cap B$ or on $B \cup \bar{A}$ are the same. This is due to the fact that like in probability theory, $N_A(B)$ is the certainty degree of the conditional event $B \mid A$ (Dubois and Prade, 1994b), viewed as the family $\{X \mid B \cap A \subseteq X \subseteq \bar{A} \cup B\}$. Indeed, $N_A(B)$ only depends on $N(A \cap B)$ and $N(\bar{A} \cup B)$.

Other, more usual forms of conditioning exist in possibility theory in the numerical or in the ordinal setting. In case of a numerical possibility scale, we can also use Dempster rule of conditioning, specialized to possibility measures, i.e., consonant plausibility measures of Shafer (1976). It leads to the definition:

$$\forall B, \, B \cap A \neq \emptyset, \, \Pi(B \mid A) = \frac{\Pi(A \cap B)}{\Pi(A)} \tag{8}$$

provided that $\Pi(A) \neq \emptyset$. The conditional necessity function is defined by $N(B|A) = 1 - \Pi(\bar{B} \mid A)$, by duality.

As already said, focusing leads to a rather uninformative conditioning process. Indeed, it leads to belief-plausibility intervals that are wider and contain the interval obtained by Dempster rule. Dempster rule leads to a shrinkening the belief-plausibility intervals, thus acknowledging that the available knowledge has been enriched. On the contrary the focusing rule loses all information (except the assumption that the class of envisaged situations lies in A), because, in the absence of revision, no mass allocated to focal sets is allowed to flow to one of its proper subsets. The particular case of possibilistic focusing (7) indicates that this loss of information is systematic since π_A is always less specific that π on the set A. In other words, if some uncertain and incomplete information is available about a class of situations described on Ω, the induced information on a subclass of it characterized by a subset of Ω can be noticeably more imprecise.

4 - Non-Numerical Settings

The distinction made in the ordinal (and logical) setting between factual evidence and generic knowledge is in good agreement with the one arising in the previous section. A set of default rules $K = \{p_i \sim> q_i, i = 1,n\}$ can be considered as being analogous to a set $\mathbb{P}(K)$ of probabilities induced by conditional probability bounds $K = \{P(B_i \mid A_i) \geq a_i, i = 1,n\}$, where B_i (resp.: A_i) is the set of models of q_i (resp.: p_i). Focusing on a reference class A gathering the models of the proposition p leads to compute bounds of $P(B \mid A)$ induced by $\mathbb{P}(K)$ for some B of interest as previously explained. Let B be the set of models of q. If the coefficients a_i are infinitesimally close to 1, that is, of the form $1 - \varepsilon$, finding that $P(B \mid A) \geq 1 - \varepsilon$ in $\mathbb{P}(K)$ is strictly equivalent to inferring the rule $p \sim> q$ from K in the preferential system of Kraus et al. (1990)). This is not surprizing because the rule $p \sim> q$ can be viewed as a conditional objects $B \mid A$, in the sense of De Finetti (1974, pp. 307-312), such that $P(B \mid A)$ is the probability of $p \sim> q$, and the logic of conditional objects (Dubois and Prade, 1994b) is a model of the preferential system of Kraus et al. (1990). This is also closely related to the approach to default reasoning in the ordinal possibility setting.

The revision problem in the ordinal approach to possibility theory is treated in details in (Dubois and Prade, 1992b). Let us briefly recall it. When revising by A, each subset B of Ω is changed into $B \cap A$, and the level of possibility $\Pi(B)$ is carried over to $B \cap A$ if $B \cap A \neq \emptyset$. Finally, all subsets $C \subseteq \Omega$ such that $\Pi(C) = \max\{\Pi(B \cap A), B \subseteq \Omega\}$ are assigned a maximal level of possibility 1. It corresponds to a notion of conditional possibility $\Pi(B \mid A)$ defined as:

$$\Pi(B \mid A) = \sup \{a \in [0,1], \Pi(A \cap B) = \min(a, \Pi(A))\}.$$

This type of ordinal conditioning has been also recently studied by Williams (1994) under the name "adjustment". Recalling that in the ordinal case, the partial ignorance function pair (N, Π) is equivalent to a subset E of possible worlds equipped with a complete ordering relation expressing plausibility, revision by subset A comes down to restricting (without changing) this ordering to $A \cap E$. Just as in the case of Bayes conditioning the ordering of elements in $A \cap E$ is left unchanged. This revision satisfies all the Alchourrón-Gärdenfors-Makinson revision postulates, and the N function coincides with what Gärdenfors (1988) calls an "*epistemic entrenchment*".

The question of representing generic knowledge in ordinal possibility theory can be addressed if possibility theory is related to conditional theories of default reasoning (Kraus et al., 1990; Lehmann and Magidor, 1992). A body of generic knowledge K is encoded as a set of default rules $p \sim> q$, each being interpreted as a nonmonotonic inference rule "if p is true then it is normal that q be true". K is also called a conditional knowledge base. As proved in (Benferhat et al., 1992), it is natural to interpret a default rule $p \sim> q$, as a constraint on possibility measures of the form $\Pi(p \wedge q) > \Pi(p \wedge \neg q)$, i.e., such a default rule can be viewed as expressing that $p \wedge q$ is more normal than $p \wedge \neg q$. A set of default rules $K = \{p_i \to q_i, i = 1,n\}$ with noncontradictory condition parts can be viewed as a family of constraints

$$\mathcal{C}(K) = \{\Pi(p_i \wedge q_i) > \Pi(p_i \wedge \neg q_i), i = 1,n\}$$

restricting a family $\Pi(K) = \{\Pi, \Pi \in \mathbb{C}(K), \Pi(p_i) > 0, i = 1,n\}$ of possibility distributions over the interpretations of a language. Given a piece of evidence encoded as a propositional sentence p and the conditional knowledge base K, the sentence q is a plausible consequence of (K,p) iff the rule p ~> q can be deduced from K. This inference has a precise meaning in the framework of possibility theory namely the (classical) inference of the strict inequality $\Pi(p \wedge q) > \Pi(p \wedge \neg q)$ for all Π in $\Pi(K)$ derived from the set of constraints $\mathbb{C}(K)$. This inference also perfectly fits with the so-called "preferential inference system" of Kraus et al. (1990) and with the logic of conditional objects (Dubois and Prade, 1995). Another more productive type of inference is the so-called "rational closure" (Lehmann and Magidor, 1992). It can be captured in the possibilistic setting by selecting a particular possibility measure Π^* in $\Pi(K)$ (the least informed one, see (Benferhat et al., 1992)) and checking the condition $\Pi^*(p \wedge q) > \Pi^*(p \wedge \neg q)$. These methods of inference of a default rule p ~> q from a body of generic knowledge K can be viewed as a focusing on the reference class pointed at by p, for the purpose of plausibly inferring q. On the contrary, a revision process consists in modifying the set of rules K by adding or deleting rules, of by restricting the set of possible worlds.

The logic of conditional objects (or any equivalent approach) has also the merit of displaying the difference between two modes of belief revision: evidence focusing and knowledge expansion that can be defined (conditional objects encode a default rule p ~> q as an entity q | p which can be symbolically processed in agreement with a 3-valued semantics; see (Dubois and Prade, 1994b)), as follows:

- *Evidence focusing*: a new piece of evidence p arrives and makes the available information on the case at hand more complete. Then E is changed into $E \cup \{p\}$ (supposedly consistent). K remains untouched. But the plausible conclusions from K and $E \cup \{p\}$, i.e., c such that $K \models c \mid E \wedge p$ may radically differ from those derived from K and E.
- *Knowledge expansion*: it corresponds to adding new generic rules tainted with possible exceptions. Insofar as the new knowledge base is consistent it is clear that due to the monotonicity of inference $\models$, all plausible conclusions derived from K can still be derived from K' since if K is a subset of K' and $K \models c \mid E$ then $K' \models c \mid E$. But more conclusions may perhaps be obtained by K'.
- *Knowledge revision*: it encompasses the situation when the result of adding new generic rules to K leads to an inconsistency. In that case some mending of the knowledge base must be carried out in order to recover consistency. Preliminary results along this line are in Boutilier and Goldszmidt (1993).

5 - Concluding remarks

We have seen that the distinction between factual evidence and generic knowledge has important consequences for a proper handling of information in flux. It may also contribute to put the Ramsey test, which is an important issue in connection with belief change, in the correct perspective.

The idea of the Ramsey test is to model belief change operations by means of conditionals expressing propositions of the form "in the context of a belief set accepting an input A leads to accept proposition B" and to assume that these conditionals belong to the belief set. A well-know result of Gärdenfors (1986) shows that putting such conditionals in the belief set is inconsistent with basic postulates of belief change, if the minimal change principle is understood as using a mere expansion when the input does not contradict the state of belief. In particular these principles imply that revision is monotonic with respect to the belief set. But is is easy to convince oneself that belief revision, because of the possible inconsistency between the input information and the belief set cannot behave in a monotonic way.

The distinction between evidence E, knowledge K and plausible consequences C, makes also clear the need for other knowledge operations. In particular, if we want to modify the set C into a set C' by either adding desirable, or deleting undesirable conclusions, for a given evidence E, we would have to modify the generic knowledge K. This operation has been called "repairing K", in (Benferhat et al., 1996) where a method is proposed for tuning the set of conclusions derivable from K by applying rational monotony postulate (Gärdenfors and Makinson, 1994), without however questioning the conclusions obtained by preferential entailment (in the sense of Kraus, Lehmann and Magidor, 1990).

Besides, the introduction of new types of conditioning corresponding to the focusing operation, in the belief functions, and numerical possibility theory settings, suggests their possible use in the modelling of default rules. Preliminary investigations can be found in Dubois and Prade (1994a, 1997).

References

Benferhat S., Dubois D., Prade H. (1992) Representing default rules in possibilistic logic", Proc. of the 3rd Inter. Conf. on Principles of Knowledge Representation and Reasoning (KR'92), Cambridge, Mass., Oct. 26-29, 673-684.

Benferhat S., Dubois D., Prade H. (1996) Beyond counter-examples to nonmonotonic formalisms: A possibility-theoretic analysis. Proc. of the 12th Europ. Conf. on Artificial Intelligence (ECAI'96) (W. Wahlster, ed.), Budapest, Hungary, Aug. 11-16, 1996, John Wiley & Sons, New York, 652-656.

Boutilier C., Goldszmidt M. (1993) Revision by conditionals beliefs. Proc. of the 11th National Conf. on Artificial Intelligence (AAAI'93), Washington, DC, July 11-15, 649-654.

De Campos L.M., Lamata M.T., Moral S. (1990) The concept of conditional fuzzy measure. Int. J. of Intelligent Systems, 5, 237-246.

De Finetti B. (1974) Theory of Probability. Wiley, New York.

Dempster A.P. (1967) Upper and lower probabilities induced by a multiple-valued mapping. Annals of Mathematical Statistics, 38, 325-339.

Domotor Z. (1985) Probability kinematics — Conditionals and entropy principles. Synthese, 63, 74-115.

Dubois D., Prade H. (1992a) Evidence, knowledge, and belief functions. Int. J. of Approximate Reasoning, 6, 295-319.

Dubois D., Prade H. (1992b) Belief change and possibility theory. In: Belief Revision (P. Gärdenfors, ed.), Cambridge Univ. Press, Cambridge, UK, 142-182.

Dubois D., Prade H. (1994a) Focusing versus updating in belief function theory. In: Advances in the Dempster-Shafer Theory of Evidence (R.R. Yager, M. Fedrizzi, J. Kacprzyk, eds.), Wiley, New York, 71-95.

Dubois D., Prade H. (1994b) Conditional objects as nonmonotonic consequence relationships. IEEE Trans. on Systems, Man and Cybernetics, 24(12), 1724-1740.

Dubois D., Prade H. (1995) Conditional objects, possibility theory, and default rules. In: Conditional Logics: From Phisolophy to Computer Science (G. Crocco, L. Fariñas del Cerro, A. Herzig, eds.), Oxford University Press, 301-336.

Dubois D., Prade H. (1996) Focusing vs. revision in possibility theory. Proc. of the 5th IEEE Inter. Conf. on Fuzzy Systems (FUZZ-IEEE'96), New Orleans, LO, Sept. 8-11, 1700-1705.

Dubois D., Prade H. (1997) Bayesian conditioning in possibility theory. Fuzzy Sets and Systems, to appear.

Dubois D., Prade H., Smets P. (1996) Representing partial ignorance. IEEE Trans. on Systems, Man and Cybernetics, 26(3), 361-377.

Fagin R., Halpern J. (1989) A new approach to updating belief. Research Report RJ 7222, IBM Research Division, San Jose, CA.

Gärdenfors P. (1986) Belief revision and the Ramsay test for conditionals. Philosophical Review, 91, 81-83.

Gärdenfors P. (1988) Knowledge in Flux — Modeling the Dynamics of Epistemic States. The MIT Press, Cambridge, MA.

Gärdenfors P., Makinson D. (1994) Non-monotonic inference based on expectations. Artificial Intelligence, 65, 197-245.

Gilboa I., Schmeidler D. (1992) Updating ambiguous beliefs. Proc. of the 4th Conf. on Theoretical Aspects of Reasoning about Knowledge (TARK'92) (Y. Moses, ed.), Monterey, CA, March 22-25, 143-162.

Huber P.J. (1981) Robust Statistics. Wiley, New York.

Jaffray J.Y. (1992) Bayesian updating and belief functions. IEEE Trans. on Systems, Man and Cybernetics, 22, 1144-1152.

Jeffrey R. (1965) The Logic of Decision. McGraw-Hill, New York. 2nd edition, University of Chicago Press, 1983.

Katsuno H., Mendelzon A.O. (1991) On the difference between updating a knowledge base and revising it. Proc. of the 2nd Inter. Conf. on Principles of Knowledge Representation and Reasoning (KR'91) Cambridge, MA, April 22-25, 387-394.

Kraus S., Lehmann D., Magidor M. (1990) Nonmonotonic reasoning, preferential models and cumulative logics. Artificial Intelligence, 44, 167-207.

Kyburg H.E., Jr. (1987) Bayesian and non-Bayesian evidential updating. Artificial Intelligence, 31, 271-293.

Lehmann D., Magidor M. (1992) What does a conditional knowledge base entail? Artificial Intelligence, 55(1), 1-60.

Levi (1980) The Enterprize of Knowledge. The MIT Press, Cambridge, MA.

Makinson D., Gärdenfors P. (1991) Relations between the logic of theory change and nonmonotonic logic. The Logic of Theory Change (Proc. of the Workshop, Konstanz, Germany, Oct. 1989) (A. Fuhrmann, M. Morreau, eds.), Lecture Notes in Artificial Intelligence, Vol. 465, Springer Verlag, Berlin, 185-205.

Moral S., De Campos L. (1991) Updating uncertain information. In: Uncertainty in Knowledge Bases (B. Bouchon-Meunier, R.R. Yager, L.A. Zadeh, eds.), Springer Verlag, Berlin, 58-67.

Shafer G. (1976) A Mathematical Theory of Evidence. Princeton University Press, Princeton, NJ.

Walley P. (1996) Measures of uncertainty in expert systems. Artificial Intelligence, 83, 1-58.

Williams M.A. (1994) Transmutations of knowledge systems. Proc. of the 4th Inter. Conf. on Principles of Knowledge Representation and Reasoning (KR'94), Bonn, Germany, 619-632.

Williams P.M. (1980) Bayesian conditionalization and the principle of minimum information. British J. for the Philosophy of Sciences, 31, 131-144.

Background and Perspectives of Possibilistic Graphical Models

Jörg Gebhardt[1] and Rudolf Kruse[2]

[1] Dept. of Mathematics and Computer Science, University of Braunschweig
38106 Braunschweig, Germany
[2] Dept. of Computer Science, Otto-von-Guericke University
39106 Magdeburg, Germany

Abstract. Graphical modelling is an important tool for the efficient representation and analysis of uncertain information in knowledge-based systems. While Bayesian networks and Markov networks from probabilistic graphical modelling are well-known for a couple of years, the field of *possibilistic graphical modelling* occurs as a new promising area of research. Possibilistic networks provide an alternative approach compared to probabilistic networks, whenever it is necessary to model uncertainty *and* imprecision as two different kinds of imperfect information. Imprecision in the sense of set-valued data has often to be considered in situations where data are obtained from human observations or non-precise measurement units. In this contribution we present a comparison of the background and perspectives of probabilistic and possibilistic graphical models, and give an overview on the current state of the art of possibilistic networks with respect to propagation and learning algorithms, applicable to data mining and data fusion problems.

1 Introduction

One major aspect concerning the acquisition, representation, and analysis of information in knowledge-based systems is the development of an appropriate formal and semantical framework for the effective treatment of uncertain and imprecise data [17]. A task that frequently appears in applications is to describe a state ω_0 of the world as specific as possible using a tuple of instantiations of a finite number of variables (attributes), based on available *generic knowledge* and application-dependent *evidential knowledge*. As an example we mention medical diagnosis, where ω_0 could be the current state of health of a person, characterized by relationships between different attributes like diseases, symptoms, and relevant attributes for observation and measurement. The aim is to describe ω_0 as specific as possible with the aid of generic knowledge (medical rules, experience of medical doctors, databases of medical sample cases) and additional evidential knowledge about the person under consideration (the instantiations of those variables or attributes that can be found by asking and examining this person).

Imprecision comes into our considerations, when generic knowledge on dependencies among variables is rather relational than functional, and when the actual information on the current state does not result in a single tuple of variable instantiations, but rather in a set of alternatives. We only know that the current state is for sure one element of this set of alternatives, but we have no preferences that help us in selecting the "true" element. In case of medical diagnosis the set of alternatives could be the set of diseases that the medical doctor regards as possible explanations for the observed symptom, without making any preferences among these diseases. Such preferences are not referred to the imprecision part of the modelling approach. They rather reflect the uncertainty about identifying the current state ω_0 within its *frame of discernment* Ω, which is the product of the domains of all variables under consideration and thus the set of all possible candidates for ω_0.

Uncertainty arises in this context from the fact that in most cases there are no functional or relational dependencies among the involved variables, but only imperfect relationships that can be quantified by degrees of confidence. If, for example, the symptom *temperature* is observed, then various disorders might explain this symptom, and the medical doctor will assign his preferences to them. The modelling of this kind of uncertainty is done in an adequate calculus, for instance with help of a probabilistic approach, with concepts taken from non-standard uncertainty calculi like possibility theory, but also on a pure qualitative way by fixing a reasonable preference relation.

For simplicity, we restrict ourselves to *finite* domains of variables. Generic knowledge on ω_0 is assumed to refer to the uncertainty about the truth value of the proposition $\omega = \omega_0$, which is specified for all alternatives ω of Ω. Such knowledge can often be formalized by a *distribution function* on Ω, for example a probability distribution, a mass distribution, or a possibility distribution, dependent of the uncertainty calculus that best reflects the structure of the given knowledge.

The consideration of evidential knowledge corresponds to *conditioning* the prior distribution on Ω that is induced by the generic knowledge. Conditioning is often based on the instantiation of particular variables. In our example of medical diagnosis we could think of instantiating the variable *temperature* by measurement with a medical thermometer. The conditioning operation leads to an inference process that has the task to calculate the posterior marginal distributions for the non-instantiated variables, and to use these distributions as the basis for decision making about the component values of ω_0.

In addition to inference and decision making aspects it is of particular interest to answer the question how generic knowledge is obtained. In the sense of an automated induction of knowledge, this problem is quite easy to handle, if there is a database of sample cases that serve as an input for justified learning algorithms which determine the distribution function on Ω that best fits the database.

Since multidimensional frames of discernment Ω in applications of practical interest tend to be intractable as a whole because of their high cardinality, the aim of efficient methods for knowledge representation and the respective inference and learning algorithms is to take advantage from *independencies* that may exist between the variables under consideration.

Independencies support decomposition techniques that reduce operating on distributions to low-dimensional subspaces of Ω. For this purpose, the field of *graphical modelling* provides useful theoretical and practical concepts for an efficient reasoning under uncertainty [4, 25, 2, 19]. Applications of graphical models can be found in a variety of areas such as diagnostics, expert systems, planning systems, data analysis, and control.

In the following, we will investigate graphical modelling from the viewpoint of the uncertainty calculus of *possibility theory*. In Section 2 we discuss some principal aspects of graphical modelling. Section 3 is for clarifying the benefits and limits of possibilistic graphical models when compared to their probabilistic counterparts. The concept of a *possibilistic graphical model* and the underlying concepts of a *possibility distribution* and *possibilistic independence* are presented in Section 4. Referred to these foundations, Section 5 deals with efficient algorithms for evidence propagation in possibilistic graphical models. In Section 6, we focus on a specific *data mining problem*, which is how to induce possibilistic graphical models from databases of sample cases. Finally, Section 7 is for some additional remarks.

2 Graphical Models

A graphical model for the representation of uncertain knowledge consists of a qualitative and a quantitative component. The *qualitative (structural)* component is a graph, for example a directed acyclic graph (DAG), an undirected graph (UG) or a chain graph (CG). In a unique way it represents the conditional independencies between the variables represented by the nodes of the graph. For this reason, it is called the *conditional independence graph* of the respective graphical model. The *quantitative* component of a graphical model is a family of distribution functions on subspaces of Ω, whose structure is determined by the conditional independence graph. The distribution functions specify the uncertainty about the values of the projections of ω_0 on the corresponding subspaces. In case of a DAG, conditional distribution functions are used in order to quantify the uncertainty relative to each component value of tuple ω_0, dependent of all possible instantiations of the direct predecessors of the node identified with this component. On the other hand, an UG with induced hypergraph structure is the appropriate choice, if non-conditional distributions are preferred, defined, for any hyperedge, on the common product domain of all variables whose identifying nodes are contained in this hyperedge.

Figure 1 shows an example of a conditional independence graph with respect to an application, where expert knowledge represented by a graphical model successfully has been used in order to improve the results of modern cattle breeding programs [21]. The purpose is primarily parentage verification for pedigree registration. The application refers to the F-blood group system which controls 4 blood group factors for blood typing. The graphical model consists of 21 attributes (variables) that are relevant for determining the genotypes and verifying the parentage of Danish Jersey cattle.

The conditional independence graph reflects the dependencies and independencies that exist between the observable variables (lysis factors, phenogroups of stated dam and stated sire), and those variables that are relevant for the intended inference process (especially genotype, dam correct, and sire correct). The involved variables have at least two, and at most eight possible values for instantiation. The genotypes, for example, are coded by $F1/F1$, $F1/V1$, $F1/V2$, $V1/V1$, $V1/V2$, and $V2/V2$, respectively.

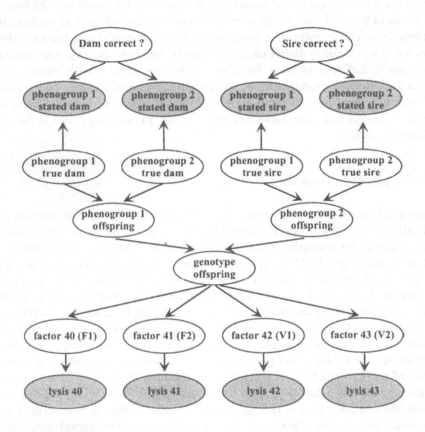

Fig. 1.: Graphical model for genotype determination and parentage verification Danish Jersey cattle in the F-blood group system

The conditional independence graph in Figure 1 can be interpreted as follows: Selecting any two nodes of the graph, their underlying variables are independent given any instantiation of all other variables whose number with respect to a topological order that agrees with the graph is not greater than the maximum of the order numbers of the two selected nodes. The *topological order* is chosen so that there is no arc from a node to another node with a asmaller order number.

Conditional independence refers to the uncertainty calculus that the graphical model is based on. In case of a probabilistic graphical model this means conditional independence of the random variables that are represented by the nodes in the graph. The common distribution function of these random variables is supposed to satisfy all independence relations expressed by the conditional independence graph. For this reason, the distribution has a factorization, where the probability of each elementary event obtained by a full instantiation of all variables equals a product of conditional probabilities in lower-dimensional subspaces. These factors are the probabilities of the single random variables given the chosen instantiation of their parent variables in the DAG [19, 25].

In the example shown in Figure 1, the decomposition provided by the mentioned factorization leads to the simplification that the user of the graphical model only needs to specify 306 conditional probabilities in subspaces of at most three dimensions, whereas the 21-dimensional frame of discernment Ω has 92.876.046.336 elements.

In the concrete application, the 306 needed conditional probabilities were specified on the base of statistical data and the experience of experts. The resulting probabilistic graphical model therefore consists of the DAG in Figure 1 as the qualitative part, and a family of 21 conditional probability distributions.

The aim of this type of graphical model is automated evidence propagation with a final step of decision making. Given, for instance, the lysis factors of a particular calf, and instantiating the corresponding variables (lysis 40, lysis 41, lysis 42, and lysis 43), the task is to calculate the posterior distributions of the other variables, especially the genotype and the parentage.

From an operational point of view, there are efficient algorithms for evidence propagation which are implemented in some commercial software tools. We will refer to this topic in more detail in Section 5.

3 Possibility Theory and Possibilistic Networks

A *possibilistic network* is a graphical model whose quantitative component is defined by a family of possibility distributions. A *possibility distribution* π is represented as a mapping from a referential set Ω to the unit interval. If π is used as an imperfect specification of a current state ω_0 of a part of the world, then $\pi(\omega)$ quantifies the degree of possibility that the proposition $\omega = \omega_0$ is true. From an intuitive point of view, $\pi(\omega) = 0$ means that $\omega = \omega_0$ is impossible, and $\pi(\omega) = 1$ says that this proposition is regarded as being possible without restriction. Any intermediary possibility degree $\pi(\omega) \in (0, 1)$ indicates that $\omega = \omega_0$ is true with restrictions, which means that there is evidence that supports this proposition as well as evidence that contradicts it.

Similar to the fact that one can find a variety of approaches to the semantics of *subjective probabilities*, there are different ways of introducing the semantics of the concept of a possibility distribution. In this context, we mention possibility distributions as the epistemic interpretation of fuzzy sets [27], the axiomatic approach to possibility theory with the aid of possibility measures [6, 7], and possibility theory based on likelihoods [5]. Furthermore, possibility distributions are also viewed as contour functions of consonant belief functions [22], or as falling shadows within the framework of set-valued statistics [24].

Introducing possibility distributions as information-compressed representations of databases of sample cases, it is convenient to interpret them as (non-normalized) one–point coverages of random sets [13], which leads to a very promising semantics of possibility distributions [8, 9]. With this approach it is quite simple to prove Zadeh's *extension principle* [26] as the adequate way of extending set-valued operations to their corresponding generalized operations on possibility distributions. It turns out that the extension principle is the only way of operating on possibility distributions that is consistent with the chosen semantical background of possibility distributions [8].

In this contribution, we confine to show an example that demonstrates how a database of sample cases is transformed into a possibility distribution. We therefore reconsider our blood group determination example of Danish Jersey cattle in Figure 1.

Table 1 represents a part of a database that is reduced to five attributes:

lysis 40	lysis 41	lysis 42	lysis 43	genotype offspring
$\{0, 1, 2\}$	6	0	6	$V2/V2$
0	5	4	5	$\{V1/V2, V2/V2\}$
2	6	0	6	*
5	5	0	0	$F1/F1$

Table 1.: Database with four sample cases

The three first rows specify imprecise sample cases, the fourth row shows a precise one. In row 1, we have three tuples $(0, 6, 0, 6, V2/V2)$, $(1, 6, 0, 6, V2/V2)$, and $(2, 6, 0, 6, V2/V2)$. They specify those states that have been regarded as possible alternatives when observing the calf that delivered the first sample case. These three tuples are the possible candidates for the state of this calf, concerning the relationship between lysis factors and genotype. In a similar way, the second row shows two possible states with respect to the second sample calf. An unknown value in the third row is indicated by *.

Assuming that the four sample cases in Table 1 are in the same way representative in order to specify possible relationships between the five selected attributes, it is reasonable to fix their probability of occurrence to 1/4.

From a probabilistic point of view, we may apply the *insufficient reason principle* to the database, stating that alternatives in set-valued sample cases are equally likely, if preferences among these alternatives are unknown. Uniform distributions on set-valued sample cases lead to a refined database of $3 + 2 + 6 + 1 = 12$ data tuples, where, for instance, $(2, 6, 0, 6, V2/V2)$ has an occurrence probability of $1/3 * 1/4 + 0 + 1/6 * 1/4 + 0 = 3/24$.

In a possibilistic interpretation of the database, we obtain for the same tuple a *degree of possibility* of $1/4 + 0 + 1/4 + 0 = 1/2$, since this tuple is considered as being possible in the first and in the third sample, but it is excluded in the two other samples. Calculating the possibility degrees for all tuples of the common domain of the five selected attributes, we get an information-compressed interpretation of the database in form of a possibility distribution. Besides the different semantical background, it is obvious that this possibility distribution is *not* a probability distribution, since the underlying possibility degrees do not sum up to 1.

Probability theory and possibility theory as the theory of possibility distributions and possibility measures [6] both provide frameworks for the formal and semantical treatment of uncertain data, but differ in their domains of appliaction. For their use in the field of knowledge-based systems with finite frames of discernment, we may emphasize the following characteristics of distinction:

Probability Theory

- serves for the *exact* modelling of uncertain, but *precise* data. There is a unique representation of a database of sample cases with the aid of a probability distribution, only if the data tuples are precise, so that all variables that are addressed within a sample case can be instantiated by observation with a particular attribute value. As a consequence, imprecise observations that lead to sets of alternative attribute values are not allowed in this modelling approach. Appropriate additional assumptions have therefore to be made in order to transform a single imprecise observation to a set of precise observations. As an example for such additional assumption we mentioned the insufficient reason principle.

- is able to model *imprecision* in set-valued sample cases by using families of probability distributions instead of single probability distributions. This kind of solution refers, for example, to *probability intervals* or *upper and lower probabilities*, respectively (cf. [17]). On the other hand, in non-trivial cases the corresponding approaches often give rise to major complexity problems, so that they are of limited interest for efficient reasoning in knowledge-based systems.

- is very suitable for applications that provide statistical data or expert knowledge that can be quantified in terms of subjective probabilities. The completeness and the normative character of probabilistic modelling for handling precision under uncertainty supports the surely most justified form of decision making.

Possibility Theory

- serves for the *approximative (information-compressed)* modelling of uncertain and/or *imprecise (set-valued)* data. For example, suppose that the available information about a sample case in a database can only be specified by a set of possible candidates for the current state ω_0. In a possibilistic approach, it is assumed that there may exist preferences among these candidates, but they are unknown, and thus it is not justified to make any particular assumptions about preferences. Under such restrictions, it is not adequate to apply concepts like the insufficient reason principle. In order to deal with the complexity problems that arise when operating on imprecise data, the information contained in the database is compressed. From a formal point of view, the information compression is due to the transformation of the random set that represents the database into its non-normalized one-point coverage. The occurring loss of information does not allow exact reasoning like in the probabilistic case, but at least approximate reasoning. On the other hand, examples from industrial practice show that many applications are not sensitive to approximate instead of exact reasoning, since the influence of data precision on the quality of decision making is often overestimated. Possibilistic reasoning can thus be established in knowledge-based systems the same way as *fuzzy control* has been established in engineering: Fuzzy control is an alternative type of approximate reasoning, namely *interpolation* in vague environments on the mathematical basis of *similarity relations* [15, 16].

- can also handle precise data as a special case of imprecise data, but as a consequence of the tolerated loss of information when dealing with possibility distributions, the reasoning process is less specific than in probabilistic approaches. The specificity of possibilistic inference mechanisms may suffice for the particular application. However, a pure probabilistic model seems to be more effective in case of precise data under uncertainty.

- has advantages in those applications of uncertain data, where a relevant part of the given information is imprecise. Sine the concept of a possibility distribution has a reasonable semantic foundation with the extension principle as a consistent way of operating on such distributions, there is a basis for well-founded possibilistic inference techniques. The establishment of methods of possibilistic decision making is topic of current research in possibility theory.

4 Theoretical Foundations of Possibilistic Networks

Graphical models take benefits of conditional independence relations between the considered variables. The basic aim is to reduce operations on distribution functions from the multidimensional frame of discernment Ω to lower-dimensional subspaces, so that the development of efficient inference algorithms is supported. Hence, the starting point of theoretical investigations for graphical models is to find an appropriate concept of *conditional independence* of variables relative to the uncertainty calculus under consideration. While the notion of (probabilistic) conditional independence is well-known for a long time, there is still some need for discussion of analogous concepts in the possibilistic setting. The main reason for it comes from the fact that possibility theory is suitable for the modelling of two different kinds of imperfect knowledge (uncertainty and imprecision), so that there are at least two alternative ways of approaching independence.

Furthermore, the various proposals for the semantics of possibility distributions allow the foundation of different concepts of independence [3]. Nevertheless, all these proposals agree to the following general definition:

Given three disjoint sets X, Y, and Z of variables (attributes), where X and Y are both not empty, X is called *independent* of Y given Z relative to a possibility distribution π on Ω, if any additional information on the variables in Y, formalized by any instantiations of these variables, the possibility degrees of all common tuples of values of the variables in X remain unchanged, if a particular instantiation of the variables in Z is chosen.

Using other words: If the Z-values of ω_0 are known, then additional restrictive information on the Y-values of ω_0 are useless in order to get more information on possible further restrictions of the X-values of ω_0.

From an operational point of view, the independence condition may be read as follows:

Suppose that a possibility distribution π is used to imperfectly specify the state ω_0. Given any crisp knowledge on the Z-values of ω_0, this distribution is conditioned with respect to the corresponding instantiations of the variables in Z. The projection of the resulting possibility distribution π on the X-values equals the possibility distribution that we obtain if we condition π with respect to the same instantiation of the variables in Z and any instantiation of the variables in Y, and then project the conditioned possibility distribution on the X-values.

How *conditioning* and *projection* have to be defined, depends on the chosen semantics of possibility distributions: If we view possibility theory as a special case of the theory of Dempster-Shafer [22], by interpreting a possibility distribution as a representation of a consonant belief function or a nested random set, then we obtain the concept of conditional indpendence of sets of variables in possibilistic graphical models from the so-called *Dempster conditioning* [22]. On the other hand, if we regard possibility distributions as (non-normalized) one-point coverages of random sets, as we used to do it for the possibilistic interpretation of databases of sample cases, then we have to choose the conditioning and the projection operation in conformity with the extension principle. The resulting type of conditional independence corresponds to *conditional possibilistic non-interactivity* [14]. For more details on this axiomatic approach to possibilistic independence, we refer to [3].

The mentioned types of conditional independence satisfy the *semi-graphoid-axioms* that have been established as the basic requirements to any reasonable concept of conditional independence in graphical models [19]. Possibilistic conditional independence in the sense of the Dempster-Shafer approach even satisfies the *graphoid-axioms* [20], in analogy with the case of probabilistic conditional independence.

With the foundations presented above, it is straight-forward to define what is called a *possibilistic independence graph*. By this we can interpret possibilistic networks, whose structural component is an UG. In a similar way it is then quite simple to introduce directed conditional independence graphs.

An undirected graph with V as its set of nodes is called *conditional independence graph* of a possibility distribution π, if for any non-empty disjoint sets X, Y, and Z of nodes from V, the separation of each node in X from each node in Y by the set Z of nodes implies the possibilistic independence of X from Y given the condition Z.

This definition makes the assumption that the so-called *global Markov property* holds for π [25]. In opposite to probability distributions, where the equivalence of global, local, and pairwise Markov property can be verified, the possibilistic setting establishes the global Markov property as the strongest of the three properties, so that it has been selected as the basis for the definition of possibilistic conditional independence graphs.

A directed acyclic conditional independence graph (see Figure 1) can be transformed to its associated undirected *moral graph*, which is obtained by eliminating the direction of all arcs in the DAG, and by "marrying" all parent nodes with the aid of additional combining edges [18]. Figure 2 is a modified representation of the DAG in Figure 1, where the new edges in the moral graph are indicated by the dotted lines.

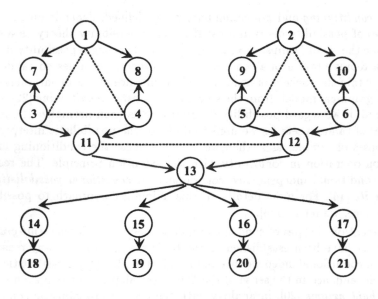

Fig. 2.: Modified representation of the DAG in Figure 1

The moral graph satisfies a subset of the independence relations of the underlying DAG, so that using the moral graph in the worst case is connected with a loss of independence information. *Triangulation* of the moral graph, which perhaps causes further loss of independence information, yields an UG whose hypergraph representation is a *hypertree*. The hyperdeges of this hypertree correspond to the nodes of the *tree of cliques* that is associated with the triangulated moral graph. The tree of cliques that may be constructed from the DAG in our blood group determination example is shown in Figure 3.

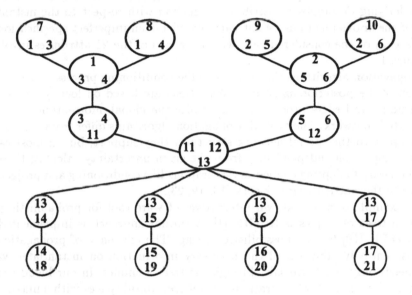

Fig. 3.: Tree of cliques of the triangulated DAG in Figure 1

The main advantage of triangulation consists in the fact that the resulting graph supports the *decomposability* properties that we need for efficient operations on distribution functions:

A possibility distribution π on Ω has a decomposition into complete irreducible components, if the decomposition is referred to a triangulated conditional independence graph G of π. The *factorization* of π relative to this decomposition is characterized in the way that π has a representation as the minimum of its projections on the maximal cliques of G. This result coincides with the decomposition of probability distributions, when the minimum operator is substituted by the product, and when degrees of possibility are changed to probabilities of elementary events.

5   Evidence Propagation in Possibilistic Networks

Dependent of the structural properties of the conditional independence graph and thus the decomposability of the particular distribution function of a graphical model, it is possible to develop efficient algorithms for evidence propagation.

In the example from Figure 1 the major task is to get information about the genotype and the correctness of assumed parentage of a given calf under consideration. Available evidential knowledge refers to the eight grey nodes in Figure 1. It is based on measuring the lysis factors and determining the phenogroups of the stated parents. Generic knowledge is coded by the family of distribution functions that specifies the quantitative part of the graphical model. Conditioning the underlying decomposed distribution function with respect to the instantiations of the above mentioned eight variables yields an imperfect specification of the state ω_0 of the considered calf, which consists of the 21 attributes involved in Figure 1.

Propagation algorithms that carry out the conditioning process and the calculation of the posterior marginal distributions are based on locally communicating node- and edge- processors that realize the global information exchange in the whole network. The typical propagation algorithms differ from each other with respect to the special network structures they support, but in most cases they are applicable independently from the given uncertainty calculus; there is only the need of adapting elementary operations like conditioning and projection relative to the particular framework [18, 19, 23].

An example of a well-known interactive software tool for probabilistic reasoning in trees of cliques is HUGIN [1]. A similar approach is implemented in POSSINFER [10] for the possibilistic setting. The efficiency of propagation is obvious, since distributing all the necessary information on instantiated variables needs only two traversals through the tree of cliques. In our blood group determination example all operations are referred to subspaces with a maximum of three dimensions.

6   Learning Possibilistic Networks from Data

If expert knowledge is not sufficient to fully specify the structural as well as the quantitative part of a possibilistic graphical model, then the question arises whether this specification can also be induced with the aid of a database of (imprecise, set-valued) sample cases. More particularly, the learning task consists of finding the best approximating decomposition of the possibility distribution that is generated from the database, relative to a chosen class of possibilistic graphical models. Structure learning can be referred to DAG structures [12] as well as hypertree structures [11]. For quantifying the approximation quality, one needs a *measure of nonspecificity* for the information represented by a possibility distribution. While probabilistic measures of uncertainty usually are based on Shannon entropy, the origin of possibilistic measures of nonspecificity is Hartley information. Theoretical considerations on this topic can be found in [11].

The addressed learning problem is NP-hard in case of non-trivial classes of graphical models, even if we restrict to n-ary relations that can be regarded as special cases of n–dimensional possibility distributions. For this reason, in analogy to learning probabilistic graphical models, heuristics are not avoidable. On the other hand, the approximation quality of the calculated graphical model can easily be calculated, so that the loss of specificity that has to be tolerated when dealing with a family of lower-dimensional possibility distributions instead of the original possibility distribution is known.

The learning algorithm in [11] approximates optimal hypertree decompositions of possibility distributions and has successfully been applied to the blood group determination example. Network induction was based on an artificial database for the 21 attributes, generated from a real database for 9 attributes and additional expert knowledge. The database consists of 747 sample cases with a large number of incomplete or imprecise cases inside.

7 Final Remarks

The concepts and methods of possibilistic graphical modelling presented in this paper have been applied within the Esprit III BRA DRUMS II (Defeasible Reasoning and Uncertainty Management Systems) and in a cooperation with Deutsche Aerospace for the conception of a data fusion tool. Furthermore, the learning algorithms are involved in the conception of a data mining tool that is developed in the research center of the Daimler-Benz AG in Ulm, Germany.

References

1. S.K. Andersen, K.G. Olesen, F.V. Jensen, and F. Jensen. HUGIN — A shell for building Bayesian belief universes for expert systems. In *Proc. 11th International Joint Conference on Artificial Intelligence*, pages 1080–1085, 1989.
2. W. Buntine. Operations for learning graphical models. *J. of Artificial Intelligence Research*, 2:159–224, 1994.
3. L.M. de Campos, J. Gebhardt, and R. Kruse. Axiomatic treatment of possibilistic independence. In C. Froidevaux and J. Kohlas, editors, *Symbolic and Quantitative Approaches to Reasoning and Uncertainty, Lecture Notes in Artificial Intelligence 946*, pages 77–88. Springer, Berlin, 1995.
4. E. Castillo, J.M. Gutierrez, and A.S. Hadi. *Expert Systems and Probabilistic Network Models*. Series: Monographs in Computer Science. Springer, New York, 1997.
5. D. Dubois, S. Moral, and H. Prade. A semantics for possibility theory based on likelihoods. Annual report, CEC–ESPRIT III BRA 6156 DRUMS II, 1993.
6. D. Dubois and H. Prade. *Possibility Theory*. Plenum Press, New York, 1988.
7. D. Dubois and H. Prade. Fuzzy sets in approximate reasoning, Part 1: Inference with possibility distributions. *Fuzzy Sets and Systems*, 40:143–202, 1991.
8. J. Gebhardt and R. Kruse. A new approach to semantic aspects of possibilistic reasoning. In M. Clarke, S. Moral, and R. Kruse, editors, *Symbolic and Quantitative Approaches to Reasoning and Uncertainty*, pages 151–159. Springer, Berlin, 1993.

9. J. Gebhardt and R. Kruse. On an information compression view of possibility theory. In *Proc. 3rd IEEE Int. Conf. on Fuzzy Systems (FUZZIEEE'94)*, pages 1285–1288, Orlando, 1994.

10. J. Gebhardt and R. Kruse. POSSINFER — A software tool for possibilistic inference. In D. Dubois, H. Prade, and R. Yager, editors, *Fuzzy Set Methods in Information Engineering: A Guided Tour of Applications*, pages 407–418. Wiley, New York, 1996.

11. J. Gebhardt and R. Kruse. Tightest hypertree decompositions of multivariate possibility distributions. In *Proc. Int. Conf. on Information Processing and Management of Uncertainty in Knowledge–Based Systems (IPMU'96)*, pages 923–927, Granada, 1996.

12. J. Gebhardt and R. Kruse. Automated construction of possibilistic networks from data. *J. of Applied Mathematics and Computer Science*, 6(3):101–136, 1996.

13. K. Hestir, H.T. Nguyen, and G.S. Rogers. A random set formalism for evidential reasoning. In I.R. Goodman, M.M. Gupta, H.T. Nguyen, and G.S. Rogers, editors, *Conditional Logic in Expert Systems*, pages 209–344. North–Holland, 1991.

14. E. Hisdal. Conditional possibilities, independence, and noninteraction. *Fuzzy Sets and Systems*, 1:283–297, 1978.

15. F. Klawonn, J. Gebhardt, and R. Kruse. Fuzzy control on the basis of equality relations with an example from idle speed control. *IEEE Transactions on Fuzzy Systems*, 3:336–350, 1995.

16. R. Kruse, J. Gebhardt, and F. Klawonn. *Foundations of Fuzzy Systems*. Wiley, Chichester, 1994.

17. R. Kruse, E. Schwecke, and J. Heinsohn. *Uncertainty and Vagueness in Knowledge Based Systems: Numerical Methods*. Artificial Intelligence. Springer, Berlin, 1991.

18. S.L. Lauritzen and D.J. Spiegelhalter. Local computations with probabilities on graphical structures and their application to expert systems. *Journal of the Royal Stat. Soc., Series B*, 2(50):157–224, 1988.

19. J. Pearl. *Probabilistic Reasoning in Intelligent Systems: Networks of Plausible Inference (2nd edition)*. Morgan Kaufmann, New York, 1992.

20. J. Pearl and A. Paz. Graphoids – A graph based logic for reasoning about relevance relations. In B.D. Boulay et al., editors, *Advances in Artificial Intelligence 2*, pages 357–363. North–Holland, Amsterdam, 1991.

21. L.K. Rasmussen. Blood group determination of Danish Jersey cattle in the F-blood group system. *Dina Research Report 8*, Dina Foulum, 8830 Tjele, Denmark, November 1992.

22. G. Shafer. *A Mathematical Theory of Evidence*. Princeton University Press, Princeton, 1976.

23. G. Shafer and P.P. Shenoy. Local computation in hypertrees. Working paper 201, School of Business, University of Kansas, Lawrence, 1988.

24. P.Z. Wang. From the fuzzy statistics to the falling random subsets, In P.P. Wang, editor. *Advances in Fuzzy Sets, Possibility and Applications*, 81–96, Plenum Press, New York, 1983.

25. J. Whittaker. *Graphical Models in Applied Multivariate Statistics*. Wiley, 1990.

26. L.A. Zadeh. The concept of a linguistic variable and its application to approximate reasoning. *Information Sciences*, 9:43–80, 1975.

27. L.A. Zadeh. Fuzzy sets as a basis for a theory of possibility. *Fuzzy Sets and Systems*, 1:3–28, 1978.

Checking Several Forms of Consistency in Nonmonotonic Knowledge-Bases*

Bertrand Mazure, Lakhdar Saïs and Éric Grégoire

CRIL – Université d'Artois
rue de l'Université SP 16
F-62307 Lens Cedex, France
{mazure,sais,gregoire}@cril.univ-artois.fr

Abstract. In this paper, a new method is introduced to check several forms of logical consistency in nonmonotonic knowledge-bases (KBs). The knowledge representation language under consideration is full propositional logic, using "Abnormal" propositions to be minimized. Basically, the method is based on the use of local search techniques for SAT. Since these techniques are by nature logically incomplete, it is often believed that they can only show that a formula is consistent. Surprisingly enough, we find that they can allow inconsistency to be proved as well. To that end, some additional heuristic information about the work performed by local search algorithms is shown of prime practical importance. Adapting this heuristic and using some specific minimization policies, we propose some possible strategies to exhibit a "normal-circumstances" model or simply a model of the KB, or to show their non-existence.

1 Introduction

Assume that a knowledge engineer has to assemble several logic-based propositional knowledge modules, each of them describing one subpart or one specific view of a complex device. In order to make fault-diagnosis possible in the future, each knowledge module describes both normal functioning conditions and faulty ones. To this end, the ontology is enriched with McCarthy's additional "Abnormal" propositions (noted Ab_i) [7] allowing default rules to be expressed together with faulty operating conditions. For instance, the rule asserting that, under normal circumstances, when the switch is on then the lights should be on is represented by the formula $switch\_on \land \neg Ab_1 \Rightarrow lights\_on$, and in clausal form by $\neg switch\_on \lor Ab_1 \lor lights\_on$. In this very standard framework, Ab_i propositions are expected to be **false** under normal operating circumstances of the device. The knowledge based system (KB) is expected to be used in a nonmonotonic way, in the sense that conclusions can be inferred when they are satisfied in some preferred models of the KB where Ab_i are **false**. Also, model-based diagnosis [9] can be performed in order to localize faulty components in the presence of additional factual data.

* This work has been supported by the Ganymède II project of the "Contrat de Plan Etat/Nord–Pas-de-Calais".

The specific issues that we want to address in this framework is the following one. How can the knowledge engineer check that the global KB is consistent? Also, how can he exhibit (or show the non-existence of) one model of the KB translating normal functioning conditions of the device? We would like these questions to be answered for very large KBs and practical, tractable, methods be proposed.

Actually, consistency checking is a ubiquitous problem in artificial intelligence. First, deduction can be performed by refutation, using inconsistency checking methods. Also, many patterns of nonmonotonic reasoning rely on consistency testing in an implicit manner. Moreover, ensuring the logical consistency of logical KBs is essential. Indeed, inconsistency is a serious problem from a logical point of view since it is global under complete (standard) logical rules of deduction. Even a simple pair of logically conflicting pieces of information gives rise to global inconsistency: every formula (and its contrary) can be deduced from it. This problem is even more serious in the context of combining or interacting several knowledge-based components. Indeed, individually consistent components can exhibit global inconsistency, due to distributed conflicting data.

Unfortunately, even in the propositional framework, consistency checking is intractable in the general case, unless P = NP. Indeed, SAT (i.e., the problem of checking the consistency of a set of propositional clauses) is NP-complete. Recently, there has been some practical progress in addressing hard and large SAT instances. Most notably, simple new methods [11] that are based on local search algorithms have proved very efficient in showing that large and hard SAT instances are consistent. However, these methods are logically incomplete in that they cannot prove that a set of clauses is inconsistent since they do not consider the whole set of possible interpretations.

However, we have discovered the following phenomenon very recently [6]. When we trace the work performed by local search algorithms when they fail to prove consistency, we extract a very powerful heuristic allowing us to locate probable inconsistent kernels extremely often. Accordingly, we proposed a new family of powerful logically complete and incomplete methods for SAT.

In this paper, we extend this previous work in order to address nonmonotonic propositional KBs, using the above "Abnormal" propositions that are expected to be false under normal circumstances. Using this preference for normal conditions, we guide the local search towards a possible "normal circumstances" model. When such a model is not found, several issues can be addressed. First, using the above heuristic and assuming that normal circumstances are satisfied, we propose a technique that allows us to prove (very often) the absence of such a model and to exhibit an inconsistent kernel. Then, dropping the special status of "Abnormal" propositions to several possible extent, we introduce various strategies for showing the consistency or inconsistency of the KB.

The paper is organized as follows. First, we recall some background about SAT, local search methods and our specific knowledge representation language. Then, we explain how the trace of local search algorithms can give rise to a powerful heuristic to locate inconsistent kernels. We then propose a family of logically complete and incomplete techniques for SAT, using a specific policy towards "Abnormal" propositions. More precisely, before relating some experimental results about the consistent combination of very large propositional KBs, we discuss some possible strategies to exhibit a normal-circumstances model or simply a model of the KB, or show their non-existence.

2 "Abnormal" propositions, SAT and local search techniques

The knowledge representation language under consideration in this paper is full propositional logic, using a finite set of propositional variables and standard logical connectives. Knowledge is assumed to be represented in clausal form. As explained in the introduction, default knowledge is represented using additional "Abnormal" propositional variables Ab_i, allowing faulty components or behaviours to be represented. Let us stress that we make a very specific use of Ab_i variables. They are intended to represent unexpected faulty conditions of the device. This means that when the KB is assembled, they could be consistently assumed **false**. This is a restricted use in the sense that we do not represent default rules where possible exceptions are expected to exist in the initial KB, like "Typical humans are right-handed".

SAT is the problem of checking whether a set of propositional clauses is consistent, i.e. whether there exists an interpretation assigning values from {**true**, **false**} to the propositional variables so that the clauses are **true** under standard compositional rules of interpretation. Such an interpretation is called a *model* of the SAT instance.

A *normal-circumstances model* of a SAT instance is defined in this paper as a model of the SAT instance where all Ab_i are interpreted false.

Although SAT is NP-complete, there has been significant progress in showing the consistency of very large and hard SAT instances, using very simple local search techniques. Let us simply recall here the most representative one, namely Selman et al.'s GSAT algorithm [11, 12]. This algorithm (see Fig. 1) performs a greedy local search for a satisfying assignment of a set of propositional clauses. The algorithm starts with a randomly generated truth assignment. It then changes ("flips") the assignment of the variable that leads to the largest increase in the total number of satisfied clauses. Such flips are repeated until either a model is found or a preset maximum number of flips (MAX-FLIPS) is reached. This process is repeated as needed up to a maximum of MAX-TRIES times.

In the following, we shall use a variant of GSAT, namely TSAT [5], a local search technique using a tabu list forbidding recurrent flips. Let us stress that all results presented in the following Sections keep holding for most members of the family of GSAT-like algorithms.

```
Procedure  GSAT
Input: a set of clauses S, MAX-FLIPS, and MAX-TRIES
Output: a satisfying truth assignment of S, if found
  Begin
    for i := 1 to MAX-TRIES do
      I := a randomly generated truth assignment
      for j := 1 to MAX-FLIPS do
          if I satisfies S then return I
          x := a propositional variable such that a change in its
              truth assignment gives the largest increase in the
              number of clauses² of S that are satisfied by I
          I := I with the truth assignment of x reversed
      done
    done
    return "no satisfying assignment found"
  End
```

Fig. 1. GSAT Algorithm: basic version

3   A heuristic to locate inconsistent kernels

In this Section, a somewhat surprising finding is presented: GSAT-like algorithms can be used to localize inconsistent kernels of propositional KBs, although the scope of such logically incomplete algorithms was normally expected to concern the proof of consistency only.

The following test[3] has been repeated very extensively, giving rise to the same result extremely often [6]. TSAT (or any other GSAT-like algorithms) is run on a SAT instance. The following phenomenon is encountered when the algorithm fails to prove that the instance is consistent. TSAT is traced and, for each clause, taking each flip as a step of time, the number of times during which this clause is falsified is updated. A similar trace is also recorded for each literal occurring in the SAT problem, counting the number of times it has appeared in the falsified clauses. Intuitively, it seemed to us that the most often falsified clauses should normally belong to an inconsistent kernel of the SAT problem if this problem is actually inconsistent. Likewise, it seemed to us that the literals that exhibit the highest scores should also take part in this kernel.

Actually, this hypothesis proved most of the time experimentally correct. This phenomenon can be summarized as follows. When GSAT-like algorithms are run on a locally inconsistent SAT problem (i.e. on a SAT problem whose smallest inconsistent kernel is a small subpart of it), then the above counters allow us to split the SAT problem into two parts: a consistent one and an unsatisfiable one.

---

[2] this number can be negative

[3] all our experimentations have been conducted on a 166 Pentium PC.

A significant gap between the scores of two parts of the clausal representation are obtained, differentiating a probable inconsistent kernel from the remaining part of the problem. Strangely enough, it appears thus that the trace of GSAT-like algorithms delivers the probable inconsistent kernel of locally inconsistent propositional KBs.

Let us stress that the validity of this experimental result depends on the size of the discovered probable kernel with respect to the size of the SAT instance. When the SAT instance tends to be globally inconsistent, i.e. when the size of the smallest inconsistent kernel converges towards the size of the SAT instance, no significant result is obtained. Fortunately, most real-life inconsistent KBs exhibit some small kernels of conflicting information.

We have proposed and experimented several ways to use the above heuristic information to prove inconsistency in a formal way [6]. The most straightforward one consists in using GSAT-like algorithms to detect the probable inconsistent kernel. Then, complete techniques like Davis and Putnam procedure (in short DP) [1] can be run on this kernel to prove its unsatisfiability and thus, consequently, the inconsistency of the global problem. Also, we can sort the clauses according to their decreasing scores and use incremental complete techniques on them until unsatisfiability is proved. In this respect, clauses outside the discovered probable kernel could also be taken into account.

More generally, the scores delivered by the GSAT-like algorithm are used to guide the search performed by the complete technique. For instance, the following (basic) procedure (seeFig. 2) combining DP with our heuristic proves extremely competitive. TSAT is run to deliver the next literal to be assigned the truth-value true by DP. This literal is selected as the one with the highest score as explained above. Such an approach can be seen as using the trace of TSAT as a heuristic in order to select the next literal to be assigned true by DP, and a way to extend the partial assignment made by DP towards a model of the SAT instance when this instance is satisfiable. Each time DP needs to select the next variable to be considered, such a call to TSAT can be performed with respect to the remaining part of the SAT instance.

4   Using "Abnormal" propositions to guide local search

Assume now that several knowledge modules need to be combined. Each module represents the knowledge about e.g. one device component or one specific view of it. "Abnormal" propositions are used to represent normal functioning circumstances and it is expected that they can be consistently interpreted as **false**. Let us make the global consistency problem more precise in this framework.

For the moment, let us assume that each module is consistent. The technique we shall develop can be used to check this basic assumption, too.

Question 1. Can we prove that the global system combining all the modules is consistent? In the negative case, can we locate the conflicting distributed information?

```
Procedure  DP+TSAT(S)
Input: a set S of clauses.
Output: a satisfying truth assignment of S if found, or a definitive
        statement that S is inconsistent.
  Begin
    S := Unit_propagate(S)
    if the empty clause is generated  then return (false)
    else if  all variables are assigned then return (true)
    else begin
        if  TSAT(S) succeeds then return (true)
        else begin
            p := the most often falsified literal during TSAT search
            return ( DP+TSAT(S∧p)∨ DP+TSAT(S∧¬p))
        end
    end
  End
```

Fig. 2. DP+TSAT Algorithm

The first question is not very satisfactory in our specific framework. Indeed, if we extract one model of the modules where some Ab_i are **true**, this will not inform us about the possible existence (or non-existence) of good (i.e. not abnormal) working conditions for the described device. However, this would be a relevant issue with respect to the consistent combination of several knowledge modules, where Ab_i are used to represent default rules where exceptions are expected to exist in the KB, like "Typical human are right-handed". Focusing on our diagnosis-oriented framework, we shall turn our attention to the following adapted consistency issues.

Question 2. Can we prove that the global system combining all the modules is normal-circumstances consistent in the sense that it exhibits at least one model where all Ab_i are **false**?

In the negative case, we can try to solve the following subsequent issues.

Question 3. Can we formally prove that no normal-circumstances model does exist, and can we locate the information that is conflicting when all Ab_i are assigned **false**?

Question 4. Can we exhibit a model where some Ab_i are **true**?

Question 5. When the answer to *Question 4.* is negative, can we locate the conflicting distributed information?

Let us adapt local search methods to the above problems. Most local search methods generate one initial interpretation randomly. However, addressing *Question 2*, the initial interpretation must be generated so that every Ab_i variable is assigned **false** while the truth value of the other variables is still selected randomly. Now, different policies should be observed with respect to the selection of the next variable to be flipped, depending on the Question in hand.

When we address *Question 2*, assigning **false** to Ab_i variables will give rise to a simplified KB. No Ab_i can be allowed to be flipped later. Running standard local search technique can deliver one model, that will obviously be a normal-circumstances model of the KB. When it fails to do so, using our heuristic, we order literals and clauses according to their scores. Clauses with highest scores form a probable inconsistent kernel (assuming all Ab_i are **false**). Running one complete method described in section 3 can solve *Question 3*.

Assume now that we want to address *Question 4*. In this case, we run a local search method, using an initial interpretation where all Ab_i variables are assigned **false** while the truth value of the other variables is still generated randomly. Indeed, we still want to guide the search towards a normal-circumstances model, preferring to exhibit a model where as many Ab_i as possible are **false**. Accordingly, we should reconsider our flipping policy. An extreme solution consists in making no distinction anymore between Ab_i variables and the other ones. We could also pounder Ab_i variables so that they are less easily flipped to **true** than the other ones. Also, we can define several policies about the initial values of Ab_i in the subsequent "tries", depending on e.g. their previous behaviour. Obviously enough, a model that is delivered by the performed local search is not necessarily a normal- circumstances one. If no model is found after a certain amount of time, we use the trace of the search to locate the most often falsified clauses and flipped variables to locate a probable inconsistent kernel. A complete method described in section 3 can be run to show this inconsistency formally and thus answer *Question 5*.

5   Experimental results

Let us now illustrate the power of our heuristic to locate inconsistency in the context of combining large KBs. We have made extensive experimentations on real-life and benchmarks problems and the same phenomenon occurred extremely often.

For example, we took several benchmarks for consistency checking [2] describing electronic devices, namely ssa7552-038, ssa7552-158, ssa7552-159 and ssa7552-160. These KBs involve 3575, 3034, 3032, 3126 (1501, 1363, 1363 and 1391) clauses (variables), respectively. Each of these KBs can be shown consistent using TSAT within 1 second CPU time. We also took bf1355-638 (4768 clauses and 2177 variables), whose concistency status cannot be settled neither by local search techniques neither using the best complete Davis and Putnam's techniques, like C-SAT [3] or DP with Jeroslow and Wang heuristic [4] (more precisely, we gave up after spending more than 24 H CPU time for each of these techniques).

We assembled these five KBs together to yield a global one. Neither local search techniques nor the above complete techniques allowed us to settle the con-

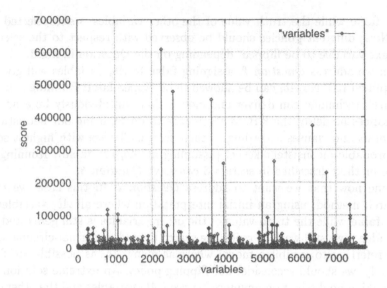

Fig. 3. scores of variables

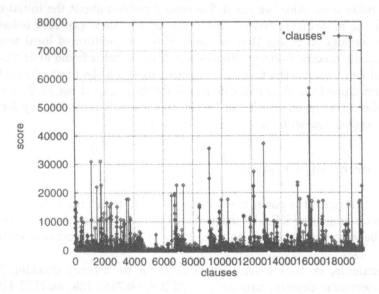

Fig. 4. scores of clauses

sistency status of this KB. Figure 3 gives the scores of the variables (*i.e.* number of times that they have been flipped) and the clauses (Fig. 4) (*i.e.* number of times that they are satisfied) obtained after running TSAT during 8 minutes. We selected the 210 clauses with the highest scores. Running a standard complete Davis and Putnam method on them, we proved the inconsistency of these clauses and thus of the whole KB in 0.03 second. Moreover, using DP+TSAT we proved in 40 seconds that the global KB without this kernel is consistent.

Actually, it is possible to prove within 32 seconds CPU time, using DP+TSAT, that, the bf problem is inconsistent.

Let us stress also that this phenomenon is also encountered when variables are shared by the combined KBs and when the inconsistent kernel is really distributed among the components. Introducing Ab_i variables and the above strategies do not alter the above results, neither.

6 Conclusions

The significance of this paper is twofold. First, we have illustrated how local search techniques can be used to prove inconsistency, which was quite unexpected. Moreover, we have illustrated the actual tractability of this approach for very large propositional KBs. Second, this work is a first step towards implementing fast nonmonotonic inference using local search techniques. Indeed, providing a new efficient complete approach for checking consistency can prove useful with respect to the direct implementation of many nonmonotonic system, like e.g. default logic [10]. By generalizing the treatment performed about Ab_i propositions in this paper, local search can be easily fine-tuned with respect to the issue of searching for specific models where some propositions are minimized.

References

1. Davis, M., Putnam, H.: A Computing Procedure for Quantification Theory. Journ. of the ACM **7** (1960) 201–215
2. Proc. of the Second DIMACS Challenge on Satisfiability Testing, Rutgers (1993)
3. Dubois, O., André, P., Boufkhad, Y., Carlier, J.: SAT vs. UNSAT, in [2].
4. Jeroslow, R.E., Wang, J.: Solving Propositional Satisfiability Problems. Ann. Maths and AI **1** (1990) 167–187
5. Mazure, B., Saïs, L., Grégoire, E.: TWSAT: a New Local Search Algorithm for SAT. Performance and Analysis. CP'95 Workshop on Studying and Solving Really Hard Problems, Cassis, France (1995) 127–130 (full version in Proc. AAAI-97)
6. Mazure, B., Saïs, L., Grégoire, E.: Detecting logical inconsistencies. Proc. AI and Maths Symposium, Fort Lauderdale (FL) (1196) 116–121
7. McCarthy, J.: Applications of circumscription for formalizing common-sense knowledge. Artificial Intelligence **28** (1986) 89–116
8. Mitchell, D., Selman, B., Levesque, H.: Hard and Easy Distributions of SAT Problems. Proc. AAAI-92 (1992) 459–465
9. Reiter, R.: A theory of diagnosis from first principles. Artificial Intelligence **32** (1987) 57–95
10. Reiter, R.: A logic for default reasoning. Artificial Intelligence **13** (1980) 81–131
11. Selman, B., Levesque, H., Mitchell, D.: A New Method for Solving Hard Satisfiability Problems. Proc. AAAI-92 (1992) 440–446
12. Selman, B., Kautz, H.A., Cohen, B.: Local Search Strategies for Satisfiability Testing. Proc. DIMACS Workshop on Maximum Clique, Graph Coloring, and Satisfiability (1993)

The α-junctions: Combination Operators Applicable to Belief Functions

Philippe SMETS

IRIDIA

Université Libre de Bruxelles.

Abstract: Derivation through axiomatic arguments of the operators that represent associative, commutative and non interactive combinations within belief function theory. The derived operators generalize the conjunction, disjunction and exclusive disjunction cases. The operators are characterized by one parameter.

1. Introduction.

In the transferable belief model (TBM), the classical and well-known combination rule is the so-called Dempster's rule of combination (for the TBM, see Smets and Kennes, 1994, for Dempster's rule of combination, see Shafer, 1976, Smets, 1990). This rule corresponds to a conjunction operator: it builds the belief induced by accepting two pieces of evidence, i.e., by accepting their conjunction. Besides there also exists a disjunctive rule of combination (Smets, 1993a). Finally, there is still a third rule, usually forgotten, that fits with the exclusive disjunction. When we noticed this third rule, we came to the idea that these three rules may be special cases of a more general combination scheme... and discovered what we will call the α-junction rules. These new rules could be extended for combining weighted sets, nevertheless our presentation is restricted to the domain covered by the TBM, i.e., to belief functions.

Conceptually what is a belief function within the TBM? It is a function that quantifies the strength of the beliefs held by a given agent, called You, at a given time t. We assume a set Ω of possible worlds, one of them is the actual world and we denote it ω_0. You, the agent, do not know exactly which world in Ω is ω_0 and all You can express is the strength of Your belief that $\omega_0 \in A$, for every $A \subseteq \Omega$. This strength is quantified by a belief function bel:$2^\Omega \rightarrow [0,1]$ with bel(A) representing the strength of Your belief that the actual world ω_0 belongs to the subset A of Ω.

These strengths result from the pieces of evidence relative to ω_0 that You have accumulated. What is a piece of evidence? Suppose a source of information, denoted S, that states that a proposition E is true and You accept at time t that S is telling the truth. We call this whole fact a piece of evidence, and we denote it $\mathscr{E}$. So a piece of evidence $\mathscr{E}$ is a triple (S, E, true) where S is a source of information, E is the proposition states by S, and true denotes that you accept as true what the source states. To be complete 'You' and 't' should also be included, but we neglect them as they stay constant all over this presentation.

This definition is not exactly the same as 'accepting E', as it will be seen once negation is introduced. Suppose S states that E is true and You accept at time t that S is telling the false (so S is lying). We define this piece of evidence as the negation of $\mathscr{E}$. It is the triple (S, E, false) and we denote it $\sim\mathscr{E}$. Its meaning will become clearer once the α-junctions will have been studied. Intuitively, it seems acceptable to defend that $\sim\mathscr{E}$ is equivalent to: S states that $\neg$E is true and You accept at time t that S is telling the truth, i.e., (S, E, false) = (S, $\neg$E, true). But if we had defended that $\mathscr{E}$ is 'accepting E', than $\sim\mathscr{E}$ would have been understood as 'not accepting E', whereas it is closer to 'accepting $\neg$E'. So using the modal operator 'accepting' is not adequate here.

Why to distinguish between $\mathscr{E}$ and E? Suppose two sources S_1 and S_2, and S_1 states E_1 and S_2 states S_2. Suppose You accept at t that at least one of S_1 or S_2 is telling the truth. This is denoted here as $\mathscr{E}_1 \vee \mathscr{E}_2$. If we had not distinguished between $\mathscr{E}$ and E, than we would have written $E_1 \vee E_2$. With such a notation, we could not distinguish the present situation with the following one: suppose the source S states that $E_1 \vee E_2$ is true and You accept at time t that S is telling the truth. In the first case, the sources are precise but You accept that maybe one of them is lying, whereas in the second case, You accept that the source tells the truth, but the source is not very precise. The first case is a problem of uncertainty (which source tells the truth), whereas the second is a case of imprecision (Smets, 1997). To furhter enhance the difference, suppose You want to better Your information. In the first case, You would worry about which source is telling the truth and collect information about the reliability of the sources. In the second case, You would worry directly about which proposition is true. In the context model, Gebhardt and Kruse (1993) also insist in taking in account the nature of the sources of information, and not only what they state.

Coming back to the two sources S_1 and S_2 where S_1 states E_1 and S_2 states E_2. They can be combined in three natural ways. (We use the same $\vee$, $\wedge$ and $\underline{\vee}$ operators for combining pieces of evidence as those used in classical logic to combine propositions. The symbol $\veebar$ denotes the exclusive disjunction operator.)

1. Suppose You accept at t that both S_1 and S_2 are telling the truth, what we denote by $\mathscr{E}_1 \wedge \mathscr{E}_2$. We call this combination a conjunctive combination or a conjunction of two pieces of evidence.

2. Suppose You accept at t that at least one of S_1 or S_2 is telling the truth, what we denote by $\mathscr{E}_1 \vee \mathscr{E}_2$. We call this combination a disjunctive combination or a disjunction of two pieces of evidence.

3. Suppose You accept at t that one and only one of S_1 or S_2 is telling the truth, what we denote by $\mathscr{E}_1 \veebar \mathscr{E}_2$. We call this combination an exclusive disjunctive combination or an exclusice disjunction of two pieces of evidence. (Note that in propositional logic, the exclusive disjunction $E_1 \veebar E_2$ is equivalent to $(E_1 \vee E_2) \wedge \neg(E_1 \wedge E_2)$)

Suppose now that $\mathscr{E}$ is the only piece of evidence that You have accumulated about which of the worlds in Ω is the actual world ω_0. $\mathscr{E}$ induces in You a belief function, denoted bel$[\mathscr{E}]$, on Ω that represents Your beliefs defined on Ω at t about the value of ω_0. The basic belief assignment (bba) related to bel$[\mathscr{E}]$ is denoted m$[\mathscr{E}]$ and m$[\mathscr{E}](A)$ denotes the basic belief mass (bbm) given to $A \subseteq \Omega$ by the bba m$[\mathscr{E}]$.

Suppose two pieces of evidence $\mathscr{E}_1$ and $\mathscr{E}_2$. Let bel$[\mathscr{E}_1]$ and bel$[\mathscr{E}_2]$ be the belief functions on Ω that they would have induced individually.

1. Suppose You accept that both sources of evidence tell the truth, then You build the belief function bel$[\mathscr{E}_1 \wedge \mathscr{E}_2]$ induced by the conjunction of $\mathscr{E}_1$ and $\mathscr{E}_2$. If we assume that this new belief function depends only on bel$[\mathscr{E}_1]$ and bel$[\mathscr{E}_2]$, what translates the idea that they are 'distinct' (Smets, 1992) or non interactive, then bel$[\mathscr{E}_1 \wedge \mathscr{E}_2]$ is obtained by Dempster's rule of combination (unnormalized in this case). The bba m$[\mathscr{E}_1 \wedge \mathscr{E}_2]$ satisfies:

$$m[\mathscr{E}_1 \wedge \mathscr{E}_2](A) = \sum_{X,Y \subseteq \Omega: X \cap Y = A} m[\mathscr{E}_1](X)\, m[\mathscr{E}_2](Y) \qquad \text{for all } A \subseteq \Omega$$

This rule is called hereafter the conjunctive rule of combination, as it results from the conjunction of the two pieces of evidence.

2. Now suppose instead that You accept that at least one source of evidence tells the truth, then You build the belief function bel$[\mathscr{E}_1 \vee \mathscr{E}_2]$ induced by the disjunction of $\mathscr{E}_1$ and $\mathscr{E}_2$. You know what would be Your beliefs if You had known which source tells the truth; they are bel$[\mathscr{E}_1]$ and bel$[\mathscr{E}_2]$, respectively. But You are not so knowledgeable about $\mathscr{E}_1$ and $\mathscr{E}_2$ and You must limit Yourself in building bel$[\mathscr{E}_1 \vee \mathscr{E}_2]$. Just as

Dempster's rule of combination fits the conjunctive case, the so-called disjunctive rule of combination solves the disjunctive case (Smets, 1993a). In that case the corresponding bba $m[\mathscr{E}_1 \vee \mathscr{E}_2]$ satisfies:

$$m[\mathscr{E}_1 \vee \mathscr{E}_2](A) = \sum_{X,Y \subseteq \Omega:\, X \cup Y = A} m[\mathscr{E}_1](X)\, m[\mathscr{E}_2](Y) \qquad \text{for all } A \subseteq \Omega$$

3. One could also imagine the case where You accept that one and only one source of evidence tells the truth, but You don't know which one is telling the truth. This is the exclusive disjunction. So we build $bel[\mathscr{E}_1 \underline{\vee} \mathscr{E}_2]$. The bba $m[\mathscr{E}_1 \underline{\vee} \mathscr{E}_2]$ satisfies:

$$m[\mathscr{E}_1 \underline{\vee} \mathscr{E}_2](A) = \sum_{X,Y \subseteq \Omega:\, X \underline{\cup} Y = A} m[\mathscr{E}_1](X)\, m[\mathscr{E}_2](Y) \qquad \text{for all } A \subseteq \Omega$$

where $\underline{\cup}$ is the symmetric difference, i.e., $X \underline{\cup} Y = (X \cap \overline{Y}) \cup (\overline{X} \cap Y)$.

These rules can in fact be extended to any number of pieces of evidence and any combination formula that states which source You accept as telling the true. So let $\mathscr{E}_1$, $\mathscr{E}_2 ... \mathscr{E}_n$ be a set of pieces of evidence, with $bel[\mathscr{E}_i]$, i=1,2...n be the belief functions induced by each piece of evidence individually. Suppose the pieces of evidence are non interactive, i.e., the belief function build from the combination of the pieces of evidence is a function of the belief functions $bel[\mathscr{E}_i]$). For instance, suppose all You accept is that $(\mathscr{E}_1 \wedge \mathscr{E}_2) \vee \mathscr{E}_3) \underline{\vee} (\mathscr{E}_4 \wedge \mathscr{E}_1)$ holds. It means You accept that one and only one of the two following cases holds: $(\mathscr{E}_1 \wedge \mathscr{E}_2) \vee \mathscr{E}_3$ or $\mathscr{E}_4 \wedge \mathscr{E}_1$. In the first case, You accept that at least one of the next two cases holds: $(\mathscr{E}_1 \wedge \mathscr{E}_2)$ or $\mathscr{E}_3$. It means that you accept that either S_1 and S_2 tell the truth or S_3 tells the truth, in a non exclusive way. In the second case, You accept that both S_4 and S_1 tell the truth. Given this complex piece of evidence, the basic belief masses related to the belief function $bel[((\mathscr{E}_1 \wedge \mathscr{E}_2) \vee \mathscr{E}_3) \underline{\vee} (\mathscr{E}_4 \wedge \mathscr{E}_1)]$ is:

$$m[((\mathscr{E}_1 \wedge \mathscr{E}_2) \vee \mathscr{E}_3) \underline{\vee} (\mathscr{E}_4 \wedge \mathscr{E}_1)](A) =$$
$$= \sum_{X,Y,Z,T \subseteq \Omega:\, ((X \cap Y) \cup Z) \underline{\cup} (T \cap X) = A} m[\mathscr{E}_1](X)\, m[\mathscr{E}_2](Y)\, m[\mathscr{E}_3](Z)\, m[\mathscr{E}_4](T) \qquad \text{for all } A \subseteq \Omega$$

This result was known for long (e.g., Dubois and Prade, 1986). It covers of course the conjunctive rule, the disjunctive rule and the exclusive disjunctive rule, three particular cases where there are only two pieces of evidence. Discovering these three cases that can be built with two pieces of evidence, we came to the idea that these three cases are nothing but special cases of a more general combination rule and we have discovered, as shown here after, the existence of a parametrized family of combination rules (with one parameter), the three special cases corresponding to special values of the parameter. We have called this new family of combination rules, the α-junction where α is the parameter of the combination rule and -junction is the common part of both the 'conjunction' and 'disjunction' words.

The concept of negation, and its meaning, came out of our developement. Suppose the pieces of evidence $\mathcal{E}$. Dubois and Prade (1986) have suggested that if bel[$\mathcal{E}$] is the belief induced by $\mathcal{E}$, then bel[~$\mathcal{E}$] could be defined so that its bba satisfies:

$$m[\sim\mathcal{E}](A) = m[\mathcal{E}](\overline{A}) \qquad \text{for all } A \subseteq \Omega$$

and where $\overline{A}$ is the complement of A relative to Ω (Dubois and Prade, 1986, have used the notation $\overline{m}$ for m[~$\mathcal{E}$]). This definition will be used later when we will study the De Morgan properties of the α-junctions.

We have thus found out the conjunction, disjunction and exclusive disjunction operators, and the negation. Readers might wonder if there are not other symmetrical junctions operators that can be built from two propositions in classical logic. In fcat, there are only eight symmetrical operators that can be built with two propositions: the tautology, the conjunction, the disjunction, and the exclusive disjunction, and their negations, the contradiction, the disjunction of the negations, the conjunction of the negations, and a junction without name. Figure 1 shows these eight operators. Two elements opposed by a diagonal are the negation of each other.

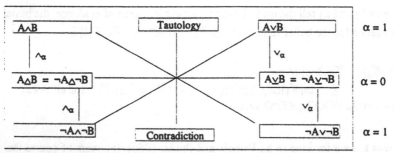

Figure 1: The eight symmetrical junctions operaotrs in classical logic, where $\underline{\vee}$ denotes the exclusive disjunction and $\triangle$ denotes its $\neg$ dual ($A\triangle B = \neg(A\underline{\vee}B) = (A\wedge B)\vee(\neg A\wedge\neg B)$). Diagonaly opposed pairs are linked by the negation operator $\neg$. The four vertical lines at left and right are the forthcoming α-conjunction and α-disjunction operaotrs.

In this paper, we present first some needed definitions and notation conventions. We proceed by studying the α-junctive rule of combination of two pieces of evidence. Then we study the disjunctive and the conjuncitve cases, and conclude.

2. Definitions and notations.

2.1. Belief functions

A basic belief assignment (bba) is defined as the function from 2^Ω to $[0,1]$, its values are the basic belief masses (bbm) and their sum over the subsets of Ω is 1. To simplify the notation we write m_1, m_2... for $m[\mathcal{E}_1]$, $m[\mathcal{E}_2]$... and even drop the reference to the underlying piece of evidence when it is irrelevant to the presentation.

In the TBM, the mass $m(A)$ for $A \subseteq \Omega$ is that part of Your belief that supports that the actual world ω_0 is in A and nothing more specific. The belief function bel is defined as
$$bel(A) = \sum_{B \subseteq \Omega: \emptyset \neq B \subseteq A} m(B).$$
It represents the total belief that supports that the actual world ω_0 is in A.

Related to m and bel, the commonality function $q:2^\Omega \to [0,1]$ is defined as :
$$q(A) = \sum_{B \subseteq \Omega: A \subseteq B} m(B)$$
and we introduce the function $b:2^\Omega \to [0,1]$ defined as:
$$b(A) = \sum_{B \subseteq \Omega: B \subseteq A} m(B) = bel(A) + m(\emptyset).$$

The meaning of q and b is essentially technical, even though $q(A)$ can be understood as the ignorance about ω_0 when You know that ω_0 belongs to A. Their major use is to be found in the combination rules. Given two bba m_i, i= 1,2, with q_i and b_i their related q- and b-functions, then $q_{1 \wedge 2}$ and $b_{1 \vee 2}$, the q and b-functions that result from their combinations, are given by:

in the conjunctive case: $\qquad q_{1 \wedge 2}(A) = q_1(A) \, q_2(A) \qquad$ for all $A \subseteq \Omega$

in the disjunctive case: $\qquad b_{1 \vee 2}(A) = b_1(A) \, b_2(A) \qquad$ for all $A \subseteq \Omega$.

Besides $b[\sim\mathcal{E}](A) = q[\mathcal{E}](\overline{A})$ and $q[\sim\mathcal{E}](A) = b[\mathcal{E}](\overline{A})$ for all $A \subseteq \Omega$, a property that fits with De Morgan law. Indeed the combination rules can be written as:
$$q[\mathcal{E}_1 \wedge \mathcal{E}_2](A) = q[\mathcal{E}_1](A) \, q[\mathcal{E}_2](A)$$
and $\quad b[\mathcal{E}_1 \vee \mathcal{E}_2](A) = b[\mathcal{E}_1](A) \, b[\mathcal{E}_2](A).$

Then: $b[\sim(\mathscr{E}_1 \wedge \mathscr{E}_2)](A) = q[\mathscr{E}_1 \wedge \mathscr{E}_2](\overline{A}) = q[\mathscr{E}_1](\overline{A}) \, q[\mathscr{E}_2](\overline{A})$

$\qquad\qquad\qquad\quad = b[\sim\mathscr{E}_1](A) \, b[\sim\mathscr{E}_2](A) = b[\sim\mathscr{E}_1 \vee \sim\mathscr{E}_2](A)$

So: $\quad b[\sim(\mathscr{E}_1 \wedge \mathscr{E}_2)](A) = b[\sim\mathscr{E}_1 \vee \sim\mathscr{E}_2](A) \quad$ for all $A \subseteq \Omega$,

Similarly,

$\qquad q[\sim(\mathscr{E}_1 \vee \mathscr{E}_2)](A) = q[\sim\mathscr{E}_1 \wedge \sim\mathscr{E}_2](A) \quad$ for all $A \subseteq \Omega$.

These two relations are the De Morgan formulas as they show that $\sim(\mathscr{E}_1 \wedge \mathscr{E}_2)$ and $\sim\mathscr{E}_1 \vee \sim\mathscr{E}_2$ induce the same bba's (and identically for $\sim(\mathscr{E}_1 \vee \mathscr{E}_2)$ and $\sim\mathscr{E}_1 \wedge \sim\mathscr{E}_2$).

2.2. Notation.

A bba m defined on Ω can be represented as a vector with $2^{|\Omega|}$ elements, so $\mathbf{m} = [m(X)]$ where X is the line index of the component of the vector $\mathbf{m}$ and $X \subseteq \Omega$. The order of the elements in $\mathbf{m}$ is arbitrary, but one order is particularly convenient as it enhances many symmetries. This order is a kind of lexico-iterated order. E.g. let $\Omega = \{a,b,c\}$, then the transpose $\mathbf{m}'$ of the vector $\mathbf{m}$ is given by:

$\mathbf{m}' = (m(\varnothing), m(\{a\}), m(\{b\}), m(\{a,b\}, m(\{c\}), m(\{a,c\}), m(\{b,c\}), m(\{a,b,c\}))$.

All matrices and vectors in this paper will be organized so that their indices obey to this order.

We use the notation 1_X to denoted a bba where all elements are null except the X'th element that equals 1: it is the bba that gives a mass 1 to the set X.

We also use the following notations:

1 is a vector where all elements equal to 1.

I is the identity matrix.

J is the matrix with elements j_{XY} where $j_{XY} = 1$ if $X = \overline{Y}$, and $j_{XY} = 0$ otherwise. With $\Omega = \{a,b\}$,

$$\mathbf{J} = \begin{bmatrix} 0 & 0 & 0 & 1 \\ 0 & 0 & 1 & 0 \\ 0 & 1 & 0 & 0 \\ 1 & 0 & 0 & 0 \end{bmatrix}.$$

The **J** matrix is the operator that transforms a bba m into its negation: $\mathbf{J}.\mathbf{m}[\mathscr{E}] = \mathbf{m}[\text{-}\mathscr{E}]$. Indeed, **J** projects m(A) on $\overline{A}$ for all $A \subseteq \Omega$. We also have $\mathbf{J}.\mathbf{J} = \mathbf{I}$, what corresponds to the involutive property of the negation operator.

Given a vector **v**, [diag v] is the diagonal matrix whose diagonal elements are the values of **v**, all other elements being 0.

2.3. Permutation.

Let P be a permutation from Ω to Ω. Let $P(X) = \{y: y = P(x), x \in X\}$. We define as L_P the permutation matrix from 2^Ω to 2^Ω obtained from the permutation P, and such that it maps the element $X \subseteq \Omega$ onto the element $P(X) \subseteq \Omega$. With $\Omega = \{a,b\}$ and P such that $P(a) = b$ and $P(b) = a$, L_P is:

$$L_P = \begin{bmatrix} 1 & 0 & 0 & 0 \\ 0 & 0 & 1 & 0 \\ 0 & 1 & 0 & 0 \\ 0 & 0 & 0 & 1 \end{bmatrix}.$$

Lemma 1: $P(A) = A$ for all P defined on Ω iff $A = \emptyset$ or $A = \Omega$.

The only subsets of Ω that are mapped onto themselves whatever the permutation are $\emptyset$ and Ω.

Permutation matrix also satisfies $(L_P)^{-1} = L_{(P^{-1})}$ (hence the parenthesis can be dropped without the risk of a typographical confusion).

2.4. Combination rules in matrix notation.

The three combinations rules can be represented under matrix forms. The conjunctive combination is introduced in Klawonn and Smets (1992). Suppose $\Omega = \{a,b\}$. Let $D(m)$ be the following matrix:

$$D(m) = \begin{bmatrix} 1 & m(\emptyset)+m(b) & m(\emptyset)+m(a) & m(\emptyset) \\ 0 & m(a)+m(\Omega) & 0 & m(a) \\ 0 & 0 & m(b)+m(\Omega) & m(b) \\ 0 & 0 & 0 & m(\Omega) \end{bmatrix}$$

Then the bba $m_1 = D(m).m_0$ is equal to the bba one would obtained by combining m_0 and m by the conjunctive rule of combination (i.e., Dempster's rule of combination, but unnormalized). In general, the A,B element of $D(m)$ is

$$m(A|B) = \sum_{X \subseteq \Omega:\, X \cap B = A} m(A \cup X),$$

i.e., the B'th column is the vector obtained by conditioning m on B, or equivalently conjunctively combining m with the bba 1_B.

Similar results are derivable for the disjunction combination described in Smets (1993a). Let $E(m)$ be the following matrix:

$$E(m) = \begin{bmatrix} m(\emptyset) & 0 & 0 & 0 \\ m(a) & m(a)+m(\emptyset) & 0 & 0 \\ m(b) & 0 & m(b)+m(\emptyset) & 0 \\ m(\Omega) & m(b)+m(\Omega) & m(a)+m(\Omega) & 1 \end{bmatrix}$$

Then the bba $m_1 = E(m).m_0$ is equal to the bba one would obtained by combining m_0 and m by the disjunctive rule of combination. In general, the A,B element of $E(m)$ is

$$\sum_{X \subseteq \Omega:\ X \cup B = A} m(A \cup X),$$ i.e., the B'th column is the vector obtained by disjunctively

combining m with the bba 1_B.

For the exclusive disjunction combination, the matrix is given by:

$$F(m) = \begin{bmatrix} m(\emptyset) & m(a) & m(b) & m(\Omega) \\ m(a) & m(\emptyset) & m(\Omega) & m(b) \\ m(b) & m(\Omega) & m(\emptyset) & m(a) \\ m(\Omega) & m(b) & m(a) & m(\emptyset) \end{bmatrix}.$$

It can be verified that the bba $m_1 = F(m).m_0$ is indeed the bba one would obtained by combining m_0 and m by the exclusive disjunction.

3. The α-junctions.

3.1. The matrix K(m).

Let m_1 and m_2 be two basic belief assignments on Ω. We assume that there exists an operator $[K(m_1)]$ induced by m_1 so that, when applied to m_2, it produces a combination m_{12} of m_1 with m_2.

$$m_{12} = [K(m_1)]\ m_2$$

The first step consists in showing why $[K(m_1)]$ is a linear operator. Suppose three bba m_0, m_1 and m_2. Let $[K(m_0)]$ be the operator induced by m_0. Suppose m_1 (m_2) is the bba that would describe Your beliefs if You accept that S_1 (S_2) tells the truth: $m_1 = m[\mathcal{E}_1]$ and $m_2 = m[\mathcal{E}_2]$. It happens You know that one of $\mathcal{E}_1$ or $\mathcal{E}_2$ will be accessible to You. Which one will be decided by a random device (such as tossing a coin). In case of success, $\mathcal{E}_1$ will be the piece of evidence You will hold, otherwise $\mathcal{E}_2$ will be the piece of evidence You will hold. Let p be the probability that a success occurs and $q = 1 - p$. Before knowing the outcome of this random experiment, Your bba is $m_{12} = p.m_1 +$

q.m_2 (for a justification of this linear relation, see Smets, 1993b). Consider the results of the combination of $[K(m_0)]$ with m_1 and m_2 individually. We postulate that before knowing the outcome of the random experiment, the result of combining $[K(m_0)]$ to m_{12} would be equal to the same linear combination of $[K(m_0)]m_1$ and $[K(m_0)]m_2$. We assume that combining and averaging commute.

Assumption A1: Linearity.

$$[K(m_0)](p.m_1 + q.m_2) = p.[K(m_0)]m_1 + q.[K(m_0)]m_1$$

This assumption is sufficient to conclude that $[K(m_0)]$ is a linear operator and can thus be represented by a matrix that we denoted by $K(m_0)$. So the operation $[K(m_0)]m_1$ is nothing but the matricial product of $K(m_0)$ with the vector m_1.

We next assume that the combination of m_1 and m_2 commute, i.e., combining m_1 with m_2 or m_2 with m_1 leads to the same result.

Assumption 2 : Commutativity.

$$K(m_0) \, m_1 = K(m_1) \, m_0$$

Theorem 1: Under assumptions A1 and A2,
$$K(m) = \sum_{X \subseteq \Omega} m(X) \, K_X.$$
where the K_X matrices are matrices which coefficients do not depend on m.

Proof:

By A1, $\quad K(m_0) \, (p.m_1 + q.m_2) = p.K(m_0) \, m_1 + q.K(m_0) \, m_2.$

By A2, $\quad K(m_0) \, (p.m_1 + q.m_2) = K(p.m_1 + q.m_2) \, m_0$

$\qquad\qquad K(m_0) \, m_1 = K(m_1) \, m_0$

$\qquad\qquad K(m_0) \, m_2 = K(m_2) \, m_0$

This being true whatever m_0, we get:

$$K(p.m_1 + q.m_2) = p \, K(m_1) + q \, K(m_2)$$

It implies that is linear in m, thus the theorem. $\hfill$ QED

From A2 we can also derive another constraint that the K_X matrices must satisfy . Let $K_X = [\, k_{AB}^X \,]$ where A, $B \subseteq \Omega$. So k_{AB}^X denotes the element of K_X at line A and column B.

Theorem 2: $k_{AY}^X = k_{AX}^Y$ for all $A, X, Y \subseteq \Omega$.

Proof: The requirement

$$K(m_1) \, m_2 = K(m_2) \, m_1 \tag{3.1}$$

becomes for $A \subseteq \Omega$,

$$\sum_{X \subseteq \Omega} m_1(X) \sum_{Y \subseteq \Omega} k_{AY}^X . m_2(Y) = \sum_{Y \subseteq \Omega} m_2(Y) \sum_{X \subseteq \Omega} k_{AX}^Y . m_1(X) \tag{3.2}$$

Being true whatever m_1 and m_2, one has $k_{AY}^X = k_{AX}^Y$ for all $A, X, Y \subseteq \Omega$.　　　　QED

So the Y-th column of K_X is equal to the X-th column of K_Y.

3.2. K_X is a stochastic matrix.

Theorem 3: For all $X \subseteq \Omega$, K_X is a stochastic matrix.

Proof: Suppose the following bba: $m_1 = 1_X$ and $m_2 = 1_Y$. Then (3.1) becomes:

$$K(m_1) \, m_2 = K(1_X) \, 1_Y = k_{AY}^X$$

So the resulting bba is the column vector with elements k_{AY}^X for $A \subseteq \Omega$. Being a bba, its elements must be non negative and add to 1.

$$k_{AY}^X \geq 0 \qquad \text{and} \qquad \sum_{A \subseteq \Omega} k_{AY}^X = 1 \qquad\qquad \text{QED}$$

Thus each column of K_X can be assimilated to a probability distribution function over 2^Ω (in fact each column is a bba).

3.3. Anonymity.

Let P be a permutation of the elements of Ω. Let L_P be the permutation matrix as defined in section 2.3. When applied to a bba m, L_P produces a new bba m_P that differs only from m by the fact that, for every A in Ω, the mass initially given to A is given after permutation to $P(A)$.

For instance let $\Omega = \{a,b\}$ and $P:\Omega \to \Omega$ so that $P(a) = b$, $P(b) = a$. Then:

$$L_P \, m = \begin{bmatrix} 1 & 0 & 0 & 0 \\ 0 & 0 & 1 & 0 \\ 0 & 1 & 0 & 0 \\ 0 & 0 & 0 & 1 \end{bmatrix} \begin{bmatrix} m(\emptyset) \\ m(a) \\ m(b) \\ m(a,b) \end{bmatrix} = \begin{bmatrix} m(\emptyset) \\ m(b) \\ m(a) \\ m(a,b) \end{bmatrix}.$$

We assume that a renaming of the elements Ω will not affect the results of the combination.

Assumption 3 : Anonymity.

Let P be a permutation of Ω to Ω and let L_P be the permutation matrix that permutes the subset A into the subset P(A). Then

$$K(L_P m_1) \, L_P m_2 = L_P K(m_1) m_2. \tag{3.3}$$

This assumption translates the following idea. Suppose we permute the elements of Ω in both m_1 and m_2, then the result of the combination is nothing but the permutation of the results of the combination of m_1 with m_2.

3.4. Symmetry.

Theorem 4: $K_{P(X)} \, L_P = L_P \, K_X$

Proof: Suppose $m_1 = 1_X$. Then $L_P 1_X = 1_{P(X)}$. Replace m_1 by 1_X in (3.3) and note that it is true for all m_2. QED

3.5. Vacuous belief.

We assume the existence of a bba (denoted m_{vac}) which combination with any bba leaves it unchanged, i.e., a neutral element for the combination.

Assumption A4. Vacuous belief.

There exists a bba m_{vac} so that for any bba m, $K(m) \, m_{vac} = m$.

Theorem 5: $K(m_{vac}) = I$.

Proof: By A2, A4 implies: $K(m_{vac}) \, m = m$ for all m, hence the theorem. QED

3.6. Associativity.

We assume that the combination is associative. This property means that the order with which the bba are combined is irrelevant.

Assumption A5: Associativity.

Let m_1, m_2 and m_3 be three bba on Ω. Then:

$$K(m_1)\,(K(m_2)m_3) = K(K(m_1)m_2)\,m_3.$$

Theorem 6: $K_X K_Y = K(K_X 1_Y)$ for all $X, Y \subseteq \Omega$.

Proof: Let $m_1 = 1_X$ and $m_2 = 1_Y$. From A5, we get:

$$K_X\,(K_Y m_3) = K(K_X 1_Y)m_3.$$

This being true for any m_3, thus the theorem. QED

Theorem 7: There exists an $X \subseteq \Omega$ so that $K_X = I$.

Proof: By theorem 5, we have: $\displaystyle\sum_{X \subseteq \Omega} m_{vac}(X)\,k_{AA}^X = 1.$

As $k_{AA}^X \in [0,1]$ (theorem 3), so $\displaystyle\sum_{X \subseteq \Omega} m_{vac}(X)\,k_{AA}^X$ is a weighted average of k_{AA}^X which

values are also in $[0,1]$. The only way to get a sum equal to 1 is:

Case 1. $k_{AA}^X = 1$ for all X (and all A), in which case $K_X = I$ for all X, and thus $K(m) = I$, a degenerated (and uninteresting) solution that will be rejected by theorem 8.

Case 2. $m_{vac}(B) = 1$ for some $B \subseteq \Omega$ and the other values of m_{vac} are null. Then $k_{AA}^B = 1$ for all $A \subseteq \Omega$. As K_B is a stochastic matrix, $k_{AC}^B = 0$ if $A \neq C$, so $K_B = I$. QED

3.7. Reversibility.

We assume that different bba induce different operators.

Theorem 8. Reversibility. Let m_1 and m_2 be two bba on Ω.

If $m_1 \neq m_2$, then $K(m_1) \neq K(m_2)$.

Proof: Let $m_1 \neq m_2$, and suppose $K(m_1) = K(m_2)$. In that case, $K(m_1)\,m_{vac} = K(m_2)\,m_{vac}$, hence, by assumption A4: $m_1 = m_2$, contrary to the initial assumption. So the theorem. QED

This is just an assumption of reversibility for the K operator. It implies that $K_X \neq K_Y$ if $X \neq Y$ (take $m_1 = 1_X$ and $m_2 = 1_Y$). It eliminates also the degenerated solution (theorem 7, case 1) for the m_{vac} determination.

Theorem 9: $m_{vac} = 1_\emptyset$ or $m_{vac} = 1_\Omega$.

Proof: Consider now the $K_B = I$ and $m_{vac} = 1_B$ (theorem 7, case 2). Let P be any permutation of the element of Ω, we have by construction

$$K_{P(B)} = L_P^{-1} K_B L_P = L_P^{-1} L_P = I.$$

Thus $K_{P(B)} = K_B$ for all P, and this means that B is either $\emptyset$ or Ω (lemma 1).   QED

We have just rediscovered the existence of two vacuous bba that are well known in the TBM. Indeed, 1_Ω is the classical vacuous belief function of the TBM, the one initially described by Shafer and the one commonly called the vacuous belief function. It is the neutral element of the conjunctive rule of combination.

The other solution $1_\emptyset$ for the vacuous bba is the negation of the previous solution. It is the neutral element of the disjunctive rule of combination (Smets, 19893a).

We call 1_Ω the and-vacuous bba and $1_\emptyset$ the or-vacuous bba. In section 4, we will study in details the or-vacuous bba, hence the familly of disjuncitve combinations. All results obtained with it will be extended to the conjunctive case by an appropriate use of the negation operator and of the De Morgan formula (section 5).

3.8. Focused bba.

Suppose a bba on Ω so that:

$m(X) \geq 0$ if $X \subseteq A$,

$m(X) = 0$ otherwise.

This bba corresponds to the case where You know (fully believe) that the actual world ω_0 belongs to A. We will say that such a bba is focused on A. In such a case, Your beliefs would be the same if You had built them on A instead of Ω.

Suppose the two bba m_1 and m_2 are focused on A, then we assume that their combination is also focused on A. Once a world is 'eliminated' by both m_1 and m_2, it stays 'eliminated' after their combination.

Assumption A6. Context preservation.

Let m_1 and m_2 be two bba's on Ω so that:

$m_1(X) = m_2(X) = 0$ for all $X \not\subseteq A$,

then $(K(m_1) m_2)(X) = 0$ for all $X \not\subseteq A$.

Theorem 10: $k_{AB}^X = 0$ for all $A \not\subseteq X$, $B \subseteq X \subseteq \Omega$.

Proof: Immediate.   QED

3.9. Summary:

In summary, we have derived that **K** must satisfy:

P1: $K(m) = \sum_{X \subseteq \Omega} m(X) \, K_X$ Linearity

P2: $k_{AY}^X = k_{AX}^Y$ Symmetry from commutativity

P3: $k_{AB}^X \geq 0, \ \sum_{A \subseteq \Omega} k_{AB}^X = 1$ Stochastic matrix.

P4: $K_{P(X)} = L_{P^{-1}} \, K_X \, L_P$ Symmetry from anonymity

P5: $K(K_X 1_Y) = K_X K_Y = K_Y K_X$ Associativity

P6: $k_{AB}^X = 0$ for all $A \not\subseteq X$, $B \subseteq X$ Context preservation

P7: $K_\emptyset = I$, $m_{vac} = 1_\emptyset$ (disjunctive case) or   or-vacuous bba

  $K_\Omega = I$, $m_{vac} = 1_\Omega$ (conjunctive case)   and-vacuous bba

4. The α-disjunctive combination.

In this section, we assume that the vacuous bba is the or-vacuous bba $1_\emptyset$. We first deduce the K_X matrices for the case $|\Omega| = 1$ and $|\Omega| = 2$, and proceed with the general case.

4.1. Case Ω = {Ω}.

Suppose Ω has only one element. We have $K_\emptyset = I$. As K_Ω is stochastic, there are $\alpha, \beta \in [0,1]$ such that:

$$K_\Omega = \begin{bmatrix} \beta & 1-\alpha \\ 1-\beta & \alpha \end{bmatrix}.$$

By P2, $\begin{bmatrix} \beta \\ 1-\beta \end{bmatrix} = \begin{bmatrix} 0 \\ 1 \end{bmatrix}$, so $\beta = 0$, and

$$K_\Omega = \begin{bmatrix} 0 & 1-\alpha \\ 1 & \alpha \end{bmatrix}.$$

These two matrices satisfy all the required properties.

4.2. Case $\Omega = \{a,b\}$.

We look now to the case $\Omega = \{a,b\}$. We have $K_\emptyset = I$. Let

$$K_a = \begin{bmatrix} 0 & 1-\alpha & x & p \\ 1 & \alpha & y & q \\ 0 & 0 & z & r \\ 0 & 0 & t & s \end{bmatrix}$$

The block of 0 results from P6. Let P be the permutation $P(a) = b$ and $P(b) = a$. Then L_P is given in section 2.3 and $L_P^{-1} = L_P$.

With $K_b = L_P^{-1} K_a L_P$ (P4) and $K_a K_b = K_b K_a$ (P5), one has:

$$K_a L_P^{-1} K_a L_P = L_P^{-1} K_a L_P K_a .$$

So $\quad (K_a L_P^{-1} K_a) = L_P^{-1} (K_a L_P K_a) L_P^{-1}.$

A (very) tedious analysis of the last equality leads to the solution to $x = y = z = 0$, $t = 1$, $p = q = 0$, $s = \alpha$. So we obtain unique solutions for K_a and K_b.

$$K_a = \begin{bmatrix} 0 & 1-\alpha & 0 & 0 \\ 1 & \alpha & 0 & 0 \\ 0 & 0 & 0 & 1-\alpha \\ 0 & 0 & 1 & \alpha \end{bmatrix} \qquad K_b = \begin{bmatrix} 0 & 0 & 1-\alpha & 0 \\ 0 & 0 & 0 & 1-\alpha \\ 1 & 0 & \alpha & 0 \\ 0 & 1 & 0 & \alpha \end{bmatrix}$$

Consider $K(K_a 1_b) = K_a K_b$ (P5). We have $K_a 1_b = 1_\Omega$, so $K(K_a 1_b) = K(1_\Omega) = K_\Omega$. Hence

$$K_\Omega = K_a K_b = \begin{bmatrix} 0 & 0 & 0 & (1-\alpha)^2 \\ 0 & 0 & 1-\alpha & \alpha(1-\alpha) \\ 0 & 1-\alpha & 0 & \alpha(1-\alpha) \\ 1 & \alpha & \alpha & \alpha^2 \end{bmatrix}.$$

We have thus obtained all the needed matrices, $K(m)$ is fully defined... and depends only on one parameter α which varies on $[0,1]$. In particular, when $\alpha = 0$, $K(m) = F(m)$, and when $\alpha = 1$, $K(m) = E(m)$ (see section 2). So $\alpha = 0$ corresponds to the exclusive disjunction and $\alpha = 1$ to the disjunction. All other values of α in $[0,1]$ correspond to new disjunctive combination operators.

4.3. The canonical decomposition of K(m).

It is worth looking at the canonical decomposition of $\mathbf{K(m)}$ into its eigenvalues - eigenvectors structure when $\Omega = \{a,b\}$, as a nice structure will emerge. Let

$$\Lambda_\emptyset = I, \quad \Lambda_a = \begin{bmatrix} 1 & 0 & 0 & 0 \\ 0 & -\alpha & 0 & 0 \\ 0 & 0 & 1 & 0 \\ 0 & 0 & 0 & -\alpha \end{bmatrix}, \quad \Lambda_b = \begin{bmatrix} 1 & 0 & 0 & 0 \\ 0 & 1 & 0 & 0 \\ 0 & 0 & -\alpha & 0 \\ 0 & 0 & 0 & -\alpha \end{bmatrix}, \quad \Lambda_\Omega = \begin{bmatrix} 1 & 0 & 0 & 0 \\ 0 & -\alpha & 0 & 0 \\ 0 & 0 & -\alpha & 0 \\ 0 & 0 & 0 & \alpha^2 \end{bmatrix}.$$

$$\text{Let } G = \begin{bmatrix} 1 & 1 & 1 & 1 \\ 1 & -\alpha & 1 & -\alpha \\ 1 & 1 & -\alpha & -\alpha \\ 1 & -\alpha & -\alpha & \alpha^2 \end{bmatrix}.$$

Then $G^{-1} \Lambda_X G = K_X$ for all $X \subseteq \Omega$.

It happens that all $\mathbf{K_X}$ for $X \subseteq \Omega$ share the same left and right eigenvectors. This decomposition allows to derive a nice representation of $\mathbf{m}_{12} = \mathbf{K(m}_1)\mathbf{m}_2$.

Lemma 2: Let $g_1 = Gm_1$, $g_2 = Gm_2$, $g_{12} = Gm_{12}$. Then :

$$g_{12}(X) = g_1(X) \, g_2(X) \qquad \text{for all } X \subseteq \Omega.$$

Proof: We have:

$$m_{12} = \sum_{X \subseteq \Omega} m_1(X) \, K_X \, m_2$$

$$= \sum_{X \subseteq \Omega} m_1(X) \, G^{-1} \Lambda_X \, G \, m_2$$

and

$$Gm_{12} = \sum_{X \subseteq \Omega} m_1(X) \, \Lambda_X \, G \, m_2.$$

With $g_1 = Gm_1$, $g_2 = Gm_2$, $g_{12} = Gm_{12}$,

so

$$g_{12} = \sum_{X \subseteq \Omega} m_1(X) \, \Lambda_X \, g_2.$$

The relation between G and Λ_X is such that

$$\sum_{X \subseteq \Omega} m_1(X) \, \Lambda_X = [\text{diag } g_1]$$

Thus $g_{12}(X) = g_1(X) \, g_2(X) \qquad$ for all $X \subseteq \Omega$. $\hfill$ QED

The lemma 2 relation is nothing but the analogous of the relation for Dempster's rule of combination when g is the commonality function. So $g = Gm$ is the analogous of the communality function within the generalized context of the α-disjunction.

The vector $g = Gm$ is nothing but the vector of eigenvalues of the matrix $K(m)$, and G is a matrix which lines are the left-eigenvectors of $K(m)$.

4.4. Extending the results to any Ω.

The generalization to any Ω is obtained by iteration. The next theorem describes the strucutre of $K(m)$.

Theorem 11: For any Ω, the K_X matrices are:

1) for $X = \emptyset$, $K_\emptyset = I$, $\Lambda_\emptyset = I$,

2) for $x \in \Omega$ and $K_{\{x\}} = [k_{AB}^x]$, we have:

if $x \notin B$,	$k_{AB}^x =$	1	if $A = B \cup \{x\}$
		0	otherwise
if $x \in B$,	$k_{AB}^x =$	α	if $B = A$
		$(1-\alpha)$	if $x \notin A$, $B = A \cup \{x\}$
		0	otherwise,

and the diagonal elements of the $\Lambda_{\{x\}}$ matrices are:

if $x \notin B$	$\lambda_x(B,B) =$	1
if $x \in B$	$\lambda_x(B,B) =$	$\alpha - 1$.

3) and for any $\emptyset \neq X \subseteq \Omega$: $\quad K_X = \prod_{x \in X} K_{\{x\}}, \quad \Lambda_X = \prod_{x \in X} \Lambda_{\{x\}}$.

4) The X'th column of G is $\Lambda_X 1$.

Proof: Obtained by iteration. Let $\Omega = \{a,b,c\}$. We have $K_\emptyset = I$. By considering bba focused on $\{a,b\}$, one obtains $k_{XY}^a = 0$ for $X \nsubseteq \{a,b\}$, $Y \subseteq \{a,b\}$. Then using bba focused on $\{a,c\}$, one gets the values of k_{XY}^a for $Y = \{c\}$ and $\{a,c\}$. The values of k_{XY}^a for $Y = \{b,c\}$ and $\{a,b,c\}$ are derived by as very tedious computation as in section 4.2. The values of $K_{\{b\}}$ and $K_{\{c\}}$ are derived through the application of L_P matrices. Finally the property P5 allows the derivation of:

$$K_X = \prod_{x \in X} K_{\{x\}} \qquad \text{for all } X \subseteq \Omega.$$

Going from spaces Ω with three to four elements and more is performed identically.

$$\text{QED}$$

The fact that the combination can be achieved by pointwise multiplications as with the commonality functions is shown in the nest theorem. This property is very usefull as the Fast Möbius Transform could be adapted to compute g and m from each other. Then the computation of the combination is obtained by transforming each bba into its corresponding g vector, combining the g vectors by pointwise multiplications, and transforming back the result into a bba (Kennes and Smets, 1990, Kennes, 1992).

Theorem 12: Let $g_1 = Gm_1$, $g_2 = Gm_2$, $g_{12} = Gm_{12}$,

then $\quad g_{12}(A) = g_1(A)g_2(A) \quad$ for all $A \subseteq \Omega$.

Proof: We still have the property that $K_X = G^{-1}\Lambda_X G$, and proof proceeds as in lemma 1. $\hspace{4cm}$ QED

5. The α-conjunctive combination.

We consider now that the vacuous bba is the and-vacuous bba 1_Ω.

In order to distinguish between the conjunctive and the disjunctive families of α-junctions, we introduce the notation $K^{\vee\alpha}(m)$ for what we had derived in the previous section. We define $K^{\wedge\alpha}(m)$ as the operator dual to $K^{\vee\alpha}(m)$ that we would have obtained if we had started with the and-vacuous belief function.

The same derivation as for the disjunctive case can be repeated using $K_\Omega = I$, instead of $K_\emptyset = I$. All results happen to be similar.

Theorem 13: For any Ω, the $K^{\wedge\alpha}_X$ matrices are:

1) for $X = \Omega$, $K^{\wedge\alpha}_\Omega = I$, $\Lambda^{\wedge\alpha}_\Omega = I$,

2) for $x \in \Omega$ and $K_{\{x\}} = [k^x_{AB}]$, we have:

if $x \in B$,	$k^{\overline{x}}_{AB}{}^{\wedge\alpha} =$	1	if $x \notin A$, $B = A \cup \{x\}$
		0	otherwise
if $x \notin B$,	$k^{\overline{x}}_{AB}{}^{\wedge\alpha} =$	α	if $A = B$
		$(1-\alpha)$	if $A = B \cup \{x\}$
		0	otherwise,

and the diagonal elements of the $\Lambda_{\{x\}}$ matrices are:

if $x \notin B$	$\lambda^{\wedge\alpha}_{\overline{x}}(B,B) =$	1
if $x \in B$	$\lambda^{\wedge\alpha}_{\overline{x}}(B,B) =$	$\alpha - 1$.

3) and for any $\Omega \neq X \subseteq \Omega$: $\quad K^{\wedge \alpha}{}_X = \prod_{x \notin X} K^{\wedge \alpha}{}_{\{\bar{x}\}}, \quad \Lambda^{\wedge \alpha}{}_X = \prod_{x \notin X} \Lambda^{\wedge \alpha}{}_{\{\bar{x}\}}.$

4) The X'th column of $G^{\wedge \alpha}$ is $\Lambda^{\wedge \alpha}{}_X 1$.

The links between $K^{\wedge \alpha}$ and $K^{\vee \alpha}$ are shown in the next theorem, that is at the core of their De Morgan nature.

Theorem 14: $\qquad K^{\wedge \alpha}(m) = J.K^{\vee \alpha}(J.m).J.$
or equivalently: for any $A, X, Y \subseteq \Omega$, $k^X_{A\ Y}{}^{\wedge \alpha} = k^{\bar{X}}_{A\ \bar{Y}}{}^{\vee \alpha}$.

This relation leads to the analogous of the De Morgan formula extended to α-junctions. We use the obvious notations:

$\qquad m[\mathscr{E}_1] \wedge_\alpha m[\mathscr{E}_2] \qquad$ for $\qquad K^{\wedge \alpha}(m[\mathscr{E}_1]) m[\mathscr{E}_2]$, and

$\qquad m[\mathscr{E}_1 \wedge_\alpha \mathscr{E}_2] \qquad$ for $\qquad m[\mathscr{E}_1] \wedge_\alpha m[\mathscr{E}_2]$.

Similarly, we define:

$\qquad m[\mathscr{E}_1] \vee_\alpha m[\mathscr{E}_2] \qquad$ for $\qquad K^{\vee \alpha}(m[\mathscr{E}_1]) m[\mathscr{E}_2]$, and

$\qquad m[\mathscr{E}_1 \vee_\alpha \mathscr{E}_2] \qquad$ for $\qquad m[\mathscr{E}_1] \vee_\alpha m[\mathscr{E}_2]$.

Then with $J.m[\mathscr{E}] = m[\sim \mathscr{E}]$, we have:

$\qquad K^{\wedge \alpha}(m[\mathscr{E}_1]) m[\mathscr{E}_2] = J.K^{\vee \alpha}(J.m[\mathscr{E}_1]) J.m[\mathscr{E}_2].$

It becomes:

$m[\mathscr{E}_1] \wedge_\alpha m[\mathscr{E}_2] = m[\mathscr{E}_1 \wedge_\alpha \mathscr{E}_2] = J.(m[\sim \mathscr{E}_1] \vee_\alpha m[\sim \mathscr{E}_2])$

$\qquad = J.m[\sim \mathscr{E}_1 \vee_\alpha \sim \mathscr{E}_2] = m[\sim (\sim \mathscr{E}_1 \vee_\alpha \sim \mathscr{E}_2)].$

So the bba induced by $\mathscr{E}_1 \wedge_\alpha \mathscr{E}_2$ and $\sim (\sim \mathscr{E}_1 \vee_\alpha \sim \mathscr{E}_2)$ are identical, what translates that $\mathscr{E}_1 \wedge_\alpha \mathscr{E}_2$ and $\sim (\sim \mathscr{E}_1 \vee_\alpha \sim \mathscr{E}_2)$ are equal, what is the De Morgan property.

In particular, when $\alpha = 1$, $K^{\vee 1}(m)$ is the disjunctive operator and $K^{\wedge 1}(m)$ is the conjunctive operator. The bba $m_1 \wedge_1 m_2$ is the one obtained by applying the conjunctive rule of combination (Dempster's rule of combination unnormalized) to m_1 and m_2.

Deriving $\alpha = 1$.

How to derive the conjunctive and disjunctive rules of combination (hence the $K^{\wedge 1}$ and $K^{\vee 1}$ operators)? Thus how to justify $\alpha = 1$? It happens that the only α-junction operator that acts as a specialization (generalization) is obtained with the 1-conjunction (1-disjunction) operator (Klawonn and Smets, 1992). So requiring that $K(m)$ acts as a specialization (generalization) on any bba implies that $\alpha = 1$, thus leads to the conjunctive rule of combination (and its disjunctive counterpart, the disjunctive rule of combination).

The case $\alpha = 0$.

Suppose two pieces of evidence E1 and E2 and their induced bba m and m. We mentioned in section 1 that:

1) the 1-conjunction ($K^{\wedge 1}$) correponds to the case where You accept that both sources tell the truth,

2) the 1-disjunction ($K^{\vee 1}$) correponds to the case where You accept that at least one source tells the truth,

3) the 0-disjunction ($K^{\vee 0}$) correponds to the case where You accept that one and only one source tells the truth, and You don't know which is which.

The $K^{\wedge 0}$ operator does not have a name: it fits with the case where You know that either none of or both sources tell the truth, a quite artifical case in practice.

The practical interest of the $\alpha = 0$ cases are limited. This might explain why they were not introduced previously. In any case, $\alpha = 0$ should not be understood as intermediate between the 1-conjunctive and 1-disjunctive rules. In fact, the $K^{\vee \alpha}$ operator is intermediate between the $K^{\vee 1}$ and the $K^{\vee 0}$ operators, and the $K^{\wedge \alpha}$ operator is intermediate between the $K^{\wedge 1}$ and the $K^{\wedge 0}$ operators.

Remark: In set theory, two operators, the joint denial and Sheffer's stroke, can be used to represent the AND, the OR and the negation with a unique symbol. We cannot extend this result to the α-junctions. Indeed their definition bears strongly on the idempotency property and $K(m)$ usually does not satisfy $K(m)m \neq m$. Hence it seems to be hopeless to find the analogous of these two special operators.

6. Some comments.

6.1. Explaining the negation $\sim\mathscr{E}$.

Suppose the bba 1_Ω defined on Ω. Then $K^{\vee 0}(1_\Omega) = J$. So $K^{\vee 0}(1_\Omega)m[\mathscr{E}] = J.m[\mathscr{E}] = m[\sim\mathscr{E}]$. As 1_Ω can be seen as the bba induced by the piece of evidence that supports nothing specific on Ω, we can define 1_Ω as $m[\mathscr{T}]$ where $\mathscr{T}$ denotes the vacuous piece of evidence, i.e., the triple (S, T, true) where T is a tautology. In particular, $\mathscr{E} \vee \mathscr{T} = \mathscr{T}$ and $\mathscr{E} \wedge \mathscr{T} = \mathscr{E}$ for any $\mathscr{E}$. Therefore $K^{\vee 0}(1_\Omega)m[\mathscr{E}] = K^{\vee 0}(m[\mathscr{T}])m[\mathscr{E}] = m[\sim\mathscr{E}]$. So we obtain an explanation of the meaning of $m[\sim\mathscr{E}]$ as being the bba induced by an exclusive disjunction between $\mathscr{E}$ and $\mathscr{T}$: $\sim\mathscr{E} = \mathscr{E}\underline{\vee}\mathscr{T}$.

In practice, $\sim\mathscr{T}$ is impossible (hopefully, if one hopes to develop a realistic model: the souce states a tautology and if $\sim\mathscr{T}$ holds, it means You accept that the source tells the false). So when $\mathscr{E}\underline{\vee}\mathscr{T}$ holds, it means You accept that the source that states E is telling the false. So $\sim\mathscr{E}$ represents the bba that would be induced if You know that the source is telling the false: whenever the source give a support that the actual world ω_0 belongs to A, You give that support to $\overline{A}$.

6.2. Spread of $m_1(X)m_2(Y)$ on Ω.

The relations $K^{\vee\alpha}(m_1)m_2$ and $K^{\wedge\alpha}(m_1)m_2$ can also be represented in such a way that one realizes that both combination operators correspond to a distribution of the product $m_1(X)m_2(Y)$ among some specific subsets of Ω. In fact the Y'th column of $K^{\vee\alpha}X$ and $K^{\wedge\alpha}X$ is the probability distribution according to which the mass $m_1(X)m_2(Y)$ is distributed on the subsets of Ω. So the terms $k_{AY}^{X}{}^{\vee\alpha}$ and $k_{AY}^{X}{}^{\wedge\alpha}$ are the proportions of $m_1(X)m_2(Y)$ that is allocated to A after the $\vee_\alpha$ and $\wedge_\alpha$ combinations of m_1 and m_2, respectively. The symmetry of the product $m_1(X)m_2(Y)$ is translated by the fact that both $k_{AY}^{X}{}^{\vee\alpha}$ and $k_{AY}^{X}{}^{\wedge\alpha}$ are both symmetric in X and Y.

6.3. Measure of the impact of $K(m)$.

A natural measure of the impact of the operator $K(m)$ is its determinant $|K(m)|$. It happens that $|K(m)| = \prod_{X \subseteq \Omega} g(X)$ where the $g(X)$ terms are the eigenvalues of $K(m)$

(see section 4.3). This relation was already obtained for the 1- conjunction (Smets, 1983) where we understood the product as a measure of the information contains in m. We think this was inappropriate and the idea of 'impact' is better.

7. Conclusions.

In conclusion, we have discovered a family of α-junction operators that include as particular cases the conjunctive rule of combination, the disjunctive rule of combination, the exclusive disjunctive rule of combination, and their negations. The operators $K^{\vee\alpha}$ (and its dual $K^{\wedge\alpha}$) generalize the classical concept of conjunction and disjunction within the context of belief function, i.e., a particular context of weighted sets. The requirements that underlie the derivation of the structure of this operator are those expected by a belief function. Their extension to other theories are not obvious. For instance, using our approach for fuzzy sets and possibility theory will probably be inadequate as the linearity requirement is not the kind of requirement assumed within these two theories.

The meaning of $K^{\vee\alpha}$ and $K^{\wedge\alpha}$ is clear with $\alpha = 0$ or 1. With other values of α, their meaning need further study. At least, we have shown that the classical conjunction and disjunction operations are just extreme cases of a general theory and that a continuum of operators can be built between the conjunction and $\triangle$ and between the disjunction and the exclisive disjunction.

Bibliography.

DUBOIS D. and PRADE H. (1986) A set theoretical view of belief functions. Int. J. Gen. Systems, 12:193-226.

GEBHARDT J. and KRUSE R. (1993) The Context Model : An Integrating View of Vagueness and Uncertainty, Int. J. of Approximate Reasoning 9, 283-314.

KENNES R. and SMETS Ph. (1990) Computational Aspects of the Möbius Transform. in Bonissone P.P., Henrion M., Kanal L.N. and Lemmer J.F. eds., Uncertainty in Artificial Intelligence 6.North Holland, Amsteram, 1991,401-416.

KENNES R. (1992) Computational aspects of the Moebius transform of a graph. IEEE-SMC, 22: 201-223.

KLAWONN F. and SMETS Ph. (1992) The dynammic of belief in the transferable belief model and specialization-generalization matrices. in Dubois D., Wellman M.P., d'Ambrosio B. and Smets Ph. Uncertainty in AI 92. Morgan Kaufmann, San Mateo, Ca, USA, 1992, pg.130-137.

SHAFER G. (1976) A mathematical theory of evidence. Princeton Univ. Press. Princeton, NJ.

SMETS Ph. (1983) Information Content of an Evidence. Int. J. Man Machine Studies, 19: 33-43.

SMETS Ph. (1990) The combination of evidence in the transferable belief model. IEEE Pattern analysis and Machine Intelligence, 12:447-458.

SMETS Ph. (1992) The concept of distinct evidence. IPMU 92 Proceedings.pg. 789-794.

SMETS Ph. (1993a) Belief functions: the disjunctive rule of combination and the generalized Bayesian theorem. Int. J. Approximate Reasoning 9:1-35.

SMETS Ph. (1993b) An axiomatic justifiaction for the use of belief function to quantify beliefs. IJCAI'93 (Inter. Joint Conf. on AI), Chambery. pg. 598-603.

SMETS Ph. (1997) Imperfect information: imprecision - uncertainty. In Motro A. and Smets Ph. eds.Uncertainty Management in Information Systems. Kluwer, Boston, pg. 225-254.

SMETS Ph. and KENNES R. (1994) The transferable belief model. Artificial Intelligence 66:191-234.

Just How Stupid is Postmodernism?

John Woods

Department of Philosophy University of Lethbridge 4401 University Drive Lethbridge, Alberta Canada T1K 3M4 Tel: (403) 329-2501 Fax: (403) 329-5109 E-mail: woods@hg.uleth.ca

Abstract. A paper prepared for presentation to the Royal Society of Canada, Vancouver, and thereafter to the University of Lethbridge, and the International Joint Conference on Qualitative and Quantitative Reasoning, Bonn.

In many ways postmodernism is a good deal less stupid than the title of this talk, which is intended as little more than a provocation. I used to think that the Greek word "logos" carried more meanings than traffic could bear. But "postmodern" displays a semantic promiscuity on a scale that makes a piker of "logos". So what is it? It is a style of thought which is suspicious of classical notions of truth, reason, identity and objectivity, of the idea of universal progress or emancipation or single frameworks, grand narratives or ultimate grounds of explanation.

Indeed [a]gainst these Enlightenment norms, it sees the world as contingent, ungrounded, diverse, unstable, indeterminate, a set of disunified cultures or interpretations which breed a degree of scepticism about the objectivity of truth, history and norms, the givenness of nature and the coherence of identities.

It may strike us at century's end that what used to be called the moral sciences are in a state that is tailor-made for postmodern summing up, although the very word for it is ignorant pomposity. (There is not a trait in the postmodern catalogue which wasn't abundantly evidenced in antiquity, with periodic recurrence ever since.) Perhaps we should not find it so striking if there were no fact of the mater about Ophelia's acquiescent sexuality or about what final moral interpretation a given body of data calls for, or about whether the id exists. Postmodern latitude is, if anything else, recognition of the slack that attends our softly scientific judgements. This is postmodernism cheaply bought, and I for one don't think much of it. Should it indeed happen that the soft sciences have nowhere to go but postmodern, the harder sciences are a harder sell; and it is to them that I should like here to turn my attention. It would be a discovery worth making a fuss over if we could show that postmodernism reposes in the very coils of the hard, which is to say, not in the history of science (which is soft) but in its essential methods, and settled practice.

I shall therefore take as my text, the mathematical theory of sets and so-called truth semantics for constantive discourse. Each is beset with paradox. With sets, it was the paradox which Russell communicated to the father of modern quantification theory in 1902.

The Russell Paradox: It may be that some sets are members of themselves. But it is certain that many sets are not self-membered, as witness the set of chairs in this room. Now there is a set of thosesets the set of all and only those sets which are not members of themselves. Call this set R. Is it a member of itself or not? If it is then, since R has only non-self-membered sets as members, it is not a member of itself. Yet if it is not a member of itself, it fulfills the conditions on membership in R, hence is a member of itself. So R is a member of itself if and only if it is not. Another difficulty arises in truth conditional semantics right at the beginning. It is the *Tarski Paradox*.

The Tarski Paradox:

> Consider the statement in the rectangle below.

> The statement in the rectangle is untrue.

Call this statement L. Is L true or not? If true, then what it says is so; hence it is not true. If it is not true, L satisfies its own predicate, i.e., L is as L says it is; hence it is true. So L is true if and only if it is not. Russell, and Frege too, thought that his paradox destroyed the concept of set; and Tarski thought that his paradox destroyed the concept of the statement.

The Tarski Paradox is an utter devastation. If valid, it destroys the concept of statement. Hence, if Tarski is a Fregean or a Russellian about concepts, there are no statements. If he is not, his lgoical classicism binds him to absolute inconsistency every statement of any natural language is true. The first alternative sounds the death-knell of constantive discourse; we lack the means even to try to say what is the case, what is so. The second alternative guarantees an alethic libertinism that also amounts to nihilism all that is is precisely what isn't, provided that Convention **T** is true. Even if **T** doesn't obtain, constantive discourse, while not impossible, is dispossessed of any rationale, since everything anyone ever says is always both true and not.

The first problem is that of constantive nihilism. The second is, depending on the status of Convention T, is ontological nihilism in a form as severe as anything dreamed up in days of yore by Gorgias, for example, or it is semantic nihilism: all and anything that is true is not.

The first is the greater problem. If the Tarski Paradox demonstrates the impossibility of statements, of constantive discourse as such then, for example, it cannot be the case that beliefs have propositional contents; i.e., it cannot be that my belief that the cat is on the mat bears any relation at all to anything identifiable as what is stated by the sentence, "The cat is on the mat", since neither that sentence nor any other states anything.

This problem about belief comes to the fore in a rather pressing way when we consider Tarski's own solution to the Paradox. If, as I am assuming, Tarski understood his strategy in the same way that Russell understood his own to the paradox of sets, then we run into vexations of a kind that I shall now describe. To do so, we need a brief detour into Russell's intellectual development. He began his work in the foundations of mathematics as an idealist. It is a commonplace of idealist thought that our ordinary concepts the concept of space, for example are inconsistent. The job of the theorist therefore is to repair the concept, to refine the inconsistency out of it. Doing so is subject to what we might call "the principle of consistent similarity", which bids the theorist to make h is new concept as similar to the original as consistency allows.

By 1903, with the publication of Principles of Mathematics, Russell had abandoned idealism for something called analysis, to which he was drawn under the forceful ministrations of G.E. Moore. On this analytical perspective, a certain view of concepts recognized a place of central importance. It was a view according to which inconsistent concepts don't exist, and since non- existent, there is nothing whatever that falls under them. As the Russell Paradox showed, the putative "concept" of set is inconsistent; hence there is no concept of set; hence no sets. It becomes quickly evident that the idealist strategy for repairing inconsistent concepts can't be applied when concepts are understood in this analytic way. The idealist strategy requires fidelity to the principle of consistent similarity the new concept must resemble the old as much as consistency allows. But on the analytic approach to concepts, there is no original concept and there are no sets. Anything proposed as the successor concept must, on the principle of consistent similarity, resemble as much as consistency allows nothing whatever. From which we have it either that no successor concept satisfies the principle or that every consistent concept whatever satisfies it, and satisfies it equally.

What was Russell to do? After much dissembling and more smoke-blowing than is quite seemly, he did the only thing he could do short of going into real estate. He stipulated. Sets were now introduced by nominal definitions, which Russell dressed up as something he dared call "mathematical analyses". Russell knew as well as anyone ever did or could that while one is free to stipulate as one pleases, no one else is required to bear them any mind. So he imposed a condition on what would count as acceptable stipulations in mathematics. A stipulation is acceptable to the extent that the right people are disposed to believe it. Thus you stipulate that p, and perhaps the community of p-enquirers come to believe it. If so, the stipulation is acceptable.

I need hardly dwell on the postmodernist skeins with which Russell's recovery of set theory is shot through: There are no facts of the matter about sets; sets are a human construct; how sets are relative to what people are prepared to think about them; sets are patches of consensus in the mathematical conversation of mankind; and so on.

Can Russell's strategy for recovery be applied to the devastation of the Tarski Paradox? Recall that Russell's strategy is stipulation supported by ellite communal belief. If Tarski's Paradox establishes the impossibility of statements, then if beliefs are propositional attitudes psychological states in some kind of a opposition to statements there are no beliefs either, and Russell's strategy fails for sets and statements alike.

If it shows anything, Tarski's Paradox establishes that the cost of persisting with the analyst's conception of concepts is the death of discourse, belief and desire (since it too is a propositional attitude).

I take it without further ado that this is too much to bear even for "the brilliant young zombies who know all about Foucault..." So what is to be done?

One option, obviously enough, is to revert to idealism and fess up about it. It is well to recur to what the reversion by us. It buys a way of recovering from paradox. Costs, if that's the word for it, are another thing. Human knowledge, whether in politics or in the foundations of mathematics, is in part at least a human artifact; and knowledge is wrought one way rather than another, for what it is wanted for. Collectively, the cost of the idealist strategy is the abandonment of realism, of the view that how the world is is independent of what we think of it, and that our beliefs are objectively true or objectively false depending on how the world is apart from what we think of it.

Naturalism offers another way of proceeding, and a more attractive one on its face for those who dislike the postmodern cachet of idealism, if the anachronism may be forgiven. Naturalism offers promise of the recovery of realism. For, unlike the old epistemology, naturalism seeks "no firmer basis for science than science itself". The naturalist "is free to use the very fruits of science in investigating its roots." It is a self-referential process, as is postmodernism itself, but no mind, since it is "a matter, as always in science, of tackling one problem with the help of our answers to others." In the case of sets, the naturalist rejigs not to preserve as much as he can of the old concept but rather with a view to facilitating the broader aims of mathematics, broadly indispensable in turn for science. Similarly, our theory of the external world will be a rational reconstruction from modest beginnings sets of triggered neural receptors at a specious present; and before long bodies will be sets of quadruples of real numbers in arbitrary co-ordinate systems. Those liking the naturalist option could do no better than to turn to Quine for instruction, for it is he more than anyone else who has given the project a commanding and detailed articulation. But caveat emptor; the raw recruit to naturalism may be unprepared for what awaits him there:

Even the notion of a cat, let alone a class or number, is a human artifact, rooted in innate disposition and cultural tradition. The very notion of an object at all, concrete or abstract, is a human contribution, a feature of our inherited apparatus for organizing the amorphous welter of neural input. Indeed, if we transform the range of objects of our science in any one-to-one fashion, by reinterpreting our terms and predicates as applying to new objects instead of the old ones, the entire evidential support of our science will remain undisturbed. How did the naturalist come to this sorry pass? And why should we not say that the strongest case ever made for the truth of postmodernism in the hard sciences has been made by him? We can say it if we like, but the irony of it all should not be lost on us (more postmodernism still). The naturalist begins his scientific account of our access to the world rooted in the realist stance. He assumes that the world is objectively there no thanks to us, and that what we come to know of it will be objectively so. Once up and running, whether in the precincts of neurological theories of perception, or in theories of the interior of the atom, or in the foundations of transfinite arithmetic; naturalism makes it clear, over and over again, that our best scientific account of how beings like us know the world reveals that we do so in ways that fulfill the canons of idealism. Here is something which deserves to be called anomalous realism. It provides that when we bring to bear the presumptions of realism upon our scientific enquiries into how we know the world it emerges that enquiry itself is idealist. In this it seems that we cannot help ourselves. The realist stance delivers the good for idealism every time, but we can't make ourselves reject the stance. We can't help being idealists while thinking that we are realists. This is what Sartre made much of under the heading of mauvaise fois bad faith. Applying the realist stance now to the persistent and pervasive phenomenon of mauvaise fois, there is little to conclude that it is naturally selected for.

One of the most recalcitrant travails of postmodernism in the arts and letters, and in the soft sciences, is postmodernism's own bad record with the question "What now?" What work is there to do in history or in literary studies if postmodernism is true and faithfully concurred with? If there are no Archimedian points, it is hard to see what the research programme could be. Not seeing where the research programme should go is like not knowing where you are. It is the kind of lostness which promotes abandonment for example, the rejection of literature in some Departments of English; or it invites intellectual rubbish, exchanged within hostile dialectical structures of a kind that Aristotle called "babbling". And it invites it positively begs for Sokal's hoax.

On this score, naturalism has the edge, not because it evades postmodern commitments but precisely because it abounds in them. The advantage given to naturalism is that it seizes upon its own postmodern consequences and lets them shape a coherent research programme.

It is to employ the best of what naturalism can offer to explain the persistence of the realist stance even in the face of the pervasive endorsement it gives to irrealism. The project in short is a naturalistic explanation of the epistemic mauvaise fois of the human condition. And that anyhow is something (and not the horary old conversation of mankind!).

Integrating Preference Orderings into Argument-Based Reasoning

Leila AMGOUD, Claudette CAYROL

Institut de Recherche en Informatique de Toulouse (I. R. I. T.)
Université Paul Sabatier, 118 route de Narbonne, 31062 Toulouse Cedex (FRANCE)
E-mail: {amgoud,testemal}@irit.fr

Abstract.
Argument-based reasoning is a promising approach to handle inconsistent belief bases. The basic idea is to justify each plausible conclusion by acceptable arguments. The purpose of this paper is to enforce the concept of acceptability by the integration of preference orderings. Pursuing previous work on the principles of preference-based argumentation, we focus here on the definition of new acceptability classes of arguments.

1 Introduction

Argument-based reasoning ([Vre91], [Pol92], [SL92], [Dun93&95], [EFK93], [EH95], [PS96]) is a promising approach for reasoning with inconsistent beliefs. It may be characterized by the following points:

- the basic principle is "inferring without revising". So, plausible inferences that follow from consistent subsets of an inconsistent belief base must be justified, with so-called arguments.

- due to the inconsistency of the available knowledge, arguments may be constructed in favor of a statement and other arguments may be constructed in favor of the opposite statement. The basic idea is to view reasoning as a process of first constructing arguments and then selecting most acceptable of them.

- the last point concerns the concept of acceptability. It has been most often defined purely on the basis of other constructible arguments. But other criteria may be taken into account for comparing arguments, such as for instance specificity [SL92], or explicit priorities on the beliefs.

The work reported here concerns preference-based argumentation. In previous work [Cay95b], we have proposed a methodological approach to the integration of preference orderings into argumentation frameworks. We have specified principles for taking into account preference orderings in the selection of acceptable arguments. Then in [ACL96], we have investigated the definition of preference relations for comparing conflicting arguments.

The topic of this paper is to report some preliminary results on the combination of these previous works.

After a brief presentation of the main principles of argument-based inference, we develop our proposal, which is focused on the concept of acceptability. Our basic idea is to account for preference in the definition of new acceptability classes of arguments. Our discussion concerns four classes and three examples of preference orderings between arguments. We show that our approach enables to recover other proposals for combining preferences and arguments, and to generalize previous results about argumentative inference.

2 Argument-Based Inference

The central notion in argument-based reasoning is the acceptability of arguments. So, we successively introduce the structure of a set of arguments and theories of acceptability.

2.1 Argumentation Framework

An argumentation framework is generally defined as a pair of a set of arguments, and a binary relation representing the "defeat" relationship between arguments.

A theory of acceptability can be defined in a general argumentation framework, without special attention paid to the internal structure of arguments, as in [Dun95]. Here, we are interested in particular argumentation frameworks where arguments are built from an inconsistent belief base, using the classical inference.

Throughout this paper L is a propositional language. $\vdash$ denotes classical entailment. K and E denote sets of formulas of L. K, which may be empty, represents a core of knowledge and is assumed consistent. Contrastedly, formulas of E represent defeasible pieces of knowledge, or beliefs. So K $\cup$ E may be inconsistent. K is the *context* and E will be referred to as the *belief base*.

We introduce the notion of argument in the framework (K, E). Similar definitions appear for instance in [SL92] and [EH95].

Definition1. An *argument* of E in the context K is a pair (H, h), where h is a formula of L and H is a subbase of E satisfying: (i) K $\cup$ H is consistent, (ii) K $\cup$ H $\vdash$ h, (iii) H is minimal (no strict subset of H satisfies i and ii). H is called the *support* and h the *conclusion* of the argument.

Definition2. The argument (H1, h1) is a *subargument* of the argument (H2, h2) iff H1 $\subseteq$ H2.

Definition3. Two arguments (H1, h1) and (H2, h2) are said *conflicting* iff h1 $\equiv$[1] $\neg$h2.

The set of all the arguments of E in the context K will be denoted by AR(K, E) (or for short AR(E) if there is no ambiguity on K).

Different definitions for the relation "defeat" lead to different argumentation frameworks.

Definition4. [EFK93] Let (H1, h1) and (H2, h2) be two arguments of E.

(H1, h1) *rebuts* (H2, h2) iff h1 $\equiv \neg$h2. (The arguments are conflicting).

(H1, h1) *undercuts* (H2, h2) iff for some h $\in$ H2, h $\equiv \neg$h1.

---

[1] $\equiv$ means logical equivalence.

See [Cay95a] for a comparative study of other families of definitions.

Example1. K = ∅, E = E1∪E2∪E3∪E4 with E1 = {p, x→o, x→t}, E2 = {o→v}, E3 = {t→¬v}, E4 = {p→x}. Let H = {p, p→x, x→o, o→v}, H' = {p, p→x, x→t, t→¬v} and H" = {p, p→x, x→t, t→¬v, o→v}. The argument (H, v) rebuts the argument (H', ¬v). The argument (H", ¬(x→o)) undercuts the argument (H, v).

Remarks

- All the arguments considered in this paper have a consistent support.

- Here, an argument is simply a pair of a set of supporting sentences and a conclusion. We do not consider structured arguments involving chains of reasons, as in [Pol92].

- A similar methodology for defining the concept of "defeat" is used in [PS96], with the same terminology but with a different structure for the arguments. In [PS96], an argument is a sequence of chained implicative rules. Each rule has a consequent part (consisting of one literal) and an antecedent part (consisting of a conjunction of literals). The consequent of each rule in a given argument is considered as a conclusion of that argument. In that sense, one argument can have contradictory conclusions.

2.2 Acceptability of Arguments

The main approaches which have been developed for reasoning within an argumentation framework rely on the idea of differentiating arguments with a notion of acceptability.

Acceptability levels may be assigned to arguments on the basis of other constructible arguments (see [EFK93] and [EH95] for various such notions of acceptability). Then, from a taxonomy of acceptability classes, consequence relations are defined.

Quite independently, Dung formalized a theory of global acceptability [Dun93&95]. The set of all the arguments that a rational agent may accept must defend itself against all attacks on it. This leads to the definition of extensions of an argumentation framework. For instance, a stable extension S of an argumentation framework (AR(E), *defeat*) is a conflict-free (no two arguments A1 and A2 in S such that A1 *defeats* A2) subset of AR(E) which *defeats* each argument which does not belong to S. A variant of Dung's formalization leads to the set of justified arguments, in [PS96].

For the purpose of the work reported here, we focus on the hierarchy of acceptability classes.

Definition5.

AR*(E) denotes the set of arguments of E with an empty support.

AR⁺(E) denotes the set of arguments of E which are not rebutted by some argument of E.

AR⁺⁺(E) denotes the set of arguments of E which are not undercut by some argument of E.

Proposition1. (H, h) belongs to AR⁺⁺(E) iff

for each element k of H, there is no argument supporting ¬k iff

for each subargument (H', h') of (H, h), (H', h') belongs to AR⁺(E).

The following inclusions hold between the so-called acceptability classes:

C4 = AR*(E) ⊆ C3 = AR⁺⁺(E) ⊆ C2 = AR⁺(E) ⊆ C1 = AR(E)

The argument (H1, h1) is said more acceptable than (H2, h2) iff there exists a class Ci $(1 \leq i \leq 4)$ containing (H1, h1) but not containing (H2, h2). For instance, arguments which are never undercut are more acceptable than arguments which are never rebutted.

Example2. $K = \emptyset$, $E = E1 \cup E2$ with $E1 = \{p, t, p \rightarrow o\}$, $E2 = \{t \rightarrow \neg o, o \rightarrow v\}$. Let $H = \{p, p \rightarrow o, o \rightarrow v\}$. The argument (H, v) $\in$ AR$^+$(E) but (H, v) $\notin$ AR^{++}(E). Indeed, (H, v) is undercut by the argument $(\{p, t, t \rightarrow \neg o\}, \neg(p \rightarrow o))$.

2.3 Argument-Based Consequence Relations

The hierarchy of acceptability classes induces a hierarchy of consequence relations as follows [EFK93]:

Definition6.

ϕ is a *certain* consequence of E iff AR*(E) contains an argument concluding ϕ.

ϕ is a *confirmed* consequence of E iff there exists $H \subseteq E$ with (H, ϕ)$\in$ AR^{++}(E).

ϕ is a *probable* consequence of E iff there exists $H \subseteq E$ with (H, ϕ)$\in$ AR$^+$(E).

ϕ is a *plausible* consequence of E iff there exists $H \subseteq E$ with (H, ϕ)$\in$ AR(E).

Example3. $K = \emptyset$, $E = \{a, a \rightarrow b, \neg a\}$. $a \rightarrow b$ is a confirmed but not a certain consequence. b is a probable but not a confirmed consequence; the argument ($\{a, a \rightarrow b\}, b\}$ belongs to AR$^+$(E) but is undercut by ($\{\neg a\}, \neg a$). a is a plausible but not a probable consequence.

A common proposal to handle multiple stable extensions is to accept a formula as a consequence when it can be classically inferred from each stable extension (conservative point of view), or when it can be classically inferred from at least one stable extension (permissive point of view).

Strong relationships (and sometimes exact correspondences) have been established between most of these consequence relations (see [Cay95a]). For instance, h is a confirmed consequence of E iff h is the conclusion of an argument which belongs to each stable extension of AR(E).

3 Integration of Preferences

3.1 Principles of Preference-Based Argumentation

The different theories of acceptability are based on the existence of defeating arguments. Other notions may be used to compare arguments: for instance specificity relations which can be syntactically extracted from the belief base [SL92] or preference orderings which are induced by a priority relation expressed on the belief base itself.

Preference orderings have recently emerged from studies in non-monotonic reasoning and belief revision as playing a crucial role. They allow for more sophisticated and more appropriate handling of conflict resolution and default information, for instance.

Previous work on prioritized coherence-based entailment [CLS95] and the tight relationships established between argument-based reasoning and coherence-based reasoning [Cay95a] indicate that preference orderings should provide promising developments for argument-based reasoning. (See also [PS96]'s proposal, where preferences between the beliefs are directly integrated in the definition of the *defeat* relation).

We have already proposed a methodological approach to the integration of preference orderings in argumentation frameworks by distinguishing two problems [Cay95b]: the definition of preference relations in order to compare conflicting arguments and the specification of principles which take into account these preference relations in order to select acceptable arguments.

The topic of this section is to connect the main results obtained in both investigation fields.

3.2 From Priorities on the Beliefs to Preferences Between Subbases

Here, we are only concerned with preference relations which are induced by priority relations expressed on the belief base itself. (See [ACL96] for a more general discussion on preferences and arguments). Moreover, the preference relations we consider are induced by preference relations defined on the supports of the arguments. Namely, let *Pref* denote a partial pre-ordering on the set of consistent subsets of E. The associated preference relation available for comparing arguments will be defined by : "(H2, h2) is preferred to (H1, h1) iff H2 is preferred to H1 w.r.t. *Pref.*"

So, we assume that the belief base E is equipped with a pre-ordering structure (partial or complete) which is not related to the semantical entailment ordering. In case of a complete pre-ordering, it is equivalent to consider that the belief base is stratified in a collection of subbases of different priority levels.

Then, the selection of preferred supports relies upon the definition of aggregation modes which extend the priority ordering (defined on the initial belief base) into a preference relation between subbases.

In [ACL96], we have presented and compared several such preference relations. Among them we have:

BDP-preference [BDP93]

This preference relation, based on certainty levels, has been introduced in the context of possibilistic logic. The belief base E is stratified in $E = E1 \cup ... \cup En$ such that beliefs in Ei have the same level of certainty, and are more reliable than beliefs in Ej where $j > i$.

The certainty level of a non-empty subbase H is defined as level(H) = max $\{j / 1 \leq j \leq n$ and $Hj \neq \varnothing\}$, where Hi denotes $H \cap Ei$.

The certainty level can be used to define a complete pre-ordering on the subsets of E. Let H, H' be two consistent subbases of E. H is preferred to H' iff level(H) $\leq$ level(H'). Note that if $H \subseteq H'$ then H is preferred to H'. We consider the associated strict ordering:

$$H \text{ is BDP-preferred to } H' \text{ iff level(H)} < \text{level(H')}.$$

Let (H, h) and (H', h') be two arguments of E. We define (H, h) $>>^{BDP}$ (H', h') iff H is BDP-preferred to H'.

ELI-preference [CRS93]

This preference relation corresponds to the well-known principle of "elitism": "everything kept must be better than something removed".

Let $\geq$ denote a (partial) pre-ordering on the belief base. The ELI-preference is the ordering defined by: H is preferred to H' iff $\forall$ k $\in$ H \ H' $\exists$k' $\in$ H' \ H, such that k > k' (i.e. k $\geq$ k' and not k' $\geq$ k).

When the belief base is stratified, the associated strict ordering can be equivalently defined in terms of levels as follows (proof in [AC96]):

H is ELI-preferred to H' iff level(H \ H') < level(H' \ H) (with level ($\varnothing$) =0).

Let (H, h) and (H', h') be two arguments of E. We define (H, h) >>^{ELI} (H', h') iff H is ELI-preferred to H'.

We have proved in [AC96] that if H is BDP-preferred to H' then H is ELI-preferred to H' and the converse holds iff level(H $\cap$ H') < level(H' \ H). This property shows that the ELI-preference may be viewed as a refinement of the BDP-preference.

Weak-BDP-preference [ACL96]

However, the consideration of examples, such as the following one, has suggested us to propose another refinement of the BDP-preference.

Example4. E = E1 $\cup$ E2 $\cup$ E3 $\cup$ E4 with E1 = {a}, E2 = {b}, E3 = {c, d}, E4 = {e}. Let H = {a, d, e} and H' = {b, c, e}. H and H' are neither comparable by the BDP-preference, nor by the ELI-preference.

We have proposed the Weak-BDP-preference, which relies upon a kind of lexicographic comparison of the levels. The idea is to extend the concept of level as follows. Let E = E1$\cup$...$\cup$En be a stratified belief base and H a consistent subbase of E. For each $1 \leq k \leq n$, we define level$_k$(H) = level(H1$\cup$...$\cup$Hk) = max {j/ $1 \leq j \leq k$ and Hj $\neq \varnothing$}. Then, level(H) = level$_n$(H).

We define: H is weakly-BDP-preferred to H' iff $\exists$ $1 \leq k \leq n$ such that level$_k$(H) < level$_k$(H'), and for each j>k level$_j$(H) = level$_j$(H').

Let (H, h) and (H', h') be two arguments of E. We define (H, h) >>^{WBDP} (H', h') iff H is weakly-BDP-preferred to H'.

The different links between these relations can be synthetized on the following schema:

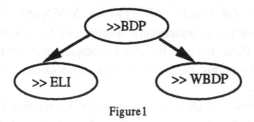

Figure1

Example1 (continued). E is stratified in E1 = {p, x$\rightarrow$o, x$\rightarrow$t}, E2 = {o$\rightarrow$v}, E3 = {t$\rightarrow$$\neg$v}, E4 = {p$\rightarrow$x}. We have: (H, v) >>^{ELI} (H', $\neg$v) and (H, v) >>^{WBDP} (H', $\neg$v) but according to the BDP-preference (H, v) is not preferred to (H', $\neg$v).

3.3 Introducing Preferences into Argumentation Frameworks

The most interesting way of combining preferences and argumentation is to account for preference in the argumentation framework itself, in the definition of new acceptability classes. Let *Pref* denote a partial pre-ordering on the set of consistent subsets of E. Following Definition5, two new classes can be defined:

Definition7.

$AR^+_{Pref}(E)$ denotes the set of arguments of E which are strictly preferred (w.r.t. *Pref*) to each rebutting argument.

$AR^{++}_{Pref}(E)$ denotes the set of arguments of E which are strictly preferred (w.r.t. *Pref*) to each undercutting argument.

Following Proposition1, two other classes can be defined:

Definition8.

(H, h) belongs to $Sub^+_{Pref}(E)$ iff for each subargument (H', h') of (H, h), (H', h') belongs to $AR^+_{Pref}(E)$.

(H, h) belongs to $Sing^+_{Pref}(E)$ iff for each element k of H, the argument ({k}, k) belongs to $AR^+_{Pref}(E)$.

The above definitions enable us to recover other proposals for combining preferences and arguments.

For instance, [Hun94] defines "the argument A supporting h *conflicts with* the argument B supporting ¬h iff B is not strictly preferred to A".

We have proved the following result [Cay95b]: there exists an argument supporting h in $AR^+_{Pref}(E)$ iff there exists an argument (H, h) such that no argument *conflicts with* (H, h).

In [SL92], we find the definition "(H1, h1) *defeats* (H2, h2) iff there exists a subargument of (H2, h2) which rebuts (H1, h1) and is not strictly preferred to (H1, h1)". We have proved the following result [Cay95b]: the argument (H, h) belongs to $Sub^+_{Pref}(E)$ iff (H, h) is never *defeated* by some argument of E.

4 Main results

The purpose of this section is to present a comparative study of the different acceptability classes: $AR^+_{Pref}(E)$, $AR^{++}_{Pref}(E)$, $Sub^+_{Pref}(E)$ and $Sing^+_{Pref}(E)$. Indeed, as in Definition6, a consequence relation can be associated to each new acceptability class. Then, we will be able to compare these consequence relations according to different points of view, such as cautiousness or coherence of the delivered conclusions.

First, we give properties which hold for any preference relation Pref respecting the minimality for set-inclusion: if H is included in H' then H must be preferred (w.r.t. Pref) to H'.

Then, we consider the particular case when Pref is based on the level concept (relations BDP, ELI, WBDP).

All the proofs can be found in [AC97].

4.1 Comparing the classes: the general case
Proposition2.

$$AR^{++}(E) \subseteq Sub^+Pref(E) \subseteq AR^+Pref(E)$$

$$AR^{++}(E) \subseteq AR^{++}Pref(E) \subseteq Sing^+Pref(E)$$

$$Sub^+Pref(E) \subseteq Sing^+Pref(E)$$

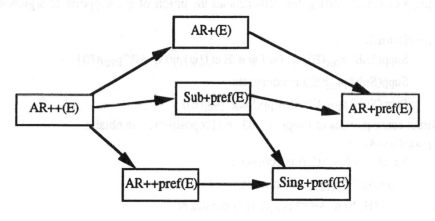

Figure2

Each arrow on the above figure can be translated by the relation "is more cautious than" on the associated consequence relations.

Generally, the different inclusions are strict as shown by the following examples.

$Sub^+_{ELI}(E)$ is strictly included in $AR^+_{ELI}(E)$:

Example5. $E = E1 \cup E2 \cup E3 \cup E4$ with $E1 = \{p, t, t \rightarrow \neg o\}$, $E2 = \{p \rightarrow o\}$, $E3 = \{o \rightarrow v\}$, $E4 = \{p \rightarrow \neg v\}$.

Let $H = \{p, p \rightarrow o, o \rightarrow v\}$. $(H, v) \in AR^+_{ELI}(E)$ and $(H, v) \notin Sub^+_{ELI}(E)$. Indeed, let $H1 = \{p, p \rightarrow o\}$ and $H2 = \{t, t \rightarrow \neg o\}$. $H1 \subseteq H$, and $(H2, \neg o) >>^{ELI} (H1, o)$.

$AR^{++}Pref(E)$ is strictly included in $Sing^+Pref(E)$:

Example6. Let $E = E1 \cup E2 \cup E3 \cup E4 \cup E5$ with $E1 = \{p, t\}$, $E2 = \{p \rightarrow o\}$, $E3 = \{t \rightarrow \neg o\}$, $E4 = \{o \rightarrow v\}$, $E5 = \{p \rightarrow \neg v\}$.

Let $H = \{p, p \rightarrow o, o \rightarrow v\}$ and $H1 = \{t, t \rightarrow \neg o, p \rightarrow o\}$. $(H, v) \in Sing^+_{BDP}(E)$ but $(H1, \neg p)$ undercuts (H, v). Since, level$(H) = 4$ and level$(H1) = 3$ $(H1, \neg p) >>^{BDP} (H, v)$. So $(H, v) \notin AR^{++}_{BDP}(E)$.

$Sub^+Pref(E)$ is strictly included in $Sing^+Pref(E)$:

Example7. Let $E = E1 \cup E2 \cup E3$ with $E1 = \{a, \neg b\}$, $E2 = \{a \rightarrow b\}$ and $E3 = \{a \wedge \neg b \rightarrow c, c \rightarrow a \wedge \neg b\}$. Let $H = \{a, \neg b, a \wedge \neg b \rightarrow c\}$. The argument $(H, c) \in Sing^+BDP(E)$ but $(H, c) \notin Sub^+BDP(E)$.

B being a subset of $AR(E)$, $Supp(B)$ denotes the union of the supports of arguments of B.

Proposition3.

$Supp(Sub^+Pref(E)) = \{\psi / \psi \in E$ et $(\{\psi\},\psi) \in AR^+Pref(E)\}$.

$Supp(Sub^+Pref(E))$ is consistent.

$Supp(Sub^+Pref(E)) = Supp(Sing^+Pref(E))$.

As direct consequences of Proposition3 and Proposition2, we obtain:

Proposition4.

$Supp(Sing^+Pref(E))$ is consistent.

$Supp(AR^{++}Pref(E))$ is consistent.

$\{h/ \exists (H, h) \in AR^{++}Pref(E)\}$ is consistent.

The results stated in Proposition4 are of great importance. Indeed, they prove that the argument-based inference defined through the classes $AR^{++}Pref(E)$ (resp. $Sub^+Pref(E)$, $Sing^+Pref(E)$) delivers safe conclusions. Moreover, since the set of consequences obtained with any of these classes is consistent, it will be possible to close it with classical deduction.

4.2 Case of level-based preference orderings

Proposition5.

For Pref in $\{BDP, ELI\}$, $AR^{++}Pref(E) \subseteq Sub^+Pref(E)$.

The converse is false as shown by:

Example8. $E = E1 \cup E2 \cup E3 \cup E4 \cup E5$ with $E1 = \{p, t\}$, $E2 = \{p \rightarrow o\}$, $E3 = \{t \rightarrow \neg o\}$, $E4 = \{o \rightarrow v\}$, $E5 = \{p \rightarrow \neg v\}$.

Let $H = \{p, p \rightarrow o, o \rightarrow v\}$. $(H, v) \in Sub^+BDP(E)$ and $(H, v) \notin AR^{++}BDP(E)$ because there exists an argument $(H1, \neg p)$ with $H1 = \{p \rightarrow o, t, t \rightarrow \neg o\}$ such that $(H1, \neg p) >>^{BDP} (H, v)$.

As a consequence of Proposition2 and Proposition5, we have:

For Pref in $\{BDP, ELI\}$, $AR^{++}Pref(E) \subseteq AR^+Pref(E)$.

The following result enables us to connect preference-based argumentation with prioritized coherence-based inference schemas (See [CLS95] for a thorough presentation of these inference schemas).

Definition9.

A consistent subbase $S = S1 \cup S2 \cup ... \cup Sn$ is an INCL-preferred subbase of E if and only if $\forall j=1..n$ $S1 \cup S2 \cup ... \cup Sj$ is a maximal (for set-inclusion) consistent subbase of $E1 \cup E2 \cup ... \cup Ej$.

INCL(E) denotes the set of INCL-preferred subbases of E.

∩ INCL(E) will denote the intersection of the INCL-preferred subbases of E.

Proposition6. For Pref in {BDP, ELI, WBDP}, we have:

If ({h}, h) ∈ $AR^+_{Pref}(E)$ then h ∈ ∩ INCL(E).

As consequences, we obtain:

For Pref in {BDP, ELI, WBDP}

$Supp(Sub^+_{pref}(E)) \subseteq \cap$ INCL(E).

If (H, h) ∈ $Sub^+_{pref}(E)$ then ∩ INCL(E) ⊢ h.

For Pref in {BDP, ELI}

If (H, h) ∈ $AR^{++}_{Pref}(E)$ then ∩ INCL(E) ⊢ h.

4.3 Related work

The above propositions enable us to generalize previous results concerning two argumentative entailment relations, proposed by [BDP93&95], in the framework of possibilistic logic.

The so-called "argued consequence" exactly corresponds to the acceptability class $AR^+_{BDP}(E)$: h is an argued consequence of E iff there exists an argument (H, h) in $AR^+_{BDP}(E)$.

The so-called "safely-supported consequence" exactly corresponds to the acceptability class $AR^{++}_{BDP}(E)$.

Moreover, [BDP95] have proved that if h is a safely-supported consequence of E then h is a classical consequence of ∩INCL(E). Proposition6 generalizes that result to other preference relations based on the concept of level.

5 Conclusion

The work reported here concerns preference-based argumentation from inconsistent belief bases. We have connected previous investigations concerning two problems: the definition of principles for integrating preferences into argumentation frameworks, and the definition of preference relations for comparing conflicting arguments.

Our proposal is focused on the notion of acceptability class and is characterized by the study of new acceptability classes. The idea is to take into account the preference relations between arguments in order to select the most acceptable arguments.

We have proposed a comparative study of the new classes. We have proved that some argumentative entailment relations defined by [BDP93&95] in the possibilistic framework can be restated using acceptability classes.

Moreover, we have obtained preliminary results concerning the relationships between preference-based argumentation and prioritized coherence-based non-monotonic inference.

We are now working in that direction, taking advantage of results obtained in the case of flat belief bases (i.e. without any priority) (see [Cay95a]). Another perspective is to study the logical properties of the consequence relations which may be associated to the acceptability classes we have defined in this paper. A first promising result concerns the consistency of the set of consequences delivered with the classes $AR^{++}Pref(E)$, $Sub^{+}Pref(E)$ and $Sing^{+}Pref(E)$.

6 References

[AC96] L. Amgoud, C. Cayrol. Etude comparative de relations de préférence entre arguments: Calcul avec un ATMS. Tech. Report n°96-33-R, IRIT, Univ. Paul Sabatier, Toulouse, Sept. 96.

[AC97] L. Amgoud, C. Cayrol. Intégration de préférences dans le raisonnement argumentatif. Tech. Report n°97-04-R , IRIT, Univ. Paul Sabatier, Toulouse, Fev. 97.

[ACL96] L. Amgoud, C. Cayrol, D. Le Berre. Comparing Arguments using Preference Orderings for Argument-based Reasoning. Proc. ICTAI'96, 400-403.

[BDP93] S. Benferhat, D. Dubois, H. Prade. Argumentative Inference in Uncertain and Inconsistent Knowledge Bases. Proc. 9° Conf. on Uncertainty in Artificial Intelligence, 411- 419, 1993.

[BDP95] S. Benferhat, D. Dubois, H. Prade. How to infer from inconsistent beliefs without revising? Proc. IJCAI'95, 1449-1455.

[Cay95a] C. Cayrol. On the relation between Argumentation and Non-monotonic Coherence-based Entailment. Proc. IJCAI'95, 1443-1448.

[Cay95b] C. Cayrol. From Non-monotonic Syntax-based Entailment to Preference-based Argumentation. In: Symbolic and Quantitative Approaches to Reasoning and Uncertainty (C. Froidevaux, J. Kohlas Eds.), LNAI 946, Springer Verlag, 99-106, 1995.

[CLS95] C. Cayrol, M.C. Lagasquie-Schiex. Non-monotonic Syntax-Based Entailment: A Classification of Consequence Relations. In: Symbolic and Quantitative Approaches to Reasoning and Uncertainty (C. Froidevaux, J. Kohlas Eds.), LNAI 946, Springer Verlag, 107-114, 1995.

[CRS93] C. Cayrol, V. Royer, C. Saurel. Management of preferences in Assumption-Based Reasoning. In: Advanced Methods in Artificial Intelligence (B. Bouchon-Meunier, L. Valverde, R.Y. Yager Eds.), LNCS 682, Springer Verlag, 13-22, 1993.

[Dun93] P. M. Dung. On the acceptability of arguments and its fundamental role in non-monotonic reasoning and logic programming. Proc. IJCAI'93, 852-857.

[Dun95] P.M. Dung. On the acceptability of arguments and its fundamental role in non-monotonic reasoning, logic programming and n-person games. Artificial Intelligence, 77: 321-357, 1995.

[EFK93] M. Elvang-Goransson, J. Fox, P. Krause. Acceptability of arguments as "logical uncertainty". Proc. ECSQARU'93, Lecture Notes in Computer Science, Springer Verlag, Vol. 747, 85 -90.

[EH95] M. Elvang-Goransson, A. Hunter. Argumentative logics: Reasoning with classically inconsistent information. Data & Knowledge Engineering, 16: 125-145, 1995.

[Hun94] A. Hunter. Defeasible reasoning with structured information. Proc. KR'94, 281-292.

[Pol92] J.L. Pollock. How to reason defeasibly. Artificial Intelligence, 57: 1-42, 1992.

[PS96] H. Prakken, G. Sartor. A System for Defeasible Argumentation, with Defeasible Priorities. Proc. FAPR'96, Lecture Notes in Artificial Intelligence, Springer Verlag, Vol.1085, 510-524.

[SL92] G.R. Simari, R.P. Loui. A mathematical treatment of defeasible reasoning and its implementation. Artificial Intelligence, 53: 125-157, 1992.

[Vre91] G. Vreeswijk. The feasibility of Defeat in Defeasible Reasoning. Proc. KR'91, 526-534.

Assumption-Based Modeling Using ABEL *

B. Anrig, R. Haenni, J. Kohlas and N. Lehmann

Institute of Informatics
University of Fribourg
CH–1700 Fribourg, Switzerland
E-Mail: rolf.haenni@unifr.ch
WWW: http://www-iiuf.unifr.ch/dss

Abstract. Today, different formalisms exist to solve reasoning problems under uncertainty. For most of the known formalisms, corresponding computer implementations are available. The problem is that each of the existing systems has its own user interface and an individual language to model the knowledge and the queries.

This paper proposes ABEL, a new and general language to express uncertain knowledge and corresponding queries. Examples from different domains show that ABEL is powerful and general enough to be used as common modeling language for the existing software systems. A prototype of ABEL is implemented in Evidenzia, a system restricted to models based on propositional logic. A general ABEL solver is actually being implemented.

1 Introduction

Today, different formalisms exist to solve reasoning problems under uncertainty. The most popular approaches are the theory of **Bayesian networks** (Lauritzen & Spiegelhalter, 1988) and the Dempster-Shafer **theory of evidence** (Shafer, 1976). For most of the known formalisms, corresponding computer implementations are available. De Kleer's idea of **assumption-based truth maintenance systems** (ATMS) proposes a general architecture for problem solvers in the domain of uncertain reasoning (de Kleer, 1986). Another general architecture is given by Shenoy's concept of **valuation networks** (Shenoy & Shafer, 1990; Shenoy, 1995). Popular software systems are Belief 1.2 (Almond, 1990), Ideal (Srinivas & Breese, 1990), MacEvidence (Hsia & Shenoy, 1989), Pulcinella (Saffiotti & Umkehrer, 1991), TresBel (Xu & Kennes, 1994), Graphical-Belief 2.0 (Almond, 1995), and Hugin (Andersen *et al.*, 1990). The problem is that each of the existing systems has its own user interface and an individual language to model the knowledge and the queries.

* Research supported by grant No. 2100–042927.95 of the Swiss National Foundation for Research.

This paper proposes ABEL[1], a general language to express uncertain knowledge and corresponding queries. In ABEL, uncertainty is expressed by **assumptions**, a special type of variables. Assumptions represent unknown circumstances, risks, or interpretations. Often, it is possible to assign probabilities to the values of an assumption. Furthermore, assumptions are the basic elements to build arguments for hypotheses. The idea of using assumptions has been introduced in de Kleer's ATMS. ABEL uses de Kleer's idea in a more general context of **assumption-based systems** (Kohlas & Monney, 1993).

Examples from different problem domains show that ABEL is powerful and general enough to be used as common modeling language for the existing software systems. This paper describes ABEL using two concrete examples. Further examples and other application domains, e.g.

- Constraint Satisfaction,
- Causal Networks,
- Failure Trees,

can be found in (Anrig *et al.*, 1997). A prototype of ABEL is implemented in Evidenzia (Lehmann, 1994; Haenni, 1996). This system is restricted to models based on propositional logic. A general ABEL solver is actually being implemented. A first running version is restricted to variables which are binary or have a set-type domain (see below). The solver can be obtained from the authors on request. We are aware of the problem, that ABEL can be used to describe problems which cannot be solved at all or which are too complex; but ABEL has no restriction in this direction.

Section 2 introduces the basic structure of ABEL statements. Section 3 and 4 present two typical examples in order to familiarize the reader with ABEL and its different constructs.

2   Basic Language Description

This section introduces the main elements of ABEL. The language is based on three other computer languages: (1) from **Common Lisp** (Steele, 1990) it adopts **prefix notation** and therewith a mess of opening and closing parentheses; (2) from **Pulcinella** (Saffiotti & Umkehrer, 1991) it uses the idea of the commands `tell`, `ask`, and `empty`; and (3) from the existing ABEL prototype (Lehmann, 1994; Haenni, 1996) it inherits the concept of **modules** and the syntax of the queries. In the sequel an informal introduction to the language will be given; for a precise language description refer to (Anrig *et al.*, 1997).

---

[1] ABEL stands for Assumption-Based Evidential Language.

Working with ABEL usually involves three sequential steps:

(1) The given information is expressed using the command `tell`. The resulting model is called **basic knowledge base**. It describes the part of the available information that is relatively constant and static in course of time such as rules, relations, or dependencies between different statements or entities. It is used to build the basic knowledge base and consists of one or several lines called **instructions**. The contents of an instruction can be a **definition** of types, variables, assumptions, or modules, a **statement**, i.e. a rule or another part of the basic knowledge base, or an application of a **module**. The sequence of instructions is interpreted as a conjunction. The syntax of a `tell`-command is the following:

```
(tell <instr-1>
      ...
      <instr-n>)
```

Every instruction can be seen as a piece of information. Therefore, the command `tell` is used to add new pieces of information to the existing basic knowledge base. Note that the instructions of a `tell`-command can be distributed among several `tell`-commands.

(2) Actual facts or observations about the concrete, actual situation are specified using the command `observe`. Usually, in order to complete a model, **observations** or **facts** are added to the basic knowledge base. Observations describe the actual situation or the concrete circumstances of the problem. Note that observations may change in course of time. It is therefore important to separate observations from the basic knowledge base. ABEL provides a command `observe` to specify observations. It expects a sequence of ABEL statements. The sequence is interpreted as a conjunction.

```
(observe <stm-1>
         ...
         <stm-n>)
```

Statements given by `observe`- and `tell`-commands are treated similarly. The difference is that `observe`-statements can be deleted or changed independently when new observations were made. To change an observation, the same statement with the new values has to be re-written. The **empty**-command can be used to delete observations.

(3) Queries about the actual knowledge base are expressed using the command `ask`. Generally, there are two different types of queries: it can be interesting to get the available information about certain variables; and it can be interesting to get symbolic or numerical arguments in favour or against certain hypotheses. In both cases, several queries can be treated at once:

```
(ask <query-1>
     ...
     <query-n>)
```

In the first case, a query is simply an ABEL expression. This type of query is useful, for example, in constraint satisfaction problems (consider (Anrig *et al.*, 1997) for examples). The second way to state queries is important in problems of assumption-based reasoning. The idea is to find arguments in favour or against hypotheses. For a given hypothesis, different types of arguments may be of interest: **support, quasi-support, plausibility,** or **doubt** (Haenni, 1996); the respective keywords in ABEL are `sp`, `qs`, `pl`, `db`. A hypothesis is an ABEL statement. Arguments are conjunctions of normal or negated ABEL constraints over assumptions. The support of a hypothesis, for example, is the set of all such conjunctions, which allow to deduce the hypothesis from the given knowledge base. It is also possible to ask for numerical arguments like **degree of support, degree of quasi-support, degree of plausibility,** or **degree of doubt** (Haenni, 1996); the respective keywords in ABEL are `dsp`, `dqs`, `dpl`, `ddb`. A numerical argument is obtained by computing the probability for the corresponding symbolic argument. This computation is based on a priori probabilities specified in the definition of the assumptions.

In step (1), **variables** and **assumptions** can be defined. Variables represent questions. For each variable, the set of possible values (answers) has to be specified, i.e. the type of the variable must be declared. An individual var-command is necessary for each type of variables. The type specifier can either be a pre-defined type, a user-defined type, or a new type specification. Pre-defined types are `integer`, `real` and `binary`. User-defined types can be built by using the keyword `type` and by either restricting a pre-defined type or specifying a so-called set-type. Assumptions are defined in a similar way. The difference between assumptions and variables is that assumptions represent uncertain events, unknown circumstances, or risks, rather than precise open questions. Assumptions are used to build arguments for hypotheses. Additionally, it is often possible to impose probabilities on the values of an assumption. The probabilities of different assumptions are assumed to be independent. In ABEL, probabilities are optional. If no probabilities are declared, then assumptions are only used to generate symbolic arguments (explanations). Examples:

```
(type weather (sun clouds rain))
(ass w1 w2 w3 weather)
(var d1 d2 d3a d3b integer[0..])
(ass ok binary 0.9)
```

Based on numerical variables (type `integer` or `real`), it is possible to build compound (algebraic) **expressions.** The syntax of ABEL expressions corresponds to LISP expressions. It uses prefix notation: the first element within a pair of opening and closing parentheses is the name of one of the pre-defined algebraic operators. The remaining elements are the operands. ABEL provides different operands. The semantic of these operators corresponds to Common LISP (Steele, 1990). Examples:

```
(+ d1 d2 d3)
(max d3a d3b)
```

Note that variables (e.g. d1, d2), assumptions (e.g. w1, w2), numbers (e.g. 17, 1/3, -23.5), symbols (e.g. sun), and sets (e.g. (sun cloud rain)) are also considered as (atomic) expressions.

ABEL expressions can be used to build so-called **constraints**. They restrict the possible values of the variables or assumptions involved to a subset of their Cartesian product. Constraints always compare two expressions. ABEL provides several operators to build constraints, e.g. =, /=, in, <, <=, >, >=. Examples:

```
(= w1 sun)
(= d3 (max d3a d3b))
(in w3 (sun clouds rain))
```

Note that variables or assumptions of type binary (e.g. ok) are also considered as (atomic) constraints.

Constraints can be used to build logical expressions called **statements**. A constraint itself is considered as (atomic) statement. ABEL supports several different types of compound statements, e.g. ->, <->, and, or, xor, not. Examples:

```
(and (>= d1 10) (<= d1 14))
(-> ok (= out (+ in1 in2)))
ok
```

The concept of **modules** has already been used in the existing ABEL prototype (Lehmann, 1994; Haenni, 1996). The idea of modules is that similar parts of the given information are modeled only once. The definition of a module can be compared with the definition of a LISP function. Every module has a name and it consists of a set of parameters and a body. The parameters of an ABEL module are variables or assumptions. The body of a module is a sequence of ABEL instructions. Therefore, it is possible to define local types, local variables, local assumptions, and local modules within the body of a module. An example for a module (as used in Section 4) is:

```
(module ADDER ((var in1 in2 out integer))
       (ass ok binary 0.97)
       (-> ok (= out (+ in1 in2)))))
```

Note that the parameters of a module have to be declared either as variables or assumptions. Additionally, types have to be specified for the parameters. The parameter list is therefore a sequence of variable and assumption definitions.

A module can be used to generate similar parts of the given information. An **instance** of a module is obtained by "calling" the module with actual parameters and a unique identifier (e.g. :A1), so an instance of the adder defined above can be generated as follows:

```
(ADDER :A1 x y f)
```

Note that the types of the actual parameters are implicitly defined by the parameter specification of the module definition. It is therefore not necessary to specify the types of the actual parameters outside the module.

3 Modeling Hints

The **mathematical theory of hints** (Kohlas & Monney, 1995) provides a formal method for building models of imprecise, uncertain, and contradictory information. The information contained in a hint always belongs to one or several **questions**. Let Θ denote the set of all possible answers for a certain question. Θ is assumed to be complete in the sense that the true (but unknown) answer is exactly one of its elements. Θ is called **frame of discernment**.

To model the uncertainty contained in a hint, another (finite) set Ω of possible **interpretations** (of the information) has to be considered. Again, exactly one of the elements of Ω is the (unknown) correct interpretation of the given hint; all other interpretations are wrong.

Each interpretation $\omega \in \Omega$ permits to restrict the possible answers to a subset $\Gamma(\omega) \subseteq \Theta$. This means that if ω turns out to be the correct interpretation of the hint, then the true answer must be within $\Gamma(\omega)$. Subsets $\Gamma(\omega)$ are called **focal sets** of the hint.

A triple $\mathcal{H} = (\Omega, \Theta, \Gamma)$ is called a **hint** relative to a frame of discernment Θ (Kohlas & Monney, 1995). A subset $H \subseteq \Theta$ represents the hypothesis that the true answer belongs to H. In order to judge a hypothesis H in the light of a hint $\mathcal{H}$, it is necessary to find arguments that allow to prove or to disprove the hypothesis.

Hints can easily be described using ABEL. The following example illustrates the use of hints and the way how to describe them in ABEL.

Suppose the construction of a new house consists of three main tasks: (1) the construction of the foundation walls, (2) the roofing, (3a) the interior decoration, and (3b) the paintings. The time needed for task (i) is denoted by d_i, $i = 1, 2, 3a, 3b$. Assume that task (2) starts as soon as task (1) is finished, whereas task (3a) and task (3b) start in parallel as soon as task (2) is finished. This scenario is illustrated in Figure 1.

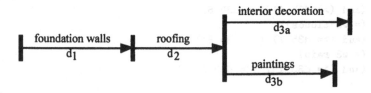

Fig. 1. A house construction time schedule.

The total time d to construct the house is the sum $d = d_1 + d_2 + d_3$ with $d_3 = max(d_{3a}, d_{3b})$. The interesting question is the duration d of the project. This is a so-called **scheduling problem**. In this type of problems, the duration of the activities are traditionally supposed to be known. In reality, they can only be estimated in dependence of particular future conditions and circumstances.

In the case of house constructing, the duration of a project depends mainly on future meteorological conditions. For example, let $\Omega_i = \{sun_i, clouds_i, rain_i\}$, $i = 1, 2, 3$, describe the possible weather conditions during phase (i). If durations are expressed in days, then Θ is the set of non-negative integers. Suppose that the durations are estimated as follows:

for d_1: $\Gamma_1(sun_1) = \{10, \ldots, 14\}$, $\Gamma_1(clouds_1) = \{12, \ldots, 16\}$,
　　　and $\Gamma_1(rain_1) = \{20, \ldots, 25\}$;
for d_2: $\Gamma_2(sun_2) = \Gamma_2(clouds_2) = \{8, 9, 10\}$, and $\Gamma_2(rain_2) = \{14, 15, 16\}$;
for d_{3a}: $\Gamma_{3a}(sun_3) = \Gamma_{3a}(clouds_3) = \Gamma_{3a}(rain_3) = \{10, \ldots, 14\}$;
for d_{3b}: $\Gamma_{3b}(sun_3) = \{4, 5, 6\}$, $\Gamma_{3b}(clouds_3) = \{7, 8\}$,
　　　and $\Gamma_{3b}(rain_3) = \{13, \ldots, 17\}$.

Obviously, the information contained in this example forms different hints, which can easily be transformed into a corresponding ABEL model:

```
(tell
  (type weather (sun clouds rain))
  (ass w1 w2 w3 weather)
  (var d1 d2 d3a d3b integer[0..])

  (-> (= w1 sun)
      (and (>= d1 10) (<= d1 14)))
  (-> (= w1 clouds)
      (and (>= d1 12) (<= d1 16)))
  (-> (= w1 rain)
      (and (>= d1 20) (<= d1 25)))
  (-> (in w2 (sun clouds))
      (and (>= d2 8) (<= d2 10)))
  (-> (= w2 rain)
      (and (>= d2 14) (<= d2 16)))
  (-> (in w3 (sun clouds rain))
      (and (>= d3a 10) (<= d3a 14)))
  (-> (= w3 sun)
      (and (>= d3b 4) (<= d3b 6)))
  (-> (= w3 clouds)
      (and (>= d3b 7) (<= d3b 8)))
  (-> (= w3 rain)
      (and (>= d3b 13) (<= d3b 17)))))
```

To complete the model, the relations between d, d_1, d_2, d_3, d_{3a}, and d_{3b} have to be added:

```
(tell
  (var d3 d integer[0..])

  (= d3 (max d3a d3b))
  (= d (+ d1 d2 d3)))
```

The model can now be used to judge hypotheses about the duration of the house construction. Examples of possible hypotheses are given in the following queries:

```
(ask
  (sp (> d 42))
  (sp (< d3b d3a))
  (pl (<= d 30)))
```

The results for these queries can be derived from the following table. It shows the minimal and maximal project durations for different weather conditions during the project phases.

Period 1	Period 2	Period 3	d_{min}	d_{max}
sun	sun/clouds	sun/clouds	28	38
clouds	sun/clouds	sun/clouds	30	40
rain	sun/clouds	sun/clouds	38	49
sun	sun/clouds	rain	31	41
clouds	sun/clouds	rain	33	43
rain	sun/clouds	rain	41	52
sun	rain	sun/clouds	34	44
clouds	rain	sun/clouds	36	46
rain	rain	sun/clouds	44	55
sun	rain	rain	37	47
clouds	rain	rain	39	49
rain	rain	rain	47	58

The support for (> d 42) will therefore be (and (= w1 rain) (= w2 rain)), the support for (< d3b d3a) will be (or (= w3 sun) (= w3 clouds)), and the plausibility for (<= d 30) will be (and (not (= w1 rain)) (not (= w2 rain)) (not (= w3 rain))).

4 Diagnostics in Technical Systems

This section discusses an instance of the problems of diagnostics in technical systems. The following example has been introduced in (Davis, 1984). Subsequently, it was used in several papers on model-based diagnostics (Reiter, 1987; de Kleer & Williams, 1987; Kohlas *et al.*, 1996). The example consist of a simple, well-known, mathematical network. In this type of problems, the aim is to find

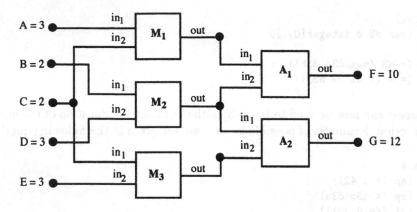

Fig. 2. An arithmetical network consisting of adders A_i and multipliers M_j.

possible explanations for an incorrect system behavior or for failure states of some components.

The network consists of three multipliers M_1, M_2, M_3, and two adders A_1, A_2. These components are connected as shown in Figure 2.

If only one failure mode is assumed, then a binary variable *ok* can be used to represent the two modes of the components. Figure 3 shows, for example, an adder component with inputs in_1, in_2 and one output *out*.

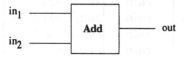

Fig. 3. A simple adder.

The behavior of the component can be expressed by a material implication:

$$ok \rightarrow (out = in_1 + in_2).$$

In ABEL, a module is built for each type of component. The reliability of a component is modeled by an assumption ok. Assume the probability 0.95 for adders and 0.97 for multipliers.

```
(tell
    (module ADDER ((var in1 in2 out integer))
        (ass ok binary 0.97)
        (-> ok (= out (+ in1 in2)))))
```

```
(module MULTIPLIER ((var in1 in2 out integer))
    (ass ok binary 0.95)
    (-> ok (= out (* in1 in2)))))
```

Next, the modules are applied following the topology of Figure 2.

```
(tell
    (MULTIPLIER :M1 a c x ok-m1)
    (MULTIPLIER :M2 b d y ok-m2)
    (MULTIPLIER :M3 c e z ok-m3)
    (ADDER :A1 x y f ok-a1)
    (ADDER :A2 y z g ok-a2))
```

The global variable ok-network reflects the state of the whole system. Observations are added as specified in Figure 2.

```
(observe
    (= a 3) (= b 2) (= c 2) (= d 3) (= e 3) (= f 10) (= g 12))
```

The observed values imply that some components must be faulty and the question is which ones. The following query can be helpful to obtain possible explanations:

```
(ask (sp tautology))
```

The result consists of four different configurations of faulty components (Kohlas & Monney, 1995):

```
(or (not M1.ok)
    (not A1.ok)
    (and (not M2.ok) (not M3.ok))
    (and (not M2.ok) (not A2.ok)))
```

The expression M1.ok denotes the assumption ok within the adder :M1.

5   Conclusion and Future Work

The actual work being done is an implementation of a general ABEL solver. The architecture of the general solver will be as follows: In a first step, the problem to be solved is decomposed and distributed over a valuation network (Shenoy & Shafer, 1990; Shenoy, 1995). The different parts of the problem are analyzed and classified in order to choose an appropriate solver. Secondly, the valuations are propagated locally through the network. The resulting marginals can be used to find arguments for queries. Future work will focus on this part of the general solver. Working with the prototype Evidenzia (Lehmann, 1994; Haenni, 1996) accumulated a lot of experience which now can be used in the implementation of a solver for ABEL. The actual version of the solver is restricted to variables which are binary or have a set-type domain.

References

Almond, R.G. 1990. *Fusion and Propagation of Graphical Belief Models: an Implementation and an Example.* Ph.D. thesis, Department of Statistics, Harvard University.

Almond, R.G. 1995. *Graphical Belief Modeling.* Chapman and Hall.

Andersen, S.K., Olesen, K.G., Jensen, F.V., & Jensen, F. 1990. HUGIN – a Shell for Building Bayesian Belief Universes for Expert Systems. *Pages 332–338 of:* Shafer, G., & Pearl, J. (eds), *Readings in Uncertain Reasoning.* Morgan Kaufmann.

Anrig, B., Haenni, R., & Lehmann, N. 1997. *ABEL - A New Language for Assumption-Based Evidential Reasoning under Uncertainty.* Tech. Rep. 97–01. University of Fribourg, Institute of Informatics.

Davis, R. 1984. Diagnostic Reasoning based on Structure and Behaviour. *Artificial Intelligence,* **24**, 347–410.

de Kleer, J. 1986. An Assumption-based TMS. *Artificial Intelligence,* **28**, 127–162.

de Kleer, J., & Williams, B.C. 1987. Diagnosing Multiple Faults. *Artificial Intelligence,* **32**, 97–130.

Haenni, R. 1996. *Propositional Argumentation Systems and Symbolic Evidence Theory.* Ph.D. thesis, Institute of Informatics, University of Fribourg.

Hsia, Y.T., & Shenoy, P.P. 1989. An Evidential Language for Expert Systems. *Pages 9–14 of:* Ras, Z.W. (ed), *Methodologies for Intelligent Systems.* North-Holland.

Kohlas, J., & Monney, P.A. 1993. Probabilistic Assumption-Based Reasoning. *In:* Heckerman, & Mamdani (eds), *Proc. 9th Conf. on Uncertainty in Artificial Intelligence.* Kaufmann, Morgan Publ.

Kohlas, J., & Monney, P.A. 1995. *A Mathematical Theory of Hints. An Approach to the Dempster-Shafer Theory of Evidence.* Lecture Notes in Economics and Mathematical Systems, vol. 425. Springer.

Kohlas, J., Monney, P.A., Anrig, B., & Haenni, R. 1996. *Model-Based Diagnostics and Probabilistic Assumption-Based Reasoning.* Tech. Rep. 96–09. University of Fribourg, Institute of Informatics.

Lauritzen, S.L., & Spiegelhalter, D.J. 1988. Local Computations with Probabilities on Graphical Structures and their Application to Expert Systems. *Journal of Royal Statistical Society,* **50**(2), 157–224.

Lehmann, N. 1994. *Entwurf und Implementation einer annahmenbasierten Sprache.* Diplomarbeit. Institute of Informatics, University of Fribourg.

Reiter, R. 1987. A Theory of Diagnosis From First Principles. *Artificial Intelligence*, **32**, 57–95.

Saffiotti, A., & Umkehrer, E. 1991. *PULCINELLA: A General Tool for Propagating Uncertainty in Valuation Networks*. Tech. Rep. IRIDIA, Université de Bruxelles.

Shafer, G. 1976. *The Mathematical Theory of Evidence*. Princeton University Press.

Shenoy, P.P. 1995. *Binary Join Trees*. Tech. Rep. 270. School of Business, University of Kansas.

Shenoy, P.P., & Shafer, G. 1990. Axioms for Probability and Belief Functions Propagation. *In:* Shachter, R.D., & al. (eds), *Uncertainty in Artificial Intelligence 4*. North Holland.

Srinivas, S., & Breese, J. 1990. IDEAL: A Software Package for Analysis of Influence Diagrams. *In: Proceedings of the Sixth Uncertainty Conference in AI, Cambridge, MA*.

Steele, G. L. 1990. *Common Lisp – the Language*. 2d edn. Digital Press.

Xu, H., & Kennes, R. 1994. Steps Toward Efficient Implementation of Dempster-Shafer Theory. *Pages 153–174 of:* Yager, R.R., Fedrizzi, M., & Kacprzyk, J. (eds), *Advances in the Dempster-Shafer Theory of Evidence*. John Wiley and Sons.

Propositional Quantification for Conditional Logic

Philippe Besnard, Jean-Marc Guinnebault, Emmanuel Mayer

IRISA
Campus de Beaulieu
35042 Rennes Cedex
France
{besnard,jguinneb,emayer}@irisa.fr

Abstract. Conditional sentences are a key issue of logical reasoning whose full scope eludes material implication. Various authors have proposed logics based on a conditional connective gearing towards specific sentences such as counterfactuals. Unfortunately, most of such conditional logics are propositional and therefore lack expressiveness. We propose in this paper to extend such logics by means of propositional quantification.

Keywords: Conditional logic.

1 Motivation

First and foremost, conditional logic (we use the generic term for any system that exhibits a connective for subjunctive implication) addresses the issue of counterfactuals sentences in natural language, that are statements of the form "if A then B" where the antecedent A is known (or expected) to be false. As a matter of example, let us assert that "If I had some oars, I could row across the river" and "I actually have no oar" [Gin86]. The truth value of the former statement need not respect the semantics of material implication, by which every counterfactual sentence is true because its antecedent is false. However, people agree that many counterfactual sentences are false. For instance, let us assume that it is actually raining and let us consider "If it wouldn't rain, flowers would speak". Although the antecedent is false, we will definitely do not conclude that this counterfactual statement is true.

Indeed, the question is: When is a counterfactual true? Conditional logic emerged from such studies in the analysis of counterfactuals. Early authors [Chi46] [Goo55] suggested that the semantics of counterfactual sentences can be captured by metalinguistic interpretation: A counterfactual "if A then B" (formally, $A \leadsto B$) can be evaluated by first conjoining A with appropriate conditions H and then checking whether A and H together make the consequent B to be true.

In continuation to the metalinguistic interpretation

"add the antecedent to your stock of knowledge, make whatever adjustments are required to maintain consistency and then consider whether or not the consequent is true" [Sta68]

the possible worlds tradition was thrown in:

"Consider a possible world in which A is true, and which otherwise differs minimally from the actual world. $A \rightsquigarrow B$ is true (false) in case B is true (false) in that possible world". [Sta68]

Generally speaking, conditional logic was investigated focusing on formulae $A \rightsquigarrow B$. However, conditional logic was mainly developed in propositional form (with a few exceptions such as [Del88], [FHK96]). Yet, the exceptions all deal with first-order quantification and even that is restrictive: We now offer a few sentences as evidence that subjunctive reasoning may involve second-order quantification.

When in doubt whether "there is a case where the President would be dismissed retroactively", it seems that we are faced with a statement that formally writes as "$\exists X \ (X \rightsquigarrow P)$" where P of course stands for "the President is dismissed retroactively".

When saying that "if I were happy whatever the situation, I would get bored" again it seems that propositional quantification is in order here that leads to the formula "$(\forall X(X \supset H)) \rightsquigarrow B$".

We claim that the above examples are no accident but can be found over and over as people often need to mention potential situations of which only some consequences are identified for instance.

There is even more to propositional quantifiers in conditional logic. The fact is that conditional logic has proven to be useful not only when it comes to subjunctive reasoning, as it was originally designed for, but also deontic reasoning and other theories. When used in to capture and encode other formal theories, the scope of propositional conditional logic is then restricted to universal theories (which can be expressed by means of formulas of the form $\forall^* A$, hence involving no existential quantifier). However, the important class of inductive theories (which can be expressed by means of formulas of the form $\forall^* \exists^* A$) then fails to come under the umbrella of conditional logic. So, introducing propositional quantifiers into conditional logic is the way to greater expressiveness, so that inductive theories can be captured while at the same time allowing for analysis of a larger class of counterfactuals (we have no idea about possible benefits in the field of deontic reasoning because we did not look at the implications of propositional quantification for conditional obligation).

By the way, propositional quantification in conditional logic is not new [Lew73]. We are simply trying to make a more general point about it than Lewis did.

2 Conditional logic

2.1 Language

The syntax of our language is based on infinite alphabets of propositional symbols together with a finite alphabet of logical symbols. We follow standard usage [Tak75] except that we call free variables "parameters", reserving "variables" for bound variables.

Symbols

The symbols of the language are:

- propositional constants P_1, P_2, ...;
- propositional parameters U_1, U_2, ...;
- propositional variables X_1, X_2, ...;
- punctuation marks, i.e. parentheses "(" and ")";
- connectives: falsum "$\perp$", verum "$\top$", negation "$\neg$", conjunction "$\wedge$", disjunction "$\vee$", material implication "$\supset$", conditional implication "$\rightsquigarrow$", necessity "$\square$" and possibility "$\diamond$";
- quantifiers "$\forall$" and "$\exists$".

Well-formed formulae

Let $A[X/Y]$ be the formula where X is substituted for every occurrence of Y in the formula A.

- P_i and U_i are atomic formulae;
- if A is a formula then $\neg A$, $\square A$ and $\diamond A$ are formulae;
- if A and B are formulae then $(A \wedge B)$, $(A \vee B)$, $(A \supset B)$ and $(A \rightsquigarrow B)$ are formulae;
- if A is a formula and the variable X_i does not occur in A then $(\forall X_i\, A[X_i/U_j])$ and $(\exists X_i\, A[X_i/U_j])$ are formulae.

As is clear from the definition, we write A, B and C to denote formulae. Other conventions we use are as follows: X is a propositional variable and U is a propositional parameter (i.e., at times we write X_1 as X, X_2 as Y, ... and U_1 as U, U_2 as V, ...). In order to improve readability, we sometimes write A_X for $A[X/U]$. Regarding abbreviation, we use $\leftrightarrow$ for material equivalence so that $A \leftrightarrow B$ is simply a shorthand for $(A \supset B) \wedge (B \supset A)$. Last, the symbol for syntactical identity is $\equiv$ (including definitional identity).

Now, we first say a few words about what symbols are taken as primitive since $\top$ and $\perp$ are not the only ones introduced for notational convenience.

Russel, who seems to be the first author to introduce a system with propositional quantifiers [Chu56], was apparently reluctant to use $\forall X\, X$ in order to define negation. It looks like he wanted to express that not everything is true, in symbols, $\neg(\forall X\, X)$, so that he rejected the definition of negation as

$\neg A \equiv A \supset \forall X\, X$ (the resulting formula expressing that not everything is true then becoming $(\forall X\, X) \supset (\forall X\, X)$ which is rather inadequate).

It does not matter whether conjunction, disjunction, and the existential quantifier are introduced by definition or not.

In contrast, necessity and possibility are defined by

$$\Box A \equiv (\neg A) \leadsto A$$

and

$$\Diamond A \equiv \neg \Box \neg A$$

as is customary [Lew73].

2.2 Proof system

The axioms and rules are those of classical logic supplemented with specific rules and axioms for conditionals. Thus, all theorems from classical logic also hold in conditional logic.

$A \supset (B \supset A)$	(axiom A1)
$(A \supset (B \supset C)) \supset ((A \supset B) \supset (A \supset C))$	(axiom A2)
$(\neg B \supset \neg A) \supset (A \supset B)$	(axiom A3)
$(\forall X\, (A \supset B_X)) \supset (A \supset (\forall X\, B_X))$	(axiom A4)
$(\forall X\, A_X) \supset A_T$	(axiom A5)
$A \leadsto A$	(axiom ID)
$(A \leadsto B) \supset (A \supset B)$	(axiom MP)
$((A \leadsto C) \wedge (B \leadsto C)) \supset ((A \vee B) \leadsto C)$	(axiom CA)
$((A \leadsto B) \wedge (A \leadsto C)) \supset (A \leadsto (B \wedge C))$	(axiom CC)
$((A \leadsto B) \wedge (B \leadsto A)) \supset ((A \leadsto C) \leftrightarrow (B \leadsto C))$	(axiom CSO)

$$\frac{A \quad A \supset B}{B} \qquad \text{(rule R1)}$$

$$\frac{A_U}{\forall X\, A_X} \qquad \text{(rule R2)}$$

$$\frac{B \supset C}{(A \leadsto B) \supset (A \leadsto C)} \qquad \text{(rule RCM)}$$

The axioms (A1), (A2), (A3), (ID), (MP), (CA), (CC) and (CSO) together with the rules (R1) and (RCM) define the conditional logic PO [KS91] which in fact is the fairly well-known conditional logic WC [Nut80]. Two derived rules are:

$$\frac{A \leftrightarrow B}{(A \rightsquigarrow C) \leftrightarrow (B \rightsquigarrow C)} \qquad \text{(rule RCEA)}$$

$$\frac{(B_1 \wedge ... \wedge B_n) \supset B}{((A \rightsquigarrow B_1) \wedge ... \wedge (A \rightsquigarrow B_n)) \supset (A \rightsquigarrow B)} \quad (n \geq 0) \quad \text{(rule RCK)}$$

As to second-order, the comprehension axiom we consider is:

$$\exists X \ (A \leftrightarrow X) \qquad \text{(axiom COMP)}$$

Finally, the replacement of equivalents applies in the form

$$\vdash (A \leftrightarrow B) \quad \Rightarrow \quad \vdash (C[A/U] \leftrightarrow C[B/U])$$

where $\vdash$ indicates deducibility from no premises.

3 Deductions

3.1 Definition of necessity from propositional quantification

The following provable formulae relate necessity to the universal quantifier.

Formula 1. $\vdash (\forall X \ (X \rightsquigarrow A)) \supset \Box A$

Formula 2. $\vdash \Box A \supset (\forall X \ (X \rightsquigarrow A))$

Therefore, we can define both modal connectives from the conditional implication and the quantifiers:

$$\Box A \equiv \forall X \ (X \rightsquigarrow A)$$

and

$$\Diamond A \equiv \neg \Box \neg A$$

which provides a nice alternative characterization of necessity (and possibility) besides Lewis' $\Box A \equiv (\neg A) \rightsquigarrow A$.

Bearing in mind that propositional quantification range over all propositions, $\forall X \ (X \rightsquigarrow A)$ makes a lot of sense as a definition of necessity in view of the possible worlds semantics for modal logic [Che80] where necessity stands for truth in all worlds (hence, truth from all propositions).

Here is a proof of $(\neg A) \rightsquigarrow A \vdash \forall X (X \rightsquigarrow A)$:

(1) $\neg A \rightsquigarrow \neg A$	(ID)
(2) $\neg A \rightsquigarrow A$	(Assumption)
(3) $\neg A \rightsquigarrow A \wedge \neg A$	((1)+(2),CC)
(4) $A \wedge \neg A \supset U \wedge \neg A$	(Tautology)
(5) $\neg A \rightsquigarrow U \wedge \neg A$	((3) + (4), RCM)
(6) $U \wedge \neg A \rightsquigarrow U \wedge \neg A$	(ID)
(7) $U \wedge \neg A \supset \neg A$	(Tautology)
(8) $U \wedge \neg A \rightsquigarrow \neg A$	((6) + (7), RCM)
(9) $(\neg A \rightsquigarrow U \wedge \neg A) \wedge (U \wedge \neg A \rightsquigarrow \neg A)$ $\supset ((\neg A \rightsquigarrow A) \leftrightarrow (U \wedge \neg A \rightsquigarrow A))$	(CSO)
(10) $U \wedge \neg A \rightsquigarrow A$	((5)+(8)+(2)+(9))
(11) $U \wedge A \rightsquigarrow U \wedge A$	(ID)
(12) $U \wedge A \supset A$	(Tautology)
(13) $U \wedge A \rightsquigarrow A$	((11)+(12), RCM)
(14) $(U \wedge A \rightsquigarrow A) \wedge (U \wedge \neg A \rightsquigarrow A)$ $\supset (((U \wedge A) \vee (U \wedge \neg A)) \rightsquigarrow A)$	(CA)
(15) $((U \wedge A) \vee (U \wedge \neg A)) \rightsquigarrow A$	((10)+(13)+(14))
(16) $U \leftrightarrow ((U \wedge A) \vee (U \wedge \neg A))$	(Tautology)
(17) $(((U \wedge A) \vee (U \wedge \neg A)) \rightsquigarrow A) \leftrightarrow (U \rightsquigarrow A)$	((16)+RCEA)
(18) $(U \rightsquigarrow A)$	((15) + (17))
(19) $\forall X (X \rightsquigarrow A)$	(18)

Careful inspection of the proof shows that the status of (2) can be turned from an assumption to the antecedent of a material implication so as to get $\vdash ((\neg A) \rightsquigarrow A) \supset \forall X (X \rightsquigarrow A)$ (warning: do *not* use the deduction theorem because it does not hold).

The proof of $(\forall X (X \rightsquigarrow A)) \supset ((\neg A) \rightsquigarrow A)$, being trivial, is omitted.

3.2 Distributivity of conditional implication over quantifiers

The provable formulae below allow us to reduce or increase the range of quantifiers w.r.t. conditional implication.

Formula 3. $\vdash (A \rightsquigarrow (\forall X \ B_X)) \supset (\forall X \ (A \rightsquigarrow B_X))$

Formula 4. $\vdash (\exists X \ (A \rightsquigarrow B_X)) \supset (A \rightsquigarrow (\exists X \ B_X))$

Both proofs are omitted, being very simple and requiring only the rule RCM as a non-classical principle.

Converse of formula 3 as well as converse of formula 4 can serve as axioms because they are of particular interest:

Formula 5 (Conditional Barcan Axiom). $(\forall X\ (A \rightsquigarrow B_X)) \supset (A \rightsquigarrow (\forall X\ B_X))$

Formula 6 (Conditional Buridan Axiom). $(A \rightsquigarrow (\exists X\ B_X)) \supset (\exists X\ (A \rightsquigarrow B_X))$

Detailed discussion of these axioms is postponed to a later section.

Here are other formulae exhibiting phenomena of distributivity of conditional implication over quantifiers.

Formula 7. $(\exists X\ (A_X \rightsquigarrow B)) \supset ((\forall X\ A_X) \rightsquigarrow B)$

Formula 8. $((\exists X\ A_X) \rightsquigarrow B) \supset (\forall X\ (A_X \rightsquigarrow B))$

In view of the rule RCEA, both formulae are equivalent. Again in view of the rule RCEA, either is equivalent to

$$((A \vee B) \rightsquigarrow C) \supset (A \rightsquigarrow C)$$

as well as

$$(A \rightsquigarrow C) \supset ((A \wedge B) \rightsquigarrow C)$$

that are themselves equivalent in propositional conditional logic. So, we have yet another example of an alternative characterization using propositional quantifiers. The four formulae below are all equivalent:

$((A \vee B) \rightsquigarrow C) \supset (A \rightsquigarrow C)$
$(A \rightsquigarrow C) \supset ((A \wedge B) \rightsquigarrow C)$
$(\exists X\ (A_X \rightsquigarrow B)) \supset ((\forall X\ A_X) \rightsquigarrow B)$
$((\exists X\ A_X) \rightsquigarrow B) \supset (\forall X\ (A_X \rightsquigarrow B))$

None is provable.

A similar result holds when considering the following related formulae.

Formula 9. $(\forall X\ (A_X \rightsquigarrow B)) \supset ((\exists X\ A_X) \rightsquigarrow B))$

Formula 10. $((\forall X\ A_X) \rightsquigarrow B) \supset (\exists X\ (A_X \rightsquigarrow B))$

In view of the rule RCEA, both formulae are equivalent and either is equivalent to

$$(A \rightsquigarrow C) \supset ((A \vee B) \rightsquigarrow C)$$

as well as

$$((A \wedge B) \rightsquigarrow C) \supset (A \rightsquigarrow C)$$

that are themselves equivalent in propositional conditional logic. To sum up, the four formulae below are all equivalent:

$(A \rightsquigarrow C) \supset ((A \vee B) \rightsquigarrow C)$
$((A \wedge B) \rightsquigarrow C) \supset (A \rightsquigarrow C)$
$(\forall X\ (A_X \rightsquigarrow B)) \supset ((\exists X\ A_X) \rightsquigarrow B))$
$((\forall X\ A_X) \rightsquigarrow B) \supset (\exists X\ (A_X \rightsquigarrow B))$

None is provable.

3.3 Barcan formula and Buridan formula

As promised, we return to Conditional Barcan (formula 5) and Conditional Buridan (formula 6).

Formula 5 [Conditional Barcan Axiom] $(\forall X\ (A \rightsquigarrow B_X)) \supset (A \rightsquigarrow (\forall X\ B_X))$

Formula 5 can be viewed as a generalization of the Barcan formula, a well-known axiom in modal logic [Bar46]:

Formula 11 (Barcan). $(\forall X\ (\Box A_X)) \supset (\Box(\forall X\ A_X))$

Importantly, the Barcan formula follows from formula 5 using only standard quantification principles.

As an axiom extending the Barcan formula in a conditional logic, formula 5 expresses that the domain of quantification is fixed (in terms of possible worlds semantics, it is the same in each world). Careful: This is so in view of the fact that the converse formula also holds. We are now going to see that such is the case.

Formula 12 (converse of Barcan). $\vdash (\Box(\forall X\ A_X)) \supset (\forall X\ (\Box A_X))$

Formula 12 can be derived from formula 3, again using only standard quantification principles.

Before dealing with the Buridan formula, we consider its converse which is easy:

Formula 13 (converse of Buridan). $\vdash (\exists X\ (\Box A_X)) \supset (\Box(\exists X\ A_X))$

Formula 13 can be derived from formula 4, once more using only standard quantification principles.

Finally, the time has come to investigate the more puzzling case.

Formula 14 (Buridan). $(\Box(\exists X\ A_X)) \supset (\exists X\ (\Box A_X))$

Similarly to the above cases, formula 14 is expected to be an immediate consequence of:

Formula 6 [Conditional Buridan Axiom] $(A \rightsquigarrow (\exists X\ B_X)) \supset (\exists X\ (A \rightsquigarrow B_X))$

Curiously, such is not the case. For the Buridan formula to follow from formula 6, an additional assumption, i.e. $\exists W\ \forall V\ \forall U((W \rightsquigarrow V) \supset (U \rightsquigarrow V))$ enters the picture. It is not too strong an assumption, though. We see again the benefits of having the existential quantifier available: Otherwise, the assumption would be stronger, either as $(\top \rightsquigarrow B) \supset (A \rightsquigarrow B)$ or the like. Anyway, here is the proof:

(1) $\Box \exists X A_X \equiv \forall Y (Y \rightsquigarrow \exists X A_X)$ (Definition)

(2) $\exists W (\forall V (W \rightsquigarrow V) \supset \forall U (U \rightsquigarrow V)$
     $\wedge (\forall Y (Y \rightsquigarrow \exists X A_X)) \supset (W \rightsquigarrow \exists X A_X))$ (Tautology)

(3) $\exists W (\forall V (W \rightsquigarrow V) \supset \forall U (U \rightsquigarrow V)$
     $\wedge ((W \rightsquigarrow \exists X A_X) \supset \exists X (W \rightsquigarrow A_X))$ (formula 6)

(4) $\exists W (\forall V (W \rightsquigarrow V) \supset \forall U (U \rightsquigarrow V)$
     $\wedge \exists X (W \rightsquigarrow A_X) \supset \exists X \forall Y (Y \rightsquigarrow A_X))$ (Tautology)

(5) $\exists X \forall Y (Y \rightsquigarrow A_X) \equiv \exists X \Box A_X$ (Definition)

(6) $\Box \exists X A_X \supset \exists X \Box A_X$ $((1) + \cdots + (5))$

3.4 Quantification as an extension of conjunction and disjunction

In modal logic, the Barcan formula can be viewed as the natural extension of $\Box A \wedge \Box B \supset \Box (A \wedge B)$ (from which $\Box A_1 \wedge \ldots \wedge \Box A_n \supset \Box (A_1 \wedge \ldots \wedge A_n)$ follows) to the case of infinite conjunction, in accordance with the view that universal quantification stands for possibly infinite conjunctions involving all entities. In semantics based on (ultra)filters, this is reflected by the fact that the Barcan formula corresponds to closure with respect to arbitrary intersections. We now look at specific formulas in light of this idea of quantification extending finite conjunction and disjunction to possibly infinite ones.

First, the rule RCK

$$\frac{(B_1 \wedge \ldots \wedge B_n) \supset B}{((A \rightsquigarrow B_1) \wedge \ldots \wedge (A \rightsquigarrow B_n)) \supset (A \rightsquigarrow B)} \quad (n \geq 0)$$

can be extended to the following quantified form

$$\frac{(\forall X \; B_X) \supset C}{(\forall X \; (A \rightsquigarrow B_X)) \supset (A \rightsquigarrow C)}$$

for which a proof exists from formula 5 using the rule RCM.
The two quantified formulae connected with conjunction are:

$$((A \rightsquigarrow B) \wedge (A \rightsquigarrow C)) \supset (A \rightsquigarrow (B \wedge C)) \qquad \text{(axiom CC)}$$

and

$$(A \rightsquigarrow (B \wedge C)) \supset ((A \rightsquigarrow B) \wedge (A \rightsquigarrow C)) \qquad \text{(axiom CM)}$$

They can also be extended to quantified forms, actually formulae 3 and 5.

Moreover, formula 7

$$(\exists X \; (A_X \rightsquigarrow C)) \supset ((\forall X \; A_X) \rightsquigarrow C)$$

can be viewed as an extension of

$$((A \rightsquigarrow C) \vee (B \rightsquigarrow C)) \supset ((A \wedge B) \rightsquigarrow C)$$

Interestingly, they are equivalent. Indeed, the latter formula is equivalent (by classical principles) to the conjunction of the formulas

$$(A \rightsquigarrow C) \supset ((A \wedge B) \rightsquigarrow C)$$

$$(B \rightsquigarrow C) \supset ((A \wedge B) \rightsquigarrow C)$$

each of which is equivalent to formula 7, as is indicated in Section 3.2.

Similarly, formula 8

$$((\exists X \; A_X) \rightsquigarrow B) \supset (\forall X \; (A_X \rightsquigarrow B))$$

can be viewed as an extension of

$$((A \vee B) \rightsquigarrow C) \supset ((A \rightsquigarrow C) \wedge (B \rightsquigarrow C))$$

and they are equivalent (as above, see Section 3.2).

3.5 Finer distinctions with quantification

Here, we try to illustrate the idea that quantification yields sharper distinction.

Propositional quantifiers leave room for another notion of possibility, maybe more appealing than Lewis' $\Diamond A \equiv \neg(A \rightsquigarrow \neg A))$:

$$\Diamond A \equiv \exists X (X \rightsquigarrow A)$$

A theorem is then

$$(\Diamond \forall X \; A_X) \supset (\forall X \; \Diamond A_X)$$

in view of the rule RCM. The converse

$$(\forall X \; \Diamond A_X) \supset (\Diamond \forall X \; A_X)$$

follows from either formula 7

$$(\exists X \; (A_X \rightsquigarrow B)) \supset ((\forall X \; A_X) \rightsquigarrow B)$$

or formula 8

$$((\exists X \; A_X) \rightsquigarrow B) \supset (\forall X \; (A_X \rightsquigarrow B))$$

in the presence of formula 5 (Conditional Barcan Axiom)

$$(\forall X\ (A \rightsquigarrow B_X)) \supset (A \rightsquigarrow (\forall X\ B_X))$$

An important difference between $\Diamond$ and $\diamondsuit$ is that $\diamondsuit\bot$ need not be a contradiction, even in relatively rich logics, whereas $\Diamond\bot$ is a contradiction in virtually all conditional logics (it only takes the axiom CN: $A \rightsquigarrow \top$ to make it a contradiction). Notice that $\Diamond A \rightarrow \diamondsuit A$ holds in any sufficiently rich logic (the axiom CEM must be available) but $\diamondsuit A \rightarrow \Diamond A$ holds in no standard conditional logic because it requires the unacceptable $\neg(A \rightsquigarrow \bot)$ (the latter formula corresponds to no reasonable principle: It states that even logically inconsistent statements induce a consistent hypothetical state of affairs).

In very weak systems of conditional logic, subtle distinction can also be found when comparing the formula $(\exists X \neg(X \rightsquigarrow \bot))$ which is less restrictive than the axiom N, i.e. $(\neg(\top \rightsquigarrow \bot))$. As another example, we may write $(\exists X \forall Y(X \rightsquigarrow Y))$ instead of $(\bot \rightsquigarrow \bot)$. Or we may distinguish between $\forall X\ \exists Y(X \rightsquigarrow Y)$ and $A \rightsquigarrow A$. In all such cases, propositional schemata are not fine enough but existential quantification is.

4 Semantics

Various semantics [Nut80] for conditional logic have been devised, each of them characterizing different variants of conditional logic. The two most popular semantics, both of the Kripke type, are the selection function semantics and the sphere semantics. The former is based on the existence of a so-called selection function f, that elicit, for each world, which worlds are relevant for a given formula. Then, a conditional formula is true if and only if its consequent is true in the relevant worlds.

The sphere semantics, which is actually less general than the selection function semantics, is based on sphere systems. Roughly speaking, a sphere system is a sequence of nested spheres (where a sphere is a set of possible worlds). A fundamental difference with the selection function semantics is that relevant worlds in a sphere do not depend on what formula is under consideration. Spheres correspond to different levels of similarity to a given world. Intuitively, the smaller a sphere is, the closer to the given world are its elements. Accordingly, a conditional formula $A \rightsquigarrow B$ is true if either (a) there exist no sphere that contains a world satisfying the antecedent A, or, (b) in all the worlds of the smallest sphere that contains a world satisfying the antecedent, $A \supset B$ holds.

We have here chosen a different semantics [KS91] which can be regarded as intermediate in generality between both previously presented semantics. Basically, it extends the sphere semantics. Each world is associated with a set of possible worlds structured by an accessibility relation. Different structures result as different properties are imposed for the accessibility relation (a sphere system is a fixed structure, actually a total pre-order).

Let W' be a subset of W. We denote by $\mathrm{Min}(W', R)$ the set of minimal elements in W' with respect to R, in symbols: $\mathrm{Min}(W', R) = \{w \in W' \mid \forall w' \in W', wRw' \Rightarrow w'Rw\}$.

We define a model M as a triple $\langle W, F, v \rangle$, where W and v are respectively a non-empty set (of worlds) and a function that maps each propositional letter and each element of W to *true* or *false*. F is a function that maps a world w to a structure $\langle W_w, R_w \rangle$ where W_w is a subset of W, and R_w is a binary relation over W_w. The relation R_w must satisfy the so-called smoothness condition, that is defined next.

Let $\|A\|_M = \{w \in W : M, w \models A\}$ be the set of worlds where A is true and let $S = \|\phi\|_M \cap W_w$ be the set of worlds in W_w satisfying ϕ.

The smoothness condition is:

$$\forall w' \in S, \exists w'' \in S \text{ s.t. } w'' \text{ is minimal in } S \text{ w.r.t. } R_{w''}$$

Then, the semantics for a conditional formula $A \rightsquigarrow B$ is:

$$M, w \models (A \rightsquigarrow B) \text{ iff } \text{Min}(\|A\|_M \cap W_w, R_w) \subseteq \|B\|_M$$

We write $M \approx_U M'$ when M and M' are two identical models except *maybe* for the assignment of the symbol U, i.e., for at least one world w, we may have $\phi_M(w, U) \neq \phi_{M'}(w, U)$.

- $M, w \models A$ if $v(w, A) = true$ (for A atomic formula);
- $M, w \models (\neg A)$ if $M, w \not\models A$;
- $M, w \models (A \wedge B)$ if $M, w \models A$ and $M, w \models B$;
- $M, w \models (A \vee B)$ if $M, w \models A$ or $M, w \models B$;
- $M, w \models (A \supset B)$ if $M, w \models (\neg A)$ or $M, w \models B$;
- $M, w \models (A \rightsquigarrow B)$ if $\text{Min}(\|A\|_M \cap W_w, R_w) \subseteq \|B\|_M$;
- $M, w \models (\Box A)$ if for all sets E in 2^W, we have $\text{Min}(E \cap W_w, R_w) \subseteq \|A\|_M$;
- $M, w \models (\forall X_i\, A)$ if for all models $M' \approx_{U_j} M$ (U_j not present in A), we have $M', w \models A[U_j/X_i]$;

We note, as syntactical equivalences, $(\Diamond A) \equiv (\neg \Box \neg A)$ and $(\exists A) \equiv (\neg \forall X \neg A)$.

In this semantics, the formulae 1 to 5 and 11 to 13 are valid. On the contrary, the formulae 7, 8 and 14 are not valid. The formulae 7 and 8 characterize the following property: $\text{Min}(S, R_w)$ increases with its first argument. In the same way, the formulae 9 and 10 are not valid and characterize the following property: $\text{Min}(S, R_w)$ decreases with its first argument.

For a better understanding of the following text, here are the conventions we use. In all the counter-example proofs, we use a set of worlds W consisting of only two elements, w_1 and w_2. For a given model M, we may thus have four derived models, which we call M_0, M_1, M_2 and M_{12}. The value of $\|U\|_{M'}$ (M' derived model) then ranges over: $\emptyset$, $\{w_1\}$, $\{w_2\}$ and W.

Finally, we use the $\oplus$ symbol for exclusive or.

As a matter of example, let us find out a counter-example to the validity of formula 6.

The semantical condition is:

$$\text{If } Min(\|A\|_M \cap W_w, R_w) \subseteq \bigcup_{M' \approx_U M} \|B_U\|_{M'}$$

then $\exists M' \approx_U M$ s.t. $Min(\|A\|_M \cap W_w, R_w) \subseteq \|B_U\|_{M'}$

To build this counter-example, let us assume that A is $\top$ and B_U is $(U \wedge C_1) \oplus (U \wedge C_2)$ The various possible sets for $\|B_U\|_{M'}$ (for M' similar to M) are:

- $\|B_U\|_{M_0} = \emptyset$;
- $\|B_U\|_{M_1} = \{w_1\}$;
- $\|B_U\|_{M_2} = \{w_2\}$;
- $\|B_U\|_{M_{12}} = \emptyset$.

We use a function $F : w \mapsto (W_w, R_w)$ such that $Min(W \cap W_w, R_w) = W$, i.e. $W_w = W$ and $R_w = Id$

$$\text{Thus } Min(\|A\|_M \cap W_w, R_w) \subseteq \bigcup_{M' \approx_U M} \|B_U\|_{M'} \text{ holds}$$

$$\text{since } \bigcup_{M' \approx_U M} \|B_U\|_{M'} = W \text{ holds.}$$

Now, there exists no similar model M' satisfying

$$\exists M' \approx_U M \text{ s.t. } Min(\|A\|_M \cap W_w, R_w) \subseteq \|B_U\|_{M'}.$$

Interpretation of quantifiers

In classical logic, evaluating a formula with propositional quantifiers only amounts to considering two extreme cases, i.e. the one where the propositional variable stands for a tautology and the one where it stands for a contradiction. In other words, when considering a formula such as $\forall X \, A_X$ or $\exists X \, A_X$, we just need to consider the cases where X is $\bot$ ($\|\bot\|_M = \emptyset$) and X is $\top$ ($\|\top\|_M = W$). Both $(A_\bot \wedge A_\top) \leftrightarrow (\forall X \, A_X)$ and $(\exists X \, A_X) \leftrightarrow (A_\bot \vee A_\top)$ are valid in classical logic. We see below that such is not the case in conditional logic.

Universal quantification The formula $(A_\bot \wedge A_\top) \supset (\forall X \, A_X)$ is not valid in conditional logic.

Existential quantification The formula $((\exists X \; A_X) \supset (A_\perp \lor A_\top)$ is not valid in conditional logic.

The semantical condition is:

$$\bigcup_{M' \approx_U M} \|A_U\|_{M'} \subseteq \|A(\perp)\|_M \cup \|A(\top)\|_M.$$

Let us use the same counter-example as for formula 6, imposing now $A_U = (U \land C_1) \oplus (U \land C_2)$.

The different possible sets for $\|A_U\|_{M'}$ (for M' similar to M) are:

- $\|A_U\|_{M_0} = \emptyset$;
- $\|A_U\|_{M_1} = \{w_1\}$;
- $\|A_U\|_{M_2} = \{w_2\}$;
- $\|A_U\|_{M_{12}} = \emptyset$.

We can easily see that the semantical condition is false in this case.

5   Conclusion

Inasmuch as the existing variants of conditional logic suffer from limited expressiveness, we have proposed a system of conditional logic with propositional quantifiers. We have investigated the simple way, which amounts to extend an existing conditional system with the classical principles of quantification. We have shown that even such a simple approach results in various interesting, sometimes unexpected, features.

In particular, we have found that several characteristic formulae of propositional conditional logic can be captured by quantified formulae. Also, we have shown that interaction of propositional quantifiers with conditional implication has a significant impact on the modal fragment of the language. Overall, introducing propositional quantifiers in conditional logic has proved to go far beyond a mere technical task with fairly predictable outcome.

In quantified modal logic, systems exist whose expressiveness allows us to capture arithmetics so that they fail to be axiomatizable. Interestingly enough, there is a right balance for that because such a system must be rich enough to enjoy closure of the simulated arithmetical operations but not too rich as the structure collapses (specifically, an ordering becomes an equivalence relation for instance). It is expected that the situation is similar with conditional logic and this is the main direction for continuation of this work.

Another direction concerns completeness, the finite model property and its relationship with decidability (that was not so obvious in modal logic [Mak69]).

References

[Bar46] Barcan (R. C.). – A functional calculus of first order based on strict implication. *Journal of Symbolic Logic* 11(1), 1946, pp. 1–16.

[Bou94] Boutilier (C.). – Conditional logics of normality: a modal approach. *Artificial Intelligence* 68, 1994, pp. 87–154.

[Chi46] Chisholm (R.). – The contrary-to-fact conditional. *Mind* 55, 1946, pp. 289–307.

[Che75] Chellas (B. F.). – Basic conditional logic. *Journal of Philosophical Logic* 4, 1975, pp. 133–153.

[Che80] Chellas (B. F.). – *Modal logic.* – Cambridge University Press, 1980.

[Chu56] Church (A.). – *Introduction to Mathematical Logic.* – Princeton University Press, Princeton, 1956.

[Del88] Delgrande (J. P.). – An approach to default reasoning based on a first-order conditional logic : revised paper. *Artificial Intelligence* 36, 1988, pp. 63–90.

[Fin70] Fine (K.). – Propositional quantifiers in modal logic. *Theoria* 36, 1970, pp. 336–346.

[FHK96] Friedman (N.), Halpern (J. Y.) and Koller (D.). – First-order conditional logic revisited. *In: Proceedings of the thirteenth National Conference on Artificial Intelligence (AAAI-96)*, 1996, pp. 1305–1312.

[Gab71] Gabbay (D. M.). – Montague type semantics for modal logics with propositional quantifiers. *Zeitschr. f. math. Logik und Grundlagen d. Math.* 17, 1971, pp. 245–249.

[Gin86] Ginsberg (M. L.). – Counterfactuals. *Artificial Intelligence* 30, 1986, pp. 35–79.

[Goo55] Goodman (N.). – *Fact, Fiction and Forecast.* – Harvard University Press, Cambridge MA, 1955.

[Kle67] Kleene (S. C.). – *Mathematical Logic.* – J. Wiley and Sons, New York, 1967.

[KS91] Katsuno (H.) and Satoh (K.). – A unified view of consequence relation, belief revision and conditional logic. *In: Proceedings of the twelfth International Joint Conference on Artificial Intelligence (IJCAI-91)*, 1991, pp. 406–412.

[Lew73] Lewis (D.). – *Counterfactuals.* – Harvard University Press, Cambridge MA, 1973.

[Mak69] Makinson (D.). – A normal modal calculus between T and S4 without the finite model property, *Journal of Symbolic Logic* 34, 1969, pp. 35–38.

[Nut80] Nute (D.). – Topics in Conditional Logic. *In: Philosophical studies series in philosophy.* – Reidel, Dordrecht, 1980.

[Nut84] Nute (D.). – Conditional logic. *In: Handbook of Philosophical Logic*, Ed. by Gabbay (D.) and Guenthner (F.), chap. 8, pp. 387–439. – Reidel, Dordrecht, 1984.

[Pri67] Prior (A. N.). – *Past, Present and Future.* – Clarendon Press, Oxford, 1967.

[Sta68] Stalnaker (R. C.). – A Theory of Conditionals. *In: Studies in Logical Theory*, Ed. by Rescher (N.), pp. 98–112. – Basil Blackwell, Oxford, 1968. Reprinted in : *Ifs*, Ed. by Harper (W. L.), Stalnaker (R. C.), and Pearce (G.), pp. 41–55. Reidel, Dordrecht, 1980.

[Sta80] Stalnaker (R. C.). – A Defense of Conditional Excluded Middle. *In: Ifs*, Ed. by Harper (W. L.), Stalnaker (R. C.), and Pearce (G.), pp. 87–104. – Basil Blackwell, Oxford, 1980.

[Tak75] Takeuti (G.). – *Proof Theory.* – North Holland, Amsterdam, 1975.

Fast-Division Architecture for Dempster-Shafer Belief Functions *

R. Bissig, J. Kohlas, N. Lehmann

Institute of Informatics
University of Fribourg
CH–1700 Fribourg
Switzerland
E-Mail: norbert.lehmann@unifr.ch
WWW: http://www-iiuf.unifr.ch/dss

Abstract. Given a number of Dempster-Shafer belief functions there are different architectures which allow to do a compilation of the given knowledge. These architectures are the Shenoy-Shafer Architecture, the Lauritzen-Spiegelhalter Architecture and the HUGIN Architecture. We propose a new architecture called "Fast-Division Architecture" which is similar to the former two. But there are two important advantages: (i) results of intermediate computations are always valid Dempster-Shafer belief functions and (ii) some operations can often be performed much more efficiently.

1 Introduction

Dempster-Shafer belief function theory (Shafer, 1976) can be used to represent uncertain knowledge. Given Dempster-Shafer belief functions $\vartheta_1, \vartheta_2, \ldots, \vartheta_m$ the problem to solve is to find the combination of these Dempster-Shafer belief functions marginalized to some interesting domain. There are three well known architectures which solve this problem:

- The Shenoy-Shafer Architecture (Shenoy & Shafer, 1990),
- The Lauritzen-Spiegelhalter Architecture (Lauritzen & Spiegelhalter, 1988),
- The HUGIN Architecture (F.V. Jensen & Olesen, 1990).

Each of these architectures has a Markov tree as underlying computational structure. A message passing scheme is used to compute the desired marginals. In a sense a compilation of the given knowledge is obtained. Queries can then be answered very fast.

We propose a new architecture called "Fast-Division Architecture". It is similar to the Lauritzen-Spiegelhalter Architecture and the HUGIN Architecture because it uses also a *division* operation. But compared to these two architectures there are two important advantages:

---

* Research supported by grant No. 2100–042927.95 of the Swiss National Foundation for Research.

(i) The result of a division operation is always a valid DS-belief function.

(ii) The division operation can often be performed much more efficiently.

In Section 2 multivariate Dempster-Shafer belief functions are introduced. Section 3 describes the Shenoy-Shafer Architecture, the Lauritzen-Spiegelhalter Architecture and the HUGIN Architecture. The Fast-Division Architecture is described in detail in Section 4.

2 Dempster-Shafer Belief Functions

In this section some basic notions of multivariate Dempster-Shafer belief functions (DS-belief functions) are recalled. More information on Dempster-Shafer belief function theory is given in (Shafer, 1976) and (Smets, 1988).

Variables and Configurations. We define Θ_x as the *state space* of a variable x, i.e. the set of values of x. It is assumed that all the variables have finite state spaces. Lower-case Roman letters from the end of the alphabet such as x,y,z,... are used to denote variables. Upper-case italic letters such as $D,E,F,...$ denote sets of variables. Given a set D of variables, let Θ_D denote the Cartesian product $\Theta_D = \times\{\Theta_x : x \in D\}$. Θ_D is called *state space* for D. The elements of Θ_D are *configurations* of D.

Projection of Sets of Configurations. If D_1 and D_2 are sets of variables, $D_2 \subseteq D_1$ and x is a configuration of D_1, then $x^{\downarrow D_2}$ denotes the *projection* of x to D_2. If A is a subset of Θ_{D_1}, then the projection of A to D_2, denoted as $A^{\downarrow D_2}$ is obtained by projecting each element of A to D_2, i.e. $A^{\downarrow D_2} = \{x^{\downarrow D_2} : x \in A\}$. Upper-case italic letters from the beginning of the alphabet such as $A,B,...$ are used to denote sets of configurations.

Extension of Sets of Configurations. If D_1 and D_2 are sets of variables, $D_2 \subseteq D_1$ and B is a subset of Θ_{D_2}, then $B^{\uparrow D_1}$ denotes the *extension* of B to D_1, i.e. $B^{\uparrow D_1} = B \times \Theta_{D_1 \setminus D_2}$.

2.1 Different External Representations

A DS-belief function φ on D assigns a value to every subset of Θ_D. There exist different ways to represent the information contained in φ. It can be represented as *mass function*, as *commonality function*, or as *belief function*. The notation $[\varphi]_m$, $[\varphi]_q$, and $[\varphi]_b$, respectively, will be used for these representations of φ. This notation is unusual but it expresses the idea of different external representations for a DS-belief function φ better than the habitual notation m_φ, q_φ, and bel_φ, respectively.

Mass Function. A *mass function* $[\varphi]_m$ on D assigns to every subset A of Θ_D a value in $[0, 1]$, that is $[\varphi]_m : 2^{\Theta_D} \to [0, 1]$. The following relations must hold:

$$[\varphi(\emptyset)]_m = 0, \tag{1}$$

$$\sum_{A \subseteq \Theta_D} [\varphi(A)]_m = 1. \tag{2}$$

If Equation (1) is not satisfied then the corresponding function can be *normalized*. Often it is more efficient to work with *unnormalized* functions.

Commonality Function. A *commonality function* $[\varphi]_q$ on D, $[\varphi]_q : 2^{\Theta_D} \rightarrow [0,1]$, can be defined in terms of a mass function:

$$[\varphi(A)]_q = \sum_{B:A \subseteq B} [\varphi(B)]_m. \tag{3}$$

Belief Function. A *belief function* $[\varphi]_b$ on D, $[\varphi]_b : 2^{\Theta_D} \rightarrow [0,1]$, can be obtained similarly in terms of a mass function:

$$[\varphi(A)]_b = \sum_{B:B \subseteq A} [\varphi(B)]_m. \tag{4}$$

Given a DS-belief function φ on D, D is called the *domain* of φ. The sets $A \subseteq \Theta_D$ for which $[\varphi(A)]_m \neq 0$ are called *focal sets*. We use $FS(\varphi)$ to denote the focal sets of φ.

Transformations. Given a DS-belief function φ on D in one of the three external representations it is always possible to derive the others:

$$[\varphi(A)]_m = \sum_{B:A \subseteq B} (-1)^{|B \setminus A|} \cdot [\varphi(B)]_q, \tag{5}$$

$$[\varphi(A)]_m = \sum_{B:B \subseteq A} (-1)^{|A \setminus B|} \cdot [\varphi(B)]_b, \tag{6}$$

$$[\varphi(A)]_q = \sum_{B:B \subseteq A} (-1)^{|B|} \cdot [\varphi(\overline{B})]_b, \tag{7}$$

$$[\varphi(A)]_b = \sum_{B:B \subseteq \overline{A}} (-1)^{|B|} \cdot [\varphi(B)]_q. \tag{8}$$

In general, transformations between different representations are computationally expensive. An optimization is obtained by using the *fast Moebius transformations* (Thoma, 1991; Kennes & Smets, 1990; Xu & Kennes, 1994) instead. But nevertheless, for every subset A of Θ_D a new value has to be calculated. Often, Θ_D contains a large number of subsets. In Subsection 4.3 a new algorithm to perform a transformation is presented. It depends on the *information* contained in φ, and not on its *structure*.

2.2 Basic Operations

The basic operations for DS-belief functions are *combination*, *marginalization*, and *extension*. Intuitively, these operations correspond to aggregation, coarsening, and refinement, respectively. Another basic operation is *division*. It is used whenever some information has to be removed from a given DS-belief function.

For every basic operation there are external representations which are more appropriate than others. Sometimes it is even necessary to perform a transformation when the requested basic operation cannot be applied directly. In the following we will focus on mass and commonality functions.

Combination. Suppose φ and ψ are DS-belief functions on D_1 and D_2, respectively. The combination of these two DS-belief functions produces a (unnormalized) DS-belief function on $D = D_1 \cup D_2$:

$$[\varphi \otimes \psi(A)]_m = \sum_{B_1, B_2} \{[\varphi(B_1)]_m \cdot [\psi(B_2)]_m : B_1^{\uparrow D} \cap B_2^{\uparrow D} = A\}, \quad (9)$$

$$[\varphi \otimes \psi(A)]_q = [\varphi(A^{\downarrow D_1})]_q \cdot [\psi(A^{\downarrow D_2})]_q. \quad (10)$$

Marginalization. Suppose φ_1 is a DS-belief function on D_1 and suppose $D_2 \subseteq D_1$. The marginalization of φ_1 to D_2 produces a DS-belief function on D_2:

$$[\varphi_1^{\downarrow D_2}(B)]_m = \sum_{A : A^{\downarrow D_2} = B} [\varphi_1(A)]_m. \quad (11)$$

Extension. Suppose φ_2 is a DS-belief function on D_2 and suppose $D_2 \subseteq D_1$. The extension of φ_2 to D_1 produces a DS-belief function on D_1:

$$[\varphi_2^{\uparrow D_1}(A)]_m = \begin{cases} [\varphi_2(B)]_m & \text{if } A = B^{\uparrow D_1}, \\ 0 & \text{otherwise}, \end{cases} \quad (12)$$

$$[\varphi_2^{\uparrow D_1}(A)]_q = [\varphi_2(A^{\downarrow D_2})]_q. \quad (13)$$

Division. Suppose ψ and φ are DS-belief functions on D. The division is formally defined by Equation (14). Note that in general the result is not a valid DS-belief function:

$$\left[\frac{\psi}{\varphi}(A)\right]_q = \begin{cases} \frac{[\psi(A)]_q}{[\varphi(A)]_q} & \text{if } [\varphi(A)]_q \neq 0, \\ 0 & \text{otherwise}. \end{cases} \quad (14)$$

3 Propagation in a Markov Tree

Given DS-belief functions $\vartheta_1, \vartheta_2, \ldots, \vartheta_m$, their corresponding domains form a *hypergraph*. For this hypergraph, first a *covering hypertree* has to be computed (Kohlas & Monney, 1995). A *Markov tree* can then be constructed. Each node N_i of the Markov tree contains a DS-belief function φ_i on the domain D_i. Finally, computations are performed on a *message passing* scheme. As a result, each node of the Markov tree will contain the global DS-belief function $\varphi := (\varphi_1 \otimes \cdots \otimes \varphi_n)$ marginalized to its domain.

In this section the axioms which allow to use a Markov tree as underlying computational structure are shown. Three existing architectures are presented. For every architecture the underlying computational structure is a Markov tree.

3.1 Local Computations

It can be shown that DS-belief functions fit perfectly well into the framework of *valuation networks*. The valuation network framework was first introduced in (Shenoy, 1989). It involves the two operations *marginalization* and *combination*, and three simple axioms which enable local computation (Shenoy & Shafer, 1990).

Axiom A1 (Transitivity of marginalization): Suppose φ is a DS-belief function on D, and suppose $F \subseteq E \subseteq D$. Then

$$\varphi^{\downarrow F} = \left(\varphi^{\downarrow E}\right)^{\downarrow F}. \tag{15}$$

Axiom A2 (Commutativity and associativity): Suppose φ_1, φ_2, and φ_3 are DS-belief functions on D_1, D_2, and D_3, respectively. Then

$$\varphi_1 \otimes \varphi_2 = \varphi_2 \otimes \varphi_1. \tag{16}$$

$$\varphi_1 \otimes (\varphi_2 \otimes \varphi_3) = (\varphi_1 \otimes \varphi_2) \otimes \varphi_3. \tag{17}$$

Axiom A3 (Distributivity of marginalization over combination): Suppose φ_1 and φ_2 are DS-belief functions on D_1 and D_2, respectively. Then

$$(\varphi_1 \otimes \varphi_2)^{\downarrow D_1} = \varphi_1 \otimes \varphi_2^{(\downarrow D_1 \cap D_2)}. \tag{18}$$

These axioms are satisfied by DS-belief functions.

3.2 Architectures

There are three well known architectures which have a Markov tree as the underlying computational structure:

- The Shenoy-Shafer Architecture (Shenoy & Shafer, 1990),
- The Lauritzen-Spiegelhalter Architecture (Lauritzen & Spiegelhalter, 1988),
- The HUGIN Architecture (F.V. Jensen & Olesen, 1990).

The Shenoy-Shafer Architecture is a very general architecture used in the framework of valuation networks. The Dempster-Shafer belief function theory fits perfectly into this framework and therefore the Shenoy-Shafer Architecture can be used for DS-belief functions. Popular software systems using the Shenoy-Shafer Architecture are Pulcinella (Saffiotti & Umkehrer, 1991) and Belief 1.2 (Almond, 1990).

The Lauritzen-Spiegelhalter Architecture and especially the HUGIN Architecture are very popular in the field of *Bayesian Networks* (Pearl, 1986). They can also be used for DS-belief functions, but it is not very common. This is due to the fact that these two architectures depend more on the *structure* of the underlying Markov tree than on the *information* contained in the Markov tree. Unfortunately, the state space for a DS-belief function can be extremely large. If, for example, the domain D_i of a node contains only 5 binary variables, there are 2^{2^5} subsets in Θ_{D_i}.

Each of the three architectures is based on a message passing scheme in a Markov tree. Each node of the Markov tree can receive and send messages from and to its neighboring nodes. Let N_i and N_j be two neighboring nodes. If node N_i has received the message from all its neighbor nodes except N_j, it computes and sends the message φ_{ij} to N_j. Later, it will receive the incoming message ψ_{ji} from node N_j. The main difference between the Shenoy-Shafer Architecture and the other two architectures is the incoming message ψ_{ji}. In the Shenoy-Shafer Architecture the message ψ_{ji} does not include the message φ_{ij} because each node keeps track of its incoming messages. For the other two architectures the message φ_{ij} is included in ψ_{ji}. Because DS-belief functions are not *idempotent* the node N_i has to remove it. The removal of the message φ_{ij} is done using the *division* operation. This operation is not needed in the Shenoy-Shafer Architecture.

There is not a big difference between the Lauritzen-Spiegelhalter Architecture and the HUGIN Architecture. For both of them, first a root node has to be selected. Then an inward propagation towards this root node begins. Secondly an outward propagation is performed. As an example Figure 1 shows the scheduling of the messages during inward propagation with node 1 selected as root node. The scheduling of the messages during outward propagation is shown in Figure 2.

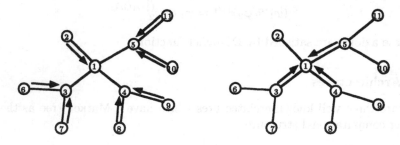

Fig. 1. Inward propagation.

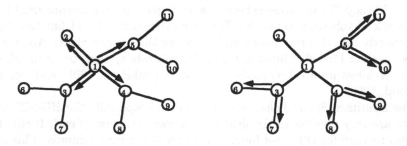

Fig. 2. Outward propagation.

During the outward propagation each node except the root computes its new DS-belief function by

$$\left(\frac{\varphi'}{\varphi_{ij}}\right) \otimes \psi_{ji} \quad \text{(Lauritzen – Spiegelhalter)} \tag{19}$$

$$\varphi' \otimes \left(\frac{\psi_{ji}}{\varphi_{ij}}\right) \quad \text{(HUGIN)} \tag{20}$$

where φ' is a DS-belief function computed during the inward propagation and N_j is the neighboring node of N_i towards the selected root node.

In the Lauritzen-Spiegelhalter Architecture the division is already performed during the inward propagation on the domain D_i. In the HUGIN Architecture the division is calculated on the smaller domain $D_i \cap D_j$. Therefore, between two connected nodes there is always a so-called *separator*. The division is performed during outward propagation on these separators.

Note that the final result of these computations is always a valid DS-belief function, although the result of the division operation is in general not a valid DS-belief function. To illustrate this consider the computation of $\left(\frac{4 \cdot 3}{6}\right)$ by $\left(\frac{4}{6}\right) \cdot 3$ or by $4 \cdot \left(\frac{3}{6}\right)$. The result is always the integer 2 but the intermediate results are not integers.

4 Fast-Division Architecture

This section describes a new architecture called "Fast-Division Architecture". This new architecture is similar to the Lauritzen-Spiegelhalter Architecture and the HUGIN Architecture described in the previous subsection, but there are two important advantages:

1. The result of a division operation is always a valid DS-belief function.
2. The division operation involved in the computations can often be performed much more efficiently.

4.1 Inward Propagation

Let $(\aleph, \Lambda)$ be a Markov tree. $\aleph = \{N_1, \ldots, N_n\}$ is a *set of nodes*. Λ is a *set of connections* between these nodes. Each node N_i contains a DS-belief function φ_i on domain D_i. First, a root node $N_r \in \aleph$ has to be selected. Then an inward propagation is started. During this inward propagation the nodes perform the following operations:

- Every node N_i except root node N_r:
  1. Combine each incoming message φ_{ki} with φ_i and store the result.
  $$\varphi_i' = \varphi_i \otimes (\otimes\{\varphi_{ki}\}). \tag{21}$$

  2. If all messages except the one from the neighbor N_j towards the root node are received:
     (a) Calculate the message φ_{ij} and send it to N_j.
     $$\varphi_{ij} = \left(\varphi_i'^{\downarrow(D_i \cap D_j)}\right)^{\uparrow D_j}. \tag{22}$$

     (b) Store the message sent to N_j on node N_i.
     $$\varphi_{ij}' = \left(\varphi_i'^{\downarrow(D_i \cap D_j)}\right)^{\uparrow D_i}. \tag{23}$$

- The root node N_r:
  Combine each incoming message φ_{kr} with φ_r and store the result.

  $$\varphi_r' = \varphi_r \otimes (\otimes\{\varphi_{kr}\}). \tag{24}$$

Note that the DS-belief functions φ_{ij} and φ_{ij}' above contain the same information. The only difference is the domain on which this information is encoded.

A node can send a message to a neighbor node when it has received a message from all other neighbor nodes. In the beginning only leaf nodes can send messages. The inward propagation is finished when the root node has received a message from all its neighbor nodes. As an example figure 1 shows the scheduling of the messages during inward propagation with node 1 selected as root node.

After the inward propagation the root node contains the global DS-belief function φ marginalized to its domain. All other nodes N_i contain the DS-belief functions φ_i' and φ_{ij}'.

4.2 Outward Propagation

After the inward propagation is finished an outward propagation is started. During this outward propagation each node performs the following operations:

- The root node N_r:
  Calculate and send a message to each of its neighbor nodes N_j

  $$\psi_{rj} = \varphi_r'^{\downarrow(D_r \cap D_j)} \tag{25}$$
  $$= (\varphi^{\downarrow D_r})^{\downarrow(D_r \cap D_j)} \tag{26}$$

- Every node N_i except root node N_r:
  If the message ψ_{ji} from neighbor node N_j is received:
  (a) Calculate the DS-belief function φ_i'' by

  $$\psi_i = \varphi_i' \otimes \psi_{ji} \tag{27}$$
  $$\varphi_i'' = \frac{\psi_i}{\varphi_{ij}'} \tag{28}$$

  (b) Calculate and send the message ψ_{ik} to each of its neighbor nodes N_k

  $$\psi_{ik} = \left(\varphi_i''^{\downarrow(D_i \cap D_k)}\right)^{\uparrow D_k} \tag{29}$$

A node can send a message to a neighbor node when it has received the message from the neighbor node towards the root node. In the beginning only the root node can send messages. The outward propagation is finished when all leaf nodes have received a message. As an example figure 2 shows the scheduling of the messages during outward propagation with node 1 selected as root node.

Theorem 1. *After inward and outward propagation every node N_i contains $\varphi^{\downarrow D_i}$, the global DS-belief function marginalized to its domain.*

Proof. See (Bissig, 1996)

4.3 Fast Division

Practical experiences show that a DS-belief function φ_i on D_i often contains only a few focal sets even if the state space for D_i is very large. The Lauritzen-Spiegelhalter Architecture and the HUGIN Architecture do not profit from this experiences. These architectures depend more on the *structure* of the corresponding Markov tree than on the *information* it contains. The new architecture presented above would also depend on the *structure* if the functions are calculated explicitly. But fortunately there exists a nice property which can speed up computations a lot.

Theorem 2 (Fast Division). *During the outward propagation each node N_i calculates its new DS-belief function in two steps as follows:*

$$\psi_i = \varphi_i' \otimes \psi_{ji} \tag{30}$$

$$\varphi^{\downarrow D_i} = \frac{\psi_i}{\varphi_{ij}'} \tag{31}$$

With ψ_i defined as above the following property holds:

$$FS(\varphi^{\downarrow D_i}) \subseteq FS(\psi_i). \tag{32}$$

Proof. See (Bissig, 1996)

Corollary 3. *For all sets $A \subseteq \Theta_{D_i}$, $A \notin FS(\psi_i)$ we have $[\varphi^{\downarrow D_i}(A)]_m = 0$. Therefore it is sufficient to calculate $[\varphi^{\downarrow D_i}(A)]_m$ using Equation (31) for sets $A \in FS(\psi_i)$.*

This is indeed a very nice property because in general the set $FS(\psi_i)$ is much smaller than the set Θ_{D_i} and a lot of unnecessary calculations can be omitted. Only for a few sets $A \in FS(\psi_i)$ it may be the case that $[\varphi^{\downarrow D_i}(A)]_m = 0$. For these sets the calculations would not have been necessary.

To calculate $[\varphi^{\downarrow D_i}(A)]_m$ for every $A \in FS(\psi_i)$ using Equation (31) it is necessary to divide two DS-belief functions. Since the mass function representation is not appropriate for the division, it is necessary to perform transformations. These transformations should of course not depend on the structure of the node N_i. The following two algorithms can be used.

Algorithm 1 (mass function to commonality function) *Suppose ϑ_i is a DS-belief function on D_i. The values $[\vartheta_i(A_k)]_m$ are known for all $A_k \in FS(\vartheta_i)$. For each of these focal sets the following formula has to be applied:*

$$[\vartheta_i(A_k)]_q = \sum_{B:A_k \subseteq B} \{[\vartheta_i(B)]_m : B \in FS(\vartheta_i)\} \tag{33}$$

Note that Equation (33) can also be used to calculate $[\vartheta_i(B)]_q$ for $B \notin FS(\vartheta_i)$.

Algorithm 2 (*commonality function* **to** *mass function*) *Suppose ϑ_i is a DS-belief function on D_i. The values $[\vartheta_i(A_k)]_q$ are known for all $A_k \in FS(\vartheta_i)$. First an ordering $\{A_1, \ldots, A_n\}$ of the set $FS(\vartheta_i)$ has to be found such that for each pair $A_k, A_l \in FS(\vartheta_i)$, whenever $A_k \supseteq A_l$ we have $k \leq l$. Such an ordering can always be found. To this ordering the following formula has to be applied sequently:*

$$[\vartheta_i(A_k)]_m = [\vartheta_i(A_k)]_q - \sum_{B:A_k \subset B} \{[\vartheta_i(B)]_m : B \in FS(\vartheta_i)\} \tag{34}$$

These two algorithms are very simple. They perform well if the set $FS(\vartheta_i)$ is relatively small. In (Dugat & Sandri, 1994) another algorithm is presented which takes advantage of the partially ordered structure of $FS(\vartheta_i)$, due to set-inclusion.

Algorithm 3 (Fast Division) *During the outward propagation Node N_i is in possession of the following DS-belief functions:*

- φ_i' : DS-belief function on Node N_i during the inward propagation

- $\left(\varphi_i'^{\downarrow(D_i \cap D_j)}\right)^{\uparrow D_i}$: Message sent to Node N_j during inward propagation

- $\left(\varphi^{\downarrow(D_i \cap D_j)}\right)^{\uparrow D_i}$: Message obtained from Node N_j during outward propagation.

Using these DS-belief functions Node N_i has to calculate the global DS-belief function marginalized to its own domain.

$$\varphi^{\downarrow D_i} = \frac{\varphi_i' \otimes \left(\varphi^{\downarrow(D_i \cap D_j)}\right)^{\uparrow D_i}}{\left(\varphi_i'^{\downarrow(D_i \cap D_j)}\right)^{\uparrow D_i}} \tag{35}$$

Therefore the following calculations have to be performed one after another:

1. *Calculate $\psi_i = \varphi_i' \otimes \left(\varphi^{\downarrow(D_i \cap D_j)}\right)^{\uparrow D_i}$*
2. *For each $A \in FS(\psi_i)$ transform*
   - *$[\psi_i(A)]_m$ into $[\psi_i(A)]_q$ using Algorithm 1.*
   - *$\left[\left(\varphi_i'^{\downarrow(D_i \cap D_j)}\right)^{\uparrow D_i}(A)\right]_m$ into $\left[\left(\varphi_i'^{\downarrow(D_i \cap D_j)}\right)^{\uparrow D_i}(A)\right]_q$ using Algorithm 1.*
3. *For each $A \in FS(\psi_i)$ calculate*
   $$[\varphi^{\downarrow D_i}(A)]_q = \frac{[\psi_i(A)]_q}{\left[\left(\varphi_i'^{\downarrow(D_i \cap D_j)}\right)^{\uparrow D_i}(A)\right]_q}$$
4. *For each $A \in FS(\psi_i)$ transform*
   $[\varphi^{\downarrow D_i}(A)]_q$ into $[\varphi^{\downarrow D_i}(A)]_m$ using Algorithm 2

The Fast Division Algorithm is shown in Figure 3. The computations have to be performed only for the focal sets of $\psi_i(A)$.

In the Fast-Division Architecture all operations except division are performed on mass functions. Commonality functions are only used to compute the division. The result of the division operation is always a valid commonality function. It will be immediately transformed back into a mass function.

$$[\psi_i(A)]_m \longrightarrow [\psi_i(A)]_q$$

$$[\varphi_i'(A)]_m \longrightarrow [\varphi_i'(A)]_q$$

$$\frac{[\psi_i(A)]_q}{[\varphi_i'(A)]_q} = [\varphi^{\downarrow D_i}(A)]_q \longrightarrow [\varphi^{\downarrow D_i}(A)]_m$$

Fig. 3. Fast Division.

5 Conclusion and Future Work

For Dempster-Shafer belief functions the Fast-Division Architecture could be an alternative to the Shenoy-Shafer Architecture. Tests based on a prototype program implemented in Common LISP were very promising. Future work will focus on comparing the Fast-Division Architecture to the Shenoy-Shafer Architecture. Until now the Fast-Division Architecture was only used for Dempster-Shafer belief functions. But perhaps it could be generalized.

References

Almond, R.G. 1990. *Fusion and Propagation of Graphical Belief Models: an Implementation and an Example.* Ph.D. thesis, Department of Statistics, Harvard University.

Bissig, R. 1996. *Eine schnelle Divisionsarchitektur für Belief-Netwerke.* M.Phil. thesis, Institute of Informatics, University of Fribourg.

Dugat, V., & Sandri, S. 1994. Complexity of Hierarchical Trees in Evidence Theory. *ORSA Journal on Computing,* **6**, 37–49.

F.V. Jensen, S.L. Lauritzen, & Olesen, K.G. 1990. Bayesian Updating in Causal Probabilistic Networks by Local Computations. *Computational Statistics Quarterly,* **4**, 269–282.

Kennes, R., & Smets, P. 1990. Computational Aspects of the Möbius Transform. *Pages 344–351 of: Proceedings of the 6th Conference on Uncertainty in Artificial Intelligence.*

Kohlas, J., & Monney, P.A. 1995. *A Mathematical Theory of Hints. An Approach to the Dempster-Shafer Theory of Evidence.* Lecture Notes in Economics and Mathematical Systems, vol. 425. Springer.

Lauritzen, S.L., & Spiegelhalter, D.J. 1988. Local Computations with Probabilities on Graphical Structures and their Application to Expert Systems. *J. R. Statist. Soc. B,* **50**(2), 157–224.

Pearl, J. 1986. Fusion, Propagation and Structuring in Belief Networks. *Artificial Intelligence, Elsevier Science Publisher B.V. (Amsterdam),* **329**, 241–288.

Saffiotti, A., & Umkehrer, E. 1991. *PULCINELLA: A General Tool for Propagating Uncertainty in Valuation Networks.* Tech. Rep. IRIDIA, Université de Bruxelles.

Shafer, G. 1976. *A Mathematical Theory of Evidence*. Princeton University Press.

Shenoy, P.P. 1989. A Valuation-Based Language for Expert Systems. *International Journal of Approximate Reasoning*, **3**, 383–411.

Shenoy, P.P., & Shafer, G. 1990. Axioms for Probability and Belief-Functions Propagation. *In:* R.D. Shachter, T.S. Levitt, L.N. Kanal (ed), *Uncertainty in Artificial Intelligence 4*. Elsevier Science Publishers B.V.

Smets, Ph. 1988. Belief Functions. *Pages 253–286 of:* Ph. Smets, A. Mamdani, D. Dubois, & Prade, H. (eds), *Nonstandard logics for automated reasoning*. Academic Press.

Thoma, H. Mathis. 1991. Belief Function Computations. *Pages 269–307 of:* I.R. Goodman, M.M. Gupta, H.T. Nguyen, & Rogers, G.S. (eds), *Conditional Logic in Expert Systems*. Elsevier Science.

Xu, H., & Kennes, R. 1994. Steps toward efficient implementation of Dempster-Shafer theory. *Pages 153–174 of:* Yager, R.R., Kacprzyk, J., & Fedrizzi, M. (eds), *Advances in the Dempster-Shafer Theory of Evidence*. Wiley.

Graduality by Means of Analogical Reasoning

J. Delechamp*°, B. Bouchon-Meunier°

* LCPC
58, Boulevard Lefebvre, 75015 Paris, France
° LAFORIA, UPMC Case 169
4, place Jussieu, 75252 Paris Cédex 05, France
bouchon@laforia.ibp.fr, delecham@inrets.fr

Abstract. There exist different types of gradual rules, involving graduality on the truth value, the possibility, the certainty associated with propositions, or dealing with the degree of similarity or the proximity of observations with reference propositions, or a graduality on the values taken by variables. In this paper, we describe the properties of gradual rules according to variations of values of the variables. Then, we extend the results to a graduality expressed through linguistic modifiers. Finally, we show that a gradual knowledge representation through modifiers can be linked with analogical reasoning.

1. Introduction

Graduality is an important concept in knowledge-based systems. First of all, many expert rules are given in a gradual form, such as "the more sunny the weather, the better the health", which is difficult to translate into classical if-then rules. Another element of graduality is involved in imprecise categories, like "a big city", for which we are able to define typical representants (Paris, London...) and, the greater the difference between a given city and the typical ones, the smallest the membership degree of this given city to the category. Human reasoning is well adapted to the processing of gradual knowledge, because of its ability to manage imprecise, qualitative, approximate information.

Analogy is an important component of human reasoning and the link between graduality and analogical reasoning is natural. Let us consider a base of already known elements, each one associated with a decision, and a new element for which we want to make a decision, according to its analogy with elements of the base. The more the new element resembles an already known element, the more similar the new decision to the decision associated with this already known element.

The capability of fuzzy sets to deal with a kind of graduality in membership degrees leads to use a fuzzy knowledge representation to work on gradual knowledge. The use of hedges or linguistic modifiers for the representation of descriptions not far from a given one is also a tool useful for the management of a kind of graduality. Further, approximate reasoning in fuzzy logic has been initiated by L.A. Zadeh as a method of automatic reasoning as close as possible to human reasoning [Zadeh 1983]. It is then natural to study gradual knowledge in a fuzzy framework and this has been extensively done in the past [Bouchon & Després 1990][Bouchon & Yao 1992][Dubois & Prade 1990].

Various types of gradual rules have been pointed out, involving graduality on the truth value, the possibility, the certainty associated with propositions, or dealing with the degree of similarity or the proximity of observations with reference propositions, or graduality on the values taken by variables. Most of the approaches are local, around a reference situation. The use of modifiers and an analogical approach have already been proposed for a non local representation of graduality.

In this paper, we choose a more global approach and we describe the properties of gradual rules according to variations of values of the variables. We extend the results to a graduality expressed through linguistic modifiers. Finally, we show that a gradual knowledge representation through modifiers can be linked with analogical reasoning.

2. Semantics of gradual rules

2.1. Main semantics of gradual rules

Various semantics have been associated with gradual rules in a fuzzy framework [Bouchon & Després 90] [Dubois & Prade 92]. For instance, there can exist a graduality on truth values, on possibility or certainty of propositions, a graduality related to similarity or proximity between propositions, or a graduality on values of variables.

Let us consider two variables X and Y defined on universes U and V, F(U) and F(V) their respective sets of fuzzy sets, and fuzzy sets A_n in F(U) and B_m in F(V). Graduality can be interpreted as a kind of link between fuzzy propositions of the form « X is A_n » and « Y is B_m ».

1) In the case of a graduality between truth values, gradual rules are of the form «the more (the less) the proposition « X is A_n » is true, the more (the less) the proposition « Y is B_m » is true [Dubois et Prade 90].

2) A graduality regarding the certainty of propositions can be expressed by means of rules «the more (the less) X is A_n , the more (the less) the proposition « Y is B_m » is certain », or «the more (the less) the proposition « X is A_n » is certain, the more (the less) the proposition « Y is B_m » is certain »

3) A graduality related to similarity is the basis of reasoning models such as AARS [Turksen & Zhao 88], which assigns to gradual rules the meaning «the more (the less) the proposition « X is similar to A_n » is true, the more (the less) the proposition « Y is similar to B_m » is true »

4) A concept of proximity, in the case of metric spaces U and V, allows to manage gradual rules such as « the more (the less) X is close to A_n, the more (the less) Y is close to B_m ». For instance, Koczy and Hirota [Koczy & Hirota 93] define the concept of closeness by means of a fuzzy distance.

5) If U and V are totally ordered, the graduality of a link between variables refers naturally to connections between variations of the variables. We can express the

graduality as « the greater (smaller) the variation of the value of X with respect to a given state in U, the greater (smaller) the variation of the value of Y with respect to a given state in V ». This means that a continuous variation of X entails a continuous variation of Y [Bouchon & Desprès 1990] or, more particularly, a variation of amplitude s_X of X with respect to the first given state entails a variation of amplitude s_Y of Y with respect to the second given state. The amplitude can be either positive or negative : for instance, for a rule of the form « the more... the more... », s_X and s_Y are positive.

These interpretations of the concept of graduality refer to different contexts, more or less local, depending on the case [Bouchon 92].
Graduality of type 1, 2 or 3 is relative to a unique description A_n of X and B_m of Y, the semantics of the rule is local with regard to A_n and B_m. For instance, for a rule such as « the more sunny the weather, the better the health », a graduality of type 1, 2 or 3 takes into account a description « sunny » of X and a description « good » of the health, while the generality of the rule refers to the global context relative to the variables « weather » and « health », which should involve several descriptions, such as « grey, rather sunny, sunny, very sunny » for the first one, « bad, mediocre, good, excellent » for the second one.
Graduality of type 4 or 5 takes into account the other descriptions of variables defined on their universes or the whole universes. It is then less local.

2.2. Properties on non local graduality
We focus on graduality of type 5 and we use gradual rules such as « the more (less) X is A_n, the more (less) Y is B_m » to describe a gradual behavior of the variables X and Y.
Let us consider real-valued universes U and V of variables X and Y, $P(U)=\{A_\alpha\}$ and $P(V)=\{B_\alpha\}$ partitions of U and V respectively, which means families of classes of U and V such that for every value x of X in U, there exists A_α such that $x \in A_\alpha$, for every value y of Y in V, there exists B_α such that $y \in B_\alpha$. We further suppose that P(U) and P(V) have the same number of classes and that partitions are ordered according to two relations $>_U$ et $>_V$. A binary operation + is supposed to be given on the set of classes of U and V respectively.
The classes of U and V are considered as descriptions of variables X and Y, they can have various forms, such as singletons corresponding to precise values, classical subsets of the universes, fuzzy subsets...

Properties of gradual rules we consider are the following :
P1 - Variations of the variables X and Y are relative to reference descriptions (respectively A_η and B_μ), and the rule « if X is A_η then Y is B_μ" is called a reference rule [Bouchon &Desprès 1990].
P2 - The graduality on U and V entails that passing from a description A_α of X to the next one (according to the order on P(U)) implies passing from a description B_β of Y

to the next one (according to the order on P(V)). This graduality is then represented through a bundle of rules settling correspondances between elements of P(U) and elements of P(V), such as « "if X is $A_{\eta \pm j}$ then Y is $B_{\mu \pm j}$". For instance, a rule of the form « the more... the more... » is expressed as : if X is $A_{\eta+j}$ then Y is $B_{\mu+j}$, with $A_{\eta+j} >_U A_{\eta}$ and $B_{\mu+j} >_V B_{\mu}$.

P3 - The gradual link between the passage of an element of the partition P(U) to the next one (or to the previous one) and an element of the partition P(V) to the next one (or to the previous one) can be expressed as an interpolation between rules of the bundles : the variation ε regarding A_{α} entails a variation ε' of B_{α} which is a continuous function of ε ($\varepsilon' = f(\varepsilon)$). More explicitely :

$\forall A_{\alpha}, A_{\beta} \in P(U),\ \forall B_{\alpha}, B_{\beta} \in P(V),$ if X is $A_{\beta} = A_{\alpha} + \varepsilon$, then Y is $B_{\beta} = B_{\alpha} + f(\varepsilon)$, with ε and $f(\varepsilon)$ two classes of U and V respectively.

2.3. Non local graduality in different frameworks

We apply the graduality described by properties P1, P2, P3 in different frameworks :
- in the case of partitions reduced to singletons of the universes,
- when classes are intervals of the real line, each one associated with a label,
- when classes are fuzzy intervals of the universes, representing linguistic descriptions of the variables

First framework : singletons

Properties P1 and P2 are naturally satisfied and P3 represents the interpolation rule between two points.

In the next frameworks, we suppose that |P(U)| and |P(V)| are finite or countable, and we note P(U)={ A_i } et P(V)={ B_i }.

Second framework : intervals of R

Let us consider $A_i = [a_i, a'_i],\ B_i = [b_i, b'_i]$ for every A_i in P(U) and B_i in P(V). We use classical operations on intervals. The gradual rule « the more (less) X is A_n, the more (less) Y is B_m » satisfies properties P1 to P3 :

- P1 : the graduality is relative to a given rule « if X is A_n then Y is B_m »,
- P2 : the bundle of rules corresponds to mappings between intervals, such as

    « if X is between a_{n+j} and a'_{n+j} then Y is between b_{m+j} and b'_{m+j} »

    (gradual rule « the more... the more... »),

    « if X is between a_{n-j} and a'_{n-j} then Y is between b_{m-j} and b'_{m-j} »

    (gradual rule « the less... the less... »),

    « if X is between a_{n+j} and a'_{n+j} then Y is between b_{m-j} and b'_{m-j} »

    (gradual rule « the more... the less... »),

    « if X is between a_{n-j} and a'_{n-j} then Y is between b_{m+j} and b'_{m+j} »

    (gradual rule « the less... the more... »)

- P3 : the interpolation between rules can be written as :

« if the value of X is in $A_i + [\varepsilon_1, \varepsilon_2]$, then the value of Y is in $B_i + [f(\varepsilon_1), f(\varepsilon_2)]$ » since f is continuous and $f([\varepsilon_1, \varepsilon_2]) = [f(\varepsilon_1), f(\varepsilon_2)]$), which is equivalent to the following rule : « if X is between $a_{n+j} + \varepsilon_1$ and $a'_{n+j} + \varepsilon_2$, then Y is between $b_{m+j} + f(\varepsilon_1)$ and $b'_{m+j} + f(\varepsilon_2)$, with ε_1 and ε_2 positive or negative.

Third framework : fuzzy intervals of R

When P(U) and P(V) contain respectively fuzzy subsets A_i and B_i of U and V, we restrict ourselves to trapezoidal fuzzy intervals $(a_i, a'_i, \delta a, \delta a')$ and $(b_i, b'_i, \delta b, \delta b')$ with kernels $[a_i, a'_i]$ and $[b_i, b'_i]$, supports $]a_i - \delta a, a'_i + \delta a'[$ and $]b_i - \delta b, b'_i + \delta b'[$, interpreted as « approximately between a_i and a'_i » and « approximately between b_i and b'_i ». The classical operations on fuzzy intervals are used [Dubois & Prade 87].

The gradual rule « the more (less) X is A_n, the more (less) Y is B_m » satisfies also the previous properties :

- P1 : the graduality is relative to a reference rule « if X is A_n then Y is B_m »
- P2 : the bundle contains rules establishing a correspondence between fuzzy intervals A_{n+j} and B_{n+j} of the form « if X is approximately between a_{n+j} and a'_{n+j}, then Y is approximately between b_{m+j} and b'_{m+j} », for a rule « the more... the more... », for instance
- P3 : the interpolation can be written through the addition $\oplus$ of fuzzy intervals chosen for the operation +, by using the notation $(\varepsilon_1, \varepsilon_2, 0, 0)$ for the precise interval $[\varepsilon_1, \varepsilon_2]$. Then, $A_j = A_i \oplus (\varepsilon_1, \varepsilon_2, 0, 0)$ and $(a_j, a'_j, \delta a, \delta a') = (a_i, a'_i, \delta a, \delta a') \oplus (\varepsilon_1, \varepsilon_2, 0, 0) = (a_i + \varepsilon_1, a'_i + \varepsilon_2, \delta a, \delta a')$, and we obtain the rules « if X is in $A_i \oplus (\varepsilon_1, \varepsilon_2, 0, 0)$ then Y is in $B_i \oplus (f(\varepsilon_1), f(\varepsilon_2), 0, 0)$ ». Equivalently, « if X is approximately between $a_{n+j} + \varepsilon_1$ and $a'_{n+j} + \varepsilon_2$, then Y is approximately between $b_{m+j} + f(\varepsilon_1)$ et $b'_{m+j} + f(\varepsilon_2)$ », with ε_1 and ε_2 positive or negative.

We can regard gradual rules involving fuzzy sets as extensions of functions defined on points or intervals and satisfying the three properties we require, and particularly the interpolation property.

In the next section, we extend the approach we propose to a graduality obtained by means of linguistic modifiers, which have already been used to describe some aspects of graduality [Bouchon & Després 90] [Bouchon 92].

3. General form of graduality through modifiers

Linguistic modifiers [Zadeh 1972] (for instance « approximately, very, really, more or less ») are useful to modulate the characterizations, by weakening or reinforcing the meaning of the characterization.

A modifier m modulates the characterization A_i (with membership function f_{A_i}) by creating a new characterization $A_j = m A_i$ with membership function f_{A_j} obtained from f_{A_i} by means of a transformation t_m as : $f_{A_j}(x) = t_m(f_{A_i})(x)$ [Bouchon 92].

We take into account the modifiers m⁺, m⁺⁺ [Bouchon 92], m*, m** [Bouchon & Yao 92] (see the annex for definitions).

The concept of graduality we use, in the case of modified characterizations, is defined by the fact that the same modifier is used for the descriptions of both variables (« if X is A_i then Y is B_i » yields « if X is mA_i then Y is mB_i ») and variations regarding membership functions are equal, i.e. :

$$\forall x \ \forall y \text{ such that} f_{B_i}(y) = f_{A_i}(x) \Rightarrow f_{B_j}(y) - f_{B_i}(y) = f_{A_j}(x) - f_{A_i}(x)$$

For all the modifiers we indicate, it can be proven that the maximum value of the variation of B_i on the left hand side is equal to $\varepsilon'_1 = f(\varepsilon_1) = \varepsilon_1 \times \dfrac{\delta b}{\delta a}$, or $\varepsilon'_1 = \varepsilon_1 \times k_1$ with $k_1 = \dfrac{\delta b}{\delta a}$, and ε_1 the maximum value of the variation of A_i on the left hand side, respectively on the right hand side $\varepsilon'_2 = f(\varepsilon_2) = \varepsilon_2 \times \dfrac{\delta b'}{\delta a'}$, or $\varepsilon'_2 = \varepsilon_2 \times k_2$ with $k_2 = \dfrac{\delta b'}{\delta a'}$, and ε_2 the maximum value of the variation of A_i on the right hand side.

In the case of m* and m**, we must have $\varepsilon_1 = \varepsilon_2 \Rightarrow \varepsilon'_1 = \varepsilon'_2$ then $k_1 = k_2$.

For the modifier m⁺, the values of variations of X or Y are not equal in any point of U or V, but we can give a general definition of the interpolation rule for all the modifiers by using α-cuts $(F)_\alpha$ of any fuzzy set F.

Let us remark that the definition of a gradual rule through α-cuts of the fuzzy sets involved in the rules has already been used, for instance by considering a collection of « α-rules » [Dubois et Prade 90] [Koczy & Hirota 93]).

If we define A_i by the family of all its α-cuts $(A_i)_\alpha$, geometric properties of membership functions yield the following results :

Property 1 : Let a rule « if X is A_i then Y is B_i » be given, with trapezoidal membership functions for A_i and B_i, the transformation of a characterization A_i into a new characterization A_j through a linguistic modifier m⁺, m⁺⁺, m*, m** can be represented by a general operation on α-cuts :

$$(A_j)_\alpha = (A_i)_\alpha + [\gamma_1 \varepsilon_1, \gamma_2 \varepsilon_2]$$

with ε_1 and ε_2 the maximum values of the variations of A_i on the left and right hand sides ($\varepsilon_1 = \varepsilon_2$ for m* and m**) and γ_1 and γ_2 the following real parameters :

for m⁺ : $\gamma_1 = \alpha - 1, \gamma_2 = 1 - \alpha$

for m⁺⁺ : $\gamma_1 = -1, \gamma_2 = 1$

for m* : $\gamma_1 = -1, \gamma_2 = -1$

for m ** : $\gamma_1 = 1, \gamma_2 = 1$

Property 2 : For the rule « if X is A_i then Y is B_i », the considered graduality implies that the transformation of B_i into B_j defined on α-cuts is analogous to the transformation of A_i into A_j in the following sense :

$$\forall \alpha \in [0, 1], \text{ if } (A_j)_\alpha = (A_i)_\alpha + [\gamma_1 \varepsilon_1, \gamma_2 \varepsilon_2] \text{ then } (B_j)_\alpha = (B_i)_\alpha + [k_1 \gamma_1 \varepsilon_1, k_2 \gamma_2 \varepsilon_2]$$

We can write $A_i \oplus E = A_j$ where the α-cuts of A_i and E are respectively $\left[a_1^\alpha, a_2^\alpha\right]$ and $[\gamma_1 \varepsilon_1, \gamma_2 \varepsilon_2]$ and A_j is the fuzzy quantity associated with them whose α-cuts are the intervals $\left[a_1'^\alpha = a_1^\alpha + \gamma_1 \varepsilon_1, a_2'^\alpha = a_2^\alpha + \gamma_2 \varepsilon_2\right]$

The interactivity of A_i and A_j does not allow us to define E from A_i and A_j by means of their α-cuts [Dubois et Prade 87]. Applying results on fuzzy relation equations [Sanchez 84] to the equation $A_i \oplus E = A_j$, we find that the inverse

equation where E is unknown has a solution if and only if $A_i \oplus (A_j \tilde{+} A_i) = A_j$,

with $A_j \tilde{+} A_i = \left\{ y / \inf(\mu_{A_i}(x) \psi \mu_{A_j}(x+y)) \right\}$ and $a \psi b \begin{cases} = 1 \text{ if } a \le b \\ = b \text{ if } a > b \end{cases}$. Then,

$E = A_j \tilde{+} A_i$ is the greatest solution.

In the general case, we can prove that $E = (\beta_1 \varepsilon_1, \beta_2 \varepsilon_2, \beta_3 \varepsilon_1, \beta_4 \varepsilon_2)$ with the following values of $\beta_1, \beta_2, \beta_3, \beta_4$ for the four modifiers :

m⁺ : $\beta_1 = 0, \beta_2 = 0, \beta_3 = -1, \beta_4 = 1$

m⁺⁺ : $\beta_1 = -1, \beta_2 = 1, \beta_3 = 0, \beta_4 = 0$

m* : $\beta_1 = -1, \beta_2 = -1, \beta_3 = 0, \beta_4 = 0$

m ** : $\beta_1 = 1, \beta_2 = 1, \beta_3 = 0, \beta_4 = 0$

The graduality represented by the modifiers can be written as :

if $\quad A_j \overset{\sim}{+} A_i \quad = \quad (\beta_1 \varepsilon_1, \beta_2 \varepsilon_2, \beta_3 \varepsilon_1, \beta_4 \varepsilon_2) \quad$ then $\quad B_j \overset{\sim}{+} B_i \quad =$
$(k_1 \beta_1 \varepsilon_1, k_2 \beta_2 \varepsilon_2, k_1 \beta_3 \varepsilon_1, k_2 \beta_4 \varepsilon_2)$

Let us remark that, for modifiers m^{++}, m*, m **, $\beta_3 = 0, \beta_4 = 0$ and the graduality can then be expressed as :

« if X is $A_i \oplus = (\beta_1 \varepsilon_1, \beta_2 \varepsilon_2, 0, 0)$, then Y is $B_i \oplus = (k_1 \beta_1 \varepsilon_1, k_2 \beta_2 \varepsilon_2, 0, 0)$ », which is equivalent to :

« if X is approximately between $a_{n+j} + \beta_1 \varepsilon_1$ and $a'_{n+j} + \beta_2 \varepsilon_2$ then Y is approximately between $b_{m+j} + k_1 \beta_1 \varepsilon_1$ et $b'_{m+j} + k_2 \beta_2 \varepsilon_2$ ».

For m$^+$, the graduality is expressed in the rule « if X is $A_i \oplus (0, 0, \beta_3 \varepsilon_1, \beta_4 \varepsilon_2)$, then Y is $B_i \oplus (0, 0, k_1 \beta_3 \varepsilon_1, k_2 \beta_4 \varepsilon_2)$ », which is equivalent to :

« if X is approximately between a_{n+j} and a'_{n+j} then Y is approximately between b_{m+j} et b'_{m+j} ».

Let us remark that, in this case, the rule « if X is A_i then Y is B_i » is linguistically equivalent to the rule « if X is m A_i then Y is m' B_i », with an only change in the representation of « approximately », which is given by the values of the difference between support and kernel of the fuzzy intervals we deal with in such a way that m' has the same form as m, with slightly different values of parameters.

We have given a general form of graduality based on the use of linguistic modifiers in the case when we can use the same modifier in the premise and in the conclusion.

The rule « if $A_j \overset{\sim}{+} A_i$ is equal to $(\beta_1 \varepsilon_1, \beta_2 \varepsilon_2, \beta_3 \varepsilon_1, \beta_4 \varepsilon_2)$ then $B_j \overset{\sim}{+} B_i$ is equal to $(k_1 \beta_1 \varepsilon_1, k_2 \beta_2 \varepsilon_2, k_1 \beta_3 \varepsilon_1, k_2 \beta_4 \varepsilon_2)$ » indicates that we can calculate $B_j \overset{\sim}{+} B_i$ as soon as we know $A_j \overset{\sim}{+} A_i$ and suggests an idea of analogy between $B_j \overset{\sim}{+} B_i$ and $A_j \overset{\sim}{+} A_i$ for the operator $\overset{\sim}{+}$. We develop this idea in the next section.

4. Graduality through modifiers and type of reasoning

Graduality has extensively been studied in the framework of generalized modus ponens processes, more particularly with regard to the use of modifiers [Bouchon 88][Bouchon 92][Bouchon & Yao 92]. We show that analogical reasoning is more suitable for this kind of gradual reasoning.

Human analogical reasoning uses the capacity of association of the memory. In a context of problem solving, cases analogous to a problem to be solved can lead us to a solution. If we restrict ourselves to descriptions relative to one variable, we denote by « X is A_i » and « Y is B_i » the propositions describing the situation and the solution of a reference case. An analogy is represented by a relationship β between X and Y, and a relation S such that, for a new situation « X is A_j » we have :

« if $A_i \beta B_i$ and $A_i S A_j$, we can find B_j such that $A_j \beta B_j$ and $B_i S B_j$ ».

In the framework of fuzzy logic, it has been proven that S can be defined from resemblance relations R on the set of fuzzy sets of U and R' on the set of fuzzy sets of V [Bouchon & Valverde 93]. We consider here a slightly different form of *analogical reasoning*, based on a dissimilarity d between fuzzy subsets of U, a dissimilarity d' between fuzzy subsets of V, as follows :

$\forall A_i \ \forall B_i$ such that $A_i \beta B_i$ and $\forall A_j$ we can find B_j satisfying

(1) $\forall s' \ \exists s$ such that, if $d(A_i ; A_j) = s$ then $d'(B_i ; B_j) = s'$.

(2) $A_j \beta B_j$.

The concept of *« dissimilarity »* is understood in a very general meaning, corresponding to the establishment of a difference betwen two fuzzy sets of the same universe. We prove that gradual reasoning defined by means of modifiers is a form of analogical reasoning, since it satisfies these two properties.

In the previous section, we have introduced an operator $\widetilde{+}$ such that

if $\qquad A_j \widetilde{+} A_i \qquad = \qquad (\beta_1 \varepsilon_1, \beta_2 \varepsilon_2, \beta_3 \varepsilon_1, \beta_4 \varepsilon_2) \qquad$ then $\qquad B_j \widetilde{+} B_i =$

$(k_1 \beta_1 \varepsilon_1, k_2 \beta_2 \varepsilon_2, k_1 \beta_3 \varepsilon_1, k_2 \beta_4 \varepsilon_2)$

which can be rewritten as :

$$\forall s' = (\beta_1 \varepsilon'_1, \beta_2 \varepsilon'_2, \beta_3 \varepsilon'_3, \beta_4 \varepsilon'_4), \ \exists s = \left(\frac{1}{k_1} \beta_1 \varepsilon'_1, \frac{1}{k_2} \beta_2 \varepsilon'_2, \frac{1}{k_1} \beta_3 \varepsilon'_1, \frac{1}{k_2} \beta_4 \varepsilon'_2 \right) \text{ such}$$

that, if $A_j \widetilde{+} A_i = s$ then $B_j \widetilde{+} B_i = s *$, We can then consider a dissimilarity $d : F(U) \times F(U) \rightarrow F(U)$ and analogously $d' : F(V) \times F(V) \rightarrow F(V)$, with $d(A,A') = A \widetilde{+} A'$, $d'(B,B') = B \widetilde{+} B'$. We have proved property (1).

The relation $A_i \beta B_i$ means that, if the value of X is A_i, the value of Y is B_i, expressed by the rule « if X is A_i, then Y is B_i » We have exhibited a transformed rule « if X is $m A_i$, then Y is $m B_i$ », with $A_j = m A_i$, $B_j = m B_i$. This rule establishes a link between A_j and B_j which can be represented by $A_j \beta B_j$, and property (2) is satisfied.

We get the following result :

Property 3 : If graduality is regarded through the use of modifiers, which means that any rule « if X is A_i then Y is B_i » is transformed into a rule « if X is $m\,A_i$ then Y is $m\,B_i$ », then graduality is a particular form of analogical resoning.

Let us remark that the form of graduality we have studied is particularly important with modifiers m* and m** corresponding to a translation of the original characterizations. The generalized modus ponens process (or compositional rule of inference) could not provide this kind of gradual reasoning, since the use of any description « X is A_i' », with a rule of the form « if X is A_i then Y is B_i » would yield a conclusion « Y is B_i' », where B_i' has a support containing (or at least equal to) the support of B_i, which is then never connected with any idea of translation of fuzzy set.

In many cases, translations are reasonable ways to represent the idea of graduality, such as « the greater X, the greater Y », for instance, expressing that a translation of the value of X to the right of the ordered universe U, entails a translation of the value of Y also to the right of the ordered universe V. The link between gradual reasoning and analogical reasoning corresponds to the fact that the relationship between variations of X and variations of Y expressed in gradual knowledge can be used to infer a value of Y from a given value of X by means of an analogy between variations of Y and variations of X.

5. Conclusion

In this paper, we have described properties of gradual rules, with regard to the variations of values of variables. Then we have extended the obtained results to a graduality defined by means of linguistic modifiers and we have given a general formalization of this kind of graduality. We have proved that gradual reasoning can be regarded as a form of analogical reasoning.

Interpolation is the main mechanism of this formalization. Relations between graduality and interpolation have already been pointed out [Buisson & Prade 89] [Koczy & Hirota 93] [Dubois & Prade & Ughetto 96], and more generally relations between approximate reasoning and interpolation [Zadeh 90][Mizumoto 94].

It will be interesting in a near future to extend this formalization to more general (non trapezoidal) fuzzy sets, to generalize the relationship between gradual reasoning and analogy, and to compare this approach to classical interpolative methods.

Annex

Transformations defining the modifiers m^+, m^{++}, m*, m** :

We consider trapezoidal membership functions f_i of fuzzy sets A_i on a given universe $U = \left[U^-, U^+\right]$ defined by four parameters $\left(m_i, n_i, a_i - m_i, n_i - b_i\right)$ which means that $f_i(x) = 0$ for every $x \in \left[U^-, m\right] \cup \left[n_i, U^+\right]$, with $m_i \geq U^-$ and $n_i \leq U^+$; $f_i(x) = \varphi'_i(x)$ for every $x \in \left[m_i, a_i\right]$, with $m_i < a_i$; $f_i(x) = 1$ for every $x \in \left[a_i, b_i\right]$, with $a_i \leq b_i$; $f_i(x) = \varphi''_i(x)$ for every $x \in \left[b_i, n_i\right]$, with $b_i < n_i$. The functions $\varphi'_i(x)$ and $\varphi''_i(x)$ are linear and such that $\varphi'_i(m_i) = \varphi''_i(n_i) = 0$, $\varphi'_i(a_i) = \varphi''_i(b_i) = 1$. We denote by $\varphi_i(x)$ the function identical with $\varphi'_i(x)$ on $\left[U^-, a_i\right]$ and identical with $\varphi''_i(x)$ on $\left[b_i, U^+\right]$.

We give real-valued parameters v, β, α, describing the range of the alteration associated with the modifier. The formal definitions of the modifier give the following membership function of $m(A_i)$ when $x \in U$.

- with m^+ : $f_{mA_i}(x) = \max(0, v\varphi_i(x) + 1 - v)$, for $v \in [0,5, \ 1]$,
- with m^{++} : $f_{mA_i}(x) = \min(1, \max(0, \varphi_i(x) + \beta))$, for $\beta \in \left]0, 0, 5\right]$,
- with m^* or m^{**} : $f_{mA_i}(x) = f_i(x + \alpha)$ if $x + \alpha \in U$, $f_{mA_i}(x) = f_i(U_i^-)$ if $x + \alpha \leq U^-$, $f_{mA_i}(x) = f_i(U_i^+)$ if $x + \alpha \geq U^+$.

References

[Bouchon 88]
Bouchon B., Stability of linguistic modifiers compatible with a fuzzy logic, in Uncertainty in Intelligent Systems, Lecture Notes in Computer Science, vol. 313, Springer Verlag, 1988 ;

[Bouchon 90]
Bouchon-Meunier B., Comment remplacer des calculs par des règles simples dans le cadre d'une logique floue, Cognitiva 90, 1990 ;

[Bouchon & Desprès 1990]
Bouchon B. and S. Desprès, Acquisition numérique/symbolique de connaissances graduelles, Actes des 3èmes journées du PRC-GDR Intelligence Artificielle, Paris, Hermès, 1990 ;

[Bouchon 92]
Bouchon-Meunier B., Fuzzy logic and knowledge representation using linguistic modifiers, in Fuzzy logic for the management of uncertainty (L.A. Zadeh, J. Kacprzyk, eds.), John Wiley & Sons, 1992, pp. 399-414

[Bouchon & Yao 92]
Bouchon-Meunier B. and Jia Yao, Linguistic modifiers and gradual membership to a category, International Journal of Intelligent Systems 7, pp 25-36, 1992 ;

[Bouchon & Valverde 93]
Bouchon-Meunier B. and L. Valverde, Analogical Reasoning and Fuzzy Resemblance, in Uncertainty in Intelligent Systems (B. Bouchon-Meunier, L. Valverde, R.R. Yager, eds.), Elsevier Science Pub, pp 247-255, 1993 ;

[Buisson & Prade 89]
Buisson B. and Prade H., Un système d'inférence pratiquant l'interpolation au moyen de règles à prédicats graduels - Une application au calcul de rations caloriques, Actes de la convention Intelligence Artificielle 88-89, Publ. Hermès, Paris, 1989 ;

[Dubois & Prade 87]
Dubois D. and H. Prade, Fuzzy numbers : an Overview, in J.C. Bezdek (dir.) Analysis of Fuzzy Information, vol. 1, Mathematics and Logic, CRC Press, Boca Raton, pp 3-39 ;

[Dubois et Prade 89]
Dubois D. and H. Prade, A typology of fuzzy "If...then...rules", in Proc. 3rd Inter. Fuzzy Systems Association Congress, Seattle, Wash., 1989, pp 782-785

[Dubois et Prade 90]
Dubois D. and H. Prade, Gradual inference rules in approximate reasoning, Information Sciences, 1990 ;

[Dubois et Prade 92]
Dubois D. and H. Prade, Fuzzy rules in knowledge-based systems -Modelling gradeness, uncertainty and preference-, in An introduction to fuzzy logic applications in intelligent systems, (R.R. Yager, L.A. Zadeh, eds.) Kluwer Academic Publishers, 1992 ;

[Dubois, Prade & Ughetto 96]
Dubois D., H. Prade and L. Ughetto, On Fuzzy Interpolation, in Systèmes de Règles Floues : cohérence, redondance et interpolation, Rapport IRIT/96-17-R, mai 1996 ;

[Koczy & Hirota 93]
Koczy L.T., Hirota K., Approximate Reasoning by Linear Rule Interpolation and General Approximation, International Journal of Approximate Reasoning, 9:197-225, 1993 ;

[Mizumoto 94]
Mizumoto M., Multifold Fuzzy Reasoning as Interpolative Reasoning, in Fuzzy Sets, Neural Networks and Soft Computing, (R.R. Yager and L.A. Zadeh, eds.) Van Nostrand Reinhold, New-York, 1994 ;

[Sanchez 84]
Sanchez E., Solution of fuzzy equations with extended operations, Fuzzy Sets Syst., 12, p237, 1984 ;

[Turksen I.B. & Zhao 88]
Turksen I.B. and Zhao Zhong, An Approximate Analogical Reasoning Approach Based on Similarity Measures, IEEE Transactions on Systems, Man, and Cybernetics, Vol. 18, n° 6, nov./dec. 1988 ;

[Ughetto,Dubois & Prade 95]
Ughetto L., Dubois D. and H. Prade, La cohérence des systèmes de règles graduelles, Rencontres francophones sur la logique floue, Cepadues Editions, 1995 ;

[Zadeh 72]
Zadeh L.A., A Fuzzy-Set Theoretic Interpretation of Linguistic Hedges, J. Cybernetics 2,2, pp. 4-34, 1972

[Zadeh 83]
Zadeh L.A., The role of fuzzy logic in the management of uncertainty in Expert Systems, Fuzzy Sets and Systems 11, pp. 199-227, 1983.

[Zadeh 90]
Zadeh L.A., Interpolative Reasoning Based on Fuzzy Logic and its Application to Control and System Analysis, invited lecture, abstract in the Proc. of the Int. Conf. on Fuzzy Logic & Neural Networks, Iizuka, Japan, 1990 .

Reasoning About Unpredicted Change and Explicit Time

Florence Dupin de Saint-Cyr and Jérôme Lang

IRIT - Université Paul Sabatier -
31062 Toulouse Cedex(France)
e-mail: {dupin, lang} @irit.fr

Abstract. Reasoning about unpredicted change consists in explaining observations by events; we propose here an approach for explaining time-stamped observations by surprises, which are simple events consisting in the change of the truth value of a fluent. A framework for dealing with surprises is defined. Minimal sets of surprises are provided together with time intervals where each surprise has occurred, and they are characterized from a model-based diagnosis point of view. Then, a probabilistic approach of surprise minimisation is proposed.

1 Introduction

Reasoning about time, action and change in full generality is a very complex task, and much research has been done in order to find solutions suited to specific subclasses of problems, obtained by making some simplificaying assumptions. Sandewall [15] provides a taxonomy for positioning the various subclasses, based on a lot of ontological and epistemological assumptions that may be present or absent in the specification of the subclass. In this paper we consider an integer representation of time and as in [15] we define *fluents* as propositions whose truth value evolves over time, *observations* as pieces of knowledge about the value of some fluents at some time points, *changes* as pairs made of a fluent f and a time point t such that the value of the fluent f at t differs from that at $t-1$. Changes may be either caused by *actions*, which are initiated by the agent, or by *events*, which are initiated by the world. This latter form of change will be called *unpredicted changes*; as they are unrelated to the performance of any action by the agent, they are totally out of the control of the agent. While many approaches in the literature focus on reasoning about change caused by actions[1], there has been significantly less attention on reasoning about unpredicted change.

In this article we propose a framework for detecting unpredicted changes from observations at different time points: the task consists thus in *explaining* observations by a set of elementary unpredicted changes. Note that it is analogous in many aspects to temporal diagnosis (an elementary change corresponding to a faulty component). We do not explicitly consider actions, since this is not our purpose; however, as we will show, our framework can handle actions as well,

---

[1] This is for instance the case for all approaches to solve the frame problem.

provided that they have a deterministic, unconditional result – thus the class of problems we consider is $\mathcal{K} - IS$ of Sandewall's taxonomy (see Section 2). Then, we propose some qualitative and quantitative criteria for ranking explanations. We propose a formal basis for handling problems in the class K-IS; namely, we argue that what has to be computed is the set of *minimal compact explanations*, where minimality means that these explanations do not contain any unnecessary change, and compactness means that they are given as concisely, as user-friendly as possible. Then we focus on computational issues and for this purpose we relate our framework to (temporal) model-based diagnosis: a problem of K-IS is translated into a system to diagnose and we show that minimal explanations correspond to minimal fault configurations. Thus, computational tools from model-based diagnosis can be reused for reasoning about unpredicted change. In Section 3, we compute a probability distribution on these explanations, which takes account of (i) the intrinsic tendency of each individual fluent to persist, and (ii) the time interval during which the change may have taken place – following the principle that the longer the time period since we last knew the value of a fluent, the fewer reasons we have to believe that this value has persisted. We illustrate this on a small example: *two cats live in a house, namely, Albert and Bart. At 6 p.m. , somebody told me that cat A. was sleeping inside the house. At 6h15 p.m., I knew that cat B. was sleeping inside the house. Now, it is 6h20 p.m., I can see a cat walking in the garden, but I can't recognize from where I am, if it is Albert or Bart.* The two intuitive minimal explanations we wish to get are the following: either *cat A. left the house between 6p.m. and 6h20 p.m.*, or *cat B. left the house between 6h15p.m. and 6h20*. Now, imagine we know the (subjective) prior probabilities for cat A. and cat B. to leave the house in a time unit (one minute); from these, and from the observations, we may compute the probabilities of all explanations. Intuitively, if the probabilities relative to cat A. and cat B. are identical, then the explanation *cat A. got outside* is preferred to *cat B. got outside*, since more time has passed between the last information we got and now – thus, a change is more likely to have occurred. However, the result may not be the same if the prior probability of cat A. getting outside is much lower than cat B's (cat A is lazier than B). What we propose in Section 3 is a simple method to compute most probable explanations in the case where probabilities of change are small within the time period considered, with a methodology similar to computing preferred diagnoses. We end up by positioning our approach w.r.t. related work and mention possible further extension.

2 Minimizing unpredicted change in K-IS

2.1 Background

From now on, the class of problems we consider coincides with the class $\mathcal{K} - IS$ in Sandewall's taxonomy. The symbol $\mathcal{K}$ means that all observations reported to the agent are correct, *i.e.*, true in the real world. I is for *inertia with integer time* and means that fluent values generally tend to persist. S is for *surprises*;

a surprise, or unpredicted change, is a change of the value of a fluent during a given interval, without any reason known by the agent; S means that surprises may occur (and are thus considered when computing completions of scenarios). Inertia implies that surprises occur with a low frequency, and should thus be *minimized* in order to be in accordance with observations. At this point we should say something about actions. In this paper we are not concerned with actions, and the general assumption we make throughout the paper is that the agent is totally passive (no action is performed)[2].

Let $\mathscr{L}$ be a propositional language built on a set of propositional variables $V = \{v_1, \ldots, v_n\}$. A *fluent* is a variable or its negation, *i.e.*, $f = v_i$ or $f = \neg v_i$. The set of all fluents is denoted by F; thus $F = \{v_1, \neg v_1, \ldots, v_n, \neg v_n\}$. Time is assumed to be discrete: the time scale T is a sequence of integers $\{0, 1, \ldots, t_{max}\}$ (it is assumed that between two consecutive time-points the system is completely inert).

If $\varphi \in \mathscr{L}$ and $t \in T$ then $[t]\varphi$ is an *elementary timed formula* (meaning that φ holds at t). (Complex) timed formulas are built from elementary timed formulas and usual connectives.

A *timed model* M is a mapping from $V \times T$ to $\{0, 1\}$: M assigns a truth value to each fluent at each time point. $M \models [t]v_i$ iff $M(v_i, t) = 1$. Satisfaction is extended to elementary timed formulas in the usual way, *i.e.*, $M \models [t]\varphi \wedge \varphi'$ iff $M \models [t]\varphi$ and $M \models [t]\varphi'$, etc., and then to complex timed formulas: $M \models [t]\varphi \wedge [t']\varphi'$ iff $M \models [t]\varphi$ and $M \models [t']\varphi'$, etc. Note that $[t]\varphi \wedge \varphi'$ is equivalent to $[t]\varphi \wedge [t]\varphi'$, $[t]\varphi \vee \varphi'$ is equivalent to $[t]\varphi \vee [t]\varphi'$, etc.

Lastly, we abbreviate $[t]\varphi \wedge [t+1]\varphi \cdots \wedge [t']\varphi$ by $[t, t']\varphi$.

Definition 1. A *scenario* Σ in K-IS consists in a set of timed-formulas

Definition 2. Let $f \in F$, $t \in T$, $t \neq 0$.

1. $\langle f, t \rangle$ is a *change* in M (notation: $C_M(f, t)$), iff $M \models [t-1]f$ and $M \models [t]\neg f$.
2. The *set of all changes in* M is $C(M) = \{\langle f, t \rangle \mid C_M(f, t)\}$.

When the exact time-point where the change occurred is not known, we affect an interval to a change occurrence and call it a *surprise*.

Definition 3. Let $f \in F$, $t, t' \in T$, $t < t'$. $\langle f, t, t' \rangle$ is a *surprise* with respect to M (notation: $S_M(f, t, t')$), iff $M \models [t]f$ and $M \not\models [t, t']f$.

Intuitively, $S_M(f, t, t')$ means that the truth value of f changed at least once between t and t' (note that it does not necessarily imply that $M \models [t']\neg f$, since f may have changed its truth value several times within $[t, t']$).

---

[2] However, the class $\mathcal{K} - IS$ does not exclude actions, but assumes that actions have deterministic and unconditional results, have only instantaneous effects, and that there is no dependency between fluents (which avoids the ramification problem). Under these assumptions, it is not hard to see that a deterministic, unconditional action can be seen as an observation: for instance, performing the action Load at time point t can be considered equivalent to observing its deterministic, unconditional result *loaded* at t.

If $t' = t + 1$, $S_M(f, t, t')$ is equivalent to $C_M(t')$; otherwise $(t' - t > 1)$ the surprise is said to be *time-ambiguous*.

The following property is straightforward:

Proposition 4. $S_M(f, t, t')$ *iff* $C_M(f, t+1)$ *or* $C_M(f, t+2)$ *or* $\ldots$ *or* $C_M(f, t')$.

2.2 Minimal explanations

The problem that we address is: given a scenario Σ, find the preferred timed-models satisfying Σ, *i.e.*, find the timed-models in which the changes are minimal: $\{M, M \models \Sigma$ and there is no M' such that $C(M') \subsetneq C(M)\}$.

Definition 5. A *pointwise explanation* PE for Σ is a set of changes $\{\langle f_i, t_i \rangle, i = 1 \cdots p\}$ such that $\Sigma \cup \{[t-1]f \rightarrow [t]f \mid f \in F, t \in T, t \neq 0, \langle f, t \rangle \notin PE\}$ is consistent.

Intuitively, PE consists in specifying which fluents change their value and when so as to be in accordance with Σ.

Example 6. Let us consider a set of pointwise observations Σ_1 such that:

$$\Sigma_1 : \begin{cases} [0] \ a \wedge b \\ [5] \ \neg a \end{cases}$$

$\{\langle a, 2 \rangle\}$ is a pointwise explanation for Σ_1: in order to explain the non-inert behaviour of the system, we must assume that the value of the fluent a changed between 0 and 5, (for instance at time point 2).

Note that $\{\langle a, 2 \rangle, \langle b, 3 \rangle\}$ is also a pointwise explanation, but, since it is not necessary to assume that b changed in order to explain Σ_1, this pointwise explanation is not minimal:

Definition 7. A pointwise explanation PE is *minimal* iff there is no pointwise explanation PE' strictly included in PE

Definition 8. An *explanation* for Σ is a set of surprises $\{< f_i, t_i, t'_i >, i = 1 \cdots p\}$ such that $\forall (t''_1, \ldots, t''_p) \in [t_1 + 1, t'_1] \times \cdots \times [t_p + 1, t'_p]$, $PE = \{\langle f_i, t''_i \rangle, i = 1 \cdots p\}$ is a pointwise explanation for Σ. We say that each of these PE is *covered* by E.

For any two explanations E and E', we say that E *covers* E' iff any pointwise explanation covered by E' is also covered by E.

Note that an explanation has to be understood as a disjunction of pointwise explanations. In the example above, $\{\langle a, 2, 5 \rangle\}$ is an explanation for Σ_1; corresponding to the pointwise explanations $\{\langle a, 2 \rangle, \langle a, 3 \rangle, \ldots, \langle a, 5 \rangle\}$.

Definition 9. An explanation E is *minimal* if any pointwise explanation which is covered by E is minimal.

The previous explanation $\{\langle a, 2, 5\rangle\}$ is minimal, but for instance $\{\langle a, 2, 5\rangle, \langle b, 1, 3\rangle\}$ or $\{\langle a, 2, 5\rangle, \langle a, 0, 5\rangle\}$ are explanations but they are not minimal.

Proposition 10. $E = \{\langle f_i, t_i, t_i'\rangle, i = 1 \ldots p\}$ *is an explanation (resp. a minimal explanation) for Σ iff*
$$\Sigma \cup \{[t]f \to [t+1]f \mid f \in F, t \in T, t \neq t_{max}, \text{ and } \nexists \langle f_i, t_i, t_i'\rangle \in E \text{ s.t. } t \in [t_i, t_i' - 1]\}$$
is consistent.

It is a minimal explanation iff the above set of formulas is maximally consistent in $\Sigma \cup \{[t]f \to [t+1]f \mid f \in F, t \in T, t \neq t_{max}\}$

Definition 11. A minimal explanation E is compact iff there is no explanation E' which strictly covers E.

In the previous example, the only compact minimal explanation for Σ_1 is $\{\langle a, 0, 5\rangle\}$.

Proposition 12. *If there is no disjunction in Σ (i.e., each pointwise observation $[t]\varphi$ is a conjunction of fluents) there is only one compact minimal explanation for Σ.*

Example 13.

$$\Sigma_2 : \begin{cases} [0] \ a \\ [5] \ a \vee c \\ [10] \ b \\ [15] \ \neg a \vee \neg b \\ [20] \ \neg c \end{cases}$$

There are 3 minimal compact explanations:
$E_1 = \{\langle a, 5, 15\rangle\}$, $E_2 = \{\langle b, 10, 15\rangle\}$ and $E_3 = \{\langle a, 0, 5\rangle, \langle c, 5, 20\rangle\}$ covering $10 + 5 + (5 \times 15)$ pointwise explanations.

- $\{\langle a, 10, 15\rangle\}$ is an explanation for Σ_2 but it is not compact since $\{\langle a, 5, 15\rangle\}$ is an explanation for Σ_2.
- $\{\langle a, 0, 15\rangle\}$ is not an explanation for Σ_2 since, for instance, $\{\langle a, 1\rangle\}$ is not a pointwise explanation for Σ_2.
- $\{\langle a, 0, 15\rangle, \langle c, 5, 20\rangle\}$ is an explanation for Σ_2 but it is not minimal since $\{\langle a, 6\rangle, \langle c, 7\rangle\}$ is covered by it and is not a minimal pointwise explanation (since $\{\langle a, 6\rangle\}$ alone is a pointwise explanation).

Definition 14. Let $Cme(\Sigma)$ be the set of all compact minimal explanations for Σ.

Proposition 15. *For each minimal pointwise explanation PE for Σ, there is a unique compact minimal explanation E which covers it.*

Thus, $Cme(\Sigma)$ is the most concise expression covering all minimal pointwise explanations.

Proposition 16. $Cme(\Sigma) = \{\varnothing\} \Leftrightarrow \Sigma \cup \{[t-1]f \to [t]f \mid f \in F, t \in T, t \neq 0\}$ *is consistent.*[3]

Intuitively, $Cme(\Sigma) = \{\varnothing\}$ means that assuming that all fluents kept their value throughout T does not lead to any inconsistency (with Σ).

2.3 Computing minimal explanations

The method consists in completing the set of observations Σ by a set of persistence assumptions in order to express that, by default, fluents persist during the interval between two consecutive observations. In order to add only "relevant" persistence assumptions, we identify, for each variable, the "relevant time points" where we have some observation regarding it.

Definition 17. The *set of relevant variables* $V(\Sigma)$ with respect to the set of observations Σ is defined by: $V(\Sigma) = \{v \in V \mid v \text{ appears in } \Sigma\}$.

The *set of relevant time points* $RT_\Sigma(v)$ to a variable $v \in V(\Sigma)$ w.r.t. Σ is defined by: $RT_\Sigma(v) = \{t \in T \mid [t]\varphi \in \Sigma \text{ and } v \text{ appears in } \varphi\}$

Let $RT_\Sigma^*(v)$ be the set of relevant time points to a variable v except the last one: $RT_\Sigma^*(v) = RT_\Sigma(v) \setminus \{max\{t \in T, t \in RT_\Sigma(v)\}\}$

In example 13: $RT_{\Sigma_2}(a) = \{0,5,15\}$, $RT_{\Sigma_2}(b) = \{10,15\}$ and $RT_{\Sigma_2}(c) = \{5,20\}$.

Definition 18. The set of persistence axioms of Σ is defined by:

$$PERS(\Sigma) = \bigcup_{v \in V(\Sigma), t \in RT_\Sigma^*(v)} \{[t]v \to [next(v,t)]v, \quad [t]\neg v \to [next(v,t)]\neg v\}$$

where $next(v,t) = \min\{t' \in RT_\Sigma(v), t' > t\}$ is the next relevant time point for v after t.

Definition 19.

- Let $MaxCons(\Sigma)$ be the set of all maximal subsets of $PERS(\Sigma)$ consistent with Σ
- let $Cmc(\Sigma) = \bigcup_{X \in MaxCons(\Sigma)}(PERS(\Sigma) \setminus X)$
- let $Candidates(\Sigma) = \{\{\langle f, t, t' \rangle \mid [t]f \to [t']f \in X\}, X \in Cmc(\Sigma)\}$.

Proposition 20.

- *Any set of surprises in $Candidates(\Sigma)$ is a minimal explanation for Σ.*
- *Every pointwise explanation for Σ is covered by a unique set of surprise in $Candidates(\Sigma)$*

---

[3] Similarly, in model-based diagnosis, when the system description and the observations are consistent with the non-failure assumptions, the preferred diagnosis is $\{\varnothing\}$ (no faulty component).

This principle of finding minimal explanations corresponds exactly to finding minimal sets of faulty components in model-based diagnosis (or candidates in ATMS [4] terminology). Hence, the computation of minimal explanation can be done by well-known algorithms from these fields.

Example 21. (Example 13 continued) The relevant time-points for Σ_2 are : $RT_{\Sigma_2}(a) = \{0, 5, 15\}$, $RT_{\Sigma_2}(b) = \{10, 15\}$ and $RT_{\Sigma_2}(c) = \{5, 20\}$.

$$
PERS(\Sigma_2) : \begin{cases}
[0]a \rightarrow [5]a \\
[5]a \rightarrow [15]a \\
[0]\neg a \rightarrow [5]\neg a \\
[5]\neg a \rightarrow [15]\neg a \\
[10]b \rightarrow [15]b \\
[10]\neg b \rightarrow [15]\neg b \\
[5]c \rightarrow [20]c \\
[5]\neg c \rightarrow [20]\neg c
\end{cases}
$$

When computing the minimal set of persistence assumptions which has to be removed from $PERS$ in order to make it consistent with Σ_2, we obtain the 3 minimal candidates $\{\langle a, 5, 15 \rangle\}$, $\{\langle b, 10, 15 \rangle\}$ and $\{\langle a, 0, 5 \rangle, \langle c, 5, 20 \rangle\}$.

Note that it is not always the case that the computed explanations are compact: consider $\Sigma_3 = \{[0]a, [5]a \vee b, [10]\neg a\}$. We find the two minimal explanation $\{\langle a, 0, 5 \rangle\}$ and $\{\langle a, 5, 10 \rangle\}$ which are not compact (since they can be compacted into $\{\langle a, 0, 10 \rangle\}$).

If we wish to obtain the *compact* minimal explanations from $Candidates(\Sigma)$, we may define a compactification procedure, where precise description is outside the scope of the paper. Intuitively, this procedure consists in compacting explanations pairwise and iterate until no pair of explanations is compactable. This operation is confluent and its result is exactly $Cme(\Sigma)$.

3 Probabilities and unpredicted change

We assume now that probabilistic information about fluents truth values and persistence is available. This will enable us to compute the probability of each explanation, and thus will help us ranking them. We will take account of 2 important factors which can influence quantitatively the persistence: time duration and the intrinsic tendency of each fluent to persist (some fluents persist longer than other, e.g., *sleeping* persists usually longer than *eating an apple*).

3.1 Markovian fluents

For the sake of simplicity, throughout this Section we assume fluents are *mutually independent* and *Markovian*[4]. We will consider stationary fluents in some cases but we will not impose this restriction to the whole section.

---

[4] These assumptions mean that timed-variables can be structured in a very simple temporal Bayesian network, which has the following form (where $\{v_1, \ldots, v_n\} \in V$

Definition 22.

- fluents are *mutually independent* iff $\forall t, t', \forall f \in F, \forall f' \in F \setminus \{f, \neg f\}$,

$$Pr([t]f \wedge [t']f') = Pr([t]f).Pr([t']f').$$

- f is *stationary* iff $\forall t, t', \forall f \in F, \quad Pr([t]f) = Pr([t']f) = p_f$.
- f is *Markovian* iff $\forall t, \quad Pr([t+1]f \mid [t]f \wedge H_{0 \to t-1}) = Pr([t+1]f \mid [t]f) = 1 - \varepsilon_f$.

Where $H_{0 \to t-1}$ is the history of the system from 0 to $t - 1$.

The independence and Markovian assumption imply that the only necessary data are, for each propositional variable v and each time-point t, a *prior probability* of v, $p_{t,v}$ and the *elementary change probabilities* ε_v and $\varepsilon_{\neg v}$. We will show later that in many cases prior probabilities are not necessary. Generally $\varepsilon_v \neq \varepsilon_{\neg v}$, *i.e.*, the persistence of a fluent may be different from the persistence of its negation (think of f =alive or f =bell-ringing).

If fluents are stationary then for each variable it is enough to consider a prior probability p_v which is independent of the time-point. Moreover, in this case, there is a relationship between the persistence of a fluent and its negation:

Proposition 23. *If f is stationary then $\varepsilon_f.p_f = \varepsilon_{\neg f}.p_{\neg f}$*

Proof. $Pr([t]f \wedge [t+1]\neg f) = Pr([t+1]\neg f \mid [t]f).Pr([t]f) = \varepsilon_f.p_f$. And $Pr([t]\neg f \vee [t+1]f) = Pr([t]\neg f) + Pr([t+1]f) - Pr([t]\neg f \wedge [t+1]f) = p_{\neg f} + p_f - \varepsilon_{\neg f}.p_{\neg f} = 1 - \varepsilon_{\neg f}.p_{\neg f}$.

Obviously, some fluents are *not* Markovian:

- either because they do not tend to persist independently of t: consider a clock which always rings around 7 a.m., then the fluent *ringing* is not Markovian (nor stationary). However, the fact that the clock is on the bedside table is usually a Markovian fluent (unless you throw it away each time it rings).
- or because their tendency to persist depend on their history: consider a bus which period is 15 minutes, the probability that the bus will come depends on the length of the interval since it last came. If you know that no bus has came for 10 minutes, you have more chance than the bus will arrive immediately than if you know that no bus has come for 5 minutes. Clearly, $Pr([t]\neg bus\_coming \mid [t-1]\neg bus\_coming) = \frac{14}{15}$, while $Pr([t]bus\_coming \mid [t-k, t-1]\neg bus\_coming) = \frac{k}{15}$. And thus *bus_coming* is not Markovian as all periodic fluents (however it may reasonably be assumed stationary without considering the rush hour) .

---

are the propositional variables):
$[0]v_1 \to [1]v_1 \to \cdots \to [t_{max}]v_1$
$[0]v_2 \to [1]v_2 \to \cdots \to [t_{max}]v_2$

$\vdots$

$[0]v_n \to [1]v_n \to \cdots \to [t_{max}]v_n$ (Dean and Kanazawa 1989 [6]).

Among all Markovian fluents, some can be distinguished:

- persistent fluents, such that $\varepsilon_f = 0$ (e.g. f =dead). A stationary and persistent fluent is a degenerate case (either it never change, or the probability of its opposite is null).
- chaotic fluents, such that: $Pr([t+1]f \mid [t]f) = Pr([t+1]f)$ (these fluents have no tendency to persist, since knowing this fluent true at t does not make it true at $t+1$ more probable). It means that $\varepsilon_f = 1 - p_{t+1,f}$, hence, chaotic fluents must be stationary.
- switching fluents (provided that the time unit is well chosen) such that $\varepsilon_f = 1$

Now, given the prior probability distribution p_f and the switch probability $Pr([t+1]\neg f \mid [t]f) = \varepsilon_f$ (which are constants, independent of t), it is possible to compute the prior probability of surprise for each stationary Markovian fluent between any time points.

Definition 24. Notation: $Pr(\langle f, t, t' \rangle) = Pr([t]f \wedge \neg[t, t']f)$ is the probability of the surprise $\langle f, t, t' \rangle$.

Proposition 25. If f is stationary and Markovian, then

$$Pr(\langle f, t, t+n \rangle) = (1 - (1 - \varepsilon_f)^n).p_f$$

Proof.

$$
\begin{aligned}
Pr(\langle f, t, t+n \rangle) &= Pr(\langle f, t, t+n \rangle) \mid [t]f).Pr([t]f) \\
&\quad + \quad Pr(\langle f, t, t+n \rangle) \mid [t]\neg f).Pr([t]\neg f) \\
&= (1 - Pr([t]\neg f \vee [t+1, t+n]f \mid [t]f)).p_f + 0 \\
&= (1 - Pr([t+2, t+n]f \mid [t+1]f).Pr([t+1]f \mid [t]f)).p_f \\
&= (1 - (1 - \varepsilon_f)^n).p_f
\end{aligned}
$$

Indeed, it looks natural that, knowing the truth value of a fluent at a given instant, the more time has passed, the more likely the fluent had its truth value changed.

3.2 Highly persistent fluents and probabilities of explanations

From proposition 25 and the independence assumption it is also possible to compute the prior probability of an explanation (minimal or not) :

$$Pr(\{\langle f_i, t_i, t'_i \rangle, i = 1 \ldots n\}) = \prod_{i=1}^{n}(1 - (1 - \varepsilon_{f_i})^{t'_i - t_i}).p_{f_i}$$

Now, these prior probabilities have to be conditioned on the observations (Σ) in order to get posterior probabilities of all explanations : if E is an explanation, then $Pr(E \mid \Sigma) = \frac{Pr(E \wedge \Sigma)}{Pr(\Sigma)}$.

In the general case these probabilities are not straightforward to compute, mainly because $Pr([t]f \wedge [t']\neg f)$ may be much lower than $Pr(\langle f, t, t'\rangle)^5$. Moreover, non-minimal explanations may have a significant probability, even in the case where $\{\varnothing\}$ is an explanation, which contradicts the principle of minimal change. This is because elementary probabilities of change (the ε_f's) may be large. Computing probabilities of change over an interval ($Pr([t]f \wedge [t']\neg f)$) requires computing the probability that an odd number of pointwise surprises occurred between t and t'. This is left outside of the scope of the paper which focuses on infrequent change (thus to be minimized). As expected intuitively, change is minimized if and only if the ε_f's are all infinitely small; more precisely, if for any considered interval $[t, t']$, $(t - t')\varepsilon_f \ll 1$.

Informally, f is *highly persistent* w.r.t. interval $[t, t']$ iff $\varepsilon_f \ll \frac{1}{t'-t}$.

Proposition 26. *Assume that all fluents are highly persistent w.r.t.* $[0, t_{max}]$, *then:*

1. $Pr(\langle f, t, t'\rangle) \approx Pr([t]f \wedge [t']\neg f) = (t' - t)\varepsilon_f.p_f + o(\varepsilon_f)$.
2. $\sum_{E \in Cme(\Sigma)} Pr(E \mid \Sigma) \approx_{\varepsilon_f \to 0, \forall f} 1$.

Item 26.2 means that non-minimal explanations for Σ have a very low posterior probability and thus they can be neglected. In particular, if $\{\varnothing\}$ is an explanation for Σ then $Pr(\{\varnothing\} \mid \Sigma) \approx 1$. Proposition 26.1 intuitively means that if f changed at least once its truth value within $[t, t']$ ($\langle f, t, t'\rangle$) then it changed exactly once (and thus $\neg f$ holds at t'). Now probabilities of minimal explanations are easy to compute (provided that all fluents are highly persistent). We start by giving two detailed examples.

Example 27. (cat A and B, pure prediction)

$$\Sigma_4 : \begin{cases} [0] \ a \\ [15] \ b \\ [20] \ \neg a \vee \neg b \end{cases}$$

This means that at 6h, cat A is sleeping inside the house, at 6h15, cat B is also sleeping in the house, and at 6h20, one cat is not sleeping inside the house. $Pr(\{\langle a, 0, 20\rangle\}) \approx 20\varepsilon_a p_a$, $Pr(\{\langle b, 15, 20\rangle\}) \approx 5\varepsilon_b p_b$.

$$Pr(\{\langle a, 0, 20\rangle \mid \Sigma_4\}) = \frac{Pr(\langle a, 0, 20\rangle \wedge \Sigma_4)}{Pr(\Sigma_4)}$$

$$Pr(\langle a, 0, 20\rangle \wedge \Sigma_4) = Pr([0]a \wedge [20]\neg a \wedge [15]b)$$
$$= Pr([0]a \wedge [20]\neg a).Pr([15]b)$$
$$\approx 20\varepsilon_a.p_a.p_b.$$

---

[5] Indeed $\langle f, t, t'\rangle$ is true if f holds at t and changed its value *at least once* between t and t', while $[t]f$ and $[t']\neg f$ is true if f changed its value *an odd number of times* between t and t'.

Similarly, $Pr(\{\langle b, 15, 20\rangle \mid \Sigma_4\}) \approx 5\varepsilon_b.p_b.p_a$

$$Pr(\Sigma_4) = Pr([0]a \wedge [15]b \wedge [20]\neg a \vee \neg b)$$
$$= Pr([0]a \wedge [15]b \wedge [20]\neg a) + Pr([0]a \wedge [15]b \wedge [20]\neg b)$$
$$- Pr([0]a \wedge [15]b \wedge [20]\neg a \wedge [20]\neg b)$$
$$\approx 20\varepsilon_a.p_a.p_b + 5\varepsilon_b.p_b.p_a - (20\varepsilon_a.p_a.p_b)(5\varepsilon_b.p_b.p_a)$$
$$\text{(the last term being negligible)}$$
$$\approx Pr(\langle a, 0, 20\rangle \wedge \Sigma_4) + Pr(\langle b, 15, 20\rangle \wedge \Sigma_4)$$

$Pr(\langle a, 0, 20\rangle \mid \Sigma_4) \approx \dfrac{20\varepsilon_a.p_a.p_b}{\varepsilon_a.p_a.p_b + 5\varepsilon_b.p_b.p_a} = \dfrac{1}{1 + \frac{\varepsilon_b}{4.\varepsilon_a}}$

and $Pr(\langle b, 15, 20\rangle \mid \Sigma_4) \approx \dfrac{1}{1 + \frac{4.\varepsilon_a}{\varepsilon_b}}$.

Remarkably, these probabilities do not depend on the prior probabilities p_a and p_b, which is natural (since a and b are known to hold at 0 and 15, respectively); this will always be the case for pure prediction problems where disjunctive observations appear only at the last time point. Note that, if $\varepsilon_a = \varepsilon_b$, we get $Pr(\langle a, 0, 20\rangle = \frac{4}{5}$ and $Pr(\langle b, 0, 20\rangle) = \frac{1}{5}$.[6]

Example 28. (pure postdiction)

$$\Sigma_5 : \begin{cases} [0]\; a \vee b \\ [5]\; \neg a \\ [20]\; \neg b \end{cases}$$

$$Pr(\langle a, 0, 5\rangle \wedge \Sigma_5) = Pr([0]a \wedge [5]\neg a \wedge [20]\neg b) \approx 5\varepsilon_a.p_a.(1 - p_b).$$
Similarly, $Pr(\langle b, 0, 20\rangle \wedge \Sigma_5) \approx 20\varepsilon_b.p_b.(1 - p_a).$
$$Pr(\Sigma_5) \approx Pr(\langle a, 0, 5\rangle \wedge \Sigma_5) + Pr(\langle b, 0, 20\rangle \wedge \Sigma_5).$$
$$Pr(\langle a, 0, 5\rangle \mid \Sigma_5) = \frac{5\varepsilon_a.p_a.(1 - p_b)}{5\varepsilon_a.p_a.(1 - p_b) + 20\varepsilon_b.p_b.(1 - p_a)}.$$

If we suppose the fluents stationary (it is natural to do so considering the usual behaviour of cats) then using proposition 23 we obtain the same kind of results as in the previous case : $Pr(\langle a, 0, 5\rangle \mid \Sigma_5) = 1/5$ and $Pr(\langle b, 0, 20\rangle \mid \Sigma_5) = 4/5$.

Example 29. Take Σ_2 of example 13, we have:
$Pr(\langle a, 5, 15\rangle) \mid \Sigma_2) = \dfrac{1}{1 + \frac{\varepsilon_b}{2\varepsilon_a} + \frac{15}{2}\varepsilon_c.\frac{p_c}{1 - p_c}}$
$Pr(\langle b, 10, 15\rangle \mid \Sigma_2) = \dfrac{1}{1 + 2\frac{\varepsilon_a}{\varepsilon_b} + 15\frac{\varepsilon_a.\varepsilon_c}{\varepsilon_b}.\frac{p_c}{1 - p_c}}$
and $Pr(\langle a, 0, 5\rangle \wedge \langle c, 5, 20\rangle \mid \Sigma_2) = \dfrac{1}{1 + \frac{2(1 - p_c)}{15(\varepsilon_c)} + \frac{\varepsilon_b}{15\varepsilon_a.\varepsilon_c}.\frac{1 - p_c}{p_c}}$

Note that if $\varepsilon_a = \varepsilon_b = \varepsilon_c$ and are very small, $Pr(\langle a, 5, 15\rangle \mid \Sigma_2) \approx \frac{2}{3}$, $Pr(\langle b, 10, 15\rangle \mid \Sigma_2) \approx \frac{1}{3}$ and $Pr(\langle a, 0, 5\rangle \wedge \langle c, 5, 20\rangle \mid \Sigma_2) \approx 0$. These results are independent of the prior probabilities p_a, p_b, p_c.

---

[6] More generally, for $\Sigma = \{[t_a]a, [t_b]b, [t']\neg a \vee \neg b\}$ and $\varepsilon_a = \varepsilon_b$, we get $Pr(\langle a, t_a, t'\rangle = \dfrac{t' - t_a}{(t' - t_a) + (t' - t_b)}$ and $Pr(\langle b, t_b, t'\rangle = \dfrac{t' - t_b}{(t' - t_a) + (t' - t_b)}$.

4 Discussion

The original aspects of our work are that (i) minimal changes (explanations) are provided together with the interval when change may have occurred, which makes the representation concise, and (ii) knowing probabilities of change of fluents from one point to the subsequent one, and fluent prior probabilities, we compute the probability of each explanation and thus we rank them accordingly. We showed that minimizing change is coherent with a probabilistic handling of change only if probabilities of change are very small (with this assumption, only minimal explanations may have significant probabilities). This probabilistic handling of change needs explicit and metric time, and is coherent with the intuitive idea that the longer the time interval during which a change may have occurred, the more likely it has actually occurred within this interval. Many related works share some features with ours. The first approach dealing with unpredicted change ("fluents that may change by themselves") is Lifschitz and Rabinov's [11]; using the situation calculus and circumscription, they minimize the set of unpredicted changes from one state to another, giving, thus, something analogous to our pointwise explanations (without the explicit temporal dimension). Our notion of "surprise" is borrowed from Sandewall's ontology [15] where surprises are defined as unpredicted changes with low frequency; in an earlier version of his book [14], he proposed to select among competing surprise sets by associating to each fluent a penalty (the higher the penalty, the more unlikely the fluent may change unpredictedly), preferred surprises sets minimizing the sum of the penalties of changing fluents. Dealing with penalties is very analogous to dealing with probabilities, especially in the case of infinitely small probabilities [7], thus our approach extends his by taking interval durations into account in defining the penalties.

A probabilistic handling of unpredicted change using explicit, metric time has been proposed for temporal projection by Dean and Kanazawa [6] and later on by Hanks and McDermott [10] in a more general setting. In [6], unpredicted change is modelled with exponential survivor functions, asserting that $Pr([t + \Delta t]f \mid [t]f) = \exp(\frac{-\Delta t}{\lambda_f})$. They do not assume that probabilities of change are small, and thus do not minimize change. We recall that the reason why we want change to be minimized is that it leads to a concise list of explanations, since only minimal explanations may have significant probabilities. This has to be related to the fact that, in [6], probabilities of change are only used for temporal projection, namely, computing the probability of given fluents at the latest time point, but they are not used to deal with explaining scenarios and the reasoning is only forward (it has for instance no postdiction capabilities). Lastly, [6] framework assumes that observations are atomic (no disjunctive observations are allowed while our approach allows the observation of any propositional formula). The same remarks apply to [10].

Other related approaches include work in the domains of probabilistic abduction and model-based diagnosis, temporal diagnosis, and also belief update. Probabilistic (and cost-based) abduction ([12], [8], [5]) attach probabilities to explanations from prior probabilities of hypotheses (faults in model-based diagnosis) and generally an independence assumption amongst them. In the domain of temporal model-based diagnosis, Friedrich and Lackinger [9] attach time intervals to fault configurations, meaning that a given component is in a failure mode during the whole interval – contrarily to our surprises which are instantaneous (recall that the interval in a surprise is understood disjunctively, not conjunctively); probabilities are attached to temporal diagnoses, the evolution of the system is modelled by Markov chains (with transition probabilities for the different modes of a component). Console et al. [2] attach time points to fault configurations; their approach is extended in a probabilistic setting by Portinale [13] who also models the evolution of the system by a Markov process. Cordier and Thiébaux's event-based diagnosis [3] consider sequences of events and compute probabilities over diagnoses from priori probabilities of events (without metric time). Boutilier's generalised update operators [1] handles unpredicted change, explaining observations by events, ranked according to their plausibility, but without explicit time (he only considers two time points). If we specialize our framework to a time scale with only two time points t_{before} and t_{now}, then it is possible to show that we obtain something very close to a generalized update operator, where the allowed events are the elementary surprises.

Further work will include the handling of dependent fluents (speciality D in Sandewall's ontology) – in the probabilistic case they may for instance be structured in a Bayesian network – and the integration of surprise minimisation together with a handling of actions with alternative effects (speciality A in Sandewall's ontology).

Acknowledgement

We would like to thank Didier Dubois and Henri Prade for their comments which helped improve this paper.

References

1. C. Boutilier. Generalised update : belief change in dynamic settings. In *Proc. of the 14th IJCAI*, volume 2, pages 1550–1556, 1995.
2. L. Console, L. Portinale, D. Theseider Dupré, and P. Torasso. Diagnostic reasoning across different time points. In *Proc. of ECAI'92*, pages 369–373, 1992.
3. M.O. Cordier and S. Thiébaux. Event-based diagnosis for evolutive systems. In *Proc. of DX'94*, 1994.
4. J. de Kleer. An assumption-based TMS. *Artificial Intelligence*, 28:127–162, 1986.
5. J. de Kleer and B. Williams. Diagnosing multiple faults. *Artificial Intelligence*, 32(1):97–130, 1987.
6. T. Dean and K. Kanazawa. Persistence and probabilistic projection. In *Proc. of IEEE Trans. on Systems, Man and Cybernetics*, volume 19(3), pages 574–585, 1989.
7. F. Dupin de Saint-Cyr, J. Lang, and T. Schiex. Penalty logic and its link with Dempster-Shafer theory. In *Proc. of the 10th Conf. on Uncertainty in Artificial Intelligence*, pages 204–211. Morgan Kaufmann, July 1994.

8. T. Eiter and G. Gottlob. The complexity of logic-based abduction. In *Proc. STACS'93*, 1993.
9. G. Friedrich and F. Lackinger. Diagnosing temporal misbehaviour. In *Proc. IJCAI'91*, pages 1116–1122, 1991.
10. S. Hanks and D. McDermott. Modelling and uncertain world i : symbolic and probabilistic reasoning about change. *Artificial Intelligence*, 66:1–55, 1994.
11. V. Lifschitz and A. Rabinov. Things that change by themselves. In *Proc. of the 11th IJCAI*, pages 864–867, Detroit, Michigan, 1989.
12. D. Poole. Representing diagnostic knowledge for probabilistic Horn abduction. In *Proc. of the 12th IJCAI*, pages 1129–1135, 91.
13. L. Portinale. Modeling uncertain temporal evolutions in model-based diagnosis. In *Proc. Uncertainty in AI'92*, pages 244–251, 1992.
14. E. Sandewall. Features and fluents: a systematic approach to the representation of knowledge about dynamical systems. Technical Report LITHIDA-R-92-30, Linköping University, 1992.
15. E. Sandewall. *Features and Fluents*. Oxford University Press, 1994.

Non-elementary Speed-Ups in Default Reasoning*

Uwe Egly and Hans Tompits

Technische Universität Wien
Abt. Wissensbasierte Systeme 184/3
Treitlstraße 3, A–1040 Wien, Austria
e-mail: [uwe,tompits]@kr.tuwien.ac.at

Abstract. Default logic is one of the most prominent formalizations of common-sense reasoning. It allows "jumping to conclusions" in case that not all relevant information is known. However, theoretical complexity results imply that default logic is (in the worst case) computationally harder than classical logic. This somehow contradicts our intuition about common-sense reasoning: default rules should help to *speed up* the reasoning process, and not to slow it down. In this paper, we show that default logic can indeed deliver the goods. We consider a sequent-calculus for first-order default logic and show that the presence of defaults can tremendously simplify the search of proofs. In particular, we show that certain sequents have only long "classical" proofs, but short proofs can be obtained by using defaults.

1 Introduction

In recent years, the complexity of non-monotonic reasoning techniques has been thoroughly investigated (e.g. [9, 10, 11, 15], an overview is given in [6]). It turned out that almost all non-monotonic systems are computationally more involving than classical logic.[2] This somehow contradicts our intuition about common-sense reasoning. One of the hopes in the development of this area was that non-monotonic rules should *speed up* the reasoning process, and not to slow it down. (Indeed, this point has recently been stressed by Reiter in his contribution for the panel at Nonmon96 [14].) The various complexity results describe how non-monotonic reasoning behaves *in the worst case*, but there are only few works which investigate how systems *can profit from non-monotonic rules*. One of the few exceptions are the works by Cadoli, Donini and Schaerf [4, 5]. Roughly speaking, they show that, unless the polynomial hierarchy collapses, propositional non-monotonic systems allow a "super-compact" representation of

---

* The authors would like to thank Thomas Eiter for his useful comments on an earlier version of this paper.

[2] The seemingly single exception of this pattern is the propositional modal logic S4. In [16] it is shown that non-monotonic S4 is at the second level of the polynomial hierarchy, whereas it is well-known that monotonic S4 is PSPACE-complete.

knowledge as compared with (monotonic) classical logic. On the other hand, by employing a *proof-theoretical* method, it is shown in [8] that, for some cases, first-order circumscription enables a non-elementary speed-up of proof length.

In this paper, we consider first-order default logic (actually, a certain undecidable fragment of it). We use a generalization, B, of Bonatti's cut-free sequent calculus [3] for propositional default logic. B consists of three parts, namely a classical LK-calculus, a decidable consistency check and certain default inference rules. The consistency check is formalized as a "complementary" sequent calculus, where "anti-sequents" of the form $\Gamma \not\vdash \Theta$ state the non-derivability of Θ from Γ.

The basic idea of our approach is the following. We compare in the calculus B the minimal proof length of "purely classical" proofs, i.e., of proofs *without* applications of default rules, with proofs where defaults are applied. More precisely, we show that there are infinite sequences $(C_k)_{k \geq 1}$, $(T_k)_{k \geq 1}$ of sequents with the following properties:

1. The minimal proof length of C_k in B is *non-elementary* in k, i.e., the proof length of C_k is of the order $s(k-1)$, where $s(0) := 1$ and $s(n+1) := 2^{s(n)}$ for all $n \geq 0$.
2. The minimal proof length of T_k in B is at most quadrupel-exponential in k.

C_k represents the fact that a certain formula G_k is "classically" derivable, whereas T_k represents the fact that G_k is proved with the help of a default. Although the derivation of G_k with the default rule involves both deriving the prerequisite of the default and checking consistency, for sufficiently large k, the length of this proof in B is much shorter than the purely classical proof. The reason is that the length of any cut-free proof of G_k is non-elementary in the size of the input formula, but short proofs can be obtained by using the cut rule – and the default simulates this instance of cut. Moreover, since for first-order cut-free sequent calculi, the size of the search space is elementarily related to the minimal proof length, a non-elementary decrease of the search space is also achieved.

A motivation of our method can be given as follows. Usually, non-monotonic techniques are applied in case a classical proof cannot be found. Although this is a reasonable procedure in decidable systems, it is not appropriate for undecidable systems like first-order logic. Indeed, if we integrate non-monotonic rules into first-order theorem provers, we have to invoke non-monotonic mechanisms *after a certain amount of time*, whenever the goal formula has not been proven classically up to this point. Accordingly, it may happen that a formula is provable both classically and with the help of non-monotonic rules. Our result shows therefore that, in certain cases, the theorem prover may easier find a proof because the presence of defaults yields a much smaller search space.

The paper is organized as follows. In Section 2 we introduce basic definitions and notations. Moreover, the generalization B of Bonatti's sequent calculus is described. In Section 3 we prove our main result, and in Section 4 we conclude with some general remarks.

2 Preliminaries

Throughout this paper we use a first-order language consisting of *variables, constants, function symbols, predicate symbols, logical connectives, quantifiers* and *punctuation symbols*. *Terms* and *formulae* are defined according to the usual formation rules. A term is *closed* if it contains no variable; a formula is closed if it contains no free variable. Furthermore, the alphabet of our language shall also include the *logical constants* $\top$ and $\bot$, representing truth and falsehood, respectively.

Let t be the function with $t(x, 0) := 2^x$ and $t(x, n + 1) := 2^{t(x,n)}$ for all $n \in \mathbb{N}_0$. In particular, let s be the non-elementary function $s(n) := t(0, n)$. Note that for each fixed $n \in \mathbb{N}_0$, the function $t(\cdot, n)$ is elementary.

2.1 Default Logic

Definition 1 [13]. Let $A, B_1, \ldots, B_n, C$ be first-order formulae. A *default d* is an expression of the form

$$\frac{A : B_1, \ldots, B_n}{C} \ .$$

A is the *prerequisite*, $B_1, \ldots, B_n$ are the *justifications* and C is the *consequent* of the default d. If $n = 0$, the default is *justification-free*, and if $A = \top$, the default is *prerequisite-free*. If $A, B_1, \ldots, B_n, C$ are closed, the default is *closed*. A set of defaults D is closed iff all defaults in it are closed.

A *default theory* is an ordered pair $T = \langle W, D \rangle$, where W is a set of closed formulae, called the *premises* of T, and D is a set of defaults. We say that T is *closed* iff D is closed, and T is *finite* iff both W and D are finite. □

Intuitively, for a default theory $T = \langle W, D \rangle$, W represents absolute (though in general incomplete) knowledge about the world ("hard facts"), while D represents *defeasible* knowledge ("rules of thumb").

For simplicity, defaults will also be written in the form $(A : B_1, \ldots, B_n/C)$.

In the following, $\text{Th}(S)$ denotes the *deductive closure* of a set S of closed formulae, i.e., $\text{Th}(S)$ is the set of all closed formulae derivable from S (in some fixed calculus).

Definition 2 [13]. Let $T = \langle W, D \rangle$ be a closed default theory. For any set S of closed formulae, let $\Gamma(S)$ be the *smallest* set K of closed formulae obeying the following conditions:

1. $K = \text{Th}(K)$;
2. $W \subseteq K$;
3. If $(A : B_1, \ldots, B_n/C) \in D$, $A \in K$ and $\neg B_1 \notin S, \ldots, \neg B_n \notin S$, then $C \in K$.

An *extension* of T is a set E of closed formulae such that $\Gamma(E) = E$, i.e., which is a *fixed point* of the operator Γ. □

A closed default theory may have none, one or several extensions. A default theory of the form $\langle W, \emptyset \rangle$ has exactly one extension, namely $\mathrm{Th}(W)$. In general, an extension of a default theory $T = \langle W, D \rangle$ is always of the form $\mathrm{Th}(W \cup G)$, for some set G such that

$$G \subseteq \mathrm{CONS}(D) := \{ C \mid (A : B_1, \ldots, B_n / C) \in D \}.$$

2.2 Classical Proof Machinery

A (*classical*) *sequent* S is an ordered tuple of the form $\Gamma \vdash \Sigma$, where Γ, Σ are finite sequences of first-order formulae. Γ is the *antecedent* of S, and Σ is the *succedent* of S. Semantically, a sequent $A_1, \ldots, A_n \vdash B_1, \ldots, B_m$ is true in an interpretation iff the formula $(\bigwedge_{i=1}^{n} A_i) \to (\bigvee_{i=1}^{m} B_i)$ is true. A sequent S is *valid* iff it is true in every interpretation.

As proof system we use the (cut-free) sequent calculus LK. Axioms (or *initial sequents*) are the two sequents $\vdash \top$ and $\bot \vdash$, and sequents of the form $A \vdash A$, where A is any first-order formula. The inference rules of LK are given below, consisting of the *logical rules*, the *quantifier rules* and the *structural rules*.

SYSTEM LK: LOGICAL RULES

$$\frac{\Gamma_1, A, \Gamma_2 \vdash \Sigma}{\Gamma_1, (A \wedge B), \Gamma_2 \vdash \Sigma} \wedge l_1 \qquad \frac{\Gamma_1, A, \Gamma_2 \vdash \Sigma}{\Gamma_1, (B \wedge A), \Gamma_2 \vdash \Sigma} \wedge l_2$$

$$\frac{\Gamma \vdash \Sigma_1, A, \Sigma_2 \quad \Lambda \vdash \Pi_1, B, \Pi_2}{\Gamma, \Lambda \vdash \Sigma_1, \Pi_1, (A \wedge B), \Sigma_2, \Pi_2} \wedge r$$

$$\frac{\Gamma_1, A, \Gamma_2 \vdash \Sigma_1 \quad \Pi_1, B, \Pi_2 \vdash \Sigma_2}{\Gamma_1, \Pi_1, (A \vee B), \Gamma_2, \Pi_2 \vdash \Sigma_1, \Sigma_2} \vee l$$

$$\frac{\Gamma \vdash \Sigma_1, A, \Sigma_2}{\Gamma \vdash \Sigma_1, (A \vee B), \Sigma_2} \vee r_1 \qquad \frac{\Gamma \vdash \Sigma_1, A, \Sigma_2}{\Gamma \vdash \Sigma_1, (B \vee A), \Sigma_2} \vee r_2$$

$$\frac{\Gamma \vdash \Sigma_1, A, \Sigma_2 \quad \Gamma_1, B, \Gamma_2 \vdash \Pi}{\Gamma, (A \to B), \Gamma_1, \Gamma_2 \vdash \Sigma_1, \Sigma_2, \Pi} \to l$$

$$\frac{\Gamma_1, A, \Gamma_2 \vdash \Sigma_1, B, \Sigma_2}{\Gamma_1, \Gamma_2 \vdash \Sigma_1, (A \to B), \Sigma_2} \to r$$

$$\frac{\Gamma_1, \Gamma_2 \vdash \Sigma_1, A, \Sigma_2}{\Gamma_1, \neg A, \Gamma_2 \vdash \Sigma_1, \Sigma_2} \neg l \qquad \frac{\Gamma_1, A, \Gamma_2 \vdash \Sigma_1, \Sigma_2}{\Gamma_1, \Gamma_2 \vdash \Sigma_1, \neg A, \Sigma_2} \neg r$$

System LK: Quantifier Rules

$$\frac{\Sigma_1, A(t), \Sigma_2 \vdash \Gamma}{\Sigma_1, \forall x A(x), \Sigma_2 \vdash \Gamma} \; \forall l \qquad\qquad \frac{\Gamma \vdash \Sigma_1, A(y), \Sigma_2}{\Gamma \vdash \Sigma_1, \forall x A(x), \Sigma_2} \; \forall r$$

$$\frac{\Sigma_1, A(y), \Sigma_2 \vdash \Gamma}{\Sigma_1, \exists x A(x), \Sigma_2 \vdash \Gamma} \; \exists l \qquad\qquad \frac{\Gamma \vdash \Sigma_1, A(t), \Sigma_2}{\Gamma \vdash \Sigma_1, \exists x A(x), \Sigma_2} \; \exists r$$

$\forall r$ and $\exists l$ must fulfill the *eigenvariable condition*: the (free) variable y must not occur in $\Gamma, \Sigma_1, \Sigma_2$. For $\forall l$ and $\exists r$, the term t must be free for x in A.

System LK: Structural Rules
WEAKENING

$$\frac{\Gamma_1, \Gamma_2 \vdash \Sigma}{\Gamma_1, A, \Gamma_2 \vdash \Sigma} \; wl \qquad\qquad \frac{\Gamma \vdash \Sigma_1, \Sigma_2}{\Gamma \vdash \Sigma_1, A, \Sigma_2} \; wr$$

CONTRACTION

$$\frac{\Gamma_1, A, \Gamma_2, A, \Gamma_3 \vdash \Sigma}{\Gamma_1, A, \Gamma_2, \Gamma_3 \vdash \Sigma} \; cl \qquad\qquad \frac{\Gamma \vdash \Sigma_1, A, \Sigma_2, A, \Sigma_3}{\Gamma \vdash \Sigma_1, A, \Sigma_2, \Sigma_3} \; cr$$

We also use the following two systems: LK_0 is LK without quantifier rules and restricted to sequents which contain only quantifier-free closed formulae, and LK_{cut} is LK together with the *cut rule*:

$$\frac{\Gamma_1 \vdash \Sigma_1, A \qquad A, \Gamma_2 \vdash \Sigma_2}{\Gamma_1, \Gamma_2 \vdash \Sigma_1, \Sigma_2} \; cut$$

Let α be a proof in LK, LK_{cut} or LK_0. We say that α is *simple* iff it has no applications of a weakening rule and its initial sequents contain atomic formulae only. Furthermore, the *length* l_α of α is the number of sequents occurring in α. The *height* h_α of α is the number of sequents occurring on the longest branch.

A formula occurrence A is *positive* (*negative*) in a formula B iff the number of implicit or explicit negation signs preceding A in B is even (odd). An occurrence Q of a quantifier is positive (negative) in a formula B iff the occurrence of the scope of Q is positive (negative) in B.[3]

In a sequent $S := A_1, \ldots, A_n \vdash B_1, \ldots, B_m$, any A_i ($1 \le i \le n$) occurs negatively and any B_j ($1 \le j \le m$) occurs positively in the sequent. A formula occurring positively (negatively) in an A_i occurs negatively (positively) in S, and

---

[3] The labels "positive occurrence" and "negative occurrence" are sometimes referred to as the *polarity* of a formula (quantifier) occurrence in the literature.

a formula occurring positively (negatively) in a B_j occurs positively (negatively) in S; the same holds for quantifier occurrences.

If $\forall x$ occurs positively (negatively) in a formula B then $\forall x$ is called a *strong* (*weak*) quantifier; if $\exists x$ occurs positively (negatively) in B then $\exists x$ is called a weak (strong) quantifier. Similarly for quantifier occurrences in sequents.

We assume that the reader is familiar with the notion of *Skolemization*. Recall that the Skolemization of a formula A eliminates *strong* quantifiers from A by introducing appropriate *Skolem functions*.[4] The Skolemization of A is denoted by $\mathsf{SK}(A)$.

For a sequent $S := A_1, \ldots, A_n \vdash B_1, \ldots, B_m$, the Skolemization $\mathsf{SK}(S)$ of S is defined as the sequent $A'_1, \ldots, A'_n \vdash B'_1, \ldots, B'_m$, where $(\bigwedge_{i=1}^n A'_i) \rightarrow (\bigvee_{i=1}^m B'_i)$ is the Skolemization of $(\bigwedge_{i=1}^n A_i) \rightarrow (\bigvee_{i=1}^m B_i)$.

Let A be a first-order formula. The *Herbrand universe of A*, symbolically $\mathcal{HU}(A)$, is the set of all closed terms constructed from the constants and function symbols occurring in A. (If A contains no constant symbol, choose an arbitrary constant from the alphabet.) The Herbrand universe of a sequent $S := A_1, \ldots, A_n \vdash B_1, \ldots, B_m$, denoted by $\mathcal{HU}(S)$, is defined to be the Herbrand universe of $(\bigwedge_{i=1}^n A_i) \rightarrow (\bigvee_{i=1}^m B_i)$.

Definition 3. Let S be a valid sequent of the form $A_1, \ldots, A_n \vdash B_1, \ldots, B_m$ containing weak quantifiers only and let $A_1^0, \ldots, A_n^0, B_1^0, \ldots, B_m^0$ be the sequent formulae without quantifiers. Furthermore, let $A_j^1, \ldots, A_j^{i_j}$ be closed substitution instances of the A_j^0 over $\mathcal{HU}(S)$, and let $B_k^1, \ldots, B_k^{p_k}$ be closed substitution instances of the B_k^0 over $\mathcal{HU}(S)$ (for $j = 1, \ldots, n$ and $k = 1, \ldots, m$). A valid sequent S' of the form

$$A_1^1, \ldots, A_1^{j_1}, \ldots, A_n^1, \ldots, A_n^{j_n} \vdash B_1^1, \ldots, B_1^{p_1}, \ldots, B_m^1, \ldots, B_m^{p_m}$$

is called a *Herbrand sequent* of S. □

If S is of the form $A_1, \ldots, A_n \vdash B_1, \ldots, B_m$, the number of formulae occurring in the sequent S is denoted by $\lambda(S)$. Here, $\lambda(S) = m + n$.

2.3 Proof Machinery for Default Logic

Bonatti [3] introduced a sequent calculus for propositional default logic formalizing *brave* reasoning, i.e., a (propositional) formula A is in some extension of a (propositional) default theory iff a certain sequent is provable in his system. This calculus consists of three parts, namely a propositional sequent calculus, a propositional anti-sequent calculus (the *complementary* system), and certain default inference rules.

Since the construction of an extension involves as an integral part a *consistency check*, a formal system for default logic must somehow take into account that certain formulae are *not provable* (to wit, in order that a default is applied, it is necessary that the negations of its justifications must not be derivable).

---

[4] Hence, Skolem functions replace *eigenvariables*.

In Bonatti's system, this feature is governed by means of the complementary sequent calculus. The formal objects of the complementary system are *anti-sequents* $\Gamma \not\vdash \Theta$ representing *invalid* statements. More exactly, an anti-sequent $\Gamma \not\vdash \Theta$ is provable in the complementary sequent calculus iff the corresponding classical sequent $\Gamma \vdash \Theta$ is invalid. In general, two logical systems are *complementary* iff objects derivable in one system are *not* derivable in the other system and vice versa.[5]

Since we use first-order formulae for our result, we must slightly generalize Bonatti's system. However, due to the undecidability of first-order logic, we cannot have a sound and complete formalization of first-order non-theorems. If one wants to construct such a sound and complete axiomatization of invalid statements, only a *decidable* subclass of first-order formulae can be used. In fact, for our purpose, it suffices to generalize only the "classical part" of Bonatti's system, but we let the "complementary part" be essentially propositional (actually, quantifier free). Bonatti's soundness and completeness results thus hold quite trivially for this construction.

We will introduce now this slightly generalized version of Bonatti's system.

In the sequel of this paper, Γ and Θ are finite sequences of quantifier-free closed formulae, Σ is a finite sequence of closed formulae (possibly *with* quantifiers), and Δ is a finite sequence of quantifier-free closed defaults.

Definition 4.

1. An *anti-sequent* is an ordered pair of the form $\Gamma \not\vdash \Theta$.
2. A (*brave*) *default sequent* is an ordered quadruple of the form $\Gamma; \Delta \mathrel{\vdash\mkern-10mu\sim} \Sigma; \Theta$.

$\square$

For a sequence s of arbitrary objects (say, formulae or defaults), $\hat{s}$ is the set of elements of s. Furthermore, for the sake of readability, the empty sequence occurring in a default sequent will be denoted by ϵ.

Definition 5.

1. An anti-sequent $\Gamma \not\vdash \Theta$ is *true* iff there is a first-order interpretation such that the classical sequent $\Gamma \vdash \Theta$ is false.
2. A default sequent $\Gamma; \Delta \mathrel{\vdash\mkern-10mu\sim} \Sigma; \Theta$ is *true* iff there is an extension E of the default theory $\langle \hat{\Gamma}, \hat{\Delta} \rangle$ such that $\hat{\Sigma} \subseteq E$ and $\hat{\Theta} \cap E = \emptyset$. $\square$

Obviously, $\Gamma \not\vdash \Theta$ is true iff $\Gamma \vdash \Theta$ is invalid.

The *complementary sequent calculus* LK_0^c is defined as follows. The axioms of LK_0^c are sequents of the form $\Phi \not\vdash \Psi$, where Φ and Ψ are finite sequences of closed atomic formulae such that $\hat{\Phi} \cap \hat{\Psi} = \emptyset$ and $(\{\bot\} \cap \hat{\Phi}) \cup (\{\top\} \cap \hat{\Psi}) = \emptyset$. The inference rules of LK_0^c comprise of the logical rules and the structural rules, which are given below.

---

[5] The term "complementary proof system" is due to Varzi [17].

SYSTEM LK$_0^c$: LOGICAL RULES

$$\frac{\Gamma_1, A, \Gamma_2, B, \Gamma_3 \not\vdash \Theta}{\Gamma_1, (A \wedge B), \Gamma_2, \Gamma_3 \not\vdash \Theta} \wedge l^c$$

$$\frac{\Gamma \not\vdash \Theta_1, A, \Theta_2}{\Gamma \not\vdash \Theta_1, (A \wedge B), \Theta_2} \wedge r_1^c \qquad \frac{\Gamma \not\vdash \Theta_1, A, \Theta_2}{\Gamma \not\vdash \Theta_1, (B \wedge A), \Theta_2} \wedge r_2^c$$

$$\frac{\Gamma_1, A, \Gamma_2 \not\vdash \Theta}{\Gamma_1, (A \vee B), \Gamma_2 \not\vdash \Theta} \vee l_1^c \qquad \frac{\Gamma_1, A, \Gamma_2 \not\vdash \Theta}{\Gamma_1, (B \vee A), \Gamma_2 \not\vdash \Theta} \vee l_2^c$$

$$\frac{\Gamma \not\vdash \Theta_1, A, \Theta_2, B, \Theta_3}{\Gamma \not\vdash \Theta_1, (A \vee B), \Theta_2, \Theta_3} \vee r^c$$

$$\frac{\Gamma_1, \Gamma_2 \not\vdash \Theta_1, A, \Theta_2}{\Gamma_1, (A \rightarrow B), \Gamma_2 \not\vdash \Theta_1, \Theta_2} \rightarrow l_1^c \qquad \frac{\Gamma_1, B, \Gamma_2 \not\vdash \Theta}{\Gamma_1, (A \rightarrow B), \Gamma_2 \not\vdash \Theta} \rightarrow l_2^c$$

$$\frac{\Gamma_1, A, \Gamma_2 \not\vdash \Theta_1, B, \Theta_2}{\Gamma_1, \Gamma_2 \not\vdash \Theta_1, (A \rightarrow B), \Theta_2} \rightarrow r^c$$

$$\frac{\Gamma_1, \Gamma_2 \not\vdash \Theta_1, A, \Theta_2}{\Gamma_1, \neg A, \Gamma_2 \not\vdash \Theta_1, \Theta_2} \neg l^c \qquad \frac{\Gamma_1, A, \Gamma_2 \not\vdash \Theta_1, \Theta_2}{\Gamma_1, \Gamma_2 \not\vdash \Theta_1, \neg A, \Theta_2} \neg r^c$$

SYSTEM LK$_0^c$: STRUCTURAL RULES
CONTRACTION

$$\frac{\Gamma_1, A, \Gamma_2, A, \Gamma_3 \not\vdash \Theta}{\Gamma_1, A, \Gamma_2, \Gamma_3 \not\vdash \Theta} cl^c \qquad \frac{\Gamma \not\vdash \Theta_1, A, \Theta_2, A, \Theta_3}{\Gamma \not\vdash \Theta_1, A, \Theta_2, \Theta_3} cr^c$$

Theorem 6 [2]. *The anti-sequent $\Gamma \not\vdash \Theta$ is provable in* LK$_0^c$ *iff it is true.*

Corollary 7 [2]. *The anti-sequent $\Gamma \not\vdash \Theta$ is provable in* LK$_0^c$ *iff the classical sequent $\Gamma \vdash \Theta$ is not provable in* LK$_0$.

The length of an LK$_0^c$-proof is defined analogously to the classical case, i.e., as the number of anti-sequents occurring in it.

Next we introduce the *default sequent calculus* B. It consists of classical sequents, anti-sequents and default sequents. Furthermore, it incorporates the systems LK for classical sequents and LK$_0^c$ for anti-sequents. Following Bonatti [3], axioms of B are of the form $\Gamma; \epsilon \vdash \epsilon; \epsilon$, for any finite sequence Γ containing only quantifier-free closed formulae. The additional inference rules of B are given below.

SYSTEM B: LOGICAL RULES

$$\frac{\Gamma \vdash A}{\Gamma; \epsilon \vdash A; \epsilon} \; l_1 \qquad\qquad \frac{\Gamma \nvdash A}{\Gamma; \epsilon \vdash \epsilon; A} \; l_2$$

$$\frac{\Gamma; \epsilon \vdash \Sigma_1; \Theta_1 \quad \Gamma; \epsilon \vdash \Sigma_2; \Theta_2}{\Gamma; \epsilon \vdash \Sigma_1, \Sigma_2; \Theta_1, \Theta_2} \; cu$$

$$\frac{\Gamma; \Delta_1, \Delta_2 \vdash \Sigma; \Theta_1, A, \Theta_2}{\Gamma; \Delta_1, (A : B_1, \dots, B_n/C), \Delta_2 \vdash \Sigma; \Theta_1, \Theta_2} \; d_1$$

$$\frac{\Gamma; \Delta_1, \Delta_2 \vdash \Sigma_1, \neg B, \Sigma_2; \Theta}{\Gamma; \Delta_1, (A : \dots, B, \dots / C), \Delta_2 \vdash \Sigma_1, \Sigma_2; \Theta} \; d_2$$

$$\frac{\Gamma_1, C, \Gamma_2; \Delta_1, \Delta_2 \vdash \Sigma_1; \Theta_1, \neg B_1, \dots, \neg B_n, \Theta_2 \quad \Gamma_1, \Gamma_2; \epsilon \vdash \Sigma_2, A; \epsilon}{\Gamma_1, \Gamma_2; \Delta_1, (A : B_1, \dots, B_n/C), \Delta_2 \vdash \Sigma_1, \Sigma_2; \Theta_1, \Theta_2} \; d_3$$

SYSTEM B: STRUCTURAL RULES

CONTRACTION

$$\frac{\Gamma_1, A, \Gamma_2, A, \Gamma_3; \Delta \vdash \Sigma; \Theta}{\Gamma_1, A, \Gamma_2, \Gamma_3; \Delta \vdash \Sigma; \Theta} \; cl_1^d$$

$$\frac{\Gamma; \Delta_1, (A : B_1, \dots, B_n/C), \Delta_2, (A : B_1, \dots, B_n/C), \Delta_3 \vdash \Sigma; \Theta}{\Gamma; \Delta_1, (A : B_1, \dots, B_n/C), \Delta_2, \Delta_3 \vdash \Sigma; \Theta} \; cl_2^d$$

$$\frac{\Gamma; \Delta \vdash \Sigma_1, A, \Sigma_2, A, \Sigma_3; \Theta}{\Gamma; \Delta \vdash \Sigma_1, A, \Sigma_2, \Sigma_3; \Theta} \; cr_1^d \qquad\qquad \frac{\Gamma; \Delta \vdash \Sigma; \Theta_1, A, \Theta_2, A, \Theta_3}{\Gamma; \Delta \vdash \Sigma; \Theta_1, A, \Theta_2, \Theta_3} \; cr_2^d$$

Theorem 8 [3]. *The default sequent $\Gamma; \Delta \vdash \Sigma; \Theta$ is derivable in B iff it is true.*

As usual, the length of a proof in B is the number of sequents occurring in it (and this includes *a fortiori* the length of the LK-proofs for classical sequents occurring as premises in applications of rule l_1, and the length of LK_0^c-proofs of anti-sequents occurring as premises in applications of rule l_2).

3 Main Result

In this section, we show how defaults can speed up proofs. We use a sequence of formulae for which Orevkov [12] showed a non-elementary lower bound on

$$C_1(\alpha, \beta, \gamma) := \exists z \; (p(\alpha, \beta, z) \wedge p(z, \beta, \gamma))$$
$$C_2(\alpha, \beta, \gamma) := \exists y \; (p(y, b_0, \alpha) \wedge C_1(\beta, y, \gamma))$$
$$C := \forall u \forall v \forall w \; (C_2(u, v, w) \rightarrow p(v, u, w))$$
$$B_0(\alpha) := \exists v_0 \; p(b_0, \alpha, v_0)$$
$$B_{i+1}(\alpha) := \exists v_{i+1} \; (p(b_0, \alpha, v_{i+1}) \wedge B_i(v_{i+1}))$$
$$A_0(\alpha) := \forall w_0 \exists v_0 \; p(w_0, \alpha, v_0)$$
$$A_{i+1}(\alpha) := \forall w_{i+1} \; (A_i(w_{i+1}) \rightarrow \overline{A}_{i+1}(w_{i+1}, \alpha))$$
$$\overline{A}_0(\alpha, \delta) := \exists v_0 \; p(\alpha, \delta, v_0)$$
$$\overline{A}_{i+1}(\alpha, \delta) := \exists v_{i+1} \; (A_i(v_{i+1}) \wedge p(\alpha, \delta, v_{i+1}))$$

Fig. 1. Abbreviations used in the following.

proof length in (cut-free) LK, but which possess short $\mathsf{LK_{cut}}$-proofs using cut. We show that these short $\mathsf{LK_{cut}}$-proofs yield short B-proofs if some quantifier-free instances of the cut formulae can be derived by applying defaults, but any such B-proof *without* defaults has a non-elementary lower bound on proof length.

Definition 9. Let F_k occur in the infinite sequence of formulae $(F_k)_{k \in \mathbb{N}}$ where

$$F_k := \forall b \; ((\forall w_0 \exists v_0 \; p(w_0, b, v_0) \wedge$$
$$\forall uvw \; (\exists y \; (p(y, b, u) \wedge \exists z \; (p(v, y, z) \wedge p(z, y, w))) \rightarrow p(v, u, w)))$$
$$\rightarrow \exists v_k \; (p(b, b, v_k) \wedge \exists v_{k-1} \; (p(b, v_k, v_{k-1}) \wedge \ldots \wedge \exists v_0 \; p(b, v_1, v_0) \ldots))). \qquad \square$$

Intuitively, $p(x, y, z)$ represents the relation $x + 2^y = z$, and F_k "computes" certain numbers using a recursive definition of this relation.

Abbreviations shown in Figure 1 are used in the following in order to simplify the notation. Using these abbreviations, F_k looks as follows:

$$F_k = \forall b \; ((A_0(b) \wedge C) \rightarrow B_k(b))$$

The formulae F_k ($k \in \mathbb{N}$) have the following properties with respect to proof length.

Proposition 10 [12]. *Let $(F_k)_{k \in \mathbb{N}}$ be the infinite sequence of formulae defined above.*

1. *There is an $\mathsf{LK_{cut}}$-proof ψ_k of $\vdash F_k$ such that $l_{\psi_k} \le c \cdot k$, for some constant c.*
2. *For any (cut-free) LK-proof α of $\vdash F_k$ it holds that $h_\alpha \ge 2 \cdot s(k) + 1$.*

Thus, eliminating the cut yields a non-elementary increase of proof length. On the other hand, each $\mathsf{LK_{cut}}$-proof using only closed cut formulae can easily be converted into a (cut-free) LK-proof of another end sequent, by simply replacing applications of the cut rule by $\rightarrow l$. Moreover, this new proof has the same length as the original proof with cut. Applying a similar method to the $\mathsf{LK_{cut}}$-proof ψ_k of $\vdash F_k$ yields an LK-proof with an end sequent S_k. For technical reasons, which will become clear later, we need a *simple* LK-proof of this sequent S_k. (Recall that a proof is simple iff it has no weakenings and its initial sequents are atomic.) The short $\mathsf{LK_{cut}}$-proof of $\vdash F_k$ presented in Orevkov [12] is not simple. It both contains weakenings and its initial sequents are not atomic. As presented in [7], the $\mathsf{LK_{cut}}$-proof ψ_k can easily be modified to a *weakening-free* $\mathsf{LK_{cut}}$-proof ψ'_k of $\vdash F_k$, whose length is also linear in k. Using a result from [1], eliminating the non-atomic sequents is straightforward and yields only an exponential increase of proof length.[6] However, we will not present the weakening-free proof of $\vdash F_k$ in all details, but sketch the proof stressing the relevant details.

If $k \geq 2$, there are two kinds of weakening-free LK-derivations, namely β_k and $\delta_k(t)$, which are relevant in the following (the slightly varying case $k = 1$ is omitted for simplicity). These cut-free derivations β_k and $\delta_k(t)$ have end sequents $A_0(b_0), C \vdash A_k(b_0)$ and $A_0(b_0), C, A_k(t) \vdash B_k(t)$, respectively. The $\mathsf{LK_{cut}}$-proof ψ'_k of $\vdash F_k$ is as follows:

$$\cfrac{\cfrac{\beta_k \qquad \delta_k(b_0)}{A_0(b_0), C \vdash B_k(b_0)} \; cut, cl, cl}{\vdash \forall b\,((A_0(b) \wedge C) \rightarrow B_k(b))} \wedge l_1, \wedge l_2, cl, \rightarrow r, \forall r$$

This derivation contains no weakenings and has one application of the cut rule, where the cut formula $A_k(b_0)$ has the free variable b_0. We eliminate the cut in this short $\mathsf{LK_{cut}}$-proof by introducing an implication into the antecedent. Since the cut formula $A_k(b_0)$ is open, an application of $\forall l$ is also added. The resulting proof, ϕ_k, looks as follows:

$$\cfrac{\cfrac{\beta_k \qquad \delta_k(b_0)}{\forall b\,(A_k(b) \rightarrow A_k(b)), A_0(b_0), C \vdash B_k(b_0)} \; \rightarrow l, cl, cl, \forall l}{\forall b\,(A_k(b) \rightarrow A_k(b)) \vdash \forall b\,((A_0(b) \wedge C) \rightarrow B_k(b))} \wedge l_1, \wedge l_2, cl, \rightarrow r, \forall r$$

Clearly, the length of ϕ_k is linearly bounded by k.

Let S_k be the end sequent of ϕ_k. Since the consistency check in B requires quantifier-free formulae, we must eliminate the quantifiers in the antecedent of S_k. We do this in two steps: first by eliminating strong quantifiers by Skolemization, and secondly by using a Herbrand sequent of the Skolemized result to get rid of the remaining quantifiers.

To that end, however, we must first make the proof ϕ_k simple.

Proposition 11 [1]. *Let α be a weakening-free LK-proof and let m be the maximal logical complexity of the formulae occurring in initial sequents of α. Then there is a weakening-free LK-proof β such that all initial sequents in β are atomic and $l_\beta < 5 \cdot l_\alpha \cdot 2^m$.*

---

[6] It is worth mentioning that certain definitions (and consequently certain proofs) of [1] are incomplete; a correct treatment can be found in [7].

Since the logical complexity of a formula occurring in an initial sequent of ψ_k is at most exponential in k, a simple calculation yields the following result.

Corollary 12. *Let ϕ_k be the short weakening-free* LK-*proof of S_k. Then there is a simple* LK-*proof ϕ'_k of S_k such that $l_{\phi'_k} < 2^{2^{d \cdot k}} = \mathrm{t}(d \cdot k, 1)$, for some constant d.*

Now we are ready to use the following result, which is a variant of Lemma 4.2 from [1].

Proposition 13 [7]. *Let γ be a simple* LK-*proof of $\vdash A$. Then there is a simple* LK-*proof γ' of $\vdash$ SK(A) and $l_{\gamma'} \leq l_{\gamma}$.*

Corollary 14. *Let ϕ'_k be the simple, cut-free proof of S_k from Corollary 12. Then there is a simple, cut-free proof ϕ''_k of* SK(S_k) *such that $l_{\phi''_k} < 2^{2^{d \cdot k}} = \mathrm{t}(d \cdot k, 1)$, for some constant d.*

Next we eliminate the remaining quantifiers in SK(S_k) by using a Herbrand sequent of SK(S_k). To estimate the length of this Herbrand sequent, we invoke yet another result from [1]. (Recall that $\lambda(\cdot)$ denotes the length of a sequent.)

Proposition 15. *Let γ be a simple* LK-*proof of a sequent S without strong quantifier introductions, and let l denote the number of initial sequents occurring in γ. Then there exists a Herbrand sequent S' of S such that $\lambda(S') \leq 2^{2 \cdot l}$.*

Since the number of initial sequents of a proof is always bounded by the length of this proof, we immediately obtain the following result.

Corollary 16. *There exists a Herbrand sequent S'_k of* SK(S_k) *such that $\lambda(S'_k) \leq 2^{2^{2^{e \cdot k}}} = \mathrm{t}(e \cdot k, 2)$, for some constant e.*

Now, the Herbrand sequent S'_k contains no quantifier whatsoever and can thus be proved by purely propositional means. By utilizing a suitable *search procedure*, each propositional (and thus each quantifier-free) sequent can be proved in LK$_0$ in an exponential number of steps.

Proposition 17. *Let S be a valid sequent containing only quantifier-free formulae. Let $\lambda(S)$ be the length of S and let m be the maximal logical complexity of the sequent formulae of S. Then there is an* LK$_0$-*proof α of S such that $l_{\alpha} \leq 2^{2^{m} \cdot \lambda(S) + m}$.*

Since each formula in S'_k is itself at most exponentially large, we obtain the following estimate:

Corollary 18. *There is an* LK$_0$-*proof ϱ_k of S'_k such that $l_{\varrho_k} < 2^{2^{2^{2^{f \cdot k}}}} = \mathrm{t}(f \cdot k, 3)$, for some constant f.*

Let us have a closer look at $\mathsf{SK}(S_k)$ and S'_k. By choosing suitable Skolem functions, $\mathsf{SK}(S_k)$ is of the form $\forall b\ (A'_k(b) \rightarrow A''_k(b)) \vdash \mathsf{SK}(F_k)$. Let D^0_k be the formula $\forall b\ (A'_k(b) \rightarrow A''_k(b))$ without quantifiers, and similarly let E^0_k be $\mathsf{SK}(F_k)$ without quantifiers. Then the Herbrand sequent S'_k of $\mathsf{SK}(S_k)$ is of the form $D^1_k, \ldots, D^{i_k}_k \vdash E^1_k, \ldots, E^{j_k}_k$ $(i_k + j_k \leq \mathsf{t}(e \cdot k, 2))$, where the D^l_k $(l = 1, \ldots, i_k)$ are closed instantiations of D^0_k and the E^m_k $(m = 1, \ldots, j_k)$ are closed instantiations of E^0_k.

It can be shown that the LK_0-proof ϱ_k of S'_k can be transformed into an LK-proof of $\bigwedge^{i_k}_{l=1} D^l_k \vdash \mathsf{SK}(F_k)$ $(i_k \leq \mathsf{t}(e \cdot k, 2))$, whose length is likewise bounded by $\mathsf{t}(g \cdot k, 3)$ for some constant g. If we denote the antecedent of this sequent by D_k, we obtain:

Lemma 19. *Let $D_k \vdash \mathsf{SK}(F_k)$ be the sequent described above. There is an LK-proof ϑ_k of this sequent whose length is bounded by $\mathsf{t}(g \cdot k, 3)$, for some constant g.*

The default sequent we are interested in is $T_k := \epsilon; (\top : D_k/D_k) \;\vdash\; \mathsf{SK}(F_k); \epsilon$. Its corresponding default theory is $T = \langle \emptyset, \{(\top : D_k/D_k)\}\rangle$; the single extension of T is $\mathrm{Th}(\{D_k\})$ and contains $\mathsf{SK}(F_k)$.

We proceed with a B-proof of T_k. The consistency check with respect to the default $(\top : D_k/D_k)$ is utilized in B by means of an LK^c_0-proof of $D_k \not\vdash \neg D_k$. First of all, it is rather straightforward to establish that D_k is satisfiable, hence $D_k \vdash \neg D_k$ is not valid. Furthermore, since D_k is quantifier-free and closed, it follows by Theorem 6 that $D_k \not\vdash \neg D_k$ must be provable in LK^c_0.

It remains to estimate the length of such an LK^c_0-proof. Recall that a suitable systematic search for an LK_0-proof of a given quantifier-free sequent S returns an exponentially large LK_0-proof, given that S is indeed LK_0-provable. However, if S is *not* provable in LK_0, *at least one branch of the systematic search tree constitutes (up to eliminating some structural rules) an LK^c_0-proof of S*. The size of such an LK^c_0-proof can be estimated as follows.

Proposition 20. *Let S be a true anti-sequent. Let $\lambda(S)$ be the length of S and let m be the maximal logical complexity of the sequent formulae of S. Then there is an LK^c_0-proof γ of S such that $l_\gamma \leq \lambda(S) \cdot (2^m - 1)$.*

Corollary 21. *There is an LK^c_0-proof ξ_k of the anti-sequent $D_k \not\vdash \neg D_k$ whose length is bounded by $2^{2^{2^{h \cdot k}}} = \mathsf{t}(h \cdot k, 2)$, for some constant h.*

Now we have all ingredients to present a (short) B-proof of the default sequent $\epsilon; (\top : D_k/D_k) \;\vdash\; \mathsf{SK}(F_k); \epsilon$:

$$
\cfrac{\cfrac{\cfrac{\overset{\displaystyle \vartheta_k}{D_k \vdash \mathsf{SK}(F_k)}}{D_k; \epsilon \vdash \mathsf{SK}(F_k); \epsilon}\, l_1 \qquad \cfrac{\overset{\displaystyle \xi_k}{D_k \not\vdash \neg D_k}}{D_k; \epsilon \vdash \epsilon; \neg D_k}\, l_2}{D_k; \epsilon \vdash \mathsf{SK}(F_k); \neg D_k}\, cu \qquad \cfrac{\cfrac{\vdash \top}{\epsilon; \epsilon \vdash \top; \epsilon}\, l_1}{}}{\epsilon; (\top : D_k/D_k) \vdash \mathsf{SK}(F_k); \epsilon}\, d_3
$$

The length of this proof is bounded by $2^{2^{2^{2^{2^{a \cdot k}}}}}$, for some constant a. On the other hand, deriving $\mathsf{SK}(F_k)$ "classically" in B is tantamount to deriving the default sequent $C_k := \epsilon; \epsilon \hspace{1mm} \vdash\!\!\!\sim \hspace{1mm} \mathsf{SK}(F_k); \epsilon$ in B. However, each such B-proof of C_k is non-elementary in k. This is a consequence of the fact that not just $\vdash F_k$ has only non-elementary long cut-free proofs (cf. Proposition 10), but so has $\vdash \mathsf{SK}(F_k)$.

Proposition 22 [12]. *Any* LK-*proof of* $\vdash \mathsf{SK}(F_k)$ *has height* $\geq r \cdot \mathsf{s}(k-1)$, *for some constant* r.

We obtain therefore our main result, which is as follows:

Theorem 23. *There is an infinite sequence of first-order formulae* $(F_k)_{k \geq 1}$ *and an infinite sequence of quantifier-free closed defaults* $(d_k)_{k \geq 1}$ *such that the following holds:*

1. *There is a* B-*proof of* $T_k := \epsilon; d_k \hspace{1mm} \vdash\!\!\!\sim \hspace{1mm} \mathsf{SK}(F_k); \epsilon$ *whose length is bounded by* $\mathsf{t}(a \cdot k, 3)$, *for some constant* a.
2. *Each* B-*proof of* $C_k := \epsilon; \epsilon \hspace{1mm} \vdash\!\!\!\sim \hspace{1mm} \mathsf{SK}(F_k); \epsilon$ *has length greater than* $b \cdot \mathsf{t}(0, k-1)$, *for some constant* b.

Thus, for sufficiently large k, proving $\mathsf{SK}(F_k)$ "classically" in B is much harder than proving $\mathsf{SK}(F_k)$ in B by applying defaults.

4 Conclusion

In most works, reasoning in default logic is studied with no relation to a concrete calculus. Instead, the deductive closure is used leaving the "implementation details" open. If default reasoning is embedded into an automated deduction system then the notion of deductive closure is not appropriate, because inferences have to be performed in the (underlying classical) calculus. But if we rely on a calculus, we have to *search* for proofs instead of taking the deductive closure for granted. Since the size of the search space is elementarily related to the length of a shortest cut-free proof, the relative efficiency of the calculus becomes a crucial property.

We considered a default sequent calculus and showed that the presence of defaults can simplify matters in the first-order case. We used a class of formulae which possess only non-elementary proofs in a cut-free sequent calculus, but short proofs can be obtained by using the cut rule. Usually, cut-free calculi are used in implementations because they have a finite branching degree. We showed that we can get a non-elementary speed-up of proof length if the "right" default rules are present. The reason for the non-elementary speed-up is the introduction of a closed implication (by applying a default) which enables a simulation of the cut rule.

Although the class of formulae used to establish our result is constructed in regard to show the best case for the speed-up, one should observe that even simpler and more natural examples may exist, which become easier to prove by considering additional (relevant) knowledge.

References

1. M. Baaz and A. Leitsch. On Skolemization and Proof Complexity. *Fundamenta Informaticae*, 20:353–379, 1994.
2. P. A. Bonatti. *A Gentzen System for Non-Theorems*. Technical Report CD-TR 93/52, Christian Doppler Labor für Expertensysteme, Technische Universität Wien, Paniglgasse 16, A–1040 Wien, 1993.
3. P. A. Bonatti. Sequent Calculi for Default and Autoepistemic Logics. In *Proceedings TABLEAUX'96*, Springer LNCS 1071, pp. 127–142, 1996.
4. M. Cadoli, F. M. Donini, and M. Schaerf. Is Intractability of Non-Monotonic Reasoning a Real Drawback? In *Proceedings of the AAAI National Conference on Artificial Intelligence*, pp. 946–951. MIT Press, 1994.
5. M. Cadoli, F. M. Donini, and M. Schaerf. On Compact Representation of Propositional Circumscription. In *Proceedings STACS '95*, Springer LNCS 900, pp. 205–216, 1995.
6. M. Cadoli and M. Schaerf. A Survey of Complexity Results for Non-Monotonic Logics. *Journal of Logic Programming*, 17:127–160, 1993.
7. U. Egly. *On Methods of Function Introduction and Related Concepts*. PhD thesis, Technische Hochschule Darmstadt, Alexanderstr. 10, D–64283 Darmstadt, 1994.
8. U. Egly and H. Tompits. Is Nonmonotonic Reasoning Always Harder? In Ilkka Niemelä, editor, *Proceedings of the ECAI'96 Workshop on Integrating Nonmonotonicity into Automated Reasoning Systems*, Fachberichte Informatik 18–96, Universität Koblenz-Landau, Institut für Informatik, Rheinau 1, D-56075 Koblenz, 1996. http://www.uni-koblenz.de/universitaet/fb/fb4/publications/GelbeReihe/RR-18-96/proceedings.html
9. T. Eiter and G. Gottlob. Propositional Circumscription and Extended Closed World Reasoning are Π_2^P-complete. *Journal of Theoretical Computer Science*, 114(2):231–245, 1993. Addendum in vol. 118, p. 315, 1993.
10. G. Gottlob. Complexity Results for Nonmonotonic Logics. *Journal of Logic and Computation*, 2:397–425, 1992.
11. W. Marek, A. Nerode and J. Remmel. A Theory of Nonmonotonic Rule Systems II. *Annals of Mathematics and Artificial Intelligence*, 5:229–264, 1992.
12. V. P. Orevkov. Lower Bounds for Increasing Complexity of Derivations after Cut Elimination. *Zapiski Nauchnykh Seminarov Leningradskogo Otdeleniya Matematicheskogo Instituta im V. A. Steklova AN SSSR*, 88:137–161, 1979. English translation in *Journal of Soviet Mathematics*, 2337–2350, 1982.
13. R. Reiter. A Logic for Default Reasoning. *Artificial Intelligence Journal* 13, p. 81, 1980.
14. R. Reiter. Nonmonotonic Reasoning: Compiled vs. Interpreted Theories. Considerations for the Panel at Nonmon96. *Proceedings of the Sixth International Workshop on Nonmonotonic Reasoning*, Timberline, Oregon, 1996.
15. J. Schlipf. Decidability and Definability with Circumscription. *Annals of Pure and Applied Logic*, 35:173–191, 1987.
16. G. Schwarz and M. Truszczyński. Nonmonotonic Reasoning is Sometimes Simpler. *Journal of Logic and Computation*, 6(2):295–308, 1996.
17. A. Varzi. Complementary Sentential Logics. *Bulletin of the Section of Logic*, 19:112–116, 1990.

A Compositional Reasoning System for Executing Nonmonotonic Theories of Reasoning

J. Engelfriet and J. Treur

Vrije Universiteit Amsterdam, Faculty of Mathematics and Computer Science
De Boelelaan 1081a, 1081 HV Amsterdam, The Netherlands
email: {joeri, jan}@cs.vu.nl, URL: http://www.cs.vu.nl

1 Introduction

An agent that is reasoning about the world often needs to draw (defeasible) conclusions that are not logically entailed by its (incomplete) knowledge about the world. Nonmonotonic reasoning systems can be used to model these reasoning patterns. Implementations of agents that can reason in a defeasible manner need to include an implemented nonmonotonic reasoning system. In different applications, agents may need different types of nonmonotonic reasoning. Therefore, it is useful to develop a generic reasoning system that covers different types of nonmonotonic reasoning (as opposed to implementations for one specific nonmonotonic formalism as in e.g., [Ni96], [RS94] or [CMT96]). Moreover, the development of such a generic reasoning system can be made in a transparent manner if a central role is played by an implementation-independent design specification based on current software engineering principles, such as compositionality and information hiding.

Reasoning can be seen as an activity of an agent taking place in time. The agent starts with a set of initial beliefs to which it applies some (nonmonotonic) rules to arrive at a new state, in which it has more knowledge. In this new state, the agent may again apply rules to arrive at a next state. Viewing the agent from the outside, we can look at the knowledge of the agent at all points in time. Thus, more formally, we can describe the behavior of this agent by a reasoning trace, or a temporal model that describes what the agent knows at the different points in time. Since (nonmonotonic) reasoning may be nondeterministic (the agent may have a choice of which rules to apply), we should allow multiple traces that start with the same set of initial beliefs. If we want to describe the reasoning of the agent exhaustively, we may incorporate traces starting with any set of initial beliefs. To specify such sets of temporal models (or traces), temporal logic can be used. This general temporal view of reasoning was put forth in [ET96], where we introduced a formal notion of temporal model, and a temporal specification language suited to specify sets of temporal models. In the current paper a design and specification of an executable nonmonotonic reasoning system to implement this generic approach to nonmonotonic reasoning is presented. For any specification of a theory of nonmonotonic reasoning, given as an input to the system, the possible reasoning traces are generated.

The system was developed using the compositional modelling environment DESIRE (framework for DEsign and Specification of Interacting REasoning components; see [LPT92], [BDJT97], [BTWW95]). This environment for the development of compositional reasoning systems and multi-agent systems has been successfully used to design various types of reasoning systems and multi-agent systems and applications in particular domains. The DESIRE software environment offers a graphical editor, an implementation generator and an execution environment to execute specifications automatically. In its use to build complex reasoning systems, DESIRE can be viewed as an advanced theorem proving environment in which both the knowledge and the control of the reasoning can be specified in an explicit, declarative and compositional manner.

In this paper, in Section 2 we briefly summarize the generic approach introduced in [ET96]. In Section 3 a very brief introduction to DESIRE is presented. In Section 4 the design of the nonmonotonic reasoning system is discussed. An example trace of the system is described in Section 5, and Section 6 gives conclusions and suggestions for further research.

2 Temporal Specification of Nonmonotonic Reasoning

In this section we will briefly review the main notions introduced in [ET96]. The basic language in which the agent expresses its beliefs will be propositional logic. Semantically, a state is a set of propositional models, representing the set of beliefs consisting of the formulas true in all models in the state.

Definition 2.1 (Temporal Model)

i) An *information state* is a non-empty set of propositional models. A propositional formula α is true in an information state M, denoted $M \vDash \alpha$, if $m \vDash \alpha$ for all $m \in M$. For two information states M, N, we say M contains more information than N, denoted $N \leq M$ if $M \subseteq N$.

i) A *temporal model* is a sequence $(\mathcal{M}_s)_{s \in \mathbb{N}}$ where each $\mathcal{M}_s$ is an information state.

ii) A temporal model $\mathcal{M}$ is *conservative* if $\mathcal{M}_s \leq \mathcal{M}_{s+1}$ for all $s \in \mathbb{N}$.

iii) The *refinement ordering* $\leq$ on temporal models is defined by:
$$\mathcal{M} \leq \mathcal{N} \quad \Leftrightarrow \quad \text{for all } s: \ \mathcal{M}_s \leq \mathcal{N}_s \text{ and } \mathcal{M}_0 = \mathcal{N}_0.$$

Note that when $N \leq M$, for all propositional formulas α we have $N \vDash \alpha \Rightarrow M \vDash \alpha$, so that indeed M contains more information than N. The conservativity of a temporal model means that the agent never forgets anything it has previously deduced. The temporal language we will introduce contains the operators H_0, F, G and C which refer respectively to knowledge at point 0, knowledge at some point in the future, at all points in the future, and to knowledge in the current time point. The formal semantics of these operators are as follows.

Definition 2.2 (Temporal Interpretation)

a) For a propositional formulae φ:

$(\mathcal{M}, s) \vDash F\varphi$	$\Leftrightarrow$	there exists $t \in \mathbb{N}$, $t > s$ such that $\mathcal{M}_t \vDash \varphi$
$(\mathcal{M}, s) \vDash G\varphi$	$\Leftrightarrow$	for all $t \in \mathbb{N}$ with $t > s$: $\mathcal{M}_t \vDash \varphi$
$(\mathcal{M}, s) \vDash C\varphi$	$\Leftrightarrow$	$\mathcal{M}_s \vDash \varphi$
$(\mathcal{M}, s) \vDash H_0\varphi$	$\Leftrightarrow$	$\mathcal{M}_0 \vDash \varphi$

b) For a temporal formula α:

$(\mathcal{M}, s) \vDash \neg\alpha$	$\Leftrightarrow$	it is not the case that $(\mathcal{M}, s) \vDash \alpha$

c) For a set A of temporal formula:

$(\mathfrak{M}, s) \vDash \bigwedge A \quad \Leftrightarrow \quad$ for all $\varphi \in A$: $(\mathfrak{M}, s) \vDash \varphi$

d) A formula φ is true in a model $\mathfrak{M}$, denoted $\mathfrak{M} \vDash \varphi$, if for all $s \in \mathbb{N}$: $(\mathfrak{M}, s) \vDash \varphi$

e) A set of formulae T is true in a model $\mathfrak{M}$, denoted $\mathfrak{M} \vDash T$, if for all $\varphi \in T$, $\mathfrak{M} \vDash \varphi$. We call $\mathfrak{M}$ a model of T.

In a specification, we allow only a certain format in the temporal formulas, which corresponds to the application of a rule by the agent:

Definition 2.3 (Reasoning Theories)

a) A formula is called a *(nonmonotonic) reasoning formula* if it is of the form

$\alpha \wedge \beta \wedge \varphi \wedge \psi \rightarrow G\gamma$, where

$\alpha = \bigwedge\{ H_0 \varepsilon \mid \varepsilon \in A \}$ for a set of propositional formulae A.

$\beta = \bigwedge\{ \neg H_0 \delta \mid \delta \in B \}$ for a set of propositional formulae B.

$\varphi = \bigwedge\{ \neg F\theta \mid \theta \in C \}$ for a set of propositional formulae C.

$\psi = \bigwedge\{ C\zeta \mid \zeta \in D \}$ for a set of propositional formulae D.

γ is a propositional formula.

b) A set Th of reasoning formulae is called a *theory of reasoning*.

In a temporal model of a theory of reasoning, all (nonmonotonic) rules which are applicable at some point in time, have actually been applied by the agent. But we also want to make sure that the agent knows *nothing more* than what it can deduce. So we will look at models of a theory (the rules have been applied) in which the knowledge of the agent over time is minimal (the agent knows nothing more).

Definition 2.4 (Minimal Temporal Model)

A temporal model $\mathfrak{M}$ is called a *minimal model* of a theory Th if it is a model of Th and for any model $\mathfrak{N}$ of Th, if $\mathfrak{N} \leq \mathfrak{M}$ then $\mathfrak{N} = \mathfrak{M}$.

The minimal models of a theory describe the reasoning process of an agent specified by this theory. The compositional reasoning system we will describe in Section 4 will find these minimal models by "executing" the theory. Our approach can therefore be seen as an executable modal (nonmonotonic) logic (see [FO95]).

We will give an example of such a theory. In [ET93] it was established that there exists a faithful translation of Reiter's default logic into the temporal language introduced above. The minimal models of the translation correspond to extensions of the theory. We will not go into the details of the translation, but rather give an example. Let the following default theory $\langle W, D \rangle$ be given: $W = \{ a, d, b \rightarrow \neg c \}$ and $D = \{ (a: b) / b, (d: c) / c, (b: \neg c) / e \}$. The atoms a and d will be initial facts. Formally, the formula $b \rightarrow \neg c$ would also be a fact, but since the generic reasoning system to be described below does not perform general propositional reasoning (it uses a subset of natural deduction called *chaining*), we will translate it into the following two rules which describe the application of the formula:

(1) $Cb \rightarrow G\neg c$,

(2) $Cc \rightarrow G\neg b$,

The default rules are translated as follows:

(1) **Cb → G¬c,**
(2) **Cc → G¬b,**
(3) **Ca ∧ ¬F¬b → Gb,**
(4) **Cd ∧ ¬F¬c → Gc,**
(5) **Cb ∧ ¬Fc → Ge**

The idea behind this translation is that we should add the conclusion of a default rule like **(b: ¬c) / e** when **b** has been derived, and **¬c** remains consistent throughout our further reasoning, which means we should never derive **c** in the future. The minimal models of the resulting theory are described by the following picture:

	M					**N**				
a	1	1	1	1	...	1	1	1	1	...
b	u	1	1	1	...	u	u	0	0	...
c	u	u	0	0	...	u	1	1	1	...
d	1	1	1	1	...	1	1	1	1	...
e	u	u	1	1	...	u	u	u	u	...

Figure 1. Minimal models

In this picture, a **1** means the corresponding atoms has been derived, a **0** means its negation has been derived, and a **u** means that neither the atoms nor its negation have been derived. The model **M** corresponds to the extension **Cn({a, b , ¬ c, d, e, b → ¬c})**, and the model **N** corresponds to the extension **Cn({a, ¬b, c, d, b → ¬c})**. The reader familiar with default logic may check that these are (the only) extensions of ⟨ **W, D** ⟩. We will use this example in Section 5, where an example trace of the generic reasoning system is given. In this same fashion one can translate a logic program to a temporal theory, yielding the stable semantics.

3 A Specification Framework for Compositional Systems

In the framework DESIRE knowledge of

(1) a task composition,
(2) information exchange,
(3) sequencing of tasks,
(4) task delegation,
(5) knowledge structures

are explicitly modelled and specified. Each of these types of knowledge is discussed below.

3.1 Task Composition
To model and specify composition of tasks, knowledge of the following types is required:

- a *task hierarchy*,
- information a task requires as *input*,
- information a task produces as a *result* of task performance
- *meta-object* relations between tasks

Within a task hierarchy *composed* and *primitive* tasks are distinguished: in contrast to primitive tasks, composed tasks are tasks for which sub-tasks are identified. Sub-tasks, in turn, can be either composed or primitive. Tasks are directly related to components: composed tasks are specified as composed components and primitive tasks as primitive components. Primitive components are executed by a simple classical deduction system, based on the inference relation chaining (modus ponens and conjunction introduction).

Information required/produced by a task is defined by *input* and *output signatures* of a component. The signatures used to name the information are defined in a predicate logic with a hierarchically ordered sort structure (*order-sorted predicate logic*). Units of information are represented by the ground *atoms* defined in the signature.

The role information plays within reasoning is indicated by the level of an atom within a signature: different (meta)levels may be distinguished. In a two-level situation the lowest level is termed *object level information*, and the second level *meta-level information*. Meta-level information contains information about object level information and reasoning processes; for example, for which atoms the values are still unknown (*epistemic information*). Often more than two levels of information and reasoning occur, resulting in meta-meta-... information and reasoning.

3.2 Information Exchange Between Tasks

Information exchange between tasks is specified as *information links* between components. Each information link relates output of one component A to input of another component B, by specifying which truth value of a specific output atom of A is linked with which truth value of a specific input atom of B. Atoms can be renamed: each component can be specified in its own language, independent of other components. The conditions for activation of information links are explicitly specified as task control knowledge: a kind of knowledge of the sequencing of tasks.

3.3 Sequencing of Tasks

Task sequencing is explicitly specified within components as *task control knowledge*. Task control knowledge includes not only knowledge of which tasks should be activated when and how, but also knowledge of the goals associated with task activation and the extent to which goals should be derived. These aspects are specified as component and link activation together with task control foci and extent to define the component's goals. Components are, in principle, black boxes to the task control of an encompassing component: task control is based purely on information about the success and/or failure of component reasoning. Reasoning of a component is considered to have been successful with respect to its task control focus if it has reached the goals specified by this task control focus to the extent specified (e.g., any or every).

3.4 Delegation of Tasks

During knowledge acquisition a task as a whole is modelled. In the course of the modelling process decisions are made as to which tasks are (to be) performed by which *agent*. This process results in the delegation of tasks to the parties involved in task execution.

3.5 Knowledge Structures

During design an appropriate structure for domain knowledge must be devised. The meaning of the concepts used to describe a domain and the relations between concepts and groups of concepts, are determined. Concepts are required to identify objects distinguished in a domain (domain-oriented ontology) , but also to express the methods and strategies employed to perform a task (task-oriented ontology). Within primitive components concepts and relations between concepts are defined in *hierarchies* and *rules* (based on *order-sorted predicate logic*). In a specification references to appropriate knowledge structures (specified elsewhere) suffice; compositional knowledge structures are composed by reference to other knowledge structures.

4 A Generic Compositional Nonmonotonic Reasoning System

The nonmonotonic reasoning task of finding minimal models of a reasoning theory is modelled as a composition of two subtasks. the first subtask generates all possible continuations of the reasoning trace, and the second subtask selects a continuation. Within the first task, first it is determined which rules are applicable. Applicable rules are the rules

$$\alpha \wedge \beta \wedge \phi \wedge \psi \rightarrow G\gamma,$$

where

$\alpha = \bigwedge\{ H_0\epsilon \mid \epsilon \in A \}$ for a set of propositional formulae **A**.

$\beta = \bigwedge\{ \neg H_0\delta \mid \delta \in B \}$ for a set of propositional formulae **B**.

$\phi = \bigwedge\{ \neg F\theta \mid \theta \in C \}$ for a set of propositional formulae **C**.

$\psi = \bigwedge\{ C\zeta \mid \zeta \in D \}$ for a set of propositional formulae **D**.

γ is a propositional formula.

for which the conditions that refer to the past and present (i.e., α, β, ψ) are fulfilled. Next, for each of the applicable rules, for the future-directed conditions ϕ two possibilities are generated:
- either the conditions that refer to the future will be fulfilled in the reasoning trace that is generated, or
- these future-directed conditions will not be fulfilled.

In the first case the rule will contribute its conclusion to the reasoning process and we have to make sure in the future that the future conditions were indeed fulfilled (we add *constraints* to ensure this). In the second case no explicit contribution will be made by the rule. However, in this second case, by the subsequent generation of the reasoning trace it will have to be guaranteed that the future-directed conditions indeed will be violated (and we again add constraints to ensure this). In this sense an implicit effect on the reasoning trace occurs: all traces that do not contradict these conditions will be rejected (see [ET97]). Note that we do not try to *execute* the future condition;

we merely guess whether it will be fulfilled. In this respect these rules are not of the form "declarative past implies imperative future" of [Ga89].

The design of the compositional nonmonotonic reasoning system has been specified in DESIRE, according to the five types of knowledge discussed in Section 3. Five levels of abstraction are distinguished in the task hierarchy (see Figure 2).

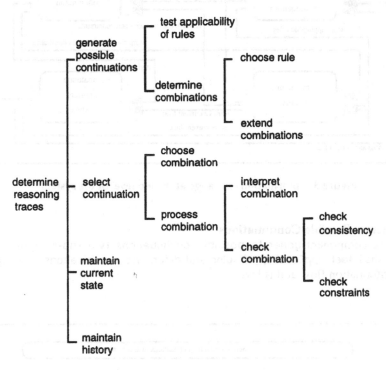

Figure 2. Complete task hierarchy of the system

4.1 Top Level of the System

At the highest level the system consists of four components. During the reasoning process, in the component maintain_current_state the facts are represented that have been derived. The component maintain_history stores relevant aspects of the reasoning process in order to perform belief revision if required. The reasoning is performed by the component generate_possible_continuations, which generates the possible next steps of the reasoning trace and select_continuation which chooses one of these possibilities. By this selection the actual next step in the reasoning trace is determined. In Figure 3 the information exchange at the top level of the system is depicted. (In this picture and the following ones, we have left out some links which go from a component to itself. For instance, such links are sometimes needed to model a closed-world assumption.)

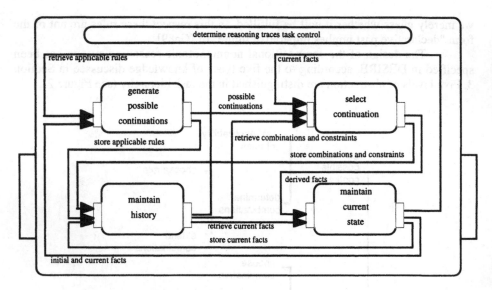

Figure 3 . Information exchange at the top level of the system

4.2 Generate Possible Continuations

Within the component generate_possible_continuations two sub-components are distinguished: test_applicability_of_rules and determine_combinations; see Figure 4 for the information flow at this level.

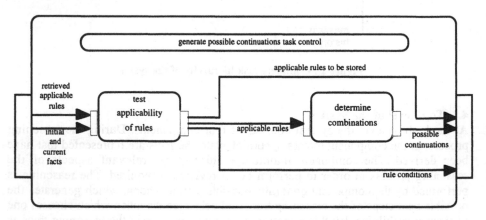

Figure 4. Information exchange within generate possible continuations

The former component is primitive and determines the rules that are applicable in the current state of the reasoning process. Its knowledge base consists of rules, for example, of the form:

```
if   posH0condition(L: Literals, R: Rules)       if   currentcondition(L: Literals, R:
Rules)                                           Rules)
and not initial_fact(L: Literals)                and not current_fact(L: Literals)
then not applicable(R: Rules)                    then not applicable(R: Rules)
```

By application of a form of the Closed World Assumption, the applicable rules are derived. The second component determines for each of the applicable rules two possibilities: it is assumed that either (+) the conditions of the rule that refer to the future will be fulfilled in the reasoning trace that is generated, or (-) these future-directed conditions will not be fulfilled.

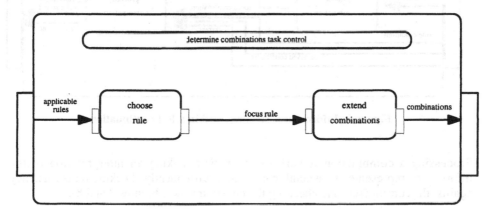

Figure 5. Information exchange within determine combinations

The component determine_combinations is rather simple (see Figure 5). The applicable rules are treated one by one and combinations are constructed. Both subcomponents are primitive. The knowledge base of the first subcomponent just consists of one rule:

```
if   applicable(R: Rules)
and not covered(R: Rules)
then infocus(R: Rules)
```

The knowledge base of the second subcomponent consists of the following two rules:

```
if   old_combination(C: combinations)       if   old_combination(C: combinations)
and infocus(R: rules)                       and infocus(R: rules)
and futuredependent(R: rules)               then new_combination(app(C:
then new_combination(app(C:                      combinations, tup(R: rules, pos)));
     combinations, tup(R: rules, neg)));
```

These rules build new combinations from old combinations and the rule in focus. The predicate app (for append) is used to build up a list, and the predicate tup (for tuple) is used to make tuples. Only rules with a part that refers to the future (futuredependent) are allowed *not* to be applied (meaning that tup(R: rules, neg) may occur in a combination).

4.3 Select Continuation

Within the component select_continuation, focus combinations are chosen one by one and processed (see Figure 6).

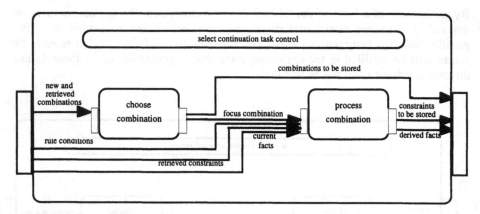

Figure 6. Information exchange within select continuation

Processing a combination is performed by first making an interpretation of the information represented by a combination, and subsequently checking on consistency against the current facts and checking the constraints (see Figures 7 and 8).

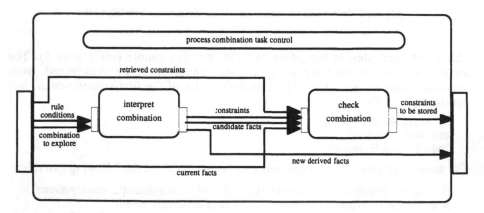

Figure 7. Information exchange within process combinations

The knowledge base of the primitive component check_consistency consists of three rules, an example of which is:

if next(A: Atoms)
and current(neg(A: Atoms))
then inconsistency

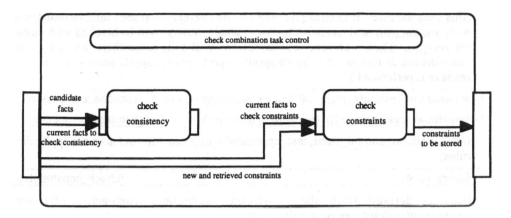

Figure 8. Information exchange within check combinations

5 Example Trace

In this section we will give an example of the execution of our generic reasoning system. The theory of reasoning is the translation of the example default theory of Section 2. When this theory is given to the generic reasoning system, the following will happen (in this list, on the left we will indicate the literals that are derived, on the right the component in which it is derived, in brackets; not everything derived in every component is listed and no information links are listed):

- a, d	(maintain_current_state)
- applicable(r3), applicable(r4)	(test_applicability_of_rules)
- new_combination(app(app(nil, tup(r3, pos)), tup(r4, pos))), new_combination(app(app(nil, tup(r3, neg)), tup(r4, pos))), new_combination(app(app(nil, tup(r3, pos)), tup(r4, neg))), new_combination(app(app(nil, tup(r3, neg)), tup(r4, neg)))	(determine_combinations)
- selectedcombination(app(app(nil, tup(r3, pos)), tup(r4, pos)))	(choose_combination)
- next(c), next(b)	(interpret_combination)
- c, b	(maintain_current_state)
- applicable(r1), applicable(r2)	(test_applicability_of_rules)
- new_combination(app(app(nil, tup(r1, pos)), tup(r2, pos)))	(determine_combinations)

(since these rules do not refer to the future, only one combination will be generated, in which both rules are applied)

- selectedcombination(app(app(nil, tup(r1, pos)), tup(r2, pos)))	(choose_combination)
- next(neg(b)), next(neg(c))	(interpret_combination)
- inconsistency	(check_consistency)

(this uses the rule **if next(neg(A: atoms)) and current(A: atoms) then inconsistency** with **next(neg(b))** and **current(b)**; now maintain_history will be invoked with target set **retrieve**. Then choose_combination will fail, since there was only one combination at this point - **app(app(nil, tup(r1, pos)), tup(r2, pos))** - so another **retrieve** is performed.)

- **selectedcombination(app(app(nil, tup(r3, neg)), tup(r4, neg)))**	(choose_combination)
- **sometimes_true(neg(c), r4)), (sometimes_true(neg(b))**	(interpret_combination)

(after maintain_current_state, test_applicability_of_rules will find no more applicable rules).

- **incorrect(r4)**	(check_constraints)

(this is derived from the constraint **sometimes_true(neg(c), r4))** and **not(current(neg(c)))**; a **retrieve** will occur)

- **selectedcombination(app(app(nil, tup(r3, pos)), tup(r4, neg)))**	(choose_combination)
- **sometimes_true(neg(c), r4), next(b), never_true(neg(b))**	(interpret_combination)
- **b**	(maintain_current_state)
- **applicable(r1), applicable(r5)**	(test_applicability_of_rules)
- **new_combination(app(app(nil, tup(r5, pos)), tup(r1, pos))))**, **new_combination(app(app(nil, tup(r5, neg)), tup(r1, pos)))**	(determine_combinations)

(**r1** has to be applied (it does not refer to the future), but for **r5** there is a choice)

- **selectedcombination(app(app(nil, tup(r5, pos)), tup(r1, pos)))**	(choose_combination)
- **next(neg(c)), next(e), never_true(c)**	(interpret_combination)
- **neg(c), e**	(maintain_current_state)

(no more applicable rules are found by test_applicability_of_rules, so check_constraints is invoked, which finds no violated constraints)

- at this point user_interaction will display the window of figure 9.

The minimal model displayed corresponds to the extension based on the literals {a, b, ¬c, d, e}. When the user clicks "no", the execution will stop. Otherwise a **retrieve** will be performed by maintain_history, after which choose_combination will choose the combination in which **r5** is not applied (but **r1** is). This will eventually lead to a violated constraint, so another **retrieve** will occur. Then choose_combination will take the last combination (for **r3** and **r4**), in which **r3** is not applied and **r4** is. Ultimately, the second minimal model will be found and displayed (corresponding to the literals {a, ¬b, c, d}). If the user wants to search for another minimal model, none will be found, and a message indicating this will be displayed, after which execution ends.

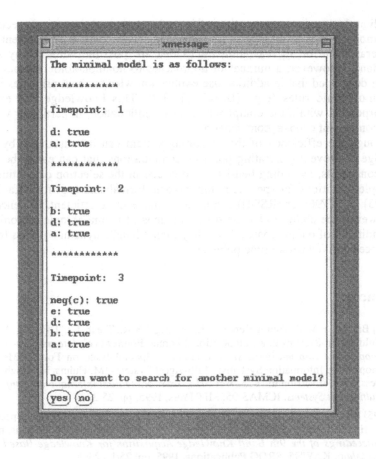

Figure 9. User interaction

6 Discussion

In this paper the framework DESIRE for the design of compositional reasoning systems and multi-agent systems was applied to build a generic nonmonotonic reasoning system. The outcome is a general reasoning system that can be used to model different nonmonotonic reasoning formalisms (see for instance [ET94]), and that can be executed by a generic execution mechanism. The main advantages of using DESIRE (compared to a direct implementation in a programming language such as PROLOG) are:

- the design is generic and has a transparent compositional structure; it is easily readable, modifiable and reusable. The generic nonmonotonic reasoning system is easily usable as a component in agents that are specified in DESIRE. For example, if an agent is designed that has default knowledge, then the generic reasoning system can be included as one of the agent's components and the representation of the agent's default knowledge can be translated to the temporal representation of the generic reasoning system.

- explicit declarative specification of both the static and dynamic aspects of the nonmonotonic reasoning processes, including their control. The current system generates one or all reasoning traces that are possible without any specific guidance. However, a number of approaches to nonmonotonic reasoning have been developed that in addition use explicit knowledge about priorities between nonmonotonic rules (e.g., [Br94], [TT92]). This knowledge can easily be incorporated within the component select_continuation, in particular within its subcomponent choose_combination.

Even though the efficiency of the reasoning system can be improved (by adding knowledge to prevent generating possible continuations that can easily be seen to violate constraints, by adding heuristic knowledge in the selection of continuations, etc.), implementations for specific nonmonotonic formalisms (e.g. for default logic, see [Ni95], [CMT96], or [RS94]) can often be made more efficient. In general they lack, however, the ability to handle different kinds of nonmonotonic reasoning, and the extendibility of our approach. Also, they cannot handle dynamic queries (e.g., has a literal been derived before time point 3).

References

[BDJT97] Brazier, F.M.T., Dunin-Keplicz, B., Jennings, N.R., Treur, J. : "DESIRE: Modelling Multi-Agent Systems in a Compositional Formal Framework", *International Journal of Cooperative Information Systems* **6** (1997) , Special Issue on Formal Methods in Cooperative Information Systems: Multi-Agent Systems (M. Huhns, M. Singh, eds.), in press. Shorter version in: V. Lesser (ed.), *Proc. of the First International Conference on Multi-Agent Systems*, ICMAS-95, MIT Press, 1995, pp. 25-32

[BTWW95] Brazier, F.M.T., Treur, J., Wijngaards, N.J.E., Willems, M.: "Formal Specification of Hierarchically (De)composed Tasks", in: B.R. Gaines, M.A. Musen (eds.), *Proceedings of the 9th Banff Knowledge Acquisition for Knowledge-Based Systems Workshop*, KAW'95, SRDG Publications, 1995, pp. 25/1 - 25/20

[Br94] Brewka, G.: "Adding Priorities and Specificity to Default Logic", in: C. MacNish, D. Pearce, L.M. Pereira (eds.), *Logics in Artificial Intelligence, Proceedings of the 4th European Workshop on Logics in Artificial Intelligence*, JELIA '94, Lecture Notes in Artificial Intelligence **838**, Springer-Verlag, 1994, pp. 247-260

[CMT96] Cholewinski, P., Marek, V.W., Truszczynski, M.: "Default Reasoning System DeReS", in: *Proceedings 5th International Conference on Principles of Knowledge Representation and Reasoning*, KR-96, Morgan Kaufmann, 1996.

[ET93] Engelfriet, J., Treur, J.: "A Temporal Model Theory for Default Logic", in: M. Clarke, R. Kruse, S. Moral (eds.), *Proceedings 2nd European Conference on Symbolic and Quantitative Approaches to Reasoning and Uncertainty*, ECSQARU '93, Lecture Notes in Computer Science **747**, Springer-Verlag, 1993, pp. 91-96

[ET94] Engelfriet, J., Treur, J. : "Temporal Theories of Reasoning", in: C. MacNish, D. Pearce, L.M. Pereira (eds.), *Logics in Artificial Intelligence, Proceedings of the 4th European Workshop on Logics in Artificial Intelligence*, JELIA '94, Lecture Notes in Artificial Intelligence **838**, Springer-Verlag, 1994, pp. 279-299. Also in: *Journal of Applied Non-Classical Logics* **5**(2), 1995, pp. 239-261

[ET96] Engelfriet, J., Treur, J. : "Specification of Nonmonotonic Reasoning", in: D.M. Gabbay, H.J. Ohlbach (eds.), *Practical Reasoning, International Conference on Formal and*

Applied Practical Reasoning, FAPR'96, Lecture Notes in Artificial Intelligence **1085**, Springer-Verlag, 1996, pp. 111-125

[ET97] Engelfriet, J., Treur, J. : "Executable Temporal Logic for Nonmonotonic Reasoning", to appear in *Journal of Symbolic Computation*, Special issue on Executable Temporal Logic

[FO95] Fisher, M., Owens, R. (eds.) : *Executable Modal and Temporal Logics, Proceedings of the IJCAI'93 workshop*, Lecture Notes in Artificial Intelligence **897**, Springer-Verlag, 1995

[Ga89] Gabbay, D. : "The Declarative Past and Imperative Future: Executable Temporal Logic for Interactive Systems", in: B. Banieqbal, H. Barringer, A. Pnueli (eds.), *Temporal Logic in Specification*, Lecture Notes in Computer Science **398**, Springer-Verlag, pp. 409-448

[LPT92] Langevelde, I.A. van, Philipsen, A.W., Treur, J. : "Formal Specification of Compositional Architectures", in: B. Neumann (ed.), *Proceedings of the 10th European Conference on Artificial Intelligence*, ECAI'92, John Wiley & Sons, pp. 272-276

[Ni95] Niemelä, I., : "Towards Efficient Default Reasoning", in: *Proceedings 14th IJCAI*, Morgan Kaufmann, 1995, pp. 312-318

[Ni96] Niemelä, I., : "Implementing Circumscription Using a Tableau Method", in: W. Wahlster (ed.), *Proceedings 12th European Conference on Artificial Intelligence*, ECAI'96, John Wiley & Sons, pp. 80-84

[RS94] Risch, V., Schwind, C.B.: "Tableau-Based Characterization and Theorem Proving for Default Logic", Journal of Automated Reasoning **13**: 223-242, 1994

[SB96] Schaub, T., Brüning, S. : "Prolog technology for default reasoning (An abridged report)", in: W. Wahlster (ed.), *Proceedings 12th European Conference on Artificial Intelligence*, ECAI'96, John Wiley & Sons, pp. 105-109

[TT92] Tan, Y.H., Treur, J. : "Constructive Default Logic and the Control of Defeasible Reasoning", in: B. Neumann (ed.), *Proceedings of the 10th European Conference on Artificial Intelligence*, ECAI'92, John Wiley & Sons, pp. 299-303

Structured Belief Bases: A Practical Approach to Prioritised Base Revision

Dov Gabbay[1] and Odinaldo Rodrigues[2]

[1] Department of Computing, Imperial College
180 Queen's Gate, London, SW7 2BZ, UK
e-mail: dg@doc.ic.ac.uk
[2] Department of Computing, Imperial College
180 Queen's Gate, London, SW7 2BZ, UK
e-mail: otr@doc.ic.ac.uk

Abstract. In this paper we present Structured Belief Bases (SBBs), a framework to reason about belief change. Structured Belief Bases can be considered as a special case of prioritised base revision, where the components of the base are allowed to be structured bases as well. This allows for the representation of complex levels of priority between sentences. Each component is resolved into a sentence via a revision operator taking the ordering into account. By adopting a right associative interpretation of the operator we avoid many of the problems with iteration faced by revision operators complying with the AGM postulates. New beliefs can be accepted by simply incorporating them into the base with highest priority.

1 Introduction

In [6], Katsuno and Mendelzon provided a semantic characterisation of all belief revision operators satisfying the AGM postulates for belief revision. They considered finite knowledge bases represented as a conjunction of sentences in propositional logic. However, when the belief base is simplified into a single sentence some information is lost, because it is not possible to distinguish derived beliefs from basic ones.

Some authors have advocated for *base revisions* instead. That is, how to change the set of basic beliefs minimally so that the new belief can be incorporated into the belief state consistently. However, even for belief bases, the problem of choice between the possible revised belief states that accomplish this minimal change is still left open. An alternative is to associate priorities to beliefs and prefer belief states that preserve the beliefs in the base with higher priority. This approach is commonly known as prioritised base revision [7].

In this work, we present structured belief bases (SBBs), which can be considered as a special kind of prioritised base revision. That is, a structured belief base is a set of components provided with an ordering representing priorities among them. We allow the components to be nested structured bases as well, so that more complex priority patterns can be represented. A revision operator provides revisions of sentences which comply with the AGM postulates for Belief Revision. This operator can be efficiently implemented for sentences represented in DNF. New beliefs can be accepted in the belief state by simply incorporating them into the base with highest priority. However, we consider a right associative interpretation of the revision of components of the base. This allows for a treatment of iteration that is simple but yet adequate for many applications.

The paper is divided as follows: in Section 2 we recap the revision operator which is going to be used in the following sections. In Section 3 we present prioritised databases a simpler mechanism than structured belief bases which are introduced in Section 4, where we also show some examples. Comparisons with prioritised base revisions are made in Section 5 and in Section 6 we show how to loosen the restrictions on the orderings of SBB. Final conclusions and future extensions to the work are outlined in Section 7.

2 Revisions of sentences

We need to recap briefly how the revision operator works. This and other operators for belief revision and reasoning about actions are presented in greater detail in [3] and in [10].

The revision operator works with formulae in disjunctive normal form (DNF). This special normal form provides great simplification of the method, since disjuncts can be seen as syntactical representations of the classes of models satisfying the sentences. A formula is said to be in DNF iff it is a disjunction of conjunctions of literals. Each component of the disjunction will be called a *disjunct*.

We assume that **L** is the language of propositional logic over some finite set of propositional variables $\mathcal{P}$[1]. A *literal* is a propositional variable or the negation of a propositional variable. The satisfaction and derivability relations ($\Vdash$ and $\vdash$, resp.) for **L** are assumed to have their usual meanings. An interpretation is a function from $\mathcal{P}$ to $\{\mathbf{tt}, \mathbf{ff}\}$, and $\mathcal{I}$ represents the set of all interpretations of **L**. $\mathrm{mod}(\psi)$ is the set of interpretations which satisfy ψ.

We will use letters from the middle of the alphabet, p, q, r, ... to represent propositional variables; capital letters from the middle of the alphabet, P, Q, R, ... to represent disjuncts of propositional logic, and the letter l (possibly subscripted), to represent literals.

Let us start with the simplest case: we have only two disjuncts P and Q, where Q has higher priority than P. If we want to accept Q, we have to keep its literals. To maintain consistency and keep some of the informative content of P, we *add* the literals in it which are consistent with Q:

Definition 1 (Prioritised Revision of Disjuncts) *Let* $P = \wedge_i l_i$ *and* Q *be two disjuncts. The* prioritised revision of P by Q, *denoted* $P \overset{*}{Q}$, *is a new disjunct*

---

[1] This is required by [6], whose results we use.

R such that

$$R = \begin{cases} Q, & \Rightarrow \text{ if } P \text{ is contradictory} \\ Q \bigwedge \{l_i | l_i \text{ (or the literal opposite in sign} \\ \quad\quad to\ l_i)\ \text{doesn't appear in } Q\}, & \Rightarrow \text{ otherwise} \end{cases}$$

The arrow in $\widehat{PQ}$ indicates that the literals in Q override (possible) complementary literals in P, where by "complementary" we mean negated literals, e.g., p and $\neg p$. It can be easily checked that $\widehat{PQ} \vdash Q$.

So far we have defined how to resolve disjuncts of sentences, but this is not enough to achieve revisions of *whole sentences*. We still need to provide a means of evaluating the degree of information change during the resolution of disjuncts. The idea is to choose those combinations of disjuncts which cause minimum change to the disjunct with lower priority.

If we think semantically, and for the case of propositional logic, this change can be measured by analysing the truth-values of propositional variables of interpretations. One possibility is to consider the "distance" between two interpretations as the number of propositional variables which are assigned different truth-values by them. This distance notion is the same used in [2]. In [10] we have proved that this distance is in fact a metric on $\mathcal{I}$.

Definition 2 *Let M and N be two elements of $\mathcal{I}$. The distance d between M and N is the number of propositional variables p_i for which $M(p_i) \neq N(p_i)$.*

Ultimately what we want is to use d to evaluate the "distance" between two sentences ψ and φ. Since each sentence is associated with a subset of $\mathcal{I}$, namely the interpretations which satisfy them, we can use the usual mechanisms of extending distance between points to distance between sets found in topology. That is, we consider the minimal distance between any two interpretations satisfying the sentences. This agrees with our intuitions for belief revision, since in a revision of ψ by φ, the goal is to choose the models of φ which are closest to models of ψ. For the special case when one of the sentences is unsatisfiable we define the distance as being arbitrarily large and represent this by ∞.

Definition 3 *Let ψ and φ, be two sentences of propositional logic. The distance dist between ψ and φ, in symbols $dist(\psi, \varphi)$, is defined as*

$$dist(\psi, \varphi) = \begin{cases} \text{minimum } d(M, N),\ for & \Rightarrow \text{ if } \mathrm{mod}(\psi) \text{ and } \mathrm{mod}(\varphi) \\ M \in \mathrm{mod}(\psi) \text{ and } N \in \mathrm{mod}(\varphi) & \quad are\ both\ non\text{-}empty \\ \infty, & \Rightarrow otherwise \end{cases}$$

By using disjunctive normal form, we have an extra advantage, because *dist* can be easily computed when only disjuncts are involved:

Proposition 4 *Let P and Q be disjuncts.*

$$dist(P, Q) = \begin{cases} \infty, & \Rightarrow \text{ if either } P \text{ or } Q \text{ is con-} \\ & \quad tradictory\ (or\ both) \\ \text{number of literals } l_i \text{ in } P \text{ whose} & \Rightarrow otherwise \\ \text{complementary literal occurs in } Q, \end{cases}$$

Proof Straightforward. □

In other words, the distance between two non-contradictory disjuncts is just the number of literals upon which they disagree. Notice that even though we evaluate distances on sentences, we are actually computing them semantically.

Now, as expected, the next step is to choose those combinations of disjuncts in ψ and φ with minimum distance. This will ultimately result in choosing the models in $\text{mod}(\psi)$ and $\text{mod}(\varphi)$ with minimum distance.

Definition 5 (Choosing disjuncts with minimum distance) *Let ψ and φ be two sentences in DNF. The disjuncts which correspond to a minimum change of information to ψ or φ, best_fit(ψ, φ), are in the following set:*

$$\text{best\_fit}(\psi, \varphi) = \{(P, Q) \mid P \in \psi, Q \in \varphi, \text{ and } dist(P, Q) \text{ is minimum}\}$$

Notice that the order of the disjuncts in each of the pairs above is important, as it identifies where each disjunct in the pair came from. Remember that in a revision, the sentence which revises has priority over the revised one. We are now in a position to define revisions of sentences:

Definition 6 (Revisions of sentences) *Let ψ and φ be two sentences in DNF. The revision of ψ by φ, $\psi \circ_r \varphi$, is a sentence γ such that:*

$$\gamma \leftrightarrow \bigvee_{(P,Q) \in \text{best\_fit}(\psi, \varphi)} \overset{*}{P Q}$$

That is, to compute the revision of sentence ψ by φ in DNF, consider the cartesian product of the disjuncts in ψ and φ, then choose those pairs with minimum distance *dist*, and finally take the disjunction of the revision of disjuncts in all such pairs.

The first condition in Definition 1 is the only consistency check used in the process. It can be easily performed since it amounts to check whether a disjunct has both positive and negative occurrences of the same propositional variable.

Remark 7 *Definition 6 does not uniquely define a sentence in DNF, since we can take the pairs in best_fit(ψ, φ) in any order. However, all such sentences will be equivalent modulo $\vdash$.*

Example 8 *Computing revisions of sentences.*

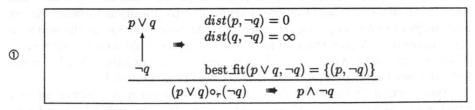

Notice that $(p \vee q) \wedge \neg q \equiv p \wedge \neg q$.

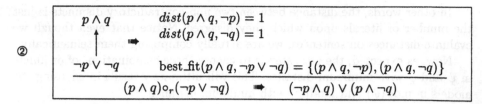

In Example ② above, we want to accept as much as possible of the informative content of $p \wedge q$ subject to $\neg p \vee \neg q$. The only way to do it is by choosing either $\neg p$ or $\neg q$ (and hence accept $\neg p \vee \neg q$) and then accept either q or p from $p \wedge q$.

Theorem 9 *The revision operator defined above verifies the AGM postulates for Belief Revision.*

Proof The proof can be done via Katsuno and Mendelzon's semantical characterisation of revision operators [6], and can be found in [3]. □

Obviously, $\circ_r$ has the properties implied by the postulates. As a consequence, the operation is commutative if and only if the two sentences are consistent with each other. The operator is not associative in general though.

Proposition 10 $\circ_r$ *is not associative.*

Proof We give a simple example as proof, which we use to discuss the subject briefly.

① $(p\circ_r p \rightarrow q)\circ_r \neg q$:

$$\boxed{\begin{array}{c} p \\ \circ_r \\ \boxed{p \rightarrow q} \\ \circ_r \\ \neg q \end{array}} \Rightarrow \boxed{\begin{array}{c} p \wedge q \\ \circ_r \\ \neg q \end{array}} \Rightarrow \boxed{p \wedge \neg q}$$

② $p\circ_r(p \rightarrow q\circ_r \neg q)$:

$$\boxed{\begin{array}{c} p \\ \circ_r \\ \boxed{\begin{array}{c} p \rightarrow q \\ \circ_r \\ \neg q \end{array}} \end{array}} \Rightarrow \boxed{\begin{array}{c} p \\ \circ_r \\ \neg p \wedge \neg q \end{array}} \Rightarrow \boxed{\neg p \wedge \neg q}$$

□

In ① above, the agent initially believes in p and $p \rightarrow q$, which results in a belief state in which he believes in both propositions p and q. When new information $\neg q$ arrives, he gives up the belief in q, but keeps the belief in p. On the other hand, in ② above, the agent reasons backwards. His priority is to believe in $\neg q$, and subject to this, in $p \rightarrow q$. He then is led to believe in $\neg p$, as the sentences are consistent with each other (if he believed in p, he would have to believe in q, which he does not). As he proceeds in his reasoning, he realises that he cannot keep the belief in p, because it contradicts the other two beliefs.

Even though in the right associative interpretation new information does require a complete recomputation of the belief state, we will advocate for its use, as it leads to a reflection upon the agent's own beliefs. One can think of the sequence of revisions as a line of reasoning from the most important sentence to the least important one. At each step only what can be consistently accepted with the current reasoning is taken in.

3 PDBs - Prioritised Databases

The more general framework of structured belief bases will be presented later. Before that, it is useful to show a simpler mechanism to illustrate the whole idea.

PDBs were devised having three principles in mind: conservativity, prioritisation and consistency. Conservativity is associated with the notion of minimal change to the belief base. However, unlike in other base revision formalisms, we require that whenever a sentence is inconsistent with sentences having higher priority, as much as possible of its informative content is kept, instead of just retracting it. Prioritisation ensures that sentences with higher priority have precedence over those with lower priority and finally Consistency requires that some mechanism is provided to solve conflicts when they arise. As a rule, we expect that the belief state associated with the PDB is consistent unless the sentence with highest priority is inconsistent. The revision operator provides a convenient and efficient way to preserve consistency and ensure conservativity at the same time.

Definition 11 *A Prioritised Database (PDB) is a finite list of sentences in DNF.*

As we have mentioned before, the order in which the sentences appear in the list is associated with their priority in the PDB. Sentences appearing later in the list have priority over those appearing earlier.

The idea is to think of a PDB as a sequence of revisions of sentences. However, as shown in the previous section, $\circ_r$ is not associative. Therefore, there are two basic ways of interpreting the sequence of sentences in a given PDB: either perceiving the revision process as a left associative operation or as a right associative one.

If Δ is a PDB, we will refer to the first interpretation as *left Delta*, in symbols, $^*\Delta$, and to the second one as *right Delta*, in symbols, Δ^*. Formally,

Definition 12 *Let $\Delta = [\varphi_1, \varphi_2, \ldots, \varphi_k]$ be a PDB.*

$$
^*\Delta =
\begin{cases}
\varphi_k & \Rightarrow \text{ if } k = 1 \\
((\varphi_1 \circ_r \varphi_2) \circ_r \ldots) \circ_r \varphi_k & \Rightarrow \text{ if } k > 1
\end{cases}
$$

$$
\Delta^* =
\begin{cases}
\varphi_k & \Rightarrow \text{ if } k = 1 \\
\varphi_1 \circ_r (\ldots \circ_r (\varphi_{k-1} \circ_r \varphi_k)) & \Rightarrow \text{ if } k > 1
\end{cases}
$$

In either case, the sentence with highest priority in a PDB is always accepted:

Proposition 13 *Let $\Delta = [\varphi_1, \varphi_2, \ldots, \varphi_k]$ be a PDB. $^*\Delta \vdash \varphi_k$ and $\Delta^* \vdash \varphi_k$.*

Proof This comes as a consequence of Theorem 9, because by the success postulate, $\varphi \circ_r \psi \vdash \psi$. $\therefore \ ^*\Delta = ((\varphi_1 \circ_r \varphi_2) \circ_r \ldots) \circ_r \varphi_k \vdash \varphi_k$. On the other hand, remember that $\varphi_{k-1} \circ_r \varphi_k \vdash \varphi_k$, $\varphi_1 \circ_r (\ldots \circ_r (\varphi_{k-1} \circ_r \varphi_k)) \vdash (\ldots \circ_r (\varphi_{k-1} \circ_r \varphi_k))$ and that $\vdash$ is transitive. □

Thus, in order to *accept* new information in a PDB Δ, all we have to do is to append the new sentence in DNF, say φ, to Δ. The sentences do not actually have to be stored in DNF, so long as they are converted to DNF before submission to the revision operator. There are efficient methods to convert sentences from conjunctive normal form to disjunctive normal form and vice-versa, e.g. the *matrix method* (see for instance [1, Chapter 2, Section 4.3]).

As mentioned in the previous section, we advocate for the right associative interpretation. This has profound consequences. First of all, the right associative interpretation results in a distinction between the *epistemic state* of the agent and the corresponding belief set. The epistemic state is associated with the PDB and the belief set with the result of applying successively the revision operator to the sentences in the PDB. Notice that the left associative interpretation would make this distinction non-existent. For instance, the epistemic state K represented by the PDB $\Delta_k = [\varphi_1, \varphi_2, \ldots, \varphi_n]$ is associated with the belief set $\varphi_1 \circ_r (\varphi_2 \circ_r \ldots (\varphi_{n-1} \circ_r \varphi_n))$. The revision of K by ψ is thus obtained by evaluating the epistemic state K', represented by the PDB $\Delta_{k'} = [\varphi_1, \varphi_2, \ldots, \varphi_n, \psi]$ which is associated with the belief set $\varphi_1 \circ_r (\varphi_2 \circ_r \ldots (\varphi_k \circ_r \psi))$.

Obviously, epistemic states represented as such carry more information than the resulting belief sets. That is, the equivalence between belief sets does not imply equivalence between epistemic states. Consider the two epistemic states $[p \rightarrow q, p]$ and $[p \wedge q]$ with equivalent belief sets $p \wedge q$. A revision of the first by $\neg p \vee \neg q$ would preserve the belief p, whereas the revision of the second would not, because a preference for p was explicitly stated in the first.

As suggested in the proof of Proposition 10, PDBs could be extended to allow for more complex structures of revisions in a given point. This is the main motivation for the definition of Structured Belief Bases which are presented next. Essentially, the extra structure of SBBs allows for the specification of subsets of the belief base where the revision of sentences should be applied first.

4 Structured Belief Bases

Definition 14 (Structured Belief Bases) *A structured belief base (SBB) is a finite list of objects each of which is either a sentence in DNF; or a finite list of sentences in DNF, or another SBB.*

Definition 15 (Level of an SBB) *The level of any sentence of propositional logic is 0. Let $\Gamma = [\Delta_1, \Delta_2, \ldots, \Delta_k]$ be a SBB. The level of Γ, in symbols $\mathrm{level}(\Gamma)$ is defined recursively as follows:*

$$\mathrm{level}(\Gamma) = \max_{i=1}^{k} \{\mathrm{level}(\Delta_i)\} + 1$$

Thus, a SBB of level 1 is just a PDB as described earlier. Naturally, an interpretation has to be given to the way the information in an SBB is perceived. Again we maintain the right associative interpretation to maintain conceptual integrity with the interpretation given to PDBs. However, since now groupings are allowed, either interpretation can be obtained (see Example 17). This allows for extra flexibility in controlling where the coherence view will be applied, since all logical consequences of each group are computed when the group is evaluated.

Definition 16 (Perceiving the information in an SBB) *Let* $\Gamma = [\Delta_1, \Delta_2, \ldots, \Delta_k]$ *be a SBB. The interpretation of the information in* Γ *is a sentence in DNF (denoted* $\mathrm{val}(\Gamma)$*). Naturally, if* φ *is a sentence, then* $\mathrm{val}(\varphi) = \varphi$*. For SBBs of higher level,* val *is defined recursively as follows:*

$$\mathrm{val}(\Gamma) = \begin{cases} \mathrm{val}(\Delta_k) & \Rightarrow if\ k = 1 \\ \mathrm{val}(\Delta_1) \circ_r (\ldots \circ_r (\mathrm{val}(\Delta_{k-1}) \circ_r \mathrm{val}(\Delta_k))) & \Rightarrow otherwise \end{cases}$$

Example 17 (Obtaining the left associative interpretation) *Let* $\Delta = [\varphi_1, \ldots, \varphi_k]$ *be a PDB. The SBB* $\Gamma = [[[[\varphi_1, \varphi_2], \ldots], \varphi_{k-1}], \varphi_k]$ *corresponds to* $*\Delta$:

$$\mathrm{val}(\Gamma) =$$
$$\mathrm{val}([[[\varphi_1, \varphi_2], \ldots], \varphi_{k-1}]) \circ_r \mathrm{val}(\varphi_k) =$$
$$(\mathrm{val}([[\varphi_1, \varphi_2], \ldots]) \circ_r \mathrm{val}(\varphi_{k-1})) \circ_r \varphi_k =$$
$$((\mathrm{val}([\varphi_1, \varphi_2]) \circ_r \ldots) \circ_r \varphi_{k-1}) \circ_r \varphi_k =$$
$$\vdots$$
$$= (((\varphi_1 \circ_r \varphi_2) \circ_r \ldots) \circ_r \varphi_{k-1}) \circ_r \varphi_k$$

The extra structure available is very useful to represent components with different priorities in a belief base, as can be seen in the following examples.

Example 18 (Extra-logical information for Belief Revision) *Suppose the belief base is represented as the SBB* $\Gamma = [Default, KB, IC]$*, where* $Default$ *represents information with least priority in* Γ*,* KB *represents ordinary beliefs in the base and* IC *contains general rules to be verified in all belief states. We omit the representation of sentences in DNF and the intermediate steps obtained by the revision operator.*

① *Initially,* $Default = [bt \rightarrow ft]$*, representing the sentence* if tweety is a bird, then it flies; $KB = [bt]$*, representing the information that tweety is a bird; and* $IC = [pt \rightarrow \neg ft]$*, representing an instance of the rule that penguins do not fly:* if tweety is a penguin, then it does not fly. *In order to have the rule verified in all belief states all is needed is to give it the highest priority in the SBB. This is the way of representing integrity constraints. Notice that multiple levels of priorities for the integrity constraints can be specified. The default information is given the lowest priority in the SBB.*

The information in KB *is subjected to the integrity constraints and then submitted to the defaults. This is to be understood as extracting from the defaults only the information compatible with the knowledge base. Also, only the information in* KB *consistent with* IC *is used in the reasoning (see figure on the right hand side).*

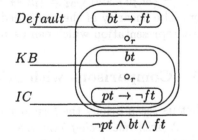

Since the agent does not have information as to whether or not tweety is a penguin, he assumes that it is not, in order to accept the information represented in the defaults. The result is

‘‘`tweety is a bird, it is not a penguin and it can fly.`’’

It can be verified that the three sentences are consistent and the result is exactly the conjunction of all of them.

② If later on the agent finds out that tweety is in fact a penguin, he can add this information to the knowledge base by appending it to the corresponding list (KB). This guarantees that it will be accepted in KB, but not necessarily in the SBB as a whole, as KB itself is subject to IC. We get the picture on the right hand side.

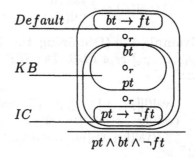

$$pt \wedge bt \wedge \neg ft$$

An alternative solution would be to reject the new information completely if it is found inconsistent with the integrity constraints. In this case, it is would not be a matter of verifying or not the principle of primacy of update, but of choosing to decline the revision process itself when appropriate.

Information conveyed by the defaults which could be consistently accepted with the rest of the belief state would also be incorporated. It is possible to avoid this, by requiring that the revision of the defaults is declined when that class is found inconsistent with the rest of KB. This can be effectively performed with sentences in DNF, since $\text{mod}(\psi) \cap \text{mod}(\varphi) \neq \emptyset$ iff there are disjuncts $P \in \psi$ and $Q \in \varphi$, such that $dist(P,Q) = 0$.

The final conclusion is

‘‘`tweety is a bird, it is a penguin and it cannot fly.`’’

SBBs can also be perceived as a belief base together with a family of total orderings one for each embedded SBB. The orderings induce equivalence classes at each level which are themselves linearly ordered. For instance, in example ②, we would have two equivalence classes w.r.t. the top ordering associated with Γ: $C_1 = \{bt \rightarrow ft\}$ and $C_2 = \{bt, pt, pt \rightarrow \neg ft\}$, where C_2 has priority over C_1. C_2 is itself associated with a second ordering whose equivalence classes are $C_{21} = \{bt, pt\}$ and $C_{22} = \{pt \rightarrow \neg ft\}$, where C_{22} has priority over C_{21} and so on. The linearity of the embedded structures makes it possible to have a simple list representation which can be promptly evaluated by the revision operator.

5 Comparisons with prioritised base revision

Let us recap briefly the formalism of prioritised base revisions.

A *prioritised belief base* is a pair $\Gamma = \langle K, \sqsubseteq \rangle$, where K is a belief base and $\sqsubseteq$ is a total pre-order on K. $\sqsubseteq$ represents priorities in K, where $x \sqsubseteq y$ denotes that

y has at least the same priority as x. If $\sqsubseteq$ is also antisymmetric, Γ is called a *linear prioritised belief base*. It is assumed that $\sqsubseteq$ always has maximal elements. The relation $\sqsubseteq$ is called an *epistemic relevance ordering* by Nebel.

K can be partitioned into a set of equivalence classes induced by $\sqsubseteq$. That is, a family of subsets of K whose elements are all at the same priority level. Let us call such partitioning $\overline{K}$. If K is finite, Γ can be represented as a list $\Gamma = [K_1, \ldots, K_n]$ where the $K_{i's}$ are the partitions of K generated by the equivalence relation induced by $\sqsubseteq$. Moreover, if Γ is linear, then each K_i is just a singleton.

In order to obtain a revision of Γ by φ, $\Gamma * \varphi$, a Levi Identity approach is used. Γ is first contracted by $\neg\varphi$ and then expanded by φ. The contraction of Γ by $\neg\varphi$ uses the epistemic relevance ordering and is called the *prioritised removal of $\neg\varphi$ from Γ*, in symbols, $\Gamma \Downarrow \neg\varphi$. It is a *family* of subsets of K each of the form:

$$X = \bigcup_{K_i \in \overline{K}} \{H_i\}$$

where each H_i is a subset of K_i. Furthermore, $X \in \Gamma \Downarrow \neg\varphi$ iff $\forall X' \subseteq K$ and $\forall i$:

$$X \cap (K_i \cup \ldots \cup K_n) \subset X' \cap (K_i \cup \ldots \cup K_n) \text{ implies } \neg\varphi \in Cn(X')$$

where K_n is the class containing the maximal elements in K w.r.t. $\sqsubseteq$. That is, starting from the most prioritised equivalence class, we consider maximal subsets of each class that together with the set being constructed do not imply $\neg\varphi$. There will possibly be a number of such X's, as there are possibly many combinations of subsets of each class that do not imply $\neg\varphi$. Therefore, the sentences in the prioritised removal of $\neg\varphi$ from Γ will be in the set $\bigcap \{Cn(X) \mid X \in \Gamma \Downarrow \neg\varphi\}$.

The prioritised base revision of Γ by φ then amounts to the following equivalence

$$\Gamma * \varphi =^{\mathrm{def}} Cn(\bigcap \{Cn(X) \mid X \in \Gamma \Downarrow \neg\varphi\} \cup \{\varphi\})$$

Thus, in the general case, it is not possible to iterate the revision process. However, if K is finite, then $\Gamma * \varphi$ can be finitely represented and the result is simply

$$Cn(\bigvee (\Gamma \Downarrow \neg\varphi) \wedge \varphi)$$

Finite prioritised belief bases and the PDBs presented in Section 3 have much in common. If the prioritised belief base is not linear, but is finite, it is possible to represent it as a PDB by taking the list of conjunctions of the sentences in each priority class. Unlike the removal of sentences that occur in the prioritised base revision, the amount of information of each class preserved depends on the distance function used. This is subject to the same criticisms about representing knowledge bases as the conjunction of their sentences. On the other hand, if there are many sentences in each class, the number of subsets in $\Gamma \Downarrow \neg\varphi$ could be considerably large.

Finite linear prioritised belief bases do not suffer such drawback. As the classes in $\Gamma\Downarrow\neg\varphi$ are simply singletons, they provide a unique result. However, it has been pointed out that the restriction of linearity imposed can be quite strong [4, page 94]. PDBs and this kind of prioritised belief bases are very similar, because the singletons can be directly associated to sentences in the PDB. The main difference is that linear prioritised belief bases operate on a all-or-nothing basis at each level. That is, whenever a sentence ψ from a class K_i yields an inconsistency with the sentence to be revised, it is ignored and the next best class is tested. On the other hand, in a PDB the sentence ψ would be *revised* by the sentences with higher priority instead. As a result, PDBs preserve more of the informative content of the belief base than prioritised linear base revisions, as can be seen in the example below.

Example 19 *Consider the following linear prioritised belief base:* $K = [\{p\wedge q\}, \{p \to r\}]$ *and suppose we want to revise it by* $\neg q$. *We first need to construct* $K\Downarrow q$. *We have to take elements from* K *starting at the highest priority level and make sure that they do not imply* q. *Thus, the result is just* $\{\{p \to r\}\}$, *since the inclusion of* $p \wedge q$ *in the set* $\{p \to r\}$ *would result in the implication of* q. *The prioritised revision of* K *by* $\neg q$ *is then* $Cn(\{p \to r\} \cup \{\neg q\})$. *However, if we had* $\{p, q\}$ *instead of* $\{p \wedge q\}$ *the result would be different, as the belief in* p *would be kept.*

By using PDBs the required revision would be obtained by simply appending $\neg q$ *to the PDB* $[p \wedge q, p \to r]$, *and computing the sequence of revisions (we write the sentences in DNF in the diagrams below):*

The belief in p is kept, and since $p \to r$, the agent also holds the belief in r. In a prioritised base revision not only would the sentence $p \wedge q$ be retracted (and consequently the belief in p), but also the belief in r previously held in the belief state. This also shows that PDBs (and SBBs in general) are not so sensitive to syntax as ordinary base revision. When each cluster in an SBB is resolved into a sentence, the revision is accomplished by the operator which complies with the AGM postulates. At that stage, the syntax becomes irrelevant, because in the revision process what is actually being taken into account are the set of models of the sentences involved.

SBBs provide a more flexible mechanism for the specification of priorities. Consequently, more complex structures can be expressed. It also has the advantage of always providing a unique result and making it possible to iterate revisions very straightforwardly. In the next section we analyse how the requirement on linearity of the ordering of an SBB could be relaxed to enhance expressibility.

6 Initial belief bases provided without ordering

In the previous sections we presented a framework to reason about belief change using priorities and a revision operator. We considered belief bases as structured sets of components. Conflicts among components are solved by the revision operator which complies with the AGM postulates for Belief Revision. The components are hierarchically ordered so that complex priorities among them can be represented.

The order which represents the hierarchy between the components is required to be linear. If we consider the evolution of the belief states of an agent through time, the linearity of the ordering does not seem to be so restrictive, as in many situations new beliefs are accepted one at a time. However, the assumption that the *initial* set of beliefs of the agent is also linearly ordered is indeed quite strong. In this section we analyse possible ways of loosening this restriction.

A flat initial sets of beliefs cannot be represented in our methodology because we can only have a single sentence at each terminal point of the structure. We have seen that taking the conjunction of the sentences in K incurs in some loss of information and we want to consider ways of implementing the revision of K by a sentence φ which support iteration. In other words, we expect the result of the revision process to be an entity of the same type of the given input.

If an agent is faced with contradictory information and is left with a number of possible states of mind which as far as his inference system is concerned are all equally plausible, he cannot simply make an arbitrary choice. The sensible thing to do seems to be to keep all possibilities open until he has gathered enough information to make a judicious choice.

Remember that revisions by a sentence φ under the partial meet approach are done by considering the maximal subsets of the base that do not imply $\neg\varphi$. If the initial base K is finite, there will be a finite number of such subsets. If it is not possible to make a choice between these subsets based on logical grounds only, one should consider that all of them are *possible belief bases* representing the current belief state of the agent, until more information has become available.

However, it is well known that it is quite difficult to handle several possible belief bases. If the base is finite, the usual approach is to take the disjunction of the conjunction of the sentences in each base, but we would like to avoid this in order to reduce the amount of information lost in the process. We should therefore somehow recombine these possible belief bases into a single one.

If $K_1, \ldots, K_n$ are the maximal subsets of K that do not imply φ, one alternative is to consider the following sets:

$$K' = \bigcap_{i=1}^{n} K_i \quad \text{and} \quad K'' = \{\vee_i (\wedge(K_i - K'))\}$$

We then proceed to take the union $K' \cup K''$. Obviously, $\varphi \notin Cn(K' \cup K'')$.

That is, we preserve the sentences that are in the intersection, but take the disjunction of the conjunction of the sentences in each possible belief base that are not in the intersection. This at least restore the sentences in the base whose belief is not being questioned. In order to fit this into the formalism of SBBs, we calculate the revision of sentences with higher priority than the initial base and then compute the sets above according to the result of the revision. Now, we can consider the conjunction of the sentences in the union $K' \cup K''$, but this time knowing that it will be consistent with the rest of the sentences in the base (the linearly ordered part). Later on, if more information w.r.t. the priorities of these sentences is available, it can be used to provide a finer tuning.

Example 20 *Let us consider an example by Nebel [7, page 57]. We make some minor modifications to illustrate the ideas suggested here.*

"During the investigation of a crime, detective Holmes decides to interrogate a suspect. The suspect tells him that he likes very much to swim, so at the time of the crime he was actually swimming at the beach":

$$\text{the suspect likes to swim} \quad \Rrightarrow \quad l$$
$$\text{swimming at the beach} \quad \Rrightarrow \quad b$$

"Occasionally, Mr. Holmes remembers that at that particular day it was quite sunny, because when he left for work it was too bright and he had to come back to fetch his sunglasses. However, it was early when he left so he is not very sure whether it stayed sunny all day. He decides to add this information anyway."

$$\text{sun is shining} \quad \Rrightarrow \quad s$$

He reckons these three sentences are roughly at the same priority level, so at this stage, he is left with the initial belief base $\{l, b, s\}$. But he is definitely sure that if the sun is shining and one goes to the beach for a swim they get a sun tan (t).

$$\begin{array}{l} \text{swimming at the beach when the sun is} \\ \text{shining results in a sun tan} \end{array} \quad \Rrightarrow \quad (b \wedge s) \to t \equiv \neg b \vee \neg s \vee t$$

Mr. Holmes gives higher priority to the information he is sure about. This results in the following extended PDB:$[\{l, b, s\}, (\neg b \vee \neg s \vee t)]$. From this information, Mr. Holmes observes that the suspect should have a sun tan (t), which he does not. So he wants to incorporate $\neg t$ into his belief base.

$$
\begin{array}{ccccc}
\left(\begin{array}{c} K = \{l, b, s\} \\ \uparrow \\ (\neg b \vee \neg s \vee t) \end{array}\right)
& \Rightarrow &
\left(\begin{array}{c} (l \wedge b \wedge s) \\ \circ_r \\ (\neg b \vee \neg s \vee t) \end{array}\right)
& \Rightarrow &
\left(\begin{array}{c} t \wedge l \wedge s \wedge b \end{array}\right)
\end{array}
$$

The maximal subsets of K that together with $\neg b \vee \neg s \vee t$ do not imply $\neg t$ are $\{l, b\}$ and $\{l, s\}$, whose intersection is $\{l\}$. The disjunction of the sentences not in the intersection is $b \vee s$. Thus, in order to restore consistency, he is left with the new belief base $K' = \{l, b \vee s\}$.

He then adds this information (¬t) to the modified PDB resulting in the following structure: $[\{l, b \vee s\}, (\neg b \vee \neg s \vee t), \neg t]$. To find out what the consequences of the PDB are, we take the conjunction of the sentences in the unordered part and revise it by the series of revisions of the linear part. This results in the sentence $(\neg t \wedge \neg s \wedge l \wedge b) \vee (\neg t \wedge \neg b \wedge l \wedge s)$, which entails the two sentences with higher priorities as well as the sentence which does not yield any contradiction, but questions whether b or s should hold, as expected.

Later on, Mr. Holmes phones the meteorology service and finds out that in fact it was sunny during that day: s, resulting in the PDB $[\{l, b \vee s\}, (\neg b \vee \neg s \vee t), \neg t, s]$. s is consistent with the information in the rest of the PDB, and now enables Mr. Holmes to conclude that the suspect was lying about being swimming at the beach that day. The final conclusion is $(\neg t \wedge \neg b \wedge l \wedge s)$.

7 Conclusions and Future Work

In this paper we presented a framework to reason about belief change. Belief bases are represented as structured sets of components. Each component is either a sentence or another structured set of sentences. This makes it possible to express complex levels of priorities between sentences. Conflicts among components are solved by a revision operator which complies with the AGM postulates for Belief Revision and can be efficiently implemented.

One of the main advantages of the methodology lies on its simplicity. The revision operator is based on simple superimposition of literals and on a distance function which is used to evaluate the degree of change caused. Iteration of the revision process can be easily achieved in the method presented.

An extension to the work may be obtained by allowing other operations among components. These might be of an epistemic nature, such as contraction, or adequate to reasoning about change, e.g. update, which would make the framework suitable to reason about a changing world as well. Some of these operations have already been defined in a similar way to the revision one shown here and can be found in [3,9]. However, a further investigation on the best ways to apply the operators still remains to be done. For instance, if an update operation is used, it seems unintuitive to consider a right associative interpretation of the sequence of sentences, as we have done for the revision one.

The distance used in the revision operator yields a total ordering of interpretations of the language w.r.t. to a given sentence. Given a sentence ψ, we can obtain the following semantical ordering $\leq_\psi$ on $\mathcal{I}$ w.r.t. to "closeness" to ψ: $M \leq_\psi N$ iff $M = N$ or $\exists I \in \text{mod}(\psi)$ s.t. $\forall J \in \text{mod}(\psi).dist(M, I) \leq d(N, J)$.

The ordering above is of course highly dependent on the function $dist$. It would be interesting to consider other distance functions and the effects they would have on the whole process. An interesting discussion on similarity and distance functions can be found in [8].

Also, the modifications proposed in the previous section represent only an initial approach to the problem of unordered belief bases. The work now being carried out focuses on how to deal with partially ordered belief bases. The idea is to use the same superimposition algorithm of Definition 1. Disjuncts will be ranked according to the partial ordering available. We now look for the combination of disjuncts from different sentences that maximizes the satisfiability of sentences in the base. Whenever there is a disagreement between two disjuncts, we try to solve it via the partial ordering and the superimposition algorithm of Definition 1, if their relative priorities are available. If not so, we split the combination in two, and include each of the disjuncts in each one of them. This is equivalent to accepting the disjunction of the two sentences and expresses an unresolved choice between them.

Acknowledgements

Odinaldo Rodrigues is supported by CNPq – Conselho Nacional de Desenvolvimento Científico e Tecnológico/Brasil. Dov Gabbay is supported by EPSRC, GR/K57268 (Proof Methods for Temporal Logics of Knowledge and Belief).

References

1. Karl Hans Bläsius and Hans-Jürgen Bürckert. *Deduction Systems in Artificial Intelligence*. Ellis Horwood Limited, 1989.
2. M. Dalal. Investigations into a theory of knowledge base revision: Preliminary report. *Proceedings of the 7th National Conference on Artificial Intelligence*, pages 475–479, 1988.
3. Dov Gabbay and Odinaldo Rodrigues. A methodology for iterated theory change. In Dov M. Gabbay and Hans Jürgen Ohlbach, editors, *Practical Reasoning - First International Conference on Formal and Applied Practical Reasoning, FAPR'96*, Lecture Notes in Artificial Intelligence. Springer Verlag, 1996.
4. P. Gärdenfors and Hans Rott. Belief revision. In C. J. Hogger Dov Gabbay and J. A. Robinson, editors, *Handbook of Logic in Artificial Intelligence and Logic Programming*, volume 4, pages 35–132. Oxford University Press, 1995.
5. Peter Gärdenfors. *Knowledge in Flux: Modeling the Dynamics of Epistemic States*. A Bradford Book - The MIT Press, Cambridge, Massachusetts - London, England, 1988.
6. Hirofumi Katsuno and Alberto O. Mendelzon. On the difference between updating a knowledge base and revising it. *Belief Revision*, pages 183–203, 1992.
7. Bernhard Nebel. Belief revision and default reasoning: Syntax-based approaches. In J. Allen, R. Fikes, and E. Sandewall, editors, *Proceedings of the Second International Conference on Principles of Knowledge Representation and Reasoning*, pages 417–428. Morgan Kaufmann, 1991.
8. Ilkka Niiniluoto. *Truthlikeness*. Kluwer Academic Publishers, 1989.
9. Odinaldo Rodrigues. Transfer report: Mphil/phd. Imperial College of Science, Technology and Medicine, February 1996. Department of Computing.
10. Odinaldo Rodrigues. *A methodology for iterated information change*. PhD thesis, Department of Computing, Imperial College, To appear.

Entrenchment Relations: A Uniform Approach to Nonmonotonicity*

Konstantinos Georgatos

Dipartimento di Informatica e Sistemistica, Università di Roma "La Sapienza", Via Salaria 113, Roma 00198, Italy, e-mail: geo@dis.uniroma1.it

Abstract. We show that Gabbay's nonmonotonic consequence relations can be reduced to a new family of relations, called entrenchment relations. Entrenchment relations provide a direct generalization of epistemic entrenchment and expectation ordering introduced by Gärdenfors and Makinson for the study of belief revision and expectation inference, respectively.

Keywords: Nonmonotonic consequence, epistemic entrenchment, belief revision.

1 Introduction

Nonmonotonicity has offered great promise as a logical foundation for knowledge representation formalisms. The reason for such a promise is that nonmonotonic logic completes in a reasonable way our (incomplete) knowledge and withdraws conclusions in the light of new information. Therefore, most approaches to central problems of Artificial Intelligence, such as belief revision, database updating, abduction and action planning, seem to rely on one way or another to some form of nonmonotonic reasoning.

There are several proposals of logical systems performing nonmonotonic inference. Among the most popular of them are: circumscription ([39]), negation as failure ([10]), default logic ([44]), (fixed points of) various modal logics ([40]) and inheritance systems ([50]). However, and despite the numerous results inter-translating one of the above systems to the other, none of the above formalisms emerged as a dominant logical framework under which all other nonmonotonic formalisms can be classified, compared and reveal their logical content. This fact signifies that our intuition on the process of nonmonotonic inference is fragmented. Although all the above mentioned logics are worth be studied and employed as a central inference mechanism, they cannot serve as a place where our basic intuitions about nonmonotonicity can finally rest.

Addressing this problem, Gabbay in [17] proposed to study nonmonotonic inference through Gentzen-like context sensitive sequents. Following this proposal, a new line of research ([17],[35],[27]) flourished by studying properties of the so-called *nonmonotonic consequence relations* leading to a semantic characterization through (a generalization of) Shoham's preferential models ([27],[36]). This line of research led to a classification of several nonmonotonic formalisms and recognized several logical properties that a nonmonotonic system should desirably satisfy such as cumulativity ([17]),

---

* Work supported by Training through Research Contract No. ERBFMBICT950324 between the European Community and Università degli Studi di Roma "La Sapienza".

distributivity ([27]) and rational monotonicity ([30]). However, there are two disadvantages of this framework:

- nonmonotonic consequence relations express the sceptical inference of a nonmonotonic proof system and therefore fail to describe nonmonotonicity in its full generality, that is, the existence of multiple extensions.
- it does not seem that there is a straightforward way to design a nonmonotonic consequence relation from existing data unless they already encode some sort of conditional information (see [30]).

These two disadvantages suggest that a nonmonotonic consequence relation is not a primitive notion but *derived* from a more basic inference mechanism.

In this paper, we shall introduce a novel framework for generating nonmonotonic inference, through a class of relations, called *entrenchment relations*. We shall see that the framework of entrenchment relations is at least as expressive as that of nonmonotonic consequence relations. In particular, nonmonotonic consequence relations can be *reduced* to entrenchment relations (in the classical case) while the inference defined through entrenchment relations admits and identifies the existence of multiple extensions. On the other hand, entrenchment relations seem to build inference easily and from the bottom up. Simple frequency data, for example, generates easily at least one class of them (rational orderings — see [2]).

Entrenchment relations are relations between single formulas and will be denoted by $\preceq$. "$\alpha \preceq \beta$" will be read as

β is at least as entrenched as α

in the sense that "α is more defeasible than β", or, in other words, "if α is accepted then so is β". For example, consider the partial description of a (transitive) entrenchment relation in Figure 1.

In that figure, a path upwards from α to β indicates that $\alpha \preceq \beta$, where $\preceq$ denotes the entrenchment relation. The entrenchment relation of Figure 1 says, for example, that $\perp$ is less entrenched than all formulas, f is less entrenched than $\neg p$, $b \to f$ and $f \to b$, and $f \to b$ is less entrenched than $p \to b$, $p \to \neg f$ and $\top$.

How will an entrenchment relation be used for inference? The idea is simple. We shall use entrenchment for *excluding* sentences.

A sentence α will infer (in a nonmonotonic way) another sentence β if α together with a sentence γ, *that is not less entrenched than* $\neg\alpha$, (classically) imply β.

The reason we exclude sentences less entrenched than $\neg\alpha$ is that if we allow such a sentence then we should also allow $\neg\alpha$. However, this brings inconsistency. For instance, using the above example and assuming p we should exclude all formulas less than $\neg p$. We are left with $\{p \to b, p \to \neg f, f \to b\}$. Adding p on those and closing under classical consequence, we get $Cn(p, b, \neg f)$. This is the nonmonotonic theory of p. With no assumptions, we should only exclude formulas less or equal to $\perp$. In this case, there are two remaining consistent sets of sentences one containing $\neg b$ and $f \to b$ and the other f and $f \to b$. Closing under consequence, we get $Cn(\neg b, \neg f, \neg p)$ and

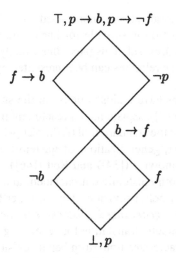

$\top, p \to b, p \to \neg f$

$f \to b$

$\neg p$

$b \to f$

$\neg b$

f

$\bot, p$

Figure 1. A (transitive) entrenchment relation.

$Cn(b, f, \neg p)$. Therefore, it is possible to have more than one possibility for extending the theory of our assumptions and that leads to the well-known phenomenon of multiple extensions.

The claim that entrenchment relations is a useful concept towards our understanding of nonmonotonicity will be substantiated by a series of representation results. We will show that Gabbay's nonmonotonic consequences relations can be expressed through entrenchment relations, and identify those classes of entrenchment relations which correspond to the classes of nonmonotonic consequence relations that have attracted special interest in the literature. In addition our framework provides more:

- *Uniformity.* Inference defined through an entrenchment relation remains the same throughout the above characterization.
- *Monotonicity.* Any strong cumulative nonmonotonic inference relies on a monotonic (on both sides) entrenchment relation.
- *Identification of multiple extensions.* The way we define inference allows the identification of multiple extensions. Therefore, both the sceptical and credulous approach towards nonmonotonicity are expressible in our framework.
- *A conceptually primitive view of nonmonotonicity.* In our framework, a nonmonotonic formalism separates into two logical mechanisms: classical logic and entrenchment.

Entrenchment relations provide a generalization of Gärdenfors-Makinson's expectation orderings introduced for the characterization of expectation nonmonotonic consequence relations ([19]). This result was later turned into a bijective correspondence and extended to rational consequence relations in [23]. In [15], incompletely specified expectation orderings were studied, but, to our knowledge, there is no study of such relations outside the non-Horn classes of nonmonotonic consequence relations. The main reason that such generalization is not immediate is that the original Gärdenfors-Makinson syntactic translation is suitable for linear entrenchment relations but inappropriate once linearity is dropped. This paper fills exactly this gap by showing that *all* nonmonotonic consequence relations can be represented by entrenchment relations.

Entrenchment relations have a close relative in the study of belief revision, called *epistemic entrenchment* ([18]). Epistemic entrenchment proved to be a very useful for belief revision and became the standard tool ([20],[52],[41],[42]) for studying the AGM postulates ([1]). Moreover, generalizations of epistemic entrenchment have been proposed by Lindström-Rabinowicz ([34]) and Rott ([46]). Lindström-Rabinowicz proposed to drop linearity from epistemic entrenchment and used such a partial ordering for the study of relational belief revision. On the other hand, Rott drops transitivity and keeps linearity for his *generalized epistemic entrenchment* and uses the original Gärdenfors-Makinson syntactic translation for generating belief revision functions. As a consequence, Rott characterizes non-Horn belief revision functions with Horn epistemic entrenchment and vice versa. Therefore, our results cannot be derived, even through a suitable translation, by the above works, although intuition and motivation should be credited to both.

Relations of orderings of sentences with other systems performing some sort of nonmonotonic reasoning are abundant ([7],[8],[28],[29],[47],[53],[13]), such as Pearl's system Z ([43]), conditional logic ([49],[31]), and possibilistic logic ([12]). It is worth mentioning that orderings appear frequently in the literature of nonmonotonic logic. Orderings of models lead to the preferential model framework ([48],[27],[25],[36],[11]), while ordering of sentences lead to *prioritization* ([14],[21],[5]). Most nonmonotonic formalisms have been enriched with priority handling ([32],[3],[9]). However, it should be made clear that entrenchment relations are *not* prioritization. Priorities are *extralogical* orderings of formulas. On the other hand, entrenchment relations is a logical formalism: a set of proof-theoretic rules that has nothing to do with arbitrary orderings of formulas (of course, an interesting question is if it can serve to encode those). This, we find, is apparent by the way inference is defined by them, and the fact that, by the results contained herein, nonmonotonic inference is reduced on entrenchment relations *alone*. Therefore, the study of entrenchment relations falls into the study of abstract consequence, as nonmonotonic consequence relations or abstract default rules ([38],[6]). However, contrary to the latter, entrenchment relations are simpler, binary relations and independent of the underlying logic, classical or four-valued. In fact, this independence is the most promising feature of entrenchment relations which, however, has not been exploited in this paper.

The further contents of this paper are as follows. In Section 2, we present entrenchment relations. We discuss their informal meaning and present various properties of them. Then, in Section 3, we define the notion of maxiconsistent and weakly maxiconsistent inference as derived from a pair of relations. Both inference schemes can generate all nonmonotonic consequence relations. In Section 4, we review nonmonotonic consequence relations and present our representation results. In Section 5, we summarize and give some directions for further work.

2 Entrenchment Relations

Gärdenfors and Makinson recently showed ([19]) that the study of a strong non-Horn class of nonmonotonic consequence relations, called *expectation inference relations*, can be reduced to the study of a particular class of linear preorders among sentences called *expectation orderings*.

The interpretation of expectation orderings which Gärdenfors and Makinson proved equivalent to expectation inference relations is the following. Assume there is an ordering $\preceq$ of the sentences of a propositional language $\mathcal{L}$, where $\alpha \preceq \beta$ means "β is at least as expected as α" or "α is at least as surprising as β". Therefore, $\preceq$ is a relation comparing degrees of defeasibility among sentences. This interpretation of $\preceq$, as well as a similar one based on possibility given in [15], although seems fit for the particular class of nonmonotonic inferences it characterizes, has in our opinion the following disadvantages. First, it has a complicated flavor by relying on notions such as expectation, defeasibility, and surprise that are far from primitive. Second, it points to a semantical interpretation by committing to a subjective evaluation of sentences and therefore is lacking the proof-theoretic interpretation meant for relations generating inference. Finally, this interpretation loses its plausibility once we weaken one of its defining properties (for example linearity) and restricts us to a unique class of orderings.

Entrenchment relations are nothing more than a generalization of the above class of orderings. We will drop first linearity from the preorder, for characterizing preferential inference, and subsequently transitivity. Note that the entrenchment interpretation is weakened once we drop transitivity: if a sentence α is less entrenched than β and β less entrenched than γ, then α should be less entrenched than γ. However, inference through an entrenchment relation remains the same, that is we still exclude sentences that relate to $\neg\alpha$, i.e. $\beta \preceq \neg\alpha$. This paper will not offer an interpretation for nontransitive entrenchment relations and the reader must rely on the syntactic constructions. Entrenchment relations define consequence and failure of transitivity indicates absence of (full) Cut.

Here are our assumptions on the language. We assume a language $\mathcal{L}$ of propositional constants closed under the boolean connectives $\vee$ (disjunction), $\wedge$ (conjunction), $\neg$ (negation) and $\rightarrow$ (implication). We use Greek letters α, β, γ, etc. for propositional variables. We also assume an underlying consequence relation $\vdash \subseteq 2^{\mathcal{L}} \times \mathcal{L}$ that will act as the underlying proof-theoretic mechanism. For the sake of simplicity, we will use classical propositional calculus. We denote the consequences of α under $\vdash$ with $\mathrm{Cn}(\alpha)$ and write $\mathrm{Cn}(X, \alpha)$ for $\mathrm{Cn}(X \cup \{\alpha\})$. We should add that nonmonotonic inference that does not contain classical tautologies is a rather rare exception.

We shall assume the following three basic properties for entrenchment relations:

1. $\alpha \preceq \alpha$ (Reflexivity)
2. If $\alpha \vdash \beta$ and $\beta \preceq \gamma$ then $\alpha \preceq \gamma$.   (Left Monotonicity)
3. If $\alpha \vdash \beta$ and $\beta \vdash \alpha$ then $\gamma \preceq \alpha$ iff $\gamma \preceq \beta$. (Logical Equivalence)

The meaning of Reflexivity is straightforward.

Left Monotonicity says that if β is less entrenched than γ so is any sentence stronger than β.

Finally, Logical Equivalence says that if a sentence is less entrenched than another sentence α then it must be less entrenched than any sentence logically equivalent to α. We summarize the above in the following definition of entrenchment frame.

Definition 2.1. An *entrenchment frame* is a triple $\langle \mathcal{L}, \vdash, \preceq \rangle$, where $\preceq$ is a relation on $\mathcal{L} \times \mathcal{L}$, called *entrenchment relation*, that satisfies the above properties, that is Reflexivity, Left Monotonicity and Logical Equivalence.

All properties of entrenchment relations mentioned in the subsequent appear on Table 1.

The following property has been considered in the framework of expectation orderings ([19])

If $\alpha \vdash \beta$ then $\alpha \preceq \beta$.　(Dominance)

In view of Reflexivity, Left Monotonicity implies Dominance. Given Dominance and Reflexivity of $\vdash$, Reflexivity of $\preceq$ follows. Dominance is a very useful property that is used abundantly in the subsequent and was in fact one of the defining properties of Gärdenfors and Makinson's entrenchment ordering and epistemic entrenchment.

The following property is derived by Left Monotonicity

$\alpha \vdash \beta$ and $\beta \vdash \alpha$ implies $\alpha \preceq \gamma$ iff $\beta \preceq \gamma$.

While Left Monotonicity allows us to strengthen arbitrarily sentences on the left, Bounded Left Disjunction and Left Disjunction allow us to weaken them. These properties amount to a disjunction property. An entrenchment relation will be called *disjunctive (weakly disjunctive)* if it satisfies Left Disjunction (Weak Left Disjunction). Similarly, for entrenchment frames. These properties are rather strong and will be used for a canonical (bijective) correspondence between entrenchment and nonmonotonic consequence relations.

Bounded Cut and Right Conjunction express our ability to strengthen the right part. Bounded Right Monotonicity (BRM), Right Monotonicity and Weak Right Monotonicity weaken the right part. It is worth noting that Right Conjunction makes a sentence, less entrenched than another sentence and its negation, less entrenched than all sentences. Right Conjunction allow us to combine the right sides using conjunction.

Here are some dependencies between properties we impose. Bounded Right Monotonicity follows from Right Monotonicity. Weak Bounded Right Monotonicity together with Bounded Cut implies Weak Left Disjunction. Weak Bounded Right Monotonicity and Weak Bounded Cut together are equivalent to Weak Equivalence. Bounded Cut and Bounded Right Monotonicity together imply Equivalence. While given Left Disjunction, Equivalence implies Bounded Cut and Bounded Right Monotonicity. Transitivity implies Right Monotonicity, and thus Bounded Right Monotonicity. Transitivity is equivalent to Right Monotonicity given Bounded Cut. Transitivity and Dominance implies Left Monotonicity.

Connectivity is the only non-Horn property among the above properties. Therefore, any class of entrenchment relations satisfying the above properties *except* Connectivity is closed under intersections. Gärdenfors and Makinson merged Connectivity and Right Conjunction into

$$\alpha \preceq \alpha \qquad \text{(Reflexivity)}$$

$$\frac{\alpha \vdash \beta \qquad \beta \preceq \gamma}{\alpha \preceq \gamma} \qquad \text{(Left Monotonicity)}$$

$$\frac{\alpha \vdash \beta}{\alpha \preceq \beta} \qquad \text{(Dominance)}$$

$$\frac{\alpha \vdash \beta \qquad \beta \vdash \alpha \qquad \gamma \preceq \alpha}{\gamma \preceq \beta} \qquad \text{(Logical Equivalence)}$$

$$\frac{\alpha \vee \beta \preceq \beta \qquad \alpha \vee \beta \preceq \alpha \qquad \alpha \vee \gamma \preceq \alpha}{\beta \vee \gamma \preceq \beta} \qquad \text{(Weak Equivalence)}$$

$$\frac{\alpha \preceq \beta \qquad \beta \preceq \alpha \qquad \gamma \preceq \alpha}{\gamma \preceq \beta} \qquad \text{(Equivalence)}$$

$$\frac{\alpha \vee \beta \preceq \alpha \qquad \alpha \vee \gamma \preceq \alpha}{\alpha \vee \beta \vee \gamma \preceq \alpha} \qquad \text{(Weak Left Disjunction)}$$

$$\frac{\beta \preceq \alpha \qquad \gamma \preceq \alpha}{\beta \vee \gamma \preceq \alpha} \qquad \text{(Left Disjunction)}$$

$$\frac{\alpha \vee \beta \vee \gamma \preceq \alpha \vee \beta \qquad \alpha \vee \beta \preceq \alpha}{\alpha \vee \gamma \preceq \alpha} \qquad \text{(Weak Bounded Cut)}$$

$$\frac{\gamma \preceq \alpha \vee \beta \qquad \beta \preceq \alpha}{\gamma \preceq \alpha} \qquad \text{(Bounded Cut)}$$

$$\frac{\alpha \vee \gamma \preceq \alpha \qquad \alpha \vee \beta \preceq \alpha}{\alpha \vee \beta \vee \gamma \preceq \alpha \vee \beta} \qquad \text{(Weak BRM)}$$

$$\frac{\gamma \preceq \alpha \qquad \beta \preceq \alpha}{\gamma \preceq \alpha \vee \beta} \qquad \text{(BRM)}$$

$$\frac{\alpha_0 \preceq \alpha_n \qquad \alpha_n \preceq \alpha_{n-1} \qquad \cdots \qquad \alpha_1 \preceq \alpha_0}{\alpha_n \preceq \alpha_0} \qquad \text{(Acyclicity)}$$

$$\frac{\alpha_0 \vee \alpha_1 \preceq \alpha_0 \qquad \alpha_1 \vee \alpha_2 \preceq \alpha_1 \qquad \cdots \qquad \alpha_n \vee \alpha_0 \preceq \alpha_n}{\alpha_0 \vee \alpha_n \preceq \alpha_0} \qquad \text{(Weak Acyclicity)}$$

$$\frac{\alpha \preceq \beta \qquad \beta \vdash \gamma}{\alpha \preceq \gamma} \qquad \text{(Right Monotonicity)}$$

$$\frac{\gamma \preceq \alpha \qquad \gamma \preceq \beta}{\gamma \preceq \alpha \wedge \beta} \qquad \text{(Right Conjunction)}$$

$$\frac{\alpha \preceq \beta \qquad \beta \preceq \gamma}{\alpha \preceq \gamma} \qquad \text{(Transitivity)}$$

$$\alpha \preceq \beta \text{ or } \beta \preceq \alpha \qquad \text{(Connectivity)}$$

Table 1. Properties for entrenchment relations

$\alpha \preceq \alpha \wedge \beta$ or $\beta \preceq \alpha \wedge \beta$. (Conjunctiveness)

and along with Dominance and Transitivity make the defining set of properties of Gärdenfors and Makinson's entrenchment orderings which is the notion we generalize.

In this paper, we will only study Horn properties. To our knowledge previous results concern only non-Horn entrenchment relations satisfying connectivity: entrenchment and rational ordering in [19] and [23], respectively.

3 Maxiconsistent and Weakly Maxiconsistent Inference

We shall now describe an inference scheme based on an entrenchment relation $\preceq$. We will define two finitary consequence relations, that is, subsets of $\mathcal{L} \times \mathcal{L}$, called *maxiconsistent* ($\vdash_{\preceq}$) and *weakly maxiconsistent* inference ($\vdash_{\preceq}^{w}$).

Our notion of inference is based on maximal consistent sets of sentences. The idea of using maximal consistent sets for inference is not new. Maximal consistent sets have been used in databases ([14]), conditional logic ([45],[51],[26],[24]), and belief revision ([1]). However, the notion of maximal consistency is already present in classical entailment. In order to compute the inferences of a formula α, one can find all maximal consistent sets that do not contain $\neg\alpha$, that is all prime filters containing α, and take their intersection. This is the filter that contains all theorems of α. Our definition of inference is similar. First, we find all maximal consistent sets whose elements do not have lower entrenchment than $\neg\alpha$. These sets do not necessarily contain α, as opposed, say, to classical logic. Next, we consider their intersection. If $\alpha \rightarrow \beta$ is contained in this intersection then α entails maxiconsistently β, that is $\alpha \vdash_{\preceq} \beta$. It is time to be more formal.

Definition 3.1. Let $\langle \mathcal{L}, \vdash, \preceq \rangle$ be an entrenchment frame. The set of *coherent sentences* for a formula $\alpha \in \mathcal{L}$ is the set

$$\text{Coh}(\alpha) = \{\beta \mid \beta \not\preceq \neg\alpha\}.$$

The *base* of α is the set

$$\mathcal{B}(\alpha) = \{U \mid U = \text{Cn}(U), U \subseteq \text{Coh}(\alpha)\}.$$

The *maximal base* of α is the set

$$\mathcal{B}_{\max}(\alpha) = \{U \mid U \in \mathcal{B}(\alpha) \text{ and if } U' = \text{Cn}(U') \text{ with } U \subset U' \text{ then } U' \notin \mathcal{B}(\alpha)\}.$$

The *extension set* of α is the set

$$e(\alpha) = \{\text{Cn}(U, \alpha) \mid U \in \mathcal{B}_{\max}(\alpha)\}.$$

The *sceptical extension* of α is the set

$$E(\alpha) = \bigcap e(\alpha),$$

and now define

$$\alpha \mathrel{\mathop{\sim}\limits_{\preceq}} \beta \quad \text{iff} \quad \beta \in E(\alpha),$$

and say that α *maxiconsistently infers* β *in the entrenchment frame* $\langle \mathcal{L}, \vdash, \preceq \rangle$.

Similarly, by using implicants of α instead of theories, we have:

The *weak base* of α is the set

$$\mathcal{B}^w(\alpha) = \{U \mid U = \mathrm{Cn}(U), U^\alpha \subseteq \mathrm{Coh}(\alpha)\},$$

where $U^\alpha = \{\alpha \to \beta \mid \beta \in U\}$, called the α-*conditionalization* of U. The *maximal weak base* of α is the set

$$\mathcal{B}^w_{\max}(\alpha) = \{U \mid U^\alpha \in \mathcal{B}^w(\alpha) \text{ and if } V = \mathrm{Cn}(V) \text{ with } U^\alpha \subset V^\alpha \text{ then } V^\alpha \notin \mathcal{B}^w(\alpha)\}.$$

The *weak extension set* of α is the set

$$e^w(\alpha) = \{U \mid U \in \mathcal{B}^w_{\max}(\alpha)\},$$

and the *sceptical weak extension* of α is the set

$$E^w(\alpha) = \bigcap e^w(\alpha).$$

Also, define

$$\alpha \mathrel{\mathop{\sim}\limits_{\preceq}^{w}} \beta \quad \text{iff} \quad \beta \in E^w(\alpha),$$

and say that α *weakly maxiconsistently infers* β *in the entrenchment frame* $\langle \mathcal{L}, \vdash, \preceq \rangle$.

Since $\mathcal{L}$ and $\vdash$ remain fixed throughout the following we shall usually drop $\langle \mathcal{L}, \vdash, \preceq \rangle$ and refer to maxiconsistent inference on an entrenchment consequence relation $\preceq$.

Note that in case $e(\alpha) = \emptyset$, we have $E(\alpha) = \bigcap \emptyset = \mathcal{L}$ and similarly for $e^w(\alpha)$. So, if $\preceq$ is $\vdash$, that is, if we equate an entrenchment relation with classical provability then both $\mathrel{\mathop{\sim}\limits_{\preceq}}$ and $\mathrel{\mathop{\sim}\limits_{\preceq}^{w}}$ collapse to classical $\vdash$.

4 Nonmonotonic Consequence Relations and their Representations

Before presenting the results of this section (and main results of this paper), we shall define a variety of classes of nonmonotonic consequence relations. The rules mentioned in the following are presented in Table 2. For a motivation of these rules see [27] and [36]. (The latter serves as an excellent introduction to nonmonotonic consequence relations.)

Definition 4.1. Following ([33],[27],[4],[30],[19]), we shall say that a relation $\sim$ on $\mathcal{L}$ is a *nonmonotonic consequence relation (based on $\vdash$)* if it satisfies Supraclassicality, Left Logical Equivalence, Right Weakening, and And. We call a nonmonotonic consequence relation $\sim$ *cumulative* if it satisfies, in addition, Cut and Cautious Monotonicity, *strongly cumulative* if it is cumulative and satisfies, in addition, Loop, *preferential* if it is cumulative and satisfies, in addition, Or, and *rational* if it is preferential and satisfies, in addition, Rational Monotonicity.

$$\frac{\alpha \vdash \beta}{\alpha \mathrel{\vdash\mkern-9mu\sim} \beta} \qquad \text{(Supraclassicality)}$$

$$\frac{\alpha \vdash \beta \quad \beta \vdash \gamma \quad \alpha \mathrel{\vdash\mkern-9mu\sim} \gamma}{\beta \mathrel{\vdash\mkern-9mu\sim} \gamma} \qquad \text{(Left Logical Equivalence)}$$

$$\frac{\alpha \mathrel{\vdash\mkern-9mu\sim} \beta \quad \beta \vdash \gamma}{\alpha \mathrel{\vdash\mkern-9mu\sim} \gamma} \qquad \text{(Right Weakening)}$$

$$\frac{\alpha \mathrel{\vdash\mkern-9mu\sim} \beta \quad \alpha \mathrel{\vdash\mkern-9mu\sim} \gamma}{\alpha \mathrel{\vdash\mkern-9mu\sim} \beta \wedge \gamma} \qquad \text{(And)}$$

$$\frac{\alpha_0 \mathrel{\vdash\mkern-9mu\sim} \alpha_1 \quad \cdots \quad \alpha_{n-1} \mathrel{\vdash\mkern-9mu\sim} \alpha_n \quad \alpha_n \mathrel{\vdash\mkern-9mu\sim} \alpha_0}{\alpha_0 \mathrel{\vdash\mkern-9mu\sim} \alpha_n} \qquad \text{(Loop)}$$

$$\frac{\alpha \mathrel{\vdash\mkern-9mu\sim} \gamma \quad \beta \mathrel{\vdash\mkern-9mu\sim} \gamma}{\alpha \vee \beta \mathrel{\vdash\mkern-9mu\sim} \gamma} \qquad \text{(Or)}$$

$$\frac{\alpha \vee \beta \mathrel{\vdash\mkern-9mu\sim} \alpha \quad \beta \vee \gamma \mathrel{\vdash\mkern-9mu\sim} \beta}{\alpha \vee \gamma \mathrel{\vdash\mkern-9mu\sim} \alpha} \qquad \text{(Weak Transitivity)}$$

$$\frac{\alpha \mathrel{\not\vdash\mkern-9mu\sim} \neg\beta \quad \alpha \mathrel{\vdash\mkern-9mu\sim} \gamma}{\alpha \wedge \beta \mathrel{\vdash\mkern-9mu\sim} \gamma} \qquad \text{(Rational Monotonicity)}$$

Table 2. Rules for Nonmonotonic Inference

4.1 From Entrenchment to Nonmonotonic Consequence Relations

The first theorem of this section shows that maxiconsistent inference in an arbitrary entrenchment frame is a nonmonotonic consequence relation. All subsequent results assume that the entrenchment frame is either disjunctive or weakly disjunctive.

Theorem 4.2. *Let $\langle \mathcal{L}, \vdash, \preceq \rangle$ be a entrenchment frame. Then its maxiconsistent inference $\mathrel{\vdash\mkern-9mu\sim}_{\preceq}$ and weakly maxiconsistent inference $\mathrel{\vdash\mkern-9mu\sim}_{\preceq}^{w}$ are a nonmonotonic consequence relation. Moreover,*

1. *If $\preceq$ satisfies Bounded Cut and Bounded Right Monotonicity then $\mathrel{\vdash\mkern-9mu\sim}_{\preceq}$ is a cumulative inference relation.*
2. *If $\preceq$ satisfies Bounded Cut and Right Monotonicity then $\mathrel{\vdash\mkern-9mu\sim}_{\preceq}$ is a strong cumulative inference relation.*
3. *If $\preceq$ satisfies Transitivity and Right Conjunction then $\mathrel{\vdash\mkern-9mu\sim}_{\preceq}$ is a preferential inference relation.*

From now on, we will assume a disjunctive or weakly disjunctive entrenchment relation because their maxiconsistent inference gives a canonical representation of non-monotonic consequence relation.

First, The following definition provide, for each nonmonotonic consequence relation, an entrenchment relation with the same maxiconsistent inference.

Definition 4.3. Given an entrenchment relation $\preceq$ then define consequence relations $\mathrel{\vdash\mkern-9mu\sim}$ and $\mathrel{\vdash\mkern-9mu\sim}'$ as follows

(N) $\alpha \hspace{1mm} \vdash \hspace{1mm} \beta$ iff $\neg\beta \preceq \neg\alpha$

($N_{\rightarrow}$) $\alpha \hspace{1mm} \vdash \hspace{1mm}'\beta$ iff $\neg\alpha \vee \neg\beta \preceq \neg\alpha$

We shall also denote $\vdash$ and $\vdash'$ with $N(\preceq)$ or and $N_{\rightarrow}(\preceq)$, respectively.

Given the above definition one can prove the following lemma

Corollary 4.4. *Let $\preceq$ be an entrenchment relation. Then*

1. *if $\preceq$ is disjunctive then $N(\preceq) = \hspace{1mm}\vdash_{\preceq}$, and*
2. *if $\preceq$ is weakly disjunctive then $N_{\rightarrow}(\preceq) = \hspace{1mm}\vdash^{w}_{\preceq}$,*

where $\vdash_{\preceq}$ and $\vdash^{w}_{\preceq}$ is the maxiconsistent and weakly maxiconsistent inference of $\preceq$, respectively.

We now have the following

Theorem 4.5. *Let $\langle \mathcal{L}, \vdash, \preceq \rangle$ be an entrenchment frame. Then*

1. *if $\preceq$ is disjunctive then the inference relation $\vdash$ defined by (N) is a nonmonotonic consequence relation such that, for all α, β in $\mathcal{L}$,*

$$\alpha \hspace{1mm} \vdash \hspace{1mm} \beta \qquad iff \qquad \alpha \hspace{1mm} \vdash_{\preceq} \beta.$$

Moreover, if $\preceq$ satisfies Bounded Cut, Bounded Right Monotonicity, Acyclicity and Conjunction then $\vdash_{\preceq}$ satisfies Cut, Cautious Monotonicity, Loop and Or, respectively.

2. *if $\preceq$ is weakly disjunctive then the inference relation $\vdash$ defined by $N_{\rightarrow}$ is a nonmonotonic consequence relation such that, for all α, β in $\mathcal{L}$,*

$$\alpha \hspace{1mm} \vdash \hspace{1mm} \beta \qquad iff \qquad \alpha \hspace{1mm} \vdash^{w}_{\preceq} \beta.$$

Moreover, if $\preceq$ satisfies Weak Bounded Cut, Weak Bounded Right Monotonicity, Weak Acyclicity and Right Conjunction then $\vdash^{w}_{\preceq}$ satisfies Cut, Cautious Monotonicity, Loop and Or, respectively.

4.2 From Nonmonotonic Consequence to Entrenchment Relations

In this section, we will show how nonmonotonic consequence relations are reduced to entrenchment relations.

We begin by defining an entrenchment relation from a nonmonotonic consequence relation. There is no unique way to generate an appropriate entrenchment relation.

Definition 4.6. Given an entrenchment relation $\preceq$ and a nonmonotonic inference relation $\vdash$, then define a consequence relation $\vdash'$ and a relation $\preceq'$ as follows

(P) $\alpha \preceq' \beta$ iff $\neg\beta \hspace{1mm} \vdash \hspace{1mm} \neg\alpha$.

($P_{\rightarrow}$) $\alpha \preceq' \beta$ iff $\neg\alpha \vee \neg\beta \hspace{1mm} \vdash \hspace{1mm} \neg\alpha$.

(P_{tr}) $\alpha \preceq'' \beta$ iff there exist $\delta_1, \ldots, \delta_n \in \mathcal{L}$ such that

$\neg\beta \hspace{1mm} \vdash \hspace{1mm} \delta_1, \delta_1 \hspace{1mm} \vdash \hspace{1mm} \delta_2, \ldots, \delta_n \hspace{1mm} \vdash \hspace{1mm} \neg\alpha$.

We shall also denote $\preceq$, $\preceq'$ and $\preceq''$ with $P(\vdash)$, $P_{\rightarrow}(\vdash)$ and $P_{tr}(\vdash)$, respectively.

Given the above definition one can prove the following lemma which shows the cases where the maps defined in Definition 4.3 and 4.6 are inverses of each other.

Lemma 4.7. *Let $\preceq$ and $\vdash\!\!\!\sim$ be an entrenchment and a nonmonotonic consequence relation, respectively. Then*

1. *$P(N(\preceq)) = \preceq$, and*
2. *$N(P(\vdash\!\!\!\sim)) = \vdash\!\!\!\sim$.*
3. *if $\preceq$ satisfies Right Monotonicity and Right Conjunction then*
   $P_\rightarrow(N_\rightarrow(\preceq)) = \preceq$,
4. *$N_\rightarrow(P_\rightarrow(\vdash\!\!\!\sim)) = \vdash\!\!\!\sim$,*
5. *if $\preceq$ is transitive then $P_{\mathrm{tr}}(N_\rightarrow(\preceq)) = \preceq$, and*
6. *if $\vdash\!\!\!\sim$ satisfies Loop then $N_\rightarrow(P_{\mathrm{tr}}(\vdash\!\!\!\sim)) = \vdash\!\!\!\sim$.*

Going from nonmonotonic consequence relations to disjunctive entrenchment relations, we have the following theorem.

Theorem 4.8. *Let $\vdash\!\!\!\sim$ be a nonmonotonic inference. Then*

1. *the relation $\preceq$ defined by (P) is a disjunctive entrenchment relation such that, for all α, β in $\mathcal{L}$,*

$$\alpha \vdash\!\!\!\sim \beta \quad \textit{iff} \quad \alpha \vdash\!\!\!\sim_{\preceq} \beta.$$

   Moreover, if $\vdash\!\!\!\sim$ satisfies Cut, Cautious Monotonicity, Loop, and Or then $\preceq$ satisfies Bounded Cut, Bounded Right Monotonicity, Acyclicity, and Conjunction, respectively.
2. *if $\vdash\!\!\!\sim$ is a preferential inference relation, then the relation $\preceq$ defined by $(P_\rightarrow)$ is a weakly disjunctive and transitive entrenchment relation satisfying Conjunction such that, for all α, β in $\mathcal{L}$,*

$$\alpha \vdash\!\!\!\sim \beta \quad \textit{iff} \quad \alpha \vdash\!\!\!\sim_{\preceq} \beta \quad \textit{iff} \quad \alpha \vdash\!\!\!\sim_{\preceq}^{w} \beta.$$

3. *if $\vdash\!\!\!\sim$ satisfies Loop, then the relation $\preceq$ defined by (P_{tr}) is a weakly disjunctive transitive entrenchment relation such that, for all α, β in $\mathcal{L}$,*

$$\alpha \vdash\!\!\!\sim \beta \quad \textit{iff} \quad \alpha \vdash\!\!\!\sim_{\preceq}^{w} \beta.$$

5 Conclusion and Further Work

In this section, we will give a summary of the correspondence between classes of entrenchment and nonmonotonic consequence relations.

Let $\mathcal{A}$ be a class of nonmonotonic consequence relations and $\mathcal{B}$ a class of entrenchment relations. Let C, C^w be maps from $\mathcal{B}$ to $\mathcal{A}$ with $C(\preceq) = \vdash\!\!\!\sim_{\preceq}$ and $C^w(\preceq) = \vdash\!\!\!\sim_{\preceq}^{w}$, respectively, where $\vdash\!\!\!\sim_{\preceq}$ an $\vdash\!\!\!\sim_{\preceq}^{w}$ are the maxiconsistent and weakly maxiconsistent inference on $\preceq$.

We will say that a class $\mathcal{A}$ of nonmonotonic consequence relations is *dual* to a class $\mathcal{B}$ of entrenchment relations and denote it with $\mathcal{A} \equiv \mathcal{B}$ if there exists a map N such that $N : \mathcal{B} \rightarrow \mathcal{A}$, $C \circ N = \mathrm{Id}_\mathcal{A}$, and $N \circ C = \mathrm{Id}_\mathcal{B}$, where Id is the identity map. Similarly,

$\mathcal{A}$ and $\mathcal{B}$ will be *weakly dual* and we denote it with $\mathcal{A} \stackrel{w}{=} \mathcal{B}$ if there exists a map N such that $N : \mathcal{B} \rightarrow \mathcal{A}$, $C^w \circ N = Id_{\mathcal{A}}$, and $N \circ C^w = Id_{\mathcal{B}}$.

We will say that a class $\mathcal{A}$ of nonmonotonic consequence relations is *a retract* of a class $\mathcal{B}$ of entrenchment relations and denote it with $\mathcal{A} \models \mathcal{B}$ if there exists a map N such that $N : \mathcal{B} \rightarrow \mathcal{A}$ and $C \circ N = Id_{\mathcal{A}}$. Similarly, $\mathcal{A}$ is *a weak retract* of $\mathcal{B}$ and we denote it with $\mathcal{A} \stackrel{w}{\models} \mathcal{B}$ if there exists a map N such that $N : \mathcal{B} \rightarrow \mathcal{A}$ and $C^w \circ N = Id_{\mathcal{A}}$.

A list of all classes of nonmonotonic and entrenchment relations mentioned in the following appear on Table 3.

NM	=	all nonmonotonic consequence relations (nmcr)
D	=	nmcr satisfying Cut
CM	=	nmcr satisfying Cautious Monotonicity
C	=	cumulative nmcr
SC	=	strong cumulative nmcr
P	=	preferential nmcr
E	=	all entrenchment relations (er)
BC	=	er satisfying Bounded Cut
BR	=	er satisfying Bounded Right Monotonicity
BCR	=	er satisfying Bounded Cut and Bounded Right Monotonicity
BA	=	er satisfying Bounded Cut, Bounded Right Monotonicity and Acyclicity
T	=	er satisfying Transitivity
TC	=	er satisfying Transitivity and Right Conjunction
d-$\mathcal{B}$	=	er satisfying the properties of $\mathcal{B}$ and Left Disjunction
wd-$\mathcal{B}$	=	er satisfying the properties of $\mathcal{B}$ and Weak Left Disjunction

Table 3. Classes of nonmonotonic and entrenchment relations

If $\mathcal{B}$ is any entrenchment relation class then d-$\mathcal{B}$ and wd-$\mathcal{B}$ are $\mathcal{B}$ augmented with Left Disjunction and Weak Left Disjunction, respectively. Clearly, d-$\mathcal{B} \subseteq$ wd-$\mathcal{B} \subseteq \mathcal{B}$.

We now have the following corollary

Corollary 5.1. *The following hold*

1. NM $\equiv$ d-E, NM $\stackrel{w}{=}$ d-E, NM $\models$ E, NM $\stackrel{w}{\models}$ E, *and* NM $\models$ wd-E.
2. D $\equiv$ d-BC *and* D $\stackrel{w}{=}$ d-BC.
3. CM $\equiv$ d-BR *and* CM $\stackrel{w}{=}$ d-BR.
4. C $\equiv$ d-BCR, C $\stackrel{w}{=}$ d-BCR, C $\models$ BCR *and* C $\stackrel{w}{\models}$ BCR.
5. SC $\equiv$ d-BA, SC $\stackrel{w}{=}$ d-BA *and* SC $\stackrel{w}{=}$ wd-T.
6. P $\models$ TC, P $\equiv$ wd-TC *and* P $\stackrel{w}{=}$ wd-TC.

What do the above results say about nonmonotonic inference? Since the characterization is uniform one can safely recognize entrenchment as a basic ingredient of nonmonotonic inference. In fact, any strong cumulative nonmonotonic consequence relation relies on the sole employment of entrenchment. However, once we look at weaker forms of nonmonotonicity, transitivity fails. We interpret this failure of transitivity as the introduction of some other mechanism which is hidden on the principal mechanisms that generate such weak forms of nonmonotonicity (e.g., extensions in default logic). Indeed, it is not the idea behind entrenchment that fails but rather the way we interpret formulas and their connectives. It seems that a formulation of entrenchment relation into a full-fledged form of a sequent calculus will help us reveal the logical content of weak forms of nonmonotonicity. Also, by allowing introduction and elimination of connectives, we can dispense with the use of classical consequence and incorporate other logics into entrenchment. This will give us a uniform way to automate the nonmonotonic process which is especially important for its applications.

Finally, it is apparent to anyone familiar with entrenchment relations in belief revision and its connections with nonmonotonicity ([37],[16]) that the above results pave the way for the study of systems weaker than the AGM set of postulates. In [22], we work out the case of preferential belief revision. Moreover, since multiple extensions can be expressed in our characterization, entrenchment becomes a tool for studying relational and multiple agent belief revision.

Acknowledgments: I thank David Makinson for his comments on an earlier draft of this paper and the two anonymous referees for their suggestions.

References

1. C. E. Alchourrón, P. Gärdenfors, and D. Makinson. On the logic of theory change: partial meet contraction and revision functions. *The Journal of Symbolic Logic*, 50:510–530, 1985.
2. G. Amati and K. Georgatos. Relevance as deduction: A logical approach to information retrieval. In *Proceedings of the 2nd International Workshop on Information Retrieval*, Glascow, United Kingdom, 1996. University of Glasgow.
3. F. Baader and B. Hollunder. How to prefer more specific defaults in terminological default logic. In R. Bajcsy, editor, *Proceedings of the 13th International Joint Conference on Artificial Intelligence (IJCAI)*, San Mateo, CA, 1993. Morgan Kaufmann.
4. J. Bell. Pragmatic logics. In *Proceedings of KR91*, pages 50–60, Cambridge, MA, 1991.
5. S. Benferhat, C. Cayrol, D. Dubois, J. Lang, and H. Prade. Inconsistency management and prioritized syntax entailment. In R. Bajcsy, editor, *Proceedings of the 13th International Joint Conference on Artificial Intelligence (IJCAI)*, pages 640–645, San Mateo, CA, 1993. Morgan Kaufmann.
6. A. Bochman. Default consequence relations as a logical framework for logic programs. In *Proceedings of the 3rd International Conference on Logic Programming and Nonmonotonic Reasoning*, number 928 in LNAI, pages 245–258, Berlin, 1995. Springer.
7. C. Boutilier. Conditional logics for default reasoning. Ph.D. Thesis, 1992.
8. C. Boutilier. Modal logics for qualitative possibility and beliefs. In *Proceedings of the 8th Workshop on Uncertainty in Artificial Intelligence*, pages 17–24, Stanford, CA, 1992.
9. G. Brewka. Adding priorities and specificity to default logic. In C. MacNish, D. Pearce, and L. M. Pereira, editors, *Logics in Artificial Intelligence (JELIA '94)*, number 838 in Lecture Notes in Computer Science, pages 247–260, Berlin, 1994. Springer-Verlag.
10. L. Clark. Negation as failure. In H. Gallaire and J. Minker, editors, *Logics and Data Bases*, pages 293–322. Plenum, New York, 1978.
11. J. P. Delgrande. A preference-based approach to default reasoning. Technical report, School of Computing Science, Simon Fraser University, 1994.
12. D. Dubois, J. Lang, and H. Prade. Possibilistic logic. In D. M. Gabbay, C. J. Hogger, and J. A. Robinson, editors, *Handbook of Logic in Artificial Intellingence and Logic Programming*, volume 3, pages 439–513. Oxford University Press, 1994.

13. D. Dubois and H. Prade. Epistemic entrenchment and possibilistic logic. *Artificial Intelligence*, 50:223–239, 1991.
14. R. Fagin, J. D. Ullman, and M. Y. Vardi. On the semantics of updates in databases. In *Proceedings of the Second ACM SIGACT-SIGMOD*, pages 352–365, 1983.
15. L. Fariñas del Cerro, A. Herzig, and J. Lang. From ordering-based nonmonotonic reasoning to conditional logics. *Artificial Intelligence*, 66:375–393, 1994.
16. M. Freund and D. Lehmann. Belief revision and rational inference. Technical Report TR 94-16, The Leibniz Center for Research in Computer Science, Institute of Computer Science,Hebrew University, July 1994.
17. D. Gabbay. Theoretical foundations for nonmonotonic reasoning in expert systems. In K. Apt, editor, *Logics and Models of Concurrent Systems*. Springer-Verlag, Berlin, 1985.
18. P. Gärdenfors and D. Makinson. Revisions of knowledge systems using epistemic entrenchment. In *Proceedings of the Second Conference on Theoretical Aspects of Reasoning about Knowledge*, pages 661–672, 1992.
19. P. Gärdenfors and D. Makinson. Nonmonotonic inference based on expectations. *Artificial Intelligence*, 65:197–245, 1994.
20. P. Gärdenfors and H. Rott. Belief revision. In D. Gabbay, editor, *Handbook of Logic in Artificial Intelligence and Logic Programming*, volume II. Oxford University Press, 1994.
21. H. Geffner and J. Pearl. Conditional entailment: Bridging two approaches to default reasoning. *Artificial Intelligence*, 53:209–244, 1992.
22. K. Georgatos. Preferential inference and revision using partial entrenchment orderings. To appear in the Annals of Pure and Applied Logic.
23. K. Georgatos. Ordering-based representations of rational inference. In J. J. Alferes, L. M. Pereira, and E. Orlowska, editors, *Logics in Artificial Intelligence (JELIA '96)*, number 1126 in Lecture Notes in Artificial Intelligence, pages 176–191, Berlin, 1996. Springer-Verlag.
24. M. L. Ginsberg. Counterfactuals. *Artificial Intelligence*, 30:35–79, 1986.
25. H. Katsuno and D. Satoh. A unified view of consequence relation, belief revision and conditional logic. In *Proceedings of IJCAI-91*, pages 406–412, Sydney, Australia, 1991.
26. A. Kratzer. Partition and revision: The semantics of conditionals. *Journal of Philosophical Logic*, 10:201–216, 1981.
27. S. Kraus, D. Lehmann, and M. Magidor. Nonmonotonic reasoning, preferential models and cumulative logics. *Artificial Intelligence*, 44:167–207, 1990.
28. P. Lamarre. S4 as the conditional logic of nonmonotonicity. In *Proceedings of KR91*, pages 357–367, Cambridge, MA, 1991.
29. P. Lamarre. From monotonicity to nonmonotonicity via a theorem prover. In *Proceedings of KR92*, pages 572–580, Cambridge, MA, 1992.
30. D. Lehmann and M. Magidor. What does a conditional knowledge base entail? *Artificial Intelligence*, 55:1–60, 1992.
31. D. Lewis. *Counterfactuals*. Harvard University Press, Cambridge, MA, 1973.
32. V. Lifschitz. Computing circumscription. In *Proceedings of the Ninth International Joint Conference on Artificial Intelligence (IJCAI-85)*, pages 121–127, 1985.
33. S. Lindström. A semantic approach to nonmonotonic reasoning: inference operations and choice. Manuscript, Department of Philosophy, Uppsala University, Uppsala, Sweden, 1990.
34. S. Lindström and W. Rabinowicz. Epistemic entrenchment with incomparabilities and relational belief revision. In A. Fuhrmann and M. Morreau, editors, *The Logic of Theory Change*, number 465 in Lecture Notes in Artificial Intelligence, pages 93–126, Berlin, 1991. Springer-Verlag.
35. D. Makinson. General theory of cumulative inference. In M. Reinfranck, editor, *Non-Monotoning Reasoning*, number 346 in Lecture Notes in Artificial Intelligence, pages 1–18. Springer-Verlag, Berlin, 1989.

36. D. Makinson. General patterns in nonmonotonic reasoning. In D. Gabbay, editor, *Handbook of Logic in Artificial Intelligence and Logic Programming*, volume III. Oxford University Press, 1994.

37. D. Makinson and P. Gärdenfors. Relations between the logic of theory change and nonmonotonic logic. In A. Fuhrmann and M. Morreau, editors, *The Logic of Theory Change*, number 465 in Lecture Notes in Artificial Intelligence, pages 185–205, Berlin, 1991. Springer-Verlag.

38. W. V. Marek, A. Nerode, and J. B. Remmel. Nonmonotonic rule systems I. *Annals of Mathematics and Artificial Intelligence*, 1:241–273, 1990.

39. J. McCarthy. Circumscription: A form of non-monotonic reasoning. *Artificial Intelligence*, 13:27–39, 1980.

40. D. McDermott and J. Doyle. Non-monotonic logic I. *Artificial Intelligence*, 13:41–72, 1980.

41. A. C. Nayak. Iterated belief change based on epistemic entrenchment. *Erkenntnis*, 41:353–390, 1994.

42. A. C. Nayak, N. Y. Foo, M. Pagnucco, and A. Satar. Changing conditional beliefs unconditionally. In *Proceedings of the TARK 96 Conference*, pages 119–135, 1996.

43. J. Pearl. System Z: a natural ordering of defaults with tractable applications to default reasoning. In *Proceedings of the 3rd Conference on Theoretical Aspects of Reasoning about Knowledge*, pages 121–135, Pacific Grove, CA, 1990.

44. R. Reiter. A logic for default reasoning. *Artificial Intelligence*, 2:147–187, 1980.

45. N. Rescher. *Hypothetical Reasoning*. North-Holland, Amsterdam, 1964.

46. H. Rott. Preferential belief change using generalized epistemic entrenchment. *Journal of Logic, Language and Information*, 1:45–78, 1992.

47. M. Ryan. Ordered presentation of theories. Ph.D. Thesis, 1992.

48. Y. Shoham. *Reasoning about Change*. MIT Press, Cambridge, 1988.

49. R. Stalnaker. A theory of conditionals. In N. Rescher, editor, *Studies in Logical Theory*. Oxford University Press, Oxford, 1968.

50. D. S. Touretzky. *The Mathematics of Inheritance Systems*. Research Notes in Artificial Intelligence. Morgan Kaufmann, Los Altos, California, 1986.

51. F. Veltman. Prejudices, presuppositions and the theory of counterfactuals. In J. Groenendijk and M. Stokhof, editors, *Amsterdam Papers in Formal Grammar*, volume I. University of Amsterdam, 1976.

52. M.-A. Williams. On the logic of theory base change. In C. MacNish, D. Pearce, and L. M. Pereira, editors, *Logics in Artificial Intelligence (JELIA '94)*, number 838 in Lecture Notes in Computer Science, pages 86–105, Berlin, 1994. Springer-Verlag.

53. W. Wobcke. On the use of epistemic entrenchment in nonmonotonic reasoning. In *Proceedings of ECAI-92*, pages 324–328, Vienna, Austria, 1992.

A Modal Logic for Reasoning about Knowledge and Time on Binary Subset Trees

Bernhard Heinemann

FernUniversität, D–58084 Hagen, Germany

Abstract. We introduce a modal logic in which one of the operators expresses properties of points (knowledge states) in a neighbourhood of a given point (the actual set of alternatives), and other operators speak about shrinking such neighbourhoods gradually (representing the change of the set of states in time). Based on a modification of the topological language due to MOSS and PARIKH [Moss and Parikh 1992] we generalize a certain fragment of propositional branching time logic to the logic of knowledge in this way. To keep the notation simple we confine ourselves to *binary* branching. Thus we define a *trimodal logic* comprising one *knowledge*–operator and two *nexttime*–operators. The formulas are interpreted in *binary ramified subset tree models*. We present an axiomatization of the set T of theorems valid for this class of semantical domains and prove its *completeness* as the main result of the paper. Furthermore, *decidability* of T is shown, and its *complexity* is determined.

1 Introduction

Subsequently we define a modal language, which is able to express properties of points and point sets: the modal operators quantify over elements of a set and over sets in a system of sets; moreover, quantification over sets is compatible with descent in the system (w.r.t. inclusion). The system of sets forms an *infinite binary tree* throughout the paper.[1] The logic originates from a knowledge-theoretic context, as it will be pointed out below. May be, however, that it is applicable in other fields of computer science as well. E.g., consider the Cantor space C of all infinite 0–1–sequences. This space plays an important part in a recently proposed model for effective analysis, in which elements of C are used as names for real numbers and computations on $\mathbb{R}$ are faithfully performed on such names [Weihrauch 1995]. It turns out that computational properties of functions and sets are closely related to topological properties of the underlying spaces (of names) and the mappings between them. E.g., continuity appears as an abstract version of computability. Concerning this model, computations are in fact *approximation procedures* which are infinite in principle. Looking at a stage of a program run the actual knowledge about the computed object x is represented by a neighbourhood of x which corresponds to the initial segment of the output already printed. Thus specifications of programs have to determine

---

[1] This assumption is made for the sake of simplicity; finite ramification would work as well.

the "admissible" sequences of neighbourhoods of x.

This knowledge–theoretic view suggests specifying by means of logical frameworks which can deal with properties of points, neighbourhoods and approximations. Especially, the *subset space logics* developed by MOSS and PARIKH [Moss and Parikh 1992] seem to be a suitable base for corresponding systems. Let us have a brief look at these modal logics. The logical language comprises two operators K and $\Box$ for *knowledge* and for *effort in time*, respectively. ($\Box$ may be viewed as an abstract program operator.) The semantical structures are, however, different from those in standard multimodal logic. Formulas are interpreted in pairs $(X, \mathcal{O})$ consisting of a non–empty set X and a distinguished set $\mathcal{O}$ of subsets of X. Such pairs $(X, \mathcal{O})$ are called *subset frames*. The elements of $\mathcal{O}$ are often called *opens*, although they need not be open sets in the sense of topology. K then varies over the elements of an open set, whereas $\Box$ describes the shrinking of an open. While K retains its S5–like character from usual logics of knowledge [Fagin et al. 1995], the modality of $\Box$ is S4–like, as it models the descent in a system of sets (w.r.t. inclusion). Moreover, the operators interact in a way which depends on the considered class of subset frames.

In the meantime the corresponding validities could be determined for several classes of subset spaces (i.e. subset frames with interpreted propositions). First we mention the class of *general subset spaces* (with no further restrictions on the set of opens), and the class of *topological spaces*, both of which are treated in [Moss and Parikh 1992] and [Dabrowski et al. 1996]. In [Georgatos 1994], the class of *tree–like spaces* has been investigated. A subset frame $(X, \mathcal{O})$ is tree–like by definition, iff $U \subseteq V$ or $V \subseteq U$ or $U \cap V = \emptyset$ holds for all $U, V \in \mathcal{O}$. In each case complete axiomatizations and decidability of the respective sets of theorems could be obtained. Finally, the subset spaces having *finite–height trees* of opens could be handled correspondingly after a slight modification concerning the semantics [Heinemann 1996b]. — One should note that the consideration of tree–like spaces is of some interest for our purposes as well, since the set of all infinite 0–1–sequences equipped with the *canonical base* of the Cantor topology belongs to this class. In fact, the resulting structure $\mathcal{C}_0$ is the standard example in this paper. Our aim is to determine the subset space logic of $\mathcal{C}_0$. This can be viewed as a first step towards a solution of the specification problem briefly mentioned above. Unfortunately we get difficulties expressing binary ramification as well as guaranteeing the real *tree* structure on the set $\mathcal{O}$ of opens because of the underlying language. As to binary ramification, one way out could be that one taken in [Heinemann 1996b] where $\Box$ is a K4–like instead of an S4–like modality. So we can give an adequate axiom scheme, but we cannot force binary ramification on the canonical model, and it is not clear how a proof of completeness can work in this case. (The same problem already appeared in ordinary modal logic; see [Gabbay 1976], Section 25).

In the given paper we take a different way to deal with the binary tree structure of the set $\mathcal{O}$. We substitute the $\Box$–operator by two modalities $\bigcirc_i$ ($i = 0, 1$), which are related to the *nexttime*–operator of usual temporal logic. In the new language called *nexttime logic for binary subset trees (NLBST)* (we omit the attribute "infinite" subsequently), specifications of knowledge concerning bounded lookahead can be formalized. — This setting is similar to that in [Heinemann 1997b], where the subset space logic of *linear* frames is examined. (A subset frame $(X, \mathcal{O})$ is linear, iff $\mathcal{O}$ is a chain w.r.t. set inclusion.)

The paper is divided into three further sections (and some conclusions). In Section 2 we introduce the syntax and the semantics of the logical language. In particular, we define the class of *binary subset tree models*. Section 3 begins with the presentation of a logical system C and is attended to proving its soundness and completeness w.r.t. binary subset tree models then. It should be mentioned that *almost canonical* completeness of C can be obtained. In the final section we treat the question of decidability for the set of C–theorems and determine the complexity of this set. — Apart from well–known notions and results from standard modal logic (see e.g. [Chellas 1980], [Goldblatt 1987]) the paper is largely self–contained. In Section 4, however, we proceed rather sketchy.

2 The Logical Language

We introduce a multimodal language, called *nexttime logic for binary subset trees (NLBST)*. The *syntax* of *NLBST* is based upon a recursively enumerable set of *propositional variables*, *PV* (denoted by upper case Roman letters). The set $\mathcal{F}$ of *NLBST–formulas* is defined recursively by the subsequent clauses:

- $PV \cup \{\top\} \subseteq \mathcal{F}$
- $\alpha, \beta \in \mathcal{F} \implies \neg\alpha, K\alpha, \bigcirc_0 \alpha, \bigcirc_1 \alpha, (\alpha \wedge \beta) \in \mathcal{F}$
- no other strings belong to $\mathcal{F}$

We omit brackets whenever possible and use the following abbreviations (besides the usual ones from sentential logic): $L\alpha$ for $\neg K \neg \alpha$ and $\boxtimes_i \alpha$ for $\neg \bigcirc_i \neg \alpha$ ($i = 0, 1$).

If X is any set, let $\mathcal{P}(X)$ denote the powerset of X. Throughout the paper the set $\{0, 1\}^*$ of all finite 0–1–strings will serve as an indexing set. If $a_i \in \{0, 1\}$ for $i = 1, \ldots, n$ and $w = a_1 \ldots a_n \in \{0, 1\}^*$, then n is the *length* of w. Moreover, let $\sqsubseteq$ denote the usual *prefix relation* on words. Using these designations we define the relevant semantical structures next.

Definition 1. (1) A *binary subset tree* is a pair $\mathcal{B} := (X, d)$ such that X is a non–empty set and

$$d : (\{0, 1\}^*, \sqsubseteq) \longrightarrow (\mathcal{P}(X), \supseteq)$$

is a monomorphism such that the following condition is satisfied:
for all $x \in X$ and $n \in \mathbb{N}$ there is exactly one word $w \in \{0, 1\}^*$ of length n such that $x \in d(w)$.
The elements of $\mathcal{O} := d(\{0, 1\}^*)$ are called *opens* of $\mathcal{B}$.
(2) $X \otimes \mathcal{O} := \{(x, d(w)) \mid w \in \{0, 1\}^*, x \in d(w)\}$ is the set of *neighbourhood situations* (of $\mathcal{B}$).

We often write U_w instead of $d(w)$ and sometimes $(X, \mathcal{O})$ instead of (X, d). In case

$$X := \{f \mid f : \mathbb{N} \longrightarrow \{0, 1\}\}$$

and $d : \{0, 1\}^* \longrightarrow \mathcal{P}(X)$ defined by

$$d(w) := \{f \in X \mid f(0) \ldots f(length(w) - 1) = w\}$$

for all $w \in \{0, 1\}^*$, we get $\mathcal{C}_0 := (X, d)$ as an important example of a binary subset tree, which we also call *the Cantor space* (by abuse of notation). Note that the above conditions on d imply that $v \sqsubseteq w$ iff $d(v) \supseteq d(w)$ for all $v, w \in \{0, 1\}^*$; in particular, $d(wa)$ is always *properly* contained in $d(w)$ for every binary subset tree (X, d).

Definition 2. Let $\mathcal{B} := (X, d)$ be a binary subset tree. A mapping $\sigma : PV \times X \longrightarrow \{0, 1\}$ is called an X-*valuation*. A triple $\mathcal{M} := (X, d, \sigma)$ is called a *binary subset tree model (based on $\mathcal{B}$)*.

Next we define *validity* of NLBST–formulas in binary subset subset tree models (X, d, σ) w.r.t. neighbourhood situations $x, U \in X \otimes \mathcal{O}$ (brackets are omitted from now on).

Definition 3 Semantics of NLBST. Let $\mathcal{M} = (X, d, \sigma)$ be a binary subset tree model based on $\mathcal{B} = (X, d)$.

(1) Let $\alpha \in \mathcal{F}$ be an NLBST–formula. Then we define by recursion on the structure of α:

$$
\begin{aligned}
&x, U_w \models_{\mathcal{M}} \top \\
&x, U_w \models_{\mathcal{M}} A &&: \Longleftrightarrow \; \sigma(A, x) = 1 \\
&x, U_w \models_{\mathcal{M}} \neg \alpha &&: \Longleftrightarrow \; x, U_w \not\models_{\mathcal{M}} \alpha \\
&x, U_w \models_{\mathcal{M}} \alpha \wedge \beta &&: \Longleftrightarrow \; x, U_w \models_{\mathcal{M}} \alpha \text{ and } x, U_w \models_{\mathcal{M}} \beta \\
&x, U_w \models_{\mathcal{M}} K\alpha &&: \Longleftrightarrow \; (\forall y \in U_w)\, y, U_w \models_{\mathcal{M}} \alpha \\
&x, U_w \models_{\mathcal{M}} \bigcirc_0 \alpha &&: \Longleftrightarrow \; x \in U_{w0} \text{ and } x, U_{w0} \models_{\mathcal{M}} \alpha \\
&x, U_w \models_{\mathcal{M}} \bigcirc_1 \alpha &&: \Longleftrightarrow \; x \in U_{w1} \text{ and } x, U_{w1} \models_{\mathcal{M}} \alpha
\end{aligned}
$$

for all neighbourhood situations x, U_w of $\mathcal{B}$, propositional variables $A \in PV$, and formulas $\alpha, \beta \in \mathcal{F}$. In case $x, U_w \models_{\mathcal{M}} \alpha$ we say that α *holds in $\mathcal{M}$ at the neighbourhood situation x, U_w*.

(2) The formula $\alpha \in \mathcal{F}$ *holds in $\mathcal{M}$* (denoted by $\models_{\mathcal{M}} \alpha$), iff it holds in $\mathcal{M}$ at every neighbourhood situation.

If there is no ambiguity, we omit the index $\mathcal{M}$ subsequently.

3 The System C

We present a list of axioms and rules, respectively, constituting a logical system C (the letter "C" indicates the relationship with the Cantor space). Our aim is to show that the theorems of this system are exactly the NLBST–formulas holding in all binary subset tree models.

Axioms

(1) All $\mathcal{F}$-instances of propositional tautologies
(2) $(A \to \boxtimes_i A) \wedge (\neg A \to \boxtimes_i \neg A)$
(3) $K(\alpha \to \beta) \to (K\alpha \to K\beta)$
(4) $K\alpha \to \alpha$
(5) $K\alpha \to KK\alpha$
(6) $L\alpha \to KL\alpha$
(7) $\boxtimes_i (\alpha \to \beta) \to (\boxtimes_i \alpha \to \boxtimes_i \beta)$
(8) $(\bigcirc_0 \top \; xor \; \bigcirc_1 \top) \wedge (L \bigcirc_0 \top \wedge L \bigcirc_1 \top)$
(9) $\bigcirc_i \alpha \to \boxtimes_i \alpha$
(10) $\bigcirc_i L\alpha \to L \bigcirc_i \alpha$
(11) $K\boxtimes_i (\alpha \to L\beta) \vee K\boxtimes_i (\beta \to L\alpha)$

for all $A \in PV$, $\alpha, \beta \in \mathcal{F}$, and $i \in \{0, 1\}$.

The following **rules** are present in the system: modus ponens, K–necessitation, and $\boxtimes_i$–necessitation ($i = 1, 2$).

Some remarks on the axioms seem to be convenient. First it should be mentioned that the schemes (2), (7), (9), (10), (11) have to be read for $i = 0, 1$, i.e. they consist of *two* schemes really. — Axiom (2) expresses the independence of the valuation from the neighbourhood component of a neighbourhood situation. (4), (5), and (6) characterize reflexivity, transitivity and the euclidean property, respectively, of the accessibility relation in ordinary modal logic. The scheme (9) corresponds to partial functionality in this sense. (10) connects knowledge and time saying that "nexttime α" is considered possible at the actual situation whenever α is possible at the succeding one. Axioms (3) and (7) first of all have a proof–theoretical meaning. Finally, (8) and (11) are related to the kind of binary ramification on the set of opens we consider in this note; *xor* means *exclusive disjunction*. — *Soundness* of the axioms w.r.t. the intended structures can easily be established.

Proposition 4. *Axioms (1)–(11) hold in every binary subset tree model.*

Proof. We only prove validity of (10) in every binary subset tree model. So let such a structure $\mathcal{M} = (X, d, \sigma)$ and any neighbourhood situation x, U_w of the underlying frame be given. W.l.o.g., let $\bigcirc = \bigcirc_0$. We omit the index 0 for the rest of the proof. Suppose that $x, U_w \models \bigcirc L\alpha$. This means that $x \in U_{w0}$ and $x, U_{w0} \models L\alpha$. Thus $y, U_{w0} \models \alpha$ for some $y \in U_{w0}$. Since $U_{w0} \subseteq U_w$, we get $y \in U_w$. Consequently, $y, U_w \models \bigcirc \alpha$. So we obtain $x, U_w \models L \bigcirc \alpha$ finally. $\square$

As to *completeness*, we will use the *canonical model* $\mathcal{M}_C$ of the system $\mathbf{C}$ extensively. Since this system is *normal* w.r.t. the modalities K and $\boxtimes_i$ ($i = 0, 1$) (i.e. it contains axioms (3), (7), and rules (2), (3), respectively), its canonical model can be built up in the usual way (see [Gol], §5). Subsequently, let the accessibility relations induced by the respective modal operators on $\mathcal{M}_C$ be denoted as follows:

$$s \xrightarrow{L} t : \iff \{\alpha \in \mathcal{F} \mid K\alpha \in s\} \subseteq t$$

$$s \xrightarrow{O_i} t :\iff \{\alpha \in \mathcal{F} \mid \boxtimes_i \alpha \in s\} \subseteq t$$

for all *maximal* C–*consistent* sets s, t from the carrier set C of $\mathcal{M}_C$ ($i = 0, 1$). (The notion of maximal C–consistency is the same as in ordinary modal logic [Chellas 1980], [Goldblatt 1987].) Note that the distinguished C–valuation of the canonical model is defined by

$$\sigma(A, s) = 1 :\iff A \in s \quad (A \in PV, s \in C).$$

Some useful properties of the canonical model are listed below. For convenience of the reader, we remind of the *truth lemma* first.

Lemma 5. *Let* $\models$ *and* $\vdash_C$, *respectively, designate satisfaction and derivability as usual. Then it holds that for all* $\alpha \in \mathcal{F}$

(a) $\mathcal{M}_C \models \alpha[s] \iff \alpha \in s$ *for all* $s \in C$;
(b) $\mathcal{M}_C \models \alpha \iff \vdash_C \alpha$.

(To distinguish satisfaction in standard modal logic from *NLBST*–satisfaction we prefer the notation as in the just stated lemma.) — In the following proposition some consequences of our special axioms are listed.

Proposition 6. *(a) The relation* $\xrightarrow{L}$ *is an equivalence relation on the set* C.

(b) For $i = 0, 1$, *the relation* $\xrightarrow{O_i}$ *is a partial function on* C.
(c) Let $s \in C$ *and assume* $O_i \alpha \in s$. *Then there exists a point* $t \in C$ *satisfying*
$$s \xrightarrow{O_i} t \text{ and } \alpha \in t \quad (i = 0, 1).$$

(d) (Analogously for $L\alpha$ *and* $\xrightarrow{L}$ *instead of* $O_i \alpha$ *and* $\xrightarrow{O_i}$.)

(e) Let $s, t, u \in C$ *be given such that* $s \xrightarrow{O_i} t \xrightarrow{L} u$. *Then there is a point* $v \in C$
satisfying $s \xrightarrow{L} v \xrightarrow{O_i} u$ ($i = 0, 1$).

Proof. The assertion stated in (a) follows in a well–known manner from axioms (4), (5) and (6); the same is true for (b) and axioms (8), (9) respectively. — The proofs of (c), (d) and (e) are similar. So we only present the proof of (e). W.l.o.g. let $O = O_0$. We consider the following set

$$S := \{\alpha \in \mathcal{F} \mid K\alpha \in s\} \cup \{O\beta \in \mathcal{F} \mid \beta \in u\}$$

We assume towards a contradiction that this set is inconsistent. Then a finite subset
$$\{\alpha_1, \dots, \alpha_n, O\beta_1, \dots, O\beta_m\}$$
of S is inconsistent as well. Using standard techniques from multimodal proof theory of normal systems we get

$$\vdash_C L O \beta \to L\neg\alpha,$$

where $\beta := \beta_1 \wedge \dots \wedge \beta_m \in u$ and $\alpha := \alpha_1 \wedge \dots \wedge \alpha_n$. With the aid of axiom (10) we obtain

$$\vdash_C O L \beta \to L\neg\alpha.$$

Since according to $s \xrightarrow{\bigcirc} t \xrightarrow{L} u$ the formula $\bigcirc L\beta$ is an element of s, we conclude that $L\neg\alpha \in s$. This contradicts $K\alpha \in s$. Consequently, the set S defined above is consistent. Hence it is contained in a maximal C–consistent set, which can serve as the desired v by the definition of the accessibility relations on the canonical model. $\qquad\square$

Next we state an immediate consequence of part (e) of the preceding proposition.

Corollary 7. *Let $n > 0$ and $s_0, \ldots, s_n, t \in C$ be given such that*

$$s_0 \xrightarrow{\bigcirc_{i_1}} s_1 \xrightarrow{\bigcirc_{i_2}} \ldots \xrightarrow{\bigcirc_{i_n}} s_n \xrightarrow{L} t,$$

where $i_1, \ldots, i_n \in \{0, 1\}$. Then there are $s_0', \ldots, s_{n-1}' \in C$ satisfying

$$s_0' \xrightarrow{\bigcirc_{i_1}} s_1' \xrightarrow{\bigcirc_{i_2}} \ldots \xrightarrow{\bigcirc_{i_{n-1}}} s_{n-1}' \xrightarrow{\bigcirc_{i_n}} t$$

and $s_j \xrightarrow{L} s_j'$ for $j = 0, \ldots, n-1$.

Subsequently, we let axiom scheme (11) come into play, which provides for the crucial property of the canonical model. — Let $[s]$ denote the $\xrightarrow{L}$–equivalence class of $s \in C$. Define the following binary relations $\succ_i$ on the set $\widetilde{C} := \{[s] \mid s \in C\}$:

$$[s] \succ_i [t] :\iff (\exists s' \in [s], t' \in [t])\, s' \xrightarrow{\bigcirc_i} t'$$

for all $[s], [t] \in \widetilde{C}$ ($i \in \{0, 1\}$). — Actually the just defined relations are *functional*.

Proposition 8. *For $i = 0, 1$, the relation $\succ_i$ on the set $\widetilde{C}$ is a partial function.*

Proof. Again, we let $\bigcirc = \bigcirc_0$ and $\succ\ =\ \succ_0$ w.l.o.g. Assume towards a contradiction that there are classes $[s], [t_1], [t_2] \in \widetilde{C}$ such that $[s] \succ [t_1]$, $[s] \succ [t_2]$, and $[t_1] \neq [t_2]$. Then there exist points $s_1', s_2' \in [s], t_1' \in [t_1], t_2' \in [t_2]$ and formulas α, β with the following properties:

$$s_1' \xrightarrow{\bigcirc} t_1' \qquad \text{and} \quad s_2' \xrightarrow{\bigcirc} t_2',$$
$$\alpha \in t_1', \neg L\alpha \in t_2' \text{ and } \beta \in t_2', \neg L\beta \in t_1'.$$

With the aid of Lemma 5(a) we get

$$\mathcal{M}_{\mathbf{C}} \models (L\bigcirc(\alpha \wedge \neg L\beta) \wedge L\bigcirc(\beta \wedge \neg L\alpha))[s],$$

contradicting Lemma 5(b) in view of axiom (11). $\qquad\square$

The first conjunct of axiom (8) implies the following assertion.

Proposition 9. *Let $s \in C$ be a point of the canonical model of the system $\mathbf{C}$. Then there exists a unique $t \in C$ such that either $s \xrightarrow{\bigcirc_0} t$ or $s \xrightarrow{\bigcirc_1} t$ holds.*

Proof. According to axiom (8), either $\bigcirc_0 \top \in s$ or $\bigcirc_1 \top \in s$. Proposition 6(c) then guarantees the existence of t with the desired properties; uniqueness follows from Proposition 6(b). □

Now we define a subset space falsifying a non–C–derivable formula which is *almost* of the type we are looking for. So let $\alpha \in \mathcal{F}$ be given and suppose that $\nvdash_C \alpha$. Choose a maximal C–consistent set $s_\alpha \in C$ containing $\neg \alpha$. For every $s \in [s_\alpha]$ define a function $f_s : \mathbb{N} \longrightarrow C$ inductively in the following way:

$$f_s(0) := s$$

$$f_s(j+1) := \text{ the unique } s' \in C \text{ satisfying } f_s(j) \xrightarrow{\bigcirc_i} s' \text{ for some } i \in \{0,1\}$$

for all $j \in \mathbb{N}$; note that f_s is well–defined because of Proposition 9. Then let the carrier of our intermediate model be the set $X := \{f_s \mid s \in [s_\alpha]\}$. In order to determine the set of opens (or, equivalently, the mapping d) we choose the standard numeration ν of $\{0,1\}^*$ counting over words by their length and over words of the same length lexicographically such that 0 precedes 1. Let $d(\nu(0)) = d(\varepsilon) := X$ (ε designates the empty word) and assume that $\nu(k)$ has already been treated for all $k \leq n$. Let $\nu(n+1) = w$ and $w = a_0 \ldots a_m$ ($a_j \in \{0,1\}$ for $j = 0, \ldots, m$). Then define

$$d(w) := \{f \in X \mid f(j-1) \xrightarrow{\bigcirc_{a_j}} f(j) \text{ for } j = 1, \ldots, m\}.$$

The non–emptiness of $d(w)$ is insured by the first part of the following lemma.

Lemma 10. *Let $w = a_0 \ldots a_m \in \{0,1\}^*$. Then there exists a function $f_w \in X$ such that*

$$f_w(j-1) \xrightarrow{\bigcirc_{a_j}} f_w(j) \text{ for } j = 1, \ldots, m.$$

Furthermore, for all $f, g \in d(w)$ it holds that

$$f(j) \xrightarrow{L} g(j) \text{ for } j = 0, \ldots, m.$$

Proof. The first assertion of the lemma is proved by induction on the length of w.

Case $w = \varepsilon$: Choose $f_\varepsilon = f_{s_\alpha}$.

Case $w = a_0 \ldots a_{m+1}$ ($a_j \in \{0,1\}$ for $j = 0, \ldots, m+1$): We only consider the case $a_{m+1} = 0$, since one proceeds analogously for $a_{m+1} = 1$. By the induction hypothesis there exists $g \in X$ such that

$$g(j-1) \xrightarrow{\bigcirc_{a_j}} g(j) \text{ for } j = 1, \ldots, m.$$

Because of the second clause of axiom (8) $L \bigcirc_0 \top \in g(m)$ holds. Thus $\bigcirc_0 \top \in s$ for some $s \in [g(m)]$ according to Proposition 6(d). Applying Corollary 7 we find a sequence

$$s_0 \xrightarrow{\bigcirc_{a_1}} s_1 \xrightarrow{\bigcirc_{a_2}} \ldots \xrightarrow{\bigcirc_{a_m}} s_m = s$$

such that $s_j \in [g(j)]$ for $j = 0, \ldots, m$. Now Proposition 6(c) yields a $t \in C$ such that $s \xrightarrow{O_0} t$. Thus $f_w := f_{s_0}$ is the wanted element of X. This ends the induction. — The second assertion of the lemma follows easily from Propositions 6(a) and 8. $\qquad\square$

What remains to be done is the definition of σ. So let $\sigma(A, f_s) = 1 :\Longleftrightarrow A \in s$ for all $A \in PV$ and $f_s \in X$. — We now prove that $\mathcal{M} := (X, d, \sigma)$ falsifies α. This will be a consequence of the subsequent theorem.

Theorem 11. *The subset space $\mathcal{M} = (X, d, \sigma)$ just constructed fulfils all binary subset tree model properties except for the condition on d stressed in Definition 1(1), possibly. Furthermore, for all $\beta \in \mathcal{F}$, $w \in \{0,1\}^*$, and $s \in [s_\alpha]$ such that $f_s \in d(w)$, we get:*

$$f_s, d(w) \models_{\mathcal{M}} \beta \Longleftrightarrow \beta \in f_s(length(w))$$

Proof. The first assertion of the theorem is clear by the construction. — Now

$$f_s, d(w) \models_{\mathcal{M}} \beta \Longleftrightarrow \beta \in f_s(length(w))$$

is proved by induction on β.

Case $\beta = A \in PV$:

$$
\begin{aligned}
f_s, d(w) \models_{\mathcal{M}} A &\Longleftrightarrow \sigma(A, f_s) = 1 &&\text{(by Definition 3)}\\
&\Longleftrightarrow A \in s &&\text{(by the definition of σ)}\\
&\Longleftrightarrow A \in f_s(length(w));
\end{aligned}
$$

axiom (2) is used repeatedly to get the latter equivalence. — We omit the other propositional cases.

Case $\beta = K\gamma$: First we state that for all words w the following holds in this case:

if $f_s \in d(w)$ and $u \in [f_s(length(w))]$, then there exists a $t \in [s_\alpha]$ satisfying $f_t \in d(w)$ and $f_t(length(w)) = u$ $(u \in C)$.

This is a consequence of Corollary 7. Then we get

$$
\begin{aligned}
f_s, d(w) \models_{\mathcal{M}} K\gamma &\Longleftrightarrow (\forall g \in d(w))\, g, d(w) \models_{\mathcal{M}} \gamma\\
&\Longleftrightarrow (\forall g \in d(w))\, \gamma \in g(length(w))\\
&\Longleftrightarrow K\gamma \in f_s(length(w)).
\end{aligned}
$$

The first equivalence holds by Definition 3. The second is valid because of the induction hypothesis. Finally, one uses the above assertion as well as the second statement in Lemma 10 to see the last one.

Case $\beta = O_0\gamma$:

$$
\begin{aligned}
f_s, d(w) \models_{\mathcal{M}} O_0\gamma &\Longleftrightarrow f_s \in d(w0) \text{ and } f_s, d(w0) \models_{\mathcal{M}} \gamma\\
&\Longleftrightarrow f_s \in d(w0) \text{ and } \gamma \in f_s(length(w0))\\
&\Longleftrightarrow O_0\gamma \in f_s(length(w))
\end{aligned}
$$

Only the last equivalence has to be commented. The direction $\Longrightarrow$ follows from standard modal proof theory, whereas one uses the definition of f_s and Proposition 6(c) for the reverse direction .

In case $\beta = \bigcirc_1 \gamma$ we argue exactly as in the previous one. This completes the proof of the theorem. $\qquad\square$

In fact, letting $w = \varepsilon$ and $s = s_\alpha$ we obtain:

Corollary 12. *Let* $\mathcal{M} = (X, d, \sigma)$ *be as in Theorem 11. Then* $\mathcal{M}$ *falsifies* α *at the neighbourhood situation* f_{s_α}, X.

The subset space $\mathcal{M} = (X, d, \sigma)$ is not yet the one we would like to have. It may still happen that two "paths" f_s, f_t eventually coincide. By means of a further construction this deficiency can be removed.

Let $X = \{f_s \mid s \in [s_\alpha]\}$ be as above. An element g_f of the Cantor space can be associated with every element $f \in X$ in a natural way: g_f is defined by

$$g_f(n) := \begin{cases} 0 \text{ if } f(n) \xrightarrow{\bigcirc_0} f(n+1) \\ 1 \text{ if } f(n) \xrightarrow{\bigcirc_1} f(n+1) \end{cases}$$

for all $n \in \mathbb{N}$. Now the carrier set $\langle X \rangle$ of the new model consists of all pairs (f, g_f) such that $f \in X$. Moreover, let $\langle d \rangle$ be defined by

$$\langle d \rangle(w) := \{(f, g_f) \mid g_f(0) \ldots g_f(length(w) - 1) = w\}$$

for all $w \in \{0, 1\}^*$. Finally, let $\langle \sigma \rangle(A, (f, g_f)) = 1 :\Longleftrightarrow \sigma(A, f) = 1$ for all $A \in PV$ and $(f, g_f) \in \langle X \rangle$. Then we get the following lemma.

Lemma 13. *Let* $\langle \mathcal{M} \rangle := (\langle X \rangle, \langle d \rangle, \langle \sigma \rangle)$. *Then* $\langle \mathcal{M} \rangle$ *is a binary subset tree model, and for all* $\beta \in \mathcal{F}$, $w \in \{0, 1\}^*$, *and* $f \in X$ *satisfying* $f \in d(w)$, *it holds that*

$$f, d(w) \models_\mathcal{M} \beta \iff (f, g_f), \langle d \rangle(w) \models_{\langle \mathcal{M} \rangle} \beta.$$

Proof. Clearly, the requirement on d previously missing is fulfilled for $\langle d \rangle$ now. The remainder of the proof is done by induction on β (not carried out here). $\square$

Combining the results of this section we get the main theorem of this paper.

Theorem 14. *A formula* $\alpha \in \mathcal{F}$ *is derivable in the system* C *iff* α *holds in all binary subset tree models.*

It should be mentioned that the Cantor space $\mathcal{C}_0$ can be embedded into the above constructed binary subset tree $(\langle X \rangle, \langle d \rangle)$. In fact, we only have to show that every 0–1–sequence g appears as some g_f, where $f \in X$. So let g be given. Consider any prefix $w = a_1 \ldots a_n \in \{0, 1\}^*$ of g. Then w induces a set

$$Y_w := \{\bigcirc_{a_1}\top, \bigcirc_{a_1}\bigcirc_{a_2}\top, \ldots, \bigcirc_{a_1}\ldots\bigcirc_{a_n}\top\}$$

of *NLBST*-formulas. Now let

$$Y := \{\beta \in \mathcal{F} \mid K\beta \in s_\alpha\} \cup \bigcup\{Y_w \mid w \text{ prefix of } g\}.$$

According to Lemma 10, Y is C–consistent. Let $\tilde{s}$ be a maximal C–consistent set containing Y. Defining $f := f_{\tilde{s}}$ we obtain $g = g_f$, as desired.

4   Decidability and Complexity

In this final section we show that the set of C–theorems is decidable and determine its complexity. Actually we deal with the set

$$SAT_C := \{\alpha \in \mathcal{F} \mid \alpha \text{ is satisfiable}\}$$

of all *NLBST*–formulas holding in some binary subset tree model at some neighbourhood situation. Because of Theorem 14, the decidability problem for the latter set is equivalent to the original one. The reason for decidability is given by the bounded "scope" of an *NLBST*–formula holding at some neighbourhood situation of a binary subset tree model. Let us explain this in a more detailed way. So, let a formula $\alpha \in \mathcal{F}$, a binary subset tree model $\mathcal{M} = (X, d, \sigma)$ and a neighbourhood situation x, U_w of (X, d) be given such that $x, U_w \models \alpha$. W.l.o.g. we may assume that $w = \varepsilon$ and $U_\varepsilon = X$ holds, as it can easily be seen. Let $rk(\alpha)$ be the *nexttime–rank* of α, i.e. the degree of nesting the $\bigcirc$–operators in α. (The exact recursive definition of $rk(\alpha)$ is evident and therefore omitted.) Let $l \in \mathbb{N}$ and $\mathcal{M}_l$ be the structure obtained from $\mathcal{M}$ by "cutting" the tree of opens after the full subtree D^l of height l rooting in $U_\varepsilon = X$ (and letting carrier X and X–valuation σ be unaltered); i.e. $\mathcal{M}_l := (X, d_l, \sigma)$, where d_l is the restriction of d to D^l. Then one can prove inductively the following *coincidence lemma.*

Lemma 15. *For all $\beta \in \mathcal{F}$, $w \in D^l$, and $y \in U_w$ it holds that $y, U_w \models_\mathcal{M} \beta \iff y, U_w \models_{\mathcal{M}_l} \beta$, if $rk(\beta) \leq l - length(w)$.*

(Strictly speaking, the validity relation $\models_{\mathcal{M}_l}$ has to be defined first; but transferring Definition 3 to such structures is obvious.) According to Lemma 15, it suffices to consider structures having a binary tree of height $\leq rk(\alpha)$ as set of opens. With the aid of the following equivalence relation on X, which depends on the considered formula $\alpha \in \mathcal{F}$, we can confine ourselves further to *finite* models:

$$x \sim y : \iff (\forall U \in \mathcal{O})(\forall A \in PV \text{ occurring in } \alpha)$$
$$[x \in U \iff y \in U] \text{ and } [\sigma(A, x) = 1 \iff \sigma(A, y) = 1]$$

Let $[x]_\sim$ denote the equivalence class of x w.r.t. the relation $\sim$. Define a subset space $[\mathcal{M}] = ([X], [\mathcal{O}], [\sigma])$ in the following manner:

- $[X] := \{[x]_\sim \mid x \in X\}$
- $[\mathcal{O}] := \{[U] \mid U \in \mathcal{O}\}$, where $[U] := \{[x]_\sim \mid x \in U\}$
- $[\sigma](A, [x]_\sim) := 1 \iff (\exists y \in [x]_\sim) \, \sigma(A, y) = 1$

for all $A \in PV$ and $[x]_\sim \in [X]$. Then the subsequent assertion can be proved by a simple induction.

Lemma 16. *For all subformulas β of α and neighbourhood situations $x, U \in X \otimes \mathcal{O}$ we have that $x, U \models_\mathcal{M} \beta \iff [x], [U] \models_{[\mathcal{M}]} \beta$.*

Going the other way round, one can "blow up" a finite structure $\mathcal{M}$ having a full finite–height tree of opens into a binary subset tree model which is equivalent to $\mathcal{M}$ w.r.t. α in a standard way.

Lemma 17. *Let $\mathcal{M} = (X, d, \sigma)$ be a structure of the type just considered. Furthermore, assume that $x, X \models_{\mathcal{M}} \alpha$ holds for some $x \in X$. Then there exists a binary subset tree model $\widetilde{\mathcal{M}} = (\widetilde{X}, \widetilde{d}, \widetilde{\sigma})$ and a point $\widetilde{x} \in \widetilde{X}$ such that $\widetilde{x}, \widetilde{X} \models_{\widetilde{\mathcal{M}}} \alpha$.*

In fact, $\widetilde{\mathcal{M}}$ is the inverse image of $\mathcal{M}$ w.r.t. an appropriate "p–morphism".

As the size of the carrier sets to be considered is bounded computably by certain invariants of α, the results of this section can be used to work out the correctness proof of a decision algorithm for the set SAT_C in a straightforward way. So we can state:

Theorem 18. *The set T of NLBST–formulas derivable in the system C is decidable.*

The above methods bound the size of the structures to be checked only to $2^{length(\alpha)}$ from above. By the next result this exponential bound can be decreased to a polynomial one. For this purpose we have to restrict the class of the relevant finite binary tree structures even more though: we must no longer require that the set indexing the opens is a *full* subtree of the set of binary strings.

Proposition 19. *Let $\alpha \in \mathcal{F}$ be satisfiable. Furthermore, let l be the number of nexttime–operators in α. Then there exists a binary subset space $(\widetilde{X}, \widetilde{d}, \widetilde{\sigma})$ such that the domain of d contains at most $l \cdot length(\alpha)$ many elements, $d(\varepsilon) = \widetilde{X}$, and the following conditions are fulfilled additionally:*

- *$x, \widetilde{X} \models \alpha$ for some $x \in \widetilde{X}$, and*
- *the size of $\widetilde{X}$ is bounded from above by $length(\alpha)$.*

In order to prove the proposition one has to look at the *generation tree* of the formula α, i.e. the tree representing the structure of α as it is built up of its subformulas. By an inductive definition going down from the root of this tree to the leaves one has to assign a suitably chosen point and a corresponding chain of subsets to every occurence of a subformula of α in the tree, which gives the bounds stated in the proposition. We omit further details.

As a consequence of the proposition, the complexity of SAT_C can be determined.

Theorem 20. *The NLBST–satisfiability problem SAT_C is NP–complete.*

By means of the theorem we easily obtain that the set of all C–theorems is co–NP–complete.

Theorem 20 should be compared with corresponding complexity results for the modal system $S5$, where NP–completeness holds as well [Ladner 1977], and for the general subset space logic, where the complexity is presumably higher: in [Heinemann 1997a] we showed $PSPACE$–hardness of the satisfiability problem of the latter system (among other things).

5 Conclusions

Looking for a language which is able to specify properties of knowledge on infinite binary branching–time structures, we were led to the topological modal logic of the Cantor space in a natural way. We changed the logical language proposed by MOSS and PARIKH [Moss and Parikh 1992] suitably, turning to the somewhat more general class of *binary subset tree models* as semantical domains. We could give a satisfactory treatment of the subset space logic of such structures then: The axiomatization of the set T of valid *NLBST*-formulas given in Section 3 was proved to be complete. Then decidability of T was shown. Moreover, the co–NP–completeness of T was established. — Future work will integrate other modal, temporal, and knowledge operators. Some work on the multi–agent case (including *common knowledge*) is in progress already [Heinemann 1996a].

References

[Chellas 1980] Chellas, B. F. 1980. *Modal Logic: An Introduction*. Cambridge: Cambridge University Press.

[Dabrowski et al. 1996] Dabrowski, A., L. S. Moss, and R. Parikh. 1996. Topological Reasoning and The Logic of Knowledge. *Ann. Pure Appl. Logic* 78:73–110.

[Fagin et al. 1995] Fagin, R., J. Y. Halpern, Y. Moses, and M. Y. Vardi. 1995. *Reasoning about Knowledge*. Cambridge(Mass.): MIT Press.

[Gabbay 1976] Gabbay, D. M. 1976. *Investigations in Modal and Tense Logics with Applications to Problems in Philosophy and Linguistics*. Dordrecht: Reidel.

[Georgatos 1994] Georgatos, K. 1994. Reasoning about Knowledge on Computation Trees. In *Proc. Logics in Artificial Intelligence (JELIA '94)*, eds. C. MacNish, D. Pearce, and L. M. Pereira, 300–315. Springer. LNCS 838.

[Goldblatt 1987] Goldblatt, R. 1987. *Logics of Time and Computation*. CSLI Lecture Notes Number 7. Stanford: Center for the Study of Language and Information.

[Heinemann 1996a] Heinemann, B. 1996a. 'Topological' Aspects of Knowledge and Nexttime. Informatik Berichte 209. Hagen: Fernuniversität, December.

[Heinemann 1996b] Heinemann, B. 1996b. 'Topological' Modal Logic of Subset Frames with Finite Descent. In *Proc. 4th Intern. Symp. on Artificial and Mathematics, AI/MATH-96*, 83–86. Fort Lauderdale.

[Heinemann 1997a] Heinemann, B. 1997a. On the Complexity of Prefix Formulas in Modal Logic of Subset Spaces. In *Logical Foundations of Computer Science, LFCS'97*, eds. S. Adian and A. Nerode. Springer. to appear.

[Heinemann 1997b] Heinemann, B. 1997b. Topological Nexttime Logic. In *Advances in Modal Logic '96*, eds. M. Kracht, M. de Rijke, H. Wansing, and M. Zakharyaschev. Kluwer. to appear.

[Ladner 1977] Ladner, R. E. 1977. The Computational Complexity of Provability in Systems of Modal Propositional Logic. *SIAM J. Comput.* 6:467–480.

[Moss and Parikh 1992] Moss, L. S., and R. Parikh. 1992. Topological Reasoning and The Logic of Knowledge. In *Proc. 4th Conf. on Theoretical Aspects of Reasoning about Knowledge (TARK 1992)*, ed. Y. Moses, 95–105. Morgan Kaufmann.

[Weihrauch 1995] Weihrauch, K. 1995. A Foundation of Computable Analysis. EATCS Bulletin 57.

How to Change Factual Beliefs Using Laws and Dependence Information

Andreas Herzig

IRIT, Université Paul Sabatier
118 route de Narbonne, F-31062 Toulouse Cedex 4, France
Email: Andreas.Herzig@irit.fr; http://www.irit.fr/~Andreas.Herzig
Tel.: +33 56155-6344, Fax: -8325

Abstract. We investigate how belief change operations can be effectively constructed. To that end we suppose given a set of laws (alias integrity constraints) together with a relation of dependence between formulas.

1 Introduction

When changing a set of beliefs K by some new piece of information (input) A, we must adjust the beliefs of K in a way such that A is believed. The result of such a belief change operation is a new set of beliefs $K \star A$.

The central question in the theory of belief change is what is meant by a minimal change of a state of belief. E.g. the well-known AGM rationality postulates for belief revision (AGM85, Gär88) partially answer that question. Another answer are the update postulates of (KM92). They all give requirements for the interplay between the (metalinguistic) belief change operator on the one hand and the classical connectives $\land$, $\lor$, $\lnot$ and the notions of consistency and theoremhood on the other.

As pointed out in (Gär90), "the criteria of minimality that have been used [in the models for belief change] have been based on almost exclusively logical considerations. However, there are a number of non-logical factors that should be important when characterizing a process of belief revision." Gärdenfors focusses on the notion of dependence (he uses the synonymous term 'relevance'), and proposes the following preservation criterion:

> *If a belief state is revised by a sentence A,*
> *then all sentences in K that are independent of the validity of A*
> *should be retained in the revised state of belief.*

This seems to be a very natural requirement for belief revision operations. But as Gärdenfors notes, "a criterion of this kind cannot be given a technical formulation in a model based on belief sets built up from sentences in a simple propositional language because the notion of relevance is not available in such a language."

Our aim is to put the notion of dependence in a formal framework, and to apply it to belief change. Such a notion should allow us to find algorithms for the effective construction of belief change operations in practical applications such as databases or robotics.

We view dependence as a weak causal connection between formulas: Given two formulas A and C, if C depends on A then changes concerning the truth value of A *might* (but need not) change the truth value of C. On the other hand, if C is independent of A then changes concerning the truth value of A will *never* influence the truth value of C. E.g. let p mean "it rains", and q "the grass is wet". Then q depends on p in the sense that there is a set of beliefs (e.g. $\{\neg q\}$) such that its revision by p makes the truth value of q change. On the other hand, if q means "the traffic light is red", then q is independent of p: Learning something about the weather (whether it started or stopped to rain) should not modify my beliefs concerning the colour of the traffic light. [1] On the other hand, if q means "the traffic light is red", then q is independent of p: Learning something about the weather (whether it started or stopped to rain) should not modify my beliefs concerning the colour of the traffic light.

As argued by Gärdenfors, dependence information enables us to construct revision operations: If we revise a set of beliefs K by some new piece of information A, then the resulting set $K \star A$ will at least contain all the formulas of K that are independent of A. (The same can be stated for updates *à la* Katsuno-Mendelzon.)

Another (and somewhat more popular) thing to integrate in a practical account of belief change are laws. Under the name integrity constraints, such priviledged formulas have been studied since a long time ago in databases and in knowledge representation. Laws are 'eternal' in the sense that they 'survive' every change of beliefs.

In this paper, we show how belief change operations based on a set of laws and a dependence relation can be effectively constructed. We suppose that a set of beliefs is represented by some finite set that is called a belief base. Nevertheless, we want to be syntax-insensitive: We want logically equivalent belief bases to be changed in the same way, and we want new pieces of information that are logically equivalent to lead to equivalent new belief bases. This will be achieved by the elimination of "inessential" atoms from formulas, via the use of prime implicates.

First we give the relevant definitions and stress the hypotheses we make. Then we introduce belief revision systems based on laws and dependence. Finally we present a reformulation and extension of the AGM-postulates that takes care of laws and dependences, and give the properties of our belief revision systems w.r.t. the latter.

2 Belief bases and syntax-sensitivity

If we want to store beliefs on a computer, we cannot represent them by deductively closed sets as done in the AGM framework, because such sets are infinite.

---

[1] Note that this may not be the case for other belief sets such as $\{q\}$.

What we need are finite *belief bases*.

Then a new problem shows up: For syntactically different but logically equivalent belief bases K_1 and K_2 it might be the case that some change operation modifies them differently. In other words, we might be sensitive to the syntax of the belief base.

This has been the starting point for the defense and study of syntax-sensitive belief change operations (Fuh91, Neb92, Han92). Despite its intuitive appeal, we think that the drawbacks of these approaches are too important: Take e.g. the four belief bases $\{p, q\}$, $\{p \wedge q\}$, $\{q \wedge p\}$, and $\{p \wedge p \wedge q\}$. Only very particular readings allow us to say that these bases do not represent the same state of affairs. Nevertheless, syntax-sensitive approaches admit operations which change these bases in four different ways.

It is for that reason we want to be syntax-insensitive. Precisely, syntax-insensitivity means two things: First, equivalent belief bases should be changed in the same way. Second, the syntactical form of the new piece of information should be irrelevant. In other words, change by logically equivalent formulas should lead to the same result. To ensure this we shall require the belief bases and the inputs to be of a particular form.

3 Classical logic

We work with the language of classical propositional logic with the connectives $\neg, \wedge, \vee, \top, \bot$. The set of atoms is denoted by ATM, and the set of formulas by FOR. For atoms we use $p, q, r \ldots$, and for formulas $A, B, C, \ldots$. For a given formula A, $atm(A)$ is the set of atoms having an occurrence in A. As usual, a *literal* is an atom or the negation of an atom, and a clause is a disjunction of literals. We do not distinguish between a clause and a finite set of literals. (Hence $\bot$ is the empty clause, and $\top$ is the empty set of clauses.) A finite set of formulas is called a *belief base*.

Given a set of formulas S, a formula A is said to be *S-consistent* iff $S \cup \{A\}$ is consistent.

Interpretations are subsets of ATM. Given a set of formulas S, an interpretation I is an *S-interpretation* iff $I \models A$ for all $A \in S$.

4 Normal forms and pure formulas

One way to get syntax-insensitive is to put formulas in some normal form. In particular, similar to (Sch93) we must avoid formulas containing atoms that can be eliminated by equivalence transformations.

Definition 1. A formula A is *pure* iff there is no equivalent formula containing less atoms.

E.g. $p \wedge q$ is pure, and $p \wedge (p \vee q)$ is not (because there exists an equivalent formula (viz. p) wherein q does not occur).

In terms of models, A is pure if and only if there is no atom p occuring in A such that for some classical p-interpretation $I \subseteq ATM$ we have $I \models A$ iff $I - \{p\} \models A$.

In classical logic, for every formula there exists an equivalent pure formula. (It is crucial here that $\top$ and $\bot$ are among our connectives, else there would be no pure formula equivalent to $p \vee \neg p$.) But how can we construct such pure formula? We need the notion of a prime implicate (Mar95).

Definition 2. A *prime implicate* of a formula A is a clause C such that

- $A \vdash C$, and
- for all clauses C' such that $A \vdash C'$ and $C \subseteq C'$ we have $C' \subseteq C$.

E.g. the clause $\{p\}$ is a prime implicate of $((p \vee q) \wedge p) \wedge (q \vee r)$, and $\{p, q\}$ is not.

We shall denote the set of all prime implicates of A by $A \downarrow$. E.g. $((p \vee q) \wedge p) \wedge (q \vee r) \downarrow = \{\{p\}, \{q, r\}\}$, and $(p \vee q) \wedge (\neg q \vee r) \downarrow = \{\{p, q\}, \{\neg q, r\}, \{p, r\}\}$.

Proposition 3. *The set of prime implicates of a formula is finite.*
The conjunction of the prime implicates of a formula A is equivalent to A.

Note that it is important for finiteness that we identify clauses such as $p \vee q, q \vee p$ and $p \vee p \vee q, \ldots$ with the set $\{p, q\}$.

This gives us the normal form we need:

Proposition 4. *If $\vdash A_1 \leftrightarrow A_2$ then $A_1 \downarrow = A_2 \downarrow$.*

Moreover we have:

Proposition 5. *The conjunction of the prime implicates of a formula is pure.*

Hence we have now at our disposal a normal form into which belief bases and inputs can be transformed in order to warrant syntax-insensitivity.

5 Dependence relations

Definition 6. An *atomic dependence relation* $\leadsto_0$ is a binary relation on ATM that is reflexive.

(Note that we neither suppose symmetry nor transitivity.)
$\leadsto_0$ induces a dependence relation $\leadsto$ on the set of formulas FOR by

$$A \leadsto C \text{ iff there exist } p \in atm(A), q \in atm(C) \text{ such that } p \leadsto_0 q.$$

Hence our dependence relations have the following properties: [2]

- $A \not\leadsto \top, A \not\leadsto \bot, \top \not\leadsto A, \bot \not\leadsto A,$

---

[2] Although we do not have symmetry, this could be added without harm.

- If $A \rightsquigarrow C$ then $A \rightsquigarrow \neg C$ and $\neg A \rightsquigarrow C$.
- If $A \rightsquigarrow C$ then $A \rightsquigarrow C \Delta C'$ and $A \Delta A' \rightsquigarrow C$ for $\Delta = \wedge, \vee, \rightarrow, \leftrightarrow$.

The complement of a dependence relation $\rightsquigarrow$ is called an *independence relation* and is noted $\not\rightsquigarrow$.

Given a dependence relation, we can define the *orthogonal* $K \perp A$ of a belief base K w.r.t. a formula A, which is the set of those elements of K that are independent of A.

Definition 7. $K \perp A = \{C : C \in K \text{ and } A \not\rightsquigarrow C\}$.

E.g. let $\rightsquigarrow$ be constructed from

$$\rightsquigarrow_0 = \{(p,p), (q,q), (r,r), (t,t), (p,q)\}$$

Then $\{p, p \vee t, r \vee t\} \perp \neg p = \{r \vee t\}$.

We have supposed that the dependence relation is induced by the atomic dependence relation. Thus we have a dependence relation with very strong properties.

This is a compromise in order to economically represent beliefs. We are aware that there is no complete theoretical justification for that: There is no theory of uncertainty whose dependence relation has such properties as ours (Fin73, Coh94, FdCH95, DFdCHP94). Indeed, it might be the case that our way of representing things is too crisp: For three atoms p, q, r it might be the case that $p \not\rightsquigarrow r$ and $q \not\rightsquigarrow r$, but nevertheless $p \wedge q \rightsquigarrow r$. [3]

Note that the notion of dependence is rather fuzzy, and hence it is unclear what the natural and intuitive properties of a dependence relation are. E.g. if one adopts probability-based dependence, dependence becomes transparent with respect to negation (in the sense that A and C are dependent iff A and $\neg C$ are dependent), but not with respect to conjunction or disjunction. (Note that this has been criticized e.g. in (Gär78) and in (Gär90)). On the contrary, things are just the other way round in the case of possibility-based dependence (DFdCHP94).

The strong properties of our dependence relation may thus make us lose some complex formulas when it comes to the construction of the new set of beliefs. But on the other hand, it gives us a particularly economic knowledge representation: If we suppose the set of atoms ATM to be finite, both the atomic dependence relation $\rightsquigarrow_0$ and the atomic independence relation $\not\rightsquigarrow_0$ contain at most $card(ATM)^2$ elements.

Note that the strong properties of dependence relations with respect to negation, conjunction and disjunction make that we are syntax-sensitive in the sense that equivalent formulas may not have the same dependencies. E.g. $p \wedge (p \vee q)$ depends on p and q, but q is inessential. It will be the use of pure formulas that will make us syntax-insensitive here.

---

[3] An intuitive example mentioned to us by D. Dubois is obtained by reading p as "I take a bath", q as "I use a hairdryer", and r as "I am going to die".

Our notion of dependence is close to that of *topics* (alias themes) as studied in (FdCL91, Lug96, DJ94). There, the themes of a formula are what the formula is about. Then one can define two formulas to be dependent if they have some theme in common. (Such a dependence is called conversational in (Coh94). The only difference is that such a notion is always symmetric (which is a property we could add as well).

We make the hypothesis of *inertia* (San93): Even factual change does not occur massively, and "almost all" of the factual beliefs survive the change. Therefore, we can expect the atomic dependence relation to be much smaller than the independence relation. This is why we chose to represent the dependence relation instead of its complement.

6 Laws and factual change

Laws (alias integrity constraints) are viewed as being "eternal" and cannot be put into question. We suppose that a law is preserved under any change. Syntactically, a law is just an arbitrary formula of FOR. We note IC the set of laws of a given domain, and we shall suppose that IC is consistent.

Hence the type of belief change that is in the scope of our approach does not allow the revision of laws. Our 'small changes' (Seg86) only concern contingent facts. We call them therefore *factual*. (As well, we shall suppose here that the dependence relation cannot be modified.)

Such a hypothesis being unsatisfactory for philosophers [4], we nevertheless think that it is reasonable and useful in the context of practical reasoning: change appears to be factual e.g. in databases or robotics.

7 Compatibility of laws and dependences

It is clear that laws and dependences are related. E.g. whenever $p \leftrightarrow q$ is in the set of laws we must have $p \leadsto q$ and $q \leadsto p$. The same holds for the law $p \to q$ (except if $IC \vdash \neg p \land q$). Such compatibilities or incompatibilities cannot be derived from the definition of dependence relations of section 5, where we did not say anything about laws yet.

A first naive requirement would be that whenever two atoms p and q both occur in some law, then they should be dependent. But this is clearly too strong. Take e.g. an electric circuit where some light is on exactly when two switches are closed (Lif86). Formally, this corresponds to a law $(p_1 \land p_2) \leftrightarrow q$, where p_i is read 'switch i is up', and q 'the light is on'. If I am told that the first of the two switches has been moved, I am prepared to change my beliefs concerning the light, but not those concerning the position of the second switch. Formally, $p_1 \leadsto q$, $p_2 \leadsto q$, but $p_1 \not\leadsto p_2$.

---

[4] Note that the evolution of laws was the motivation of the founders of the AGM-theory: Alchourrón and Makinson were interested in the derogation of (legal) norms, and Gärdenfors in the evolution of scientific theories.

Intuitively, the dependence relation must contain enough dependences of p in order to allow any change of p respecting the laws. The following condition is central and warrants what we need.

Definition 8. A set of laws IC and an atomic dependence relation $\leadsto_0$ are *compatible* iff for every IC-interpretation I and every IC-consistent atom p there exists an $(IC \wedge p)$-interpretation I^+ and there exists an $(IC \wedge \neg p)$-interpretation I^- such that all atoms in $(I - I^+) \cup (I - I^-)$ depend on p. [5]

E.g. let $IC = \{p \to q\}$. If $p \not\leadsto_0 q$ or $q \not\leadsto_0 p$ then IC and $\leadsto$ are incompatible. Contrarily, if $\leadsto$ is constructed from $\leadsto_0 = \{(p,p), (q,q), (p,q), (q,p)\}$ then IC and $\leadsto$ are compatible.

This gives us the following property for general formulas.

Proposition 9. *Let IC a set of laws and $\leadsto$ a dependence relation such that IC and $\leadsto$ are compatible. Let A be an IC-consistent formula. Then for every IC-interpretation I there is an $(IC \wedge A)$-interpretation I' such that all atoms in $I - I'$ depend on A.*

Suppose the language is finite, and $\leadsto$ and IC are given. It is straightforward to check compatibility of $\leadsto$ and IC: we must instantiate A by all maximally consistent conjunctions of literals. Such compatibility checks are exponential, but note that they can be done once for ever when a particular system for reasoning about actions and plans is designed. Now a way to generate all minimally compatible dependence relations is to incrementally apply compatibility tests.

8 Dependence-based belief change

Now we are ready to formally define dependence-based belief change operations.

Definition 10. A belief change system is a couple $(IC, \leadsto_0)$, such that

- IC is a finite set of formulas (the laws),
- $\leadsto_0$ is a dependence relation, and
- IC and $\leadsto_0$ are compatible.

Now for every belief change system we can construct a unique revision operation $\star$.

Definition 11. Let $(IC, \leadsto_0)$ be a belief change system. The associated revision operation $\star$ is such that for every belief base K

- $K \star A = A{\downarrow} \cup IC$ if $\emptyset \in K{\downarrow}$
- $K \star A = K \cup \{A\} \cup IC$ if $K \cup IC \not\vdash \neg A$
- $K \star A = ((K{\downarrow}) \bot (A{\downarrow})) \cup A{\downarrow} \cup IC$ if $K \cup IC \vdash \neg A$ and $\emptyset \notin K{\downarrow}$.

---

[5] i.e. $p \leadsto_0 q$ for all atoms $q \in I - I^+$, and $p \leadsto_0 q$ for all atoms $q \in I - I^-$, and

In other words, in order to revise a belief base K by A we first check whether K is IC-inconsistent. If this is the case then $K \star A$ is $A{\downarrow} \cup IC$. Else we check whether K and A are IC-consistent. If this is the case, we simply add A and IC to K. Else we put K and A in normal form by computing the respective sets of prime implicates $K{\downarrow}$ and $A{\downarrow}$, and then compute the orthogonal $(K{\downarrow}){\perp}(A{\downarrow})$. The new belief base results from adding $A{\downarrow}$ and IC to the latter.

E.g. let $IC = \emptyset$, and let $\leadsto$ be constructed from

$$\leadsto_0 = \{(p,p),(q,q),(r,r),(t,t),(p,r)\}$$

Let $K = \{p, q \vee t, r \vee t\}$. Then $K \star \neg p = \{q \vee t, \neg p\}$.

Clearly, computing prime implicates is expensive. But it seems to be the only way to avoid syntax-sensitivity.

9 Postulates

In the AGM-framework (Gär88), nothing is said about laws. Therefore, some postulates must be reformulated in terms of IC-consistence in order to take them into account. We have adapted the formulation in (KM92). Only the postulate R0 does not appear in the original set.

R0$_{IC}$ $K \star A$ is a belief base such that $K \star A \vdash IC$.
R1$_{IC}$ $K \star A \vdash A$.
R2$_{IC}$ If $K \cup \{A\}$ is IC-consistent then $K \star A \vdash K$.
R3$_{IC}$ If A is IC-consistent then $K \star A$ is IC-consistent.
R4$_{IC}$ If $K_1 \dashv\vdash K_2$ and $A_1 \dashv\vdash A_2$ then $K_1 \star A_1 \dashv\vdash K_2 \star A_2$. [6]
R5$_{IC}$ $(K \star A_1) \cup \{A_2\} \vdash K \star (A_1 \wedge A_2)$.
R6$_{IC}$ If $(K \star A_1) \cup \{A_2\}$ is IC-consistent then $K \star (A_1 \wedge A_2) \vdash (K \star A_1) \wedge A_2$.

(Remember that we do not distinguish between a finite set and the conjunction of its elements.)

But we not only have things to say about laws, but also about dependences. The first and main postulate involving dependence that we can formulate is

R1$_\leadsto$ If $A \not\leadsto C$ and $K \vdash C$ then $K \star A \vdash C$.

This is just the preservation criterion of (Gär92).

But we can also say more subtle things about the interplay between dependence and the classical connectives. We only give the following two postulates. The first one deals with conjunction.

R2$_\leadsto$ If $A_1 \not\leadsto A_2$ then $K \star (A_1 \wedge A_2) \vdash (K \star A_1) \star A_2$.

This means that revision by the conjunction of independent formulas can be done sequentially.

The next postulate is about disjunction.

---

[6] $A \dashv\vdash B$ is a shorthand for '$A \vdash B$ and $B \vdash A$'.

R3. If $A_1 \not\leadsto A_2$, $A_2 \not\leadsto A_1$ and $K \vdash \neg(A_1 \vee A_2)$ then $K \star (A_1 \vee A_2) \vdash \neg(A_1 \wedge A_2)$.

In other words, if both $\neg A_1$ and $\neg A_2$ are in the belief set and A_1 and A_2 are independent of each other, then the revision by $A_1 \vee A_2$ leads to the exclusive disjunction.

To illustrate that, suppose a murder has occurred, and the butler and the gardener (who were both initially supposed to be innocent) get suspected: Either the butler was it, or the gardener, *or both*. Suppose my opinion about the butler's guiltiness does not depend on that about the gardener, and vice versa. Then R3. says that we should suppose that exactly one of them is the murderer (i.e. we suppose they did not act together). Under the independence hypothesis this fits nicely in Gärdenfors' principle of informational economy. In turn, if this hypothesis cannot be made, it is somewhat adventurous to exclude the possibility of both being murderers.

Note that such a principle of minimal change (together with the underlying independence hypothesis) is implicit in several formal approaches to the dynamics of belief such as circumscription (EGG93) or Winslett's Possible Models Approach (Win88).

10 Properties

Proposition 12. *Every dependence-based belief change operations satisfies the AGM revision postulates $R0_{IC}$ - $R5_{IC}$.*

It is easy to prove that our belief change operation satisfies $R0_{IC}$, $R1_{IC}$, $R2_{IC}$, and $R5_{IC}$. In order to establish that equivalent formulas give us the same result ($R4_{IC}$) we use the properties of prime implicates (proposition 4). It is more difficult to establish that the result of a revision by IC-consistent formulas is IC-consistent ($R3_{IC}$). Here we must exploit that formulas and belief sets correspond to prime implicates, and that dependences and constraints are compatible: For the main case, suppose K and A are consistent. Hence there is a classical model I of K. As $K \leftrightarrow K\!\downarrow$ and $(K\!\downarrow)\bot(A\!\downarrow) \subseteq K\!\downarrow$, I is a model of $(K\!\downarrow)\bot(A\!\downarrow)$ as well. As IC and $\leadsto$ are compatible, by proposition 9 there exists a classical $IC \wedge A$-interpretation I' such that all atoms in $I - I'$ depend on A. As $(K\!\downarrow)\bot(A\!\downarrow)$ does not contain any of these atoms, I' is a $(K\!\downarrow)\bot(A\!\downarrow)$-interpretation as well. Hence $((K\!\downarrow)\bot(A\!\downarrow)) \cup A\!\downarrow \cup IC$ is satisfiable.

Proposition 13. *There are dependence-based belief change operations which do not satisfy the postulate $R6_{IC}$.*

In the proof we give IC and $\leadsto_0$ such that the associated revision operation does not satisfy $R6_{IC}$.

Note that $R6_{IC}$ has been criticized by means of a counterexample (Sta92).

Proposition 14. *Every dependence-based belief change operations satisfies the postulates $R1.$, $R2.$, and $R3.$.*

11 Conclusion

We have showed how one can construct a belief change operation from a set of laws together with a dependence relation. We have shown that all AGM-postulates except one are satisfied, and we have stated several new dependence-based postulates. It would be nice to find a set of postulates completely characterizing our family of operations.

One can proceed in the same way in the case of updates à la Katsuno-Mendelzon (KM92). There, given a set of laws IC and an atomic dependence relation $\leadsto_0$ the definition of the update operation $\leftrightarrow\!\!\diamond$ is even simpler:

- $K \leftrightarrow\!\!\diamond A = ((K\!\downarrow)\bot(A\!\downarrow)) \cup A\!\downarrow \cup IC$

Then the only KM-postulate which does not immediately follow is

(U8) $(B_1 \vee B_2)\leftrightarrow\!\!\diamond A \leftrightarrow . B_1\leftrightarrow\!\!\diamond A \vee B_2\leftrightarrow\!\!\diamond A$

This must be investigated further.

Finally, another continuation of this paper is to look for other natural postulates for revisions involving dependence.

12 Acknowledgements

Some of the ideas in this paper have been presented at the workshop on formal epistemology that has been organized by Hans Rott at the European Summer School on Logic, Language and Information in Barcelona (August 1995).

Thanks to Daniel Lehmann for several email exchanges on the interpretation of disjunction in revision. I am indebted to Pierre Marquis for discussions on prime implicates, and to Luis Fariñas del Cerro and Marcos A. Castilho for stimulating comments on preceding drafts.

References

Carlos Alchourrón, Peter Gärdenfors, and David Makinson. On the logic of theory change: Partial meet contraction and revision functions. *J. of Symbolic Logic*, 50:510–530, 1985.

D. Cohen. Some steps towards a general theory of relevance. *Synthese*, 101:171–185, 1994.

Didier Dubois, Luis Fariñas del Cerro, Andreas Herzig, and Henri Prade. An ordinal view of independence with applications to nonmonotonic reasoning. In Ramon Lopez de Mantaras and David Poole, editors, *Proc. Int. Conf. on Uncertainty in AI (UAI'94)*, pages 1855–203, Seattle, 1994.

Robert Demolombe and Andrew Jones. A logic for reasoning about "is about". Technical report, ONERA-CERT, Toulouse, 1994.

Thomas Eiter, Georg Gottlob, and Yuri Gurevich. Curb your theory! a circumscriptive approach for inclusive interpretation of disjunctive information. In Ruzena Bajcsy, editor, *Proc. 13th Int. Joint Conf. on Artificial Intelligence (IJCAI'93)*, pages 640–645. Morgan Kaufmann Publishers, 1993.

Luis Fariñas del Cerro and Andreas Herzig. Possibility theory and independence. In Bernadette Bouchon-Meunier, Ronald R. Yager, and Lotfi A. Zadeh, editors, *Advances in Intelligent Computing - IPMU'94, Selected Papers*, number 945 in LNCS, pages 292–301. Springer-Verlag, 1995.

Luis Fariñas del Cerro and Valérie Lugardon. Sequents for dependence logics. *Logique et Analyse*, 133-134:57–71, 1991.

Terrence Fine. *Theories of probability*. Academic Press, New York, 1973.

André Fuhrmann. Theory contraction through base contraction. *J. of Philosophical Logic*, 20:175–203, 1991.

Peter Gärdenfors. On the logic of relevance. *Synthese*, 37:351–367, 1978.

Peter Gärdenfors. *Knowledge in Flux: Modeling the Dynamics of Epistemic States.* MIT Press, 1988.

Peter Gärdenfors. Belief revision and irrelevance. *PSA*, 2:349–356, 1990.

Peter Gärdenfors, editor. *Belief revision*. Cambridge University Press, 1992.

Sven-Ove Hansson. In defence of the ramsey test. *J. of Philosophy*, pages 522–540, 1992.

Hirofumi Katsuno and Alberto O. Mendelzon. On the difference between updating a knowledge base and revising it. In Gärdenfors Gär92, pages 183–203. (preliminary version in Allen, J.A., Fikes, R., and Sandewall, E., eds., Principles of Knowledge Representation and Reasoning: Proceedings of the 2nd International Conf., pages 387–394. Morgan Kaufmann Publishers, 1991).

Vladimir Lifschitz. Frames in the space of situations. *Artificial Intelligence J.*, 46:365–376, 1986.

V. Lugardon. *Sur les fondements de la notion de dépendance et de son application à la théorie de l'action*. PhD thesis, Institut de recherche en informatique de Toulouse (IRIT), Université Paul Sabatier, 1996.

Pierre Marquis. Knowledge compilation using theory prime implicates. In *Proc. 14th Int. Joint Conf. on Artificial Intelligence (IJCAI'95)*, pages 837–843, 1995.

Bernhard Nebel. Syntax-based approaches to belief revision. In Gärdenfors Gär92, pages 247–275.

Erik Sandewall. The range of applicability of nonmonotonic logics for the inertia problem. In *Proc. 13th Int. Joint Conf. on Artificial Intelligence (IJCAI'93)*, 1993.

Karl Schlechta. Some completeness results for propositional conditional logics. *J. of the IGPL*, 3(1), 1993. available from http://www.mpi-sb.mpg.de/igpl/Journal/V3-1/.

Krister Segerberg. On the logic of small changes in theories, I. *Auckland Philos. Papers*, 1986.

Robert Stalnaker. What is a nonmonotonic consequence relation? In *(Informal) Working Notes of the 4th Int. Workshop on Nonmonotonic Reasoning*, Plymouth, Vermont, 1992.

M. Winslett. Reasoning about action using a possible models approach. In *Proc. 7th Conf. on Artificial Intelligence (AAAI'88)*, pages 89–93, St. Paul, 1988.

Using Default Logic for Lexical Knowledge

Anthony Hunter

Department of Computer Science
University College London
Gower Street
London WC1E 6BT, UK

a.hunter@cs.ucl.ac.uk

Abstract. Lexical knowledge is knowledge about the morphology, grammar, and semantics of words. This knowledge is increasingly important in language engineering, and more generally in information retrieval, information filtering, intelligent agents and knowledge management. Here we present a framework, based on default logic, called Lexica, for capturing lexical knowledge. We show how we can use contextual information about a given word to identify relations such as synonyms, antinyms, specializations, and meronyms for the word. We also show how we can use machine learning techniques to facilitate engineering a Lexica knowledgebase.

1 Introduction

Lexical knowledge is knowledge about the semantics, morphology, and usage, of words. Handling words is central to many reasoning activities, and as a result lexical knowledge is increasingly important in language engineering, and more generally in information retrieval, information filtering, intelligent agents and knowledge management. Lexical knowledge can facilitate in the resolution of ambiguity in information.

For example, if we know that a newspaper article is about `oil`, it is usually reasonable to derive that it is about `petroleum`, with exceptions such as in contexts about `cooking`.

Lexical knowledge can also facilitate in the identification of synonyms, related terms, antinyms, and specializations for a word. It can also be used to identify meronymic relations, such as `engine` is `part-of` a `car`, and parts-of-speech such as relating actors with actions: For example, for the actor `terrorist` an appropriate action is `terrorism`.

In this paper, we briefly review the need for lexical knowledge — with particular emphasis on intelligent agents — and then show the need for a new approach to providing lexical knowledge. To address this need, we present a framework called Lexica for providing context-dependent lexical knowledge. The Lexica framework is based on default logic. We show how grammatical and semantic knowledge can be captured in Lexica. We also show how machine learning can be used to build Lexica knowledgebases.

1.1 An example of the need for lexical knowledge

Rapidly increasing amounts of information, particularly textual information, is being made available electronically, though the Internet, newsfeed, electronic databases, etc. This has created a pressing need to develop intelligent agents to search these sources for information that meets a user's needs.

Current search engines for the Internet (for example Yahoo, Lycos, and Alta Vista) use keywords given by the user to locate items that may be of interest to the user. Unfortunately, these search engines are limited in their ability to use background knowledge. They incorporate little knowledge on the meanings of keywords, morphology of words, nor on related terms for particular contexts. In particular, there is no way for users to provide background knowledge that could improve the precision and recall.

Statistical techniques in information retrieval and filtering offer some solutions for the Internet, but lack a systematic means for using background information, and are lacking in facilities for users to specify background knowledge. Statistical techniques are well suited to repositories of information where comprehensive statistical analyses can be undertaken (see for example [Cro93]), but are less well suited to heterogeneous distributed sources, such as the Internet, where the topics are so diverse, locally managed, and constantly evolving.

An increasingly important alternative is the approach of intelligent agents (for example [Mae94,GLC+95,GF95,Lie95,BPK+96,FJ96,Mou96]). Some incorporate a limited amount of background knowledge and allow users to input some background knowledge. However, they lack a systematic means for using large amounts of complex background knowledge. This raises the need to incorporate default knowledgebases as a repository for background knowledge. Using explicit defaults offers a lucid representation for users, and it aids maintainability and incrementality.

1.2 Lexicons

What is a lexicon? According to Trask [Tra93], a lexicon within the study of grammar has traditionally been used as a repository of miscellaneous facts, with little in the way of generalizations. This view has shifted and recent theories of language are using lexicons for significant proportions of linguistic knowledge [Bri91]. A wide variety of machine-readable lexicons have been developed (for reviews see [WSG96,GPWS96]), though many are oriented to specific approaches or tasks, such as EDR [Yok95] and Acquilex [Bri91] for machine translation.

Yet there is a need to general purpose information about words, such as for intelligent agents, information retrieval, information filtering, and information extraction. Perhaps the most significant example of such a general purpose system is WordNet [BFGM91,Mil95]. This is a semantic network containing lexical knowledge on over 90,000 word senses and it is now found to be an increasingly important resource on synonyms, generalizations, and specializations of words, for applications in information systems. In WordNet, each set of words that are strict synonyms (i.e. the words can be interchanged in a sentence) is called a synset. The following is an example of a synset.

{Molotov-cocktail, petrol-bomb, gasoline-bomb}

Whilst WordNet separates different meanings of the same word by putting the same word in more than one synset, there is no explicit machinery for determining in which context a particular wordsense should be used. Moreover, there is no logical reasoning with the relations in the semantic network.

Another kind of problem with WordNet is that the knowledge is very general. WordNet has been applied to information retrieval [Voo94] and information filtering [Eng96] and in these studies the utility of WordNet was limited by this generality. Furthermore, WordNet was limited by the inability of the user to be able to add context-dependent knowledge appropriate for the application domain.

1.3 Lexical knowledgebases

In order to build on the success of systems such as WordNet, and address some shortcomings, we need to develop more sophisticated systems that incorporate richer, more complex, knowledge about words. Essentially, we require knowledgebases containing structured knowledge about words. This calls for knowledgebased systems technology that can extend the approach of lexicons by allowing for automated identification of contexts for a word, by selecting knowledge on a context-dependent basis, and by supporting automated reasoning.

This need for lexical knowledge raises significant knowledge representation and reasoning questions. Lexical knowledge is default knowledge. This is knowledge that is usually correct, though can have some exceptions. Representing and reasoning with default knowledge in computing is difficult, and in anything other than small examples, it is necessary to adopt a logical approach in order to minimize these difficulties.

We therefore require a formal approach to knowledge representation and reasoning that can handle the context-dependent default knowledge. In this work, we explore the use of non-monotonic logics for building these more sophisticated lexical knowledgebases.

2 Contextual information

A context is a setting for a word. If a word is polysemous, then the word is a member of more than one context. Different contexts can denote different word senses for a word. In this way, a context can be viewed as a boundary on the meaning of a word. For example, the word bank can be described as being a member of contexts including river and financial-institution.

In language, the words surrounding a particular word can indicate the context for the word. For example, for a word in some text, the words in the same paragraph can usually indicate the context for the word.

In this section, we show how we can use a classification, or decision, tree to test whether a set of words is in a particular context. We then show how we can use machine learning techniques to generate such classification trees.

2.1 Context classification trees

Each classification tree is developed to test for a single classification. In this work, each classification is a context. Given a set of words, presence or absence of particular words in the set of words, is used by a classification tree to classify the set of words as either a positive or negative example for the classification. Hence, a classification tree determines whether the set of words is in a particular context.

A classification tree for a context C is a binary tree, where each node is a word, except the leaves which are labelled either POSITIVE or NEGATIVE. Given a set of words, start at the root: If the root is in S, then take the left subtree, otherwise take the right subtree. Upon taking the subtree, repeat the process, until reaching a leaf. If the leaf is POSITIVE then S is in context C.

Consider the example of a classification tree in Fig 1 for the classification aircraft-accident. Given the set $S = \{$crash, boeing, engine, runway$\}$, the tree classifies S as being in the context aircraft-accident.

Given a set of words S, there might be a number of classification trees with different contexts, and the set S is found to be in each of the contexts using the trees. To handle this see section 2.3.

2.2 Learning classification trees

In this work, we have used the ID3 inductive learning algorithm developed by Ross Quinlan [Qui86]. ID3 is an approach to machine learning based on constructing a classification, or decision, tree for a set of training examples. Training examples are presented as a table — each row is an example and each column is an attribute of the examples. The last attribute is the classification of the example. For example, in learning a decision tree for determining whether a patient has a particular disorder, we use a table of patients — some who have the disorder and some who do not. Each column refers to particular symptoms or tests, and the final column states whether the patient has the disorder. Once a decision tree has been constructed, and then tested successfully with examples not used in the training, it can be used to classify further examples.

We have used ID3 to classify textual information. The methodology involves taking an item of text, removing the stop words[1], and then using the remaining

---

[1] Stop words are words that usually offer relatively little semantic information in a sentence, such as for example, the, a, because, and what. They normally constitute about 50% of the words in a sentence.

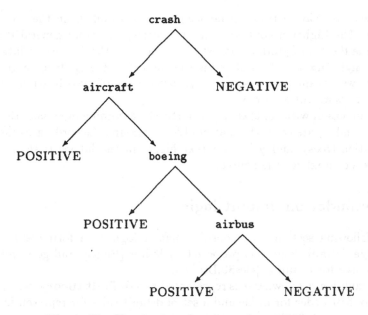

Fig. 1. Classification tree for `aircraft-accident`

words as a training (learning) example. Each item of text is a about at least one topic. Topics are used as the classifications for the examples. Each attribute in the table of training examples is a word. If a training example contains that word, then "yes" is entered into the corresponding position in the table, and "no" otherwise. Tables up to 236 attributes have been constructed containing up to 60 examples (rows) [Sha96]. We equate the notion of topic with that of context. So we can use these classification trees as described in section 2.1.

2.3 Reasoning with contextual information

We assume a set of atomic contexts. These are disjoint and exhaustive. They are the most focussed contexts we consider. We also allow contexts that are Boolean combinations of atomic contexts. These non-atomic contexts may have alternative names for lucidity. A classification tree can be trained for any Boolean combination of the atomic contexts. For example, if the language includes the non-atomic contexts `personal-finance` and `business-finance`, then the language includes the Boolean combinations such as

$$\text{personal-finance} \land \text{business-finance}$$

$$\text{personal-finance} \lor \text{business-finance}$$

The first is a more specialized context, whereas the second is a more generalized context. The second might have an alternative name such as `finance`.

In this way, we have a Boolean lattice of contexts built from the set of atomic contexts. The higher a context is in the lattice, the more general it is. Since we assume the set of atomic contexts is exhaustive, the top of the lattice is the context **anything** and the bottom is the context **nothing**. If for a given set of words S, we obtain contexts $x_1, ..., x_n$, using a set of classification trees, then $x_1 \wedge ... \wedge x_n$ is a context for S.

We can reason with contexts in the lattice by assuming inferences that hold: If x_1 and ... and x_n are contexts that hold for the source (according to the context classification trees), and y is a context higher in the lattice than $x_1 \wedge ... \wedge x_n$, then y is a context for the source.

3 Reminder on default logic

In the following sections, we consider default logic as a formalism for lexical knowledge. Default logic was proposed by Reiter [Rei80], and good reviews are available (see for example [Bes89,Bre91]).

In default logic, knowledge is represented as a **default theory**, which consists of a set of first-order formulae and a set of default rules for representing default information. A **default rule** is of the following form, where α, β and γ are classical formulae,

$$\frac{\alpha : \beta}{\gamma}$$

The inference rules are those of classical logic plus a special mechanism to deal with default rules: Basically, if α is inferred, and $\neg\beta$ cannot be inferred, then infer γ. For this, α is called the pre-condition, β is called the justification, and γ is called the consequent. Informally, an **extension** is a maximally consistent set of inferences (classical formulae) that follow from a default theory.

Basing the framework on default logic brings advantages. Default logic provides an efficient representation for context-dependent reasoning and handling of exceptions, and it is a well-understood formalism for representing uncertain information. In addition, there are prototype implementations of inference engines for default logic that can be used for developing default logic knowledge-bases [Nie94,LS95,Sch95].

4 Outline of Lexica framework

The Lexica framework is based on default logic. In this framework we represent morphological, grammatical, and semantic relations using default logic. The user queries the system to find information about a word. The information can include synonyms, generalizations, specializations, meronyms, related terms, different lexical categories of the word, and so on.

A key feature of the Lexica framework is the identification of the context for a query. The context is identified from the input that the user provides. The interaction between a user and the system can be summarized as follows:

Input: A query word plus a source, defined as follows.

Query word. A word for which further information is required.

Source. A set of words used to help identify the contexts for a query word. A source may be obtained in a number of ways. For example, it could be given directly by a user seeking information about a query word, or it could be obtained from a sentence containing the query word.

Output: Set of relations providing further information about the query word.

A Lexica knowledgebase is composed of the following three sets of knowledge that are used to provide the output from the system.

1. **A set of context classification trees.** Given a source S and a query word q, the context classification trees are used to identify contexts that hold for $S \cup \{q\}$. Consider a tree T that tests whether $S \cup \{q\}$ is in context **x**: If the test is positive, then $S \cup \{q\}$ is in context **x**.
2. **A context lattice.** A Boolean lattice generated from a set of contexts. These atomic contexts are disjoint and exhaustive. Every context that can be identified by the set of context classification trees is present in the context lattice.
3. **A set of default rules.** These default rules represent context-dependent lexical knowledge. Given a query word and a set of context propositions derived from the source, these default rules are used to provide further information about the query word.

From input to output, reasoning with a Lexica knowledgebase is a three-stage process.

1. From the source and query word, contexts are found using the set of context classification trees. These contexts are called **primary contexts**.
2. From the primary contexts, further contexts are inferred from the context lattice. These further contexts are called **inferred contexts**. By reflexivity the inferred contexts include the primary contexts.
3. From inferred contexts, semantic relations that hold for the query word are identifed by reasoning with the default rules.

As an example, consider the following sentence.

> **The bank of a river in a flood plain is usually low.**

Suppose the query word is **bank**, and the set of stop words in this sentence is the following.

$$\{\text{The, of, a, in, is }\}$$

This leaves the following set as the source.

$$\{\text{river, flood, plain, usually, low }\}$$

Assuming that `river` can be identified as a context by a context classification tree, and that `valley` can be identified as a context by a context classification tree, then `river` and `valley` are primary contexts containing `bank`. As a result, the following is an inferred context. We may choose to use an alternative name such as `river-bank` for it.

$$\text{river} \wedge \text{valley}$$

In the next subsection, we show how we represent the input to the system. We then show how we can represent lexical knowledge using default logic. Finally, we show how we query the system.

4.1 Representing inputs

The input to a Lexica system is a query word q and a source S. Let C denote the set of inferred contexts that hold for $S \cup \{q\}$. From the inputs, we form a set Q of formulae that we use as part of the default theory to derive the output. We now show how we form Q from the inputs.

$S \cup \{q\}$ is in context x
iff context(x) is in Q.

We can restate this as follows, where in a Boolean lattice, the downset of x is the set of all elements less than or equal to x in the lattice.

context(x) is in Q
iff the least element for the inferred contexts of $S \cup \{q\}$
is in the downset of x in the context lattice for $S \cup \{q\}$.

We also require the complement for the context relation.

¬context(x) is in Q
iff the least element for the inferred contexts of $S \cup \{q\}$
is not in the downset of x in the context lattice for $S \cup \{q\}$.

The query word is the word for which further information is sought. Via the relations that hold for the query word, we also seek information about further words. For example, if the query word is `bank`, and the following relation holds, we then seek further information about `river-bank`.

synonym(bank,river-bank)

These words for which we seek further information are called **focus words**, and we denote this by the relation **focus**.

If q is the query word,
then we represent this as focus(q) in Q.

We propagate focus words by axioms of the following form. These capture the transitivity of focus for particular relations such as **synonym**, **related-term**, and **meronym**.

$$\mathtt{focus(x) \wedge synonym(x, y) \rightarrow focus(y)}$$

$$\mathtt{focus(x) \wedge related\text{-}term(x,y) \rightarrow focus(y)}$$

$$\mathtt{focus(x) \wedge meronym(x, y) \rightarrow focus(y)}$$

The exact combination of axioms required in Q depends on which relations are used in the knowledgebase.

4.2 Representing semantic and grammatical relations

We assume a semantic relation is a binary relation between a pair of words. Types of relation include synonymy, antinymy, specialization, and meronymy. We qualify semantic relations according to context. For example, in the context of **river**, **bank** is a synomyn of **river-bank**, whereas in the context of **corporate-finance**, **bank** is a synonym of **merchant-bank**.

$$\frac{\mathtt{focus(bank) \; : \; context(river)}}{\mathtt{synonym(bank, river\text{-}bank)}}$$

$$\frac{\mathtt{focus(bank) \; : \; context(corporate\text{-}finance)}}{\mathtt{synonym(bank, merchant\text{-}bank)}}$$

We now consider some defaults for finding synonyms for **car**. The first says that **synonym(car,automobile)** holds if **context(road)** holds. The second says that in the more general situation where **context(transport)** holds, we also need **context(rail)** to not hold. The third rule is a weaker alternative to the second option: In the general situation where **context(transport)** holds, we also need to check ¬**context(road)** does not hold. In the framework, we have freedom as to whether we require a particular context (or negation of a context) is needed as a precondition or justification.

$$\frac{\mathtt{focus(car) \wedge context(road) \; : \; \top}}{\mathtt{synonym(car, automobile)}}$$

$$\frac{\mathtt{focus(car) \wedge context(transport) \; : \; \neg context(rail)}}{\mathtt{synonym(car, automobile)}}$$

$$\frac{\mathtt{focus(car) \wedge context(transport) \; : \; context(road)}}{\mathtt{synonym(car, automobile)}}$$

In some situations, `automobile` is not an appropriate synonym for `car`, such as in the case of `wagon`.

$$\frac{\texttt{focus(car)} \wedge \texttt{context(rail)} \; : \; \neg\texttt{context(road)}}{\texttt{synonym(car,wagon)}}$$

Another word sense for `car` is in the context of `lisp`. Here we consider the **specialization** relation as consequent.

$$\frac{\texttt{focus(car)} \wedge \texttt{context(lisp)} \; : \; \top}{\texttt{specialization(car,lisp-function)}}$$

If the context `lisp` cannot be determined, then the following default may be appropriate.

$$\frac{\texttt{focus(car)} \wedge \texttt{context(computing)} \; : \; \neg\texttt{context(transport)}}{\texttt{specialization(car,lisp-function)}}$$

As another example, consider the polyseme **case**. Here we a provide default for the **baggage** word sense.

$$\frac{\texttt{focus(case)} \; : \; \texttt{context(transport)} \wedge \neg\texttt{context(legal)}}{\texttt{synonym(case,baggage)}}$$

We now consider other semantic relations, including `located` and `made-of`, that can hold for a given word.

$$\frac{\texttt{focus(knife)} \; : \; \texttt{context(cooking)}}{\texttt{located(knife,kitchen)}}$$

$$\frac{\texttt{focus(hull)} \; : \; \texttt{context(ship)}}{\texttt{made-of(hull,steel)}}$$

$$\frac{\texttt{focus(hull)} \; : \; \texttt{context(sailing-ship)}}{\texttt{made-of(hull,wood)}}$$

We can draw on a richer taxonomy of meronymic relations, in particular [Cru86,WCH87], in order to develop further semantic relations. For example, "member/collection", "portion/mass", "place/area", and "component/integral-object".

Semantic information is also important in applying morphological and grammatical rules. Consider, for example, the following rules.

$$\frac{\texttt{focus(bank)} \; : \; \texttt{context(finance)}}{\texttt{category(bank, verb)} \vee \texttt{category(bank, noun)}}$$

$$\frac{\texttt{focus(bank)} \; : \; \texttt{context(river)}}{\texttt{category(bank, noun)}}$$

Since many morphological and grammatical rules are context-dependent, these can also be usefully presented in a Lexica knowledgebase.

4.3 Obtaining output

We now consider how we can reason with a Lexica knowledgebase in order to derive lexical information about a query word. First we need to assume some general Lexica knowledge, represented as a set of classical formulae, denoted G. This includes formulae such as the following for generating further useful semantic relations.

$$\text{synonym}(x, y) \land \text{synonym}(y, z) \rightarrow \text{synonym}(x, z)$$

$$\text{synonym}(x, y) \rightarrow \text{synonym}(y, x)$$

$$\text{specialization}(x, y) \land \text{specialization}(y, z) \rightarrow \text{specialization}(x, z)$$

A Lexica knowledgebase is a default theory (D, W), where D is a set of semantic rules (discussed in section 4.2), and W is the union of G (discussed above) and Q (discussed in section 4.1). If E is an extension of (D, W), then E contains a set of semantic relations concerning the query word.

Given an extension E, we need to extract the semantic relations concerning E. Let R denote the set of semantic and grammatical relations in E. So for example, if E includes synonym(happy,joyous), then synonym(happy,joyous) is in R.

4.4 Using the Lexica framework

Reasoning with a Lexica knowledgebase is non-monotonic with respect to the source: Taking a superset of the source may cause lexical inferences to be retracted. This gives the context-dependent reasoning that is necessary for lexical knowledge.

The definition of the Lexica framework does not exclude multiple extensions. For a given query word and source, the generation of multiple extensions implies that with respect to the source, the query word is ambiguous. This may be because the context is underdetermined.

Abstracting from a Lexica knowledgebase (D, W), we can obtain a semantic network — where the nodes are words and the arcs are semantic relations. We obtain this semantic network by taking the consequents of all the default rules in D. Call this network G. So $G = (N, A)$ is a directed graph where N is a set of nodes and A is a set of directed arcs. G is not necessarily a connected graph. For example, consider the following set of three defaults:

$$\frac{\text{focus}(\text{car}) : \text{context}(\text{road})}{\text{synonym}(\text{car}, \text{automobile})}$$

$$\frac{\text{focus}(\text{automobile}) : \text{context}(\text{road})}{\text{synonym}(\text{automobile}, \text{motor-car})}$$

$$\frac{\text{focus}(\text{road}) : \neg\text{context}(\text{sea})}{\text{synonym}(\text{road}, \text{street})}$$

By abstracting from this Lexica knowledge, we obtain the semantic network composed of the following arcs. This network does not form a connected graph.

```
synonym(car,automobile)
synonym(automobile,motor-car)
synomym(road,street)
```

However, observe that given a source and a query word, an extension E of the corresponding Lexica knowledgebase will be such that the set of semantic relations R in E form a connected subgraph. This results from the propagation of the **focus** relation using the axioms in the Q subset of the default theory. To continue the above example, suppose the source just contains the word **car** and so the query word is **car**, then the extension contains the semantic relations **synonym(car,automobile)** and **synonym(automobile,motor-car)**, but not **synomym(road,street)**.

5 Discussion

In this paper, we have presented an important problem — reasoning with lexical knowledge — where default logic has much to offer. We have shown how we can use classification trees to identify contexts for a word, and shown how we can use the identified contexts to reason with lexical knowledge about the word in default logic. We have also shown that machine learning techniques can be used to generate context classification trees.

Our immmediate goal in developing the Lexica framework is to develop the Lexica framework for goal-directed reasoning. For simplicity, we chose Reiter's version of default logic. But, for efficiency, a goal-directed form of default reasoning is more appropriate. In particular, we are investigating the use of the XRay query answering system for default logics [Sch95].

For an application, it is possible that a relatively large number of default rules would be required for an acceptable level of performance. To address this viability problem, we aim to investigate a number of avenues: (1) Using the framework in restricted domains that require a limited number of default rules; (2) Using inductive logic programming ([Mug92]) to generate default rules for a domain; (3) Using co-locational data for knowledge engineering; and (4) Using machine-readable dictionaries and thesauri for knowledge engineering [Mei93].

The Lexica framework is complementary to formalizations of the notion of "aboutness" such as [BH94,Buv95,Hun96]. A Lexica knowledgebase could potentially be used in such frameworks to allow identification of, and reasoning with, relations such as "article A is about topic T".

Finally, we can consider the Lexica approach as a move towards reusable knowledgebases or general knowledge systems. CYC is perhaps the best known, and certainly the most intensively developed example of a general knowledge system [Len95]. There are many problems with reuse of knowledge, as highlighted during the development of CYC. It is likely that initial success will eminate from more highly structured systems with constrained querying, such as in the Lexica approach, than in more general systems, such as CYC, that have a wider range of knowledge and querying.

Acknowledgements

I would like to thank an anonymous referee for some helpful suggestions.

References

[Bes89]   Ph Besnard. *An Introduction to Default Logic*. Springer, 1989.

[BFGM91] R Beckworth, C Fellbaum, D Gross, and G Miller. WordNet: A lexical database organized on psycholinguistic principles. In U Zernik, editor, *Lexical Acquisition: Exploiting On-line Resources to Build a Lexicon*, pages 211–226. Lawrence Erlbaum Associates, 1991.

[BH94]   P Bruza and T Huibers. Investigating aboutness axioms using information fields. In *Proceedings of the 18th ACM SIGIR Conference on Research and Development in Information Retrieval (SIGIR'94)*, pages 112–121. Springer, 1994.

[BPK+96] U Borghoff, R Pareschi, H Karch, M Nohmeier, and J Schlichter. Constraint-based information gathering for a network publication system. In *Proceedings of the First International Conference on the Practical Application of Intelligent Agents and Multi-agent Technology*. Pratical Applications Company, 1996.

[Bre91]   G Brewka. *Common-sense Reasoning*. Cambridge University Press, 1991.

[Bri91]   T Briscoe. Lexical issues in natural language processing. In E Klein and F Veltman, editors, *Natural Language and Speech*, pages 39–68. Springer, 1991.

[Buv95]   S Buvac. Resolving lexical ambiguity using a formal theory of context. In K van Deemter and S Peters, editors, *Semantic Ambiguity and Underspecification*, pages 101–124. CSLI Publications, 1995.

[Cro93]   B Croft. Knowledge-based and statistical approaches to text retrieval. *IEEE Expert*, pages 8–12, 1993.

[Cru86]   D Cruse. *Lexical Semantics*. Cambridge University Press, 1986.

[Eng96]   B Engleder. *Filtering News Articles*. MSc Thesis, Department of Computing, Imperial College, London, 1996.

[FJ96]   A Falk and I Jonsson. PAWS: An agent for WWW-retrieval and filtering. In *Proceedings of the First International Conference on the Practical Application of Intelligent Agents and Multi-agent Technology*. Practical Applications Company, 1996.

[GF95]   B Grosof and D Foulger. Globenet and RAISE: Intelligent agents for networked newsgroups and customer service support. Technical report, IBM Research Division, T J Watson Research Center, New York, 1995.

[GLC+95] B Grosof, D Levine, H Chan, C Parris, and J Auerbach. Reusable architecture for embedding rule-based intelligence in information agents. Technical report, IBM Research Division, T J Watson Research Center, New York, 1995.

[GPWS96] L Gutherie, J Pustejovsky, Y Wilks, and B Slator. The role of lexicons in natural language processing. *Communications of the ACM*, 39(1):63–72, 1996.

[Hun96]   A Hunter. Intelligent text handling using default logic. In *Proceedings of the IEEE Conference on Tools with Artificial Intelligence*, pages 34–40. IEEE Computer Society Press, 1996.

[Len95]   D Lenat. CYC:a large-scale investment in knowledge infrastructure. *Communications of the ACM*, 38(11):33–38, 1995.

[Lie95]   H Lieberman. Letizia: An agent that assists web browsing. In *Proceedings of the Fourteenth International Joint Conference on Artificial Intelligence*. Morgan Kaufmann, 1995.

[LS95]   T Linke and T Schaub. Lemma handling in default logic theorem provers. In *Symbolic and Qualitative Approaches to Reasoning and Uncertainty*, volume 946 of *Lecture Notes in Computer Science*, pages 285–292. Springer, 1995.

[Mae94]   P Maes. Agents that reduce work and information overload. *Communications of the ACM*, 37(7):31–40, 1994.

[Mei93]   W Meijs. Exploring lexical knowledge. In C Souter and E Atwell, editors, *Corpus-based Computational Linguistics*, pages 249–260. Rodopi, 1993.

[Mil95]   G Miller. WordNet: A lexical database for English. *Communications of the ACM*, 38(11):39–41, 1995.

[Mou96]   A Moukas. Amalthaea: Information discovery and filtering using a multiagent evolving ecosystem. Technical report, MIT Media Laboratory, Cambridge MA, 1996.

[Mug92]   S Muggleton. *Inductive Logic Programming*. Academic Press, 1992.

[Nie94]   I Niemelä. A decision method for non-monotonic reasoning based on autoepistemic reasoning. In *Proceedings of the Fourth International Conference Principles of Knowledge Representation and Reasoning*, pages 473–484. Morgan Kaufmann, 1994.

[Qui86]   J Quinlan. Induction of decision trees. *Machine Learning*, 1:81–106, 1986.

[Rei80]   R Reiter. Default logic. *Artificial Intelligence*, 13:81–132, 1980.

[Sch95]   T Schaub. A new methodology for query-answering in default logics via structure-oriented theorem proving. *Journal of Automated Reasoning*, 15:95–165, 1995.

[Sha96]   A Shaikh. *Data Mining Using Inductive Logic Programming*. MSc Thesis, Department of Computing, Imperial College, London, 1996.

[Tra93]   R Trask. *A Dictionary of Grammatical Terms in Linguistics*. Routledge, 1993.

[Voo94]   E Voorhees. Query expansion using lexical-semantic relations. In W Croft and C van Rijsbergen, editors, *Proceedings of the Seventeenth International ACM-SIGIR Conference on Research and Developement in Information Retrieval*, pages 61–69, 1994.

[WCH87]   M Winston, R Chaffin, and D Herrman. A taxonomy of part-whole relations. *Cognitive Science*, 11:417–444, 1987.

[WSG96]   Y Wilks, B Slator, and L Guthrie. *Electric Words: Dictionaries, Computers, and Meanings*. MIT Press, 1996.

[Yok95]   T Yokoi. The EDR Electronic Dictionary. *Communications of the ACM*, 38(11):42–44, 1995.

A Layered, Any Time Approach to Sensor Validation

Pablo H. Ibargüengoytia[1], Sunil Vadera[1] and L. Enrique Sucar[2]

[1] University of Salford
Department of Mathematics and Computer Science
Salford, M5 4WT,
{P.Ibar / S.Vadera}@cms.salford.ac.uk
[2] ITESM - Campus Morelos
AP C-99, Cuernavaca, Mor., 62020, Mexico
esucar@campus.mor.itesm.mx

Abstract. Sensors are the most usual source of information in many automatic systems such as automatic control, diagnosis, monitoring, etc. These computerised systems utilise different models of the process being served which usually, assume the value of the variables as a correct reading from the sensors. Unfortunately, sensors are prone to failures. This article proposes a layered approach to the use of sensor information where the lowest layer validates sensors and provides the information to the higher layers that model the process. The proposed mechanism utilises *belief networks* as the framework for failure detection, and uses a property based on the Markov blanket to isolate the faulty sensors from the apparently faulty sensors. Additionally, an any time version of the sensor validation algorithm is presented and the approach is tested on the validation of temperature sensors in a gas turbine of a power plant.

Keywords: *Uncertainty, Belief networks, sensor validation.*

1 INTRODUCTION

Current applications of artificial intelligence (AI) in real domains include different functions like automation and process control, diagnosis, monitoring etc. However, these applications require an overall process model where usually, its inputs are mainly sensors. In general, all possible functions of a computerised system require the use of reliable information in order to take the right decisions. This article proposes a mechanism for intelligent, real time sensor validation which can be utilized as a separate module that works together with other functions in industrial plants. In other words, it is assumed that a layered scheme is used in which the lowest level concentrates on validating the signals transmitted by sensors as presented in Fig. 1 [12]. Faults are detected in a decentralised and hierarchical approach, so they can be easily isolated and repaired. Additionally, suppose that the higher layers of the system represent other important and critical functions, e.g., the fault diagnosis of a nuclear plant. The intermediate

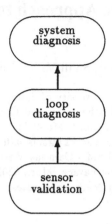

Fig. 1. Layered diagnosis architecture.

layer (loop diagnosis) may be using *model-based reasoning* to diagnose a control loop in the plant, whereas the system diagnosis layer may be utilizing a different approach. Irrespective of the approach used by the diagnosis layer, it needs to assume that the information it utilizes is accurate. This is the goal of the proposed mechanism: to provide reliable information from sensors to higher layers of a system.

This article is organized as follows. Section 2 describes the proposed sensor model and also, summarizes related work on sensor validation. Section 3 describes the architecture of the proposed model and describes the probabilistic reasoning model for fault detection. Section 4 presents an any time extension of the validation algorithm more suitable for a real time environment. Section 5 presents the results of applying the approach to a gas turbine of the Gomez Palacio power plant in México. Finally, section 6 gives the conclusions and future work.

2   THE SENSOR AND PROCESS MODELS

The input of a sensor is the value V_s which is considered unknown and inaccessible, and the output is the measurement V_m (Fig. 2). A sensor is declared *faulty* if the output measurement V_m gives an incorrect representation of the V_s [12]. A detection of a fault is made when the output of the sensor V_m exceeds some threshold, or a non permitted deviation from a characteristic trend. But, what exactly is a characteristic trend?

This question is being answered differently by many investigators. However, in all the approaches, the central idea is to *estimate* the value that a sensor must deliver based on its environment. Some examples of these environments are the following:

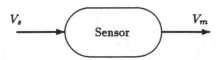

V_s → Sensor → V_m

Fig. 2. Basic model of a sensor.

- history of a single signal in time,
- history of the state of the process in time,
- state of all related signals at a specific time

This estimation process is what makes the various validation approaches different. Specifically, this project uses probabilistic methods for estimation based on all the related signals at specific time instants.

As an aid to relating these different approaches, the rest of the paper will use the following terminology:

variable state (V_s): this refers to the measurand, i.e., the input to the sensor. This is the physical parameter being measured.

variable measure (V_m): this refers to the measurement, i.e., the output of the sensor. This is a numerical representation of the physical parameter state.

variable estimated (V_e): this refers to the estimated value of the variable.

sensor state (S): this refers to the condition of the sensor. This parameter has one of the binary values {correct, faulty}.

Based on these definitions, Fig. 3 shows some simplified models that can be used to represent sensor information in a physical process. These models are dependency models indicating *causality*. In (a), either the variable state *causes* the variable measure,or V_m *depends* on the value of V_s. This is the most obvious and basic model of a single sensor. Figure 3(b) shows a model including three nodes: the measure V_m depends on the variable state V_s and on the sensor state S, i.e., V_m displays a realistic representation of the variable state if the sensor is working properly $(S = correct)$. Finally, since V_s is unknown and inaccessible, it is replaced with its estimation V_e in Fig. 3(c). Here, the inference on the sensor state S is dependant on the measure and the estimation. In fact, Fig. 3(c) represents the goal of this project, namely to obtain the state of the sensor based on the reading and the estimated value. In other words, this model makes explicit the conditional probability of a fault in a sensor, given the measure and the estimation, i.e., $P(S \mid V_m, V_e, \delta)$, where δ represents previous knowledge about the sensor. For example, δ might represent the mean time between failures reported by the manufacturer, the physical location of the sensor in the plant, time between the last maintenance, etc.

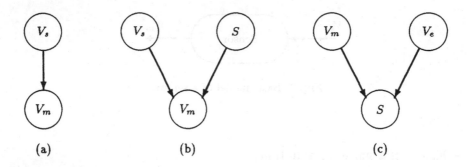

Fig. 3. Basic model of a sensor performance. (a) represents that V_m depends on V_s. (b) represents an enhanced model where V_m depends on V_s and the state of the sensor S. (c) displays the proposed basic approach, namely, S can be inferred with the values of the measure and the estimated real value.

2.1 Related work on sensor validation

Since digital computers have been accepted in industrial and specially in critical applications, the validation of sensors has been a concern for manufacturers and users. Several approaches have been proposed to detect inconsistencies in measures. The literature reports various survey papers from different application domains (e.g. [1], [12]). Recent work on this area includes the SEVA project [7] and the TIGER project [9]. The SEVA project concentrates on the design of self validating sensors using emerging techniques from digital communications for field devices. Based on microprocessors, sensors are able to detect their own failures and report the uncertainty or quality measure of their own readings. The TIGER project utilizes model-based reasoning, i.e., the system possesses a mathematical model of the process and runs a simulator in order to compare the observed output with the one estimated by the simulator. Its main function is the monitoring of the complete gas turbine, including all its parts and processes.

This article concentrates only on sensor validation utilizing probabilistic methods. Approaches proposed by Dean [3] and Horvitz [6] also have a similar focus and are summarized below.

The approach proposed by Dean presents a probabilistic model for estimating the current value of a sensor based on previous states of all sensors in the system. This approach utilises knowledge about sensor errors in terms of conditional probabilities. If x represents the system state vector, and z represents the measurement vector, then Dean represents knowledge about the performance of the sensors that produced z as a conditional probability density function, $p(x(k) \mid z(1), \ldots, z(k))$, indicating the probability that x is the true state of the system given that z has been observed at time k. This, represented graphically at a certain instant of time in the scalar case, corresponds to the simplified model of Fig. 3(a). Dean assumed that the estimated value can be determined by the *mean* of the conditional probability density function which has *white Gaussian* noise. However, this assumption is made in order to make the computation tractable even if it may not hold in the real environment.

Horvitz and co-workers described the utilisation of Belief networks for the diagnosis of gas turbines for an auxiliary power unit of a commercial aircraft. They aimed to model the whole process by using a belief network that includes sensor validation as well as the fault diagnosis process. Further, the sensor validation parts of the belief network utilizes a mixture of the different validation models outlined in Fig. 3. For example, in a fragment of their model, they represent three instances where sensors play a role. Figure 4 presents these cases. In (a), they utilise directly the value V_m as the unique source of information. In (b), they use a simplified model where V_s and S cause V_m (i.e., the model in Fig. 3(b)). Finally, in Fig. 4(c) they use a sensor state S that must obtain its value ({correct, false}) from another source (human or computerised).

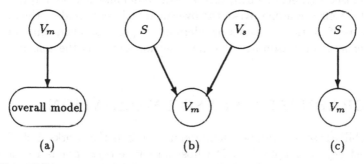

Fig. 4. Three different uses of sensor related nodes in a overall process model. In (a), the measure is utilized by the model assuming it is correct. (b) V_m depends on V_s and the state of the sensor S. In (c), the measure provided by the sensor depends only on its state.

Such an approach can be appropriate when it is not necessary to validate all the sensors with the same reliability. However, the inclusion of different sensor validation models within the full diagnosis model can result in a larger and more complex network. Hence, this article proposes a layered approach that first concentrates on sensor validation and then, on the fault diagnosis process.

2.2 Proposed Approach

Consider a system whose inputs are sensors. Utilising a simplified model of a sensor shown in Fig. 3(c), this paper proposes a general model which *separates* the sensor validation mechanism and the process diagnosis mechanism (or controlling, or monitoring etc.) as shown in Fig. 5.

The signals are validated in the first module, utilizing a sensor validation model proposed in section 3. It transmits the same signal V_m but produces a signal S which indicates the degree of belief in the measurements to the process

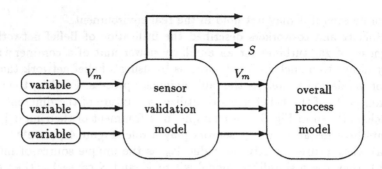

Fig. 5. General proposed model for sensor validation.

model layer [see Fig. 3(c)]. Then, this overall model may decide how and whether to utilize the sensor reading. Also, the overall model may decide if the failure is in the sensor itself, or there exists a problem in the process. Then, this corresponds to a higher level of reasoning. The next section explains the sensor validation model.

3   PROBABILISTIC SENSOR VALIDATION

The probabilistic sensor validation model utilizes belief networks. Belief networks are directed acyclic graphs (DAG) whose structure corresponds to the dependency relations of the set of variables represented in the model [10]. The nodes in this application represent the measures while the arcs represent the conditional probabilities between nodes. The structure of the network makes explicit the dependence and independence relations between the variables. The nodes generally represent discrete variables although recent work on belief networks proposes continuous representations of variables when the application requires it [5]. Alternatively, continuous variables can be discretized into some intervals according to the precision required and depending on the computational cost that is acceptable. In the current implementation, discretization is done by simply dividing the range into 10 intervals although more sophisticated approaches are available [4].

The validation algorithm considers every sensor as suspicious and obtains, one by one, the estimated value V_e of each sensor using probability propagation based on its most related variables. In belief networks, the most related variables consist of a *Markov blanket* of a node. A Markov blanket is defined as the set of variables that makes a variable independent from the others. For example, Fig. 6 shows a reduced, simplified model of a gas turbine where m represents the readings of the megawatts generated, t represents the temperature, p the pressure, g and a represent the gas and air supplied for the combustion respectively. Then, the Markov blanket of m is formed by its children t and p, and the Markov blanket of t is formed by its parent m and its children g and a. Utilizing these concepts, if a fault exists in one of the sensors, it will be revealed in all the sensors in its Markov blanket. On the contrary, if a fault exists outside a sensor's Markov blanket, it will not affect the estimation V_e of that sensor. Then, the Markov blanket of a sensor acts as its protection against others faults, and also protects others from its own failure [8].

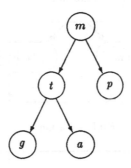

Fig. 6. A reduced Bayesian network of a turbine.

Figure 7 describes the complete process for discriminating faulty and correct sensors. The *sensors* block represents a list with all sensors to be validated. These are processed one by one by the *validation* module utilizing the following algorithm:

1. Read the value of the variable provided by the sensor.
2. Read the value of all variables that appear in the Markov blanket of the selected variable.
3. Propagate the probabilities and obtain the posterior probability distribution for the selected variable.
4. If the probability (obtained in 3) of the value acquired in step 1 is lower than a specified value (described below), return *failure*; else return *success*

The result of the validation module consists of a posterior probability distribution as shown in Fig. 8. Then, an error is detected if the difference between the real value read and the highest probability interval's value exceeds some threshold.

This algorithm provides the two lists of the potential level shown in Fig. 7, i.e., the *apparently correct* and the *apparently faulty* sensors. Then, the probabilistic reasoning can only tell if a sensor has a *potential* (real or apparent) fault, but, without considering other sensors, it can not tell if the fault is real or apparent. Thus, the next step is to distinguish between real and apparent faults, considering that one or more sensors may fail at the same time. This is performed by the *fault isolation* module described next.

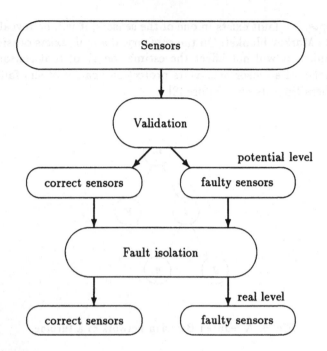

Fig. 7. Block diagram of the proposed mechanism for sensor validation.

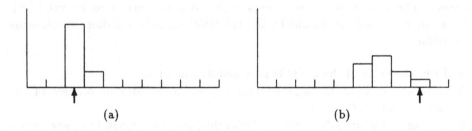

(a)   (b)

Fig. 8. Description of some possible results. The arrows indicate the interval of the real value of a sensor. (a) shows a narrow distribution where the real value coincides with the estimated. (b) shows a wider probability distribution where the real value has a low probability to be correct.

4 ANY TIME FAULT ISOLATION

The presence of a faulty sensor causes a constrained area of manifestation which forms a context. This constrained area is called the *extended Markov blanket* (EMB) of a variable. It is formed by the Markov blanket of a variable plus the variable itself. For example, in Fig. 6 the EMB of m is the set $\{m, t, p\}$, while the EMB of t is formed by $\{m, t, g, a\}$.

Suppose that there is a failure in t, and the sensors are validated following the order m, t, p, g, a. First, since t provides a wrong reading, the validation of m will report a potential fault. Then, the validation of t will also report a fault. Next, p will be validated as correct, and finally, the validation of g and a will report potential faults. At the end of the cycle, the potential fault list will contain the sensors $\{m, t, g, a\}$. This corresponds exactly with t's EMB. In other words, a sensor's Markov blanket acts like a signature that it exhibits when it is faulty [8].

This property can be used to formulate a sensor validation procedure which works in batch mode where all the sensors are examined and a faulty sensor identified. In real time applications, it is sometimes better to have some answers instead of waiting for the most accurate answer which may be too late. This problem has been addressed by a number of authors by using the concept of any time algorithms [2]. This property of an algorithm implies an incremental quality of the response with respect to time. In the sensor validation problem, the quality of the response consists of the number of sensors considered potentially correct and faulty.

An any time version of the sensor validation approach can be based on the following observations. If a sensor is found to be apparently faulty, then the Markov blanket property implies that there must be a faulty sensor in its EMB. However, if a sensor is apparently correct, there is no guarantee that all the sensors in its EMB are correct. Nevertheless, one would expect that most of the time, sensors in its EMB will also be correct. Hence, this information and the size of the Markov blanket can be used to rank the order in which the sensors are examined and leads to the following any time sensor validation algorithm.

1. Let S be a list of sensors to be validated, C be a list of validated sensors initialized to the empty list, F be a list of faulty sensors initialized to the empty list, PC and PF be lists of potentially correct and potentially faulty sensors initialized to empty lists, but where sensors that have already been tested are marked with an asterisk.
2. Select the first sensor s_i from S.
3. While S is not empty, PC and PF have sensors not yet tested do
   - apply the validation module on s_i (as given in section 3)
   - if s_i is correct,
     - then [move s_i to C, move all the sensors in its Markov blanket to PC].
     - else move all sensors in its EMB to PF.
   - if the tested sensors in PF corresponds to the EMB of a sensor, then move that sensor to F and its MB to PC (since one can expect that their failure was due to the isolated fault).
   - select the next sensor: if PF has untested sensors then select it from PF, otherwise if S is not empty then select it from S, otherwise if PC has untested sensors then select it from PC.

To illustrate the algorithm, consider the model of Fig. 6. The upper part of Table 1 shows the operation of the algorithm when there are no failures, and the lower part shows a situation when the sensor g is faulty. In the table, the final column gives a tuple that indicates the number of faulty sensors, the number of correct sensors, and the potentially faulty and correct sensors. This tuple can be viewed as a measure of the quality of the response provided by the any time algorithm at the end of each cycle. The manner in which this information is utilized will depend on the higher layer that may be application dependent. For example, a typical quality function Q may take the following form:

$$Q(F, C, PF, PC) = \alpha F + \beta C + \gamma PF + \delta PC \tag{1}$$

where $\alpha, \beta, \gamma, \delta$ are weights given to the number of sensors in each of the lists F, C, PF, PC. Suppose for presentation purposes, $(\alpha, \beta, \gamma, \delta) = (10,10,2,2)$, then Fig 9 gives a graphical representation of the assigned quality of the response against time for the examples in Table 1. This is known as the performance profile of the system.

Table 1. Example of the algorithm of the model of Fig. 6. Rows represent steps of validation whereas columns represent the content of the different lists.

validate	result	S	C	PC	F	PF	Q
		[t,m,p,g,a]	[]	[]	[]	[]	(0,0,0,0) = 0
t	OK	[p]	[t]	[m, g, a]	[]	[]	(0,1,0,3) = 16
p	OK	[]	[t, p]	[m, g, a]	[]	[]	(0,2,0,3) = 26
m	OK	[]	[t, p, m]	[g, a]	[]	[]	(0,3,0,2) = 34
g	OK	[]	[t, p, m, g]	[a]	[]	[]	(0,4,0,1) = 42
a	OK	[]	[t, p, m, g, a]	[]	[]	[]	(0,5,0,0) = 50
		[t,m,p,g,a]	[]	[]	[]	[]	(0,0,0,0) = 0
t	Fault	[p]	[]	[]	[]	[m, g, a, t^*]	(0,0,4,0) = 8
m	OK	[]	[m]	[t^*, p]	[]	[g, a, t^*]	(0,1,3,2) = 20
g	Fault	[]	[m]	[t^*, p]	[]	[a, t^*, g^*]	(0,1,3,2) = 20
a	OK	[]	[m, a]	[t^*, p]	[]	[t^*, g^*]	(0,2,2,2) = 28
g is	isolated	[]	[m, a]	[t^*, p]	[g]	[]	(1,2,0,2) = 34
p	OK	[]	[m, a, p]	[t^*]	[g]	[]	(1,3,0,1) = 42

5 THE CASE STUDY: A GAS TURBINE

The techniques presented in this paper are being applied to the validation of temperature sensors in a gas turbine of a combined cycle power plant. The temperature is the most important parameter in the operation of a turbine since the optimal performance requires the operation at the maximum permitted values. However, a little increase in the temperature, over a permitted value, may cause severe damage. For this reason, the above sensor validation process is executed

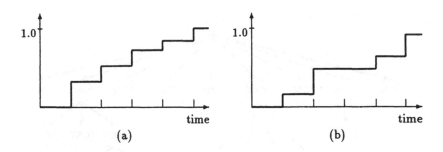

Fig. 9. Performance profile of the examples: (a) without failure, (b) with a failure simulated in sensor g.

over the set of temperature sensors across the turbine. The sensors were grouped into several measurements categories as follows:

- 6 beadings (CH1 - CH6)
- 7 disk cavities (CA1 - CA7)
- 1 cavity air cooling (AEF)
- 2 exciter air (AX1 - AX2)
- 3 blade path (EM1 - EM3)
- 2 lub oil (AL1 - AL2)

A dependency model for these temperatures was obtained utilizing an automatic learning program which, based on data, produces the optimal tree [11]. In total, the data set has 21 variables and 870 instances of the readings. These readings were taken during the start up phase of the plant. The 21 variables are continuous and were discretized for building the model, and for performing probability propagation. Figure 10 shows the probabilistic tree obtained with this data set. A tree was chosen since the inference algorithm for obtaining the posterior probabilities is faster since it depends only on the depth of the tree [10].

Notice that the dependencies can be explained as the heat *propagation* from the centre of the turbine (CH4) to the extremes. CH4 is the measure of the beading temperature which is closer to the combustion chamber, and can be modelled as the tree's root, i.e., the variable which *causes* the other variable's heating. This explanation is intuitive since, based on this data set, it is very difficult for an expert to modify it according to his experience, or to include other variables that could represent other aspects of the process.

There are two sets of experiments that are of interest in this case study. First, the overall sensor validation algorithm needs to be evaluated, and second, the performance profile of the any time algorithm needs to be determined for this application. Some initial experiments have been conducted to evaluate the validation algorithm and the results are described in the following subsection.

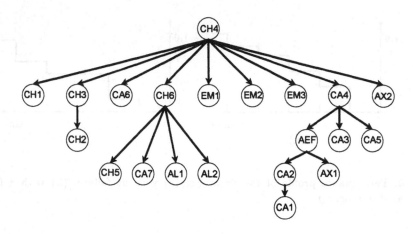

Fig. 10. Belief network of the temperature sensors in a gas turbine of a power plant.

The experiments to determine the performance profile are incomplete at the time of writing but will be available prior to the conference.

5.1 Evaluation of the sensor validation algorithm

This section presents the experimental results obtained when the algorithm is applied to the validation of the temperature sensors. The experiments were carried out in two parts: (i) to check its operation when there are no faults, and (ii) to check its operation when there is a simulated failure.

Table 2 shows the posterior probabilities of several intervals for some variables (only 6 intervals are shown, the other ones have zero probability). The interval of the variable's actual value (V_m as defined in section 2) is underlined.

Table 2. Posterior probabilities per variable per interval, with no faults.

CH4	0.00	<u>1.00</u>	0.00	0.00	0.00	0.00
CH6	0.00	0.35	<u>0.37</u>	0.28	0.00	0.00
CH5	0.00	0.00	0.00	0.08	<u>0.92</u>	0.00
CA7	<u>1.00</u>	0.00	0.00	0.00	0.00	0.00
AL1	0.00	<u>1.00</u>	0.00	0.00	0.00	0.00
AL2	0.00	<u>0.88</u>	0.12	0.00	0.00	0.00

The results of a simulated fault in CH6 are indicated in Table 3. Notice that all actual values are in intervals where the posterior probabilities are zero. That is, in all cases, the real value of the readings has 0% probability of being correct. For example, the line for the CH5 results shows that the real value, which corresponds to the first interval, has 0% probability. However, the posterior probability distribution shows that the correct value has a 40% probability of being in the fifth interval, and a 60% in the sixth. The results shown in the table are the sensors that constitute the CH6's EMB. The fault simulated was the reading of CH6 with its maximum value when the correct value is close to its minimum.

Table 3. Posterior probabilities per variable per interval, with a simulated fault.

CH4	0.00	0.00	0.00	0.00	0.00	0.00
CH6	1.00	0.00	0.00	0.00	0.00	0.00
CH5	0.00	0.00	0.00	0.00	0.40	0.60
CA7	0.00	0.00	0.00	0.00	0.00	1.00
AL1	0.00	0.00	0.00	0.27	0.32	0.42
AL2	0.00	0.00	0.00	0.00	0.22	0.78

These results show that: (i) in the fault free results, the real value coincides always with a highest probability interval, (ii) in the simulated fault results, the real value indicates a zero probability in and only in all the variables of the faulty sensor's EMB. However, these results are based on simulated faults on just one sensor.

6 CONCLUSIONS AND FUTURE WORK

This article has presented a layered approach to sensor validation. In comparison to other approaches, such as that of Horvitz [6], the main benefit of using a layered approach is that it enables the construction of models in a modular fashion. That is, it is easier to construct a model for sensor validation and then a model for the process than it is to construct an overall model in one step. This separation of the sensor validation layer can also result in simpler higher layer models and leave the higher layers to utilize other AI techniques.

The lowest layer, that of sensor validation, is based on the use of Bayesian networks. A Bayesian network is used to define the relationships between variables and to estimate the expected value of a sensor. The expected value is then compared with the actual reading obtained. If these measures differ then a faulty sensor is suspected. A faulty sensor is then distinguished from apparently faulty sensors by the use of a property based on the Markov blanket.

An any time version of the validation algorithm, that improves the quality of its answer incrementally, has also been presented. This any time algorithm uses the expected outcomes of a sensor together with the Markov property as a basis for ranking the order in which sensors are tested.

The approach has been implemented and is being tested on the validation of temperature sensors in a gas turbine of a combined cycle power plant. Some preliminary experimental results have been presented and suggest that the approach is promising.

In addition to further experiments, future research will attempt to use the approach on different application domains.

Acknowledgments

Special thanks to Professor F.A. Holland who provided invaluable advice in the presentation of this paper. Thanks also to the anonymous referees for their comments which improved this article. This research is supported by a grant from CONACYT and IIE under the In-House IIE/SALFORD/CONACYT doctoral programme.

References

1. M. Basseville. Detecting changes in signals and systems. *Automatica*, 24(3):309–326, 1988.
2. T. Dean and M. Boddy. An analysis of time dependent planning. In *Proc. Seventh Natl. Conf. on AI*, St. Paul, MN, U.S.A., 1988.
3. T. Dean and M.P. Wellman. *Planning and control*. Morgan Kaufmann, Palo Alto, Calif., U.S.A., 1991.
4. J. Dougherty, R. Kohavi, and M. Sahami. Supervised and unsupervised discretization of continuous features. In A. Prieditis and S. Russell, editors, *Machine Learning, Proceedings of the Twelfth International Conference*, San Francisco, CA, U.S.A., 1995. Morgan Kaufmann.
5. E. Driver and D. Morrell. Implementation of continuous bayesian networks using sums of weighted gaussians. In *Proc. Eleventh Conference on Uncertainty in Artificial Intelligence*, Montreal, Quebec, Canada, 1995.
6. M. Henrion, J.S. Breese, and E.J. Horvitz. Decision analysis and expert systems. *AI Magazine*, Winter:64–91, 1991.
7. M.P. Henry and D.W. Clarke. The self-validating sensor: rationale, definitions and examples. *Control Engineering Practice*, 1(4):585–610, 1993.
8. P.H. Ibargüengoytia, L.E. Sucar, and S. Vadera. A probabilistic model for sensor validation. In *Proc. Twelfth Conference on Uncertainty in Artificial Intelligence*, pages 332–339, Portland, Oregon, U.S.A., 1996.
9. R. Milne and C. Nicol. Tiger: knowledge based gas turbine condition monitoring. *AI Communications*, 9:92–108, 1996.
10. J. Pearl. *Probabilistic reasoning in intelligent systems*. Morgan Kaufmann, Palo Alto, Calif., U.S.A., 1988.
11. L.E. Sucar, J. Pérez-Brito, and J.C. Ruiz-Suarez. Induction of dependence structures from data and its application to ozone prediction. In G.F. Forsyth and M. Ali, editors, *Procedings Eight International Conference on Industrial and Engineering Applications of Artificial Intelligence and Expert Systems (IEA/AIE)*, pages 57–63, DSTO:Australia, 1995.
12. S.K. Yung and D.W. Clarke. Local sensor validation. *Measurement & Control*, 22(3):132–141, 1989.

treeNets: A Framework for Anytime Evaluation of Belief Networks

N. Jitnah and A. Nicholson

Department of Computer Science, Monash University, Clayton, VIC 3168, Australia,
{njitnah,annn}@cs.monash.edu.au

Keywords: uncertainty, Bayesian Networks, anytime algorithms, practical reasoning, graphical models.

Abstract. We present a new framework for evaluation of belief networks (BNs). It consists of two steps: (1) transforming a belief network into a tree structure called a **treeNet** (2) performing anytime inference by searching the treeNet. The root of the treeNet represents the query node. Whenever new evidence is incorporated, the posterior probability of the query node is re-calculated, using a variation of the polytree message-passing algorithm. The treeNet framework is geared towards anytime evaluation. Evaluating the treeNet is a tree search problem and we investigate different tree search strategies. Using a best-first method, we can to increase the rate of convergence of the anytime result.

1 Introduction

Belief (or Bayesian) networks [18] have become a popular representation for reasoning under uncertainty, as they integrate a graphical representation of causal relationships with a sound Bayesian foundation. Belief network evaluation involves the computation of posterior probability distributions, or beliefs, of query nodes, given evidence about other nodes. Belief networks (BNs) have been used in a range of applications, in particular real-time decision systems such as medical diagnosis [1] and automated vehicle control [13]. A useful characteristic of such real-time decision systems is to ensure that the quality of the result gradually improves with computation time; these are called "anytime" algorithms [9]. Much research has been done to develop efficient inference algorithms for BNs; both exact and approximate inference in general BNs are NP-hard [6, 7].

The treeNet framework allows incremental evaluation of a BN. A treeNet is a representation of a BN; a given BN is transformed into a tree structure with a query node at the root of the tree. Estimates for the belief of the query node are iteratively computed by traversing the tree. This incremental evaluation results in a form of "anytime" algorithm.

Other researchers have developed anytime algorithms for BN evaluation [5, 20, 14] and proposed the use of tree representations for efficient representation of evaluation of a BN [8, 4] (Sect. 2); in this paper we combine these approaches. The representation, construction and tree traversal evaluation of a treeNet is described in Sect. 3. A "best-first" tree traversal strategy requires an assessment of the relative importance of various nodes in terms of contributing the most towards the query node. In Sect. 4 we look at the concept of connection strength in a BN, and describe candidate measures, including a new measure for the strength of an arc linking multi-state nodes. We present performance results showing the anytime aspects of our evaluation method, and comparisons of different tree traversal strategies in Sect. 5. Advantages and limitations of the treeNet method are discussed in Sect. 6. We outline possible extensions to our framework in Sect. 7.

2   Background and Related Work

Belief networks are directed acyclic graphs where nodes correspond to random variables. Directed arcs specify independence assumptions between nodes; this allows a more compact representation of the joint probability distribution of all the state variables. A *conditional probability distribution* (CPD) is associated with each node. The CPD gives the probability of each node value for all combinations of the values of its parents. The probability distribution for a node with no parents is its prior distribution. Given these priors and the CPDs, we can compute posterior probability distributions for all the nodes in a network, which represent *beliefs* about the values of these nodes. Observation of specific values for nodes is called *evidence*. Beliefs are updated by re-computing the posterior probability distributions given the evidence. Both exact and approximate inference of general BNs has been shown to be NP-hard [6, 7].

Anytime algorithms [22] are algorithms where the quality of the result gradually improves with computation time. Their capability to trade accuracy of result for computation time makes them very useful tools, especially in real-time systems, where it is critical to obtain a quick estimate for planning and decision purposes, but the exact computation may be too lengthy. Usual characteristics of anytime algorithms include: measurable quality of the result, monotonicity and interruptibility. In Sect. 6, we discuss the treeNet framework in terms of these requirements. For BN evaluation, the quality or accuracy of the result is measured in terms of the distance between the exact and estimated beliefs of query nodes. If this distance decreases as computation time increases, the algorithm has an anytime flavour. The rate at which this distance decreases is obviously important. Simulation approaches [5] are classified as anytime algorithms because the accuracy of results improves as the sample size grows.

Other anytime algorithms for BN evaluation have been developed [14, 20] and compared [17]. In Wellman and Liu's state-Space abstraction [20], the states of selected nodes are merged; the subsequent size reduction of the CPDs allows faster evaluation of the BN. By iteratively refining the merged states and re-evaluating the BN, a progressively more accurate result is obtained. Empirical comparisons between other anytime BN evaluation algorithms and the treeNet anytime evaluation have yet to be conducted. However, since the anytime performance of the treeNet evaluation depends on the tree traversal, for which many existing strategies can be applied, it appears more promising than state-space evaluation, where the issue is choosing which nodes to abstract and which states within a given node to merge.

Our method of evaluating a treeNet is related to the Variable Elimination algorithm of [21]. The structure of the treeNet can be regarded as a factorisation of the posterior probability of the designated query node, where each term indicates the contribution of a different variable to the final result. Every step of the anytime evaluation will include a new term in the factorisation. The strategy used for searching the treeNet will then correspond to a chosen ordering over the factors. Our best-first strategy is an optimal ordering which achieves fastest convergence of the result towards the exact answer. BN evaluation algorithms based on variable elimination are discussed more generally in [11].

Recent research in the area of BNs has given rise to alternative representations and evaluation schemes. The main motivation behind these new approaches is efficiency in encoding the network and in the method of evaluation. Here we review briefly two approaches that use a tree representation: Query DAGs and Context-Specific Independence.

Query DAGs (Q-DAG) are introduced in [8]. This approach consists of firstly compiling a BN into an arithmetic expression called a Q-DAG, then answering queries by evaluating the expression. Each leaf of the Q-DAG corresponds to the belief of a designated query node and each other node of the Q-DAG represents an arithmetic operation or a value. Constructing the Q-DAG is an off-line process and is only done once. When evidence is added or retracted, the Q-DAG is evaluated on-line. We borrow the idea of turning a BN into a tree from the Q-DAG method. Our approach is also query-oriented: the treeNet is a framework within which the belief of a designated query node is computed by an anytime algorithm. Like a Q-DAG, a treeNet is also built off-line and evaluated on-line when we want to update the belief of our query node.

Context-Specific Independence (CSI), proposed in [4], refers to independencies that are encoded in the CPDs of a BN. They only hold within certain contexts, i.e. if specific evidence values are assigned to some nodes. This approach also uses the tree representation: a CPD in which CSI has been identified is encoded as a tree that supports efficient BN evaluation algorithms. CSI can be used to simplify the evaluation of a treeNet by identifying branches to be pruned: this situation occurs when a node taking on a certain value renders one or more of its neighbours independent.

We discuss related work on connection strength in Sect. 4.1.

3 Basic Framework

The treeNet is a framework for incremental evaluation of a BN. We first transform a given BN into a tree structure with a query node Q at the root of the tree. We then iteratively compute estimates for the belief of Q by traversing the tree. At every treeNet node N visited, we update the belief of Q by recomputing the term corresponding to N. Thus, each iteration of the algorithm takes into account the influence of a new node on the query node Q. We obtain the exact belief when all the nodes of the treeNet are visited. The process of constructing and evaluating a treeNet is described in this section and we discuss performance issues in Sect. 5. We first explain the procedure for transforming a BN into a treeNet. Then we describe how weights are assigned to the arcs of a BN. Finally we present the algorithm used to evaluate a treeNet.

3.1 Structure of a treeNet

Our starting point is a BN, with vertices $\mathbf{V}$, directed edges $\mathbf{E}$ and a conditional probability distribution (CPD) attached to each vertex to encode the probability of each state of the variable given each state combination of its parents. Assume the BN has no underlying loops [1]. Given a query node $Q \in \mathbf{V}$, root nodes $\mathbf{R} \subset \mathbf{V}$ and leaf nodes $\mathbf{L} \subset \mathbf{V}$, we build a treeNet by traversing the BN in breadth-first fashion, starting from Q. The treeNet construction algorithm is given in Fig. 1. The result is a tree with Q at the root and $\mathbf{R}$ and $\mathbf{L}$ at the leaves. Thus, each treeNet node N corresponds to a vertex $V \in \mathbf{V}$ and each treeNet branch B corresponds to an edge $E \in \mathbf{E}$.

Input: a BN $= (\mathbf{V}, \mathbf{E})$ with query node $Q \in \mathbf{V}$
Output: a treeNet $\mathbf{T}$
1. Let Q be the root N_0 of $\mathbf{T}$
2. Mark Q with 0
3. $i = 0$
4. For each N_i in $\mathbf{T}$, loop:
        Find the corresponding V_i in the BN
        For each unmarked neighbour W of V_i
            add a treeNet node N_{i+1}
            add a branch from N_i to N_{i+1}
            mark W with $i + 1$
5. $i = i + 1$
6. Repeat from step 4 until all $V \in \mathbf{V}$ are marked

Fig. 1. Algorithm for constructing treeNet

In addition to the CPDs from the original BN, each N stores a real-valued vector P. If N corresponds to $R \in \mathbf{R}$, the P represents the prior probabilities. If N corresponds to $L \in \mathbf{L}$, P is a unit vector. Otherwise, P holds the probabilities of the node, averaged over all the state combinations of the parents. Evidence is entered into a node by replacing its P vector by one consisting of a 1 for the evidence state and 0's for all other states. Each N also stores information needed for belief propagation, such as the number of states and the number of parents. For a given query node, the structure of a treeNet is fixed. However the

---

[1] We discuss the relaxation of this assumption in Sect. 7.

structure can be simplified if we use d-separation and/or CSI to prune the tree (see Sect. 3.4).

Let us consider an example of the construction of a treeNet from a BN. Suppose that the original BN is that shown in Fig. 2(a). If node A is taken as the query node, we obtain the treeNet shown in Fig. 2(b). Alternatively, if the query node is O, the resulting treeNet is that shown in Fig. 2(c).

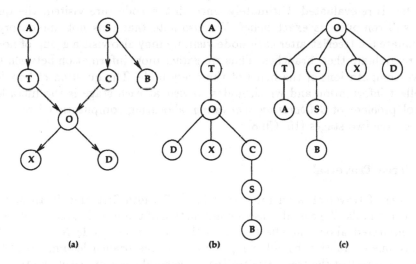

(a)　　　　　　(b)　　　　　　(c)

Fig. 2. Example showing how (a) a BN (a modified polytree version of the Asia network [15]), is transformed into different treeNets depending on the query node: (b) query node A (c) query node O.

3.2 Evaluating a treeNet

Exact Evaluation. The algorithm for evaluating a treeNet is a variation of the message-passing polytree algorithm [18]. The main modification is that messages are only passed in one direction, towards the query node, instead of two-way propagation. Since we are only interested in the beliefs of the query node, two-way propagation is not necessary. As in the polytree algorithm, we have π and λ messages which are real-valued vectors representing probabilities of states. In the treeNet, node N sends a π message to its parent X if there was an arc $N \to X$ in the original BN. But if there was an arc $N \leftarrow X$ in the original BN, then node N sends a λ message to its treeNet parent X. The query node computes its beliefs by combining incoming messages from its children in the treeNet. Other treeNet nodes combine their incoming messages and compute an outgoing message using their stored CPD.

Anytime Evaluation. Now, let us see how we can use the treeNet framework to develop an anytime version of the polytree algorithm. We obtain an initial estimate of the beliefs of our query node by collecting messages from all its immediate children in the treeNet. Other nodes are ignored, hence the query node's children have "vacuous" incoming messages [2] which are simply the vectors P described in Sect. 3.1.

We then traverse the treeNet. As each node is visited, it sends to its parent a message to be propagated towards the query node Q, which recomputes a more accurate estimate of its own beliefs. Only the branch on which the newly visited node lies is re-evaluated. Ultimately, once all the nodes are visited, the query node will compute its exact belief. We also note that it is not mandatory to re-evaluate the treeNet after each node visit: we may also visit a group of nodes, then re-evaluate their branches. Thus we gather more information between successive computations of the belief of the query node. The question of how long to collect information and which nodes to visit at each stage is the meta-level control problem of *deliberation scheduling*, allocating computational resources between the two stages [10, Ch.8].

3.3   Tree Traversal

The order of traversal is an important issue. Breadth-first and depth-first are obvious methods. We can also use a combination of these. For example, if we are more interested about the effect of a particular evidence node N, we can start the anytime evaluation by following depth-first the branch leading to N then traverse the rest of the tree breadth-first. In general, we can orient the traversal by focusing on branches leading to nodes of interest.

The use of arc weights provide a flexible way of evaluating the treeNet. Weights of arcs at a node give an estimate of how much the node can be affected by each neighbour. The neighbour at the heaviest arc will have the greatest influence on the node of interest. Traversing the treeNet by visiting the node connected by the heaviest arc at each iteration results in a faster convergence of the result. Details of possible measures for determining such an arc weight are given in Sect. 4.

3.4   Pruning the treeNet and Approximate Evaluation

Pruning branches is an obvious way of simplifying the treeNet. We want to prune all branches leading to "irrelevant" nodes, i.e. nodes from which the posterior of our query node Q is independent. D-separation [18] and CSI [4] provide criteria by which independent nodes, given certain evidence assignments, are identified. Both are straightforwardly applicable to the treeNet scheme.

For example, given the BN of Fig 2(a), suppose node A is the query node, (see Fig 2(b)) and O is the evidence node, D and X can be pruned. If O is the query node (see Fig 2(c)) and T is the evidence node, A is d-separated from O by T and can be pruned. As for CSI, if we can determine that an arc of the BN

---

[2] The term "vacuous message" is borrowed from [12]

is irrelevant in the given context, we simply prune the corresponding treeNet branch.

If an exact result is not ultimately required, we can use arc weight to decide whether some branches of the treeNet may be pruned. In time-critical applications, deliberation scheduling in combination with arc weights give us a way to decide whether it is worth visiting a particular node at a certain point in time. We might decide to ignore a node, and hence prune the branches starting there, if the gain in accuracy of our result would be insignificant.

3.5 An Example

This example shows how we evaluate the treeNet of Fig. 2(c), with evidence in node $X = 0$. The π and λ messages for the treeNet evaluation, and the initial P vectors, are shown in Fig. 3. The CPDs of the original BN are given in Fig. 4. The treeNet evaluation results are shown in Fig. 5. We start by visiting the immediate children of O, to get an initial estimate. Then we visit the rest of the tree in depth-first order. The final result for the belief of O is $(.639, .361)$, which is the exact answer.

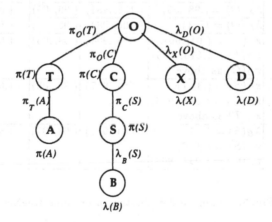

$$\pi(A) = (.01, .99) \quad \text{prior for } A$$
$$\pi(S) = (.3, .7) \quad \text{prior for } S$$
$$\pi(T) = (.03, .97) \quad \tfrac{1}{|\Omega(A)|} \sum_{a \in \Omega(A)} p(T \mid A = a)$$
$$\pi(C) = (.055, .945) \quad \tfrac{1}{|\Omega(S)|} \sum_{s \in \Omega(S)} p(C \mid S = s)$$
$$\lambda(B) = (1, 1) \quad \text{unit vector since } B \text{ is a leaf}$$
$$\lambda(D) = (1, 1) \quad \text{unit vector since } D \text{ is a leaf}$$
$$\lambda(X) = (1, 0) \quad \text{the evidence}$$

Fig. 3. The π and λ messages used for evaluation are shown on the treeNet above, with initial P vectors on each node of the treeNet below.

$$p(A) = (.01, .99) \qquad p(S) = (.3, .7)$$

$$p(T \mid A) = \begin{pmatrix} .05, .95 \\ .01, .99 \end{pmatrix} \quad p(C \mid S) = \begin{pmatrix} .1, .9 \\ .01, .99 \end{pmatrix}$$

$$p(B \mid S) = \begin{pmatrix} .6, .4 \\ .3, .7 \end{pmatrix} \qquad p(O \mid T, C) = \begin{pmatrix} .81, .19 \\ .57, .43 \\ .8, .2 \\ .05, .95 \end{pmatrix}$$

$$p(X \mid O) = \begin{pmatrix} .98, .02 \\ .05, .95 \end{pmatrix} \quad p(D \mid O) = \begin{pmatrix} .8, .2 \\ .45, .55 \end{pmatrix}$$

Fig. 4. CPDs for example BN.

Node Visited	Messages passed	$\pi(O)$	$\lambda(O)$	Belief of node O
T	$\pi_O(T) = (.03, .97)$	$(.866, 1.134)$	$(1, 1)$	$(.433, .567)$
C	$\pi_O(C) = (.055, .945)$ $\pi_O(T)$ as above	$(.106, .984)$	$(1, 1)$	$(.106, .894)$
X	$\lambda_X(O) = (.98, .05)$	$(.106, .984)$	$(.98, .05)$	$(.699, .301)$
D	$\lambda_D(O) = (1, 1)$	$(.106, .984)$	$(.98, .05)$	$(.699, .301)$
A	$\pi_T(A) = (.01, .99)$ $\pi_O(T) = (.010, .990)$ $\pi_O(C)$ as above	$(.096, .904)$	$(.98, .05)$	$(.676, .324)$
S	$\pi_C(S) = (.3, .7)$ $\pi_O(C) = (.037, .963)$ $\pi_O(T)$ as above	$(.083, .917)$	$(.98, .05)$	$(.639, .361)$
B	$\lambda_B(S) = (1, 1)$ $\pi_C(S) = (.3, .7)$ $\pi_O(C) = (.037, .963)$	$(.083, .917)$	$(.98, .05)$	$(.639, .361)$

Fig. 5. Incremental evaluation of example treeNet.

4 Measures for Weight of Node Connections

The crucial aspect of the incremental treeNet evaluation described above is the order of tree traversal. We would like to add those nodes which have the greatest influence on the query node first. Influence is determined by a complex combination of location in relation to the query node and the CPDs. Since we are adding nodes by traversing the tree, we are looking for a measure describing the strength of the connections between adjacent nodes. In this context, if a node has several neighbours, a best-first strategy means we should first consider the neighbour linked by the strongest connection.

4.1 Connection Strengths

The concept of a connection strength (CS) between two adjacent nodes in a binary BN is introduced in [3]; this measure was proposed for use when graphically displaying BNs, with a thicker arc indicating higher connection strength. A method for approximate inference based on link strengths is also described. The CS from parent node A to child node B, both binary nodes, is defined as the difference in the resulting belief of B as the state of A changes. The distance between the beliefs of B is measured as the difference of log-odds:

$$CS(A, B) = \left| log \frac{p(b \,|a)}{p(\neg b \,|a)} - log \frac{p(b \,|\neg a)}{p(\neg b \,|\neg a)} \right| \tag{1}$$

Thus, a high CS indicates that the parent highly influences the child. If B has other parents, the CS is taken as the maximum of the log-odds difference, over all state combination of the other parents.

4.2 Arc Weights

Based on this idea of CS, we propose a new measure for the strength of an arc linking multi-state nodes. We use the term *arc weight* for our measure. The weight of an arc provides an estimate of how much a change in the parent of the arc can affect the child. Given a node Y with parent X and a set of other parents $\mathbf{Z}$, we calculate the weight $w(X, Y)$ of the arc $X \rightarrow Y$ as follows:

$$w(X,Y) = \frac{1}{|\Omega(\mathbf{Z})|} \sum_{\mathbf{k} \in \Omega(\mathbf{Z})} \sum_{i,j \in \Omega(X)} D(p(Y \,|X = i, \mathbf{Z} = \mathbf{k}), p(Y \,|X = j, \mathbf{Z} = \mathbf{k}))$$

$$\tag{2}$$

where $\Omega(N)$ is the set of states of node N, and D is a distance between the two probability distributions, in terms of any suitable distance measure. The weight is computed locally at each arc because only the values stored in the associated CPD are required. In general, for nodes X and Z where X is an ancestor of Z but not necessarily a direct parent, we define the arc weight as:

$$W(X, Z) = \frac{w(Z', Z)}{l(X, Z)} \tag{3}$$

where $l(X, Z)$ is the length of the path between X and Z, and Z' is the direct parent of Z along the branch from X to Z. By taking into account the distance from the query node, we are incorporating the impact of topological proximity and relevance described in [19]. Since messages are also sent from child to parent of the original BN, a reverse measure of the weight of an arc is also needed. It is obtained by computing a CPD of the reversed arc $Y \rightarrow X$, using Bayes' Law and repeating the calculation.

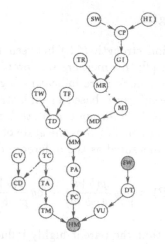

Fig. 6. 22 node polytree, adapted from the example in [3], used to obtain experimental results.

Bhattacharyya Distance. An appropriate distance measure for multi-state nodes is the Bhattacharyya [2] distance between two distributions P and Q is given by:

$$B(P, Q) = -log \sum_i \sqrt{P_i Q_i}. \tag{4}$$

Another distance measure which can be used is the average Kullback-Leibler (KL) distance [18] between two distributions P and Q. However, since the KL measure is normally used to calculate errors between an exact distribution and an estimate, we deem it less appropriate. In the formula for arc weight given in Eq 2, we take the average over state combinations of the other parents. An alternative (used in [3]) is to take the maximum. The formula then gives a maximum bound on how much a change in the parent can affect the child.

When using arc weights to traverse the treeNet, each branch of the treeNet is assigned a weight as defined in Eq. 2. To choose the best treeNet node to visit next, we scan all unvisited nodes directly connected to visited ones and select the node where the ratio of the branch weight to the path length from query node is largest. This corresponds to the value calculated by Eq. 3. In this way, we are choosing the node which most largely influences our query node.

5 Experimental Results

The experimental results given in this section involve a polytree network of 22 nodes, shown in Fig. 6. It is adapted from the example in [3]. We constructed a treeNet for this network with query node HM.

We evaluate the treeNet using breadth-first, depth-first and best-first strategies. We calculate an initial estimate of the posterior of the query node HM by visiting its direct children in the treeNet. At each subsequent node visit, we update the posterior of HM and calculate the error in the result in terms of the Bhattacharyya distance between the current estimate and the exact value [3]. The following graphs show how the error decreases as more nodes of the treeNet are included.

Comparison of Tree Traversal Strategies. In Fig. 7 we compare the performances of breadth-first, depth-first and best-first (arc weights) strategies. The evidence node is FW, shaded in Fig. 6. The major dip in the curves occurs when node PA is visited. This large fall in the error shows that PA has a great influence on our query node. Because arc PA $\rightarrow$ PC has a large weight, our best-first strategy chooses to visit PA first, hence the early drop in the arc weights curve. All three curves eventually fall to zero but at different times.

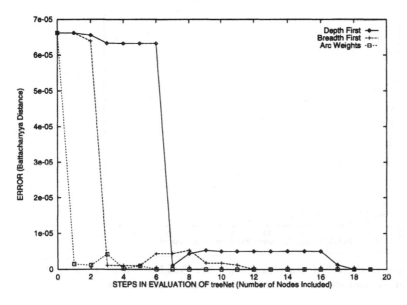

Fig. 7. Comparison of tree traversal strategies: Breadth-First, Depth-First and Best-First (using Arc Weights).

Comparison of Best-First Measures. For such a small example network, there is little difference in the performance between using the arc weight and the connection strength. Similarly, the difference when using the maximum or average arc weight is minimal. So we do not show the graphs here due to space constraints.

---

[3] To obtain the exact answer, we initially evaluated the network using a join-tree algorithm.

Comparison of Arc Weights and Breadth-First with Evidence in Roots of BN. In [17] we showed that the structure of the network, and the location of evidence nodes, affect the performance of various BN inference algorithms. In Fig. 8 we compare the performances of breadth-first and best-first (arc weights) strategies with evidence in the root nodes of the BN (CV, TC, TW, TW, TF, TR, SW, HT). The same experiment is repeated by separately instantiating a different root node each time and calculating the errors as before. Then we average the errors over all the different instantiations. The curves follow a similar trend to Fig. 7 but they are smoother since individual peaks are eliminated in the averaging process.

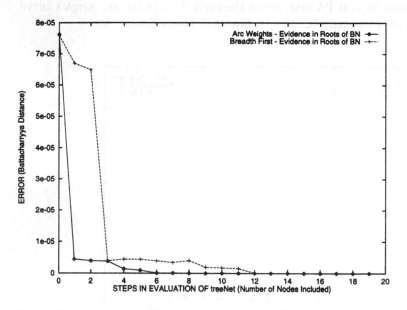

Fig. 8. Comparison of treeNet evaluation using Arc Weights and Breadth First. One root node is set as evidence for each run, with average results shown.

6 Discussion

Traditional BN evaluation algorithms rely on the complete network, that is its associated data structures, as well as the evaluation code, to be available on-line. This is often costly in storage resources. Traditional algorithms are also not geared towards answering particular queries. However, in many applications, the nodes to be queried are pre-determined in the specifications of the system. The query-oriented characteristic of the treeNet formulation simplifies BN evaluation: only messages relevant to the belief of the query node are computed and transmitted. A treeNet can also be more efficiently stored on-line than a complete BN since we only want to have immediate access to the branch of the tree under investigation. Other branches may be swapped out and back when needed.

Evaluating a treeNet is essentially a tree search problem, therefore, many tree search optimisation techniques can be used. Biasing the search in favour of arcs of heavier weights is only one of the possible variations. In general, we can guide the search using criteria relevant to the specific requirements of different systems, hence leading to fast, customised, on-line evaluation procedures. We can also develop efficient techniques for the off-line construction of the tree, but since this process is only carried out once, it is not a foremost concern.

As already mentioned in Sect. 3.4, the structure of the treeNet facilitates identification of independencies induced by specific evidence assignments, using d-separation and CSI. D-separation is often costly to verify exhaustively in BNs because of the complexity of finding paths between a query node and a potentially d-separated node. However, in a treeNet, it is straightforward to follow the path from any node to the query, which is fixed. For a given set of conditioning nodes, namely the evidence nodes, it becomes easy to determine which nodes are d-separated from the query.

In real-time BN applications, where there is a continuous inflow of information, we have to add new evidence or retract old evidence to keep up with the evolution of the system. We must be able to update the belief of our query node at the rate of arrival of new information. In a treeNet, new information is easy to incorporate since we only re-evaluate the branch of the tree where the updated evidence lies. The new evidence is propagated to the query node along a direct path; other branches are ignored and can even be moved to off-line storage. If it can be determined in advance which nodes will receive new evidence most frequently, then we can store only the relevant branches on-line and load other parts of the treeNet intermittently. Like Q-DAGS, since tree structures are simple to conceptualise and widely accepted, users unfamiliar with BNs may find a treeNet easier to work with [8].

The treeNet incremental evaluation does not strictly satisfy the ideal requirements of anytime algorithms [22]. We can obtain a measurable quality of result only by comparing the output to the exact result which will not be available at run-time when the system is installed. While we can show empirically that in most cases, on average, the treeNet algorithm produces more accurate results over time, we cannot guarantee monotonicity. For example, in Fig. 7 the graph is non-monotonic. The treeNet evaluation algorithm is fully interruptible.

The main limitation of the treeNet paradigm is that it is only applicable to polytree structures. However, since most BNs with underlying loops can be converted to clique-trees [18] or join-trees [16] the treeNet construction and evaluation process can be adapted to suit a larger category of BNs. We discuss this possibility further in Sect. 7. Also, we have only constructed treeNets containing a single query node. Multiple query nodes are the obvious extension which we plan to implement in further work.

7 Extensions

So far, we have presented the application of the treeNet formulation to polytree BNs. Realistic problems are almost invariably encoded as BNs with underlying loops, so we now outline how to adapt treeNet construction and evaluation to cater for the larger category of DAGs. Given a BN that contains underlying loops, we first transform it to a clique-tree [18]. A treeNet can then be constructed for the clique-tree using the same algorithm as for polytrees. The root of the treeNet will then correspond to the clique which contains the designated query node. If the query node is present in more than one clique, it is preferable to choose the smallest of the candidate cliques to minimise the size of messages collected by the root during evaluation. The other nodes of the treeNet will correspond to other cliques of the clique-tree. Instead of the original CPDs of the BN, the nodes of the clique will now store clique probabilities.

This treeNet can be evaluated incrementally using a chosen order of visiting the nodes. When a node is visited, it sends a message to its treeNet parent to be propagated towards the query clique. To extract the belief of our query node, we perform a marginalisation operation on the updated beliefs of the query clique. The beliefs of any other member of the query clique can similarly be extracted if necessary. If we extend our measure of arc weights to clique-tree arcs, we can use the weights to orient the traversal towards the next most influential clique at each iteration. This problem will be investigated in further work.

Approximate models of BNs, which are more efficient to encode and evaluate, are often preferred in situations where an approximate result is acceptable. Instead of using the full BN as starting point, we can also use an approximate model, for example one from which the weak links have been removed [15]. In general, any model approximation technique can be applied to the BN prior to constructing a treeNet.

8 Conclusion

We have presented a new framework for implementing evaluation of belief networks (BNs), which consists of two steps: (1) transforming a belief network into a tree structure called a **treeNet** (2) performing anytime inference by searching the treeNet. We have described a measure for arc weights which we have shown is suitable for a best-first tree traversal strategy, that performs better than standard breadth-first or depth-first methods. While the approach is described and implemented for polytrees, we have outlined possible extensions to handle general BNs.

Acknowledgements: we thank the anonymous reviewers for helpful comments.

References

1. I. Beinlich, H. Suermondt, R. Chavez, and G. Cooper. The alarm monitoring system: A case study with two probabilistic inference techniques for belief networks. In *Proc. of the 2nd European Conf. on AI in medicine*, pages 689–693, 1992.

2. A. Bhattacharyya. On a measure of divergence between two statistical populations defined by their probability distributions. *Bulletin of the Calcutta Mathematics Society*, 35:99–110, 1943.

3. B. Boerlage. Link strengths in bayesian networks. Master's thesis, Dept. of Computer Science, U. of British Columbia, 1995.

4. C. Boutilier, N. Friedman, M. Goldszmidt, and D. Koller. Context-specific independence in bayesian networks. In *Proc. of 12th Conf. on UAI*, pages 115–123, 1996.

5. Homer L. Chin and Gregory F. Cooper. Bayesian belief network inference using simulation. In *Uncertainty in Artificial Intelligence 3*, pages 129–147, 1989.

6. G.F. Cooper. The computational complexity of probabilistic inference using bayesian belief networks. *Artificial Intelligence*, 42:393–405, 1990.

7. P. Dagum and M. Luby. Approximating probabilistic inference in belief networks is NP-hard. *Artificial Intelligence*, pages 141–153, 1993.

8. A. Darwiche and G. Provan. Query-dags: A practical paradigm for implementing belief-network inference. In *Proc. of 12th Conf. on UAI*, pages 203–210, 1996.

9. T. Dean and M. Boddy. An analysis of time-dependent planning. In *Proc. AAAI-88*, pages 49–54, 1988.

10. T. Dean and M. P. Wellman. *Planning and control.* Morgan Kaufman Publishers, San Mateo, Ca., 1991.

11. R. Dechter. Bucket elimination: A unifying framework for probabilistic inference. In *Proceedings of the Twelfth Conference on Uncertainty in AI*, pages 211–219, 1996.

12. D. L. Draper and S. Hanks. Localized partial evaluation of a belief network. In *Proc. of UAI 94*, pages 170–177, 1994.

13. Jeff Forbes, Tim Huang, Keiji Kanazawa, and Stuart Russell. The batmobile: Towards a bayesian automated taxi. In *Proc. of the 14th Int. Joint Conf. on AI (IJCAI'95)*, pages 1878–1885, 1995.

14. E.J. Horvitz, H.J. Suermondt, and G.F. Cooper. Bounded conditioning: Flexible inference for decisions under scarce resources. In *Proc. of the 5th Workshop on Uncertainty in AI*, pages 182–193, 1989.

15. Uffe Kjaerulff. Reduction of computation complexity in bayesian networks through removal of weak dependencies. In *Proc. of 10th Conf. on UAI*, pages 374–382, 1994.

16. S.L. Lauritzen and D.J. Spiegelhalter. Local computations with probabilities on graphical structures and their applications to expert systems. *Journal of the Royal Statistical Society*, 50(2):157–224, 1988.

17. A.E. Nicholson and N. Jitnah. Belief network algorithms: a study of performance based on domain characterisation. Technical Report 96/249, Department of Computer Science, Monash University, 1996.

18. J. Pearl. *Probabilistic Reasoning In Intelligent Systems: Networks of Plausible Inference*. Morgan Kaufmann, 1988.

19. K. L. Poh and E Horvitz. Topological proximity and relevance in graphical decision models. In *Proc. of 12th Conf. on UAI*, pages 427–435, 1996.

20. M. Wellman and C. Liu. State-space abstraction for anytime evaluation of probabilistic networks. In *Proc. of 10th Conf. on UAI*, pages 567–574, 1994.

21. N. Zhang and D. Poole. Exploiting causal independence in bayesian network inference. *Journal of Artificial Intelligence Research*, 5:301–328, 1996.

22. S. Zilberstein. Composition and monitoring of anytime alforithms. In *IJCAI 95: Anytime Algorithms and Deliberation Scheduling Workshop*, pages 14–21, 1995.

A Logically Sound Method for Uncertain Reasoning with Quantified Conditionals

Gabriele Kern-Isberner

FernUniversität Hagen, Fachbereich Informatik
P.O. Box 940, D-58084 Hagen, Germany
e-mail: gabriele.kern-isberner@fernuni-hagen.de

Abstract. Conditionals play a central part in knowledge representation and reasoning. Describing certain relationships between antecedents and consequences by "if–then–sentences" their range of expressiveness includes commonsense knowledge as well as scientific statements. In this paper, we present the principles of maximum entropy resp. of minimum cross-entropy (*ME-principles*) as a logically sound and practicable method for representing and reasoning with quantified conditionals. First the meaning of these principles is made clear by sketching a *characterization* from a completely conditional-logical point of view. Then we apply the techniques presented to derive *ME-deduction schemes* and illustrate them by examples in the second part of this paper.

1 Introduction

Knowledge is often expressed by "if A then B"–statements (*conditionals*, written as A → B), where the *antecedent* A describes a precondition which is known (or assumed) to imply the *consequence* B. Usually, a conditional supposes a special relationship between its antecedent and its consequence. The most characteristic property of a conditional is that its meaning is restricted to situations where its antecedent is true. Therefore to represent conditionals appropriately, the need for a third logical value *u*, interpreted as *undetermined* or *undefined*, arises, thus making it necessary to leave the area of classical two-valued logic (cf. [15]).

A lot of different approaches to a *logic of conditionals* have been made (for a survey, cf. [16]), also aiming at reflecting more general relationships between antecedent and consequence so as to capture the manifold meanings of commonsense conditionals. In general, conditionals are used to describe relationships that are assumed to hold *mostly* or *possibly*, or that are only *believed* to hold. Beside such qualitative approaches, the validity of conditionals may be quantified by *degrees of trueness* or *certainty* (cf. e.g. [2]). Cox [3] argued that a logically consistent handling of quantified conditionals is only possible within a probabilistic framework, where the degree of certainty associated with a conditional is interpreted as a conditional probability. In fact, probability theory provides a sound and convenient machinery to be used for knowledge representation and automated reasoning (cf. e.g. [5],[4], [14], [18], [23]). But its clear semantics and strict rules require a lot of knowledge to be available for an adequate modelling of problems.

In contrast to this, usually only relatively few relationships between relevant variables are known, due to incomplete information. Or maybe, an abstractional representation is intended, incorporating only fundamental relationships. In both cases, the knowledge explicitly stated is not sufficient to determine uniquely a probability distribution. One way to cope with this indetermination is to calculate upper and lower bounds for probabilities (cf. [23], [5]). This method, however, brings about two problems: Sometimes the inferred bounds are quite bad, and in any case, one has to handle intervals instead of single values.

An alternative way that provides best expectation values for the unknown probabilities and guarantees a logically sound reasoning is to use the *principle of maximum entropy* resp. the *principle of minimum cross entropy* to represent all available probabilistic knowledge by a unique distribution. If P is a probability distribution, its *entropy* is defined as $H(P) = -\sum_{\omega} p(\omega) \log p(\omega)$, and if Q is another distribution, $R(Q, P) = \sum_{\omega} q(\omega) \log \frac{q(\omega)}{p(\omega)}$ gives the *cross-* or *relative entropy of Q with respect to P*. Entropy is a numerical value for the uncertainty inherent to a distribution, and relative entropy measures the information-theoretic distance between P and Q (cf. e.g. [20], [13], [9], [8]). So if $\mathcal{R}$ is a set of conditionals, each associated with a probability, then the "best" distribution to represent $\mathcal{R}$ – and only $\mathcal{R}$ – is the one which fulfills all conditionals in $\mathcal{R}$ and has maximum entropy. By an analogous argument, if prior knowledge given by a distribution P has to be adjusted to new probabilistic knowledge $\mathcal{R}$, the one distribution should be chosen that satisfies $\mathcal{R}$ and has minimum relative entropy to P. In the sequel, the abbreviation *ME* will indicate representation of resp. adjustment to probabilistic knowledge at optimum (*M*inimum or *M*aximum) *E*ntropy.

There are three articles [21],[17],[11] that characterize optimum entropy distributions as sound bases for logically consistent inferences. In [11], the author chose a completely conditional-logical environment for this characterization, proving that optimum entropy representations are most appropriate for probabilistic conditionals. Together with the results of Cox [3], this establishes reasoning at optimum entropy as a most fundamental inference method in the area of quantified uncertain reasoning.

The characterization given in [12] rests on only four very basic and intelligible axioms for conditional reasoning, and it is developed step by step from a conditional-logical argumentation. In the first part of this paper, we will sketch this characterization for a deeper understanding of ME-reasoning. The representation of the ME-distribution central to the argumentation given then turns out to be not only of theoretical but also of practical use: Due to it, some deduction schemes for ME-inferring will be presented in the second part of this paper. For instance, we will show how knowledge is propagated transitively and give explicit ME-probability values for *cautious cut* and *cautious monotony*. These deduction schemes, however, are global, not local, i.e. all knowledge available has to be taken into account in their premises to give sound results. But they provide useful insights into the practice of ME-reasoning. Furthermore, a few examples will be given.

2 A probabilistic conditional language

Let $V = \{V_1, V_2, V_3, \ldots\}$ be a finite set of binary propositional variables, and let $\mathcal{L}(V)$ be the propositional language over the alphabet V with logical connectives $\vee, \wedge$ resp. juxtaposition and $\neg$. For each V_i, $\dot{v}_i \in \{v_i, \bar{v}_i\}$ stands for one of the two possible outcomes of the variable (where negation is indicated by a bar). The set of all *atoms* (*complete conjunctions, elementary events*) over the alphabet V is denoted by Ω: $\Omega = \{\omega | \omega = \dot{v}_1 \dot{v}_2 \ldots, \dot{v}_i \in \{v_i, \bar{v}_i\}\}$. Each probability distribution P over V induces a probability function p on Ω and vice versa. Thus, given a distribution P, a probability can be assigned to each propositional formula $A \in \mathcal{L}(V)$ via $p(A) = \sum_{\omega : A(\omega)=1} p(\omega)$, where $A(\omega) = 1$ resp. $= 0$ means that A is true resp. false in the world described by ω.

We extend $\mathcal{L}(V)$ to a *probabilistic conditional language* $\mathcal{L}^*$ by adding a *conditional operator* $\rightsquigarrow$: The wff in $\mathcal{L}^*$ are defined to have syntax $A \rightsquigarrow B[x]$ with $A, B \in \mathcal{L}(V)$, $x \in [0,1]$, and they are called *probabilistic conditionals* or *probabilistic rules*. Their semantics is based on P via conditional probabilities: We write $P \models A \rightsquigarrow B[x]$ iff $p(A) > 0$ and $p(B|A) = \dfrac{p(AB)}{p(A)} = x$. By a simple probabilistic calculation, we see $p(B|A) = x$ iff $\dfrac{p(AB)}{p(A\overline{B})} = \dfrac{x}{1-x}$. *Probabilistic facts* are probabilistic rules $\top \rightsquigarrow B$ with tautological antecedent. Thus the classical language $\mathcal{L}(V)$ may be embedded into $\mathcal{L}^*$ by identifying the classical propositional formula B with the probabilistic conditional $\top \rightsquigarrow B[1]$.

3 Probabilistic representation and adjustment at optimum entropy

The *problem of adapting new probabilistic knowledge* may now be adequately formalized as follows, using the notation of the previous section:

(*) Given a (prior) distribution P and a set
$\mathcal{R} = \{A_1 \rightsquigarrow B_1[x_1], \ldots, A_n \rightsquigarrow B_n[x_n]\}$ of probabilistic conditionals, which (posterior) distribution P^* should be chosen such that it satisfies $P^* \models \mathcal{R}$ and that it is — in some sense — "nearest" to P?

To avoid inconsistencies between prior knowledge and new information, we assume P to be positive. The *representation problem*

(**) Given a set $\mathcal{R} = \{A_1 \rightsquigarrow B_1[x_1], \ldots, A_n \rightsquigarrow B_n[x_n]\}$ of probabilistic conditionals, which distribution P^* that satisfies $P^* \models \mathcal{R}$ should be chosen to represent $\mathcal{R}$ best?

is a special case of (*) with a uniform distribution as prior distribution, starting from complete ignorance.

Let P_e denote the ME-solution to (*), that is P_e has minimum cross-entropy to P among all distributions Q with $Q \models \mathcal{R}$. Then P_e may be represented as

$$p_e(\omega) = \alpha_0 p(\omega) \prod_{\substack{1 \le i \le n \\ A_i B_i(\omega)=1}} \alpha_i^{1-x_i} \prod_{\substack{1 \le i \le n \\ A_i \overline{B_i}(\omega)=1}} \alpha_i^{-x_i}, \tag{1}$$

for $\omega \in \Omega$ (cf. e.g [9], [11]). In particular, P_e satisfies $\mathcal{R}$, so $p_e(B_i|A_i) = x_i$, which is equivalent to postulating $\dfrac{p_e(A_i B_i)}{p_e(A_i \overline{B_i})} = \dfrac{x_i}{1-x_i}$ for all i, $1 \le i \le n$. Regarding (1), this last equation yields

$$\alpha_i = \frac{x_i}{1-x_i} \frac{\displaystyle\sum_{\omega: A_i \overline{B_i}(\omega)=1} p(\omega) \prod_{\substack{j \ne i \\ A_j B_j(\omega)=1}} \alpha_j^{1-x_j} \prod_{\substack{j \ne i \\ A_j \overline{B_j}(\omega)=1}} \alpha_j^{-x_j}}{\displaystyle\sum_{\omega: A_i B_i(\omega)=1} p(\omega) \prod_{\substack{j \ne i \\ A_j B_j(\omega)=1}} \alpha_j^{1-x_j} \prod_{\substack{j \ne i \\ A_j \overline{B_j}(\omega)=1}} \alpha_j^{-x_j}}, \quad 1 \le i \le n, \tag{2}$$

as a necessary (and sufficient) condition for any distribution of type (1) to satisfy $\mathcal{R}$. α_0 may be chosen adequately to ensure that p_e is a probability function, i.e. that $\sum_{\omega \in \Omega} p_e(\omega) = 1$. The next section will show that equations (1) and (2) are crucial for the understanding of ME-adjustment (cf. Theorem 8). All proofs may be found in [11].

4 Characterizing the principle of minimum cross entropy within a conditional-logical framework

4.1 The principle of conditional preservation

Following Calabrese [2], a conditional $A \rightsquigarrow B$ can be represented as an *indicator function* $(B|A)$ on elementary events, setting

$$(B|A)(\omega) = \begin{cases} 1 & : \ \omega \in AB \\ 0 & : \ \omega \in A\overline{B} \\ u & : \ \omega \notin A \end{cases}$$

where u stands for *undefined*. This definition captures excellently the non-classical character of conditionals within a probabilistic framework. According to it, a conditional is a function that polarizes AB and $A\overline{B}$, leaving $\overline{A}$ untouched.

As we saw earlier in section 2, conditional probabilities are ratios, the most fundamental of which being *elementary ratios*, i.e. ratios of probabilities of elementary events. They can be regarded as the atoms of the conditional-logical structure of a distribution, and products of elementary ratios provide a suitable means to investigate conditional structures. In statistics, logarithms of such expressions are used to measure the interactions between the variables involved (cf. [7], [24]).

Informally, the *principle of conditional preservation* is to state that all products of elementary ratios shall remain unchanged if there is no indication for change found in the set $\mathcal{R}$, representing new conditional information. As a consequence, all (statistical) interactions between variables are preserved as far as possible.

To make the idea of conditional preservation concrete, we have to devise a method to compare the elementary events involved in such products on the base of the information in $\mathcal{R}$. To this end, we will formalize the notion of *conditional structures* of elementary events resp. of multi-sets of elementary events by means of a group-theoretical representation.

Let Ω denote the set of all elementary events, and let $\mathcal{R} = \{A_1 \rightsquigarrow B_1[x_1], \ldots, A_n \rightsquigarrow B_n[x_n]\}$ be a set of probabilistic conditionals. To each conditional $A_i \rightsquigarrow B_i[x_i]$ in $\mathcal{R}$ we associate two symbols a_i, b_i. Let $F_{\mathcal{R}} = \langle a_1, b_1, \ldots, a_n, b_n \rangle$ be the free abelian group with generators $a_1, b_1, \ldots, a_n, b_n$, i.e. $F_{\mathcal{R}}$ consists of all elements of the form $a_1^{\mu_1} b_1^{\nu_1} \ldots a_n^{\mu_n} b_n^{\nu_n}$ with integers $\mu_i, \nu_i \in \mathcal{Z}$ (the ring of integers), and each element can be identified by its exponents, so that $F_{\mathcal{R}}$ is isomorphic to $\mathcal{Z}^{2n}$. The commutativity of $F_{\mathcal{R}}$ corresponds to the fact that the conditionals in $\mathcal{R}$ shall be effective all at a time, without assuming any order of application.

For each $i, 1 \leq i \leq n$, we define a function $\sigma_i : \Omega \to F_{\mathcal{R}}$ by setting

$$\sigma_i(\omega) = \begin{cases} a_i & \text{if} \quad (B_i|A_i)(\omega) = 1 \\ b_i & \text{if} \quad (B_i|A_i)(\omega) = 0 \\ 1 & \text{if} \quad (B_i|A_i)(\omega) = u \end{cases}.$$

$\sigma_i(\omega)$ represents the manner in which the conditional $A_i \rightsquigarrow B_i[x_i]$ affects the elementary event ω. The neutral element 1 of $F_{\mathcal{R}}$ corresponds to the non-applicability of $A_i \rightsquigarrow B_i[x_i]$ in case that the antecedent A_i is not satisfied.

The function $\sigma = \sigma_{\mathcal{R}} : \Omega \to F_{\mathcal{R}}$, $\sigma(\omega) = \prod_{1 \leq i \leq n} \sigma_i(\omega)$ describes the all-over effect of $\mathcal{R}$ on ω. $\sigma(\omega)$ contains at most one of each a_i or b_i, but never both of them because each conditional affects ω in a well-defined way.

Definition 1. (i) Let $\omega \in \Omega$ be an elementary event. $\sigma(\omega) = \prod_{1 \leq i \leq n} \sigma_i(\omega) \in F_{\mathcal{R}}$ is called the *conditional structure* of ω with respect to $\mathcal{R}$.

(ii) Let $\Omega_1 = \{\omega_1 : r_1, \ldots, \omega_{m_1} : r_{m_1}\}$ denote a multi-set of elementary events, with each r_j indicating the multiplicity of the corresponding ω_j. The element $\Omega_1^{\sigma} := \prod_{1 \leq j \leq m_1} \sigma(\omega_j)^{r_j} \in F_{\mathcal{R}}$ is called the *conditional structure of Ω_1* .

Thus the conditional structure Ω_1^{σ} of a multi-set $\Omega_1 = \{\omega_1 : r_1, \ldots, \omega_{m_1} : r_{m_1}\}$ is represented by a group element which is a product of the generators a_i, b_i of $F_{\mathcal{R}}$, with each a_i occurring with exponent $\sum_{k:\sigma_i(\omega_k)=a_i} r_k = \sum_{k:(B_i|A_i)(\omega_k)=1} r_k$, and each b_i occurring with exponent $\sum_{k:\sigma_i(\omega_k)=b_i} r_k = \sum_{k:(B_i|A_i)(\omega_k)=0} r_k$ (note that both sums may be zero in case that the corresponding conditional is not touched at all). So the exponent of a_i in Ω_1^{σ} indicates the number of elementary events in Ω_1 which *confirm* the conditional $A_i \rightsquigarrow B_i$, each event being counted with its multiplicity, and in the same way the exponent of b_i indicates the number of elementary events that *refute* $A_i \rightsquigarrow B_i$. Ω_1^{σ} encodes this information in an elegant manner. Making use of the group structure allows to form products and thus provides a convenient representation and handling of conditional structures.

Definition 2. Two multi-sets $\Omega_1 = \{\omega_1 : r_1, \ldots, \omega_{m_1} : r_{m_1}\}$ and $\Omega_2 = \{\nu_1 : s_1, \ldots, \nu_{m_2} : s_{m_2}\}$ of elementary events with equal cardinalities $\sum_{1 \leq k \leq m_1} r_k = \sum_{1 \leq l \leq m_2} s_l$ are $\mathcal{R}$-*equivalent* iff $\Omega_1^\sigma = \Omega_2^\sigma$, i.e. iff their conditional structures with respect to $\mathcal{R}$ are identical.

We are now in the position to formalize exactly what is to be understood by conditional preservation. Adaptations following this principle will be called *c-adaptations*.

Definition 3. Let P^* be a distribution which fulfills $\mathcal{R}$.
P^* is called a *c-adaptation of P to $\mathcal{R}$* iff it satisfies the following two conditions:

(i) For any $\omega \in \Omega$, $p^*(\omega) = 0$ if and only if there is a conditional $A_i \rightsquigarrow B_i[x_i]$ in $\mathcal{R}$ such that either $x_i = 1$ and $A_i \overline{B_i}(\omega) = 1$, or $x_i = 0$ and $A_i B_i(\omega) = 1$;

(ii) the $\mathcal{R}$-equivalence of any two multi-sets $\Omega_1 = \{\omega_1 : r_1, \ldots, \omega_{m_1} : r_{m_1}\}$ and $\Omega_2 = \{\nu_1 : s_1, \ldots, \nu_{m_2} : s_{m_2}\}$ of elementary events implies

$$\frac{p^*(\omega_1)^{r_1} \ldots p^*(\omega_{m_1})^{r_{m_1}}}{p^*(\nu_1)^{s_1} \ldots p^*(\nu_{m_2})^{s_{m_2}}} = \frac{p(\omega_1)^{r_1} \ldots p(\omega_{m_1})^{r_{m_1}}}{p(\nu_1)^{s_1} \ldots p(\nu_{m_2})^{s_{m_2}}}.$$

Thus a c-adaptation is completely based both on the conditional structure of P and of $\mathcal{R}$, using the structure of P as a reference point and that of $\mathcal{R}$ as a guideline for changes. It realizes perfectly a conditional-logical approach to the adaptation problem (*):

Postulate (P1): conditional preservation
The solution P^* to (*) is a c-adaptation.

The next theorem provides a catchy and easy characterization of c-adaptations:

Theorem 4. *Let P^* be a distribution that satisfies $\mathcal{R} = \{A_1 \rightsquigarrow B_1[x_1], \ldots, A_n \rightsquigarrow B_n[x_n]\}$; let P be a positive prior distribution.*
P^* is a c-adaptation of P to $\mathcal{R}$ iff for each conditional $A_i \rightsquigarrow B_i[x_i]$ in $\mathcal{R}$, there is a pair of non-negative factors (α_i^+, α_i^-) associated with it such that

$$p^*(\omega) = \alpha_0 p(\omega) \prod_{\substack{1 \leq i \leq n \\ A_i B_i(\omega)=1}} \alpha_i^+ \prod_{\substack{1 \leq i \leq n \\ A_i \overline{B_i}(\omega)=1}} \alpha_i^- \tag{3}$$

for all elementary events ω, where $\alpha_0 > 0$ is a normalization factor.

From now on, we will use an operational notation to distinguish posterior distributions that satisfy Postulate (P1), i.e. the results of c-adaptations of P to $\mathcal{R}$: $P * \mathcal{R}$ shall denote any distribution of the form (3), where the factors $\alpha_i^+, \alpha_i^-, 1 \leq i \leq n$, are solutions to

$$\frac{\alpha_i^+}{\alpha_i^-} = \frac{x_i}{1 - x_i} \frac{\displaystyle\sum_{\omega : A_i \overline{B_i}(\omega)=1} p(\omega) \prod_{\substack{j \neq i \\ A_j B_j(\omega)=1}} \alpha_j^+ \prod_{\substack{j \neq i \\ A_j \overline{B_j}(\omega)=1}} \alpha_j^-}{\displaystyle\sum_{\omega : A_i B_i(\omega)=1} p(\omega) \prod_{\substack{j \neq i \\ A_j B_j(\omega)=1}} \alpha_j^+ \prod_{\substack{j \neq i \\ A_j \overline{B_j}(\omega)=1}} \alpha_j^-} \tag{4}$$

with $\alpha_i^- = 0$ for $x_i = 1$. (4) ensures that the resulting $P * \mathcal{R}$ fulfills all of the conditionals in $\mathcal{R}$.

4.2   The functional concept

In general, there will be many different posterior c-adaptations, corresponding
to different solutions of (4). Of course, this situation is not very satisfactory because we intuitively feel that there are "good" solutions and "bad" solutions. For
instance, if the prior distribution P already satisfies the information contained in
$\mathcal{R}$ it would be reasonable to expect that the scheme does not alter any probability, yielding the prior as the posterior distribution. In general, the question for a
"best" solution and the need to determine it uniquely arises. There should best
be a function which calculates an appropriate posterior distribution from prior
knowledge P and new conditional information $\mathcal{R}$.

The factors α_i^+ and α_i^- crucial for a c-adaptation are solutions to the equations
(4) which reflect the complex logical interactions between the rules in $\mathcal{R}$ and the
dependency from the P. Their common quotient $\alpha_i := \dfrac{\alpha_i^+}{\alpha_i^-}$ symbolizes the core
of the impact the conditionals in $\mathcal{R}$ are to have on the new distribution, as
well as it takes into account the influence of the prior P. Therefore it plays a
key role, and we assume the factors α_i^+ and α_i^- to be functionally dependent
on it: $\alpha_i^+ = F_i^+(\alpha_i) = F^+(x_i, \alpha_i)$ resp. $\alpha_i^- = F_i^-(\alpha_i) = F^-(x_i, \alpha_i)$. The
functions F^+ and F^- are supposed to be sufficiently regular and are to follow a
pattern independent of the specific form of a rule $A_i \rightsquigarrow B_i[x_i]$, thus realizing a
fundamental inference pattern. We state these assumptions as

Postulate (P2): functional concept
The factors α_i^+ and α_i^- in (3) are determined by two regular real functions
$F^+(x, \alpha)$ and $F^-(x, \alpha)$, defined for $x \in [0, 1]$ and non-negative real α,
such that

$$\alpha_i^+ = F^+(x_i, \alpha_i) \quad \text{and} \quad \alpha_i^- = F^-(x_i, \alpha_i),$$

where $\alpha_i = \dfrac{\alpha_i^+}{\alpha_i^-}$ denotes their common quotient.

To symbolize the presence of the functional concept, we will use $*_F$ instead of $*$.

4.3   Logical consistency and representation invariance

Surely, the adaptation scheme (3) will be considered sound only if the resulting
posterior distribution can be used as a prior distribution for further adaptations.
This is a very fundamental meaning of *logical consistency*.

Postulate (P3): logical consistency
For any positive distribution P and any sets $\mathcal{R}_1, \mathcal{R}_2 \subset \mathcal{L}^*$, the (final)
posterior distribution which arises from a two-step process of adjusting
P first to $\mathcal{R}_1$ and then adjusting this intermediate posterior to $\mathcal{R}_1 \cup \mathcal{R}_2$ is
identical to the distribution resulting from simply adapting P to $\mathcal{R}_1 \cup \mathcal{R}_2$
(provided that all adaptation problems are solvable).

More formally, the operator $*$ satisfies (P3) iff the following equation holds:

$$P * (\mathcal{R}_1 \cup \mathcal{R}_2) = (P * \mathcal{R}_1) * (\mathcal{R}_1 \cup \mathcal{R}_2)$$

Theorem 5. *If the adjustment operator $*_F$ satisfies the postulate (P3) of logical consistency then there is a regular function $c(x)$ such that*

$$F^- (x, \alpha) = \alpha^{c(x)} \quad and \quad F^+ (x, \alpha) = \alpha^{c(x)+1}$$

for any positive real α and any $x \in (0, 1)$.

We surely expect the result of our adjustment process to be independent of the syntactic representation of probabilistic knowledge in $\mathcal{R}$:

Postulate (P4): representation invariance
If two sets of probabilistic conditionals $\mathcal{R}$ and $\mathcal{R}'$ are probabilistically equivalent the posterior distributions $P * \mathcal{R}$ and $P * \mathcal{R}'$ resulting from adapting a prior P to $\mathcal{R}$ resp. to $\mathcal{R}'$ are identical.

Here, as in general, two sets of rules $\mathcal{R}$ and $\mathcal{R}'$ are called *probabilistically equivalent* iff each rule in one set is derivable from rules in the other by elementary probabilistic calculation. Using the operational notation, we are able to express (P4) more formally: The adjustment operator $*$ satisfies (P4) iff

$$P * \mathcal{R} = P * \mathcal{R}'$$

for any two probabilistically equivalent sets $\mathcal{R}$ and $\mathcal{R}'$.

Theorem 6. *If the operator $*_F$ is to meet the fundamental demands for logical consistency (P3) and for representation invariance (P4), then F^- and F^+ necessarily must be*

$$F^+ (x, \alpha) = \alpha^{1-x} \quad and \quad F^- (x, \alpha) = \alpha^{-x}.$$

4.4 Uniqueness

So far we have proved that the demands for logical consistency and for representation invariance uniquely determine the functions which we assumed to underly the adjustment process in the way we described in Subsection 4.2. We recognize that the posterior distribution necessarily is of the same *type* (1) as the ME-distribution if it is to yield sound and consistent inferences. But - are there possibly several different solutions of type (1), only one of which being the ME-distribution? The uniqueness which makes the characterization complete is now affirmed by the next theorem. This unique posterior distribution of type (1) must be the ME-distribution, $*_F$ then corresponds to ME-inference, and ME-inference is known to fulfill (P3) and (P4) as well as many other reasonable properties, cf. [10], [17], [21], [22].

Theorem 7. *There is at most one solution $P * \mathcal{R}$ of the adaptation problem of type (1).*

The following theorem summarizes our results in characterizing ME-adjustment within a conditional-logical framework:

Theorem 8. *Let $*_e$ denote the ME-adjustment operator, i.e. $*_e$ assigns to a prior distribution P and some set $\mathcal{R} = \{A_1 \rightsquigarrow B_1[x_1], \ldots, A_n \rightsquigarrow B_n[x_n]\}$ of probabilistic conditionals the one distribution $P_e = P *_e \mathcal{R}$ which has minimal cross-entropy with respect to P among all distributions that satisfy $\mathcal{R}$ (provided the adjustment problem is solvable at all). Then $*_e$ yields the only adaptation of P to $\mathcal{R}$ that obeys the principle of conditional preservation (P1), realizes a functional concept (P2) and satisfies the fundamental postulates for logical consistency (P3) and for representation invariance (P4). $*_e$ is completely described by (1) and (2).*

Thus a characterization of the ME-principle within a conditional-logical framework is achieved, and its implicit logical mechanisms have been revealed clearly.

5 Some ME-deduction rules

After having proved ME-adjustment and ME-representation to be logically sound methods for uncertain reasoning, we will now leave the abstract level of argumentation and turn to concrete inference patterns. It must be emphasized, however, that ME-inferring is a *global*, not a *local* method: Only if *all* knowledge available is taken into account, the results of ME-inference are reliable to yield best expectation values. Thus it is not possible to use only partial information for reasoning, and then continue the process of adjusting from the obtained intermediate distribution with the information still left. It is important that in the two-step adjustment process $(P * \mathcal{R}_1) * (\mathcal{R}_1 \cup \mathcal{R}_2)$ dealt with in the consistency postulate (P3) (cf. section 4.3) the second adaptation step uses full information $\mathcal{R}_1 \cup \mathcal{R}_2$. In fact, the distributions $(P *_e \mathcal{R}_1) *_e (\mathcal{R}_1 \cup \mathcal{R}_2)$ and $(P *_e \mathcal{R}_1) *_e \mathcal{R}_2$ differ in general.

For this reason, the deduction rules to be presented in the sequel do not provide a convenient (and complete) calculus for ME-reasoning. But they illustrate effectfully the reasonableness of that technique by calculating explicitly inferred probabilities of rules in terms of given prior probabilities. In contrast to this, the inference patterns for deriving lower and upper bounds for probabilities presented in [5] and [23] are local, but they are afflicted with all problems typical to methods for inferring intervals, not single values (cf. section 1).

It must be pointed out that in principle, ME-reasoning is feasible for all consistent probabilistic representation and adaptation problems by iterative propagation, realized e.g. by the expert system shell SPIRIT (cf. [19]) far beyond the scope of the few inference patterns given below (also cf. Example 3).

We will use the following notation:

$$\frac{\mathcal{R} \; : \; A_1 \rightsquigarrow B_1[x_1], \ldots, A_n \rightsquigarrow B_n[x_n]}{A_1^* \rightsquigarrow B_1^*[x_1^*], \ldots, A_m^* \rightsquigarrow B_m^*[x_m^*]}$$

iff $\mathcal{R} = \{A_1 \rightsquigarrow B_1[x_1], \ldots, A_n \rightsquigarrow B_n[x_n]\}$ and $P_0 *_e \mathcal{R} \models \{A_1^* \rightsquigarrow B_1^*[x_1^*], \ldots, A_m^* \rightsquigarrow B_m^*[x_m^*]\}$, where P_0 is a uniform distribution of suitable size.

5.1 Chaining rules

Proposition 9 (Transitive Chaining). *Suppose A, B, C to be propositional variables, $x_1, x_2 \in [0, 1]$. Then*

$$\frac{\mathcal{R} \ : \ a \leadsto b[x_1], \ b \leadsto c[x_2]}{a \leadsto c[\frac{1}{2}(2x_1x_2 + 1 - x_1)]} \tag{5}$$

Proof. According to equations (1) and (2), the posterior distribution $P_0 *_e \{a \leadsto b[x_1], b \leadsto c[x_2]\}$ may be calculated as shown in the following table:

ω	$P_0 * \mathcal{R}$
abc	$\alpha_0 \alpha_1^{1-x_1} \alpha_2^{1-x_2}$
$ab\bar{c}$	$\alpha_0 \alpha_1^{1-x_1} \alpha_2^{-x_2}$
$a\bar{b}c$	$\alpha_0 \alpha_1^{-x_1}$
$a\bar{b}\bar{c}$	$\alpha_0 \alpha_1^{-x_1}$
$\bar{a}bc$	$\alpha_0 \alpha_2^{1-x_2}$
$\bar{a}b\bar{c}$	$\alpha_0 \alpha_2^{-x_2}$
$\bar{a}\bar{b}c$	α_0
$\bar{a}\bar{b}\bar{c}$	α_0

with $\alpha_1 = \dfrac{x_1}{1-x_1} \dfrac{2}{\alpha_2^{1-x_2} + \alpha_2^{-x_2}} = \dfrac{x_1}{1-x_1} \alpha_2^{x_2} \dfrac{2}{\alpha_2 + 1}$ and $\alpha_2 = \dfrac{x_2}{1-x_2}$.

Now the probability of $a \leadsto c$ may be calculated in a straightforward manner:

$$p^*(c|a) = \frac{p^*(ac)}{p^*(a)} = \frac{p^*(abc) + p^*(a\bar{b}c)}{p^*(abc) + p^*(ab\bar{c}) + p^*(a\bar{b}c) + p^*(a\bar{b}\bar{c})} =$$

$$= \frac{\alpha_1 \alpha_2^{1-x_2} + 1}{\alpha_1 \alpha_2^{-x_2}(\alpha_2 + 1) + 2} = \frac{1}{2}(2x_1x_2 + 1 - x_1), \text{ as desired.}$$

Example 1. Suppose the propositional variables A, B, C are given the meanings $A=$*Being young*, $B=$*Being single*, and $C=$*Having children*, respectively. We know (or assume) that young people are usually singles (with probability 0.9) and that mostly, singles do not have children (with probability 0.85), so that $\mathcal{R} = a \leadsto b[0.9], b \leadsto \bar{c}[0.85]$. Using (5) with $x_1 = 0.9$ and $x_2 = 0.85$, ME-reasoning yields $a \leadsto \bar{c}[0.815]$ (the negation of C makes no difference). Therefore from the knowledge stated by $\mathcal{R}$ we may conclude that the probability of an individual not to have children if (s)he is young is best estimated by 0.815.

In many cases, however, rules must not be simply connected transitively as in Proposition 9 because definite exceptions are present. Let us consider the famous "Tweety the penguin"-example.

Example 2. Birds are known to fly mostly, i.e. *bird* $\leadsto$ *fly*$[x_1]$ with a probability x_1 between 0.5 and 1, penguins are definitely birds, *penguin* $\leadsto$ *bird*$[1]$, but no one has ever seen a flying penguin, so *penguins* $\leadsto$ *fly*$[x_2]$ with a probability x_2 very close to 0. What may be inferred about Tweety who is known to be a bird *and* a penguin?

The crucial point in this example is that two pieces of evidence apply to Tweety, one being more specific than the other. The next proposition shows that ME-reasoning is able to cope with *categorical specificity*.

Proposition 10 (Categorical Specificity). *Suppose A, B, C to be propositional variables, $x_1, x_2 \in [0, 1]$. Then*

$$\frac{\mathcal{R} \;:\; a \rightsquigarrow b[x_1], \; c \rightsquigarrow b[x_2], \; c \rightsquigarrow a[1]}{ac \rightsquigarrow b[x_2]}$$

Proof. Let $\alpha_1, \alpha_2, \alpha_3$ be the factors associated with the probabilistic conditionals $a \rightsquigarrow b[x_1], c \rightsquigarrow b[x_2], c \rightsquigarrow a[1]$. Using equation (2) we obtain

$$\alpha_3 = \frac{1}{0} \frac{\alpha_2^{1-x_2} + \alpha_2^{-x_2}}{\alpha_1^{1-x_1}\alpha_2^{1-x_2} + \alpha_1^{-x_1}\alpha_2^{-x_2}}, \text{ thus, by convention, } \alpha_3^0 = 1 \text{ and } \alpha_3^{-1} = 0. \text{ This}$$

implies $\alpha_1 = \dfrac{x_1}{1-x_1} \dfrac{\alpha_2^{-x_2} + 1}{\alpha_2^{1-x_2} + 1}$ and $\alpha_2 = \dfrac{x_2}{1-x_2} \dfrac{\alpha_1^{-x_1} + 0}{\alpha_1^{1-x_1} + 0} = \dfrac{x_2}{1-x_2}\alpha_1^{-1}$, thus

$\alpha_1\alpha_2 = \dfrac{x_2}{1-x_2}$. According to (1), the posterior probability of the conditional $ac \rightsquigarrow b$ can now be calculated as follows:

$$p^*(b|ac) = \frac{p^*(abc)}{p^*(abc) + p^*(a\bar{b}c)} = \frac{\alpha_1^{1-x_1}\alpha_2^{1-x_2}}{\alpha_1^{1-x_1}\alpha_2^{1-x_2} + \alpha_1^{-x_1}\alpha_2^{-x_2}} = \frac{\alpha_1\alpha_2}{\alpha_1\alpha_2 + 1} = x_2.$$

Proposition 10 states that specific information dominates more general information, as it is expected. In its proof, however, we used essentially that the specificity relation $c \rightsquigarrow a[1]$ is categorical. If its probability lies somewhere in between 0 and 1, the equational systems determining the α_i's become more complicated. But it can be solved at least by iteration, e.g. by the aid of SPIRIT (cf. [19]), if the conditional probabilities involved are numerically specified. Within a qualitative probabilistic context, Adam's ϵ-semantics [1] presents a method to handle exceptions and to take account of subclass specificity. Goldszmidt, Morris and Pearl [6] showed how reasoning based on infinitesimal probabilities may be improved by using ME-principles.

Example 3. A knowledge base is to be built up representing "Typically, students are adults", "Usually, adults are employed" and "Mostly, students are not employed" with probabilistic degrees of uncertainty $0.99(< 1)$, 0.8 and 0.9, respectively. Let A, S, E denote the propositional variables $A = $ Being an *Adult*, $S = $ Being a *Student*, and $E = $ Being *Employed*. The quantified conditional information may be written as $\mathcal{R} = \{s \rightsquigarrow a[0.99], a \rightsquigarrow e[0.8], s \rightsquigarrow \bar{e}[0.9]\}$. From this, SPIRIT calculates $p^*(\bar{e}|as) = 0.8991 \approx 0.9$. So the more specific information s dominates a clearly, but not completely.

5.2 Cautious monotony and cautious cut

Obviously, ME-logic is nonmonotonic: conjoining the antecedent of a conditional with a further literal may alter the probability of the conditional dramatically (cf. Example 3). But a weak form of monotony is reasonable and can indeed be proved:

Proposition 11 (Cautious Monotony). *Suppose A, B, C to be propositional variables, $x_1, x_2 \in [0, 1]$. Then*

$$\frac{\mathcal{R} \; : \; a \rightsquigarrow b[x_1], \; a \rightsquigarrow c[x_2]}{ab \rightsquigarrow c[x_2]} \tag{6}$$

Proof. Let α_1 be the ME-factor belonging to the first conditional, and α_2 that of the second one. Then immediately $\alpha_1 = \dfrac{x_1}{1 - x_1}$ and $\alpha_2 = \dfrac{x_2}{1 - x_2}$, by (2), so that

$$p^*(c|ab) = \frac{\alpha_1^{1-x_1} \alpha_2^{1-x_2}}{\alpha_1^{1-x_1} \alpha_2^{1-x_2} + \alpha_1^{1-x_1} \alpha_2^{-x_2}} = \frac{\alpha_2}{\alpha_2 + 1} = x_2.$$

(6) illustrates how ME-propagation respects conditional independence (cf. [22]): $p^*(c|ab) = p^*(c|a) = x_2$.

The monotony inference rule deals with *adding* information to the antecedent. Another important case arises if literals in the antecedent have to be *deleted*. Of course we cannot expect the classical cut rule to hold. But, as in the case of monotony, a *cautious cut rule* may be proved:

Proposition 12 (Cautious Cut). *Suppose A, B, C to be propositional variables, $x_1, x_2 \in [0, 1]$. Then*

$$\frac{\mathcal{R} \; : \; ab \rightsquigarrow c[x_1], \; a \rightsquigarrow b[x_2]}{a \rightsquigarrow c[\frac{1}{2}(2x_1 x_2 + 1 - x_2)]}$$

Proof. Let α_1, α_2 be the ME-factors associated with the conditionals $ab \rightsquigarrow c[x_1]$, $a \rightsquigarrow b[x_2]$ in $\mathcal{R}$.

Again by using (2), we see $\alpha_1 = \dfrac{x_1}{1 - x_1}$ and $\alpha_2 = \dfrac{x_2}{1 - x_2} \alpha_1^{x_1} \dfrac{2}{\alpha_1 + 1}$. According to (1), the probability of the conditional in question may be calculated as follows:

$$p^*(c|a) = \frac{\alpha_1^{1-x_1} \alpha_2^{1-x_2} + \alpha_2^{-x_2}}{\alpha_1^{1-x_1} \alpha_2^{1-x_2} + \alpha_1^{-x_1} \alpha_2^{1-x_2} + 2\alpha_2^{-x_2}} = \frac{\alpha_1^{1-x_1} \alpha_2 + 1}{\alpha_1^{-x_1} \alpha_2 (\alpha_2 + 1) + 2} =$$

$$= \frac{\alpha_1 \dfrac{x_2}{1 - x_2} \dfrac{2}{\alpha_1 + 1} + 1}{2 \dfrac{x_2}{1 - x_2} + 2} = \frac{1}{2}(2x_1 x_2 + 1 - x_2).$$

5.3 Conjoining literals in antecedent and consequence

The following deduction schemes deal with various cases of inferring probabilistic conditionals with literals in antecedents and consequences being conjoined. Three of them – *Conjunction Left, Conjunction Right, (ii) and (iii)* – are treated in [23] under similar names, thus allowing a direct comparison of ME-inference to probabilistic local bounds propagation. Cautious monotony (6) may be found in that paper, too, where it is denoted as *Weak Conjunction Left*. We will omit the straightforward proofs.

Proposition 13 (Conjunction Right). *Suppose* A, B, C *to be propositional variables,* $x_1, x_2 \in [0, 1]$. *Then the following ME-inference rules hold:*

(i)
$$\frac{\mathcal{R} \; : \; a \rightsquigarrow b[x_1], \; a \rightsquigarrow c[x_2]}{a \rightsquigarrow bc[x_1 x_2]}$$

(ii)
$$\frac{\mathcal{R} \; : \; a \rightsquigarrow b[x_1], \; ab \rightsquigarrow c[x_2]}{a \rightsquigarrow bc[x_1 x_2]}$$

(iii)
$$\frac{\mathcal{R} \; : \; a \rightsquigarrow b[x_1], \; b \rightsquigarrow c[x_2]}{a \rightsquigarrow bc[x_1 x_2]}$$

Proposition 14 (Conjunction Left). *Suppose* A, B, C *to be propositional variables,* $x_1, x_2 \in [0, 1]$. *Then*

$$\frac{\mathcal{R} \; : \; a \rightsquigarrow b[x_1], \; a \rightsquigarrow bc[x_2]}{ab \rightsquigarrow c[\frac{x_2}{x_1}]}$$

5.4 Reasoning by cases

The last inference scheme presented in this paper will show how probabilistic information obtained by considering exclusive cases is being processed at maximum entropy:

Proposition 15 (Reasoning by cases). *Suppose* A, B, C *to be propositional variables,* $x_1, x_2 \in [0, 1]$. *Then*

$$\frac{\mathcal{R} \; : \; ab \rightsquigarrow c[x_1], \; a\bar{b} \rightsquigarrow c[x_2]}{a \rightsquigarrow b[(1 + \frac{x_1^{x_1}(1 - x_1)^{1 - x_1}}{x_2^{x_2}(1 - x_2)^{1 - x_2}})^{-1}],}$$

$$a \rightsquigarrow c[x_1(1 + \frac{x_1^{x_1}(1 - x_1)^{1 - x_1}}{x_2^{x_2}(1 - x_2)^{1 - x_2}})^{-1} + x_2(1 + \frac{x_2^{x_2}(1 - x_2)^{1 - x_2}}{x_1^{x_1}(1 - x_1)^{1 - x_1}})^{-1}]$$

Proof. The ME-factors α_1 and α_2 associated with the conditionals in $\mathcal{R}$ (in order of appearance above) are computed to be $\alpha_1 = \dfrac{x_1}{1 - x_1}$ and $\alpha_2 = \dfrac{x_2}{1 - x_2}$. Following (1) we thus obtain

$$p^*(b|a) = \frac{\alpha_1^{1-x_1} + \alpha_1^{-x_1}}{\alpha_1^{1-x_1} + \alpha_1^{-x_1} + \alpha_2^{1-x_2} + \alpha_2^{-x_2}} = \frac{\alpha_1^{-x_1}(\alpha_1 + 1)}{\alpha_1^{-x_1}(\alpha_1 + 1) + \alpha_2^{-x_2}(\alpha_2 + 1)} =$$

$$= \frac{1}{1 + \dfrac{x_1^{x_1}(1 - x_1)^{1 - x_1}}{x_2^{x_2}(1 - x_2)^{1 - x_2}}}, \text{ as desired. The probability of the second conditional}$$

$a \rightsquigarrow c$ is proved by applying the fundamental probabilistic equality $p^*(c|a) = p^*(c|ab)p^*(b|a) + p^*(c|a\bar{b})p^*(\bar{b}|a)$, and using the information given by $\mathcal{R}$.

6 Concluding remarks

We showed that using the principles of optimum entropy provides a powerful and sound machinery for probabilistic reasoning. The presentation of an ME-distribution given by (1) and (2) is crucial for understanding the logical mechanisms that underlie ME-adjustment, as it is important for practical reasoning. The inference patterns presented in Section 5 only make use of the principle of maximum entropy to represent conditional knowledge, and perhaps, this will be a major field of application for ME-reasoning. But, as the axiom (P3) necessary for the characterization shows, the process of adjusting prior probabilistic knowledge to new information in a logically consistent way – and thus using the more general cross-entropy – is indispensable for the whole principle.

One of the most striking features of ME-reasoning is its thoroughness: In fact, only one method – that of minimizing cross-entropy – is used to realize representation, adaptation and instantiating of probabilistic knowledge. The last statement arises from the fact that instantiating a distribution P with respect to evidence A means taking the conditional distribution $P(\cdot|A)$, and this amounts obviously to calculating the ME-distribution $P *_e \{A[1]\}$.

Considering the principle of maximum entropy as subordinate to the principle of minimum cross-entropy apparently means to accept that total ignorance is represented by uniform distributions. This has been disputed for decades if not even for centuries (cf. e.g. [9]). The calculations in this paper, however, based on the formulas (1) and (2), show clearly how ignorance is handled by ME-reasoning: *Non-knowledge* is realized as *non-occurrence*. In fact, the normalizing factor α_0 which includes the prior uniform probabilities is always cancelled in inferring posterior *conditional* probabilities. Only if *facts* are to be deduced, α_0 affects the posterior probabilities, but merely representing the probabilistic convention $P^*(\Omega) = 1$.

References

1. E.W. Adams. *The Logic of Conditionals*. D. Reidel, Dordrecht, 1975.
2. P.G. Calabrese. Deduction and inference using conditional logic and probability. In I.R. Goodman, M.M. Gupta, H.T. Nguyen, and G.S. Rogers, editors, *Conditional Logic in Expert Systems*, pages 71–100. Elsevier, North Holland, 1991.
3. R.T. Cox. Probability, frequency and reasonable expectation. *American Journal of Physics*, 14(1):1–13, 1946.
4. D. Dubois and H. Prade. Conditional objects and non-monotonic reasoning. In *Proceedings 2nd Int. Conference on Principles of Knowledge Representation and Reasoning (KR'91)*, pages 175–185. Morgan Kaufmann, 1991.
5. D. Dubois, H. Prade, and J.-M. Toucas. Inference with imprecise numerical quantifieres. In Z.W. Ras and M. Zemankova, editors, *Intelligent Systems - state of the art and future directions*, pages 52–72. Ellis Horwood Ltd., Chichester, England, 1990.
6. M. Goldszmidt, P. Morris, and J. Pearl. A maximum entropy approach to nonmonotonic reasoning. In *Proceedings AAAI-90*, pages 646–652, Boston, 1990.

7. I.J. Good. Maximum entropy for hypothesis formulation, especially for multidimensional contingency tables. *Ann. Math. Statist.*, 34:911–934, 1963.

8. A.J. Grove, J.Y. Halpern, and D. Koller. Random worlds and maximum entropy. *J. of Artificial Intelligence Research*, 2:33–88, 1994.

9. E.T. Jaynes. *Papers on Probability, Statistics and Statistical Physics*. D. Reidel Publishing Company, Dordrecht, Holland, 1983.

10. R.W. Johnson and J.E. Shore. Comments on and correction to "Axiomatic derivation of the principle of maximum entropy and the principle of minimum cross-entropy". *IEEE Transactions on Information Theory*, IT-29(6):942–943, 1983.

11. G. Kern-Isberner. Characterizing the principle of minimum cross-entropy within a conditional logical framework. Informatik Fachbericht 206, FernUniversität Hagen, 1996.

12. G. Kern-Isberner. Conditional logics and entropy. Informatik Fachbericht 203, FernUniversitaet Hagen, 1996.

13. S. Kullback. *Information Theory and Statistics*. Dover, New York, 1968.

14. S.L. Lauritzen and D.J. Spiegelhalter. Local computations with probabilities in graphical structures and their applications to expert systems. *Journal of the Royal Statistical Society B*, 50(2):415–448, 1988.

15. N.Rescher. *Many-Valued Logic*. McGraw-Hill, New York, 1969.

16. D. Nute. *Topics in Conditional Logic*. D. Reidel Publishing Company, Dordrecht, Holland, 1980.

17. J.B. Paris and A. Vencovská. A note on the inevitability of maximum entropy. *International Journal of Approximate Reasoning*, 14:183–223, 1990.

18. J. Pearl. *Probabilistic Reasoning in Intelligent Systems*. Morgan Kaufmann, San Mateo, Ca., 1988.

19. W. Rödder and C.-H. Meyer. Coherent knowledge processing at maximum entropy by spirit. In E. Horvitz and F. Jensen, editors, *Proceedings 12th Conference on Uncertainty in Artificial Intelligence*, pages 470–476, San Francisco, Ca., 1996. Morgan Kaufmann.

20. J.E. Shore. Relative entropy, probabilistic inference and AI. In L.N. Kanal and J.F. Lemmer, editors, *Uncertainty in Artificial Intelligence*, pages 211–215. North-Holland, Amsterdam, 1986.

21. J.E. Shore and R.W. Johnson. Axiomatic derivation of the principle of maximum entropy and the principle of minimum cross-entropy. *IEEE Transactions on Information Theory*, IT-26:26–37, 1980.

22. J.E. Shore and R.W. Johnson. Properties of cross-entropy minimization. *IEEE Transactions on Information Theory*, IT-27:472–482, 1981.

23. H. Thöne, U. Güntzer, and W. Kiessling. Towards precision of probabilistic bounds propagation. In D. Dubois, M.P. Wellmann, B. D'Ambrosio, and P. Smets, editors, *Proceedings 8th Conference on Uncertainty in Artificial Intelligence*, pages 315–322, San Mateo, Ca., 1992. Morgan Kaufmann.

24. J. Whittaker. *Graphical models in applied multivariate statistics*. John Wiley & Sons, New York, 1990.

Belief Functions with Nonstandard Values

Ivan Kramosil *

Institute of Computer Science
Academy of Sciences of the Czech Republic
Pod vodárenskou věží 2
182 07 Prague 8, Czech Republic
e–mail: kramosil@uivt.cas.cz,   fax: 42-2-85 85 789

Abstract. The notions of basic probability assignment and belief function, playing the basic role in the Dempster–Shafer model of uncertainty quantification and processing often called Dempster–Shafer theory, are generalized in such a way that their values are not numbers from the unit interval of reals, but rather infinite sequences of real numbers including those greater than one and the negative ones. Within this extended space it is possible to define inverse probability assignments and, consequently, to define the dual operation to the Dempster combination rule, also to assignments ascribing, to the whole space of discourse, the degree of belief "smaller than any positive real number" or "quasi-zero", in a sense; the corresponding inverse assignments than take "quasi-infinite" values. This approach extends the space of invertible, or non-dogmatic, in the sense introduced by Ph. Smets, basic probability assignments and belief functions, when compared with the other approaches suggested till now.

1 Introduction: Invertibility Problem for Belief Functions

Belief functions play the role of the main numerical characteristics, or quantitative degrees, of uncertainty in the so called Dempster–Shafer (D. S.) theory, or rather D. S.approach to uncertainty quantification and processing. Their combinatorial definitions, sufficient for our purposes, read as follows. Let S be a nonempty finite set (of possible actual states of a system, of possible answers to a question or solutions to a problem,...). *Basic probability assignment* (BPA) on S is a probability distribution on the power-set $\mathcal{P}(S)$ of all subsets of S, i. e., BPA on S is a mapping m which ascribes to each subset A of S a real number $m(A)$ from the unit interval $\langle 0, 1 \rangle$ of real numbers ($m : \mathcal{P}(S) \to \langle 0, 1 \rangle$, in symbols) in such a way that $\sum_{A \subset S} m(A) = 1$. The *belief function*, defined or induced by the BPA m, is the mapping $bel_m : \mathcal{P}(S) \to \langle 0, 1 \rangle$ such that, for each $A \subset S$

$$bel_m(A) = (1 - m(\emptyset))^{-1} \sum_{\emptyset \neq B \subset A} m(B), \qquad (1)$$

---

* This work has been sponsored by the grant no. A1030504 of the Grant Agency of the Acad. of Sci. of the CR.

supposing that $m(\emptyset) < 1$ holds for the empty subset $\emptyset$ of S. If $m(\emptyset) = 1$, bel_m is not defined.

Let m_1, m_2 be two BPA's on the same finite set S. The BPA m_3 on S, defined for each $A \subset S$ by

$$m_3(A) = \sum_{B \subset S,\, C \subset S,\, B \cap C = A} m_1(B)\, m_2(C) \tag{2}$$

is called the (non-normalized) *Dempster product* of the BPA's m_1, m_2, and denoted by $m_1 \oplus m_2$. As can be easily verified, m_3 is in fact a BPA on S, i.e., a probability distribution on $\mathcal{P}(S)$. The binary operation $\oplus$, ascribing a new BPA to each pair of BPA's, is called the (non-normalized) *Dempster combination rule*. The belief function defined by the BPA $m_1 \oplus m_1$ is then denoted by $bel_{m_1} \oplus bel_{m_2}$ and it is also called the Dempster product of the belief functions bel_{m_1} and bel_{m_2}.

The problem how to define an inverse operation to the Dempster combination rule for BPA's and for belief functions, i.e., an operation $\ominus$ such that $((m_1 \oplus m_2) \ominus m_2)(A) = m_1(A)$ would hold for (at least, some) BPA's m_1 on S and for each $A \subset S$, possesses a natural motivation and an intuitive interpretation. Or, the Dempster rule reflects, under some conditions of stochastical independence definable by the stochastical independence of the corresponding set-valued (generalized) random variables, a modification of one's system of degrees of beliefs when the subject in question becomes familiar with the degrees of beliefs of another subject and accepts the arguments on which the degrees of beliefs of the second subject are based. The inverse operation would enable, then, to erase the impact of this modification and to return back to one's original degrees of beliefs in the case when the reliability of the second subject is put into serious doubts.

In what follows, we shall limit ourselves to the Dempster products of BPA's, defining the corresponding operation for belief function in the same way as in (1) with m replaced by the Dempster product of the two (possibly generalized) BPA's generating the two belief functions to be combined together. As can be easily seen for the BPA's, and as it will be proved for the generalizations of BPA's which will be introduced and investigated below, there is a one-to-one relation between BPA's and belief functions, so that such a definition of operations over belief functions will be correct.

It is almost obvious, or it can be easily proved, that the operation $\oplus$ is commutative and associative, and the *vacuous* BPA m_V, defined by $m_V(S) = 1$, hence, $m_V(A) = 0$ for each $A \subset S$, $A \neq S$, is the (only) unit element in the space of BPA's with respect to $\oplus$. In symbols, the equalities

$$(m_1 \oplus (m_2 \oplus m_3))(A) = ((m_1 \oplus m_2) \oplus m_3)(A) \tag{3}$$
$$(m \oplus m_V)(A) = (m_V \oplus m)(A) = m(A),$$
$$(m_1 \oplus m_2)(A) = (m_2 \oplus m_1)(A)$$

are valid for all $BPA's$ m_1, m_2, m_3, for which the Dempster products in question are defined and for all $A \subset S$. We shall write $m_1 \equiv m_2$ for BPA's m_1, m_2, if $m_1(A) = m_2(A)$ holds for each $A \subset S$.

Given BPA's m_1, m_2, m_3 on S, let us suppose, for a while, that there exists a BPA m_2^{-1} such that $m_2 \oplus m_2^{-1} \equiv m_V$, hence, that there exists an inverse BPA to m_2 with respect to the operation $\oplus$ and to the unit element m_V in the space of BPA's. Then, due to (3)

$$(m_1 \oplus m_2) \oplus m_2^{-1} \equiv m_1 \oplus (m_2 \oplus m_2^{-1}) \equiv m_1 \oplus m_V \equiv m_1, \qquad (4)$$

so that we could define $m_1 \ominus m_2$ by $m_1 \oplus m_2^{-1}$. So, the problem of a dual operation to the Dempster combination rule transforms or reduces to the problem, whether and how it is possible to define, given a BPA m, its inverse BPA m^{-1} such that $m \oplus m^{-1} \equiv m_V$ would hold.

2 Finite Basic Signed Measure Assignments

As can be almost immediately seen, the only BPA to which there exists an inverse BPA, is the vacuous BPA m_V. Trivially, $m_V \oplus m_V \equiv m_V$, so that $m_V^{-1} \equiv m_V$ according to the common intuition behind the unit elements in multiplicative structures. If $m \not\equiv m_V$, them $m(S) < 1$ holds, so that, for each BPA m_1, (2) yields that

$$(m \oplus m_1)(S) = \sum_{B,C \subset S,\, B \cap C = S} m(B)\, m_1(C) = m(S)\, m_1(S) < 1, \qquad (5)$$

hence, $m \oplus m_1 \not\equiv m_V$.

An outcome from this situation consists in an appropriate extension of the notion of BPA. This pattern has been already several times followed in the history of mathematics: the notion of integer extends the space of natural numbers in such a way that substraction can be defined as a total operation over the space of integers. In the same way, the notion of rational number enables to extend the domain of the operation of division, even if this extension is not total – rational numbers with zero divisors cannot be defined.

The extension of the notion of BPA can be achieved at least in the two following ways. The first one, introduced and developed by Ph. Smets in [5], may be called *algebraic*. Let $A \subset S$, let $0 \le w \le 1$ be a real number. Then A^w denotes the so called *simple support* BPA, defined by $A^w(S) = w$, $A^w(A) = 1 - w$, $A^w(B) = 0$ for each $B \subset S$, $B \ne S$, $B \ne A$. As proved in [3], if m is a BPA over a finite set S such that $m(S) > 0$ holds ("non-dogmatic BPA", in the terms of [5]), the induced belief function bel_m can be defined by the Dempster product $\oplus_{A \subset S} A^{w(A)}$ for appropriate reals $w(A)$, here $\oplus_{A \subset S} A^{w(A)}$ denotes $A_1^{w(A_1)} \oplus A_2^{w(A_2)} \oplus \cdots \oplus A_m^{w(A_m)}$ for $m = 2^{\text{card}(S)}$ and for an ordering $A_1, A_2, \ldots, A_m$ of all subsets of S; due to (3) the choose of the ordering does not matter. Consequently, also the "belief functions" such that $w(A) > 1$ holds for some $A \subset S$ are defined, investigated, and interpreted, and the invertibility problem is solved for non-dogmatic BPA's.

Another approach can be called the *measure-theoretic* one. Each BPA over a finite set S can be defined by a set-valued (generalized) random variable U

defined as a measurable mapping which takes a fixed probability space $\langle \Omega, \mathcal{A}, P \rangle$ into the measurable space $\langle \mathcal{P}(S), \mathcal{P}(\mathcal{P}(S)) \rangle$. Let us recall that probability space consists of a nonempty space Ω, a nonempty σ-field $\mathcal{A}$ of subsets of Ω, and a σ-additive probability measure P on $\mathcal{A}$ (so that $P(\Omega) = 1$ and $P(\bigcup_{i=1}^{\infty} A_i) = \sum_{i=1}^{\infty} P(A_i)$ for each sequence $A_1, A_2 \ldots$ of mutually disjoint sets from $\mathcal{A}$. Such a set-valued random variable can be defined by

$$U(\omega) = \{s \in S : \rho(s, X(\omega)) = 1\} \tag{6}$$

for each $\omega \in \Omega$, where $X(\cdot)$ is a random variable defined on $\langle \Omega, \mathcal{A}, P \rangle$ and taking its values in the space of values of empirical data and/or observations (hence, X reflects the empirical data charged by uncertainty of stochastical nature), and ρ is a *compatibility relation* taking the values 0 or 1, i.e. $\rho(s, x) = 1$ means that $s \in S$ can be actual state of the system (the true answer to the question or the solution to the problem) supposing that x is the observed empirical value. Within this formal framework, a BPA m on a finite set S is defined by a set-valued random variable U, if the equality

$$m(A) = P(\{\omega \in \Omega : U(\omega) = A\}) \tag{7}$$

holds for each $A \subset S$. The induced belief function bel_m is then defined by

$$bel_m(A) = P(\{\omega \in \Omega : U(\omega) \subset A\} / \{\omega \in \Omega : U(\omega) \neq \emptyset\}) \tag{8}$$

for each $A \subset S$, supposing that this conditional probability is defined, i.e., supposing that $P(\{\omega \in \Omega : U(\omega) \neq \emptyset\})$ $(= 1 - m(\emptyset))$ is positive.

This definition of BPA and belief function can be extended to infinite sets S, if a nonempty σ-field $\mathcal{S} \in \mathcal{P}(\mathcal{P}(S))$ is fixed and the measurability of U with respect to the measurable space $\langle \mathcal{P}(S), \mathcal{S} \rangle$ is assured (cf. [2] for more detail). However, let us generalize the notions of BPA and belief function in a different way, replacing the probability measure in (7) and (8) by the so called *signed measure*. Signed measure defined on a measurable space $\langle \Omega, \mathcal{A} \rangle$ is a mapping μ which takes the σ-field $\mathcal{A}$ into the extended line of real numbers, in symbols, $\mu : \mathcal{A} \to R^* = (-\infty, \infty) \cup \{\infty\} \cup \{-\infty\}$, and possessing the following properties:

(i) μ is σ-additive, i.e. $\mu(\bigcup_{i=1}^{\infty} A_i) = \sum_{i=1}^{\infty} \mu(A_i)$ for each sequence $A_1, A_2, \ldots$ of mutually disjoint sets from $\mathcal{A}$ including the case when $\sum_{i=1}^{\infty} \mu(A_i) = \pm\infty$, the usual conventions for sums with infinite members being accepted;

(ii) $\mu(\emptyset) = 0$ for the empty subset $\emptyset$ of Ω;

(iii) μ takes at most one of the infinite values $\pm\infty$, so that no expressions like $\infty - \infty$ can occur, consequently, the definition of σ-additivity in (i) is correct.

As easily follows, the notion of BPA is generalized to that of *basic signed measure assignment* (BSMA); BSMA on a finite set S is a mapping $m : \mathcal{P}(S) \to R^*$ taking at most one of the infinite values $\pm\infty$. If $m(\emptyset) = 0$, then *signed belief function* bel_m, induced by m, is defined by $bel_m(A) = \sum_{B \subset A} m(B)$ for each $A \subset S$. When comparing this definition with that of BPA, we have strenghened the condition $m(\emptyset) < 1$ (positive probability of consistence) to that of $m(\emptyset) = 0$

(almost sure consistence) because we want to escape from the necessity to define conditional signed measures and because of the different nature of the sets of zero signed measure when compared with the case of probability measure (a set of zero signed measure can be the union of two or more disjoint sets of non-zero signed measure).

The following notions will help us to present the next statement. A BSMA m on a finite set S is called *finite*, if $-\infty < m(A) < \infty$ holds for each $A \subset S$. *Commonality function* induced by BSMA (by a BPA, in particular) m on a finite set S is a mapping $q : \mathcal{P}(S) \to R^*(\mathcal{P}(S) \to \langle 0, 1 \rangle$, in the case of BPA) such that $q(A) = \sum_{B \supset A} m(B)$ for each $A \subset S$. A BSMA m is called *invertible*, if $q(A) \neq 0$ holds for each $A \subset S$. This notion generalizes that of *non-dogmatic BPA*, introduced in [5], as for a BPA m the inequality $q(A) \neq 0$ (in fact, $q(A) > 0$) holds for all $A \subset S$ iff $m(S) > 0$. Let us recall, finally, that m_V is the vacuous BPA (hence, also BSMA) on S defined by $m_V(S) = 1$, $m_V(A) = 0$ for all $A \subset S$, $A \neq S$.

Theorem 1. Let m be a finite invertible BSMA on a finite set S. Let m^{-1} be the BSMA on S defined recurrently in this way:

$$m^{-1}(S) = (m(S))^{-1}, \tag{9}$$

$$m^{-1}(A) = (q(A))^{-1} \left(- \sum_{B,C \subset S,\, B \cap C = A,\, B \neq A} m^{-1}(B)\, m(C) \right), \tag{10}$$

if $A \subset S$, $A \neq S$. Then $m \boxplus m^{-1} \equiv m_V$ holds, where $\boxplus$ generalizes the Dempster combination rule to BSMA's in such a way that

$$(m_1 \boxplus m_2)(A) = \sum_{B,C \subset S,\, B \cap C = A} m_1(B)\, m_2(C) \tag{11}$$

for each finite BSMA's m_1, m_2 on a finite set S and for each $A \subset S$; $\equiv$ denotes the equality for each $A \subset S$. $\qquad \Box$

Proof. First of all, we have to prove that the definitions (9) and (10) are correct. Applying the condition $\sum_{B \supset A} m(B) \neq 0$ to the case $A = S$ we obtain that $m(S) \neq 0$ for invertible BSMA's. Hence, $(m(S))^{-1}$ is defined. The summation on the right-hand side of (10) goes over the sets B such that $B \cap C = A$ and $B \neq A$, hence, over the sets B, $A \subset B \subset S$, $\mathrm{card}(B) > \mathrm{card}(A)$. Consequently, $m^{-1}(A)$ is correctly defined in the recurrent way with respect to the decreasing cardinality of A. For the whole set S we obtain

$$(m \boxplus m^{-1})(S) = m(S)\, m^{-1}(S) = 1. \tag{12}$$

Let $A \subset S$, $A \neq S$. The set $\{\langle B, C \rangle : B, C \subset S,\ B \cap C = A\}$ can be decomposed into four disjoint subsets:

$$\{\langle A, A \rangle\},\ \{\langle A, B \rangle : A \subset B \subset S,\ A \neq B\},\ \{\langle B, A \rangle : A \subset B \subset S,\ A \neq B\}, \tag{13}$$
$$\{\langle B, C \rangle : B \subset S,\ C \subset S,\ B \neq A,\ C \neq A,\ B \cap C = A\}.$$

Using the relation (11) we obtain that

$$(m \boxplus m^{-1})(A) = m(A) m^{-1}(A) + \sum_{B \subset S, B \supset A, B \neq A} m(B) m^{-1}(A) + \qquad (14)$$

$$+ \sum_{B \subset S, B \supset A, B \neq A} m(A) m^{-1}(B) + \sum_{B, C \subset S, B \neq A, C \neq A, B \cap C = A} m(B) m^{-1}(C) =$$

$$= m^{-1}(A) \left[m(A) + \sum_{B \subset S, B \neq A, B \supset A} m(B) \right] + \sum_{B, C \subset S, B \supset A, B \neq A, C = A} m(C) m^{-1}(B) +$$

$$+ \sum_{B, C \subset S, B \neq A, C \neq A, B \cap C = A} m^{-1}(B) m(C),$$

as the last sum contains with each product $m(B) m^{-1}(C)$ also the product $m(C) m^{-1}(B)$. Or, if $\langle B, C \rangle$ is such that $B \neq A$, $C \neq A$ and $B \cap C = A$, the same holds for the pair $\langle C, B \rangle$. Consequently,

$$(m \boxplus m^{-1})(A) = m^{-1}(A) \left[\sum_{B \subset S, \ B \supset A} m(B) \right] + \qquad (15)$$

$$+ \sum_{B \subset S, \ C \subset S, B \supset A, B \neq A, C \supset A, B \cap C = A} m^{-1}(B) m(C).$$

Combining (15), (10) and the definition of $q(A)$ we obtain that $(m \boxplus m^{-1})(A) = 0$, if $A \subset S$, $A \neq S$, so that $m \boxplus m^{-1} \equiv m_V$ holds. The theorem is proved. $\square$

As an illustration, let us consider the conditioning or restriction of a BPA or BSMA m to a nonempty proper subset A of S. From the formal point of view it is defined by the Dempster product $m \boxplus m_{A,1}$, where $m_{A,1}$ is the simple support BPA A^0 in the notation introduced in [5], so that $m_{A,1}(A) = 1$, $m_{A,1}(B) = 0$ for each $B \subset S$, $B \neq A$. As $m_{A,1}(S) = 0$ for all $A \neq S$, $m_{A,1}$ is not invertible, so that the problem of elimination of the conditioning operation is unsolvable even in the space of finite BSMA's. However, taking an $\varepsilon > 0$ and defining the ε-quasiconditioning of m to A by the Dempster product $m \boxplus m_A^\varepsilon$, where m_A^ε is the BPA A^ε, i.e. $m_A^\varepsilon(S) = \varepsilon$, $m_A^\varepsilon(A) = 1 - \varepsilon$, and $m_A^\varepsilon(B) = 0$ for all $B \subset S$, $B \neq A$, $B \neq S$, we can easily see that m_A^ε is invertible, as $\sum_{B \supset A} m(B) \geq m(S) > 0$ holds for each $A \subset S$. Namely, $(m_A^\varepsilon)^{-1}(S) = 1/\varepsilon$, $(m_A^\varepsilon)^{-1}(A) = -(1 - \varepsilon)/\varepsilon$, and $(m_A^\varepsilon)^{-1}(B) = 0$ for each $B \subset S$, $B \neq A$, $B \neq S$. An easy calculation yields that $m_A^\varepsilon \boxplus (m_A^\varepsilon)^{-1} \equiv m_V$ holds, so that $(m \boxplus m_A^\varepsilon) \boxplus (m_A^\varepsilon)^{-1} \equiv m$ holds for each finite BSMA m on a finite set S.

As the following example proves, the invertibility problem for dogmatic BPA's cannot be, in general, solved neither when taking into account the BSMA's with infinite values and when defining the values of the expressions like $0 \cdot \infty$ and $0 \cdot (-\infty)$ in a no matter which way consistent with the product operation on finite real numbers. Let m be a BPA on a finite set S such that $m(S) = 0$ and $m(A) = 0$ for at least one $A \subset S$ such that $A = S - \{s\}$ for some $s \in S$.

Now, the equality $(m \boxplus m^{-1})(S) = m(S) m^{-1}(S) = 1$ can be satisfied only when taking $m^{-1}(S) = \infty$ and setting $0 \cdot \infty = \infty \cdot 0 = 1$. Consequently, we must set $0 \cdot (-\infty) = -1$, to be consistent with $(-1) \infty = -\infty$ and with the associativity of multiplication. For the subset A of S we obtain:

$$(m \boxplus m^{-1})(A) = \sum_{B,C \subseteq S, B \cap C = A} m(B) m^{-1}(C) = \tag{16}$$

$$= m(A) m^{-1}(A) + m(A) m^{-1}(S) + m^{-1}(S) m(A) =$$

$$= 0 \cdot m^{-1}(A) + 0 \cdot \infty + \infty \cdot 0 = 2 + 0 \cdot m^{-1}(A).$$

The demand that $(m \boxplus m^{-1})(A) = 0$ cannot be satisfied by no matter which value of $m^{-1}(A)$. If $m^{-1}(A)$ is finite, $0 \cdot m^{-1}(A) = 0$ and $(m \boxplus m^{-1})(A) = 2$, if $m^{-1}(A) = \infty$, then $0 \cdot m^{-1}(A) = 1$ and $(m \boxplus m^{-1})(A) = 3$, if $m^{-1}(A) = -\infty$, then $0 \cdot m^{-1}(A) = -1$ and $(m \boxplus m^{-1})(A) = 1$. Hence, $m \boxplus m^{-1} \not\equiv m_V$ in all cases.

The solution consisting in abandoning the deterministic character of algebraic operations over R^* and defining the value of $0 \cdot \infty$ and expressions like this not by a single value from R^* but rather by a subset of R^* would bring us very far beyond the intended scope of this paper and will not be investigated here.

3 Nonstandard Spaces of Values for BSMA's

In the rest of this paper we shall generalize the approximative solution to the invertibility problem for BPA's illustrated above by the ε-quasiconditioning approach. The weak point of this approximation consists in the fact that it introduces a new and ontologically independent parameter ε, the actual value of which cannot be justified only within the framework of the used mathematical formalism. Consequently, the subject (user) must choose some value on the grounds of her/his subjective opinion taking into consideration, e. g. the intended field of application and other extra-mathematical circumstances. Below, we shall present a model which enables to invert also BPA's ascribing to the whole space S a value "greater than 0 but smaller than any positive ε", in other words said, a "quasi-zero value", both these notions being given a correct mathematical sense. The corresponding inverse "generalized BPA" then will take "quasi-infinite values" smaller than ∞ but greater than any finite real number. Our approach will use some elementary ideas on which nonstandard (models of) analysis are based, but we prefer the "neutral" adjective "dynamic" to "nonstandard", in order not to involve associations, analogies, and perhaps expectations, going much far than our approach enables and justifies. Of course, no preliminaries concerning nonstandard analysis are assumed.

Let $\mathcal{R} = X_{i=1}^{\infty} R_i$, $R_i = (-\infty, \infty)$ for each $i \in \mathcal{N}^+ = \{1, 2 \ldots\}$, be the space of all infinite sequences of real numbers. For each $x \in \mathcal{R}$, x_i denotes its i-th member, so that $x = \langle x_i \rangle_{i=1}^{\infty}$. Given $x \in \mathcal{R}$, set $w(x) = \lim_{i \to \infty} x_i$ supposing that this limit value is defined and including the cases when $w(x) = \pm\infty$. Let $\mathcal{R}_c = \{x \in \mathcal{R} : w(x) \text{ is defined}\}$ be the space of all convergent infinite sequences of real numbers, let $\mathcal{R}_{cf} = \{x \in \mathcal{R}_c : -\infty < w(x) < \infty\}$ be the subspace of convergent infinite sequences with finite limit values. Let us define three following binary relations in $\mathcal{R}$:

(i) *identity*: given $x, y \in \mathcal{R}$, $x = y$ iff $x_i = y_i$ for each $i \in \mathcal{N}^+$;
(ii) *strong equivalence* (s. e.): given $x, y \in \mathcal{R}$, $x \approx y$ iff there exists $i_0 \in \mathcal{N}^+$ such that $x_i = y_i$ for each $i \geq i_0$;
(iii) *weak equivalence* (w. e.): given $x, y \in \mathcal{R}$, $x \sim y$ iff $w(x) = w(y)$.

Hence, $=$ and $\approx$ are relation on $\mathcal{R}$, $\sim$ is a relation on $\mathcal{R}_c$, undefined, if x or y are in $\mathcal{R} - \mathcal{R}_c$. As can be easily proved, $=$ and $\approx$ are *equivalence relations* on $\mathcal{R}$, $\sim$ is an *equivalence relation* on $\mathcal{R}_c$, so that the factor-spaces $\mathcal{R}/ = $ (it can be identified with $\mathcal{R}$), $\mathcal{R}_\approx = R/\approx$, and $R_{c\sim} = \mathcal{R}_c/\sim$ are well-defined. The strong equivalence is a coarsening of the identity relation and the weak equivalence is a coarsening of the strong equivalence on $\mathcal{R}_c$, hence $x = y$ implies $x \approx y$ for each $x, y \in \mathcal{R}$, and $x \approx y$ implies $x \sim y$ for each $x, y \in \mathcal{R}_c$.

Arithmetical operations in $\mathcal{R}$ will be defined in the pointwise way, so that $x + y = \langle x_i + y_i \rangle_{i=1}^{\infty}$ and $xy = \langle x_i y_i \rangle_{i=1}^{\infty}$ for each $x, y \in \mathcal{R}$. It follows easily that $\sum_{j=1}^{n} x^j = \left\langle \sum_{j=1}^{n} x_i^j \right\rangle_{i=1}^{\infty}$ and $\prod_{j=1}^{n} x^j = \left\langle \prod_{j=1}^{n} x_i^j \right\rangle_{i=1}^{\infty}$ holds for each finite sequence $x^1, x^2, \ldots, x^n$ of sequences from $\mathcal{R}$. Both the addition and multiplication operations can be uniquely extended to equivalence classes from $\mathcal{R}_\approx$ or $\mathcal{R}_{cf\approx}$. Let $[x]_\approx = \{y \in \mathcal{R} : y \approx x\}$, let $[x]_\sim = \{y \in \mathcal{R}_{cf} : y \sim x\}$. Setting $[x]_\approx + [y]_\approx = (x + y)_\approx$ and $[x]_\sim + [y]_\sim = [x + y]_\sim$, in the first case for all $x, y = \mathcal{R}$, in the other one for all $x, y \in \mathcal{R}_{cf}$, we obtain a correct extension of this arithmetical operation to $\mathcal{R}_\approx$ and $\mathcal{R}_{cf\sim}$, as the definitions do not depend on the chosen representants of the classes $[x]_\approx, [y]_\approx, [x]_\sim$, and $[y]_\sim$. Or, if $x_1 \approx x$ and $y_1 \approx y$, then $x_1 + y_1 \approx x + y$ obviously holds, and if $x_1 \sim x$, $y_1 \sim y$ holds for $x, y \in \mathcal{R}_{cf}$, then $x_1 + y_1 \sim x + y$ holds as well. The extension to finite sums is obvious. For $[x]_\approx \cdot [y]_\approx$ defined by $[xy]_\approx$, if $x, y \in \mathcal{R}$, and for $[x]_\sim \cdot [y]_\sim$ defined by $[xy]_\sim$, if $x, y \in \mathcal{R}_{cf}$, as well as for finite products, the situation is analogous. However, if $w(x) = \pm\infty$ or $w(y) = \pm\infty$, the definition of $[x]_\sim + [y]_\sim$ and $[x]_\sim \cdot [y]_\sim$ are not correct. As a counterexample, take simply $x = \langle n \rangle_{i=1}^{\infty}$, $y = \langle -\sqrt{n} \rangle_{n=1}^{\infty}$, and $y_1 = \langle -n \rangle_{n=1}^{\infty}$. Then $w(y) = w(y_1) = \infty$, so that $y \sim y_1$, but $w(x + y) = \infty \neq 0 = w(x + y_1)$. Similarly, for x as before, $y = \langle 1/n \rangle_{n=1}^{\infty}$, $y_1 = \langle 1/\sqrt{n} \rangle_{n=1}^{\infty}$, $w(y) = w(y_1) = 0$, but $w(xy) = 1 \neq \infty = w(xy_1)$. Also the extension of both the operations to infinite sums and products is impossible, take $x^j = \langle x_i^j \rangle_{i=1}^{\infty}$, such that $x_j^i = 1$, $x_i^j = 0$ for each $i \neq j$. Then $w(x^j) = 0$ for each $j \in \mathcal{N}^+$, but $w\left(\sum_{j=1}^{\infty} x^j\right) = 1$. We use intentionally the same symbols $+$ and $\cdot$ for operations in $R, \mathcal{R}, \mathcal{R}_\approx$, and $\mathcal{R}_{cf\sim}$, to emphasize their analogous role in all the cases. It should be always clear from the context, in which space these operations work.

Also the ordering relation can be extended from R to other spaces under consideration. If $x, y \in \mathcal{R}$, we write $x > y$, if $x_i > y_i$ holds for each $i \in \mathcal{N}^+$, and we write $x > .y$, if there exists $i_0 \in \mathcal{N}^+$ such that $x_i > y_i$ holds for each $i \geq i_0$, if $x, y \in \mathcal{R}_c$, we write $x \succ y$, if $w(x) > w(y)$ holds. The inequalities $\geq$ and $\geq$.

on $\mathcal{R}$, and $\succeq$ on $\mathcal{R}_c$, are defined as above, just replacing $x_i > y_i$ by $x_i \geq y_i$ and $w(x) > w(y)$ by $w(x) \geq w(y)$. The relations $>$. and $\geq$. can be extended to $\mathcal{R}_\approx$ and $\succ$, $\succeq$ to $\mathcal{R}_{\mathrm{cf} \sim}$, setting $[x]_\approx > .[y]_\approx$, if $x > .y$, and setting $[x]_\sim \succ [y]_\sim$, if $x \succ y$; both these definitions are correct, i.e., independent of the representants of the equivalence classes in question. Inequalities $\geq$ and $\succeq$ are extended to $\mathcal{R}_\approx$ and $\mathcal{R}_{\mathrm{cf} \sim}$ in the same way.

Let $a \in R$ be a real number, let $a^* = \langle a_i \rangle_{i=1}^\infty$, $a_i = a$ for each $i \in \mathcal{N}^+$ be the corresponding constant sequence from $\mathcal{R}_{\mathrm{cf}}$. The real line can be embedded into $\mathcal{R}_\approx$, when identifying each $a \in R$ with the equivalence class $[a^*]_\approx$, and R can be embedded into $\mathcal{R}_{\mathrm{cf} \sim}$, when identifying a with $[a^*]_\sim$. However, there is an important difference. The second mapping takes R *onto* $\mathcal{R}_{\mathrm{cf} \sim}$, as for each $x \in \mathcal{R}_{\mathrm{cf}}$ there is a real number a, namely, $a = w(x)$, which is mapped on $[x] \in \mathcal{R}_{\mathrm{cf} \sim}$. Contrary to this fact, there are many $x \in \mathcal{R}$ such that no $a \in R$ is mapped onto $[x]$, take, e.g. $x = \langle n \rangle_{n=1}^\infty$ or $y = \langle 1/n \rangle_{n=1}^\infty$. Denoting the classes from $\mathcal{R}_\approx$ which are images of real numbers from R, as *standard real numbers*, $[x]$ and $[y]$ defined above are examples of *nonstandard real numbers*. In particular, x is an example of *quasi-infinite nonstandard real number*, as $x > .[a^*]$ obviously holds for each $a \in R$, and y can serve as an example of *quasi-zero nonstandard real number*, as $y > .[0^*]$, but also $[a^*] > .y$ for all $a \in R$, $a > 0$, can be easily verified.

In what follows, when defining a generalization of the notion of basic signed measure assignment, using the apparatus just developed, we have to choose between two ways. Either, we can take $\mathcal{R}$ as the space of values and to prove some results up to the equivalence relation $\approx$ in $\mathcal{R}$. Or, we can take $\mathcal{R}_\approx$ as the space of values and to prove identity of some values in the factor-space $\mathcal{R}_\approx$ (Taking $\mathcal{R}_{\mathrm{cf} \sim}$ as the space of values, we would have to abandon the desirable enrichment of the real line R). Even if the second way is more elegant from the purely mathematical point of view, we shall prefer the first approach in order to minimize the technical difficulties involved by our considerations and argumentations below, and to keep the intuition behind as immediate and lucid as possible.

4 Basic Dynamic Assignments

Definition 1. Let S be a nonempty finite set. *Basic dynamic assignment* (BDA) on S is a mapping $m : \mathcal{P}(S) \to \mathcal{R} = \mathbb{X}_{i=1}^\infty R_i$, $R_i = (-\infty, \infty)$, $i = 1, 2, \ldots$. *Dynamic non-normalized belief function* induced by BDA m on S is a mapping $bel_m : \mathcal{P}(S) \to \mathcal{R}$ such that $bel_m(A) = \sum_{B \subset A} m(B)$ for each $A \subset S$. BDA m on S is called *convergent* BDA (CBDA), if $m(A) \in \mathcal{R}_c$ for each $A \subset S$. If m is a BPA or a finite BSMA on S, then m^* is the BDA on S defined by $m^*(A) = (m(A))^*$ for each $A \subset S$, if m is a BDA on S, then $m^* \equiv m$, i.e., $m^*(A) = m(A)$ for each $A \subset S$. Let us recall that $a^* = \langle a, a, \ldots \rangle$ for each $a \in R = (-\infty, \infty)$. Let m_1, m_2 be BPA's, BSMA's or BDA's. Then m_1 is *strongly equivalent* to m_2, $m_1 \approx m_2$ in symbols, if $m_1^*(A) \approx m_2^*(A)$ holds for each $A \subset S$; m_1 is *weakly equivalent* to m_2, $m_1 \sim m_2$ in symbols, if $m_1^*(A) \sim m_2^*(A)$ holds for each $A \subset S$.

Let m_1, m_2 be BDA's on a finite set S. Their *Dempster product* $m_1 \boxplus m_2$ is the BDA on S defined by the relation

$$(m_1 \boxplus m_2)(A) = \sum_{B,C \subseteq S,\, B \cap C = A} m_1(B)\, m_2(C) \tag{17}$$

for each $A \subset S$. If m is BDA on S, then m_i is the finite BSMA on S defined by $(m_i)(A) = (m(A))_i$ for each $i \in \mathcal{N}^+$. Let us recall that x_i is the i-th member of a sequence $x = \langle x_i \rangle_{i=1}^\infty \in \mathcal{R}$. Vacuous BDA on S is such m_V that $m_V(S) = 1^*$ and $m_V(A) = 0^*$ for each $A \subset S$, $A \neq S$, obviously, $(m_V)_i$ is the vacuous BSMA (and the vacuous BPA) on S for each $i \in \mathcal{N}^+$. The *dynamic commonality function* induced by a BPA on a finite set S is the mapping $q_m : \mathcal{P}(S) \to \mathcal{R}$ defined by $q_m(A) = \sum_{B \supset A} m(B)$ for each $A \subset S$. □

The following assertion deals with the particular case of BSMA's, but it will play the role of an important auxiliary statement in the main assertion (Theorem 3) below.

Theorem 2. For each finite BSMA m on a finite set S, and for each $\varepsilon > 0$, there exists a BSMA m_0 on S such that m_0^{-1} is defined and $|m(A) - m_0(A)| < \varepsilon$ holds for each $A \subset S$. □

Proof. Let m be a finite BSMA over a finite set S, let $\varepsilon > 0$ be given, let $0 < \varepsilon_1 < \varepsilon$. If the inequality $\sum_{B \supset A} m(B) \neq 0$ is satisfied for all $A \subset S$, then m^{-1} is defined and for $m_0 \equiv m$ the assertion trivially holds. If this is not the case, set

$$i_1 = \max\left\{ n \in \mathcal{N}^+ : \exists A \subset S,\, \mathrm{card}(A) = n,\, \sum_{B \supset A} m(B) = 0 \right\}. \tag{18}$$

Set also, for each $A \subset S$,

$$\begin{aligned}
m_1(A) &= m(A) + \varepsilon_1, \text{ if } \mathrm{card}(A) = i_1 \text{ and } \sum_{B \supset A} m(B) = 0, \\
m_1(A) &= m(A) \qquad \text{otherwise.}
\end{aligned} \tag{19}$$

In particular, (19) implies that $m_1(A) = m(A)$ for all $A \subset S$ such that $\mathrm{card}(A) > i_1$ or $\mathrm{card}(A) < i_1$ holds.

Let $A \subset S$ be such that $\mathrm{card}(A) > i_1$ holds. Then $\sum_{B \supset A} m_1(B) = \sum_{B \supset A} m(B) \neq 0$ is valid, due to the definition of i_1, as for each $B \supset A$ the relations $\mathrm{card}(B) > i_1$ and $m_1(B) = m(B)$ follow. Let $A \subset S$ be such that $\mathrm{card}(A) = i_1$ and $\sum_{B \supset A} m(B) \neq 0$. Then, again, $m_1(B) = m(B)$ for all $B \supset A$, but now including also the case $A = B$ due to (19), so that $\sum_{B \supset A} m_1(B) = \sum_{B \supset A} m(B) \neq 0$. Finally, let $A \subset S$ be such that $\mathrm{card}(A) = i_1$ and $\sum_{B \supset A} m(B) = 0$. Then $m_1(B) = m(B)$ for each $B \supset A$, $B \neq A$, but $m_1(A) = m(A) + \varepsilon_1$, so that

$$\sum_{B \supset A} m_1(B) = \sum_{B \supset A,\, B \neq A} m_1(B) + m_1(A) = \sum_{B \supset A,\, B \neq A} m(B) + m(A) + \varepsilon_1 = \tag{20}$$

$$= \sum_{B \supset A} m(A) + \varepsilon_1 = \varepsilon_1 \neq 0.$$

Consequently, for each $A \subset S$, $\operatorname{card}(A) \geq i_1$, the inequality $\sum_{B \supset A} m_1(B) \neq 0$ holds.

By induction, let us apply the same modification to the BSMA m_1. Set

$$i_2 = \max\left\{ n : \exists A \subset S,\ \operatorname{card}(A) = n,\ \sum_{B \supset A} m_1(B) = 0 \right\}, \qquad (21)$$

and define m_2 by (19), just with m replaced by m_1. Obviously, $i_2 < i_1$ and $\sum_{B \supset A} m_2(B) \neq 0$ for all $A \subset S$, $\operatorname{card}(A) \geq i_2$, by the same way of reasoning as above. Moreover, $m_2(B) = m_1(B)$ for all $B \subset S$ such that $\operatorname{card}(B) > i_2$, so that $m_2(B)$ for all $B \subset S$ with $\operatorname{card}(B) > i_1$. So, repeating this induction step n_0-times for an appropriate $n_0 \leq \operatorname{card}(S)$, we arrive at a BSMA m_{n_0} such that $\sum_{B \supset A} m_{n_0}(B) \neq 0$ holds for each $A \subset S$, consequently, $m_{n_0}^{-1}$ is defined. For each particular $A \subset S$ the original value $m(A)$ is changed at most once during the procedure leading from m to m_{n_0}, so that either $m_{n_0}(A) = m(A)$, or $m_{n_0}(A) = m(A) + \varepsilon_1$ for each $A \subset S$. Hence, for each $A \subset S$ the inequality $|m(A) - m_{n_0}(A)| \leq \varepsilon_1 < \varepsilon$ immediately follows, so that, setting $m_{n_0} = m_0$, we can conclude the proof. $\qquad\square$

Theorem 3. Let m be a CBDA on a finite set S. Then there exist BDA's m_1 and m_1^{-1} on S such that $m \sim m_1$ and $m_1 \boxplus m_1^{-1} \equiv m_V$. In other words: for each CBDA m there exists an invertible BDA weakly equivalent to m. $\qquad\square$

Proof. Let m be CBDA on a finite set S. For each $i \in \mathcal{N}^+$, let m_i^0 be such a BSMA on S, that $|m_i(A) - m_i^0(A)| < 1/i$ and $q_i(A) \neq 0$ holds for each $A \subset S$, such m_i^0 exists due to Theorem 2. Let m^0 be the BDA on S such that $m^0(A) = \langle m_i^0(A) \rangle_{i=1}^\infty$ for each $A \subset S$. As $w(m(A)) = \lim_{i \to \infty} m_i(A)$ exists for each $A \subset S$, $w(m^0(A)) = \lim_{i \to \infty} m_i^0(A) = w(m(A))$ exists as well, so that $m \sim m_0$ holds. Let $(m_i^0)^{-1}$ be the BSMA on S defined by (9) and (10) (cf. Theorem 1 above) for the BSMA m_i^0. Then $(m_i^0 \boxplus (m_i^0)^{-1}) \equiv m_V$ (the vacuous BSMA on S), so that, setting $(m^0)^{-1}(A) = \langle (m_i^0)^{-1}(A) \rangle_{i=1}^\infty$ we obtain, that $m^0 \boxplus (m^0)^{-1} \equiv m_V$, here m_V is the vacuous CBDA on S. The assertion is proved. $\qquad\square$

Let us illustrate this statement by the example of the single support BPA m_A defined by $m_A(A) = 1$, $m_A(B) = 0$ for each $B \subset S$, $B \neq A$, where A is a subset of S; this BPA is used in the conditioning operation. Here we can define $m_A^0 \sim m_A$ in such a way that $(m_A^0)_i(A) = 1 - (1/i)$, $(m_A^0)_i(S) = 1/i$, $(m_A^0)(B) = 0$ for each $B \subset S$, $B \neq S$, $B \neq A$. Let $(m_A^0)^{-1}$ be the BDA defined by $((m_A^0)^{-1})_i(S) = i$, $((m_A^0)^{-1})_i(A) = -(i-1)$, $((m_A^0)^{-1})_i(B) = 0$ for all $B \subset S$, $B \neq S$, $B \neq A$, $i \in \mathcal{N}^+$. An easy calculation then yields that $m_A \sim m_A^0$ and $m_A^0 \boxplus (m_A^0)^{-1} \equiv m_V$ hold.

Items [3] and [4] in the list of references below can serve as good introductory surveys into the domain of Dempster–Shafer theory. In [1] the reader can find the basic information concerning measures, probability measures and signed measures.

5 Conclusions

As we already introduced and briefly discussed in the very end of Chapter 1, the motivation for our effort originated from the idea to define an operation dual to the Dempster combination rule in order to be able to eliminate the impact of beliefs, later proved as unreliable, to the combined degrees of beliefs. The solution proposed here consists in an appropriate extension of the space of basic probability assignments and corresponding belief functions and copies the paradigm used already many times in mathematics. E. g., negative integers have been introduced in order to enable to define substraction as total operation and analogous motivations led to the notions of rational, real and complex numbers. The particular problem of deconditionalization cannot be solved neither within this extended space of basic signed measure assignments and signed belief functions, so that we have followed the way similar to that used in nonstandard analysis and enabling to define basic signed measure assignments and signed belief functions with "almost zero" or "almost infinite" values which still enable to solve the invertibility and, consequently, the deconditionalization problem. The author believes that at least within the frameworks of mathematical formalisms such results can be seen as interesting and useful applications of one mathematical theory into another one, sheding perhaps a new light to both of them.

References

1. P. R. Halmos. *Measure Theory*. D. Van Nostrand, New York – Toronto – London, 1950.
2. I. Kramosil. Believability and plausibility functions over infinite sets. *International Journal of General Systems*, 23(2):173–198, 1994.
3. G. Shafer. *A Mathematical Theory of Evidence*. Princeton University Press, Princeton, New Jersey, 1976.
4. P. Smets. The representation of quantified belief by the transferable belief model. Technical Report TR/IRIDIA/94–19.1, Institut de Récherches Interdisciplinaires et de Dévelopments en Intelligence Artificielle, Université Libre de Bruxelles, 1994, 49 pp.
5. P. Smets. The canonical decomposition of a weighted belief. In *IJCAI–95 – Proceedings of the 14th International Joint Conference on Artificial Intelligence*, Vol. 2, pp. 1896–1901, Montréal, Québec, Canada, August 20 – 25 1995.

Error Tolerance Method in Multiple-Valued Logic

Soowoo Lee

Technische Hochschule Darmstadt, FB Informatik, FG Intellektik
Alexanderstr. 10, 64283 Darmstadt, Germany
email: lee@intellektik.informatik.th-darmstadt.de

Abstract. Because standard logic is based on only two truth values, it is not suitable for reasoning with uncertainty or vague knowledge. Such knowledge requires nonstandard logics, for example fuzzy logic, multiple-valued logic, probabilistic and possibilistic logic. We provide a logical environment and a proof procedure for representing and reasoning about this kinds of knowledge.

1 Introduction

We give a new proof procedure *error tolerance method* (*ETM*) in multiple-valued logic which is based on the connection method [1] for classical logic. The formalism of *ETM* is a family of signed logics which has been developed independently by Hähnle [4, 5] and Murray [10, 11]. Not only sharp information but also vague formulas can be proved in the *ETM*.

While other truth value spaces may be of interest as well, in this paper we consider the unit interval [0,1] as a set of truth value. Information that is given by percentage numbers assigned to propositions, like in the probabilistic logic can be represented in the unit interval, as well as the value range of membership functions in the sense of fuzzy logic. Truth values can be assigned to both sides of an implication (that is premise and conclusion), and differences between connected literals in a formula determine the validity grade of this formula.

Some basic definitions are given in Section 2. In Section 3 we give a characterization theorem which provides a formal criterion for the validity grade of a formula. Different kinds of proof procedures can be developed with the help of this characterization theorem. We provide one such proof procedure (named Δ_1^0) which is sound and complete wrt. the characterization theorem. In Section 4 we introduce widely studied related works, namely signed logic, annotated logic [6, 3] and fuzzy operator logic [12, 13]. With a simple example we illustrate that our approach has advantages over these logics in this truth value space.

2 Basic Definition

Let $\mathcal{A}$ be a set of atoms, the unit interval [0,1] be a set of truth values, and let I be an Interpretation that assigns to each atom a truth value in [0,1]. *Sign* is a regular subset of the unit interval [0,1]. A regular subset is an upset or its complement. For example, [i,1] is the upset of i, where $i \in [0,1]$ (written ↑i) and [0,i) is the complement of [i,1] (written (↑i)′). We write ↑i:a to denote a signed atom, for $a \in \mathcal{A}$. Formulas can be built in the usual way from the set of signed atoms with connectives ¬, ∧ and ∨.

Definition 2.1 Let $a \in \mathcal{A}$ and F, G Formulas. An interpretation I is said to *satisfy*
1. the signed atom ↑i:a iff $I(a) \in$ ↑i;
2. the signed literal ¬(↑i:a) iff I does not satisfy ↑i:a;
3. the disjunction $F \vee G$ if it satisfies F or G;
4. the conjunction $F \wedge G$ if it satisfies both F and G.

The logical consequence $F \rightarrow G$ can be interpreted $\neg F \vee G$, like in classical logic. "I satisfies ¬(↑i:a)" is equivalent to "I satisfies (↑i)′:a", because "I does not satisfy ↑i:a" means $I(a) \notin$ ↑i and therefore $I(a) \in$ (↑i)′. Analogously "I satisfies ¬(¬(↑i:a))" is equivalent to "I satisfies ↑i:a".

Definition 2.2 ↑s (respectively (↑s)′) is called *support set* of ↑i (respectively (↑i)′) iff ↑i $\subseteq$ ↑s $\subset$ [0,1] (respectively (↑i)′ $\subseteq$ (↑s)′ $\subset$ [0,1]).

For example ↑0.6 is a support set of ↑0.8 and (↑0.8)′ a support set of (↑0.6)′. In the following we define only the upset (↑i), but its complement ((↑i)′) can be defined analogously.

Definition 2.3 Let W be a set of regular subsets, then a *distance function* $\delta : W \times W \rightarrow [0,1]$ is defined as $\delta(↑x,↑y) = |x - y|$.

We call *support distance* the distance between an upset and its support set. For example, $\delta(↑0.6,↑0.8) = 0.2$.

Definition 2.4 Let ↑i be a sign and α a value, then a set of support sets can be defined by the function ω: $\omega(\alpha,↑i) = \{$ ↑s | ↑s support set of ↑i, and $\delta(↑s,↑i) \le \alpha/2$ $\}$. This set of support sets is called a *support set space*.

For example, if $i = 0.8$ and $\alpha = 0.4$, then the support set space is $\omega(0.4,↑0.8) = \{$ ↑s | $0.6 \le s \le 0.8$ $\}$. Based on the support set space we extend the base case of definition 2.1 as follows:

Definition 2.5 An interpretation I satisfies ↑i:a with *error tolerance* α iff there is a support set ↑s in the support set space $\omega(\alpha,↑i)$ such that I satisfies ↑s:a.

If an interpretation I satisfies ↑i:a with α, but I does not satisfies ↑i:a with γ, for any $\gamma < \alpha$, then we call α *minimal error tolerance* and denote this value by α_{min}. The following lemma shows this property. Instead of "with error tolerance α" we often say just "with α", if it causes no confusion.

Lemma 2.1 Suppose an Interpretation I satisfies ↑i:a with α, then I satisfies ↑i:a with β, for all $\beta \geq \alpha$.

<u>Proof</u> : I satisfies ↑i:a with α means that we can find a support set ↑s of ↑i in a support set space $\omega(\alpha, \uparrow i)$, so that I satisfies ↑s:a. This support set ↑s can always be found in the support set space $\omega(\beta, \uparrow i)$, because

$$\omega(\beta, \uparrow i) \supseteq \omega(\alpha, \uparrow i)$$

Therefore I satisfies ↑i:a with β.

Definition 2.1 can then be extended to error tolerance as follows:

Definition 2.6 Let F and G be formulas.
I satisfies $(F \vee G)$ with α iff I satisfies F with α or I satisfies G with α.
I satisfies $(F \wedge G)$ with α iff I satisfies F with α and I satisfies G with α.
I satisfies $\neg F$ with α iff I does not satisfies F with α.
I satisfies $(F \rightarrow G)$ with α iff I satisfies $(\neg F \vee G)$ with α.

Definition 2.7 A formula F is *valid* with α iff for all interpretation I, I satisfies F with α.

For "F is valid with α" we say also "F is $(1 - \alpha)$-valid".

The following definitions have to be introduced, because we will use the direct proof technique of matrix form which is oriented on the connections. For more details we refer to [1, 2]. Let F be a *matrix* (a matrix is a set of clauses) of the form $\{C_1, \ldots, C_n\}$ with $C_i = \{L_{i1}, \ldots, L_{im_i}\}$. Concerning *direct proof* the following representation applies:
• a literal stays a literal,
• a clause is the conjunction of literals and
• a formula the disjunction of clauses.
In this case we speak of *positive representation* (for the *negative representation* we mean the analogous inverse representation). Within the positive representation Horn clause is defined as follows:

Definition 2.8 A clause is called a *Horn clause* if it contains at most one negation. A matrix is called a *Horn matrix* if all its clauses are Horn clauses.

Definition 2.9 Let $\{C_1, \ldots, C_n\}$ be a matrix, then each set $\{L_1^{C_1}, \ldots, L_n^{C_n}\}$ is a *path* through the matrix (L^C denotes a literal L in a clause C).

So the matrix

$$\begin{bmatrix} \neg a & \neg b & \neg c \\ a & b & c \end{bmatrix}$$

has the following four paths

$$\{a, \neg a, \neg b, \neg c\}, \{a, \neg a, c, \neg c\}, \{a, b, \neg b, \neg c\}, \{a, b, c, \neg c\}.$$

Definition 2.10 A *connection* (or a complementary literal) in a matrix is a subset of the form $\{a, \neg a\}$ of paths through the matrix. A set of connections in a matrix is called a *mating*. A mating is called *spanning*, if each path contains at least one of its connection.

The mating in the following matrix is spanning.

$$\begin{bmatrix} & \neg a & \neg b & \neg c \\ a & b & c & \end{bmatrix}$$

Based on the above definitions we extend the notion of the connection as follows.

Definition 2.11 A Δ-*connection* in a matrix is a subset of the form $\{\uparrow i{:}a, (\uparrow j)'{:}a\}$ of paths through the matrix. An *error distance* (*ED*) of a Δ-connection is the function $E\delta : W \times W \to [0,1]$ with

$$E\delta(\uparrow i, (\uparrow j)') = \begin{cases} |i - j| & \text{if } i > j \\ 0 & \text{otherwise.} \end{cases}$$

For example, $\{\uparrow 0.8{:}a, (\uparrow 0.6)'{:}a\}$ and $\{\uparrow 0.5{:}b, (\uparrow 0.9)'{:}b\}$ are Δ-connections. The error distance of the first connection is $E\delta(\uparrow 0.8, (\uparrow 0.6)') = 0.2$ and the second $E\delta(\uparrow 0.5, (\uparrow 0.9)') = 0$.

Definition 2.12 A set of Δ-connections in a matrix is called a Δ-*mating* of the matrix.

Definition 2.13 A Δ-mating of a matrix is called Δ-*spanning*, if each path through the matrix contains at least one of its Δ-connection.

The Δ-mating in the following matrix is Δ-spanning.

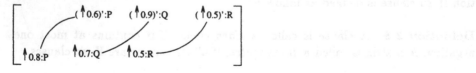

3 Characterization Theorem of *ETM* and Proof Procedure

The connection method is a family of proof procedures which is based on the characterization theorem for classical logic. This theorem states that a matrix is valid iff its mating is spanning. A path which does not contain any connection, can be interpreted as a falsefiable interpretation (in the sense of a counter example). In this paper we introduce the *error tolerance method* (*ETM*). We generalize the characterization theorem (named Δ-*characterization theorem*) that enable us to determine the validity degree of a matrix which can be used to represent uncertain propositions. We also provide a sound and complete proof procedure (named Δ_1^0) for Horn matrices which is based on the Δ-characterization theorem.

At first we explain the meaning of $max_F(min_{path}(ED))$, where $min_{path}(ED)$ is the smallest error distance of a path if this path contains Δ-connections. For example, $min_{path}(ED) = 0.1$ for the path $\{\uparrow0.9{:}a,(\uparrow0.7)'{:}a,(\uparrow0.8)'{:}a\}$, since there exist two Δ-connections $\{\uparrow0.9{:}a,(\uparrow0.7)'{:}a\}$ and $\{\uparrow0.9{:}a,(\uparrow0.8)'{:}a\}$ with the error distances 0.2 and 0.1. There are many such values $(min_{path}(ED))$ in a matrix F, if F has many paths. We write $max_F(min_{path}(ED))$ to denote the largest value of them.

Theorem (Δ-characterization theorem)[1] A Δ-mating of a matrix F is Δ-spanning and $max_F(min_{path}(ED)) = \alpha$ iff F is valid with α_{min}.

Just like in the classical characterization theorem it is understandable that if Δ-mating of F is Δ-spanning then F is valid and vice versa. The Δ-characterization theorem additionally states that the validity degree of F depends on the error distance $(max_F(min_{path}(ED)))$. Do we have to represent all paths in a matrix to obtain this value? If we must do that, it has no sense for the application because of computational complexity. The proof procedure Δ_1^0 is similar to the extension proof procedure for classical logic [1, 2]. The extension calculus checks paths through a matrix systematically for the occurrence of connections. If there is a path without a connection then the matrix is invalid, otherwise valid. We consider the following matrix:

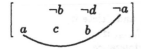

We don't need to represent all four paths in this matrix, but only the connection between a and $\neg a$. This is the basic idea of the extension calculus. Δ_1^0 works the same way but computes error distances for each connection, and gives the largest error distance of all connections as a result. If more than one proof exists in the matrix, then Δ_1^0 gives the smallest proof as a result. Once more, Δ_1^0 returns the largest error distance of a proof, and if alternative proofs exists, then the smallest value among them. The algorithm of Δ_1^0 is given as follows.

————————————< Begin of the algorithm Δ_1^0 > ————————————

F: Input matrix;

---

[1] The proof for this theorem is given in [7]

WAIT: Sub goal (path to proof) that is still open;
ALT_k: Alternative connections set with configuration at that time;
ALT_s: Alternative start clauses set with configuration at that time;
HIS: Smallest error tolerance value with proof step (for representing);
ED_{mas}: Larger error distance between now and previous error distances;
$X \leftarrow Y$: the value assignment of Y to X;
push(*STACK*, *item*): the value of which is *STACK* extended by *item* at the end;
pop(*STACK*): the value of which is (*STACK*, *item*) where *item* denotes the previous entry
at the end, and *STACK* denotes the previous value of *STACK* but with this *item* deleted;
{ }: denotes the empty list.

$D \leftarrow F$; $WAIT \leftarrow \{\}$; $ALT_k \leftarrow \{\}$; $ALT_s \leftarrow \{\}$; $HIS \leftarrow \{\}$; $ED_{mas} = 0$.

If $D = \{\}$
Then Return: D is valid 0.
save possible start clauses set in ALT_s;
choose clause $C \in ALT_s$; delete C from ALT_s

So long as $D \neq \{\}$

choose literal $L \in C$; delete L from C
If $C \neq \{\}$
Then $WAIT \leftarrow$ push($WAIT$, (C, D))
If exists a clause $K \in D$ and a literal $L' \in K$ that L' and L can be connected
Then delete K from D;
    delete a literal from K that is with L connected;
    update the ED_{mas} with the error distance between L and L', if this is lager;
    save alternative connections in ALT_k
    If $K = \{\}$
    Then If $WAIT = \{ \}$
        Then If $ALT_k = \{ \}$
            Then Return: D is valid with α
                (α is the smallest value between ED_{mas}
                and errortolerance value from *HIS*).
            Else choose a new configuration from ALT_k and delete it;
                update errortolerance in *HIS* with ED_{mas}, if
                this is smaller then the previous errortolerance
        Else ($WAIT$, (C, D)) $\leftarrow$ pop($WAIT$)
    Else $C \leftarrow K$
Else If $ALT_k \neq \{\}$
Then choose a new configuration from ALT_k and delete it
Else If $ALT_s = \{ \}$
    Then If $HIS = \{ \}$
        Then Return: invalid.
        Else Return: valid with α

(α is the smallest value between ED_{max}
and errortolerance value from HIS).
Else choose a new configuration from ALT_i and delete it

is $D = \{\}$ **Return:** D is invalid.

────────────────$<$ End of algorithmus Δ_1^0 $>$ ────────────────

Assume that Δ_1^0 is applied to a matrix, and we have "valid with α" as a result. For the soundness we only need to prove that the mating of F is Δ-spanning and $max_F(min_{Path}(KD)) = \alpha$. Then the Δ characterization theorem guarantees that F is valid with error tolerance α. Analoguesly we can argument for the completeness.

Theorem (Soundness and Completeness of Δ_1^0) [2] Let F be a Horn matrix. Δ_1^0 is applied to the matrix F, then return "valid with α" iff the mating of the matrix F is Δ-spanning and $max_F(min_{Path}(ED)) = \alpha$.

The inference mechanism of Δ_1^0 is explained with the following examples. For the sake of simplicity, we write only i for the upset $\uparrow$i: in these examples.

Example 3.1

$$Fact : \quad 0.7a$$
$$Rule : \quad 0.8b \leftarrow a$$
$$Rule : \quad 0.9c \leftarrow 0.9b$$
$$Query : ?c$$

Δ_1^0 returns the value 0.3 as error tolerance because 0.3 is the largest error distance (ED) between 0.7a and a.

Example 3.2

$$Fact : \quad 0.9a$$
$$Fact : \quad 0.7b$$
$$Fact : \quad 0.8c$$
$$Rule : \quad d \leftarrow a, 0.7b, c$$
$$Query : ?d$$

Δ_1^0 returns the value 0.2 (ED between 0.8c and c) as error tolerance because all propositions in the premise (a, 0.7b and c) of the last rule must be satisfied (conjunction).

────────
[2] The proof for this theorem is given in [7]

Example 3.3

$$Fact: \quad 0.7a$$
$$Fact: \quad 0.8b$$
$$Rule: \quad c \leftarrow 0.8a$$
$$Rule: \quad c \leftarrow b$$
$$Query: ?c$$

In this case there are two proofs, one error tolerance with 0.2 (ED between b and 0.8b) and the other 0.1 (ED between 0.8a and 0.7a). Then Δ_1^0 returns the smaller value 0.1 as result. This inference is intuitively plausible because people are usually interested in the better solution.

The following example shows that Δ_1^0 can be used for the classical proof procedure. In this case a certain proof with error tolerance value 0 exists.

Example 3.4

$$Fact: \quad r$$
$$Rule: \quad q \leftarrow r$$
$$Rule: \quad p \leftarrow q, r$$
$$Query: ?p, q$$

4 Related Work

In this section we introduce inference mechanism of widely studied related works, fuzzy operator logic, annotated logic and signed logic.

4.1 Fuzzy Operator Logic

The unit interval [0,1] is also the truth value space in the fuzzy operator logic (FOpL). The fuzzy operator (designated λ) ranges over the interval [0,1]. Formulas in FOpL can be enhanced with this fuzzy operator as follows.

Let a be an atom, and F and G fuzzy formulas. Then the following are also fuzzy formulas:

1. $\lambda a, \lambda F, \lambda G$
2. $\lambda(F \wedge G)$
3. $\lambda(F \vee G)$
4. $\forall_x F$
5. $\exists_x F$

For example, a fuzzy formula F is the same as fuzzy formula $1F$ (with fuzzy operator 1). Let F be a fuzzy formula of the form λF, then the negation of the formula (written $\neg F$) is $(1 - \lambda)F$. If F and G are fuzzy formulas, then $F \rightarrow G = \neg F \vee G$. Example for fuzzy formulas are: $F = 0.2a \wedge 0.7(0.8b \vee 0.9c)$ and $G = 0.6\forall_x(M(x) \rightarrow H(x))$.

Fuzzy operator is used to define the level of uncertainty for a particular proposition or formula e.g. let a be the statement "A client is old." and b be "a client's risk tolerance is low." Each statement can be described with the help of the fuzzy operator. For example, $0.8a$ might express a level of uncertainty of 0.8 for the claim that the client is old. The formula $a \rightarrow 0.9b$ describes the inference that when a client is old, then he or she has low risk tolerance with a level of uncertainty of 0.9. The semantics of the fuzzy operator is given via a kind of "fuzzy" product:

Definition 4.1 Let λ_1, $\lambda_2 \in [0, 1]$, then
$$\lambda_1 \otimes \lambda_2 = (2\lambda_1 - 1) \cdot \lambda_2 - \lambda_1 + 1.$$

The truth value of a fuzzy formula λF is given via the confidence function C_I, defined below.

Definition 4.2 Let a be an atom, I an interpretation in the sense of classical logic, F and G fuzzy formulas, λ a fuzzy operator, then:
1. $C_I(a) = I(a)$.
2. $C_I(\lambda F) = \lambda \otimes C_I(F)$.
3. $C_I(F \wedge G) = \min\{C_I(F), C_I(G)\}$.
4. $C_I(F \vee G) = \max\{C_I(F), C_I(G)\}$.
5. $C_I(\forall x\ F) = \min\{C_I(F^x/t)|\ t$ denotes a member of D, where D is the domain of I$\}$.
6. $C_I(\exists x\ F) = \max\{C_I(F^x/t)|\ t$ denotes a member of D, where D is the domain of I$\}$.

Example 4.1 Let $F = 0.2a \wedge 0.7(0.8b \vee 0.9c)$ and
$C_I(a) = 0$, $C_I(b) = 1$, $C_I(c) = 1$. Then
$C_I(F) = \min\{C_I(0.2a), 0.7 \otimes \max\{C_I(0.8b), C_I(0.9c)\}\}$
$\qquad = \min\{0.8, 0.66\} = 0.66$

The notion of satisfiability and validity are as follows:

Definition 4.3 Let I be an interpretation of fuzzy formula F, then
1. I Λ-*satisfies* F, if $C_I(F) \geq \Lambda$.
2. I Λ-*refute* F, if $C_I(F) < \Lambda$.

Definition 4.4 Let I be an interpretation of fuzzy formula F.
F is said to be Λ-*valid* iff for any interpretation I, I Λ-satisfies F. F is Λ-*inconsistent* iff for any interpretation I, I Λ-refute F. F is Λ-*satisfiable* iff F is not Λ-inconsistent.

G is said to be *logical consequence* of F, if for every interpretation such that $C_I(F) \geq \Lambda$, we have $C_I(G) \geq \Lambda$. The notion of complementarity is defined with respect to the threshold Λ. Two literals $\lambda_1 a$ and $\lambda_1 a$ are said to be Λ-*complementary* if $\lambda_1 \leq 1 - \Lambda$ and $\lambda_2 \geq \Lambda$.

Definition 4.5 Let C_1, C_2 be two clauses, and let L_1, L_2 be Λ-complementary literals in C_1 and C_2 respectively. Let R_1, R_2 be two clauses such that $C_1 = R_1 \vee L_1$ and $C_2 = R_2 \vee L_2$. Then the Λ-*resolvent* of C_1 and C_2 on the literals L_1 and L_2 is the clause $R_1 \vee R_2$.

We illustrate the inference mechanism in FOpL by the following example.

Example 4.2 Let $\Theta = \{a, a \rightarrow b, 0.6(0.1c \rightarrow d)\}$, $F = b \wedge d \wedge 0.43c$. We show that F is logical consequence of Θ. First, the skolem form of $\Theta \cup \{0F\}$ is the following clauses:

$$(1)\ a$$
$$(2)\ 0a \vee b$$
$$(3)\ 0.42c \vee 0.6d$$
$$(4)\ 0b \vee 0d \vee 0.57c$$

Let $\Lambda = 0.57$, then applying Λ-resolution yields

$$(5)\ b \qquad \text{from } (1), (2)$$
$$(6)\ 0d \vee 0.57c \text{ from } (4), (5)$$
$$(7)\ \square \qquad \text{from } (3), (6)$$

From this refutation we can conclude that F is 0.57-valid. The following example illustrates a drawback of the above method.

Example 4.3 Let a be the statement "A client is old", and b be "A client's risk tolerance is low". We consider a small logic program, written in PROLOG notion:

$$Fact: \quad (1)\ 0.8a$$
$$Rule: \quad (2)\ b \leftarrow a$$
$$Query: (3)\ ?b$$

The answer of the question is: "a client's risk tolerance is low" is 100 % certain, if the statement "a client is old" is 100 % certain. Unfortunately, the statement "a client is old" is only 80 % certain. Which answer might be expected for this question? If one thinks strictly, one may answer that nothing can be inferred about b, because the premise of the rule $(b \leftarrow a)$ has a level of uncertainty 1, and therefore this rule can not be applied. Others may say that the statement b can be inferred with 80 %, because $0.8a$ is a fact. What is the conclusion FOpL allows? The logic program is transformed into the clause form:

$$(C1)\ 0.8a$$
$$(C2)\ 0a \vee b$$
$$(C3)\ 0b$$

With maximal threshold value $\Lambda = 0.8$ the empty clause can be derived . We can then infer 0.8-validity of b, which is plausible. But the problem is the following case.

$$Fact: \quad (1)\ a$$
$$Rule: \quad (2)\ b \leftarrow 0.8a$$
$$Query: (3)\ ?b$$

In this case it suffices if statement a has a certainty degree of 0.8 (80 %), if one wants to infer the statement b. We have the statement a as a certain fact. Of course we expect a different conclusion as in the first case. But we obtain the same result:

$$(C1)\ a$$
$$(C2)\ 0.2a \lor b$$
$$(C3)\ 0b$$

A maximal threshold value to derive the empty clause is still $\Lambda = 0.8$. So we claim that the inference mechanism of FOpL is insufficient.

4.2 Annotated Logic

Annotated logic is a family of multiple-valued logics which has complete lattice as a truth value space. We write $\mathcal{T}$ to denote the complete lattice under some ordering $\leq$, which has $\perp$ as the least and $\top$ as the greatest element. Let $\mathcal{A}$ be a set of atoms and $a \in \mathcal{A}$, then $a{:}\mu$ is called an *annotated atom*, and an *annotated literal* is either an annotated atom or the negation of an annotated atom. The formulas are those that are constructed from annotated literals using the connectives $\land$ and $\lor$. As usual, an interpretation I is a mapping from $\mathcal{A}$ to $\mathcal{T}$.

Definition 4.6 Let F and G be formulas, an interpretation I is said to satisfy
- the annotated atom $a{:}\mu$ iff $I(a) \geq \mu$;
- the annotated literal $\sim(a{:}\mu)$ iff I does not satisfy $a{:}\mu$;
- the disjunction $F \lor G$ if it satisfies either F or G;
- the conjunction $F \land G$ if it satisfies both F and G.

Two annotated literals L_1 and L_2 are said to be *complementary* if they have the respective forms $a{:}\mu$ and $\sim(a{:}\rho)$, where $\mu \geq \rho$. The annotated logic contains the following two inference rules as a deduction method.

Definition 4.7 The *annotated resolvent* of two given clauses $(L_1 \lor R_1)$ and $(L_2 \lor R_2)$ is $R_1 \lor R_2$, if the literals L_1 and L_2 are complementary.

Definition 4.8 The *annotated reductant* of two given clauses $(a{:}\mu_1 \lor D_1)$ and $(a{:}\mu_2 \lor D_2)$ is $(a{:}Sup\{\mu_1, \mu_2\}) \lor D_1 \lor D_2$, in which μ_1 and μ_2 are incomparable.

To illustrate the same example as in FOpL (example 4.3) for the unit interval $[0,1]$, we don't need the second inference rule because the unit interval offers the total ordering.

Example 4.4 Let a and b be statements. We consider the same logic program, written in the syntax of annotated logic.

$$Fact:\quad (1)\ a{:}0.8$$
$$Rule:\quad (2)\ b{:}1 \leftarrow a{:}1$$
$$Query:\ (3)\ ?\ b{:}1$$

The program is transformed into the clause form of annotated logic.

$$(C1)\ a{:}0.8$$
$$(C2)\ \sim(a{:}1) \lor b{:}1$$
$$(C3)\ \sim(b{:}1)$$

The resolution can not be applied to $C1$ and $C2$, because $0.8 \not\geq 1$. Therefore we know that $b{:}1$ is not a logical consequence of (1) and (2). This inference is plausible, if one think strictly. But by *ETM* we can get more information; $b{:}1$ is a logical consequence of (1) and (2), if we allow the error tolerance 0.2. We consider the second case:

$$Fact: \quad (1)\ a{:}1$$
$$Rule: \quad (2)\ b{:}1 \leftarrow a{:}0.8$$
$$Query: (3)\ ?\ b{:}1$$

This program is also transformed into the clause form of annotated logic.

$$(C1)\ a{:}1$$
$$(C2) \sim (a{:}0.8) \vee b{:}1$$
$$(C3) \sim (b{:}1)$$

We apply the annotated resolution, then the empty clause can be derived as follows:

$$(R1) \sim (a{:}0.8) \text{ (from } C2 \text{ and } C3)$$
$$(R2)\ \Box \qquad \text{(from } C1 \text{ and } R1)$$

We can conclude that $b{:}1$ is the logical consequence of (1) and (2). *ETM* yields the same result in this case.

In contrast to FOpL, annotated logic can infers different result (see example 4.3). We claim that this is an advantage of annotated logic over FOpL.

4.3 Signed Logic

More then two truth values can be represented in signed logic, in this sense the signed formula logic is also a family of multiple-valued logic. We consider signed logic with the unit interval $[0,1]$ as a set of truth values, regular signs, and inference rules at the atomic level. Under this condition syntax and semantics of signed logic are analog to definition 2.1 in section 2. It is not a severe restriction to consider the atomic level because many inference techniques deal with atomic formulas. In general there are no restrictions for the truth value space in signed formula logic. Therefore it is not very surprising that signed logic captures the inference rule of annotated logic and fuzzy operator logic [8, 9]. Due to the usage of set operations by signed logic, SLD-like proof can be implemented which is not always possible by annotated logic. Recently for signed logic a total ordering has been considered as the truth value space [5]. We now give definitions of resolution for the signed logic.

Definition 4.9 Two signed formulas $S_1{:}a$ and $S_2{:}a$ are said to be *complementary*, if $S_1 \cap S_2 = \emptyset$.

Definition 4.10 Let $S_1{:}a$ and $S_2{:}a$ be complementary ($S_1 \cap S_2 = \emptyset$), then $(D \vee E)$ is called *signed resolvent* of clauses $\{S_1{:}a \vee D\}$ and $\{S_2{:}a \vee E\}$.

The following example shows that the signed resolution yields the same result as by the annotated logic (example 4.4).

Example 4.5 Let a and b be statements. We consider the signed logic program as follows:

$$Fact: \quad (1) \uparrow 0.8{:}a$$
$$Rule: \quad (2) \uparrow 1{:}b \leftarrow \uparrow 1{:}a$$
$$Query: (3) \ ? \uparrow 1{:}b$$

The program is transformed into the clause form of signed logic.

$$(C1) \uparrow 0.8{:}a$$
$$(C2) (\uparrow 1)'{:}a \vee \uparrow 1{:}b$$
$$(C3) (\uparrow 1)'{:}b$$

Since the conjunction ($\cap$) of signs of the literals $\uparrow 0.8{:}a$ and $(\uparrow 1)'{:}a$ is not an empty set, so deriving the empty clause is not possible. We conclude that $b{:}1$ is not a logical consequence of (1) and (2). We consider the second case:

$$Fact: \quad (1) \uparrow 1{:}a$$
$$Rule: \quad (2) \uparrow 1{:}b \leftarrow \uparrow 0.8{:}a$$
$$Query: (3) \ ? \uparrow 1{:}b$$

This program is also transformed into the clause form of signed logic.

$$(C1) \uparrow 1{:}a$$
$$(C2) (\uparrow 0.8)'{:}a \vee \uparrow 1{:}b$$
$$(C3) (\uparrow 1)'{:}b$$

The empty clause can be derived as follows:

$$(R1) (\uparrow 0.8)'{:}a \text{ (from } C2 \text{ and } C3)$$
$$(R2) \ \square \qquad \text{(from } C1 \text{ and } R1)$$

Although annotated logic and signed logic have different kinds of syntax and semantics both logics return the same result for this truth value space. *ETM* is also capable of this kinds of reasoning, but provides more information, namely the error tolerance.

5   Conclusion

In this paper we showed that the error-tolerance-method (ETM) is an adequate inference mechanism for vague informations. We have also showed that automated reasoning in classical propositional logic is a special case of the proof procedures in ETM. Another difficult task for this proof procedure is the large search space, because we have to check all alternative proof paths even if a proof has already been found. But not only ETM has this difficulty; most proof procedures in multiple-valued logics (for example fuzzy prolog) do have the same difficulties. We have implemented the proof procedure (algorithm Δ_1^0) in the program language PROLOG. One possible direction of future work is lifting the ETM to first order logic.

References

1. W. Bibel. *Automatic theorem proving.* Vieweg Verlag, Braunschweig, 1987.
2. W. Bibel. *Deduktion – Automatisierung der Logik.* Handbuch der Informatik. Oldenbourg, München, 1992.
3. N.C.A. da Costa, J.J. Lu, and V.S. Subrahmanian. Automatic theorem proving in paraconsistent logics. pages 72–86. Foundations and Experimantal Result. Proc. 10th CADE, LNCS, Springer Verlag, 1989.
4. R. Hähnle. Uniform notation of tableaux rules for multiple-valued logics. In *Proc. interational Symposium on multiple-valued logics*, pages 238–245. IEEE Press, 1991.
5. R. Hähnle. Exploiting data dependencies in many-valed logics. pages 48–69. Journal of applied Non-Classiccal Logics, 1996.
6. M. Kiefer and V.S. Subrahmanian. Theory of generalized annotated logic programming and its applications. pages 335–367. the journal of Logic Programming 12, 1992.
7. S.W. Lee. *Deduktive Realisierung von Vagheit mittels einer mehrwertigen Logik.* Erscheinen als Dissertation, Darmstadt, 1997.
8. J.J. Lu, N.V. Murray, and E. Rosenthal. Signed formulas and annotated logics. pages 48–53. Proc. of the 23rd IEEE International Symposium on Multiple-Valued Logic, 1993.
9. N.V. Murray, J.J. Lu, and E. Rosenthal. Signed formulas and fuzzy operator logics. pages 75–84. Proc. of the 8th International Symposium on Methodologies for Intelligent Systems, 1994.
10. N.V. Murray and E. Rosenthal. Resolution and path dissolution in multiple-valued logics. pages 570–579. Proc. of the International Symposium on Methodologies for Intelligent Systems, 1991.
11. N.V. Murray and E. Rosenthal. Adapting classical inference technique to multiple-valued logics using signed formulas. pages 237–253. Fundamenta Informaticae Journal 21, 1994.
12. T. Weigert, X.H. Liu, and J.P. Tsai. λ-implication in fuzzy operator logic. pages 259–278. Information Science, 1991.
13. T.J. Weigert, J.P. Tsai, and X.H. Liu. Fuzzy operator logic and fuzzy resolution. pages 59–78. Journal of Automated Reasoning 10, 1993.

Representing and Reasoning with Events from Natural Language

Miguel Leith and Jim Cunningham

Department of Computing
Imperial College of Science, Technology and Medicine
180 Queen's Gate, London, England
E-mails: *mfl@doc.ic.ac.uk* and *rjc@doc.ic.ac.uk*

Abstract. Linguistic categories such as progressives, perfectives, tense and temporal adverbials are at the heart of our ability to describe events in natural language. Following on from the work of Moens and Steedman and the later work of Kent, we have identified a fragment of an interval tense logic of Halpern and Shoham that is expressive enough to represent the temporal readings of many simple sentences involving the linguistic categories listed above, and computable enough for entailment checking to be manageable in a reasonable time scale. We show how one can model the semantics of formulae from the fragment using simple timelines and how one can support entailment checking by comparing timelines using a simple algorithm.

Keywords: temporal reasoning, natural language understanding.

1 Introduction

Linguistic categories such as progressives, perfectives, tense and temporal adverbials are at the heart of our ability to describe events in natural language. Through these categories, which we will loosely call *temporal*, we express the relative positions of events in time, characterized by temporal relations such as *later than, adjoins, overlaps, begins, ends* or *includes*. Even from the simplest sentences involving temporal categories people are able to understand something of the relative positioning of events in time. For example the sentence *Max stopped running* might give rise to the informal temporal reading in Fig. 1.

The analysis of events in natural language is complicated by the existence of sentences which embody a sense of culmination, for example *accomplishments* like *Jill ate the apple*, and sentences which involve various types of temporal adverbial phrase such as *in five minutes* or *at three o'clock*. Even for quite simple sentences it is hard to find a straightforward formalism that is suitable for representing their temporal readings and checking their entailments while being applicable in wider problems such as analysing narrative texts. Our work here focuses on providing such a formalism.

The logical representation of tense and aspect in natural language has considerable tradition and many open issues which will not be discussed here. The

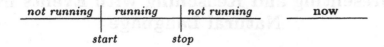

Fig. 1. Time periods for the sentence *Max stopped running*

reader is referred to the recent book by Verkyl [9]. We take as our point of departure a thesis by Kent [4]. Kent draws on work of Moens [6], Lascarides [5], and Moens and Steedman [7] in order to develop a formal account of linguistic behaviour in a complex interval logic.[1] In doing so he provides substantial evidence for the adequacy of interval models in explicating the entailments between simple related sentences.[2] However despite the use of representational conventions which extend modal style tense logics, from a computational perspective the absence of a proof theory for Kent's logic is a hindrance to exploitation.

We build on certain aspects of Kent's work but adopt a different choice of formalism and characterization of linguistic behaviour. We show how we may use formulae in the temporal modal logic known as HS, described by Halpern and Shoham in [3], to represent the temporal readings of various sentences involving temporal linguistic categories while retaining the implicit temporality and economy of expression of tense language. We then show that by using these temporal readings and the properties of HS logic we can demonstrate some of the entailments that exist between sentences. Our aims are made achievable by defining a fragment of HS called HSF1. The fragment is sufficiently expressive to represent the temporal readings of the kinds of sentences we are interested in, while the semantic models that we build for its formulae (which we refer to as *timelines*) are quite simple, allowing entailments to be checked by carrying out comparisons with a straightforward algorithm. Examples of the types of sentences we will be looking at are given in Fig. 2. They include certain uses of the present progressive, past tense, present perfective and past progressive, with both activity and accomplishment verb phrases. Some sentences also include simple temporal adverbial expressions. Examples of the types of entailments we can account for are given in Fig. 3. Within the later sections of this paper we will give examples of how we model the listed sentences and entailments.

The structure of this paper, then, is as follows. In Sect.2 we give a summary of the syntax and semantics of the logic HS and discuss the modal operators which are of immediate use to us. In Sect.3 we describe the logic fragment HSF1, show how we can model the semantics of its formulae as timelines and

---

[1] Kent's logical notation is complex partly because time is allowed to branch in both directions and intervals can be either open or closed. Logical operators are defined as mappings between interval sets, his proofs are then external mathematical arguments.

[2] There is an additional point to be made regarding the expressiveness of interval logics when compared to point logics. Certain properties may hold at intervals in time but not at points within those intervals, as was commented by Allen and Ferguson in [2].

1 - *Max is running.*
2 - *Max has run.*
3 - *Max stopped running.*
4 - *Jill ate the apple in five minutes.*
5 - *Jill is eating an apple.*
6 - *Fred was playing the piano for an hour.*
7 - *Mark started racing at two o'clock.*

Fig. 2. Example sentences for which we wish to model semantic readings

1 - *Max is running* $\Rightarrow$ *Max has run*
2 - *Jill is eating an apple* $\not\Rightarrow$ *Jill has eaten the apple*
3 - *Jill stopped eating an apple* $\not\Rightarrow$ *Jill ate the apple*
4 - *Fred walked to the station* $\Rightarrow$ *Fred was walking to the station*

Fig. 3. Example entailments which we wish to model

illustrate some of our temporal readings. In Sect.4 we describe an entailment checker which can carry out comparisons between timelines in a reasonable time scale, and demonstrate its use in modelling linguistic entailments. Finally in Sect.5 we make our conclusions.

2 The Logic HS

HS is the interval tense logic of Halpern and Shoham, which was first described in [3] and further investigated by Venema in [8]. In HS the set of possible worlds is the set of closed intervals in a temporal frame which may be dense or discrctc. Its six basic modal operators express relative positional interval relationships, and are described below.

$\langle B \rangle \varphi$: φ *holds at a beginning interval of the current interval.*
$\langle \underline{B} \rangle \varphi$: φ *holds at an interval begun by the current interval*
$\langle E \rangle \varphi$: φ *holds at an end interval of the current interval.*
$\langle \underline{E} \rangle \varphi$: φ *holds at an interval ended by the current interval.*
$\langle A \rangle \varphi$: φ *holds at an interval adjoining the current interval at its end.*
$\langle \underline{A} \rangle \varphi$: φ *holds at an interval adjoining the current interval at its beginning.*

Another six may be expressed by derived modal operators[3], namely $\langle L \rangle \varphi \equiv \langle A \rangle \langle A \rangle \varphi$ (later), $\langle \underline{L} \rangle \varphi \equiv \langle \underline{A} \rangle \langle \underline{A} \rangle \varphi$ (earlier), $\langle D \rangle \varphi \equiv \langle B \rangle \langle E \rangle \varphi$ (during), $\langle \underline{D} \rangle \varphi \equiv \langle \underline{B} \rangle \langle \underline{E} \rangle \varphi$ (included by), $\langle O \rangle \varphi \equiv \langle E \rangle \langle \underline{B} \rangle \varphi$ (overlapping) and $\langle \underline{O} \rangle \varphi \equiv \langle B \rangle \langle \underline{E} \rangle \varphi$ (overlapped by).

---

[3] Those familiar with Allen's interval logic, described in [1], may notice a similarity between the HS modal operators and Allen's interval relations. Specifically, DURING(i1, i2) is similar to $\langle D \rangle$ and $\langle \underline{D} \rangle$, STARTS(i1, i2) to $\langle B \rangle$ and $\langle \underline{B} \rangle$, FINISHES(i1, i2) to $\langle E \rangle$ and $\langle \underline{E} \rangle$, BEFORE(i1, i2) to $\langle L \rangle$ and $\langle \underline{L} \rangle$, OVERLAP(i1, i2) to $\langle O \rangle$ and $\langle \underline{O} \rangle$, MEETS(i1, i2) to $\langle A \rangle$ and $\langle \underline{A} \rangle$

There are also derived modal operators for the beginning point and end point of an interval. Since the proposition [B]false only holds at point intervals in HS logic, Halpern and Shoham define the beginning point modal operator [[BP]] and the end point modal operator [[EP]][4] as:

$$[[BP]]\varphi \equiv ((\varphi \wedge [B]false) \vee (\langle B \rangle (\varphi \wedge [B]false))$$
$$[[EP]]\varphi \equiv ((\varphi \wedge [B]false) \vee (\langle E \rangle (\varphi \wedge [B]false))$$

The correspondence of the fourteen modal operators to fourteen possible relative positions between intervals is illustrated in the diagram Fig. 4.

current interval ——————

$\langle \underline{L} \rangle \varphi$	——	1
$\langle \underline{A} \rangle \varphi$	——	2
$\langle \underline{O} \rangle \varphi$	————	3
$\langle \underline{E} \rangle \varphi$	————	4
$\langle \underline{D} \rangle \varphi$	————	5
$\langle B \rangle \varphi$	——	6
$\langle \underline{B} \rangle \varphi$	————	7
$\langle D \rangle \varphi$	—	8
$\langle E \rangle \varphi$	——	9
$\langle O \rangle \varphi$	————	10
$\langle A \rangle \varphi$	——	11
$\langle L \rangle \varphi$	—	12
$[[BP]]\varphi$		13
$[[EP]]\varphi$		14

Fig. 4. Pictorial illustration of relative interval positions and their HS formulae

2.1   Syntax of HS

Given Φ_0 as the set of atomic propositions, made up of the propositional constants p_0, p_1, ..., the syntax of HS is defined inductively as follows.

All members of Φ_0 are well formed formulae (wffs).
false is a wff.
$\varphi_1 \rightarrow \varphi_2$ is a wff iff φ_1 is a wff and φ_2 is a wff.
$\langle B \rangle \varphi$, $\langle E \rangle \varphi$, $\langle A \rangle \varphi$, $\langle \underline{B} \rangle \varphi$, $\langle \underline{E} \rangle \varphi$ and $\langle \underline{A} \rangle \varphi$ are all wffs iff φ is a wff.

The connectives ¬, ∧, ∨ and ↔ and the truth constant *truth* are defined in terms of → and *false* in the usual manner. The necessitation operators [B], [E] etc. are defined as usual as $\neg \langle B \rangle \neg$, $\neg \langle E \rangle \neg$ etc.

---

[4] We will tend to look at points simply as special cases of intervals.

2.2 Semantics of HS

A *(temporal) frame* is a pair $F = (T, <)$, where T is a set of time points and $<$ is a strict partial order on T. The *interval set* of a frame F is called $INT(F)$ and is defined as the set of all closed intervals $[s, t] = \{x \in T \mid s \leq x \leq t\}$ in T. A model M is a pair (F, V), where F is a frame and V is a *valuation*, i.e. a map $\Phi_0 \mapsto 2^{INT(F)}$. Next we inductively define the truth relation $\models$ as follows:

$\models_{[s,t]}^M P$ iff $[s, t] \in V(P)$

$\not\models_{[s,t]}^M false$

$\models_{[s,t]}^M (\varphi_1 \rightarrow \varphi_2)$ iff $\models_{[s,t]}^M \varphi_1$ implies $\models_{[s,t]}^M \varphi_2$

$\models_{[s,t]}^M \langle B \rangle \varphi$ iff there is a point u such that $s \leq u < t$ and $\models_{[s,u]}^M \varphi$

$\models_{[s,t]}^M \langle \underline{B} \rangle \varphi$ iff there is a point u such that $t < u$ and $\models_{[s,u]}^M \varphi$

$\models_{[s,t]}^M \langle E \rangle \varphi$ iff there is a point u such that $s < u \leq t$ and $\models_{[u,t]}^M \varphi$

$\models_{[s,t]}^M \langle \underline{E} \rangle \varphi$ iff there is a point u such that $u < s$ and $\models_{[u,t]}^M \varphi$

$\models_{[s,t]}^M \langle A \rangle \varphi$ iff there is a point u such that $t < u$ and $\models_{[t,u]}^M \varphi$

$\models_{[s,t]}^M \langle \underline{A} \rangle \varphi$ iff there is a point u such that $u < s$ and $\models_{[u,s]}^M \varphi$

The concepts of *validity* and *satisfiability* of formulae with respect to models, frames and classes of frames are defined in the usual manner.

2.3 Operators of HS That Are of Immediate Use to Us

It has been shown by Halpern and Shoham in [3] that the validity problem in HS logic is undecidable in all but the simplest classes of frames such as finite sets of integers. However, as is pointed out by Venema in [8], there may be natural and useful fragments of the logic that behave better. Such a fragment that interests us is one where we restrict the set of modal operators to all of the possibility operators plus one of the necessitation operators, [D]. The [D] operator is of immediate use to us, as it captures a highly important principle in temporal logics known as *homogeneity*. A proposition is said to be homogeneous if it is true at all sub-intervals of any intervals at which it holds. Homogeneity is used by Kent in [4] as a property on states. If someone was running for five minutes, for instance, then they were running at all periods of time within that five minutes, unless an exception is specified, in which case the proposition may be split into two or more homogeneous periods instead. By having a formula [D](*Max_run*) hold at an interval we are saying that Max is running at all intervals/points during that interval. Using the [D] operator to characterize homogeneity appears to be more elegant than in classical temporal logics, where one defines an axiom such as $HOLDS(p, T) \Leftrightarrow (\forall t.IN(t, T) \Rightarrow HOLDS(p, t))$.

3 HSF1: A Fragment of HS logic

The HSF1 fragment contains all of the possibility operators from HS and one of the necessitation operators, [D]. We begin this section by defining the syntax and semantics of the fragment. We then describe how we represent the semantics of formulae from the fragment using specialised models which we refer to as *timelines*. The algorithm by which timelines are constructed is described. This is followed by a description of our treatment of states and culminations, and lastly by some examples of how we represent temporal readings of simple sentences as HSF1 formulae.

3.1 Syntax of HSF1

The syntax of HS is defined using grammar rules as follows.

HSF1 : Exp2 | PossOp HSF1
PossOp : $\langle D \rangle$ | $\langle \underline{D} \rangle$ | $\langle E \rangle$ | $\langle \underline{E} \rangle$ | $\langle B \rangle$ | $\langle \underline{B} \rangle$ | $\langle L \rangle$ | $\langle \underline{L} \rangle$ | $\langle A \rangle$ | $\langle \underline{A} \rangle$ | $\langle O \rangle$ | $\langle \underline{O} \rangle$
Exp2 : Exp3 | HSF1 $\wedge$ HSF1 | HSF1 $\vee$ HSF1
Exp3 : ExtendedAtomic | [[*BP*]] Atomic | [[*EP*]] Atomic
ExtendedAtomic : Atomic | [D] Atomic
Atomic : PropConstant | $\neg$ PropConstant
PropConstant : p_1 | p_2 | p_3 ...

3.2 Semantics of HSF1

Semantics of the modal operators in HSF1 are the same as those given in the semantics of HS. However, we impose a restriction on frames by requiring that the set of points T in a temporal frame is dense, linear and unbounded.

3.3 Timelines

Proposition 1. *Any HSF1 formula may be modelled semantically by a finite number of alternative data structures which we will refer to as timelines.*

Definition 2. A timeline is a pair $\langle Ordering, Bindings \rangle$ where *Ordering* is a sequence of point labels in linear dense unbounded time and *Bindings* is a set of associations $\langle (Pt1, Pt2): P \rangle$ where *Pt1* and *Pt2* are point labels (from the ordering) signifying the bounds of an interval and P is an extended atomic formula.

Definition 3. A timeline is said to be inconsistent iff any two of its bindings take the following forms or their converses:

$\langle (Pt1, Pt2) : p \rangle$ and $\langle (Pt1, Pt2) : \neg p \rangle$
$\langle (Pt1, Pt2) : [D](p) \rangle$ and $\langle (Pt1, Pt2) : [D](\neg p) \rangle$
$\langle (Pt1, Pt2) : [D](p) \rangle$ and $\langle (Pt3, Pt4) : \neg p \rangle$ where $Pt1 < Pt3$ and $Pt4 < Pt2$
$\langle (Pt1, Pt2) : [D](p) \rangle$ and $\langle (Pt3, Pt4) : [D](\neg p) \rangle$
       where $(Pt3 < Pt1$ and $Pt1 < Pt4)$ or $(Pt3 < Pt2$ and $Pt2 < Pt4)$

Proposition 1 is justified because in the case of the HSF1 fragment, the semantics of an HSF1 formula can be interpreted by a procedure which builds specialised semantic models. The models, which we refer to as *timelines*, comprise an ordering relation represented by a series of point labels and a truth valuation represented by a set of associations between pairs of point labels (intervals) and extended atomic formulae. In general there may be more than one possible timeline for a given HSF1 formula but each temporal reading of a sentence in our examples is represented by a formula which has only one timeline.

We will give here an informal explanation of the procedure by which timelines are built, which is known as the Timeline Construction Algorithm. The algorithm is a recursive traversal of an HSF1 formula that builds an ordering relation and truth valuation along the way. At each stage of the recursive traversal we are at a particular *current interval* denoted by a pair of point labels e.g. (*Pt1*, *Pt2*). Each modal operator (except for [*D*]) is a possibility operator in its positive form (e.g. $\langle E \rangle (\varphi)$ etc.) and effectively asserts the existence of an interval in given temporal relation to the current interval. Thus the effect of encountering a possibility operator in the algorithm is to change the current interval and update the ordering relation to reflect the points of the new current interval in their correct positions. If the algorithm encounters an extended atomic formula it will create a *binding* between the current interval and the formula. For instance if it encounters a formula [*D*](*p*) at current interval (*Pt3*, *Pt4*) it will create a binding $\langle (Pt3, Pt4) : [D](p) \rangle$ and add it to the list of bindings that the algorithm is building.

By the preceding argument a timeline modelling an HSF1 formula of depth n (where we count the starting interval and each possibility modality as adding 1 to the depth) has at most $2n$ time points in its ordering. Also a timeline modelling an HSF1 formula that contains m extended atomic formulae has at most m bindings.

3.4 Single-timeline HSF1 Formulae

Generally an HSF1 formula may have more than one timeline that may model its semantics. However those formulae which may be modelled by only one timeline are of special interest. Our temporal reading formulae all share this property, which is advantageous since entailments between them may be checked very quickly. It is only certain pairs of HSF1 operators which, when appearing as sub-formulae, result in alternative timelines. A simple example is the formula $\langle B \rangle \langle \underline{B} \rangle (p)$. At the stage where we analyse $\langle \underline{B} \rangle$ the current interval is, say, (*Pt1*, *Pt3*) and the ordering is [... *Pt1* ... *Pt3* ... *Pt2* ...]. The new current interval will be (*Pt1*, *Pt4*) where *Pt4* is a new point that may be inserted anywhere after *Pt3* in the ordering. This means it can go in any of three places: before, at or after *Pt2*. We call pairs of HSF1 operators that behave this way *cross-directional*. The three basic cross-directional pairs are: $\langle B \rangle$ and $\langle \underline{B} \rangle$, $\langle E \rangle$ and $\langle \underline{E} \rangle$, $\langle A \rangle$ and $\langle \underline{A} \rangle$[5]. Expansions of derived operators result in further cross-directional pairs e.g. since $\langle L \rangle \equiv \langle A \rangle \langle A \rangle$, the pair $\langle \underline{A} \rangle$ and $\langle L \rangle$ is cross-directional.

---

[5] Venema used the term *mirror images* in [8] in reference to these pairs, because they are reflections of each other in a northwest plane correspondence.

3.5 The Treatment of Start, Stop and Culmination

In our temporal readings of natural language sentences great importance is given to three special types of time point: a state's start point, stop point and culmination point. Here we define what we mean by a state and illustrate each of the three point types in terms of the HSF1 formulae that hold at them.

Definition 4. An HSF1 state is a formula of the form $[D](P)$ where P may be any atomic formula. The state derived from an atomic proposition p is on at any interval/point at which $\langle \underline{D} \rangle ([D](p))$ holds and off at any interval/point at which $\langle \underline{D} \rangle ([D](\neg p))$ holds.

A start point of a state is a point of transition from off to on. This means that $[D](\neg p)$ must hold at an interval adjoining the point on the left, and $[D](p)$ must hold at an interval adjoining the point on the right.

Definition 5. A start point of a state $[D](p)$ is any point at which the formula $\langle \underline{A} \rangle ([D](\neg p)) \wedge \langle A \rangle ([D](p))$ holds.

| a | b |

Index
a - $[D](\neg p)$ b - $[D](p)$

Fig. 5. Start point of a state

A stop point of a state is a point of transition from on to off. This means that $[D](p)$ must hold at an interval adjoining the stop point on the left, and $[D](\neg p)$ must hold at an interval adjoining the stop point on the right.

Definition 6. A stop point of a state $[D](p)$ is any point at which the formula $\langle \underline{A} \rangle ([D](p)) \wedge \langle A \rangle ([D](\neg p))$ holds.

| a | b |

Index
a - $[D](p)$ b - $[D](\neg p)$

Fig. 6. Stop point of a state

A culmination point of a state is a point at which the state is said to be *finished* or *completed*. Culmination points are a critical factor in our understanding of accomplishment expressions such as *Max walked to the station* or *Jill ate the apple* where the progression of an event is *directed* towards a particular goal whose status is relevant to our perception of proceedings. In the case of activity expressions such as *Max worked* or *Fred played the piano* a directed progression is not apparent. We define a culmination of a state as a stop point at which the state's atomic proposition is true.

Definition 7. A culmination point of a state $[D](p)$ is any stop point of $[D](p)$ at which p holds.

$$| \quad a \quad |_b \quad c \quad |$$

Index
a - $[D](p)$ b - p c - $[D](\neg p)$

Fig. 7. Culmination point of a state

If we know that a state definitely did not culminate (i.e. it was broken off) we may represent this as a *break point*.

Definition 8. A break point of a state $[D](p)$ is any stop point of $[D](p)$ at which $\neg p$ holds.

$$| \quad a \quad |_b \quad c \quad |$$

Index
a - $[D](p)$ b - $\neg p$ c - $[D](\neg p)$

Fig. 8. Break point of a state

3.6 Temporal Readings as HSF1 Formulae

In this subsection we show how formulae from the HSF1 fragment may be used to represent the temporal readings of the sentences given in Fig. 2.

HSF1 formula:

$\langle \underline{L}\rangle ([D](\neg run(max)) \wedge \langle \underline{A}\rangle ([D](run(max)) \wedge \langle \underline{A}\rangle ([D](\neg run(max)))))$

Timeline

| | a | | b | | c | | | * | |

Timeline Index

a - $[D](\neg run(max))$ c - $[D](\neg run(max))$

b - $[D](run(max))$ * - now

Fig. 9. A temporal reading of the sentence *Max stopped running*

In the above reading of *Max stopped running* one sees a timeline similar to the informal example given in the introduction of this paper, but now expressed by an HSF1 formula. Looking back into the past from **now** we see an interval *c* where the state of Max running is <u>off</u> at all sub-intervals following the point at which Max stops running. Adjoining *c* on the left is the interval *b* for which the state is <u>on</u> at all sub-intervals, preceded by another interval *a* for which the state is <u>off</u> prior to his starting to run.

HSF1 formula:

$\langle \underline{L}\rangle ([D](\neg eat(jill, apple)) \wedge \langle \underline{A}\rangle ([[EP]](eat(jill, apple)) \wedge [D](eat(jill, apple)) \wedge$ *dura-*
tion(minutes(5)) $\wedge \langle \underline{A}\rangle ([D](\neg eat(jill, apple)))))$

Timeline

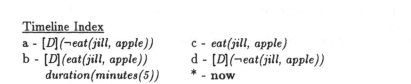

Timeline Index

a - $[D](\neg eat(jill, apple))$ c - *eat(jill, apple)*

b - $[D](eat(jill, apple))$ d - $[D](\neg eat(jill, apple))$

     duration(minutes(5))   * - now

Fig. 10. A temporal reading of the sentence *Jill ate the apple in five minutes*

In the above reading of *Jill ate the apple in five minutes*, we can see two things of fresh interest. The basic timeline structure is similar to that of *Max stopped running* in Fig. 9 but there is a culmination point at the end of the state for which Jill is eating an apple (point *c*) which indicates that Jill has finished eating the apple, and a duration proposition holding at interval *b*, which is our interpretation of the adverbial phrase *in five minutes* in its durational sense. Hence the proposition *duration(minutes(5))* holds only at the outermost interval of the eating state.

HSF1 formula:

$\langle \underline{D} \rangle ([D](eat(jill, apple)) \wedge \langle A \rangle ([D](\neg eat(jill, apple))) \wedge \langle \underline{A} \rangle ([D](\neg eat(jill, apple))))$

Timeline

	a			b			c	

	*	

Timeline Index

a - $[D](\neg eat(jill, apple))$ * - now

b - $[D](eat(jill, apple))$ c - $[D](\neg eat(jill, apple))$

Fig. 11. A temporal reading of the sentence *Jill is eating an apple*

The temporal reading of the present progressive sentence *Jill is eating an apple* takes a form where the eating state includes the current interval. Present progressive readings are the same for activity verb phrases such as *run*. The distinction in behaviours occurs is that where one can entail *Max has run* from *Max is running* or *Max stopped running*, one may not entail *Jill has eaten the apple* from *Jill is eating an apple* or *Jill stopped eating an apple*. In the past tense and perfective usage of accomplishment verb phrases such as *eat an apple* there is a definite sense of having culminated the eating event. In contrast activity verb phrases such as *run* which involve generic undirected states do not necessarily culminate in perfective and past tense readings.

HSF1 formula:

$\langle \underline{L} \rangle ([D](race(mark)) \wedge [[BP]](clock(hour(2))) \wedge \langle \underline{A} \rangle ([D](\neg race(mark))))$

Timeline

	a		b	c				*	

Timeline Index

a - $[D](\neg race(mark))$ c - $[D](race(mark))$

b - $clock(hour(2))$ * - now

Fig. 12. A temporal reading of the sentence *Mark started racing at 2 o'clock*

In our last example we show a reading of a sentence involving an at-adverbial phrase. We read the use of *at* with a chronological complement as signifying that a proposition representing the clock time, *2 o'clock* in this case, holds at a particular point, in this case the point at which Mark started racing. The index *b* appears as a subscript because the proposition it indexes holds at a point rather than a durative interval.

4 An Entailment Checker for HSF1 formulae

An entailment $F1 \models F2$ where $F1$ and $F2$ are HSF1 formulae holds if and only if for each timeline that may be *constructed* for $F1$ a timeline may be *constructed* for $F2$ such that the latter is a *substructure* of the former.

4.1 The Substructure Fitting Algorithm

The Substructure Fitting Algorithm establishes whether a timeline is a substructure of another timeline by attempting to *fit* the former into the latter. A successful fit means that a new timeline which represents the match of the two timelines can be built. We will use the terms *substructure timeline, superstructure timeline* and *matched timeline* to distinguish between the operative timelines in this algorithm. Note that for an entailment $F1 \models F2$ a superstructure timeline is a model of $F1$ and a substructure timeline is a model of $F2$.

The principle of the algorithm is to attempt to build a new timeline (the matched timeline) which has a set of bindings that pairs each binding of the substructure timeline with a binding from the superstructure timeline such that the resultant ordering is consistent with the orderings of the substructure timeline and superstructure timeline. At each stage of the recursive algorithm, a binding (SubBd) is removed from the set of substructure bindings (SubBds) and an attempt is made to match its formula (SubBdF) with a formula (SupBdF) of some binding (SupBd) from the set of superstructure bindings (SupBds) such that the formulae match in such a way that the new ordering (MatchOrd) can be updated to remain consistent with the superstructure ordering (SupOrd) and substructure ordering (SubOrd). This is repeated until either SubBds has been emptied (success) or it has been established that SubBds cannot possibly be emptied by the algorithm (failure).

The formulae in a pair of bindings may be matched in any of the ways depicted in Fig. 13.

Exact match, atomic case

    [*prop*]

    [*prop*]

Exact match, homogeneous case

    [$[D](prop)$]

    [$[D](prop)$]

Right homogeneous match

[$[D](prop)$]

[$[D](prop)$]

Left homogeneous match

[$[D](prop)$]

[$[D](prop)$]

Internal match, homogeneous case

[$[D](prop)$]

[$[D](prop)$]

Internal match, atomic case

[$[D](prop)$]

[*prop*]

Fig. 13. How formulae in a pair of bindings may be matched

4.2 Entailment Examples

Entailment check:

$\langle \underline{D} \rangle ([D](run(max)) \wedge$
$\quad \langle A \rangle ([D](\neg run(max))) \wedge$
$\quad \langle \underline{A} \rangle ([D](\neg run(max))) \models$
$\langle \underline{L} \rangle ([D](run(max)) \wedge$
$\quad \langle \underline{A} \rangle ([D](\neg run(max)))$

Superstructure (case 1 of 1):

```
        |  a  |          b          |  c  |
                |  *  |
```

Timeline Index
a - $[D](\neg run(max))$
b - $[D](run(max))$
* - now
c - $[D](\neg run(max))$

Substructure (case 1 of 1):

```
        |  d  |  e  |        |  *  |
```

Timeline Index
d - $[D](\neg run(max))$
e - $[D](run(max))$
* - now

Match:

```
        | a/d |          b          |  c  |
            |  e  |     |  *  |
```

Fig. 14. Entailment check for: *Max is running* $\Rightarrow$ *Max has run*

<u>Entailment check:</u>

$$\begin{array}{l} \langle \underline{L} \rangle ([D](\neg eat(jill,\ apple)) \wedge \\ \quad \langle \underline{A} \rangle ([D](eat(jill,\ apple)) \wedge \\ \qquad \langle \underline{A} \rangle ([D](\neg eat(jill,\ apple))))) \models \\ \langle \underline{L} \rangle ([D](\neg eat(jill,\ apple)) \wedge \\ \quad \langle \underline{A} \rangle ([[EP]](eat(jill,\ apple)) \wedge \\ \qquad [D](eat(jill,\ apple)) \wedge \\ \qquad \langle \underline{A} \rangle ([D](\neg eat(jill,\ apple))))) \end{array}$$

Superstructure (case 1 of 1):

| a | b | c | | * |

<u>Timeline Index</u>

a - $[D](\neg eat(jill,\ apple))$
b - $[D](eat(jill,\ apple))$
c - $[D](\neg eat(jill,\ apple))$
* - **now**

Substructure (case 1 of 1):

| d | e |$_f$ g | | * |

<u>Timeline Index</u>

d - $[D](\neg eat(jill,\ apple))$
e - $[D](eat(jill,\ apple))$
f - $eat(jill,\ apple)$
g - $[D](\neg eat(jill,\ apple))$
* - **now**

Match:
None.

Fig. 15. Entailment check for: *Jill stopped eating an apple* $\not\Rightarrow$ *Jill ate the apple*

5 Conclusion

We believe that the contribution of this paper is in demonstrating a route to simple and precise temporal representations of natural language events. It is pleasing to explicate prior formal analyses using interval semantics by utilising an expressive interval logic, interesting to find that only a tractable fragment seems to be needed, and satisfying to discover that timelines are sufficient to model the semantics of the fragment's formulae.

It is not evident to us that natural language should require excessively complex temporal models. AI approaches to understanding natural language, in our view, properly place emphasis on representing complex propositions and coping with the dynamic effect of actions. Nevertheless, the problem of temporal representation arises both in the local understanding and generation of elegant sentences and in the wider problems of scenario understanding and planning. A forthcoming thesis will discuss work on the understanding of narrative texts by analysing them into timelines using a process of refinement and consistency checking. Rather than a monolithic integrated solution, the need to mixing different logics and representations may be important for human and machine understanding of natural language.

References

1. James Allen. Towards a general theory of action and time. *Artificial Intelligence*, 23(2):123–154, 1984.
2. James Allen and George Ferguson. Actions and events in interval temporal logic. Technical Report 521, University of Rochester, 1994.
3. Joseph Halpern and Yoav Shoham. A propositional modal logic of time intervals. In *Proceedings of Symposium on Logic in Computer Science*, pages 279–292, Cambridge, Massachusetts, June 16-18, 1986.
4. Stuart Kent. *Modelling Events from Natural Language*. PhD thesis, Imperial College, London, 1993.
5. Alex Lascarides. *A Formal Semantic Theory of the Progressive*. PhD thesis, University of Edinburgh, 1988.
6. Marc Moens. *Tense, Aspect and Temporal Reference*. PhD thesis, University of Edinburgh, 1987.
7. Marc Moens and Mark Steedman. Temporal ontology and temporal reference. *Computational Linguistics 14*, pages 15–28, 1988.
8. Yde Venema. Expressiveness and completeness of an interval tense logic. *Notre Dame Journal of Formal Logic*, 31(4):529–547, 1990.
9. Henk Verkuyl. *A theory of aspectuality*. Cambridge University Press, 1993.

Reasoning About Security: A Logic and a Decision Method for Role-Based Access Control

Fabio Massacci*

Computer Laboratory,
University of Cambridge (UK)

Abstract. Role-based access control (RBAC) is one of the most promising techniques for the design and implementation of security policies and its diffusion may be enhanced by the development of formal and automated method of analysis.

This paper presents a logic for practical reasoning about role based access control which simplifies and adapts to RBAC the calculus developed at Digital SRC. Beside a language and a formal semantics, a decision method based on analytic tableaux is also given. Analytic tableaux make it possible to reason about logical consequence, model generation and consistency of a formalised role-based security policy.

1 Introduction

Access control is a key issue for the security of distributed systems (see [28] for an introduction) and plays an important role in the security policies of many organisations such as the military [4], banking [7] or health services [2].

Recently, new formal models for access control have emerged besides the traditional access matrix [21] for the design of more sophisticated policies [5, 11, 15, 25, 26]. Logic and formal reasoning techniques are also used for describing and reasoning about security policies and access control [1, 8, 16, 22, 29]

Among the new paradigms, Role-Based Access Control (RBAC) [11, 12, 15, 26] has received particular attention. It stems from an extensive field study [13] which pointed out that in practice permissions are assigned to a user according his/her roles in the organisation. Moreover, the explicit representation of roles makes also possible to simplify the system design and to use engineering principles such as least privilege, separation of duties, etc. [11, 26].

As languages become more expressive and systems more complex, formal methods of analysis (possibly automated) are necessary: the simple table (or list) lookup of the access matrix [18, 27] must be replaced by careful reasoning for complex and highly expressive systems such as [1, 5, 8, 11, 16, 22, 26].

This work aims at practical but formal reasoning methods for RBAC:

1. *define a logic* (a language and a semantics) to express RBAC policies in a simple and (hopefully) natural way;

---

* Current address: Dip. Informatica e Sistemistica, Univ. di Roma I, "La Sapienza", via Salaria 113, I-00198 Roma, massacci@dis.uniroma1.it.

2. *find a decision method* (a terminating inference procedure) which can be used to verify in a direct and automatic way the consistency or the logical consequences of an RBAC policy.

The logic we propose is a simplified and enhanced version of the logic for access control developed in [1, 22]. The language is tailored to express directly role hierarchies, privileges attribution and user attributions as requested in many formal models of RBAC [11, 26]. We also developed a model theoretic semantics — the relevance for security analysis of formal semantics has been pointed out in [29] — and have improved the semantics proposed in [1].

A decision method is equally important: suppose we formalised the key RBAC features of a health care system, we also want to "do something" with this formalisation. For instance checking whether it is consistent, or whether a particular (un)desired property is a logical consequence of our model or not. To this extent axiom systems, which are often used to define formal models based on logics [8, 16, 19, 20], are inadequate: axioms systems require people to be good logicians and cannot be automated at all. The exclusive focus on axiomatisation, without calculi for automated reasoning, may condemn logics for practical reasoning to be unused in practice and therefore fail in their very first aim.

Therefore we developed a decision method based on analytic tableaux for modal logics [14] and in particular on Simple Step Tableaux [23, 9]. Tableaux have many advantages: they can be used for both logical consequence and satisfiability; deduction steps follows the semantics of the logic and the calculus can be easily automated along the lines of "lean theorem proving" [3].

We view our calculus as a building block of a more general framework such as authentication logics for cryptographic protocols [6, 30], where the issue of jurisdiction of agents is a key problem. In this perspective access control can be seen as a problem of jurisdiction in complex and distributed systems.

Notice that we *do not* advocate the use of logic and semantic tableaux for actually implementing RBAC systems. In the same way that a civil engineer use mathematical instruments to design a bridge but concrete and iron to build it, we believe that logic and decision methods based on logic should be used in the *design and verification phase* of RBAC systems.

In this paper we present a logical language for RBAC and describe how RBAC features can be captured (§2), together with the desired reasoning capabilities (§3). Next we present the formal semantics (§4) and the tableaux-based decision method (§5). The discussion on related works (§6) concludes the paper.

2   A Logic for RBAC: Language and Features

The main actors of RBAC (see [11, 26] for an introduction) are *users* and *roles*, denoted respectively by U and R. We refer to either of them as *atomic principals A*. We use also composed principals (denoted by P) which are user under a certain roles i.e. "U as R". Intuitively by U as R we denote a principal U who has restricted his privileges to those allowed by the role R; for instance

$fm205$ as $WebMaster$. Notice that U may also be claiming a role R which has not been assigned to him by the organisation.

For *role hierarchies* [11, 26] we use "R_1 isa R_2", that is role R_1 has at least all privileges of R_2; for instance $SwAdm$ isa $User$. For *users assignments* to roles we use the construct "U has R"; for instance maj has $SysAdm$.

The role of operations over objects is played by *atomic requests* which are uninterpreted operations [11, 26], e.g. *login(ftp)*. The intuition is that they are requests from a user and can be granted or not. Therefore we interpret them as propositional letters which can be true in a particular state of the system (granted) or false (not granted). We admit more complicated requests with boolean connectives $\land$, $\lor$, $\supset$: for instance *login(telnet)* $\supset$ *login(ftp)* whose intuitive (and formal) interpretation is "if telnet has been granted so has ftp". Requests are denoted with φ, ψ.

User requests of the form "P req φ" corresponds to the intuition that P requests φ to be executed (or tries to execute φ);e.g. maj req ($fm205$ has $User$). For the meaning of R req φ, where R is a role, we follow [1] and interpret it as "somebody, who has role R, requests φ" (see [1, 22] for discussion). Note that "P req φ" is itself a request and that an user may claim that a request comes from somebody else e.g. $fm205$ req ($kw10009$ req $exec(emacs)$).

To visualize the intuition behind the formulae we may compare a request such as $fm205$ req $rm(test.c)$ to a command typed by the user on a terminal: fm205{ely}: rm test.c. As we press enter, it is evaluated according our access privileges and either executed or not (i.e. $rm(test.c)$ may be true or false).

Notice the difference with the logics of knowledge and obligations for computer security (see for instance [8, 16, 20]) whose main concern is usually confidentiality, characterized by the following axiom schema:

$$\text{"}A \text{ knows } \varphi\text{"} \supset \text{"}A \text{ is allowed to know } \varphi\text{"} \tag{1}$$

This work on access control follows [1] and focuses more on a theory of actions, to verify that the actions executed by a system respect its security policy. In this sense the modal operator A req φ is closer to the Do operator of Kanger in [19] or the "A sees to it that X" in [20], where X is a action. The key observation, from the viewpoint of access control, is that we should rather use an operator for "A *tries* to see to it that X", since A may not be allowed to do X. This is the underlying intuition behind this work and also [1] and is crucial to understand the subsequent properties that we may find desirable.

The link between principals and requests (or privileges) is given by *privileges attributions* [26]. The construct "P controls φ" denotes that principal P has access control over φ. Loosely speaking, P can (is allowed to) make φ happen. In the literature on authentication this is usually referred as *jurisdiction* of a principal [6, 30] and it is characterised by the axiom

$$\text{"}A \text{ says } \varphi\text{"} \land \text{"}A \text{ has jurisdiction on } \varphi\text{"} \supset \varphi \tag{2}$$

Our aim is to to replace it by following one (more complex yet more realistic):

$$A \text{ req } \varphi \land B \text{ controls } \varphi \land \text{"some relation between } A \text{ and } B\text{"} \supset \varphi \tag{3}$$

$$RH = \begin{cases} SysAdm \text{ isa } SwAdm \\ SysAdm \text{ isa } PostMaster, \\ SwAdm \text{ isa } User, \\ PostMaster \text{ isa } User, \\ User \text{ isa } RemUser, \end{cases} \quad UA = \begin{cases} maj \text{ has } SysAdm \\ pb \text{ has } PostMaster \\ gt \text{ has } SwAdm \\ gt \text{ has } DepPostMaster \\ kw10009 \text{ has } User \\ fm205 \text{ has } RemUser \end{cases}$$

$$PA = \begin{cases} SysAdm \text{ controls } (fm205 \text{ has } User) \\ SwAdm \text{ controls } write(emacs - source) \\ User \text{ controls } login(telnet) \\ User \text{ controls } run(emacs) \\ RemUser \text{ controls } login(ftp) \\ (fm205 \text{ as } User) \text{ controls } rm(test.c) \end{cases}$$

Fig. 1. A Logical Formalisation of an RBAC

where the relation between A and B is specified according RBAC policies. Again this interpretation makes sense in our framework where φ is a "potential action" and not a state of affairs. This can be rephrased as follows: if A tried to execute φ and A is allowed to execute φ then φ is indeed executed.

In "pure" RBAC only roles are connected to privileges [26] so we should only have formulae like R controls s. We don't want to force this commitment on the logic because (i) it is not necessary and (ii) there are applications where privileges may be attached to individual users or to particular users in particular roles. For instance in the UNIX file system $fm205$ can write his own files (when he logs in as a legitimate user) and those permission are linked to him as an individual user rather than a role (which may be assigned to many users) e.g. $(fm205$ as $User)$ controls $rm(test.c)$.

Roles hierarchics, users and privileges attribution are the major requirements for RBAC [12, 11, 26]. An *RBAC Security Policy* is characterised as follows:

Role Hierarchy (RH) $\bigwedge_{i,j}\{R_i \text{ isa } R_j\}$,
Privileges Attribution (PA) $\bigwedge_{i,j}\{P_i \text{ controls } \varphi_j\}$,
Users Attribution (UA) $\bigwedge_{i,j}\{U_i \text{ has } R_j\}$.

For a particular *realization* or run of the system we simply add the following:

User Requests (UR) $\bigwedge_{i,j}\{P_i \text{ req } \varphi_j\}$

For sake of simplicity, we freely exchange a set of formulae with the conjunction of its elements. An example is shown in Fig. 1.

Example 1. With the RBAC model in Fig. 1 when maj req $(fm205 \text{ has } User)$ together with $fm205$ req $(login(telnet) \wedge run(emacs))$ should the system grant $login(telnet) \wedge run(emacs)$?

Finally we introduce cryptographic keys and the standard notation $\{s\}_K$ to denote a request s encrypted (or signed) with the key K [6]. The construct

$$
\begin{array}{ll}
UA = \left\{
\begin{array}{l}
Alice \text{ has } Account \\
Bob \text{ has } Marketing \\
Charlie \text{ has } Store \\
Sam \text{ has } Server
\end{array}
\right.
&
PA = \left\{
\begin{array}{l}
Marketing \text{ controls } order \\
Account \text{ controls } pay \\
Store \text{ controls } recvd \\
Alice \text{ controls } (Bob \text{ req } order) \\
Server \text{ controls } (K_A \text{ speaks-for } Alice) \\
Server \text{ controls } (K_B \text{ speaks-for } Bob)
\end{array}
\right.
\end{array}
$$

$$
UR = \left\{
\begin{array}{l}
\{Bob \text{ req } order\}_{K_A} \\
\{recvd\}_{K_C} \\
\{K_A \text{ speaks-for } Alice\}_{K_S} \\
\{K_C \text{ speaks-for } Charlie\}_{K_S}
\end{array}
\right.
\qquad
\begin{array}{l}
KA = \left\{ K_S \text{ speaks-for } Sam \right. \\[1em]
Constr = \left\{ Account \text{ req } (order \wedge recvd \supset pay) \right.
\end{array}
$$

Fig. 2. RBAC Models with Keys for Example 2

K **speaks-for** P indicates that K is associated to a principal P or, more intuitively, that "a key speaks for a principal". We use composed principals because one key may be used by *maj* and another by *maj* as *User*.

Since we are interested in access control and not in authentication or confidentiality we assume that the problems of associating keys with principals or finding whether keys are good etc. has been tackled already (see [6, 22, 30]). It is possible to associate a key to more than one principal (for shared key encryption). In this case the logic has been tailored to assume that the request came from the principal with most privileges (to avoid a denial of service).

Encrypted Requests (ER) $\bigwedge_{i,j} \{\varphi_j\}_{K_i}$
Key Assignments (KA) $\bigwedge_{i,j} \{K_i \text{ speaks-for } P_j\}$

Example 2. *Alice* is responsible for the accounts. *Bob* for marketing and ordering of goods and *Charlie* is the storekeeper, in charge of receiving and shipping of goods. *Alice* is trusted in forwarding *Bob* orders. *Sam* is the security office responsible for the authentication server. The account department pays only after the store has certified goods' arrival. Messages and certificates are shown in Fig. 2. Should the department *pay*?

In logical term we are asking whether

$$
UA \wedge PA \wedge KA \wedge UR \wedge Constraints \models^? pay
$$

where the symbol $\models$ denotes traditional logical consequence.

In full fledged RBAC models [11, 15, 26] there are also other *constraints*. Most static constraints, such as separation of duties, can be easily captured:

$$
\neg (U \text{ has } R_j \wedge U \text{ has } R_i) \quad \text{for mutually exclusive } R_i, R_j
$$

We can also express the fact that roles hierarchies must not yield cycles with $\neg (R_1 \text{ isa } R_2 \wedge R_2 \text{ isa } R_1)$ for distinct R_1 and R_2.

Once we have defined the language we can briefly sketch the properties we are interested to model. The basic properties are *transitivity and cumulativity* for role hierarchies and user attributions:

$$(Tr) \quad R_1 \text{ isa } R_2 \land R_2 \text{ isa } R_3 \supset R_1 \text{ isa } R_3$$
$$(Cm) \quad U \text{ has } R_1 \land R_1 \text{ isa } R_2 \supset U \text{ has } R_2$$

Other desired properties concern the relation between users and roles which follow from the meaning of R req φ as "somebody with role R requested φ":

$$(R) \qquad U \text{ has } R \land U \text{ req } \varphi \supset R \text{ req } \varphi$$
$$(U/R) \quad U \text{ has } R \land (U \text{ as } R) \text{ req } \varphi \supset R \text{ req } \varphi$$

Some simplification properties for users and roles are also useful, in particular *idempotence*[2]: it does not make sense to have "logically stammering" principals for whom U req $(U$ req $\varphi)$ is logically different from U req φ. *Awareness* is also important: principals should be aware of the consequences of what they requested (or didn't request):

$$(Id) \quad U \text{ req } (U \text{ req } \varphi) \supset U \text{ req } \varphi$$
$$(K) \quad U \text{ req } (\varphi \supset \psi) \land U \text{ req } \varphi \supset U \text{ req } \psi$$
$$(4) \quad U \text{ req } \varphi \supset U \text{ req } (U \text{ req } \varphi)$$
$$(5) \quad \neg U \text{ req } \varphi \supset U \text{ req } \neg(U \text{ req } \varphi)$$

and equally for roles. In this way we have simplifications such as U req $(\varphi \land U$ req $\psi)$ implies U req $(\varphi \land \psi)$.

As in other system for reasoning about security [1, 8, 16] we avoid reflexivity, i.e. A req $\varphi \supset \varphi$, since the request of an agent may not be granted. Indeed the whole point of using **controls** is exactly to denote explictly the case where reflexivity holds, for a particular principal and a particular request. This idea is also used in [1] for their access control system and in [30].

$$(Ctrl) \quad P \text{ controls } \varphi \equiv [(P \text{ req } \varphi) \supset \varphi]$$

The intuition may become clearer by rephrasing it as "P controls φ means that if P tries to execute φ then φ will indeed be executed by the system".

In further development of the logic it may be desirable to replace material implication with an intuitionistic or a relevant implication. With $(Ctrl)$ a principal may pretend to control whatever she does not explicitly request i.e. $\neg P$ req $\varphi \supset P$ controls φ. Such situation may also be a feature if we are interested in modelling "bluffing" principals. In any case this is still sufficient for modelling a wide range of situations.

Simplification properties are not imposed on cryptographic keys since it is open whether it would make sense to logically flatten multiple encryption:

$$\{p \land \{q\}_K\}_K \equiv^? \{p \land q\}_K$$

---

[2] Idempotence has been advocated also in [1] and corresponds to K4 axioms of modal logic and a weak form of the axiom T [10, 17].

Therefore keys have different logical properties from users (in contrast with [1] where keys share the same properties of other principals). We only require that

$$(KKey) \ \{\varphi \supset \psi\}_K \wedge \{\varphi\}_K \supset \{\psi\}_K$$
$$(SF) \quad K \ \text{speaks-for} \ P \wedge \{\varphi\}_K \supset P \ \text{req} \ \varphi$$

Notice that even if we don't flatten encryption, we still want to derive from K speaks-for U and $\{p \wedge \{q\}_K\}_K$ the request U req $(p \wedge q)$.

3 The Reasoning Services

Once we have formalised a RBAC system with a set G of role hierarchies, users attributions etc. we expect a number of *reasoning services*:

Consistency Check: check whether G has a model. Typically verify the compatibility of role hierarchies, users attributions etc. with the constraints.

Logical Consequence: decide whether a property φ logically follows from G. It could be used to decide whether a certain request should be granted or whether R_1 isa R_2 is entailed even if is not explicitly written in G.

Model Generation: in case a property φ does not follows from G (it is not a logical consequence) derive a counter model for it.

Satisfiability: if logical consequence is too strong, decide whether ψ is compatible with G i.e. there is a model for G where a state satisfies ψ.

Of course we also expect *soundness, completeness and termination*.

The *proof search* of tableaux methods is based on *satisfiability* [14]. Suppose we have a G and we want to prove that a request ψ is satisfiable, then we try to construct a model for the initial formula by assuming that ψ holds in the actual world and breaking connectives or adding formulae from G and breaking them too. New states may also be introduced. If we always end up with contradictions then the initial formula is unsatisfiable.

To prove *logical consequence* i.e. $G \models \varphi$, we apply refutational theorem proving: start with the negation of φ and try to derive a countermodel. If we fail then the formula is valid otherwise we have directly the counterexample. *Consistency* of G can be proved by starting with $True$ and verifying that it is satisfiable. *Model generation* comes for free in this approach.

Notice that if we fail an axiomatic proof we don't know whether we (or our programs) are just bad logician or a proof does not actually exists. Tableaux methods yield a proof (no counter model found) or a counter model (a proof does not exists).

4 Formal Syntax and Semantics

The language, already described in §2, is the following:

$$P ::= U \mid R \mid P \ \text{as} \ R \quad \varphi, \psi ::= r \mid \neg \varphi \mid \varphi \wedge \psi \mid P \ \text{req} \ \varphi \mid \{\varphi\}_K \mid$$
$$R_1 \ \text{isa} \ R_2 \mid U \ \text{has} \ R \mid K \ \text{speaks-for} \ P$$

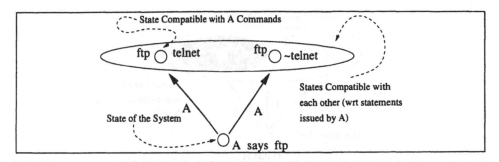

Fig. 3. A Simple Kripke Model

Other connectives are abbreviations, such as material implication $\varphi \supset \psi \equiv \neg(\varphi \wedge \neg\psi)$. We also use $P\,\text{controls}\,\varphi$ as a shortcut (see axiom (*Ctrl*)).

The semantics is based on Kripke models (see [10, 17] for an introduction). A *model* is a pair $\langle W, \mathcal{I} \rangle$ where W is a non empty set of states and $\mathcal{I}$ an interpretation such that for every atomic principal A it is $A^{\mathcal{I}} \subseteq W \times W$ and for every atomic request r it is $r^{\mathcal{I}} \subseteq W$.

Intuitively $r^{\mathcal{I}}$ represents the states where request r has been granted while the relation $A^{\mathcal{I}}$ models the compatibility between states: if $\langle w, w^* \rangle \in U^{\mathcal{I}}$ this means that all requests made by user U in state w has been granted in state w^*. An example is shown in Fig. 3

There are *further requirements on roles and users* i.e. the corresponding relations $U^{\mathcal{I}}$ and $R^{\mathcal{I}}$ must be K45 relations [10, 17] i.e. transitive and euclidean[3] Intuitively a K45 is an "almost equivalence relation" which is not reflexive and thus we have $A^{\mathcal{I}} = (\{w_0\} \cup W^*) \times W^*$ for $\{w_0\} \cup W^* \subseteq W$ and $w_0 \notin W^*$. as in Fig. 3. A K45 relation satisfies the properties of idempotence and awareness we have required in §2.

Keys K are also such that $K^{\mathcal{I}} \subseteq W \times W$ but we do not impose any constraint on the compatibility relation. However, the semantics of the **speaks-for** operator has been crafted to satisfy the properties described in §2.

The interpretation $\mathcal{I}$ is extended as follows:

$$
\begin{aligned}
(\varphi \wedge \psi)^{\mathcal{I}} &= \varphi^{\mathcal{I}} \cap \psi^{\mathcal{I}} \\
(\neg\varphi)^{\mathcal{I}} &= W - \varphi^{\mathcal{I}} \\
(P\,\text{req}\,\varphi)^{\mathcal{I}} &= \left\{ w \mid \forall w' \in W \text{ if } \langle w, w' \rangle \in P^{\mathcal{I}} \text{ then } w' \in \varphi^{\mathcal{I}} \right\} \\
(R_1\,\text{isa}\,R_2)^{\mathcal{I}} &= \left\{ w \mid \forall w' \in W \text{ if } \langle w, w' \rangle \in R_2^{\mathcal{I}} \text{ then } \langle w, w' \rangle \in R_1^{\mathcal{I}} \right\} \\
(U\,\text{as}\,R)^{\mathcal{I}} &= \left\{ \langle w, w' \rangle \mid \exists w'' \langle w, w'' \rangle \in U^{\mathcal{I}} \text{ and } \langle w'', w' \rangle \in R^{\mathcal{I}} \right\} \\
(U\,\text{has}\,R)^{\mathcal{I}} &= \left\{ w \mid \forall w' \in W \text{ if } \langle w, w' \rangle \in R^{\mathcal{I}} \text{ then } \langle w, w' \rangle \in U^{\mathcal{I}} \right\} \\
(\{s\}_K)^{\mathcal{I}} &= \left\{ w \mid \forall w' \in W \text{ if } \langle w, w' \rangle \in K^{\mathcal{I}} \text{ then } w' \in \varphi^{\mathcal{I}} \right\} \\
(K\,\text{speaks-for}\,P)^{\mathcal{I}} &= \left\{ w \mid \forall w' \in W \text{ if } \langle w, w' \rangle \in P^{\mathcal{I}} \text{ then } \langle w, w' \rangle \in K^{\mathcal{I}} \right\}
\end{aligned}
$$

In the sequel we write $w \|\!\!-\varphi$ for $w \in \varphi^{\mathcal{I}}$.

The graphical representation of the semantics of "U has R" is given in Fig. 4. The basic intuition is that $R\,\text{req}\,\varphi$ is interpreted as "somebody who has R said

---

[3] A relation R is euclidean if wRw' and wRw'' implies $w'Rw''$.

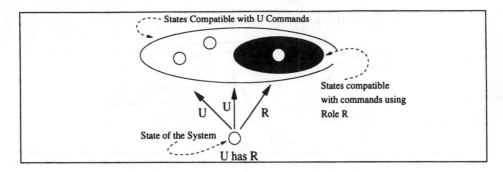

Fig. 4. A User with a Role

φ". So if U **has** R then all possible commands by U are also valid for R since that "somebody" could be indeed U. Therefore the set of possible states for R is contained in that of U.

Definition 1. A request φ is *satisfiable* iff there is a model $\langle W, \mathcal{I} \rangle$ where $(\varphi)^{\mathcal{I}}$ is not empty; it is *valid* iff for every model $\langle W, \mathcal{I} \rangle$ it is $(\varphi)^{\mathcal{I}} = W$.

To define logical consequence we use the set of *Global axioms* G which holds in every possible state [14, 23]. Global axioms are used in this framework for role hierarchies, user attributions etc.

$$G = RH \wedge PA \wedge UA \wedge KA \wedge UR \wedge Contraints$$

Definition 2. A request φ is a *logical consequence* of global axioms G iff for every model $\langle W, \mathcal{I} \rangle$ if $\forall v \in W : v \|\!\!-G$, then $\forall w \in W : w \|\!\!-\varphi$.

Example 3. Consider the hierarchy top-secret TS, secret S, classified C and unclassified U. The users are *Alice* with TS privileges, *Bob* with S and *Charlie* who has none. The system received a message and some delegation certificates (Fig. 5). Should $read(file) \wedge distrib(file)$ be granted with a write-up policy [28]?

In other terms is $read(file) \wedge distrib(file)$ a logical consequence of the RBAC in Fig. 5? Even in this simple case is not easy to see that it is *not*. Hence the need of a decision method.

5 A Decision Method Based on Tableaux

Prefixed tableaux use *prefixed requests*, i.e. pairs $\langle \sigma : \varphi \rangle$ where φ is a requests and σ is prefix, formally defined as $\sigma ::= 1 \mid \sigma.U.n \mid \sigma.R.n$ where n is an integer.

The labelling follows the intuition behind the model: "1" is the actual (real) state. If σ "names" a state then $\sigma.Alice.n$ (for $n = 2, 3 \ldots$) encodes in the name itself the property that the state "called" $\sigma.Alice.n$ is compatible with state σ according *Alice*'s requests. After "giving names to things" we can speak of

$$RH = \left\{ \begin{array}{l} TS \text{ isa } S \\ S \text{ isa } C \\ C \text{ isa } U \end{array} \right. \qquad PA = \left\{ \begin{array}{l} TS \text{ controls } distrib(file) \\ TS \text{ controls } read(file) \\ (Bob \text{ as } S) \text{ controls } read(file) \\ Charlie \text{ controls } write(file) \\ Server \text{ controls } (K_A \text{ speaks-for } Alice) \\ Server \text{ controls } (K_B \text{ speaks-for } Bob) \\ Server \text{ controls } (K_C \text{ speaks-for } Charlie) \end{array} \right.$$

$$UA = \left\{ \begin{array}{l} Alice \text{ has } TS \\ Bob \text{ has } S \\ Charlie \text{ has } U \end{array} \right.$$

$$KA = \left\{ \begin{array}{l} K_S \text{ speaks-for } Server \\ K_A \text{ speaks-for } Alice \end{array} \right. \qquad UR = \left\{ \begin{array}{l} \{K_B \text{ speaks-for } Bob\}_{K_S} \\ \{Alice \text{ req } (read(file) \wedge distrib(file))\}_{K_B} \end{array} \right.$$

Fig. 5. Global Description of an RBAC Model

$$\alpha: \dfrac{\sigma : \varphi \wedge \psi}{\begin{array}{c} \sigma : \varphi \\ \sigma : \psi \end{array}} \qquad \beta: \dfrac{\sigma : \neg(\varphi \wedge \psi)}{\sigma : \neg \varphi \mid \sigma : \neg \psi} \qquad dn: \dfrac{\sigma : \neg\neg\varphi}{\sigma : \varphi} \qquad G: \dfrac{\vdots}{\sigma : \varphi} \quad \begin{array}{l} \text{with } \varphi \in G \text{ and} \\ \sigma \text{ present} \end{array}$$

$$K: \dfrac{\sigma : A \text{ req } \varphi}{\sigma.A.n : \varphi} \qquad 4: \dfrac{\sigma : A \text{ req } \varphi}{\sigma.A.n : A \text{ req } \varphi} \qquad 5: \dfrac{\sigma.A.n : A \text{ req } \varphi}{\sigma : A \text{ req } \varphi} \qquad \pi: \dfrac{\sigma : \neg(A \text{ req } \varphi)}{\sigma.A.m : \neg\varphi}$$

with $\sigma.A.n$ present, $\sigma.A.m$ new, and $A \in \{U, R\}$

Fig. 6. Rules for K45 multi-modal logic

their properties. For instance $1.Alice.2 : Bob \text{ controls } login(ftp)$ tells that the privilege attribution $Bob \text{ controls } login(ftp)$ holds in the state $1.Alice.2$.

A *tableau* is a rooted dyadic tree where nodes are labelled with requests, and a *branch* B is path from the root to a leaf [14, 23]. A prefix is *present* in a branch if there is a prefixed formula with that prefix already in the branch, and it is *new* if it is not already present. Intuitively a branch is a (tentative) model for the initial formula. The rules for K45 are standard [14, 23] and shown in Fig. 6.

The rules for users under certain roles are simple and recall sequential composition for PDL [9]. They are shown in Fig. 7 together with the rules for keys. The latter require a set of propositions C to represent unused cleartexts.

Roles hierarchy (RH) require more work due to the K45 properties on roles used to get idempotence. The difficulty — and therefore the need of extra rules w.r.t. [24] — is due to the interaction of K45 requirements with the isa requirements for the relation $R^{\mathcal{I}}$. The rules are shown in Fig. 8. Rule (IK) is not necessary given rule Tr and rules $K^*, 4^*, 5^*$. However its application followed by "traditional" K45 rules (Fig. 6) is sufficient in most cases.

For user attribution it is enough to replace "R_1" with "U" and "R_1 isa" with "U has" in all rules in Fig. 8.

Notice that if isa (or has) are not nested with req (as it is normally the case for RBAC) then rules (IK) and $(I\pi)$ are the only necessary ones.

To define a tableaux proof we follow [9]. If B is the branch of a tableau, B/σ

$$\langle as \rangle : \frac{\sigma : \neg(U \text{ as } R)\,\text{req}\,\varphi}{\sigma : \neg(U \text{ req } (R\,\text{req}\,\varphi))} \qquad\qquad [as] : \frac{\sigma : (U \text{ as } R)\,\text{req}\,\varphi}{\sigma : U \text{ req } (R\,\text{req}\,\varphi)}$$

$$[Key] : \frac{\sigma : \{s\}_K \quad \sigma.K.n \text{ present}}{\sigma.K.n : \varphi} \qquad \langle Key \rangle : \frac{\sigma : \neg\{s\}_K \quad \sigma.K.m \text{ new}}{\sigma.K.m : \neg\varphi}$$

$$[sf] : \frac{\sigma : K \text{ speaks-for } P \quad \sigma : \{s\}_K}{\sigma : P \text{ req } \varphi} \qquad \langle sf \rangle : \frac{\sigma : \neg(K \text{ speaks-for } P)}{\sigma : \{c_h\}_K \quad c_h \text{ new}}{\sigma : \neg(P \text{ req } c_h)}$$

Fig. 7. Users in Roles and Keys

$$IK : \frac{\sigma : R_1 \text{ isa } R_2 \quad \sigma : R_1 \text{ req } \varphi}{\sigma : R_2 \text{ req } \varphi} \qquad I\pi : \frac{\sigma : \neg(R_1 \text{ isa } R_2)}{\sigma : R_1 \text{ req } x_i \quad x_i \text{ new}}{\sigma : \neg(R_2 \text{ req } x_i)}$$

$$IC : \frac{\sigma.R_2.n : R_1 \text{ isa } R_2}{\sigma : R_2 \text{ req } (R_1 \text{ isa } R_2)} \qquad I4 : \frac{\sigma : R_1 \text{ isa } R_2}{\sigma.R_2.n : R_1 \text{ isa } R_2}$$

$$Tr : \frac{\sigma : R_1 \text{ isa } R_2 \quad \sigma : R_2 \text{ isa } R_3}{\sigma : R_1 \text{ isa } R_3}$$

$$K^* : \frac{\sigma : R_1 \text{ req } \varphi \quad \sigma : R_1 \text{ isa } R_2}{\sigma.R_2.n : \varphi} \qquad 4^* : \frac{\sigma : R_1 \text{ req } \varphi \quad \sigma : R_1 \text{ isa } R_2}{\sigma.R_2.n : R_1 \text{ req } \varphi}$$

$$5^* : \frac{\sigma.R_2.n : R_1 \text{ req } \varphi \quad \sigma : R_1 \text{ isa } R_2}{\sigma : R_1 \text{ req } \varphi} \qquad C^* : \frac{\sigma.R_2.n : R_1 \text{ req } \varphi \quad \sigma.R_2.n : R_1 \text{ isa } R_2}{\sigma : R_2 \text{ req } (R_1 \text{ req } \varphi)}$$

Fig. 8. Roles Hierarchy

indicates the set of prefixed requests in $\mathcal{B}$ labelled with the prefix σ, i.e.:

$$\mathcal{B}/\sigma = \{\varphi \mid \langle \sigma : \varphi \rangle \in \mathcal{B}\}.$$

A prefix σ is *reduced* if π-rules are the only rules which have not been applied to requests of $\mathcal{B}/\sigma$. It is *fully reduced* if all rules have been applied. It is a *copy* of a prefix σ' if $\mathcal{B}/\sigma = \mathcal{B}/\sigma'$. In case two x_i, x_j are present in both prefix we assume them equal if they are generated by the same $R_1 \text{ isa } R_2$ (similarly for y_k and c_h). The intuition is that we should use π-rule to create a new state only if we have not seen it before.

Definition 3. A branch $\mathcal{B}$ is *π-completed* if (i) all prefixes are reduced, and (ii) for every σ' which is not fully reduced there is a fully reduced copy σ.

Definition 4. A branch $\mathcal{B}$ is *closed* if it contains both $\sigma : \varphi$ and $\sigma : \neg\varphi$, for some request φ and some prefix σ. A branch is *open* if it is π-completed and not closed. A tableau is closed if all branches are closed; it is open if at least one branch is open.

Definition 5. A *validity tableaux proof* for request φ with axioms G is the closed tableau starting with $\langle 1 : \neg\varphi \rangle$.

Theorem 6 (Strong Soundness). *If request φ has a validity proof for G then φ is a logical consequence of G.*

Completeness has been obtained with a restriction: an RBAC policy is *stable* iff the `isa`-operator occurs positively[4] only in the Role Hierarchy. This is quite natural since roles are determined by the organisation [13, 11, 26] and therefore should not be changed by the principals at run time. On the contrary user attributions and privileges can be changed and thus we do not constrain them.

Theorem 7. *Let G with $\{\neg\varphi\}$ be a stable RBAC policy, if φ is a logical consequence of G then it has a validity proof.*

Theorem 8 (Termination). *Let G with $\{\neg\varphi\}$ be a stable RBAC policy, the construction of a closed or a π-completed tableau for G and $\neg\varphi$ can be terminated by loop checking if the role hierarchy is acyclic.*

The *computational complexity* of this method needs to be investigated with more details: multi-modal logics with global axioms are EXPTIME complete [17] and therefore we cannot go below this threshold in general. Indeed the termination proof constructs a (worst case) exponential model using a concept similar to the Fisher-Ladner closure of dynamic logic [9, 10, 17]. Note in comparison that the DEC-SRC calculus is undecidable [1].

Yet, hard cases requires ad hoc constructions and are extremely unlikely to appear in global axioms for RBAC. For instance the simple restriction to user requests of the form U `req` φ, $(U$ as $R)$ `req` φ or $\{\varphi\}_K$ where φ does *not* contains `req` (that is no nested `req`) leads to a collapse down to NP since the individual relation for users and roles is a K45 relation [17]. Therefore it is realistic to expect the calculus to be tractable for practical reasoning.

As a practical example we show the deduction for Example 1 (page 4) in Fig. 9. For simplicity we suppose to have direct rules for `controls` rather than transforming it into its definition. We abbreviate *login(telnet)* in *telnet* and similarly for *run(emacs)*.

6 Related Works and Conclusions

To place this work among the various approaches to access control we can loosely classify them in two main streams[5]: transition and policy analysis.

---

[4] Under the scope of an even number of negations.

[5] Of course the very same article may deal with both of them.

$$1 : \neg(telnet \wedge emacs)$$
$$1 : maj\,\text{req}\,(fm205\,\text{has}\,User)$$
$$1 : maj\,\text{has}\,SysAdm$$
$$1 : SysAdm\,\text{req}\,(fm205\,\text{has}\,User)$$
$$1 : SysAdm\,\text{controls}\,(fm205\,\text{has}\,User)$$

$$\swarrow \quad \searrow$$

$$1 : \neg SysAdm\,\text{req}\,(fm205\,\text{has}\,User) \qquad fm205\,\text{has}\,User$$
$$\bot \qquad\qquad\qquad\qquad\qquad\qquad fm205\,\text{req}\,(telnet \wedge emacs)$$
$$\qquad\qquad\qquad\qquad\qquad\qquad User\,\text{req}\,(telnet \wedge emacs)$$
$$\qquad\qquad\qquad\qquad\qquad\qquad (*)$$

$$(*)$$
$$\swarrow \quad \searrow$$

$$1 : \neg telnet \qquad\qquad\qquad\qquad 1 : \neg emacs$$
$$1 : User\,\text{controls}\,telnet \qquad\qquad 1 : User\,\text{controls}\,emacs$$

$$\swarrow \quad \searrow : telnet \qquad\qquad\qquad \swarrow \quad \searrow$$

$1 : \neg User\,\text{req}\,telnet$	$1 : telnet$	$1 : \neg User\,\text{req}\,emacs$	$: emacs$
$1.User.2 : \neg telnet$	$\bot$	$1.User.3 : \neg emacs$	$\bot$
$1.User.2 : telnet \wedge emacs$		$1.User.3 : telnet \wedge emacs$	
$1.User.2 : telnet$		$1.User.3 : telnet$	
$1.User.2 : emacs$		$1.User.3 : emacs$	
$\bot$		$\bot$	

Fig. 9. Deduction for the request of Example 1

Transition analysis studies the evolution of the system and its *safety properties*: given some privileges and some operations to change them, prove that the system will not evolve in an undesirable state. For instance this problem can be tackled in the form of transition between access matrices [18, 27]. *Policy analysis* verifies that security mechanisms do actually fulfill the desired requirements and can be used to define decision procedure for granting access. This operation is a simple task for access matrices but it is more involved for expressive models [1, 5, 11, 22, 26]. The approaches based on deontic logic [8, 16] can also be placed within this latter form of analysis.

At present, most works on RBAC are concentrated on the aspect of policy analysis and this work is no exception to it. This is also due to the novel features of the model which has been fully characterised only recently [11, 26].

Among the features described in [11, 26] the practical reasoning mechanism proposed here captures the key points such as role hierarchies, user and privileges attributions. Static constraints (such as mutually exclusive roles) are also tackled. On the contrary dynamic separation of duties is not provided here, because the notion of session and the related dynamic attribution of roles [11, 26] have not been considered so far. In comparison with the model proposed by [15] we have decoupled the notions of privileges and sub-roles (through the operators `isa` and `controls`) which are still merged together within and-roles in [15]. Our method should make it possible a more modular construction.

The approach proposed in this paper is closer to the logic for access control developed in [1] of which is at the same time an enhancement and a restriction. In particular we have incorporated into the semantics some features advocated by [1, 22] such as idempotence of principals and refined the notions of the "speaks-for" operator of [1], which eliminate its undesired properties. For this logic it has been possible to provide soundness, completeness and decidability results, whereas the DEC-SRC logic is undecidable. However this work is also a restriction wrt [1, 22] since we do not consider delegation. Extension to incorporate a full fledged notion of delegation are planned.

Beside the definition of a logic and a semantics for RBAC, one of the contribution of this paper is the development of a decision method which extends to a logic for RBAC the tableaux methods used for the logics of knowledge and belief [14, 23] and the DEC-SRC calculus [24].

The combination of logical languages and decision methods based on logic may be a step further towards the use of formal methods for verification and design of RBAC with practical reasoning tools in computer security.

Acknowledgments

I would like to thank L. Paulson and the Computer Laboratory for their hospitality in Cambridge, M. Abadi, the Computer Security group (Cambridge), and the anonymous referees for many suggestions which helped to improve this paper. This research has been partly supported by ASI, CNR and MURST 40% and 60% grants and by EPSRC grant GR/K77051 "Authentication Logics".

References

1. M. Abadi, M. Burrows, B. Lampson, and G. Plotkin. A calculus for access control in distributed systems. *ACM Trans. on Programming Languages and Systems*, 15(4):706–734, 1993.
2. R. Anderson. A security policy model for clinical information systems. In *Proc. of the Symp. on Security and Privacy*. IEEE Press, 1996.
3. B. Beckert and R. Goré. Free variable tableaux for propositional modal logics. In *Proc. of TABLEAUX-97*, LNAI. Springer-Verlag, 1997. To appear.
4. D. Bell and L. La Padula. Secure computer systems: unified exposition and MULTICS. Report ESD-TR-75-306, The MITRE Corporation, March 1976.
5. E. Bertino, S. Jajodia, and P. Samarati. Supporting multiple access control policies in database systems. In *Proc. of the Symp. on Security and Privacy*, pp. 94–109. IEEE Press, 1996.
6. M. Burrows, M. Abadi, and R. Needham. A logic for authentication. *ACM Trans. on Comp. Sys.*, 8(1):18–36, 1990. Also as research report SRC-39, DEC - System Research Center, 1989.
7. D. Clark and D. Wilson. A comparison of commercial and military computer security policies. In *Proc. of the Symp. on Security and Privacy*, pp. 184–194. IEEE Press, 1987.
8. F. Cuppens and R. Demolombe. A deontic logic for reasoning about confidentiality. In *3rd Int. Workshop on Deontic Logic in Computer Science*, Portugal, 1996.
9. G. De Giacomo and F. Massacci. Tableaux and algorithms for propositional dynamic logic with converse. In *Proc. of the 13th Int. Conf. on Automated Deduction (CADE-96)*, LNAI 1104 , pp. 613–628. Springer-Verlag, 1996.

10. R. Fagin, J. Halpern, Y. Moses, and M. Vardi. *Reasoning about Knowledge*. The MIT Press, 1995.
11. D. Ferraiolo, J. Cugini, and K. Richard. Role-based access control (RBAC): Features and motivations. In *Proc. of the Annual Computer Security Applications Conf.*. IEEE Press, 1995.
12. D. Ferraiolo and R. Kuhn. Role based access control. In *Proc. of the NIST-NCSC Nat. (U.S.) Comp. Security Conf.*, pp. 554–563, 1992.
13. D. Ferraiolo, D. Gilbert, and N. Lynch. An examination of federal and commercial access control policy needs. In *Proc. of the NIST-NCSC Nat. (U.S.) Comp. Security Conf.*, pp. 107–116, 1993.
14. M. Fitting. *Proof Methods for Modal and Intuitionistic Logics*. Reidel, 1983.
15. L. Giuri and P. Iglio. A formal model for role based access control with constraints. In *Proc. of the Computer Security Foundations Workshop*, pp. 136–145. IEEE Press, 1996.
16. J. Glasgow, J. MacEwen, and P. Panangaden. A logic for reasoning about security. In *Proc. of the Symp. on Security and Privacy*, pp. 2–13. IEEE Press, 1990.
17. J. Halpern and Y. Moses. A guide to completeness and complexity for modal logics of knowledge and belief. *Artificial Intelligence*, 54:319–379, 1992.
18. M. Harrison, W. Ruzzo, and J. Ullman. Protection in operating systems. *Comm. of the ACM*, 19(8):461–471, 1976.
19. S. Kanger. Law and logic. *Theoria*, 38(3):105–132, 1972.
20. C. Krogh. Obligations in multiagent systems. In *Scandinavian Conf. on Artificial Intelligence (SCAI-95)*, pp. 29–31. ISO Press, 1995.
21. B. Lampson. Protection. *ACM Operating Sys. Reviews*, 8(1):18–24, 1974.
22. B. Lampson, M. Abadi, M. Burrows, and E. Wobber. Authentication in distributed systems: Theory and practice. *ACM Trans. on Computer Systems*, 10(4):265–310, 1992.
23. F. Massacci. Strongly analytic tableaux for normal modal logics. In *Proc. of the Int. Conf. on Automated Deduction (CADE-94)*, LNAI 814, pp. 723–737. Springer Verlag, 1994.
24. F. Massacci. Tableaux methods for access control in distributed systems. In *Proc. of TABLEAUX-97*, LNAI. Springer-Verlag, 1997. To appear.
25. C. McCollum, J. Messing, and L. Notargiacomo. Beyond the pale of MAC and DAC - defining new forms of access control. In *Proc. of the Symp. on Security and Privacy*, pp. 190–200, IEEE Press, 1990.
26. R. Sandhu, E. Coyne, H. Feinstein, and C. Youman. Role-based access controls models. *IEEE Computer*, 29(2), February 1996.
27. R. Sandhu. The typed access matrix model. In *Proc. of the Symp. on Security and Privacy*, pp. 122–136. IEEE Press, 1992.
28. R. Sandhu and P. Samarati. Access control: Principles and practice. *IEEE Communications Magazine*, pp. 40–48, September 1994.
29. P. Syverson. The use of logic in the analysis of cryptographic protocols. In *Proc. of the Symp. on Security and Privacy*, pp. 156–170. IEEE Press, 1991.
30. P. Syverson and P. van Oorschot. On unifying some cryptographic protocols logics. In *Proc. of the Symp. on Security and Privacy*. IEEE Press, 1994.

Process Modeling with Different Qualities of Knowledge

Wolfgang May*

Institut für Informatik, Universität Freiburg
Am Flughafen 17, 79110 Freiburg, Germany
may@informatik.uni-freiburg.de

Abstract. For modeling structured processes with different levels of atomic actions, classical linear or branching time Kripke structures with first-order states are insufficient: They do not provide any means for modeling independent parallel threads of activity which have to be joined at some point, action refinement, or procedure concepts. In all three cases, it is important to distinguish facts resp. knowledge derived in the current thread of activity from facts which are not concerned in this activity. This problem is also closely related to the frame problem.

In this paper, hierarchical Kripke structures are introduced for modeling hierarchically structured processes, also coping with the different qualities of knowledge arising in this context: every state consists of a total first-order interpretation, which gives the state as-is, and a partial first-order interpretation containing all procedure knowledge which has been derived in the current thread of activity.

A correct and complete set of axioms is presented for reasoning about hierarchical Kripke structures.

1 Introduction

Variants of Kripke structures are widely used as a model-theoretic base for reasoning about processes, for instance with state-oriented logics as CTL or Dynamic Logic [Har79]. Furthermore, the more process-oriented Hennessy-Milner Logic [HM85] is interpreted over Kripke structures. In some contexts, for example Transaction Logic [BK94], logics refer to path-structures which also can be regarded as variants of Kripke structures.

In any case, modeling independent parallel threads of activity which are re-joined at some synchronization point causes problems. Additionally, action abstraction or refinement is not supported in those models. The problems arising by those features are closely related to the frame problem [MH69]: At the beginning of a (sub)process, there is no local knowledge, the state as-is is the same as the state where the process has been started from. Every thread of activity reads and derives its own facts, making up its local knowledge about the global process. By carrying out actions, a process gains local knowledge from examining the state as-is and modifying it. When joining different threads of activity, the local knowledge after the join consists of the local knowledge of the predecessor state modified by the accumulated knowledge of all incoming threads. The state

---

* Supported by grant no. GRK 184/1-97 of the Deutsche Forschungsgemeinschaft.

as-is is obtained from the predecessor state, overwritten by the sum of the local knowledge of all participating threads.

From the model-theoretic point of view, the problem is solved by modeling different qualities of knowledge in the states: every state consists of a total first-order interpretation, giving the state as-is, and a partial first-order interpretation containing all procedure knowledge which has been derived in the current thread of activity.

The paper is structured as follows: This section closes with a short glance on related work on structured process modeling. Section 2 briefly reviews of first-order Kripke structures. In Section 3, the algebra of actions used in this paper is introduced. In a first step, in Section 4, the modeling of parallel actions is introduced which is extended in Section 5 with procedures to the complete scope of hierarchical processes. Section 6 contains an illustrating example, Section 7 closes with some concluding remarks.

Related Work. In [BK94], a concept for defining transactions as sequences of elementary actions in a logic-programming style is defined, based on the model-theoretic notion of *path structures*. States and transitions are represented by an abstract concept of a theory and a transition oracle which can be instantiated to any specific formal framework. Parallelism is modeled via interleaving.

From the process-algebraic point of view, action refinement has been dealt with in [DG91, DG95, AH93]. In those works, only the dynamic aspects are considered, thus there is no notion of *states*.

In [AH96], an abstract framework for reactive modules is presented which permits parallel composition and abstraction from the internal behaviour of modules. There, the focus is on the observable behaviour of communication variables, the transitions performed by composed modules are given in a declarative style, similar to the transition oracle of [BK94].

From the software engineering point of view, the concept can further be extended with the usual modularization concepts of import, export, visible signature, renaming, and hiding (cf. [BHK90]).

Specialized instances of the concept have been used for modeling nested transactions in the state-oriented deductive database language Statelog [LML96], and for extending the Evolving Algebra specification method [May97]. Here, the concept is presented from the model-theoretic point of view in its full generality.

2 First-Order Kripke Structures

Every language of first-order logic includes the symbols) and (, the boolean connectives $\neg$ and $\vee$, the quantifier $\exists$, and an infinite set of variables $\mathsf{Var} := \{x_1, x_2, \dots\}$. A particular language is given by its *signature* Σ consisting of function symbols and predicate symbols with fixed arities $\mathrm{ord}(f)$ resp. $\mathrm{ord}(p)$. Terms, first-order formulas, and the notions of bound and free variables are defined as usual.

A *first-order interpretation* $\mathfrak{I} = (I, \mathfrak{U})$ over a signature Σ consists of a non-empty set $\mathfrak{U}$ (*universe*) and a mapping I which maps every function symbol $f \in \Sigma$ to a function $I(f) : \mathfrak{U}^{\mathrm{ord}(f)} \to \mathfrak{U}$ and every predicate symbol $p \in \Sigma$ to a function $I(p) : \mathfrak{U}^{\mathrm{ord}(f)} \to \{\mathrm{T}, \mathrm{F}\}$. For *partial* interpretations, $I(f)$ and $I(p)$ are partial functions/mappings. The set of interpretations (partial interpretations) over a fixed signature Σ is denoted by $\mathbb{I}_\Sigma$ ($\mathbb{I}_\Sigma^p$).

Definition 1 A first-order *Kripke structure* over a signature Σ is a triple $\mathfrak{K} = (\mathfrak{G}, \mathfrak{R}, \mathfrak{M})$ where $\mathfrak{G}$ is a set of states, $\mathfrak{R} \subseteq \mathfrak{G} \times \mathfrak{G}$ an *accessibility relation*, and for every $g \in \mathfrak{G}$, $\mathfrak{M}(g) = (M(g), \mathfrak{U}(g))$ is a first-order interpretation of Σ with universe $\mathfrak{U}(g)$. $\mathfrak{G}$ and $\mathfrak{R}$ are called the *frame* of $\mathfrak{K}$. In some cases, for a set $\mathfrak{X}$, the accessibility relation $\mathfrak{R} \subseteq \mathfrak{G} \times \mathfrak{X} \times \mathfrak{G}$ is labeled[2].

A *path* p in a Kripke structure $\mathfrak{K} = (\mathfrak{G}, \mathfrak{R}, \mathfrak{M})$ is a sequence $p = (g_0, g_1, g_2, \ldots)$, $g_i \in \mathfrak{G}$ with $\mathfrak{R}(g_i, g_{i+1})$ holding for all i. □

3 The Action Algebra

Allowing composite actions, the set A of actions is given as an algebra, based on a set Σ_A of action symbols (analogous to function and predicate symbols), the underlying data universe $\mathfrak{U}$ and a set op_A of operators for composing actions.

Since for modeling distributed processes, parallelism is a crucial point and modeling of parallelism by interleaving is not always satisfactory, a truly parallel composition is needed. *Parallel composition* of actions is denoted by $\|$. $\in_\|$ denotes membership in parallel composition: For an elementary action a, $a \in_\| b :\Leftrightarrow a = b \vee (\exists c, d : b = (c \| d) \wedge (a \in_\| c \vee a \in_\| d))$, i.e. if b is already elementary, or a is a parallel component of some parallel component of b.

Actions consisting of different alternative subactions which can be executed in a single step are constructed by the *alternative composition* $\vee$.

Although sequential composition is modeled naturally by successive steps, it should be possible to combine actions by sequential composition which should be executed atomically in a single step. *Sequential composition* to an *atomically executing* action is denoted by $\otimes$.

For iterative computations, *finite iteration* a^ω and *iterative closure* $a^\star$ of an action a are defined. On the respective level of abstraction, a^ω and $a^\star$ are regarded as atomic actions.

Also, for implementing abstraction and action refinement, procedure definitions are provided, altogether summarized in the following definition:

Definition 2

- For an n-ary action symbol a which does not occur on the left side of a procedure definition (see third item of this definition) and $u_1, \ldots, u_n \in \mathfrak{U}$, $a(u_1, \ldots, u_n)$ is an *elementary action*.
- For actions $a, a_1, a_2, a_1 \| a_2, a_1 \otimes a_2, a \vee b, a^\omega$ and $a^\star$ are composite actions.
- For an action a with arguments $(u_1, \ldots, u_n)$ and an action symbol p, $p = p(u_1, \ldots, u_n) := a$, is a *procedure definition*. p is called a *procedure*.

Since procedure symbols are also action symbols, procedures are also actions, thus, A is a non-free algebra. With this, action refinement via procedure definitions fits easily into the concept: from an abstract point of view, an action

---

[2] cf. Dynamic Logic [Har79], Hennessy-Milner Logic [HM85], or Transaction Logic [BK94]; in these cases, $\mathfrak{X}$ corresponds to a set of actions.

symbol p is used for an elementary action, whereas from a more detailed level, it is used as a procedure symbol.

The sets of elementary resp. composite actions are denoted by A_e resp. A_c, the set of procedures is denoted by A_p. □

Definition 3 For an interpretation $\mathfrak{I}$ and an action a, the state reached by executing a in $\mathfrak{I}$ is denoted by $a(\mathfrak{I})$. For nondeterministic a, $a(\mathfrak{I})$ is set-valued. □

3.1 Specification of Actions

The modeling of actions directly leads to the frame problem [MH69]: it has to be specified, which facts – dependent on the current state – *must* hold in the state reached by execution of a particular action, and which facts are simply not effected. The former are reflected in the specification of the action, whereas the latter are assumed to remain unchanged which is normally incorporated into frame axioms. Thus, the *specification* of an action *alone* should not give a *total* interpretation of the successor state – since then there would be no room for parallel composition. Instead, the specification formalism should talk about *partial* interpretations, which can be combined easily.

For composing (partial) interpretations, semantic counterparts to terms and atomic formulas are needed:

Definition 4 Given an interpretation $\mathfrak{I}$ with universe $\mathfrak{U}$ over a signature Σ, a symbol $v \in \Sigma$ and $u_1, \dots, u_{\text{ord}(v)} \in \mathfrak{U}$, the tuple $(v; u_1, \dots, u_{\text{ord}(v)})$ is called a *function application* over Σ and $\mathfrak{U}$ if v is a function symbol, and it is called a *predicate application* if v is a predicate symbol. □

Definition 5 The (semantical) *domain* of a (partial) interpretation $\mathfrak{I} = (I, \mathfrak{U})$ over a signature Σ is the (mathematical) domain of the mapping I:

$$\text{dom}(\mathfrak{I}) := \{P \mid P = (p; u_1, \dots, u_m) \text{ a predicate application over } \Sigma \text{ and } \mathfrak{U} \text{ and}$$
$$(I(p))(u_1, \dots, u_m) \text{ is defined}\} \ \cup$$
$$\{F \mid F = (f; u_1, \dots, u_n) \text{ a function application over } \Sigma \text{ and } \mathfrak{U} \text{ and}$$
$$(I(f))(u_1, \dots, u_n) \text{ is defined}\} \ . \qquad \square$$

Note, that these sets are subsets of the Herbrand base and the Herbrand universe.

Definition 6 In the following, let all interpretations be over a fixed signature Σ and universe $\mathfrak{U}$. Let $\sqcup$ denote the least upper bound of two elements in the flat lattices given by ($\bot < u < \top$ for all $u \in \mathfrak{U}$) resp. ($\bot < t < \top$ for $t \in \{\top, F\}$). For two (partial) interpretations $\mathfrak{I}$ and $\mathfrak{I}'$, the *union* $\mathfrak{J} := \mathfrak{I} \cup \mathfrak{I}'$ is defined as

$$(J(p))(u_1, \dots, u_{\text{ord}(p)}) := \sqcup((I(p))(u_1, \dots, u_{\text{ord}(p)}), (I'(p))(u_1, \dots, u_{\text{ord}(p)}))$$
$$(J(f))(u_1, \dots, u_{\text{ord}(f)}) := \sqcup((I(f))(u_1, \dots, u_{\text{ord}(f)}), (I'(f))(u_1, \dots, u_{\text{ord}(f)})) \ .$$

$\mathfrak{J} = \mathfrak{I} \cup \mathfrak{I}'$ is *consistent*, if $J(p)(\bar{u}) \neq \top$ and $J(f)(\bar{u}) \neq \top$ for all $\bar{u} \in \mathfrak{U}^\omega$. In this case, $\mathfrak{J}$ is a partial interpretation over Σ with universe $\mathfrak{U}$.

For an interpretation $\mathfrak{I}$ over Σ and sets F of function applications and P of predicate applications over Σ and $\mathfrak{U}$, the restriction $\mathfrak{I}|_R = (I|_R, \mathfrak{U})$ of $\mathfrak{I}$ to $R := F \cup P$ is defined as

$$(I|_R \ (p))(u_1, \dots, u_n) := \begin{cases} (I(p))(u_1, \dots, u_n) & \text{if } (p; u_1, \dots, u_n) \in P \\ \bot & \text{otherwise, and} \end{cases}$$

$$(I|_R \ (f))(u_1, \dots, u_n) := \begin{cases} (I(f))(u_1, \dots, u_n) & \text{if } (f; u_1, \dots, u_n) \in F \\ \bot & \text{otherwise} \end{cases}$$

For modifying an interpretation in some parts with a (partial) interpretation, an "overwriting" operator is needed: given two interpretations $\Im$ and $\Im'$ over Σ and sets F of function applications and P of predicate applications over Σ and $\mathfrak{U}$, the *superposition* $\Im = (J, \mathfrak{U}) = \Im \uplus_R \Im'$ of $\Im$ with $\Im'$ on $R := F \cup P$ is defined as

$$(J(p))(u_1, \ldots, u_n) := \begin{cases} (I'(p))(u_1, \ldots, u_n) \text{ if } (p; u_1, \ldots, u_n) \in P \\ (I(p))(u_1, \ldots, u_n) \text{ otherwise, and} \end{cases}$$

$$(J(f))(u_1, \ldots, u_n) := \begin{cases} (I'(f))(u_1, \ldots, u_n) \text{ if } (f; u_1, \ldots, u_n) \in F \\ (I(f))(u_1, \ldots, u_n) \text{ otherwise} \end{cases}.$$

For two interpretations $\Im$, $\Im'$, $\Im \uplus \Im' := \Im \uplus_{\mathrm{dom}(\Im')} \Im'$. □

The specification of an action a can be given by partial interpretations as

$$\mathrm{Spec}(\mathsf{a}) := \{(\mathfrak{P}, \mathfrak{Q}) \in \mathbb{I}_\Sigma^p \times \mathbb{I}_\Sigma^p \mid \text{for every } \Im \in \mathbb{I}_\Sigma \text{ with } \Im \supseteq \mathfrak{P},$$
$$\mathsf{a} \text{ is executable in } \Im \text{ and } \Im \uplus_{\mathrm{dom}(\mathfrak{Q})} \mathfrak{Q} \in \mathsf{a}(\Im)\} .$$

A specification in this way allows a constructive computation of the successor state, e.g. it can be used to define a partial transition oracle [BK94].

Definition 7 For an action a and an interpretation $\Im$,

$$\mathsf{a}^{part}(\Im) := \{\mathfrak{Q} \mid (\mathfrak{P}, \mathfrak{Q}) \in \mathrm{Spec}(\mathsf{a}) \text{ and } \mathfrak{P} \subseteq \Im\} .$$ □

Proposition 1 *For an interpretation $\Im$ and an action a,*

$$\mathsf{a}(\Im) = \{\Im \mid \exists \mathfrak{Q} \in \mathsf{a}^{part}(\Im) : \Im = \Im \uplus \mathfrak{Q}\} .$$ □

For composite actions, it is convenient to orient the interpretation on their representation in the action algebra. Elementary actions and operators are interpreted separately (usually the semantics of the operators is predefined on a meta-level while the actions are interpreted according to the application domain).

Then, the specifications of composed actions are composed from the specifications of their elementary subactions, i.e. the operators also act as operations on specifications, making specifications homomorphic and the diagram at the right commutative.

$$\begin{array}{ccc} & \mathrm{Spec} & \\ \mathbb{A} & \longrightarrow & \mathbb{I}^p \times \mathbb{I}^p \\ {\scriptstyle\mathrm{op}_\mathbb{A}}\downarrow & & \downarrow{\scriptstyle\mathrm{op}_{\mathrm{Spec}}} \\ \mathbb{A} & \longrightarrow & \mathbb{I}^p \times \mathbb{I}^p \\ & \mathrm{Spec} & \end{array}$$

Definition 8 For a given set $\mathbb{A}_e$ of elementary actions and a specification $\mathrm{Spec}(\mathbb{A}_e)$ formulated by relations of partial interpretations,

$$\mathrm{Spec}(\mathsf{a} \parallel \mathsf{b}) = \{(\mathfrak{P}, \mathfrak{Q}) \mid \mathfrak{P}, \mathfrak{Q} \in \mathbb{I}_\Sigma^p \text{ and there are } (\mathfrak{P}_a, \mathfrak{Q}_a) \in \mathrm{Spec}(\mathsf{a}),$$
$$(\mathfrak{P}_b, \mathfrak{Q}_b) \in \mathrm{Spec}(\mathsf{b}) \text{ s.t. } \mathfrak{P} = \mathfrak{P}_a \cup \mathfrak{P}_b \text{ and } \mathfrak{Q} = \mathfrak{Q}_a \cup \mathfrak{Q}_b\} ,$$

$$\mathrm{Spec}(\mathsf{a} \otimes \mathsf{b}) = \{(\mathfrak{P}, \mathfrak{Q}) \mid \mathfrak{P}, \mathfrak{Q} \in \mathbb{I}_\Sigma^p \text{ and there are } (\mathfrak{P}_a, \mathfrak{Q}_a) \in \mathrm{Spec}(\mathsf{a}),$$
$$(\mathfrak{P}_b, \mathfrak{Q}_b) \in \mathrm{Spec}(\mathsf{b}) \text{ s.t. } \mathfrak{Q}_a \cup \mathfrak{P}_b \text{ is consistent and}$$
$$\mathfrak{P} = \mathfrak{P}_a \cup \mathfrak{P}_b|_{\mathrm{dom}(\mathfrak{P}_b)\backslash\mathrm{dom}(\mathfrak{Q}_a)} \text{ and } \mathfrak{Q} = \mathfrak{Q}_a \uplus \mathfrak{Q}_b\} ,$$

$$\mathrm{Spec}(\mathsf{a} \vee \mathsf{b}) = \mathrm{Spec}(\mathsf{a}) \cup \mathrm{Spec}(\mathsf{b}) \cup \mathrm{Spec}(\mathsf{a} \parallel \mathsf{b}) ,$$

$$\mathrm{Spec}(\mathsf{a}^\omega) = \{(\mathfrak{P}, \mathfrak{Q}) \mid \exists n : (\mathfrak{P}, \mathfrak{Q}) \in \mathrm{Spec}(\underbrace{\mathsf{a} \otimes \ldots \otimes \mathsf{a}}_{n-\text{times}})\} ,$$

$$\mathrm{Spec}(\mathsf{a}^\star) = \{(\mathfrak{P}, \mathfrak{Q}) \mid \exists n : (\mathfrak{P}, \mathfrak{Q}) \in \mathrm{Spec}(\underbrace{\mathsf{a} \otimes \ldots \otimes \mathsf{a}}_{n-\text{times}}) \text{ and}$$

$$\text{there is no } (\Im, \Im) \in \mathrm{Spec}(\mathsf{a}) \text{ s.t. } \mathfrak{Q} \cup \Im \text{ is consistent}\} .$$

(the additional condition in $\mathsf{a}^\star$ states that a cannot be applied any more.)□

Example 1 Let the two actions copy and inc be given by pairs of partial interpretations as follows: $\mathsf{Spec}(\mathsf{copy}) = \{((\{x = N\}, \{x.bak = N\}))\}$, $\mathsf{Spec}(\mathsf{inc}) = \{((\{x = N\}, \{x = N+1\}))\}$.

Then, $\mathsf{Spec}(\mathsf{copy} \parallel \mathsf{inc}) = \{((\{x = N\}, \{x.bak = N, x = N+1\}))\}$, $\mathsf{Spec}(\mathsf{copy} \otimes \mathsf{inc}) = \{((\{x = N\}, \{x.bak = N, x = N+1\}))\}$, and $\mathsf{Spec}(\mathsf{inc} \otimes \mathsf{copy}) = \{((\{x = N\}, \{x.bak = N+1, x = N+1\}))\}$. □

Proposition 2 *With the above specification, the operators $\parallel$ and $\vee$ are commutative, all three binary operators are associative, and $*$ is idempotent.* □

4 Modeling Parallel Composite Actions with Kripke Structures

In this section, only parallel composite actions are considered, i.e., $A = \langle \parallel, A_e \rangle$. The state space is augmented with auxiliary states which result from application of a single action from another state. The states emerging from parallel execution of several actions are computed from those auxiliary states. In addition to the temporal successor relation $\mathfrak{R}$, an accessibility relation $\mathfrak{T}$ of auxiliary states and a relation $\mathfrak{S}$ denoting the return of results are introduced. All accessibility relations are labeled with the actions executed in the corresponding transition. Furthermore, the semantics of processing the results has to be specified.

Definition 9 A *Kripke structure with elementary parallel changes* is a tuple $\mathfrak{K} = (\mathfrak{G}, \mathfrak{G}_p, \mathfrak{A}, \mathfrak{T}, \mathfrak{R}, \mathfrak{S}, \mathfrak{M})$ (cf. Figure 1):

$\mathfrak{G}$ is the set of *main states*,

$\mathfrak{G}_p$ is the set of *auxiliary states*,

$\mathfrak{A}$ is a set of *labels*, which are used to represent actions.

$\mathfrak{R} \subseteq \mathfrak{G} \times \mathfrak{A} \times \mathfrak{G}$ is the *temporal successor relation*,

$\mathfrak{T} \subseteq \mathfrak{G} \times A_e \times \mathfrak{G}_p$ is the *accessibility relation of auxiliary states from main states*,

$\mathfrak{S} \subseteq \mathfrak{G}_p \times A_e \times \mathfrak{G}$ is the *return relation from auxiliary states to main states*,

$\mathfrak{M}$ is a mapping $\mathfrak{G} \to \mathbb{I}$ and $\mathfrak{G}_p \to \mathbb{I}^p$, giving an interpretation for every state.

The tuple $(\mathfrak{G}, \mathfrak{G}_p, \mathfrak{A}, \mathfrak{T}, \mathfrak{R}, \mathfrak{S})$ is called a *Kripke frame with elementary parallel changes*.

For $h \in \mathfrak{G}$ and $a \in A_e$ and $h \in \mathfrak{G}$, let $\mathfrak{S}^{-1}(a, h) := \{g \mid \mathfrak{S}(g, a, h)\}$, denoting the set of $\mathfrak{S}$-input states for an action a and a main state h. □

The labels must provide a detailed description of the transitions, containing sufficient information to allow a reasonably strong axiomatization of the possible models of a given specification of actions.

The auxiliary states reached by elementary actions can be axiomatized directly from the specification Spec_{A_e} of elementary actions:

$$\mathfrak{T}(g, a, h) \;\Rightarrow\; \mathfrak{M}(h) \in a^{part}(\mathfrak{M}(g)) \quad .$$

For the main states, for every transition $g \xrightarrow{a} h$, it has to be determined from the labeling of $\mathfrak{R}$, which $\mathfrak{S}$-inputs have to be considered:

Definition 10 For every transition $g \xrightarrow{a} h$ modeling execution of an action a, every elementary parallel component action $b \in_{\parallel} a$ is modeled by $\mathfrak{Q}(g, b, i)$ and $\mathfrak{S}(i, b, h)$ for some auxiliary state $i \in \mathfrak{G}$. Then,

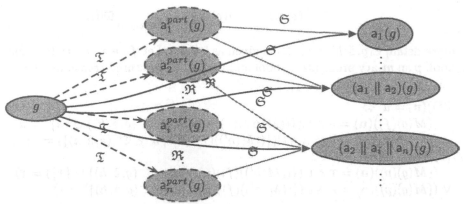

▶ represents the computation of the union of the procedure knowledge of the
$\mathfrak{S}$-inputs corresponding to an $\mathfrak{R}$-transition and superposing it to the $\mathfrak{R}$-input.
(Here, $a(\_)$ denotes some element of $a(\_)$ as defined in Def. 3 resp. Def. 7.)

Fig. 1. Fragment of a Kripke Structure with elementary parallel changes

---

- $b[i]$ is the *share* of b of $g \xrightarrow{a} h$, and
- if $a = a_1 \| \ldots \| a_n, a_{n+1}, \ldots, a_m \in A_e$ and for every a_i, $a_i[g_i]$ is the share of
  a_i of $g \xrightarrow{a} h$, then $\tilde{a} := a_1[g_1] \| \ldots \| a_n[g_n]$ is the *parallel decomposition* of
  $g \xrightarrow{a} h$. $(g, \tilde{a}, h)$ is called the *structured* transition.   □

A *consensus specification* $\mathfrak{C}$ is a function which maps a structured transition
$(g, \tilde{a}, h)$ to a first-order interpretation (of the target state) over a suitable signature by evaluating the respective $\mathfrak{S}$-input states:

Definition 11 For a structured transition $(g, \tilde{a}, h)$,
$$\mathfrak{C}^{\leftarrow}(g, \tilde{a}, h) := \{g' \mid \exists b : b[g'] \in_\| \tilde{a}\}$$
is the set of auxiliary states which have to be considered as results of component
actions for determining $\mathfrak{M}(h)$.   □

The definition shows that setting $\mathfrak{A}$ (the set of labels of $\mathfrak{R}$) to the set of parallel decompositions of transitions encodes sufficient information into the Kripke
frame: For every modeled transition $g \xrightarrow{a} h$, $\mathfrak{C}^{\leftarrow}$ can be determined from $\tilde{a}$. Then,
$\mathfrak{R}$ yields exactly the structured transitions modeled by a Kripke structure. From
this, additional requirements on Kripke frames arise:

Definition 12 A Kripke frame with elementary parallel changes is

- *complete*, if for every $\mathfrak{R}(g, \tilde{a}, h)$ and $b \in A_e$ with $b \in_\| a$, there is a h' s.t.
  $\mathfrak{S}(h', b, h)$, i.e., for every elementary component action, there is some state
  which is queried as its result.
- *consistent* if for all $g, h \in \mathfrak{G}$, $h' \in \mathfrak{G}_p$, $a, b \in A$: $(\mathfrak{R}(g, \tilde{a}, h) \wedge \mathfrak{S}(h', b, h) \wedge$
  $b \in_\| a) \Rightarrow \mathfrak{T}(g, b, h')$, i.e., only states are evaluated as results of an action
  which actually arise from executing this action.   □

Now, for complete and consistent Kripke frames (and only those are reasonable),
the target states of $\mathfrak{R}$ can be axiomatized:

Theorem 3 *The following consensus function $\mathfrak{C}$ implements the specification of*
$\|$ *given in Definition 8:*

$$\mathfrak{C}(g,\tilde{a},h) = \mathfrak{M}(g) \uplus \bigcup\nolimits_{h' \in \mathfrak{C}^{\leftarrow}(g,\tilde{a},h)} \mathfrak{M}(h') .$$

□

In more detail, $\mathfrak{C}(g,\tilde{a},h) = (C,\mathfrak{U})$ is given as follows: Let f be an n-ary function symbol, p an m-ary predicate symbol and $(g,\tilde{a},h)$ a structured transition. Then, $\mathfrak{C}(g,\tilde{a},h) = (C,\mathfrak{U})$ where

$(C(f))(\mathbf{u}) = u \Leftrightarrow$
$\qquad ((M(g)(f))(\mathbf{u}) = u \wedge \sqcup(\{((M(h'))(f))(\mathbf{u}) \mid h' \in \mathfrak{C}^{\leftarrow}(g,\tilde{a},h)\} \cup \{u\}) = u)$
$\qquad \vee ((M(g)(f))(\mathbf{u}) = v \neq u \wedge \sqcup(\{((M(h'))(f))(\mathbf{u}) \mid h' \in \mathfrak{C}^{\leftarrow}(g,\tilde{a},h)\}) = u)$
$(C(p))(\mathbf{u}) = \text{T} \Leftrightarrow$
$\qquad ((M(g)(p))(\mathbf{u}) = \text{T} \wedge \sqcup(\{((M(h'))(f))(\mathbf{u}) \mid h' \in \mathfrak{C}^{\leftarrow}(g,\tilde{a},h)\} \cup \{\text{T}\}) = \text{T})$
$\qquad \vee ((M(g)(p))(\mathbf{u}) = \text{F} \wedge \sqcup(\{((M(h'))(f))(\mathbf{u}) \mid h' \in \mathfrak{C}^{\leftarrow}(g,\tilde{a},h)\}) = \text{T})$

Definition 13 A labeled Kripke structure with parallel changes is a model of a specification Spec_A and a consensus specification $\mathfrak{C}$, if its frame is complete and consistent, and for all $g,h \in \mathfrak{G}$ and $a \in \mathfrak{A}$,

$$\mathfrak{T}(g,a,h) \Rightarrow a \in A_e \wedge \mathfrak{M}(h) \in a^{part}(\mathfrak{M}(g)) \quad , \text{ and}$$
$$\mathfrak{R}(g,\tilde{a},h) \Rightarrow \mathfrak{M}(h) = \mathfrak{C}(g,\tilde{a},h) .$$

□

Example 2 Imagine four fields, x_1, x_2 and y_1, y_2 and some boxes which can be positioned on them, and two nondeterministic actions a_1: if a ball lies on x_1, put it on y_1 or y_2, and similarly a_2 for x_2.
Let g be a state s.t. two identical boxes are positioned on x_1 and x_2, and y_1, y_2 are empty, and additional states as specified:

State	x_1	x_2	y_1	y_2
g	B	B	$\neg B$	$\neg B$
g_1	$\neg B$	u	B	u
g_2	$\neg B$	u	u	B
g_3	u	$\neg B$	B	u
g_4	u	$\neg B$	u	B
h	$\neg B$	$\neg B$	B	B

Obviously, $a_1(g) = \{g_1,g_2\}$, $a_2(g) = \{g_3,g_4\}$ and $(a_1 \parallel a_2)(g) = \{h\}$, which is reached by the structured transitions $\mathfrak{R}(g, a_1[g_1] \parallel a_2[g_4], h)$ and $\mathfrak{R}(g, a_1[g_2] \parallel a_2[g_3], h)$. Then, $\mathfrak{C}^{\leftarrow}(g, a_1[g_1] \parallel a_2[g_4], h) = \{g_1, g_4\}$ and $\mathfrak{C}^{\leftarrow}(g, a_1[g_2] \parallel a_2[g_3], h) = \{g_2, g_3\}$.

□

Furthermore, if for every state and every elementary action there is at most one possible auxiliary state to be used as $\mathfrak{S}$-input, it is sufficient to set the set of labels $\mathfrak{A} := A$:

Definition 14 A Kripke frame with elementary parallel changes is *unambiguous*, if for every $\mathfrak{S}(g,a,h)$ and $\mathfrak{S}(g',a,h)$, $g = g'$, i.e. for each component action there is at most one $\mathfrak{S}$-input.

□

Proposition 4 *For an unambiguous frame, the set of (partial) states to be considered for establishing consensus can be determined by looking only at the accessibility relations:*

$$\mathfrak{C}^{\leftarrow}(g,\tilde{a},h) = \mathfrak{C}^{\leftarrow}(g,a,h) := \{h' \mid \exists b \in A_e : b \in_{\parallel} a \wedge h' \in \mathfrak{S}^{-1}(b,h)\} \subset \mathbb{I}^p$$

□

Remark: By state-splitting, every Kripke frame can be made unambiguous by preserving all safety, guarantee, and fairness properties.

5 Hierarchical Kripke Structures: Complex Actions, Procedures

Until now, only relatively elementary computations can be performed atomically by single transitions. Thus, it is desirable to perform complex computations as atomic steps or to refine actions by more detailed actions. With respect to the Kripke structures developed in the preceding section, this means to compute the auxiliary states not only by application of single actions (via $\mathfrak{T}$), but also as results of complex computations.

The model-theoretic treatment of the hierarchical concept requires a structured state space: internal computations of complex actions have to be hidden from the superordinate computation. Parallel actions which have to be regarded as atomic see and work on different views of the world. Only when the subactions commit, their results are joined and become visible in the common successor state. This leads to a modular concept with hidden data. To allow a homogenous treatment of procedures and actions – which is also important for incorporating action-refinement –, updates raised by procedures have to be handled in the same way as updates resulting from elementary actions. Thus, besides the *temporal* aspect, also the *hierarchical* aspect has to be modeled by the accessibility relations.

Inherited and Derived "Knowledge". In order to describe complex subcomputations – possibly consisting of several levels – , all states have to be total interpretations since also states on lower levels can be starting points for subcomputations. Thus, using partial interpretations as presented in the preceding section, fails. On the other hand, it must be possible to distinguish between facts which are derived in the current subcomputation, and facts which are inherited from the state where this subcomputation started. The problem is solved by introducing explicit "qualities" of facts into the model. The interpretation of predicate and function symbols is divided into knowledge which is derived in the current subcomputation and knowledge which is only known from the context.

Definition 15 A triple $\mathfrak{I} = (\langle I \rangle_1, \langle I \rangle_2, \mathfrak{U})$ is a *2-structure* if

- $\langle \mathfrak{I} \rangle_1 := (\langle I \rangle_1, \mathfrak{U})$ (the state "as is") is a total first-order interpretation, and
- $\langle \mathfrak{I} \rangle_2 := (\langle I \rangle_2, \mathfrak{U})$ (procedure knowledge) is a partial first-order interpretation,

and $\langle I \rangle_1 \supseteq \langle I \rangle_2$ holds. □

The 2-specification of actions can be obtained from their standard specification: the precondition queries the state as-is, its effect changes the procedure knowledge (and also the state as-is):

$$\text{Spec}^2(\mathsf{a}) := \{((\mathfrak{P}, \emptyset), (\emptyset, \mathfrak{Q})) \mid (\mathfrak{P}, \mathfrak{Q}) \in \text{Spec}(\mathsf{a})\}$$

(if read-access should be included into procedure knowledge, require $\text{dom}(\mathfrak{Q}) \supseteq \text{dom}(\mathfrak{P})$, this decision depends on the intended semantics).

Theorem 5 *For a state $\mathfrak{I}$ and an action a which are described by a 2-interpretation $(\langle I \rangle_1, \langle I \rangle_2, \mathfrak{U})$ and a 2-specification $\text{Spec}^2(\mathsf{a}) = ((\mathfrak{P}, \emptyset), (\emptyset, \mathfrak{Q}))$,*

$$\mathsf{a}(\langle I \rangle_1, \langle I \rangle_2, \mathfrak{U}) = (\langle I \rangle_1 \uplus \mathfrak{Q}, \langle I \rangle_2 \uplus \mathfrak{Q}, \mathfrak{U})$$

is the 2-interpretation obtained by executing a in $\mathfrak{I}$.

Particularly, $a(\langle I\rangle_1, \emptyset, \mathfrak{U}) = (\langle J\rangle_1, \langle J\rangle_2, \mathfrak{U})$ *where for the 1-specification of* a, $(\langle J\rangle_1, \mathfrak{U}) = a(\langle I\rangle_1, \mathfrak{U})$ *and* $(\langle J\rangle_2, \mathfrak{U}) = a^{part}(\langle I\rangle_1, \mathfrak{U})$. □

PROOF: The projection on the total states is equivalent to Proposition 1: $\langle a(\mathfrak{I})\rangle_1 = a(\langle\mathfrak{I}\rangle_1)$, and the derived knowledge is accumulated.

Definition 16 A *hierarchical labeled Kripke structure* is a tuple $\mathfrak{K} = (\mathfrak{G}, \mathfrak{A}, \mathfrak{T}, \mathfrak{Q}, \mathfrak{R}, \mathfrak{S}, \mathfrak{M})$ (cf. Figure 2) where

$\mathfrak{G}$ is the set of *states*,

$\mathfrak{A}$ is a set of *labels*,

$\mathfrak{R} \subseteq \mathfrak{G} \times \mathfrak{A} \times \mathfrak{G}$ is the *temporal successor relation*,

$\mathfrak{T} \subseteq \mathfrak{G} \times \mathbb{A}_e \times \mathfrak{G}$ is the *accessibility relation for elementary actions*,

$\mathfrak{Q} \subseteq \mathfrak{G} \times \mathbb{A}_p \times \mathfrak{G}$ is the *accessibility relation for procedure calls*,

$\mathfrak{S} \subseteq \mathfrak{G} \times (\mathbb{A}_p \cup \mathbb{A}_e) \times \mathfrak{G}$ is the *labeled return relation* from states reached by elementary actions or executions of procedures.

$\mathfrak{M} : \mathfrak{G} \to \mathbb{I} \times \mathbb{I}_p$ maps states to 2-structures (i.e. $\langle\mathfrak{M}\rangle_1 : \mathfrak{G} \to \mathbb{I}, \langle\mathfrak{M}\rangle_2 : \mathfrak{G} \to \mathbb{I}_p$). The tuple $(\mathfrak{G}, \mathfrak{A}, \mathfrak{T}, \mathfrak{Q}, \mathfrak{R}, \mathfrak{S})$ is a *hierarchical Kripke frame*. □

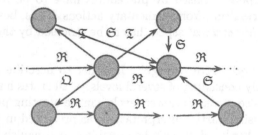

Fig. 2. Fragment of a Hierarchical Kripke Structure

Definition 17 Transitions are also hierarchically structured, i.e. the execution of a procedure is not represented by a single auxiliary state but by a sequence of states. Additionally to Definition 10,

- if for some transition $g \xrightarrow{a} h$, execution of a procedure $p \in \mathbb{A}_p$ s.t. $p \in_{\|} a$ is modeled by a procedure call $\mathfrak{Q}(g, p, g_0)$, several subsequent structured transitions $\mathfrak{R}(g_i, \tilde{a}_{i+1}, g_{i+1})$ for $0 \leq i < n$, and a return $\mathfrak{S}(g_n, p, h)$, then $p[g_0, \tilde{a}_1, g_1, \dots, \tilde{a}_n, g_n]$ is the share of p of $g \xrightarrow{a} h$.

- For a transition $g \xrightarrow{a} h$ s.t.
  - $a = p_1 \| \dots \| p_n \| a_{n+1} \| \dots \| a_m$, $p_1, \dots, p_n \in \mathbb{A}_p$, $a_{n+1}, \dots, a_m \in \mathbb{A}_e$,
  - for every $n < i \leq m$, $a_i[g_i]$ is the share of a_i of $g \xrightarrow{a} h$, and
  - for every $1 \leq i \leq n$, $p_i[g_{i,0}, \tilde{a}_{i,1}, g_{i,1}, \dots, \tilde{a}_{i,k_i}, g_{i,k_i}]$ is the share of p_i of $g \xrightarrow{a} h$,

$$\tilde{a} := (p_1[g_{1,0}, \tilde{a}_{1,1}, g_{1,1}, \dots, \tilde{a}_{1,k_1}, g_{1,k_1}] \| \dots \| p_n[g_{n,0}, \tilde{a}_{n,1}, g_{n,1}, \dots, \tilde{a}_{n,k_n}, g_{n,k_n}]$$
$$\| a_{n+1}[g_{n+1}] \| \dots \| a_m[g_m])$$

is the *hierarchical decomposition* of $g \xrightarrow{a} h$. Also, $(g, \tilde{a}, h)$ is called the *structured* transition. □

Similar to Section 4, setting $\mathfrak{A}$ to the set of hierarchical decompositions of transitions encodes sufficient information into the Kripke frame.

Consistency requirements. Additional consistency requirements on the frame develop from the given procedure definitions: if some parallel components of a transition are procedures, they are described more detailed by the hierarchical structure. The share of a procedure of a transition has to be consistent with the procedure definition:

For relating actions regarded as atomic on some level of abstraction with their more detailed "implementation" on the next lower level, complex actions have to be decomposed into their component actions. In general, there is no unique decomposition. From the point of view of verification, stating properties of composite actions in any logic defines subproblems which have to be solved by subproofs. Those proofs are also based on the decompositions of actions.

Definition 18 For an action a, $[a] \subset A$ is the set of *traces*, i.e. the set of actions containing neither $\vee$, $^\omega$, nor *, which represent possible executions of a:

- $[a(u_1, \dots, u_n)] := a(u_1, \dots, u_n),$
- $[a \parallel b] := \{a' \parallel b' \mid a' \in [a] \text{ and } b' \in [b]\},$
- $[a \vee b] := [a] \cup [b],$
- $[a \otimes b] := \{a' \otimes b' \mid a' \in [a] \text{ and } b' \in [b]\},$
- $[a^\omega] := \{a_1 \otimes a_2 \otimes \dots \otimes a_n \mid n \in \mathbb{N} \text{ and } a_i \in [a]\},$
- $[a^\star] := \{a_1 \otimes a_2 \otimes \dots \otimes a_n \otimes a^\perp \mid n \in \mathbb{N} \text{ and } a_i \in [a]\},$
  where $a^\perp$ is the action which is executable iff a is not executable and which does not make any changes on the data.
- For a procedure definition $p := a$, $[p] := \{p[a'] \mid a' \in [a]\}$. In this definition, the procedure name is included since it is used when describing the semantics on different abstraction levels. □

Proposition 6 $[\,]$ *is the "inverse" to the $\vee$-operator on actions: For every action* a *and state* $\mathfrak{J}$,

$$a \equiv \bigvee_{\{b \mid b \in [a]\}} b \quad , \ i.e. \quad a(\mathfrak{J}) = \bigcup_{\{b \mid b \in [a]\}} b(\mathfrak{J}) \,. \qquad \square$$

The shares and decompositions have to agree with the respective definitions of procedures:

Definition 19 The share $p[g_0, a_1[\dots], g_1, \dots, a_n[\dots], g_n]$ of a procedure p of a transition is consistent with the definition of p iff $p[a_1 \otimes \dots \otimes a_n] \in [p]$. □

Extending Definitions 12 and 14, the consistency of hierarchical Kripke frames is defined as follows (the definitions of completeness and unambiguity are retained):

Definition 20 A hierarchical Kripke frame is *consistent* iff for all $g, h, h' \in \mathfrak{G}$, $a, b \in \mathfrak{A}$:

- $\mathfrak{R}(g, \bar{a}, h) \wedge \mathfrak{S}(h', b, h) \wedge b \in_\parallel a \Rightarrow$
  $((b \in A_e \Rightarrow \mathfrak{T}(g, b, h')) \wedge (b \in A_p \Rightarrow \exists g' : \mathfrak{Q}(g, b, g') \wedge \mathfrak{R}^*(g', h')))$,
  i.e. only states are evaluated as results of an action if they actually arise from executing this elementary action or procedure, and
- all shares of procedures are consistent. □

Proposition 7

- For a structured transition $\mathfrak{R}(g, \tilde{a}, h)$, where for each $b \in_{\|} a$, $b[\ldots, g_b]$ is the share of b of this transition,

$$\mathfrak{C}^{\leftarrow}(g, \tilde{a}, h) = \{g_b \mid b \in_{\|} a\} .$$

- If a hierarchical Kripke frame is unambiguous,

$$\mathfrak{C}^{\leftarrow}(g, \tilde{a}, h) = \mathfrak{C}^{\leftarrow}(g, a, h) := \{h' \mid \exists b \in \mathbb{A}_e \cup \mathbb{A}_p : b \in_{\|} a \wedge h' \in \mathfrak{S}^{-1}(b, h)\} .$$

□

Additionally to the interpretation $\mathsf{Spec}_{\mathbb{A}}$ of the action algebra $\mathbb{A}$, the semantics of procedure calls and the processing of their results have to be specified. A procedure call consists of copying the current state as is to the initial state for executing the procedure. When starting the execution of a procedure, there is no procedure knowledge. For the computation of the target state of a transition, the consensus function is adapted to 2-interpretations and procedures:

Theorem 8 *The following 2-consensus function implements the specification of $\|$ given in Definition 8:*

$$\langle\mathfrak{C}\rangle_2(g, \tilde{a}, h) = \langle\mathfrak{M}\rangle_2(g) \uplus \bigcup_{h' \in \mathfrak{C}^{\leftarrow}(g, \tilde{a}, h)} \langle\mathfrak{M}\rangle_2(h') ,$$
$$\langle\mathfrak{C}\rangle_1(g, \tilde{a}, h) = \langle\mathfrak{M}\rangle_1(g) \uplus \langle\mathfrak{C}\rangle_2(g, \tilde{a}, h).$$

Assume $\langle\mathfrak{M}\rangle_1(g)$ to be consistent. Then, for all h s.t. there are $g, \tilde{a}$ with $\mathfrak{R}(g, \tilde{a}, h)$, $\langle\mathfrak{C}\rangle_1(g, \tilde{a}, h)$ is total and consistent iff $\bigcup_{h' \in \mathfrak{C}^{\leftarrow}(g, \tilde{a}, h)} \langle\mathfrak{M}\rangle_2(h')$ is consistent, i.e. there are no contradicting results of procedures or actions. □

Definition 21 A hierarchical Kripke structure with constant universe $\mathfrak{U}$ is a model of a 2-specification $\mathsf{Spec}_{\mathbb{A}}$ and a 2-consensus function $\mathfrak{C}$ if for all $g, h \in \mathfrak{G}$ and $a \in \mathbb{A}$,

$$\begin{aligned}
\mathfrak{R}(g, \tilde{a}, h) &\Rightarrow \mathfrak{M}(h) = \mathfrak{C}(g, \tilde{a}, h) \\
\mathfrak{T}(g, a, h), a \in \mathbb{A}_e &\Rightarrow \mathfrak{M}(h) = a(\langle M(g)\rangle_1, \emptyset, \mathfrak{U}) \\
\mathfrak{Q}(g, a, h), a \in \mathbb{A}_p &\Rightarrow \mathfrak{M}(h) = (\langle M(g)\rangle_1, \emptyset, \mathfrak{U}) .
\end{aligned}$$

□

This directly leads to an extension of $\mathfrak{T}$ to procedures: $\mathfrak{T}' : \mathfrak{G} \times \mathfrak{A}_p \times \mathfrak{G}$ is defined as an equivalent to $\mathfrak{T} : \mathfrak{G} \times \mathfrak{A}_e \times \mathfrak{G}$ for procedures:

$$\mathfrak{T}' = \{(g, \mathsf{p}, h') \mid \mathsf{p} \in \mathbb{A}_p, (g, \mathsf{p}, h_0) \in \mathfrak{Q} \text{ and there are } a_1, h_1, \ldots, a_n, h_n = h' \text{ with } (h_{i-1}, a_i[\ldots], h_i) \in \mathfrak{R} \text{ and } \mathsf{p}[a_1 \otimes \ldots \otimes a_n] \in [\mathsf{p}]\}.$$

Proposition 9

For every $g, h' \in \mathfrak{G}$, $\mathsf{p} \in \mathbb{A}_p$, $\mathfrak{T}'(g, \mathsf{p}, h') \Rightarrow \mathfrak{M}(h') = \mathsf{p}(\langle M(g)\rangle_1, \emptyset, \mathfrak{U}) .$ □

For 2-structures, a sound and complete set of axioms can be given, based on the idea of inheritance of knowledge:

Let φ be a function or predicate application, i.e. of the form $f(u_1, \ldots, u_n)$ or $p(u_1, \ldots, u_n)$ for n-ary $f \in \Sigma$ resp. $p \in \Sigma$.

$$\frac{\mathfrak{R}(g, \_, h) \quad , \quad \langle\mathfrak{M}(g)\rangle_1 \vdash \varphi}{\langle\mathfrak{M}(h)\rangle_1 \vdash \varphi} \qquad \frac{\mathfrak{R}(g, \_, h) \quad , \quad \langle\mathfrak{M}(g)\rangle_2 \vdash \varphi}{\langle\mathfrak{M}(h)\rangle_2 \vdash \varphi}$$

$$\frac{\mathfrak{Q}(g, \_, h) \quad , \quad \langle\mathfrak{M}(g)\rangle_1 \vdash \varphi}{\langle\mathfrak{M}(h)\rangle_1 \vdash \varphi}$$

$$\frac{\mathfrak{T}(g, \_, h) \quad , \quad \langle\mathfrak{M}(g)\rangle_1 \vdash \varphi}{\langle\mathfrak{M}(h)\rangle_1 \vdash \varphi} \qquad \frac{\mathfrak{T}(g, a, h) \quad , \quad \mathsf{Spec}(a), \mathfrak{M}(g) \vdash \varphi}{\langle M(h)\rangle_2 \vdash \varphi}$$

$$\frac{\mathfrak{S}(g,\_,\mathsf{a},h) \quad , \quad \mathsf{a} \in \mathfrak{A}_e \quad , \quad \langle \mathfrak{M}(g)\rangle_2 \vdash \varphi}{\langle \mathfrak{M}(h)\rangle_2 \vdash \varphi}$$

$$\frac{\langle \mathfrak{M}(g)\rangle_2 \vdash \varphi \quad , \quad \text{not } \langle \mathfrak{M}(g)\rangle_2 \vdash \neg\varphi}{\langle \mathfrak{M}(g)\rangle_2 \vdash \varphi}$$

$$\frac{\langle \mathfrak{M}(g)\rangle_2 \vdash \varphi}{\langle \mathfrak{M}(g)\rangle_1 \vdash \varphi} \qquad \frac{\langle \mathfrak{M}(g)\rangle_1 \vdash \varphi \quad , \quad \text{not } \langle \mathfrak{M}(g)\rangle_1 \vdash \neg\varphi}{\langle \mathfrak{M}(g)\rangle_1 \vdash \varphi}$$

In the initial state $\mathfrak{E}$, $\langle \mathfrak{E}\rangle_1$ is a total first-order interpretation.

Since procedure knowledge is obtained by actions, it is an important decision, which data should be regarded as procedure knowledge: in the above axiom system, only the data written by an action make up procedure knowledge. Alternatively, one could regard all data read or written by an action as procedure knowledge, i.e., requiring $\Omega \supseteq \mathfrak{P}$ for every 2-specification $((\mathfrak{P},\emptyset),(\emptyset,\Omega))$.

6   Example

Assume a company with a very competitive internal policy: employees get bonuses for good work; if an employee gets a bonus and stays in the same team, his manager also gets the same bonus since he seems to be a good manager. Employees can be transferred to another team in the company. An internal rule of the company is, that every manager should have at least \$1000 more than each of his subordinates. If this condition is not satisfied in some team by the manager mgr and some employee emp, obviously emp did not reach this salary by a bonus internal to the team, but came from outside the team. Thus, the manager did not educate his team-members well to fit the demands of this position. In this case, the manager is degraded to an ordinary team member with the same salary as emp, and the team is directly subordinated to the next higher manager (until a new team manager is employed). If there is a position to be filled, and there is no employee of the same team capable of doing this, some external candidate has to be employed.

The following elementary actions are given by 2-specifications:

$employ(emp, team, sal)$:

$\{(((\{\neg\mathsf{isEmp}(emp)\},\emptyset)(\emptyset,\{\mathsf{isEmp}(emp),\mathsf{team}(emp)=team,\mathsf{sal}(emp)=sal\}))\}$,

$award(emp, amt)$:$\{(((\{\mathsf{sal}(emp)=Sal,\neg\mathsf{pres}(emp)\},\emptyset),(\emptyset,\{\mathsf{sal}(emp)=Sal+amt\}))\}$,

$awardP(emp, amt)$:$\{(((\{\mathsf{sal}(emp)=Sal,\mathsf{pres}(emp)\},\emptyset),(\emptyset,\{\mathsf{sal}(emp)=Sal+amt\}))\}$,

$transfer(emp, team', sal')$:

$\{(((\{\mathsf{isEmp}(emp)\},\emptyset),(\emptyset,\{\mathsf{team}(emp)=team',\mathsf{sal}(emp)=sal'\}))\}$,

$check(emp)$:

$\{(((\{\mathsf{team}(emp)=Team,\mathsf{sal}(emp)=Sal,\mathsf{mgr}(Team)=Mgr,\mathsf{sal}(Mgr)=Sal',$

$\qquad \mathsf{team}(Mgr)=Team',\mathsf{mgr}(Team')=Mgr',Sal'<Sal+1000\},\emptyset),$

$\quad (\emptyset,\{\mathsf{team}(Mgr)=Team,\mathsf{mgr}(Team)=Mgr',\mathsf{sal}(Mgr)=Sal\})),$

$\quad ((\{\mathsf{team}(Emp)=Team,\mathsf{sal}(Emp)=Sal,\mathsf{mgr}(Team)=Mgr,\mathsf{sal}(Mgr)=Sal',$

$\qquad \mathsf{team}(Mgr)=Team',\mathsf{mgr}(Team')=Mgr',Sal'\geq Sal+1000\},\emptyset),$

$\quad (\emptyset,\emptyset))\}$.

Now, there are some interactional dependencies, corresponding to atomic execution of a sequence of actions:

- if someone gets a bonus and stays within the same team, his manager has to get a bonus before he is probably degraded (both actions can be applicable). This corresponds to having a state which is not visible at the upper level.

- if someone is newly employed or changes the team, perhaps his new manager must be degraded. Here, the second action enforces an external consistency condition. In databases, this is normally enforced by *triggers*, which seem to be a natural application area for this concept.

Thus, there are the following procedures:

reward(emp) := (award(emp, amt)⊗reward(mgr(team(emp))))∨awardP(emp, amt),
hire($emp, team, sal$) := employ($emp, team, sal$) ⊗ check(emp), and
move($emp, team, sal$) := transfer($emp, team, sal$) ⊗ check(emp).

Employee	Salary	Team	MgrOf
John	2000	I	
Paul	2000	I	
George	2000	I	
Ringo	2000	II	
Frank	3200	III	I
Mike	3200	III	II
Bob	4500	Pres.	III

Consider the state ℑ as given in the table on the left. In the following, names are abbreviated by their initials, and a short notation for facts is employed. Parallel execution of move(P,II,2300) and reward(J,500) in state ℑ leads to the state space shown in Figure 3 (for every state, only the procedure knowledge is given).

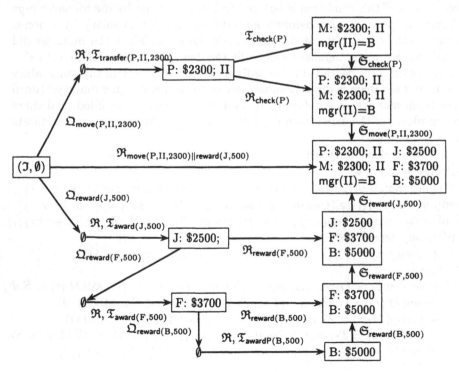

Fig. 3. Example State Space

7 Concluding Remarks

The hierarchical state space with different qualities of knowledge allows a flexible treatment of several interesting features of processes. It also provides a model theoretic base for reasoning about hierarchical transaction concepts.

- Safety conditions on processes can be stated for different levels of abstraction: While a subprocess is running, those conditions can be temporarily violated.
- Checking the admissibility of changes and blocking inadmissible ones: for any fact $p(\bar{x})$ that should be guaranteed, derive it in some subprocess.
- Hypothetical updates: if import and export declarations are added, every subprocess can work on infomation which is imported but not exported without having any effect to the calling process. By this it can create a hypothetical scenario, check the outcome and report the consequences. This can be used to evaluate several alternatives in parallel.

The model is also suitable for verifying action refinements wrt. abstract specifications and proving correctness properties of complex systems.

Acknowledgements.
The author thanks GEORG LAUSEN and BERTRAM LUDÄSCHER for many fruitful discussions and their help with improving the presentation of this paper.

References

[AH93] L. Aceto and M. Hennessy. Towards Action-Refinement in Process Algebras. *Information and Computation*, 103, 1993.

[AH96] R. Alur and T. Henzinger. Reactive Modules. In *Proc. 11th Symp. on Logic in Computer Science (LICS)*, 1996.

[BHK90] J. Bergstra, J. Heering, and P. Klint. Module Algebra. *Journal of the ACM*, 37(2):335–372, 1990.

[BK94] A. J. Bonner and M. Kifer. An Overview of Transaction Logic. *Theoretical Computer Science*, 133(2):205–265, 1994.

[DG91] P. Degano and R. Gorrieri. Atomic refinement in Process Description languages. In *16th Symp. on Mathematical Foundations of Computer Science*, Springer LNCS 520, pp. 121-130, 1991.

[DG95] P. Degano and R. Gorrieri. A Causal Operational Semantics of Action Refinement. *Information and Computation*, 122(1):97–119, 1995.

[Har79] D. Harel. *First-Order Dynamic Logic*, Springer LNCS 68, 1979.

[HM85] M. Hennessy and R. Milner. Algebraic Laws for Non-determinism and Concurrency. *Journal of the ACM*, 32:137–161, 1985.

[LML96] B. Ludäscher, W. May, and G. Lausen. Nested Transactions in a Logical Language for Active Rules. In *Proc. Intl. Workshop on Logic in Databases (LID)*, San Miniato, Italy, Springer LNCS 1154, pp. 197-222, 1996.

[May97] W. May. Specifying Complex and Structured Systems with Evolving Algebras. In *Proc. TAPSOFT'97*, To appear in Springer LNCS Series, 1997.

[MH69] J. McCarthy and P. Hayes. Some Philosophical Problems from the Standpoint of Artificial Intelligence. In B. Meltzer and D. Michie, editors, *Machine Intelligence 4*, pages 463–502. Edinburgh University Press, 1969.

A Fuzzy Analysis of Linguistic Negation of Nuanced Property in Knowledge-Based Systems

Daniel Pacholczyk

Angers University, LERIA, 2 Boulevard Lavoisier, F-49045 Angers Cedex, France

Abstract. In this paper, we present a Pragmatic Model dealing with Linguistic Negation in Knowledge-Based Systems. It can be viewed as a Generalization of Linguistic Approach to Negative Information within a Fuzzy Context. We define the Linguistic Negation with the aid of a Similarity Relation between Nuanced Properties. We propose a Choice Strategy allowing the User to explain the intended Meaning of Linguistic Negations. This Model improves the abilities in Management of a Knowledge Base, since Information can refer to Linguistic Negations either in Facts or in Rules.

1 Introduction

In this paper, we present a Model dealing with Negative Information within a Fuzzy Context. The Imprecise Information being generally evaluated in a Numerical way, our model is based upon the Fuzzy Set Theory proposed by Zadeh in [24]. Moreover, this Model has been conceived in such a way that the User deals with statements expressed in Natural Language, that is to say, referring to a Graduation Scale containing a finite number of Linguistic expressions. In many Discourse Universes, a part of Knowledge is represented by facts, denoted as « x is A », and its Management is founded upon classical Rules, denoted as if « x is A » then « y is B ». So, their Representation can be handled with Fuzzy Sets associated with respective Properties.

In previous papers (see [5], [6] and [17]) we have introduced Fuzzy Linguistic Operators f_α and Modifiers m_β allowing the User to express Knowledge having the following form « x is f_α m_β A ». Property, denoted as « f_α m_β A », requiring for its expression a list of linguistic terms, or a list of *Nuances* (ex. : (f_α, m_β)), is called *Nuanced Property*. So, the User can introduce in his Knowledge Description an assertion like « basketball players are *really very* tall ». This point is briefly presented in Section 2.

Moreover, the User also wishes to include in Knowledge Base following fact « the number of spectators is not high » or a particular rule if « an individual is not tall » then «it is invisible in a crowd». More formally, we have to propose a formal Representation of « x is not A » and of « if « x is not A » then « y is B » ». In other words, the basic Problem is then the explicit Interpretation of « x is not A ».

It is obvious that the explicit interpretation of « x is not A » is not easy. The linguists Ducrot and Schaeffer in [7], Muller in [14], Culioli in [4], Ladusaw in [12] and Horn in [11] have proposed methods generally dealing with precise Properties. In Section 3, we present the fundamental ideas concerning these Approaches to Negation in Linguistics. At this point, we can present the linguistic Analysis of fact like « the number of spectators is not high ». This Negation implies *1)* the *rejection* of

« the number of spectators is high» and *2) a reference to another Nuanced Property* like « very low » in order to translate the previous Negation into the fact « the number of spectators is very low ». It can be noted that « very low » is in weak *Agreement*, or, has a weak *Similarity* with « high ». Similarity Relations, conceived as a Reflexive, Symmetric but not Transitive Relation, have been proposed by Tversky in [23], Baldwin and Pilsworth in [1], Zadeh in [26], Ruspini in [20] and Pacholczyk in [15]. Our main idea has been to approach symbolically the Negation by using the weakly transitive Similarity Relation proposed by Pacholczyk in [15].

In this paper, we study Linguistic Negation of Nuanced Property. More precisely, we *i) complete* the results obtained by Desmontils and Pacholczyk (see [5] and [6]) and *ii) improve* the Model proposed by Pacholczyk in [17] in order to deal with the previous Facts or Rules. First, in Section 3, we present the main ideas of Negation Analysis in Linguistics. The basic notions leading to the symbolic Concept of the θ_i-*similarity of Fuzzy Sets* are defined in Sect. 4.1 and 4.2. The Section 5 is devoted to the Interpretation of Linguistic Negation in a Fuzzy Context. In Sect. 5.4, we generate a set of ρ-*plausible Linguistic Negations* in the domain. Then, we propose to the User a *Choice Strategy* of its interpretation of « x is not A » in Sect. 5.5. Moreover, in Sect. 5.6 we note that our interpretation leads to Properties intuitively connected with Negation in Common sense Reasoning. Finally, in Sect. 5.7, we present the interpretation of some Rules containing Linguistic Negations.

2 Brief Description of Nuanced Properties in Terms of Fuzzy Modifiers and Operators

In this Section, we briefly describe the Symbolic Representation of Nuanced Properties proposed by Desmontils and Pacholczyk (see [5] and [6]). Let us suppose that our Discourse Universe is characterized by a finite number of concepts C_i. A set of Properties P_{ik} is associated with each C_i, whose Description Domain is denoted as D_i. The Properties P_{ik} are said to be the *basic Properties* connected with concept C_i. As an example, basic Properties such as « thin », « big » and « enormous » can be associated with the particular « weight » concept (*Cf.* Fig. 3). This Universe description can be connected with some concepts proposed in linguistics by Brondal (see [2] and [3]), Hjelmslev (see [9] and [10]), Ducrot and Schaeffer in [7] and Ladusaw in [12]. Indeed, our Concepts and their associated Properties stand for *semantic Categories* and *semantic Units*.

A finite set of *Fuzzy Modifiers* m_α allows us to define *Nuanced Properties*, denoted as « $m_\alpha P_{ik}$ », whose membership L-R function simply results from P_{ik} by using a translation and a contraction. We can select the following Modifiers set (*Cf.* Fig. 1): $M_7 = \{$extremely little, very little, rather little, moderately ($\varnothing$), rather, very, extremely $\}$.

In order to modify the Precision or Imprecision of each « $m_\alpha P_{ik}$ », we use a finite set of *Fuzzy Operators* f_α defining new *Nuanced Properties* « $f_\alpha m_\beta P_{ik}$ ». Their membership L-R functions simply result from the ones of « $m_\beta P_{ik}$ ». The following set F_6 gives us a possible choice of Fuzzy Operators (*Cf.* Fig. 2): $F_6 = \{$vaguely, neighboring, more or less, moderately ($\varnothing$), really, exactly $\}$.

Remark. Please, note that graduation scales presented here have been chosen by experts working on Project « Filoforms » in an Image Synthesis field. Another application can lead to other linguistic terms without modifying our methodology.

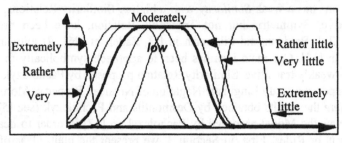

Fig. 1. A set of Fuzzy Modifiers

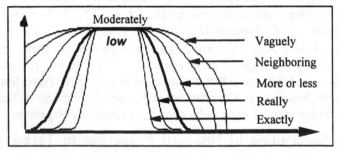

Fig. 2. A set of Fuzzy Operators

Remark. Note that the membership L-R functions of Operators and Modifiers have been constructed in such a way that we obtain, a satisfactory interpretation of the associated linguistic terms, and, a plausible linguistic Negation.

Remark. It is obvious that more than one Concept can be connected with a particular Domain. As an example, « height » concerns either « humans » or « buildings ». So, strictly speaking, it must be necessary to define for any concept its Dependence on other Concepts in order to express its particular Graduation Scale. This point not being studied in this paper, we simply denote as D_i the Domain of the Concept C_i.

3 Negation in Linguistics

In this Section we present some aspects resulting from the Linguistic Analysis of Negation proposed by Muller in [14], Culioli in [4], Horn in [11], Ducrot and Schaeffer in [7] and by Ladusaw in [12]. Note that our Argumentation is restricted here to a statement having the form « x is not A », where A denotes a Fuzzy Property.

3.1 Negation Characterization

Linguists have pointed out that Negation must be defined within a *Pragmatic Context.* More precisely, saying that « x is not A », the Speaker characterizes as Negation *i)* the *judgement of rejection* and *2)* the *semiologic means* exclusively used to notify this rejection. So, « x is A » is not in Adequation with Discourse Universe, but « x is not A » does not necessarily imply its Adequation with this Universe. It can only be a step in the outcome of this precise Adequation by the Speaker. Another point has also been pointed out by Linguists. Generally, Common sense Reasoning prefers affirmative statements to negative ones. In other words, Reasoning Process is based upon Rules consisting of Affirmative statements. So, a statement like « x is not A » has to be translated as an equivalent affirmative statement « x is P ».

3.2 Possible Interpretations of « x is not A »

We have chosen some characteristic cases to present the basic argumentation of linguists leading to the interpretations of « x is not A ». Note that we have extended these interpretations by using Operators or Modifiers applied to Fuzzy Properties.

Linguistic Negation based upon A. Saying that « John is not tall », the User (or Speaker) does not deny a certain height, he simply denies that his height can be high. So, his precise Adequation to Reality can be « John is *extremely little* tall ». In this case, the Speaker refers to the same Property and expresses a weak agreement between « tall » and « *extremely little* tall ».

Marked and not marked Properties
a - Let us suppose that three Properties « thin », « big » and « enormous » are associated with the basic concept « weight » (*Cf.* Fig. 3). Then, « x is not thin » can mean that the speaker 1) rejects « x is thin » and 2) refers to « x is enormous » or « x is *really* big », but not to « x is *vaguely* big ».
It can be noted that « big » and « enormous » have a weak agreement with « thin ». On the other hand, asserting that « x is not big » generally means that « x is thin », that is to say, the affirmative Interpretation is precise and unique. So, linguists distinguish a *marked* property like « thin » from a *not marked* property such as « big ». This distinction is important *since the Negation of not marked Property is explicitly defined.*

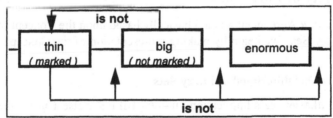

Fig. 3. Negations of Properties associated with the concept « weight »

b - The same analysis with two Properties « unkind » and « nice » associated with the basic concept « friendliness », leads to results collected in Fig. 4.

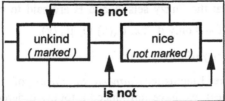

Fig. 4. Negations of Properties associated with the concept « friendliness »

A new property denoted as « not-A ». When the Speaker asserts that « Mary is not ugly », in this particular case, this semblance of Negation is in fact an affirmative reference to the new Property « not-ugly ». So, in some cases, Negation of « A » introduces *a new Basic Property* denoted as « not-A ».

Reference to logical ¬ A. In this case, the interpretation of the assertion « this head-scarf is not green » is « this head-scarf has another color than green ». So, « x is not A » means that « x is P (color) » with P ≠ A. The Property P is exactly ¬ A. We can note that between P and A, no similarity can exist.

A simple rejection of « x is A ». Saying that « John is not guilty », the Speaker only rejects the fact « John is guilty ». So, we progress weakly with the adequation of « x is not guilty » to Reality.

Remark. We can summarize the main aspects of Negation denoted as « x is not A » :
- It means at least the judgement of rejection of « x is A ».
- A « weak agreement », or « weak similarity », exists between the Fuzzy Properties « A » and « not A ».
- Some Negations refer to the same Property.
- Some Properties are not marked. In other words, the linguistic Negations are precisely defined.
- Some Properties are marked. In this case, the linguistic Negation refers to other Properties defined in the same domain.
- Some Negations refer to new Properties.
- Linguistic Negation can simply refer to Logical Negation.

In the following parts of this paper, we propose a *Generalization of the previous Linguistic Approach* within the context of a *Nuanced Property* denoted as A in the assertion « x is not A ».

4   Similarity of Fuzzy Sets

Our Approach to « Agreement » being basically based upon the Concept of Similarity degree of Fuzzy Sets, we begin by making a presentation of the Concept *of Nuanced Similarity*.

4.1   Nuanced Neighborhood of Fuzzy Sets

Let us recall Lukasiewicz's Implication : $u \rightarrow v = 1$ if $u \leq v$ else $1-u+v$, where u and v belong to [0, 1]. The *neighborhood* relation υ_α is defined in [0, 1] as follows.

Definition 1. u and v are *α-neighboring*, denoted as $u \upsilon_\alpha v$ if, and only if, Min $\{u \rightarrow v, v \rightarrow u\} \geq \alpha$.

Definition 2. Given that the Fuzzy sets A and B are said to be *α-neighboring*, we denote this as $A \approx_\alpha B$, if and only if $\forall x, \mu_A(x) \upsilon_\alpha \mu_B(x)$.

4.2   Nuanced Similarity of Fuzzy Sets

In order to define the Linguistic *Nuanced Similarity* of Fuzzy Sets (or their corresponding Properties), we have introduced a totally ordered partition of [0, 1] : $\{I_1, I_2,..., I_7\} = [0] \cup]0, 0.25] \cup]0.25, 0.33] \cup]0.33, 0.67[\cup [0.67, 0.75[\cup [0.75, 1[\cup [1]$. We have defined a one to one correspondence between these intervals and the following totally ordered set of linguistic expressions LS = $\{\theta_1,...,\theta_7\}$:

LS ={not at all, very little, rather little, moderately ($\emptyset$), rather, very, entirely}.

Remark. It is obvious that these choices are arbitrary. But, note that they lead to results corresponding in a satisfactory way to those intuitively expected.

Now, we can put the linguistic definition of θ_i similar Fuzzy Sets.

Definition 3. Given $\alpha = \text{Max}\{\delta \,|\, A \approx_\delta B\}$. Then, the Fuzzy Sets A and B are said to be θ_i *similar* if and only if the subscript i is such that $\alpha \in I_i$.

Remark. The partial similarity of Nuanced Properties is defined through the partial Similarity of their corresponding Fuzzy Sets.

Proposition 1. *For any Fuzzy Property A :*
- A and $m_\alpha A$ are at least $\emptyset$ similar iff $m_\alpha \in$ {rather little, moderately, rather}.
- A and $f_\alpha A$ are at least $\emptyset$ similar iff $f_\alpha \in$ {more or less, moderately, really}.

Proof. The Fuzzy Operators and Modifiers have been defined by Desmontils and Pacholczyk (see [5] and [6]) in such a way that any Basic Property satisfies these Properties.   □

In the next sections, we illustrate our argumentation by using the following example.

Example. The concept being « the number of intersection points », its associated properties will be « low », « average » and « high ». Then, a plausible interpretation of these properties is given in Fig. 5.

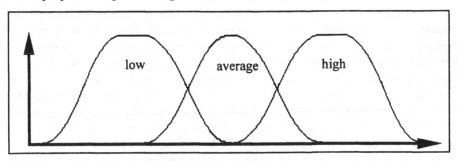

Fig. 5. A set of Basic Properties

5 Towards an Interpretation of Linguistic Negation

5.1 Linguistic Interpretation of « x is A »

As already pointed out by Scheffe in [21], linguistically speaking, the statement « x is A » implies some statements. So, applying this result to Nuanced Properties, « x is A » may be interpreted as one of the following statements : « x is $\emptyset_f$ A », « x is really A » or « x is more or less A ». As an example, the statement « x is low » corresponds to one of the following statements : « x is $\emptyset_f$ low », « x is really low » and « x is more or less low ».

In the set of Fuzzy operators F_6, we can put : $G_1 = $ {more or less, $\emptyset_f$, really} and $G_2 = $ {vaguely, neighboring, exactly}. So, linguistically speaking we put :
« x is A » $\Leftrightarrow$ {« x is f_α A » with $f_\alpha \in G_1$}.

Remark. Within the Fuzzy Context, this equivalence requires a Generalization of the linguistic Analysis of « x is A » presented in Section 3.

5.2 Linguistic Interpretation of « x is not A »

Within a given field, the Linguistic Negation of « x is A », denoted as « x is not A », can receive the following meaning : the speaker (1), *rejects* all previous interpretations of « x is A », and (2), *refers in many cases* to a Nuanced Property P having a « weak agreement » with A, and, in such a way that « x is P » is equivalent to the statement « x is not A ».

By using our previous « Nuances », P is « f_β A » with $f_\beta \in G_2$, *or* « m_δ A » with $m_\delta \neq \varnothing$, *or* « f_χ B » with B $\neq$ A defined in the same domain and $f_\chi \in F_6$. The main difficulty is then due to the fact that P is not explicitly given.

In the following, we propose a Choice Method based upon the notion Nuanced Similarity of Fuzzy Sets, which can be viewed as a Generalization to Nuanced Properties of the Linguistic Approaches presented in Section 3.

5.3 Intuitive Approach to Linguistic Negation

In order to define a plausible Axiomatics of Linguistic Negation, we present here an intuitive approach to Linguistic Negation of Common Sense within the Nuanced Property Context.

[I1] : Saying that « x is not A », the Speaker *rejects all the implicit meanings involved in « x is A »*. As an example, saying that « x is not low », we reject « x is $\varnothing$ low », « x is more or less low » and « x is really low ».

[I2] : Saying that « x is not A », *in many cases* the Speaker *refers globally* to a property P, based upon a *basic Property* defined in the same domain as A, « strongly » different from A (or « weakly » similar to A), and chosen in such a way that his Linguistic Negation means that « x is P ». Using the previous example, his implicit reference can be « average » or « high » or « extremely low ».

[I3] : Among the possible interpretations resulting from the previous global condition, the wanted interpretation must be *locally Neighboring*. More explicitly, when x satisfies « x is P » to a relatively high degree, then it satisfies « x is A » to a degree weakly neighboring from it, and conversely. Moreover, if need be this local condition of weak neighborhood must allow the speaker to define P as a Nuanced Property based upon A. As an exemple, he can refer to the nuanced property : « really extremely low ».

[I4] : Let us suppose that more than one Property is defined in the Domain. If the *Fuzzy Property A is not marked*, then its Linguistic Negation refers to a Nuanced Property based *upon one precise basic Property* P defined in this domain. If the *Fuzzy Property A is marked*, then its Linguistic Negation is a Nuanced Property constructed upon *at least one basic Property P* (different from A) defined in this domain. In both cases, P is « weakly » similar to A. As an example (*Cf.* Fig. 3), « x is not thin » refers to « big » or « enormous ». Finally, the Fuzzy Property being *neither marked nor not marked*, then the speaker refers in many cases *to all the basic Properties* defined in the domain. As an example (*Cf.* Fig. 5), if we suppose that « low » satisfies this condition, then asserting that « x is not low », the Speaker can refer to « high », « average », and also to « low ».

[I5] : In order to define the Nuanced Negation P, the Speaker uses a *Simplicity Rule*: its construction requires a small number of « Nuances ». As an example, he interprets « x is not low » as « x is rather high» rather than « x is more or less very little high ».

[I6] : In some particular cases, semblance of Negation is in fact an affirmative reference to the specific Nuance of a new basic Property denoted as « not-A » defined in the same domain. In other words, asserting « x is not A » the Speaker implicitly refers to « x is f_α not-A » with $f_\alpha \in G_1$. As an example, « Mary is not ugly » is equivalent to « Mary is $\varnothing$ not-ugly », « Mary is more or less not-ugly » or « Mary is really not-ugly ».

[I7] : In particular cases, a *Simplicity Combination Rule* is necessary to explicit Linguistic Negations. As an example, « x is not average » can be translated by « « x is very low » *or* « x is rather high » ».

In order to satisfy these intuitive properties associated with Linguistic Negation, we have formally defined a Concept of *ρ-plausible Negation* and constructed a *Choice Strategy of Linguistic Negation* among the all ρ-plausible solutions.

5. 4   ρ-plausible Linguistic Negations

Definition 4. For a given Fuzzy Property A, let ρ be a real number such that $0.33 \geq \rho \geq 0$. If P, defined in the same domain as A, satisfies the conditions :

[C1] : $\forall f_\alpha \in G_1, P \neq f_\alpha A$ (Initial Rejection)

[C2] : P and A are θ_i-similar with $\theta_i <$ moderately (or $\varnothing$), (Global Condition)

[C3] : $\forall x, ((\mu_A(x) = \xi \geq 0.67 + \rho) \Rightarrow (\mu_P(x) \leq \xi - 0.67))$, (Local Condition)

[C4] : $\forall x, ((\mu_P(x) = \xi \geq 0.67 + \rho) \Rightarrow (\mu_A(x) \leq \xi - 0.67))$, (Local Condition)

then « x is P » is said to be a *ρ-plausible Linguistic Negation* of « x is A ».

Remark. We can point out the fact that values 0.33 and 0.67 are not arbitrarily chosen in the partition of [0, 1] defined in Sect. 4.2. More precisely, the Neighborhood degree is low if its value is less than 0.33, and strong if its value is greater than 0.67.

Proposition 2. *Given ρ such that $0.33 \geq \rho \geq 0$, each ρ-plausible Linguistic Negation P is globally less than moderately similar to A, and locally less than 0.33-neighboring with A for the more significant values.*

Proof. We can verify the Adequation of this definition with the previous intuitive properties. Condition [C1] (translation of [I1]) results from the Linguistic Interpretation of « x is A » proposed in Sect. 5.1. The Similarity of Fuzzy Sets, as defined in Sect. 4.2, has been conceived in such a way that a *weak global Similarity* leads to a degree less than moderately ([C2]-[I2]). Conditions [C3]-[C4] translate a *weak local Neighborhood* of membership degrees (conditions [I3]-[I4]). Indeed, [C3] gives us:$\mu_P(x) \rightarrow \mu_A(x) = 1$. This hypothesis leads to: $\mu_A(x) \rightarrow \mu_P(x) \leq \xi \rightarrow (\xi - 0.67) = 1 - \xi + (\xi - 0.67) = 0.33$. So, we have: $\mu_A(x) \cup_\alpha \mu_P(x)$, with $\alpha \leq 0.33$. In other words, these conditions define a weak global Similarity and a weak Neighborhood for the more significant values. We can note that: $\mu_A(x) = 1 \Rightarrow \mu_P(x) \leq 0.33$. □

Remark. As noted before, the Fuzzy Operators have been defined in such a way that : $\forall f_\alpha \in G_1$, P and f_α A will be θ_i similar with $\theta_i \geq$ moderately. So, [C2] implies [C1]. This condition has only been preserved to copy exactly the intuitive approach.

Example. Figures 6 and 7 give us illustrations of definition with the Property « low ».

Remark. For any Property A, the function of ρ is to increase (or not) the number of Nuanced Negations and to accept (or not) some Negations based upon A.

In the following, we suppose that a lower value of ρ has been chosen in order to fulfil the previous remark. So, in the next sections our examples will be used with ρ = 0.3. In other words, we accept solutions based upon A.

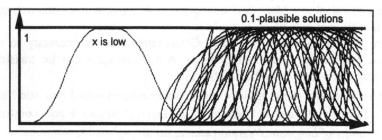

Fig. 6. 0.1-plausible Negations of « x is low »

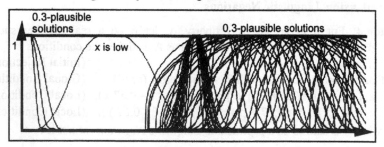

Fig. 7. 0.3-plausible Negations of « x is low »

Proposition 3. *For any Property A, there exists a value of ρ such that the set of ρ-plausible Linguistic Negations is not empty.*

Proof. We can distinguish two cases :

a : A is the only property defined in the domain. As noted before, Fuzzy Operators and Modifiers have been defined in such a way that some nuanced properties based upon A satisfy previous conditions. More precisely, it is the case when ρ ≥ 0.3.

b : At least two different Fuzzy Properties have been defined in the domain. Then, the translations and contractions have been defined in such a way that some Nuanced Properties based upon them fulfil all the conditions for any ρ ≥ 0.1. □

5.5 A Choice Strategy

It is obvious that the number of ρ-plausible Negations can be high (*Cf.* Fig. 6 and 7). This is due to the fact that our model accepts a great number of Operator and Modifier combinations. In order to allow the User to make his Choice from a limited number of solutions, we have constructed a Strategy essentially based upon a *Simplicity Rule* : he uses a very small number of Nuances (generally at most two) to define a Nuanced Property. So, too often Nuanced ρ-plausible Solutions will be rejected. We can now go into all the details of the Strategy leading to an explicit interpretation of the User or Speaker Linguistic Negation.

5.5.1 Construction of ρ-plausible solution Sets

First, we ask the User for Possibility or not of Negation based upon A. So, we can determinate a value of ρ satisfying this condition. If he does not make a Choice, we include Possibility of Negation based upon A by putting ρ = 0.3.

This being so, we can define the set of ρ-plausible Linguistic Negations of « x is A » by using successively the following Rules.

[R1] : *Simplicity Principle.* Among the previous ρ-plausible solutions, we define the set S of Nuanced Properties P based upon *at most two Nuances of a Basic Property.*

[R2] : *Increasing Similarity.* For each Similarity degree θ_i with θ_i < moderately, we define the subsets S_i of S whose elements P are θ_i *similar* to A.

[R3] : *Increasing Complexity.* We constitute a partition of each S_i in subsets $S_{i\ P}$ where P is defined in the same Domain as A. The last subset will be $S_{i\ A}$. Moreover, each $S_{i\ P}$ is reorganized in such a way that its elements appear ordered to an increasing *Complexity* extend, that is to say, the number of Nuances (different from moderately) required in their construction.

Example. The concept being « the number of intersection points », and the properties « low », « average » and « high » (*Cf.* Fig. 3), the Choice Strategy suggests the set S of 0.3-plausible Linguistic Negations of « x is low » collected in Fig. 8.

Remark. The Choice is initially made among approximately 20 interpretations of « x is not low » when ρ= 0.3, and among 10 interpretations when ρ= 0.1. So, the User can choose, for any ρ, solutions as « high », « extremely high », « exactly average », « really average », and in addition, « extremely low », « really extremely low » and « exactly extremely low » for ρ= 0.3.

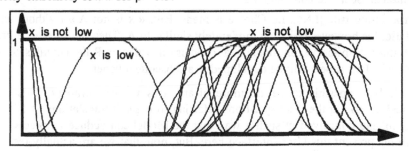

Fig. 8. The 0.3-plausible Linguistic Negations having 1 as Complexity

5.5.2 User or System Choice of the Linguistic Negation

Using the previous decomposition of S, we can construct a Choice Strategy based upon the following Rules denoted [LNi]. This Strategy allows the User to explain his interpretation of a *particular occurrence* of the assertion « x is not A ». In other words, this procedure must be applied to each previous occurrence appearing in a Knowledge Base. So, each occurrence of « x is not A » being explained with the aid of a particular Rule [LNi], we associate the *mark* i to this particular occurrence. We have generalized the notion of the *marked or not marked* Property proposed in linguistics, in such a way that we can manage the Fuzzy Property.

An interactive Process proposes to the User the following ordered Rules.

[LN0] : If the Negation is in fact the *logical Fuzzy Negation* denoted as ¬A, then :

- if ¬A exists as a Property defined in the same domain as A, we propose ¬A,
- if not, then for any P such that P⊂(¬A), we propose a Choice among the solutions of $S_{i\,P}$, i=1,2, ...

[LN1] : If the Negation *is based upon* A, we propose a Choice among the plausible solutions of $S_{i\,A}$, i=1,2, ...

[LN2] : If the Negation *is based upon only one P≠A*, we ask the User for its basic Negation P. Then, his Choice has to be made among the plausible solutions of the previous sets $S_{i\,P}$, i=1,2, ...

[LN3] : If the Negation can be *based upon B* and C *different from A*, then the User has to retain his interpretation among the solutions of $S_{i\,B}$ and $S_{i\,C}$ where i=1,2, ...

[LN4] : If the Negation can be based upon *one of all the basic properties*, his interpretation is one of the plausible solutions of $S_{i\,P}$ where i=1,2, ... , for any P.

[LN5] : If the Negation *is in fact a New basic Property denoted as not-A*, we must ask the User for its explicit meaning, and for its membership L-R function.

[LN6] : If the Negation is *simply the rejection* of « x is A », then any element of S is a potential solution but he does not choose one of them.

[LN7] : If the Negation requires a *Combination based upon two basic Properties B and C*, we ask the User for the adequate Linguistic Operator. So, we propose a Choice among the Simple Combination solutions of $S_{i\,B}$ and $S_{i\,C}$, i=1, 2, ...

[LN8] : If no Choice is made with rules [LN0], [LN1], [LN2], [LN3], [LN4] or [LN7] then the System can propose a *Default Choice*.

Remark. The solutions are always proposed by increasing Complexity

Remark. This Choice Strategy recovers all interpretations of Linguistic Negations presented in Sect. 3.2 and Sect. 5.3.

Remark. Using rule [LN6], no Choice is made. But, « x is not A » (without explicit translation) is however less than moderately similar to A. Then, an Inference Process based upon Similarity can deduce a conclusion when a Rule contains such an hypothesis. This aspect will be developed in a subsequent paper.

Remark. It is possible to define more clearly this Choice Strategy. This can be achieved by asking the User questions referring to explicit translations of Linguistic Negation, and conceived in such a way that the Model can deduce its precise Mark. We do not present this User Interface here. But, note that we are actually testing the efficiency of a convivial Query System in the Project « Filoforms ».

Remark. Note that a lot of works on Linguistic Negation have already been proposed. Among these, the old papers by Trillas, Lowen, Ovchinnokov and Esteva are not quoted here since very different points of view are considered. In the recent paper by Torra in [22], one will find a closer point of view. But, our Definition and our Treatment of Linguistic Negation of Nuanced Property lead to a different Model. We propose a Generalization of the Linguistic Approach to Negation within a Pragmatic Context based upon the Judgement of Rejection and the Semiologic means exclusively used to notify this Rejection. Our Definition of the ρ-plausible Linguistic Negation is basically founded upon the Notions of weak Similarity and weak Neighborhood. Moreover, the Basic properties associated with a given Concept are not supposed to be totally ordered.

5.5.3   The Model in Action

We present here some results obtained by using the previous Choice Strategy.

Example 1 (Cf. Fig. 5 and 8). The concept and its associated Properties being the ones presented in § 5.1, we search for the interpretation of Negation « x is not low ».

a : If its mark is 1, then 3 solutions are proposed, that is to say, « x is extremely low », « x is really extremely low » and « x is exactly extremely low ».

b : If its mark is 4, then the Choice is made among the following solutions :

- Complexity = 1 : « x is m_α high » with $\alpha=2..7$; « x is f_β high » with $\beta=1..3, 5, 6$; « x is exactly average » and « x is extremely low ».

- Complexity = 2 : « x is really m_α average » with $\alpha=1, 2, 6, 7$;« x is exactly m_α average » with $\alpha=1..3, 5.. 7$; « x is really extremely low » and « x is exactly extremely low » ; « x is f_β m_α high » with $\alpha=1..3, 5..7$ and $\beta=1..3, 5, 6$.

Example 2. Properties « thin », « big » and « enormous » are associated with the basic concept « weight » (*Cf.* Fig. 3). We study now some Negations.

a : « x is not big » has received the mark 2. So, the User can construct its negation with « thin ». We first propose the solutions having a Complexity equal to 1 : « x is m_α thin » with $\alpha=4, 5, 6, 7$; Solutions having 2 as a complexity are obtained as before for Property « thin ».

b : « x is not thin » has received the mark 3. The Linguistic Negation is then based upon basic Properties « big » and « enormous ». So, we can propose the successive solutions :

- Complexity = 1 : « x is m_α enormous » with $\alpha=2..7$; « x is f_β enormous » with $\beta=1..3, 5, 6$; « x is exactly big ».

- Complexity = 2 : « x is really m_α big » with $\alpha=1, 2, 6, 7$; « x is exactly m_α big» with $\alpha=1..3, 5..7$; « x is f_β m_α enormous » with $\alpha=1..3, 5..7$ and $\beta=1..3, 5, 6$.

Example 3. Given the fact « x is really small » and the rule « if « x is not tall » then « x is invisible in a crowd » », suppose that the occurrence of « x is not tall » has received the mark 8. Since « x is very small » belongs to S_1, this solution can be a Default Choice. In this case, the hypothesis implies the rule « if « x is very small » then «x is invisible in a crowd», and we can deduce that « x is invisible in a crowd».

Remark. The implementation of this Choice Strategy is actually in progress within Project « Filoforms » in Image Synthesis. We can point out that Designers generally refer to Negations having 1 as a complexity.

Remark. It is obvious that previous results must be applied to any instance of « x is not A », denoted as « a is not A ». Indeed, two different instances require two different applications of the Choice Strategy. As an example, saying that « John is not tall » can be translated « John is very small », and, on the other hand, the translation of « Jack is not tall » can be « Jack is really very tall ».

5.6   General Properties of previous Linguistic Negation

Finally, we can point out the fact that this Linguistic Negation satisfies some Common sense properties of Negation.

Proposition 4. *Given a Fuzzy Property A, « x is A » does not automatically define the Knowledge about « x is not A ».*

Proof. This property results directly from our construction Process of Linguistic Negation. Knowing exactly A does not imply, as does the Logical Negation, precise Knowledge of its Negation, since most of them require complementary information, as a mark of the property, and a Choice among possible interpretations. □

Proposition 5. *Given a Fuzzy Property A, its double Negation does not generally lead to A.*

Proof. Using Figure 3, the User can choose « x is thin » as the interpretation of « x is not big », and « x is enormous » as the negation of « x is thin ». □

Proposition 6. *If « x is P » is a ρ-plausible Linguistic Negation of « x is A », then « x is A » is a ρ-plausible Linguistic Negation of « x is P ». Moreover, « x is A » is a ρ-plausible double Negation of « x is A ».*

Proof. This properties result from the definition of ρ-plausible Negation. □

Proposition 7. *Given the Rule « if « x is A » then « y is B » », we can deduce that « if « y is not B » then « z is A' » » where A' is a ρ-plausible Negation of A.*

Proof. This property results from the definition of ρ-plausible Linguistic Negation. □

Remark. Strictly speaking, the previous Approach to Linguistic Negation gives us a *Pragmatic Model* leading to results consistent with those intuitively expected, and this, within the Context of Nuanced Fuzzy Properties. Its formal integration in Many-valued Predicate Logics is actually being examined. Our main objective is to define a new Adequate Fuzzy Negation Operator having Properties like the one proposed in a logical Approach to the Negation of Precise assertions by Lenzen in [13], Gabbay in [8], Pearce and Wagner in [19] and by Pearce in [18]. At this point, we can point out that Propositions 4-7 give us the first results as the Basis of a possible formalization within Many-valued Logics.

5.7 Deductive Process dealing with Linguistic Negation

As pointed out in Section 1, the User expresses some Facts or Rules with the use of Linguistic Negations. It results from previous analysis that our interactive Choice Strategy allows him to explain affirmatively their intended meanings. That is to say, the presence of such Linguistic Negations in the Facts or Rules of a Knowledge Base does not generally modify the use of the existing Deductive Process, since, in many cases, Negative assertions can be translated as Affirmative ones. In other words, the User can generally explain initial Rules in the following equivalent form: if « x is A » then « y is B ». We now illustrate this point through the Analysis of Rules initially containing Linguistic Negations.

Example. In following Figures 9 and 10, we have collected Fuzzy Properties associated with concepts « height » and « appearance ». Let us now analyse some Rules containing Linguistic Negations.

1 : if « x is not tall » then « x is not visible in a crowd ». It results from User interpretation that we obtain the following equivalent Rule : if « x is small » then « x is invisible in a crowd ».

2 : if « x is not small » then « x is visible in a crowd». Let us suppose that the User has selected « x is very tall » as translation of « x is not small ». So, the Rule is equivalent to the following one : if « x is very tall » then « x is visible in a crowd ».

3 : if « x is not small » then « x is not invisible in a crowd ».The previous Choice gives us : if « x is very tall » then « x is not invisible in a crowd ». If « x is not invisible in a crowd » receives as an affirmative translation « x is rather visible in a crowd », then we obtain : if « x is very tall » then « x is rather visible in a crowd ».

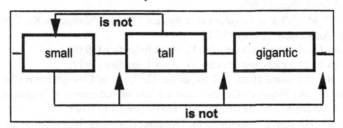

Fig. 9. Nuanced Properties associated with the concept « height »

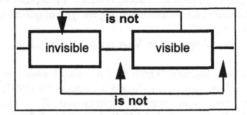

Fig. 10. Nuanced Properties associated with the concept « appearance »

Remark. It is obvious that information does not generally correspond with the Premise of Rules. So, a Deductive Process must be founded upon some Generalized Modus Ponens Rule, like the Zadeh one proposed in [25]. But, such a Process creates difficulty in making an explicit Deduction in terms of the basic Properties of the Discourse Universe. We are actually achieving a variant of the Zadeh one via the Similarity of Properties.

6   Conclusion

We have presented an Approach to Linguistic Negation of Nuanced Properties based upon a Similarity Relation. We have proposed a Choice Strategy improving the abilities in the Management of a Knowledge Base. Indeed, the User can refer to Linguistic Negations either in Facts or in Rules, since this Interactive Strategy allows him to explain their intended meanings, and also to exploit this particular Knowledge by using new Rules based upon Affirmative Information.

References

1. Baldwin J. F., Pilsworth B. W.: Axiomatic approach to implication for approximate reasoning with fuzzy logic. Fuzzy Sets and Syst. 3 (1980) 193-219
2. Brondal V.: « Essais de linguistique générale ». (1943) Copenhage
3. Brondal V.: Les parties du discours : étude sur les catégories linguistiques. EINAR MUNKSGAARD (1948) Copenhague

4. Culioli A.: Pour une linguistique de l'énonciation : Opérations et Représentations. Tome 1. (1991) Ophrys 2ds. Paris

5. Desmontils E., Pacholczyk D.: Apport de la théorie des ens. flous à la modélisation déclarative en Synthèse d'images. Proc. LFA'96 (1996) 148-149

6. Desmontils E., Pacholczyk D.: Modélisation déclarative en Synthèse d'images: traitement semi-qualitatif des propriétés imprécises ou vagues. Proc. AFIG'96 (1996) 173-181

7. Ducrot O ., Schaeffer J.-M. et al.: Nouveau dictionnaire encyclopédique des sciences du langage. Seuil, Paris (1995)

8. Gabbay D. M.: What is Negation in a System ? Logic Colloquium'86, F.R. Drake & J. K. Truss Amsterdam (1988) 95-112

9. Hjelmslev L .: La catégorie des cas (1). Acta Jutlandica (1935)

10. Hjelmslev L .: La catégorie des cas (2). Acta Jutlandica (1937)

11. Horn L. R : A Natural History of Negation. The Univ. of Chicago Press Chicago (1989)

12. Ladusaw W. A.: Negative Concord and « Made of Judgement ». Negation, a notion in Focus. H. Wansing, W. de Gruyter eds. Berlin (1996) 127-144

13. Lenzen W.: Necessary Conditions for Negation Operators. Negation, a notion in Focus. H. Wansing, W. de Gruyter eds. Berlin (1996) 37-58

14. Muller C .: La négation en français. Public. romanes et françaises Genève (1991)

15. Pacholczyk D.: Contribution au traitement logico-symbolique de la connaissance. Thèse d'Etat Part. C, Paris 6 (1992)

16. Pacholczyk D.: A new Approach to Vagueness and Uncertainty. CC-AI 9 (4) (1992) 395-436

17. Pacholczyk D.: About Linguistic Negation of Nuanced Property in Declarative Modeling in Image Synthesis. Proc. of « Fuzzy Days'5 in Dortmund » (1997) (to appear)

18. Pearce D.: Reasoning with negative information II : Hard Negation, Strong Negation and Logic Programs. LNAI 619 Berlin (1992) 63-79

19. Pearce D. ; G. Wagner G.: Reasoning with negative information I : Hard Negation, Strong Negation and Logic Programs. Language, Knowledge, and Intentionality : Perspectives on the Philosophy of J. Hintikka. Acta Philosophica Fennica 49 Helsinki (1990) 430-453

20. Ruspini E. H.: The Semantics of vague knowledge. Revue Internationale de Systémique 3 (4) (1989) 38 -420

21. Scheffe P.: On foundations of reasoning with uncertain facts and vague concepts. Fuzzy Reasoning and its Applications (1981) 189-216

22. Torra V.: Negation Functions Based Semantics for Ordered Linguistic Labels. Int. Jour. of Intelligent Systems (1996) 975-988

23. Tversky A.: Features of Similarity. Psychological Review (1977) 4 -84

24. Zadeh L. A.: Fuzzy Sets. Inform. and Control 8 (1965) 338-353

25. Zadeh L. A.: PRUF-A meaning representation language for natural languages. Int. J. Man-Machine Studies 10 (4) (1978) 395-460

26. Zadeh L. A.: Similarity relations and Fuzzy orderings. Selected Papers of L.A. Zadeh 3 (4) (1987) 387-420

Normative Argumentation and Qualitative Probability

Simon Parsons

Department of Electronic Engineering, Queen Mary and Westfield College,
Mile End Road, London E1 4NS, United Kingdom.
S.D.Parsons@qmw.ac.uk

Abstract. In recent years there has been a spate of papers describing systems for plausible reasoning which do not use numerical measures of uncertainty. Some of the most successful of these have been systems for argumentation, and there are advantages in considering the conditions under which such systems are normative. This paper discusses an extension to previous work on normative argumentation, exploring the properties of a particular normative approach to argumentation and suggesting some uses of it.

1 Introduction

In the last few years there have been a number of attempts to build systems for reasoning under uncertainty that are of a qualitative nature—that is they use qualitative rather than numerical values, dealing with concepts such as increases in belief and the relative magnitude of values. Between them, these systems address the problem of reasoning in situations in which knowledge is uncertain, but in which there is a limited amount of numerical information quantifying the degree of uncertainty. Three main classes of system can be distinguished—systems of abstraction, infinitesimal systems, and systems of argumentation.

In systems of abstraction, [5, 14, 19, 22], the focus is often, though not always [3], on modelling how the probability of hypotheses changes when evidence is obtained and never commits to exact probability values. They thus provide an abstract version of probability theory which ignores the actual values of individual probabilities but which is nevertheless sufficient for planning and design [13] tasks. Infinitesimal systems [10, 23] deal with beliefs that are very nearly 1 or 0, providing formalisms that handle order of magnitude probabilities. Infinitesimal systems may be used for diagnosis [4] as well as providing a general model of default reasoning. Systems of argumentation [1, 2, 6, 11, 12, 21] are based on the idea of constructing logical arguments for and against propositions, establishing the overall validity of such propositions by assessing the persuasiveness of the individual arguments.

However, unlike other qualitative systems, systems of argumentation do not always have a clear semantics. In particular, it is not always clear what it means to have an argument for or against something. As a result, in a previous paper [17] I made the suggestion that it might be beneficial to investigate whether systems of argumentation might be made normative in a probabilistic sense, thus combining the expressiveness of argumentation with the clear semantics of probability. This paper takes that work further, building a more expressive system, exploring its properties, providing soundness and completeness results and suggesting ways in which the system might be used.

2 Introducing systems of argumentation

In classical logic, an argument is a sequence of inferences leading to a conclusion. If the argument is correct, then the conclusion is true. In the system of argumentation proposed by Fox, Krause and colleagues [11] this traditional form of reasoning is extended to allow arguments to indicate support and doubt in propositions, as well as proving them. This is done by assigning two labels to every sentence that is deduced. One label, the grounds, identifies the argument for that sentence by indicating the formulae used in the deduction. The other label, the sign, indicates the force of the argument. The force can be taken to be the change in belief in the proposition warranted by the argument. This form of argumentation may be summarised by the following schema:

$$\text{Database } \vdash_{ACR} (\text{Sentence}, \text{Grounds}, \text{Sign})$$

where $\vdash_{ACR}$ is a suitable consequence relation. This approach to argumentation is one of many, but it is the one which will be adopted in this paper. Having chosen to work with this form of argumentation, we then choose a particular formalisation of this form of argumentation to work with, creating a system which we will call $\mathcal{SA}''$. Again there are many possibilities of which this formalisation is just one, particularly convenient, one. Others are explored in [8, 9, 11, 16]. Since this formalisation is expressed in very general terms, $\mathcal{SA}''$ can actually be refined to produce a number of different systems depending upon the precise interpretation one puts on the signs. Eventually a particular interpretation is selected, generating a system $\mathcal{NA}''$, and the bulk of the paper investigates the properties and uses of this system.

We start the definition of $\mathcal{SA}''$ with a set of atomic propositions $\mathcal{L}$. We also have a set of connectives $\{\rightarrow, \wedge, \neg\}$, and the following set of rules for building the well-formed formulae (wffs) of the language.

- If $l \in \mathcal{L}$ then l is a simple well-formed formula ($swff$).
- If l is an $swff$, then $\neg l$ is an $swff$.
- If l and m are $swff$s then $l \wedge m$ is an $swff$.
- If l and m are $swff$s then $l \rightarrow m$ is an implicational well-formed formula ($iwff$).
- The set of all wffs is the union of the set of $swff$s and the set of $iwff$s.

The reason for distinguishing the $swff$s and the $iwff$s is that whilst $\neg$ and $\wedge$ have their usual logical meaning, $\rightarrow$ does not represent material implication but a connection between the signs of antecedent and consequent. Thus there is a

$$\text{Ax} \frac{}{\Delta \vdash_{ACR} (St, \{l\}, Sg)} \ (l : St : Sg) \in \Delta$$

$$\land\text{-E1} \frac{\Delta \vdash_{ACR} (St \land St', G, Sg)}{\Delta \vdash_{ACR} (St, G, \mathsf{conj_{elim}}(Sg))}$$

$$\land\text{-E2} \frac{\Delta \vdash_{ACR} (St \land St', G, Sg)}{\Delta \vdash_{ACR} (St', G, \mathsf{conj_{elim}}(Sg))}$$

$$\land\text{-I} \frac{\Delta \vdash_{ACR} (St, G, Sg) \qquad \Delta \vdash_{ACR} (St', G', Sg')}{\Delta \vdash_{ACR} (St \land St', G \cup G', \mathsf{comb_{conj}}(Sg, Sg'))}$$

$$\rightarrow\text{-E} \frac{\Delta \vdash_{ACR} (St, G, Sg) \qquad \Delta \vdash_{ACR} (St \rightarrow St', G', Sg')}{\Delta \vdash_{ACR} (St', G \cup G, \mathsf{comb_{imp}}(Sg, Sg'))}$$

$$\rightarrow\text{-I} \frac{\Delta, (St, \emptyset, Sg) \vdash_{ACR} (St', G, Sg')}{\Delta \vdash_{ACR} (St \rightarrow St', G, \mathsf{comb_{imp}^{-1}}(Sg, Sg'))}$$

$$\neg\text{-E1} \frac{\Delta \vdash_{ACR} (\neg St, G, Sg)}{\Delta \vdash_{ACR} (St, G, \mathsf{neg_{form}}(Sg))}$$

$$\neg\text{-I} \frac{\Delta \vdash_{ACR} (St, G, Sg)}{\Delta \vdash_{ACR} (\neg St, G, \mathsf{neg_{form}}(Sg))}$$

Fig. 1. Argumentation Consequence Relation

fundamental difference between the *swff*s and the *iwff*s which will become clear later in the paper. The set of all *wff*s that may be defined using $\mathcal{L}$, may then be used to build up a database Δ where every item $d \in \Delta$ is a triple $(i : l : s)$ in which i is a token uniquely identifying the database item (for convenience we will use the letter 'i' as an anonymous identifier), l is a *wff*, and s is a statement about belief in l. With this formal system, we can take a database and use the argumentation consequence relation $\vdash_{ACR}$ defined in Figure 1 to build arguments for propositions that we are interested in[1].

Typically we have several arguments for a given proposition, and so *flatten* them to get a single sign. Thus we have a function $\mathsf{Flat}(\cdot)$ from a set of arguments **A** for a proposition l from a particular database Δ to the pair of that proposition and some overall measure of validity:

$$\mathsf{Flat}(\mathbf{A}) = \langle l, v \rangle$$

---

[1] The change from the triple (*Indentifier, Sentence, Sign*) in the database to the argument (*Sentence, Set of Indentifiers, Sign*) is to make a distinction between database items and derived sentences.

where $\mathbf{A} = \{(l, G, Sg) \mid \Delta \vdash_{ACR} (l, G, Sg)\}$, and v is the result of a suitable combination of the Sg that takes into account the structure of the the arguments. The value of v is calculated by another function flat:

$$v = \mathsf{flat}\Big(\{\langle G_i, Sg_i\rangle \mid (l, G_i, Sg_i) \in \mathbf{A}\}\Big)$$

Together $\mathcal{L}$, the rules for building the formulae, the connectives, and $\vdash_{ACR}$ define $\mathcal{SA}''$. In fact, $\mathcal{SA}''$ is really the basis of a family of systems of argumentation, because one can define a number of variants of $\mathcal{SA}''$ by using different sets of signs. Each set will have its own set of functions $\mathsf{conj}_{\mathsf{elim}}$, $\mathsf{comb}_{\mathsf{conj}}$, $\mathsf{comb}_{\mathsf{imp}}$, $\mathsf{comb}_{\mathsf{imp}}^{-1}$ and $\mathsf{neg}_{\mathsf{form}}$, and its own means of flattening arguments, flat. The meanings of the signs, flattening functions, and combination functions delineate the semantics of the system of argumentation. The purpose of this paper is to suggest a way in which these functions may be given a specific probabilistic interpretation. The reason for doing this is to provide a system of argumentation, $\mathcal{NA}''$, which is normative in the sense that it accords to the norms of probability theory.

3 A probabilistic semantics

The idea behind the semantics is to provide a precise characterisation of the intuitive idea that constructing an argument which supports a proposition is a reason for one to increase one's belief in that proposition and that constructing an argument against a proposition is a reason for decreasing one's belief in that proposition. Taking the position that a degree of belief may be expressed as a probability it is possible to modify the notion of a probabilistic influence in qualitative probabilistic networks (QPNs) [22] to give the signs of $\mathcal{NA}''$ a probabilistic interpretation.

3.1 The meaning of formulae

In particular we take triples $(i : l : \uparrow)$ to denote the fact that $\Pr(l)$ increases, and similar triples $(i : l : \downarrow)$, to denote the fact that $\Pr(l)$ decreases. Triples $(i : l : \leftrightarrow)$, denote the fact that $\Pr(l)$ is known to neither increase nor decrease. It should be noted that the triple $(i : l : \uparrow)$ indicates that the change in value of $\Pr(l)$ either goes up, or does not change—this inclusive interpretation of the notion of "increase" is taken from QPNs—and of course a similar proviso applies to $(i : l : \downarrow)$. Since we want to reason about changes in belief which equate to the usual logical notion of proof, we also consider changes in belief to 1 and decrease in belief to 0, indicating these by the use of the symbols $\Uparrow$ and $\Downarrow$. The meaning of a proposition $(i : l : \Uparrow)$ is that the probability of l becomes 1, while $(i : l : \Downarrow)$ means that the probability of l becomes 0. We also have triples $(i : l : \updownarrow)$ which indicate that the change in $\Pr(l)$ is unknown.

Implications, by which we mean *iwff*s, can be given a probabilistic interpretation by making the triple $(i : a \to c : +)$ denote the fact that:

$$\Pr(c \mid a, X) \geq \Pr(c \mid \neg a, X)$$

for all $X \in \{x, \neg x\}$ for which there is a triple $(i : x \to c : s)$ (where s is any sign) or $(i : \neg x \to c : s)$, while the triple $(i : a \to c : -)$ denotes the fact that:

$$\Pr(c \,|\, a, X) \leq \Pr(c \,|\, \neg a, X)$$

again for all $X \in \{x, \neg x\}$ for which there is a triple $(i : x \to c : s)$ or $(i : \neg x \to c : s)$. As a result an implication $(i : a \to c : +)$ means that there is a probability distribution over the formulae c and a such that an increase in the probability of a makes c more likely to be true, and an implication $(i : a \to c : -)$ means that there is a probability distribution over the propositions c and a such that an increase in the probability of a makes c less likely to be true. We do not make much use of triples such as $(i : c \to a : 0)$ since they have no useful effect but include them for completeness—$(i : c \to a : 0)$ indicates that $\Pr(c)$ does not change when $\Pr(a)$ changes. We also have implications such as $(i : a \to c : ?)$ which denotes the fact that the relationship between $\Pr(c \,|\, a, X)$ and $\Pr(c \,|\, \neg a, X)$ is not known, so that if the probability of a increases it is not possible to say how the probability of c will change. With this interpretation, implications between atomic propositions correspond to qualitative influences in QPNs. As a result of this link to QPNs we require that implications are causally directed, by which we mean that the antecedent is a cause of the consequent. This is the usual restriction imposed in probabilistic networks [20].

It should be noted that the effect of declaring that there is an implication $(i : a \to c : +)$ is to create considerable constraints on the probability distribution over a and c. In fact we have:

Theorem 1. *A consequence of the probabilistic semantics is that an implication* $(i : a \to c : +)$ *places the same restrictions over the probabilities of a and c as* $(i : a \to \neg c : -)$, $(i : \neg a \to c : -)$, $(i : \neg a \to \neg c : +)$ *and* $(i : c \to a : +)$, $(i : c \to \neg a : -)$, $(i : \neg c \to a : +)$, $(i : \neg c \to \neg a : +)$.

Proof: The condition $\Pr(c \,|\, a, X) \geq \Pr(c \,|\, \neg a, X)$ is exactly that for $(i : \neg a \to c : -)$, and implies $1 - \Pr(c \,|\, a, X) \leq 1 - \Pr(c \,|\, \neg a, X)$ so that $\Pr(\neg c \,|\, a, X) \leq \Pr(\neg c \,|\, \neg a, X)$. This is the condition for $(i : a \to \neg c : -)$ and $(i : \neg a \to \neg c : +)$. This takes care of the restrictions on reasoning from a to c. For the reverse cases we recall that $\Pr(c \,|\, a, x) \geq \Pr(c \,|\, \neg a, x)$ implies $\Pr(a \,|\, c, y) \geq \Pr(a \,|\, \neg c, y)$ [22]. $\square$

Clearly analogous restrictions are imposed by implications like $(i : a \to c : -)$. We also have *categorical* implications which allow propositions to be proved true or false. In particular, an implication $(i : a \to c : ++)$ indicates that when a is known to be true, then so is c. Thus it denotes a constraint on the probability distribution across a and c such that if $\Pr(a)$ becomes 1, so does $\Pr(c)$. This requires that:

$$\Pr(c \,|\, a, X) = 1$$

for all $X \in \{x, \neg x\}$ for which there is a triple $(i : x \to c : s)$ or $(i : \neg x \to c : s)$ [15]. Note that this type of implication also conforms to the conditions for implications labelled with $+$, and that if $\Pr(c \,|\, \neg a, x) = 1$ then $\Pr(c)$ is always equal to $\Pr(a)$. Similarly, a probabilistic interpretation of an implication $(i : a \to c : --)$ which denotes the fact that if a is true, c is false requires that:

$$\Pr(c \mid a, X) = 0$$

for all $X \in \{x, \neg x\}$ for which there is a triple $(i : x \to c : s)$ or $(i : \neg x \to c : s)$. There are two more types of categorical implication which are symmetric to those already introduced. The first is of the form $(i : a \to c : -+)$ which denotes the constraint:

$$\Pr(c \mid \neg a, X) = 1$$

for all $X \in \{x, \neg x\}$ for which there is a triple $(i : x \to c : s)$ or $(i : \neg x \to c : s)$. The second is of the form $(i : a \to c : +-)$ which denotes the constraint:

$$\Pr(c \mid \neg a, X) = 0$$

for all $X \in \{x, \neg x\}$ for which there is a triple $(i : x \to c : s)$ or $(i : \neg x \to c : s)$. Unsurprisingly, the introduction of categorical implications imposes restrictions on other implications involving the same formulae.

Theorem 2. *A consequence of the probabilistic semantics is that an implication $(i : a \to c : ++)$ places the same restrictions over the probabilities of a and c as $(i : a \to \neg c : --)$, $(i : \neg a \to c : -+)$ and $(i : \neg a \to \neg c : +-)$, and implies that for all other implications $(i : x \to c : s)$ where $x \neq a$, it is the case that $s \in \{++, +, -, -+\}$.*

Proof: The implication $(i : a \to c : ++)$ requires $\Pr(c \mid a, X) = 1$ which is exactly the constraint imposed by $(i : a \to \neg c : -+)$. $\Pr(c \mid a, X) = 1$ implies $\Pr(\neg c \mid a, X) = 0$ which is the condition imposed by $(i : a \to \neg c : --)$ and $(i : \neg a \to \neg c : +-)$. The restriction on implications such as $(i : x \to c : s)$ follows directly from the mutual incompatibility of the constraints on the conditional probabilities imposed by the categorical implications [15]. $\square$

This property is exactly what we should expect. If $\Pr(a)$ is related to $\Pr(c)$ then it is also related to $\Pr(\neg c)$, and its negation is related to both $\Pr(c)$ and $\Pr(\neg c)$. Furthermore, if the occurrence of a proves c to be true, then no other evidence can change this conclusion so implications labelled $+-$ or $--$ may not have c as their consequent. Similar results hold for other types of categorical implication.

Theorem 3. *A consequence of the probabilistic semantics is that an implication $(i : a \to c : ++)$ or $(i : a \to c : -+)$ places the same restrictions over the probabilities of a and c as $(i : c \to a : +)$, $(i : c \to \neg a : -)$, $(i : \neg c \to a : -)$ and $(i : \neg c \to \neg a : +)$.*

Proof: An categorical implication $(i : a \to c : ++)$ or $(i : a \to c : -+)$ is just a more extreme version of $(i : a \to c : +)$, and while it won't necessarily reverse to give a categorical implication, it will reverse just like $(i : a \to c : +)$ so the result follows from Theorem 1. $\square$

Again, analogous results hold for the other kinds of categorical implications.

	⇑	↑	↔	↓	⇓	↕
⇑	⇑	↑	↑	↕	↕	↕
↑	↑	↑	↑	↕	↕	↕
↔	↑	↑	↔	↓	↓	↕
↓	↕	↕	↓	↓	↓	↕
⇓	↕	↕	↓	↓	⇓	↕
↕	↕	↕	↕	↕	↕	↕

(a)

	++	+-	+	0	-	-+	--	?
⇑	⇑	↑	↑	↔	↓	⇓	↓	↕
↑	↑	↑	↑	↔	↓	↓	↓	↕
↔	↔	↔	↔	↔	↔	↔	↔	↔
↓	↓	↓	↓	↔	↑	↑	↑	↕
⇓	↓	⇓	↓	↔	↑	↑	⇑	↕
↕	↕	↕	↕	↔	↕	↕	↕	↕

(b)

Table 1. Conjunction introduction comb$_{conj}$ (a) and implication elimination comb$_{imp}$ (b)

3.2 The proof rules

We now turn to providing a probabilistic interpretation for the functions used to build arguments starting with conjunction introduction and elimination. When introducing conjunction we have:

Definition 4. The function comb$_{conj}$: $Sg \in \{⇑, ↑, ↔, ↓, ⇓, ↕\} \times Sg' \in \{⇑, ↑, ↔, ↓, ⇓, ↕\} \mapsto Sg'' \in \{⇑, ↑, ↔, ↓, ⇓, ↕\}$ is specified by Table 1(a) where, as with all combinator tables in this paper, the first argument is taken from the first column and the second argument is taken from the first row.

When eliminating a conjunction with sign Sg we assign both conjuncts the sign conj$_{elim}(Sg)$.

Definition 5. The function conj$_{elim}$: $Sg \in \{⇑, ↑, ↔, ↓, ⇓, ↕\} \mapsto Sg' \in \{⇑, ↑, ↔, ↓, ⇓, ↕\}$ is as follows:

$$\text{conj}_{elim}(Sg) = \begin{cases} ⇑ & \text{if } Sg = ⇑ \\ ↕ & \text{otherwise} \end{cases}$$

What this means is that most of the time it is not possible to determine how the probability of the individual conjuncts change. This is an unfortunate but unavoidable property of probability theory.

To deal with implication we need two further functions, comb$_{imp}$ to establish the sign of a formula generated by the rule of inference →-E, and comb$_{imp}^{-1}$ to establish the sign of an implication generated by →-I. This means that the main use of comb$_{imp}$ is to combine the change in probability of a formula a, say, with the constraint that the probability of a imposes upon the probability of another formula c. Since this constraint is expressed in exactly the same way as qualitative influences are in QPNs, comb$_{imp}$ performs the same function as ⊗ [22], and is merely an extension of it.

Definition 6. The function comb$_{imp}$: $Sg \in \{⇑, ↑, ↔, ↓, ⇓, ↕\} \times Sg' \in \{++, +-, +, 0, -, -+, --, ?\} \mapsto Sg'' \in \{⇑, ↑, ↔, ↓, ⇓, ↕\}$ is specified by Table 1(b).

Note the asymmetry in the table which stems from the definition of the categorical implications. If the asymmetry did not exist, categorical implications would be close to logical bi-implication. As one might guess from its name, the function $comb_{imp}^{-1}$ is merely the inverse of $comb_{imp}$:

Definition 7. The function $comb_{imp}^{-1} : Sg \in \{⇑, ↑, ↔, ↓, ⇓, ↕\} \times Sg' \in \{⇑, ↑, ↔, ↓, ⇓, ↕\} \mapsto Sg'' \in \{++, +-, +, 0, -, -+, --, ?\}$ is specified by Table 2(a). Blank spaces represent impossible combinations.

The rules for handling negation are applicable only to *swffs* and permit the negation to be either introduced or eliminated by reversing the direction of change of the sign, for example allowing $(i : \neg a : ↑)$ to be rewritten as $(i : a : ↓)$. This leads to the definition of neg_{form}:

Definition 8. The function $neg_{form} : (i : \neg a : s), s \in \{⇑, ↑, ↔, ↓, ⇓, ↕\} \mapsto (i : a : s'), s' \in \{⇑, ↑, ↔, ↓, ⇓, ↕\}$ relates s to s' by Table 2 (b).

Note that neg_{form} is not defined over the values $++, +-, +, 0, -, -+,$ and $--$. Although an implication $(i : a \to b : +)$ has a kind of inverse relation with $(i : a \to b : -)$, there is no such relation with $(i : \neg(a \to b) : s)$—indeed, $(i : \neg(a \to b) : s)$ is not even an implication, since its main connective is $\neg$—and because of this it is not possible to apply neg_{form} to an implication.

Having defined the combination of formulae to build arguments, we need to specify suitable function flat to be used to assess the overall strength of several arguments for the same formula. Now, in general, the overall strength of several arguments will be influenced by the interaction between steps in the arguments, and it will be necessary to take into account the grounds of each argument when flattening them in order to correctly handle any dependencies between the values of the formulae concerned. This is why flat is defined over the grounds as well as the signs of the arguments it operates on. However, in $\mathcal{NA}''$, because the effect of each implication is defined to occur whatever other arguments are formed (this is a result of the constraint imposed on the conditional probabilities by the implications), all combinations are completely local, and the structure of the

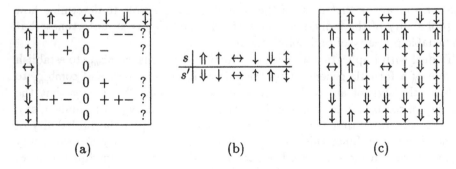

(a)

	⇑	↑	↔	↓	⇓	↕
⇑	++	+	0	-	--	?
↑		+	0	-		?
↔			0			
↓		-	0	+		?
⇓	-+	-	0	+	+-	?
↕			0			?

(b)

s	⇑	↑	↔	↓	⇓	↕
s'	⇓	↓	↔	↑	⇑	↕

(c)

	⇑	↑	↔	↓	⇓	↕
⇑	⇑	⇑	⇑	⇑		⇑
↑	⇑	↑	↑	↕	⇓	↕
↔	⇑	↑	↔	↓	⇓	↕
↓	⇑	↕	↓	↓	⇓	↕
⇓	⇓	⇓	⇓	⇓	⇓	⇓
↕	⇑	↕	↕	↕	⇓	↕

Table 2. Implication introduction $comb_{imp}^{-1}$ (a), negation of *swffs* neg_{form} (b), and flattening flat (c).

arguments may be disregarded when flattening (for exactly the same reason as it is in QPNs [5]). As a result, flat is simply the iterated application of the function ⊕—an extended version of the qualitative addition function used by QPNs:

$$\mathsf{flat}\Big(\{\langle G_i, Sg_i\rangle\}\Big) = \bigoplus_i Sg_i$$

where the correct way to combine the changes in probability using ⊕ is as follows:

Definition 9. The function ⊕ : $Sg \in \{\Uparrow, \uparrow, \leftrightarrow, \downarrow, \Downarrow, \updownarrow\} \times Sg' \in \{\Uparrow, \uparrow, \leftrightarrow, \downarrow, \Downarrow, \updownarrow\} \mapsto Sg'' \in \{\Uparrow, \uparrow, \leftrightarrow, \downarrow, \Downarrow, \updownarrow\}$ is specified by Table 2 (c). Blank spaces represent impossible combinations.

4 Example

For an illustration of some of the reasoning possible in $\mathcal{N}\mathcal{A}''$, consider the following database. This encodes the information that three events have a bearing on whether or not I lose my job—being ill makes it more likely I will lose my job, doing good research makes it less likely I will lose my job, and embezzling money makes it certain I will lose my job—while being ill makes it more likely I will go to hospital:

$$r1 : good\_research \rightarrow lose\_job : - \qquad \Delta_1$$
$$r2 : ill \rightarrow lose\_job : +$$
$$r3 : embezzle\_money \rightarrow lose\_job : ++$$
$$r4 : ill \rightarrow hospital : +$$
$$f1 : good\_research : \Uparrow$$
$$f2 : ill : \Uparrow$$

It also becomes known that I am ill and do good research (this is a fictional example). From this information we can build the arguments:

$$\Delta_1 \vdash_{ACR} (hospital, \{f2, r4\}, \uparrow)$$
$$\Delta_1 \vdash_{ACR} (lose\_job, \{f1, r1\}, \downarrow)$$
$$\Delta_1 \vdash_{ACR} (lose\_job, \{f2, r2\}, \uparrow)$$

The first argument means that I am increasingly likely to go to hospital. The second two arguments will flatten to give $\updownarrow$, indicating that it is impossible to say how the probability of losing my job will change. These are exactly the conclusions that would be drawn by the equivalent QPN illustrating the fact that $\mathcal{N}\mathcal{A}''$ is capable of representing binary QPNs (as discussed in [16]) and can be regarded as a mechanism for the construction of binary QPNs. If we now add the fact that I am known to embezzle money from my employer:

$$(f4 : embezzle\_money : \Uparrow)$$

to the database, we can build a new argument:

$$\Delta_3 \vdash_{ACR} (lose\_job, \{f4, r3\}, \Uparrow)$$

meaning that the arguments about my loss of job will now flatten to give ⇑, so that my loss of job is assured. Thus $\mathcal{N}\mathcal{A}''$ is capable of a form of defeasible reasoning in which certain information can outweigh previously known uncertain information.

5 Soundness and completeness

Armed with the interpretation introduced in Section 3, $\mathcal{N}\mathcal{A}''$ has a probabilistic semantics. However, the results presented so far have a rather baroque appearance, and so might seem *ad hoc* to the sceptical reader. However, they are not. The proof mechanism given above is provably sound for the propagation of changes in probability, as shown by the following result:

Theorem 10. *The construction and flattening of arguments in $\mathcal{N}\mathcal{A}''$ is sound with respect to probability theory.*

Proof: The soundness of $\mathcal{N}\mathcal{A}''$ follows immediately from the soundness of the the way in which changes in probabilities are propagated and flattened and thus from the soundness of the proof rules and combination tables.

(**Conjunction introduction**): Consider the probabilities $\Pr(a)$ and $\Pr(b)$ of the two propositions being conjoined. $\Pr(a \wedge b) = \Pr(a).\Pr(b \mid a) = \Pr(a \mid b).\Pr(b)$. Thus if at least one of $\Pr(a)$ and $\Pr(b)$ increases and the other does not decrease, then $\Pr(a \wedge b)$ will increase. If one increases and one decreases, then the change in $\Pr(a \wedge b)$ cannot be determined. If both increase to 1, $\Pr(a \wedge b)$ increases to 1. Similar reasoning completes the proof.

(**Conjunction elimination**): There are two parts to the proof. One for the part of the function that gives ⇑ and one for the part that gives ↕. For the first, the following suffices—the only way in which $Pr(a \wedge b)$ can increase to 1 is if both $\Pr(a)$ and $\Pr(b)$ increase to 1 (since the notion of 'increasing to 1' takes into account the fact values may have been 1 all along). For the second part we need the following argument. Giving any sign as ↕ is always sound (since it means that nothing at all is being said about the relevant probability). However, it is also possible to prove that no more precise rule can be proposed. This is done by considering what the probability of $\Pr(a \wedge b)$ should be if $\Pr(a) = \uparrow$ and $\Pr(b) = \downarrow$. The answer is that $\Pr(a \wedge b)$ can either increase, decrease, or not change depending on the relative magnitudes of the changes in $\Pr(a)$ and $\Pr(b)$. Turning this around, it is clear that no firm conclusions about changes in $\Pr(a)$ and $\Pr(b)$ can be drawn from particular changes in $\Pr(a \wedge b)$ other than $\Pr(a \wedge b) = \Uparrow$.

(**Implication elimination**): First consider implications labelled with $+$. From the definition of such implications it is clear that combining any increase in probability with an implication labelled $+$ will generate a possible increase in probability, in other words ↑. Similarly combining any decrease in probability with an implication labelled $+$ will generate ↓, combining no change in probability with such an implication will generate ↔, and combining a change of ↕ with such an implication will generate ↕. An implication labelled $-$ will also give

↔ when combined with no change in probability and ↕ when combined with ↕, but otherwise will have the opposite behaviour to that of an implication labelled +. From its definition it is clear that an implication labelled ++ will behave like an implication labelled + except when combined with a change ⇑ when it will generate a change of ⇑. The results for other implications can be obtained analogously.

(Implication introduction): The soundness of Table 2(a) follows directly from that of Table 1 (b). Where only one possible type of implication can give Sg' the relevant sign is given by the table, where two or more can give Sg', for instance when $Sg = \uparrow$ and $Sg' = \uparrow$ when the sign of the implication could be $++$, $-+$ or $+$, the most inclusive sign is given, $+$ in the case of our example since it subsumes $++$ and $-+$.

(Negation elimination and introduction): Consider a proposition a. If $\Pr(a)$ increases to 1 then clearly $\Pr(\neg a)$ decreases to zero, and if $\Pr(\neg 1)$ increases to 1 then clearly $\Pr(a)$ decreases to zero. This takes care of the function for ⇓ and ⇑. The other cases are handled similarly.

(Flattening): Table 2(c) follows directly from qualitative addition [22] and the fact that categorical changes in probability cannot be altered by non-categorical changes—the latter follows from the definition of categorical implications [15]. The spaces in the table follow from Property 2. □

The other thing that might worry the sceptical reader is the completeness of $\mathcal{N}\mathcal{A}''$. However, since we only allow the initial database to contain implications $(i : St \to St' : Sg)$ where St is a direct cause of St' we have the following result:

Theorem 11. *The construction and flattening of arguments in $\mathcal{N}\mathcal{A}''$ is complete with respect to probability theory for reasoning in a causal direction.*

Proof: Immediate from the definition of $\vdash_{ACR}$ and the causal direction of the implications—all possible causally directed inferences can be made by the application of the appropriate proof rules. □

Causal completeness has its limitations, since from a database consisting of an implication $(r1 : a \to b : +)$ and a fact $(f1 : b : \uparrow)$ no arguments may be built using $\vdash_{ACR}$, yet using probability theory one can infer that $\Pr(a)$ increases. Extending $\mathcal{N}\mathcal{A}''$ to cope with this kind of evidential reasoning is straightforward and is discussed in [18].

6 Discussion

The probabilistic semantics of $\mathcal{N}\mathcal{A}''$ gives us two things. Firstly, it gives us a precise probabilistic notion of what it means to have an argument for something. If we accept the probabilistic interpretation of *wff*s then given a database of causal relations (implications) and evidence which leads to some changes in belief we can infer what changes in belief are implied. Only those propositions supported by sound arguments will undergo a change in belief and only those propositions

If	and	and	then
$St = w$	$Sg = \Uparrow$	$\Pr(w)_{initial} = p$	$\Pr(w)_{final} = 1$
$St = w$	$Sg = \uparrow$	$\Pr(w)_{initial} = p$	$p \leq \Pr(w)_{final} \leq 1$
$St = w$	$Sg = \leftrightarrow$	$\Pr(w)_{initial} = p$	$\Pr(w)_{final} = p$
$St = w$	$Sg = \downarrow$	$\Pr(w)_{initial} = p$	$p \geq \Pr(w)_{final} = 0$
$St = w$	$Sg = \Downarrow$	$\Pr(w)_{initial} = p$	$\Pr(w)_{final} = 0$
$St = w$	$Sg = \updownarrow$	$\Pr(w)_{initial} = p$	$0 \leq \Pr(w)_{final} \leq 1$

If	and	then
$St = v \to w$	$Sg = ++$	$\Pr(w \mid v, x) = 1$
$St = v \to w$	$Sg = +-$	$\Pr(w \mid \neg v, x) = 0$
$St = v \to w$	$Sg = +$	$\Pr(w \mid v, x) \geq \Pr(w \mid \neg v, x)$
$St = v \to w$	$Sg = 0$	$\Pr(w \mid v, x) = \Pr(w \mid \neg v, x)$
$St = v \to w$	$Sg = -$	$\Pr(w \mid v, x) \leq \Pr(w \mid \neg v, x)$
$St = v \to w$	$Sg = -+$	$\Pr(w \mid \neg v, x) = 1$
$St = v \to w$	$Sg = --$	$\Pr(w \mid v, x) = 0$
$St = v \to w$	$Sg = ?$	The relationship between $\Pr(w \mid v, x)$ and $\Pr(w \mid \neg v, x)$ is unknown.

Table 3. What a derived formula means.

for which an argument may be built have a change of belief warranted in them. Secondly, the semantics gives us a means of determining how changes in probability are propagated. If we encode our probabilistic knowledge of the world by writing down *swff*s and then build arguments for and against propositions using $\vdash_{ACR}$, we can identify the changes in probability of those propositions. Either way, if after building arguments and flattening we have an pair (St, Sg) where St is any *wff* then Sg indicates the change in probability of St. If, on the other hand we have (St, Sg) where St is an *iwff* $St' \to St''$ then Sg indicates the constraint between $\Pr(St')$ and $\Pr(St'')$. The full denotation of any pair (St, Sg) is given by Table 3. Since reasoning with probability is normative in the sense that it accords to the norms of probability theory (which can be justified by the usual Dutch book argument) this makes $\mathcal{N}\mathcal{A}''$ normative.

Another advantage of $\mathcal{N}\mathcal{A}''$ is that it makes it possible to give a probabilistic analysis of different styles of argumentation. This section gives examples of three such analyses using small examples (more extensive analysis may be found in [18]). In some systems of argumentation [7], arguments are flattened by counting the number of arguments for and against a proposition and giving it the sign of the majority. Consider the following example:

$$f1 : embezzle\_funds : -. \qquad \Delta_2$$
$$f2 : good\_tutor : -.$$
$$f3 : good\_research : +.$$
$$r1 : embezzle\_funds \to lose\_job : +.$$

$$r2 : good\_tutor \rightarrow lose\_job : -.$$
$$r3 : good\_research \rightarrow lose\_job : -.$$

there are three arguments affecting the proposition "*lose_job*"

$$\Delta_2 \vdash_{ACR} (lose\_job, (f1, r1), -).$$
$$\Delta_2 \vdash_{ACR} (lose\_job, (f2, r2), +).$$
$$\Delta_2 \vdash_{ACR} (lose\_job, (f3, r3), -).$$

Counting arguments suggests we should conclude that belief in "*lose_job*" decreases since there are two arguments against it and only one for it. Considering using $\mathcal{N}\mathcal{A}''$ here shows that concluding $\langle lose\_job, - \rangle$ involves the assumption:

$$\Delta \Pr(good\_tutor).(\Pr(lose\_job \,|\, good\_tutor) - \Pr(lose\_job \,|\, \neg good\_tutor))$$
$$\leq \Delta \Pr(embezzle).(\Pr(lose\_job \,|\, embezzle) - \Pr(lose\_job \,|\, \neg embezzle))$$
$$+ \Delta \Pr(research).(\Pr(lose\_job \,|\, research) - \Pr(lose\_job \,|\, \neg research))$$

when the size of the changes in probability propagated across each implication are taken into account [19]. Another kind of flattening is that based on directness of argument [12]. With the database:

$$f1 : good\_research : + \qquad\qquad \Delta_3$$
$$r1 : good\_research \rightarrow good\_tutor : -$$
$$r2 : good\_tutor \rightarrow lose\_job : -.$$
$$r3 : lose\_job \rightarrow no\_money : +.$$
$$r4 : good\_research \rightarrow job\_in\_industry : +.$$
$$r5 : job\_in\_industry \rightarrow no\_money : -.$$

we get the arguments:

$$\Delta_3 \vdash_{ACR} (no\_money, (f1, r1, r2, r3), +).$$
$$\Delta_3 \vdash_{ACR} (no\_money, (f2, r4, r5), -).$$

Flattening by directness leads us to conclude $\langle no\_money, - \rangle$. since the argument for it is the shorter of the two. This time the probabilistic semantics of $\mathcal{N}\mathcal{A}''$ expose the assumption as being:

$$(\Pr(industry \,|\, research) - \Pr(industry \,|\, \neg research))$$
$$.(\Pr(no\_money \,|\, industry) - \Pr(no\_money \,|\, \neg industry))$$
$$\geq (\Pr(good\_tutor \,|\, research) - \Pr(good\_tutor \,|\, \neg research))$$
$$.(\Pr(lose\_job \,|\, good\_tutor) - \Pr(lose\_job \,|\, \neg good\_tutor))$$
$$.(\Pr(no\_money \,|\, lose\_job) - \Pr(no\_money \,|\, \neg lose\_job))$$

which is similar to that underlying counting arguments, but without placing constraints on the change in the probability of the initial facts. Finally, several authors [11, 21] have considered the use of rebuttal and undercutting to flatten arguments—a proposition is rebutted when it is directly argued against and is undercut when one of the steps in the argument for it is argued against. Consider the following example:

$$f1 : good\_tutor : -. \qquad \Delta_4$$
$$f2 : high\_research\_output : +.$$
$$r1 : good\_tutor \rightarrow lose\_job : -.$$
$$r2 : lose\_job \rightarrow no\_money : +.$$
$$r3 : high\_research\_output \rightarrow lose\_job : -.$$

From Δ_4 we can build the arguments:

$$\Delta_4 \vdash_{ACR} (lose\_job, (f1, r1), (+)).$$
$$\Delta_4 \vdash_{ACR} (no\_money, (f1, r1, r2), (+)).$$
$$\Delta_4 \vdash_{ACR} (lose\_job, (f2, r3), (-)).$$

and find that the third argument rebuts proposition $lose\_job$ and undercuts the the proposition no_money. The usual interpretation of this is that the $lose\_job$ is more affected by the third argument than no_money, and this is exactly what one would conclude from $\mathcal{N}\mathcal{A}''$ since the changes associated with the third argument will be no smaller for $\Pr(lose\_job)$ than for $\Pr(no\_money)$.

7 Summary

This paper has discussed a means of giving a probabilistic semantics to a system of argumentation. It is thus in some senses an extension of previous work on such systems of argumentation [11] as well as probabilistic systems of argumentation which use a more restricted base logic [17]. With a solid basis in probability theory, the system can be used to combine the advantages of a sound means of handling uncertainty with the expressiveness of a logical method of knowledge representation, an expressiveness that will be increased with the planned extension to a first order system which includes disjunction. Amongst other things the system is capable of defeasible reasoning and knowledge-based model construction for qualitative probabilistic networks. Furthermore, because of its qualitative nature, the system may be used when probabilistic knowledge of a domain is incomplete and the fact that it is soundly based on probability theory makes it a useful basis for a qualitative decision theory [8]. Finally, it should be noted that $\mathcal{N}\mathcal{A}''$ has similarities with Neufeld's system for default reasoning [14]. Some of these are explored in [16].

References

1. S. Benferhat, D. Dubois, and H. Prade. Argumentative inference in uncertain and inconsistent knowledge bases. In *Proceedings of the 9th Conference on Uncertainty in Artificial Intelligence*, 1993.
2. A. Darwiche. Argument calculus and networks. In *Proceedings of the 9th Conference on Uncertainty in Artificial Intelligence*, 1993.

3. A. Darwiche and M. Ginsberg. A symbolic generalisation of probability theory. In *Proceedings of the 10th National Conference on Artificial Intelligence*, 1992.
4. A. Darwiche and M. Goldszmidt. On the relation between kappa calculus and probabilistic reasoning. In *Proceedings of the 10th Conference on Uncertainty in Artificial Intelligence*, 1994.
5. M. J. Druzdzel. *Probabilistic reasoning in Decision Support Systems: from computation to common sense*. PhD thesis, Carnegie Mellon University, 1993.
6. P. M. Dung. On the acceptability of arguments and its fundamental role in non-monotonic reasoning and logic programming. In *Proceedings of the 13th International Conference on Artificial Intelligence*, 1993.
7. J. Fox. A unified framework for hypothetical and practical reasoning (2): lessons from clinical medicine. In *Proceedings of the Conference on Formal and Applied Practical Reasoning*, 1996.
8. J. Fox and S. Parsons. On using arguments for reasoning about actions and values. In *Proceedings of AAAI Spring Symposium on Qualitative Preferences in Deliberation and Practical Reasoning*, 1997.
9. J. Fox, S. Parsons, P. Krause, and M Elvang-Gørannson. A generic framework for uncertain reasoning. In *Proceedings of the IMACS III International Workshop on Qualitative Reasoning and Decision Technologies*, 1993.
10. M. Goldszmidt and J. Pearl. Qualitative probabilities for default reasoning, belief revision and causal modelling. *Artificial Intelligence*, 84:57–112, 1996.
11. P. Krause, S. Ambler, M. Elvang-Gøransson, and J. Fox. A logic of argumentation for reasoning under uncertainty. *Computational Intelligence*, 11:113–131, 1995.
12. R. Loui. Defeat among arguments: a system of defeasible inference. *Computational Intelligence*, 3:100–106, 1987.
13. N. F. Michelena. *Monotonic influence diagrams: application to optimal and robust design*. PhD thesis, University of California at Berkeley, 1991.
14. E. Neufeld. A probabilistic commonsense reasoner. *International Journal of Intelligent Systems*, 5:565–594, 1990.
15. S. Parsons. Refining reasoning in qualitative probabilistic networks. In *Proceedings of the 11th Conference on Uncertainty in Artificial Intelligence*, 1995.
16. S. Parsons. Comparing normative argumentation with other qualitative probabilistic systems. In *Proceedings of the Conference on Information Processing and the Management of Uncertainty*, 1996.
17. S. Parsons. Defining normative systems for qualitative argumentation. In *Proceedings of the Conference on Formal and Applied Practical Reasoning*, 1996.
18. S. Parsons. Normative argumentation and qualitative probability. Technical report, Department of Electronic Engineering, Queen Mary and Westfield College, 1996.
19. S. Parsons. *Qualitative approaches to reasoning under uncertainty*. MIT Press, (to appear), Cambridge, MA, 1997.
20. J. Pearl. *Probabilistic reasoning in intelligent systems; networks of plausible inference*. Morgan Kaufmann, San Mateo, CA., 1988.
21. J. L. Pollock. Justification and defeat. *Artificial Intelligence*, 67:377–407, 1994.
22. M. P. Wellman. *Formulation of tradeoffs in planning under uncertainty*. Pitman, London, 1990.
23. N. Wilson. An order of magnitude calculus. In *Proceedings of the 11th Conference on Uncertainty in Artificial Intelligence*, 1995.

Towards a Formalization of Narratives: Actions with Duration, Concurrent Actions and Qualifications

Anna Radzikowska

Institute of Mathematics, Warsaw University of Technology
Plac Politechniki 1, 00–661 Warsaw, Poland
Email: annrad@im.pw.edu.pl

Abstract. The paper addresses the qualification problem in the context of narratives where actions with duration, causal chains of actions and concurrent actions occur. In order to represent these scenarios, we define an *action description language* $\mathcal{AL}_1$ which enables to represent different types of system constraints and guarantees correct interactions between actions in simple cases of concurrency. A simple linear discrete model of time is assumed. Two preferential methods of reasoning are proposed which amount to global and chronological maximization of executable actions, respectively. We provide a translation from $\mathcal{AL}_1$ into circumscription and show that this translation is sound and complete relative to the semantics of $\mathcal{AL}_1$.

1 Introduction and Motivations

This paper concerns narrative reasoning, that is reasoning about actions which occur at particular times, which are actually executed during specific periods of time as well as about properties of the world that hold at different times. The importance of narratives has been recognized in, for example, [11], [3], [6]. When we deal with the notion of flow of time, the narrative-based formalism seems especially adequate.

We focus on the qualification problem in the context of a broad class of scenarios where actions with duration, nondeterministic actions, causal chains of actions and concurrent and/or overlapping actions occur.[1] Roughly speaking, the qualification problem is that of determining conditions under which actions are executable. When concurrency is involved, substantially new problems arise. This work is partially inspired by the well-known examples traditionally studied in the theory of concurrency. The intention is to show *how these problems may be embedded in the context of action scenarios*. We essentially concentrate on the mutual exclusion problem and show how to avoid undesired interactions between actions. It seems obvious that dynamic systems involving concurrency should be able to properly deal with mutual exclusion. In fact, this problem is the instance of the qualification problem: to ensure intended synchronization and cooperation

---

[1] We say that actions overlap iff time intervals of their duration are *not* disjoint.

of processes, some actions should be blocked – consequently, executions of some actions contribute to qualifications for other actions.

In general, several actions may affect the same fluent at a time (clearly, if it does not cause inconsistency). As we know, such events are often undesirable or even impossible. To cope with this problem, we introduce the notion of *protected fluents*. Intuitively, a fluent is protected by an action A iff no action except A may affect this fluent during the duration of A. This way actions are guaranteed to gain exclusive access to specific fluents.

An action execution usually causes twofold changes: those that necessarily occur when the action terminates and those that occur during the duration of the action, but not necessarily when the action ends. The former may be called *final results*, the latter *intermediate results* of the action. When overlapping actions are allowed, intermediate effects may affect executability conditions of actions. We restrict ourselves to simple cases of intermediate results. Specifically, many actions make use of some resources when being executed, some of which cannot be shared by different actions (e.g. printing a document sets the printer's mode to "printing" and no other document may be printed meanwhile). In order to represent such cases, we suggest introducing fluents that correspond to modes of the system resources: when an action makes use of a resource, the corresponding fluent is specified as *reserved* and made true throughout this action; when the action ends, it is made false (the resource is released), if no action may change its value at this time.

Traditionally, fluents are assumed to persist by default. Similarly, it seems natural to assume that they are unprotected (resp. unreserved) unless otherwise is explicitly specified.

Reserved fluents combined with the protection device introduce a specific type of system constraints. To be sure, it would be also possible to explicitly specify which actions are mutually exclusive. The advantage of our approach is that mutual exclusion is left as the subject of inference.

State constraints may invoke indirect effects of actions or may prevent them from being carried out. Since the ramification problem is out of the scope of this paper, all constraints are assumed to function as qualification constraints only.

Most of the existing approaches involving concurrency (e.g. [1], [13], [6], [9]) focus on reasoning about effects of concurrently executed actions. For simplicity, we assume that the effect of concurrently executed actions is the joint outcome of the particular actions achievable when these actions are performed separately.

Occurrences of actions need not imply their performances, e.g. because their preconditions do not hold, their effects violate observations[2] or system constraints, they are invoked in states in which they cannot start, or they require access to resources currently used by other actions. However, it seems intuitively justified to assume that an action execution immediately follows its occurrence unless there are reasons which make the action unexecutable. Yet this default rule does not determine precisely which actions are actually executed, so some preferential strategy is needed. Actually, this problem has not been discussed in

---

[2] Notice that in most formalisms observations do not affect actions.

the literature so far (at least, as far as we know). We suggest two such methods: *chronological* and *global maximization of executable actions*. The first method, CHM, amounts to prefer actions which occur at earlier points in time (it realizes in fact the FIFO strategy), the second one, GLM, selects all maximal sets of executable actions (relative to set inclusion).

The actual reasoning process proceeds in two stages. First, for every possible configuration of potentially executable actions we determine all courses of events constrained by the inertia law. This step amounts to solve the frame problem and may be viewed as a simulation of possible behaviors of the system. Next, from among the remaining "histories" the preferred ones are selected according to either of the proposed preferential policies.

Following the methodology proposed by Gelfond and Lifschitz ([2]), we first define an *action description language* $\mathcal{AL}_1$ (as the extension of the language $\mathcal{AL}_0$ introduced in [8]) for representing commonsense knowledge about action scenarios, and then translate this language into circumscriptive theories. As a final result, we establish soundness and completeness of this translation relative to the semantics of $\mathcal{AL}_1$.

The paper is organized as follows. Section 2 provides general characteristics of domains under consideration. The action language $\mathcal{AL}_1$ is defined in Section 3. In Section 4, we outline a propositional fluent logic PFL-1 and provide circumscriptive schemes syntactically characterizing the preferential methods GLM and CHM. In Section 5 we present a translation from $\mathcal{AL}_1$ into circumscription and show that it is sound and complete relative to the semantics of $\mathcal{AL}_1$. The paper is completed with concluding remarks and options for future work.

2 Domains of Actions: Underlying Assumptions

In this paper we consider action scenarios where all actions are assumed to be known together with all their effects (i.e. all results are explicitly specified).

Narrative-based formalisms enable to represent facts that are *observed* to hold at particular times. We assume that all observations are fully reliable.

Background knowledge is traditionally represented by *state constraints* reflecting facts that must hold in every state of the world. We generalize these constraints by admitting rules which specify *temporal* (not necessarily *permanent*) persistence of some properties of the world (e.g. since an agent has lost his job, he is worried as long as he has no money). These rules are also useful to express irreversible changes in the world (e.g. whenever a turkey is killed, it remains dead forever).

Considering actions with duration we assume that the exact time of changes each action brings about is unknown – what is only guaranteed is that a postcondition of the action holds when the action terminates.

Executability conditions of actions are usually expressed by their preconditions reflecting facts which, when satisfied, make the action leading to particular (final) results. However, successful course of an action depends also on what is going on throughout its duration (e.g. the shooting action makes a turkey killed, provided that it does not manage to hide meanwhile). Following the terminology proposed in [10], these conditions will be called *prevail conditions*.

Causal chains of actions[3] reflect sequences $< A_1, A_2, \ldots >$ of actions such that an action A_i results from executing some previous action(s) A_j, $j < i$. In particular, A_j may either invoke the resulting action A_i or it may cause a situation which triggers an occurrence of A_i.[4] For simplicity, invoked actions are assumed to occur a fixed number of time units after termination of causal actions (similarly, triggered actions occur immediately in causal situations).

Finally, we assume that only the state in which an action occurs implies how long it may be performed.

3 An Action Language $\mathcal{AL}_1$

3.1 Syntax of $\mathcal{AL}_1$

Likewise the well-known language A ([2]), an action language $\mathcal{AL}_1$ is a family of languages each one of which is characterized by three nonempty sets of symbols: a set $\mathcal{A}$ of *actions*, a set $\mathcal{F}$ of *fluents* and a set $\mathcal{T}$ of *timepoint values*. For concreteness, the set $\mathcal{T}$ will be identified with the set $\mathbb{N}$ of all natural numbers.

A formula (denoted by α, π or φ) is a propositional combination of fluents. There are seven types of statements in $\mathcal{AL}_1$.

A *performance statement* is an expression of the following two forms

$$\textbf{A performs from } s \textbf{ to } t; \tag{1}$$

$$\textbf{A fails at } t. \tag{2}$$

Intuitively, (1) means that the action A is executed from time s to time t, whereas (2) says that A that occurs at time t cannot be performed at this time.

A *value statement* is an expression of the form

$$\alpha \textbf{ at } t \tag{3}$$

which intuitively means that the formula α holds at time t. If $t = 0$ then (3) is written **initially** α.

An *occurrence statement* is an expression of the form

$$\textbf{A occurs at } t \tag{4}$$

which informally says that the action A occurs at time t.

A *state constraint statement* is an expression of the form

$$\alpha \textbf{ since } \pi \textbf{ until } \varphi \tag{5}$$

which says that when the condition π holds then since then the formula α holds as long as the condition φ does not (e.g. *Worried* **since** $\neg HasJob$ **until** $HasMoney$). If $\varphi \equiv \bot$ (resp. $\pi \equiv \top$) then the part **until** (resp. **since**) is omitted.[5] If $\varphi \equiv \bot$ and $\pi \equiv \top$ then (5) is abbreviated to **always** α.

---

[3] We have previously considered these events in [7] and [8].

[4] Notice that invoked (resp. triggered) actions may be viewed as delayed, dynamic effects of causal actions.

[5] $\top$ and $\bot$ stand for truth constants *True* and *False*, respectively.

Next, we define an *evoke statement* of the following two forms:

$$\pi \text{ triggers A;} \tag{6}$$

$$\text{A invokes B after } d \text{ if } \pi. \tag{7}$$

The statement (6) means that any situation satisfying π immediately causes the occurrence of the action A. The statement (7) says that the execution of the action A causes the occurrence of the action B d units of time after A completes if π holds when A begins (we drop the part **if**, provided that $\pi \equiv \top$).

An *action unexecutability statement* is of the following forms

$$\text{impossible A since } \pi \text{ until } \varphi; \tag{8}$$

$$\text{B disables A until } \varphi \text{ if } \pi. \tag{9}$$

Intuitively, (8) says that since the time when π holds A cannot begin as long as φ is not satisfied. If $\varphi \equiv \neg\pi$ then (8) is abbreviated to **impossible A if** π.
The statement (9) asserts that after executing the action B that starts in a state satisfying π, the action A cannot begin as long as the condition φ does not hold.

Furthermore, we have four types of *action law statements*. For convenience, we drop the part **if** (resp. **while**) if $\pi \equiv \top$ (resp. $\varphi \equiv \top$).

An *effect statement* is an expression of the following two forms

$$\text{A causes } \alpha \text{ if } \pi \text{ while } \varphi; \tag{10}$$

$$\text{A releases } f \text{ if } \pi \text{ while } \varphi. \tag{11}$$

Intuitively, (10) says that the action A makes the formula α true at the end of its performance if its precondition π holds when A starts and during its duration the prevail condition φ holds too (e.g. **Fire causes** $\neg Alive$ **if** $Loaded$ **while** $\neg Hidden$). The statement (11) informally states that the fluent f is exempt from the inertia law when the action A is executed starting in a state where π holds and φ is satisfied during the whole performance of A (e.g. **Toss releases** $Heads$).

A *reserve statement* is of the form

$$\text{A reserves } f \text{ if } \pi \tag{12}$$

which says that the action A reserves the fluent f (i.e. f is true in all intermediate points of the duration of A) if it starts when π is satisfied. For example, **Print reserves** $Printing$.

Next, we have a *protect statement* of the form

$$\text{A protects } f \text{ if } \pi \tag{13}$$

which intuitively means that the action A protects the fluent f if π holds when A begins. For example, **Print protects** $Printing$.

Finally, we introduce a *duration statement* of the form

$$\text{A has duration } d \text{ if } \pi \tag{14}$$

which states that whenever the action A begins in a state satisfying π, it is executed during d time units.

Statements of the forms (3)–(14) will be called *domain statements*, whereas statements of the forms (1)–(4) will be referred to as *atomic queries*. A *query* is a propositional combination of atomic queries.

A set Υ of domain statements in $\mathcal{AL}_1$ is a *domain description*. The language of Υ will be denoted by $\mathcal{L}_\Upsilon$. For a domain description Υ, the set of all fluents and the set of all actions occurring in Υ will be denoted by $\mathcal{F}_\Upsilon$ and $\mathcal{A}_\Upsilon$, respectively.

A domain description Υ is *finite* iff it is a finite set of statements and the sets $\mathcal{F}_\Upsilon$ and $\mathcal{A}_\Upsilon$ are finite also.

3.2 Semantics of $\mathcal{AL}_1$

This section provides the semantics of the language $\mathcal{AL}_1$. Let us start with some preliminary notation and terminology.

For a formula α, by $fl(\alpha)$ we denote the set of all fluents occurring in α.

Given two functions $\Phi, \Psi : \mathcal{X} \to 2^\mathcal{Y}$, we write $\Phi \preceq \Psi$ to denote that for every $x \in \mathcal{X}$, $\Phi(x) \subseteq \Psi(x)$. By $\Phi \prec \Psi$ we denote that $\Phi \preceq \Psi$ but not $\Psi \preceq \Phi$.

A *valuation* is a function $v : \mathbb{N} \times \mathcal{F} \to \{0, 1\}$. This function can be easily extended for the set $Fm_\mathcal{F}$ of all formulae. A *history function* $H : \mathbb{N} \times Fm_\mathcal{F} \to \{0, 1\}$ is defined according to the truth tables in propositional logic.

The semantics of $\mathcal{AL}_1$ is centered upon the notion of a *structure*.

Definition 3.1 A *structure* is a tuple $\Sigma = (H, I, R, O, E, P, C)$, where

- $H : \mathbb{N} \times Fm_\mathcal{F} \to \{0, 1\}$ is a *history function*;
- $I : \mathbb{N} \times \mathcal{A} \times \mathbb{N} \to 2^\mathcal{F}$ is an *influence function* such that $I(s, \mathsf{A}, t)$ is the set of fluents which may change after executing the action A between time s and t;
- $R : \mathbb{N} \times \mathcal{A} \times \mathbb{N} \to 2^\mathcal{F}$ is a *reservation function* such that $R(s, \mathsf{A}, t)$ is the set of fluents reserved by the action A from time s to t; this function is such that for any $s, t \in \mathbb{N}$ and any $\mathsf{A} \in \mathcal{A}$,

$$\text{for any } f \in R(s, \mathsf{A}, t), \ H(t', f) = 1 \text{ for all } s < t' < t; \tag{15}$$

- $O : \mathbb{N} \to 2^\mathcal{A}$ is an *occurrence function* such that $O(t)$ yields the set of actions occurring at time t;
- $E \subseteq \mathbb{N} \times \mathcal{A} \times \mathbb{N}$ is a *performance relation* such that $(s, \mathsf{A}, t) \in E$ yields that the action A is executed from time s to t; this relation is such that for any $\mathsf{A} \in \mathcal{A}$ and any $s, t \in \mathbb{N}$,

$$(s, \mathsf{A}, t) \in E \text{ implies } (s, \mathsf{A}, t') \notin E \text{ for all } t' \neq t; \tag{16}$$

- $P : \mathbb{N} \times \mathcal{A} \times \mathbb{N} \to 2^\mathcal{F}$ is a *protection function* such that $P(s, \mathsf{A}, t)$ is the set of fluents protected by executing the action A from time s to t;
- $C : \mathbb{N} \to 2^\mathcal{A}$ is a *concession function* such that for any $t \in \mathbb{N}$, $C(t)$ is the set of actions that are allowed to start at time t. □

The functions I and R determine regions of action influences, so only within these areas changes are allowed. The condition (15) specifies that each reserved fluent f is true throughout the duration of the action that reserves f, whereas (16) captures the natural requirement that every executable action has the uniquely defined duration.

For a given structure $\Sigma = (H, I, R, O, E, P, C)$ and a statement σ of $\mathcal{AL}_1$, we define that Σ is a *structure for* σ, written $\Sigma \bowtie \sigma$, as follows:

- $\Sigma \bowtie$ A **performs from** s **to** t iff $(s, A, t) \in E$.
- $\Sigma \bowtie$ A **fails at** t iff $A \in O(t)$ & $A \notin C(t)$.
- $\Sigma \bowtie$ A **occurs at** t iff $A \in O(t)$.
- $\Sigma \bowtie \alpha$ **at** t iff $H(t, \alpha) = 1$.
- $\Sigma \bowtie \alpha$ **since** π **until** φ iff
  $$\forall t, d \in \mathbb{N} \; \Big[H(t, \pi) = 1 \; \& \; \big(\forall t \leq t' \leq t + d \; H(t', \varphi) = 0 \big) \Rightarrow H(t + d, \alpha) = 1 \Big].$$
- $\Sigma \bowtie \pi$ **triggers** A   iff $\forall t \in \mathbb{N} \; \Big[H(t, \pi) = 1 \Rightarrow A \in O(t) \Big]$.
- $\Sigma \bowtie$ A **invokes** B **after** d **if** π iff
  $$\forall s, t \in \mathbb{N} \; \Big[(s, A, t) \in E \; \& \; H(s, \pi) = 1 \Rightarrow B \in O(t + d) \Big].$$
- $\Sigma \bowtie$ **impossible** A **since** π **until** φ iff
  $$\forall t, d \in \mathbb{N} \; \Big[H(t, \pi) = 1 \; \& \; \big(\forall t \leq t' \leq t + d \; H(t', \varphi) = 0 \big) \Rightarrow A \notin C(t + d) \Big].$$
- $\Sigma \bowtie$ B **disables** A **until** φ **if** π iff $\forall s, t, d \in \mathbb{N} \; \Big[(s, B, t) \in E \; \& \; H(s, \pi) = 1$
  $$\& \; \big(\forall t \leq t' \leq t + d \; H(t', \varphi) = 0 \big) \Rightarrow A \notin C(t + d) \Big].$$
- $\Sigma \bowtie$ A **causes** α **if** π **while** φ iff $\forall s, t \in \mathbb{N} \; \Big[(s, A, t) \in E \; \& \; H(s, \pi) = 1 \; \&$
  $$\& \; \big(\forall s \leq t' \leq t \; H(t', \varphi) = 1 \big) \Rightarrow H(t, \alpha) = 1 \; \& \; \bigwedge_{f \in fl(\alpha)} f \in I(s, A, t) \Big].$$
- $\Sigma \bowtie$ A **releases** f **if** π **while** φ iff $\forall s, t \in \mathbb{N} \; \Big[(s, A, t) \in E \; \& \; H(s, \pi) = 1 \; \&$
  $$\& \; \big(\forall s \leq t' \leq t \; H(t', \varphi) = 1 \big) \Rightarrow f \in I(s, A, t) \Big].$$
- $\Sigma \bowtie$ A **reserves** f **if** π iff
  $$\forall s, t \in \mathbb{N} \; \Big[(s, A, t) \in E \; \& \; H(s, \pi) = 1 \Rightarrow f \in R(s, A, t) \Big].$$
- $\Sigma \bowtie$ A **protects** f **if** π iff
  $$\forall s, t \in \mathbb{N} \; \Big[(s, A, t) \in E \; \& \; H(\pi, s) = 1 \Rightarrow f \in P(s, A, t) \Big].$$
- $\Sigma \bowtie$ A **has duration** d **if** π iff
  $$\forall t \in \mathbb{N} \; \Big[H(\pi, t) = 1 \; \& \; A \in O(t) \cap C(t) \Rightarrow (t, A, t + d) \in E \Big].$$

We say that an atomic query ξ is *true* in a structure Σ iff $\Sigma \bowtie \xi$. For non-atomic queries the truth conditions are defined as usual, namely $\Sigma \bowtie \neg \xi$ iff $\Sigma \not\bowtie \xi$ and $\Sigma \bowtie \xi_1 \wedge \xi_2$ iff $\Sigma \bowtie \xi_1$ and $\Sigma \bowtie \xi_2$.

Notice that the interpretation of the effect statement (11) indicates only that the action A affects the fluent f, so f has an arbitrary value when A terminates. Similarly, the interpretation of the reserve (resp. protect) statement just indicates that the action reserves (resp. protects) the corresponding fluent.

Observe also that regions of possible changes are determined by effect, release and reserve statements, since only those statements determine values of functions I and R. Then constraint statements cannot invoke indirect effects of actions. When a constraint is violated due to results of some action, it means that this action cannot be performed. Consequently, state constraints in $\mathcal{AL}_1$ function as qualification constraints only.

Given a domain description Υ, let us now formalize the process of simulation of all possible courses of events constrained by the inertia law which additionally satisfy the assumption under which every fluent is *normally* unprotected (resp. unreserved). First, we are actually interested in structures for all statements in Υ with minimal regions of action influences (since only within these areas changes are allowed). Secondly, it is required that both protected and reserved areas are as minimal as possible also. Since all possible courses of events are to be determined, every concession function should be taken into account.

Definition 3.2 Let Υ be a domain description and $\Sigma = (H, I, R, O, E, P, C)$ be a structure. We say that Σ is a *structure for* Υ iff

(S1) Σ is a structure for every statement in Υ;

(S2) there is no structure $\Sigma' = (H', I', R', O', E', C', P')$ satisfying **(S1)** such that $H' = H$, $C' = C$, $I' \preceq I$, $R' \preceq R$, $P' \preceq P$ and either $I' \prec I$ or $R' \prec R$ or $P' \prec P$. $\square$

Structures for Υ still admit spurious changes since there is no interconnection between their history and influence functions. It is required that each fluent is allowed to change only if it is affected by some action. Also, for an arbitrary concession function C, sets of executable actions are not uniquely defined: there are structures for Υ where the relation E contains tuples (s, A, t) such that A has in fact no effect over the time interval $[s, t]$ (i.e. $I(s, \mathrm{A}, t) \cup R(s, \mathrm{A}, t) = \emptyset$), since its precondition (or prevail condition) fails. Each action should be then viewed as executable if and only if it affects some fluent. Next, structures for Υ do not take into account the protection device, so protected fluents may be still affected by several actions at a time. Then the class of structures should be further restricted in order to guarantee that every protected fluent can be affected only the action that protects it. Finally, values of reserved fluents are still undefined at termination points of reserving actions. By our assumption, such a fluent f is made false, provided that at this time no actions affect f except those that reserve it and simultaneously terminate at this very time.

Given a structure Σ for Υ, two timepoints $s, t \in \mathbb{N}$ and an action $\mathrm{A} \in \mathcal{A}$, by $\mathrm{INFL}(s, \mathrm{A}, t)$ we denote the set $I(s, \mathrm{A}, t) \cup R(s, \mathrm{A}, t)$, i.e. the set of all fluents affected by executing A from time s to t.

Definition 3.3 Let $\Sigma = (H, I, R, O, E, P, C)$ be a structure for a domain description Υ. We say that Σ is a *frame for* Υ iff

(F1) for any $t \in \mathbb{N}$ and any $f \in \mathcal{F}$, if $H(t, f) \neq H(t+1, f)$ then $f \in \mathrm{INFL}(s, \mathrm{A}, t)$ for some $\mathrm{A} \in \mathcal{A}$ and $s', t' \in \mathbb{N}$ such that $s' \leq t < t'$;

(F2) for any $\mathrm{A} \in \mathcal{A}$ and any $s, t \in \mathbb{N}$, $(s, \mathrm{A}, t) \in E$ iff $\mathrm{INFL}(s, \mathrm{A}, t) \neq \emptyset$;

(F3) for any $\mathrm{A}, \mathrm{B} \in \mathcal{A}$ and any $s, t, s', t' \in \mathbb{N}$ such that $(s, t] \cap (s', t'] \neq \emptyset$, the condition $P(s, \mathrm{A}, t) \cap \mathrm{INFL}(s', \mathrm{B}, t') \neq \emptyset$ implies $\mathrm{A} = \mathrm{B}$ and $s = s'$;

(F4) for any $\mathrm{A} \in \mathcal{A}$ and any $s, t \in \mathbb{N}$, if $f \in R(s, \mathrm{A}, t)$ then $H(t, f) = 0$, provided that for any $\mathrm{B} \in \mathcal{A}$ and any $s', t' \in \mathbb{N}$

   (a) $s' < t < t'$ implies $f \notin R(s', \mathrm{B}, t')$;
   (b) $s' < t \leq t'$ implies $f \notin I(s', \mathrm{B}, t')$. $\square$

Frames for Υ are defined for arbitrary concession functions. As a result, some actions may be blocked, though there is no reason for them not to be executed. While formalizing the CHM and GLM policies, those frames for Υ that admit such a behavior will be rejected.

Let Σ be a frame for Υ. For a timepoint t, we write $\mathrm{B}(\Sigma, t)$ to denote the set $\mathrm{B}(\Sigma, t) = \{\mathtt{A} \in \mathcal{A} \mid (t, \mathtt{A}, t') \in E \text{ for any } t' \in \mathbb{N}\}$. Intuitively, $\mathrm{B}(\Sigma, t)$ is the set of actions that start at time t (relative to the frame Σ).

Let $\Sigma = (H, I, R, O, E, P, C)$, $\Sigma' = (H', I', R', O', E', P', C')$ be frames for Υ. We say that Σ is

- *CHM-preferred* over Σ', written $\Sigma \ll_{\mathrm{CHM}} \Sigma'$, iff there is a timepoint $t \in \mathbb{N}$ such that $\mathrm{B}(\Sigma', t) \subset \mathrm{B}(\Sigma, t)$ and for every $s < t$, $\mathrm{B}(\Sigma, s) = \mathrm{B}(\Sigma', s)$;

- *GLM-preferred* over Σ', written $\Sigma \ll_{\mathrm{GLM}} \Sigma'$, iff $E' \subset E$.

Definition 3.4 We say that a frame Σ for a domain description Υ is a *P-model of* Υ, $P \in \{\mathrm{GLM}, \mathrm{CHM}\}$, iff there is no frame Σ' for Υ such that $\Sigma' \ll_P \Sigma$. □

The set of all CHM-models (resp. GLM-models) of Υ will be denoted by $\mathrm{ChMOD}(\Upsilon)$ (resp. $\mathrm{GMOD}(\Upsilon)$).

Let $P \in \{\mathrm{CHM}, \mathrm{GLM}\}$. A domain description Υ is called *P-consistent* iff it has a P-model. We say that Υ *P-entails* a query ξ, in symbols $\Upsilon \models_P \xi$, iff ξ is true in every P-model of Υ.

Proposition 3.5 $\mathrm{ChMOD}(\Upsilon) \subseteq \mathrm{GMOD}(\Upsilon)$ *for any domain description Υ in $\mathcal{AL}_1$.*

Proof. Suppose that $\Sigma = (H, I, R, O, E, P, C)$ is not a GLM-model of a domain description Υ. Therefore there is a structure $\Sigma' = (H', I', R', O', E', P', C')$ such that $\Sigma' \ll_{\mathrm{GLM}} \Sigma$. It means that $E \subset E'$. Put $\overline{E} = E' \setminus E$. Let $(s_0, \mathtt{A}_0, t_0) \in \overline{E}$ be such that for every $(s, \mathtt{A}, t) \in \overline{E}$, $s_0 \leq s$. So $\mathrm{B}(\Sigma, s') = \mathrm{B}(\Sigma', s')$ for every $s' < s_0$, and $\mathrm{B}(\Sigma, s_0) \subset \mathrm{B}(\Sigma', s_0)$. Hence $\Sigma' \ll_{\mathrm{CHM}} \Sigma$, so $\Sigma \notin \mathrm{ChMOD}(\Upsilon)$. □

From Proposition 3.5 it immediately follows:

Corollary 3.6 *For any domain description Υ and any query $\xi \in \mathcal{L}_\Upsilon$, $\Upsilon \models_{\mathrm{GLM}} \xi$ implies $\Upsilon \models_{\mathrm{CHM}} \xi$.* □

3.3 Examples

The following example illustrates that in $\mathcal{AL}_1$ state constraints contribute to qualifications for actions only.

Example 3.7[6] In an ancient kingdom there are two blocks, A and B, and either of which may be yellow. Due to order of the emperor at most one block may be yellow at a time. Initially the block A is yellow and at time 1 a painter tries to paint the other block yellow which takes him 3 time units.
Below is the corresponding domain description Υ_1

initially $YellowA$;	**PaintB has duration** 3;
always $\neg(YellowA \wedge YellowB)$;	**PaintB occurs at** 1.
PaintB causes $YellowB$;	

---

[6] This is a slightly modified version of an example from [5].

Minimization of regions where changes are allowed leads to structures for Υ_1, where $YellowA \notin I(t, \text{PaintB}, t')$ for any t and t'. **(F1)** implies that $YellowA$ cannot be changed at all. Thus, in view of the constraint statement, $YellowB$ cannot be made true anywhere. Hence painting the block B is impossible (regardless of the applied preferential method). □

The next examples show how mutual exclusion can be encoded in $\mathcal{AL}_1$.

Example 3.8 Consider a simplified version of the *Producer-Consumer Problem*. Suppose that two agents, A and B, want to get an item from the buffer at time 1 and 2, respectively. An agent A needs 3 time units to get the item, for B it takes 2 units of time. When the buffer is empty, it is fulfilled immediately in a single unit of time. Getting the item makes the buffer empty.

initially ¬*Empty*;	*Empty* **triggers** Put;
GetA **causes** *Empty* **if** ¬*Empty*;	GetA **occurs at** 1;
GetB **causes** *Empty* **if** ¬*Empty*;	GetB **occurs at** 2;
Put **causes** ¬*Empty*;	GetA **has duration** 3;
GetA **protects** *Empty*;	GetB **has duration** 2;
GetB **protects** *Empty*;	Put **has duration** 1.

Notice that by protecting the fluent *Empty* both getting actions cannot overlap. Intuitively, the agent A should manage to get the item since he demands it earlier (so B fails to get it). It is easy to check that we obtain exactly one CHM-model Σ of Υ_2, where $E = \{(1, \text{GetA}, 4), (4, \text{Put}, 5)\}$. Notice also that there are exactly two GLM-models of Υ_2: Σ and Σ', where $E' = \{(2, \text{GetB}, 4), (4, \text{Put}, 5)\}$. □

Example 3.9 Consider a simplified version of the *Readers-Writers Problem*. Assume that the reading actions occur at time 1 and 2, writing actions occur at time 2 and 3, and each reading (resp. writing) action takes 2 (resp. 3) time units.

initially ¬*Reading* ∧ ¬*Writing*;	Read **occurs at** 1;
Read **protects** *Writing*;	Read **occurs at** 2;
Write **protects** *Writing*;	Write **occurs at** 2;
Write **protects** *Reading*;	Write **occurs at** 3;
Read **reserves** *Reading*;	Write **has duration** 3
Write **reserves** *Writing*;	Read **has duration** 2.

The fluent *Writing* is protected by the action Read, so no writing action can be executed when Read is in progress. This action does not protect the fluent *Reading*, since several reading actions may be executed at a time. Writing actions protect both fluents, so such an action never overlaps with any other one.

This scenario has exactly two CHM-models, namely Σ_1 and Σ_2, such that $E_1 = \{(1, \text{Read}, 3), (2, \text{Read}, 4)\}$ and $E_2 = \{(1, \text{Read}, 3), (3, \text{Write}, 6)\}$, respectively. Hence the writing action occurring at time 2 is unexecutable.

If action performances are maximized globally then we obtain exactly three GLM-models: Σ_1, Σ_2 and Σ_3; the last one is such that $E_3 = \{(2, \text{Write}, 5)\}$. Then we cannot infer which action actually takes place and which one fails. □

There are cases where chronological maximization of action executions leads to counterintuitive results, but the global maximization provides desired ones.

Example 3.10 There are two entrance doors to a hall, A and B, and both are initially opened. An agent is supposed to close the door A and B at time 1 and at time 3, respectively, which takes him a single unit of time. However, at time 5 at least one of the doors is observed opened.

initially $OpenA \wedge OpenB$;	**CloseA occurs at** 1;
$OpenA \vee OpenB$ **at** 5;	**CloseB occurs at** 3;
CloseA causes $\neg OpenA$;	**CloseA has duration** 1;
CloseB causes $\neg OpenB$;	**CloseB has duration** 1.

Intuitively, exactly one of the doors was closed. The GLM policy leads to the very conclusion. However, maximizing executions of actions chronologically we obtain exactly one GLM-model where the door A was closed at time 1. □

4 Propositional Fluent Logic PFL-1

In this section we briefly present the propositional fluent logic PFL-1 based on the logic PFL proposed in [7] and [8]. PFL-1 is a three-sorted classical logic with equality including sorts for actions, temporal entities and truth-valued fluents. We assume a discrete linear model of time containing all natural numbers. The language of PFL-1 will be denoted by $\mathcal{L}(\text{PFL-1})$.

An alphabet of PFL-1 involves three denumerable sets of object constants: a set $\mathcal{F}$ of *fluent constants*, a set $\mathcal{A}$ of *action constants* and the set $\mathbb{N}$ of timepoints. $\mathcal{L}(\text{PFL-1})$ has the predicate $<$ and two functions $+$ and $-$ interpreted as usual on natural numbers. We have a predicate $Holds$ with a timepoint and a fluent as arguments. $Holds(t, f)$ says that the fluent f is true at time t. Also, there are two *occlusion* predicates $Occlude_F$ and $Occlude_R$[7] which take an action, two timepoints and a fluent as arguments. $Occlude_F(A, s, t, f)$ states that the fluent f is affected by executing the action A from time s to t and f occurs in *final* results of this action (so f may change at any time $s < t' \leq t$); $Occlude_R(A, s, t, f)$ says that the action A *reserves* f from time s to t. Next, we introduce predicates $Occurs$ and $Disabled$ which take an action and a timepoint as arguments. $Occurs(A, t)$ says that the action A occurs at time t, $Disabled(A, t)$ means that the action A is blocked at time t. Finally, we have a predicate $Action$ with an action and two timepoints as arguments. $Action(s, A, t)$ says that the action A is executed from time s to t.

An action scenario represented in $\mathcal{L}(\text{PFL-1})$ is called a *scenario description* and denoted by Γ. Within this theory we distinguish the following disjoint sets of axioms: *observation axioms* Γ_{OBS} representing usually partial states of the system at particular points in time, *action axioms* Γ_{ACS} representing action occurrences or action descriptions, *unique name axioms* Γ_{UNA} for actions and

---

[7] The predicate $Occlude$ (with a timepoint and a fluent as the only arguments) was proposed by Sandewall ([11]) to explicitly indicate where changes are allowed.

fluents and *constraint axioms* Γ_{CONS} representing constraint rules together with the following axiom Γ_{ECON}

$$\forall a, s, t.\ Action(s, a, t) \rightarrow \big[\forall t'.\ t \neq t' \rightarrow \neg Action(s, a, t')\big] \wedge \\ \wedge \big[\exists f.\ Occlude_F(a, s, t, f) \vee Occlude_R(a, s, t, f)\big] \tag{17}$$

which specifies that each executable action has uniquely defined duration and it affects some fluent.

4.1 Circumscriptive Policy

In this section we provide syntactic characterizations of the CHM and GLM preferential policies. Our approach is based on the PMON preferential entailment[8] introduced by Sandewall ([11]). For a given scenario description Γ, this method amounts to determine all potential histories with minimal regions where changes are allowed, but relative to action axioms only. Then histories which contradict observations and those where spurious changes occur are rejected.

Introducing the reservation tool does not essentially change this idea: now *both* occlusion predicates are jointly minimized in order to determine regions where any changes (i.e. final and intermediate) are allowed. This policy will be called *Generalized PMON* (GPMON, for short).

Following Sandewall's idea, we introduce the following *nochange axioms* Γ_{NCH}

$$\forall t, f.\ \neg\big[Holds(t, f) \equiv Holds(t+1, f)\big] \rightarrow \\ \rightarrow \exists a, s', t'.\ (s' \leq t < t') \wedge \big(Occlude_F(a, s', t', f) \vee Occlude_R(a, s', t', f)\big)$$

specifying that a fluent may change only when it is *occluded* by some action.

Moreover, we add the following *reservation constraint axioms* Γ_{RCA}:

$$\forall a, s, t, f.\ Occlude_R(a, s, t, f) \rightarrow (\forall t'.\ s < t' < t \rightarrow Holds(t', f))$$

$$\forall t, f.\ \big[\exists a, s.\ Occlude_R(a, s, t, f)\big] \wedge \big[\forall s', t'.\ s' < t \leq t' \rightarrow (\forall a.\ \neg Occlude_F(a, s', t', f))\big] \wedge \\ \wedge \big[\forall a, s', t'.\ (s' < t \leq t') \wedge Occlude_R(a, s', t', f) \rightarrow t' = t\big] \rightarrow \neg Holds(t, f)$$

to specify that each reserved fluent is true throughout its reservation period and at the end of this period it is made false, provided that all actions affecting this fluent at this time simultaneously terminate then and reserve this fluent too.

Given a scenario description Γ, the following *GPMON-Circumscription of* Γ, in symbols $\text{CIRC}_{\text{GPMON}}(\Gamma)$, realizes the GPMON policy

$$\Gamma_{\text{RCA}} \wedge \Gamma_{\text{NCH}} \wedge \Gamma \wedge \text{CIRC}_{SO}\big(\Gamma_{\text{ACS}};\ Occlude_R, Occlude_F;\ Occurs, Action\big),$$

where $\text{CIRC}_{SO}(\Gamma_{\text{ACS}};\ Occlude_F, Occlude_R;\ Occurs, Action)$ is the second-order circumscription of the occlusion predicates within the theory Γ_{ACS} with $Occurs$ and $Action$ allowed to vary.

The GLM policy selects models of Γ where the inertia holds and as many as possible actions are executed. This is realized by global minimization of the predicate *Disabled*.

---

[8] PMON stands for *Pointwise Minimization of Occlusion with Nochange Premises*.

Definition 4.1 (GLM-Circumscription) Let Γ be a scenario description. The *GLM-Circumscription of Γ*, written $\text{CIRC}_{\text{GLM}}(\Gamma)$, is the formula

$$\text{CIRC}_{SO}\big(\widetilde{\Gamma}\,;\,Disabled\,;\,Holds, Occlude_F, Occlude_R, Occurs, Action\big),$$

where $\widetilde{\Gamma}$ is $\text{CIRC}_{\text{GPMON}}(\Gamma)$. $\qquad\square$

The CHM policy amounts to select models of Γ (satisfying the inertia law) where earlier occurring actions are preferred. To syntactically characterize this strategy, the following circumscription is proposed.

Definition 4.2 (CHM-Circumscription) Let Γ be a scenario description. The *CHM-Circumscription of Γ*, written $\text{CIRC}_{\text{CHM}}(\Gamma)$, is the sentence

$$\text{CIRC}_{PW}\big(\widetilde{\Gamma}\,;\,Disabled/\Omega\,;\,Holds, Occlude_F, Occlude_R, Occurs, Action\big),$$

where $\text{CIRC}_{PW}(\widetilde{\Gamma}\,;\,Disabled/\Omega\,;\,Holds, Occlude_F, Occlude_R, Occurs, Action)$ is the pointwise circumscription formula ([4]) equivalent to

$$\widetilde{\Gamma}(Disabled, Holds, Occlude_F, Occlude_R, Occurs, Action)\wedge$$
$$\wedge \forall t, a\, \forall \Phi\, \overline{\Psi}.\ \neg\Big[\widetilde{\Gamma}(\Phi, \overline{\Psi}) \wedge Disabled(a, t) \wedge \neg\Phi(a, t)\wedge$$
$$\wedge\big(\forall t', b.\ \neg\Omega(t, a, t', b) \rightarrow Disabled(b, t') \equiv \Phi(b, t')\big)\Big]$$

with $\Omega(t, a, t', b) = \big((t, a) = (t', b) \vee (t' > t)\big)$ and $\widetilde{\Gamma} = \text{CIRC}_{\text{GPMON}}(\Gamma)$. $\qquad\square$

5 Translating $\mathcal{AL}_1$ into Circumscription

In this section we provide a two-step translation from $\mathcal{AL}_1$ into circumscriptive theories. We restrict our attention to finite domain descriptions. For a domain description Υ, the first step determines the corresponding scenario description Γ^Υ in $\mathcal{L}(\text{PFL-1})$, whereas the next one amounts to apply the GLM-Circumscription (resp. the CHM-Circumscription) of the theory obtained in the first step.

Let Υ be a finite domain description and let $\mathcal{F}_\Upsilon, \mathcal{A}_\Upsilon$ be sets of all fluents and all actions in $\mathcal{L}_\Upsilon$. Let $\mathcal{L}(\text{PFL-1})$ be a language, where $\mathcal{F}_\Upsilon$ and $\mathcal{A}_\Upsilon$ are sets of fluent constants and action constants, respectively.

For every statement σ in Υ we define a corresponding first-order formula $\Theta(\sigma)$ of the language $\mathcal{L}(\text{PFL-1})$ in the following way:

- If σ is a performance statement (1)–(2) then $\Theta(\sigma)$ is

$$Action(s, A, t),$$
$$Occurs(A, t) \wedge Disabled(A, t).$$

- If σ is a value statement (3) then $\Theta(\sigma)$ is

$$Holds(t, \alpha),$$

  where $Holds(t, \neg f)$ stands for $\neg Holds(t, f)$ and $Holds(t, \beta_1 \wedge \beta_2)$ stands for $Holds(t, \beta_1) \wedge Holds(t, \beta_2)$.

- If σ is an occurrence statement (4) then $\Theta(\sigma)$ is $Occurs(A, t)$.

- If σ is a state constraint statement (5) then $\Theta(\sigma)$ is

$$\forall t, d.\ Holds(t, \pi) \wedge \big(\forall t'.\, t \leq t' \leq t+d \to \neg Holds(t', \varphi)\big) \to Holds(t+d, \alpha)\big].$$

- If σ is an evoke statement (6)–(7) then $\Theta(\sigma)$ is

$$\forall t.\ Holds(t, \pi) \to Occurs(A, t),$$
$$\forall s, t.\ Action(s, A, t) \wedge Holds(s, \pi) \to Occurs(B, t+d).$$

- If σ is an action unexecutability statement (8)–(9) then $\Theta(\sigma)$ is

$$\forall t, d.\ Holds(t, \pi) \wedge \big(\forall t'.\, t \leq t' \leq t+d \to \neg Holds(t', \varphi)\big) \to Disabled(A, t+d)\big),$$
$$\forall s, t, d.\ Action(s, B, t) \wedge Holds(s, \pi) \wedge \big(\forall t'.\, t \leq t' \leq t+d \to \neg Holds(t', \varphi)\big) \to$$
$$\to Disabled(A, t+d)\big].$$

- If σ is an effect statement (10)–(11) then $\Theta(\sigma)$ is

$$\forall s, t.\ Action(s, A, t) \wedge Holds(s, \pi) \wedge \big(\forall t'.\, s \leq t' \leq t \to Holds(t', \varphi)\big) \to$$
$$\to Holds(t, \alpha) \wedge \bigwedge\nolimits_{f \in fl(\alpha)} Occlude_F(A, s, t, f)\big],$$
$$\forall s, t.\ Action(s, A, t) \wedge Holds(s, \pi) \wedge \big(\forall t'.\, s \leq t' \leq t \to Holds(t', \varphi)\big) \to Occlude_F(A, s, t, f).$$

- If σ is a reserve statement (12) then $\Theta(\sigma)$ is

$$\forall s, t.\ Action(s, A, t) \wedge Holds(s, \pi) \to Occlude_F(A, s, t, f)$$

- If σ is a protect statement (13) then $\Theta(\sigma)$ is

$$\forall s, t.\ Action(s, A, t) \wedge Holds(s, \pi) \to \big[\forall a, s', t'.\ \big(\exists t''.\, (s < t'' \leq t) \wedge (s' < t'' \leq t')\big) \wedge$$
$$\wedge \big(Occlude_F(a, s', t', f) \vee Occlude_R(a, s', t', f)\big) \to (a = A) \wedge (s = s')\big].$$

- Finally, if σ is a duration statement (14) then $\Theta(\sigma)$ is

$$\forall t.\ Holds(t, \pi) \wedge Occurs(A, t) \wedge \neg Disabled(A, t) \to Action(t, A, t+d).$$

Let $\Gamma^{\Upsilon}_{\text{OBS}}$ be the set of first-order formulae in $\mathcal{L}(\text{PFL-1})$ obtained by translating all value statements in Υ, $\Gamma^{\Upsilon}_{\text{CONS}}$ be the set that contains the axiom Γ_{ECON} (17) and all formulae resulting from this translation of all state constraint and protect statements[9] in Υ and let $\Gamma^{\Upsilon}_{\text{ACS}}$ be the sets of formulae obtained by translating all the remaining statements in Υ. Then $\Gamma^{\Upsilon} = \Gamma^{\Upsilon}_{\text{OBS}} \cup \Gamma^{\Upsilon}_{\text{ACS}} \cup \Gamma^{\Upsilon}_{\text{CONS}} \cup \Gamma^{\Upsilon}_{\text{UNA}}$ is a scenario description corresponding to Υ.

5.1 Soundness and Completeness

The following theorem relates the notion of entailment in the sense of classical logic and the P-entailment relation ($P \in \{\text{CHM}, \text{GLM}\}$) for the language $\mathcal{AL}_1$.

Theorem 5.1 *Let Υ be a finite domain description, Γ^{Υ} be a scenario description corresponding to Υ, $P \in \{\text{CHM}, \text{GLM}\}$, and let $\Theta_P(\Upsilon)$ denote $\text{CIRC}_P(\Gamma^{\Upsilon})$. For any query ξ in $\mathcal{L}_{\Upsilon}$, $\Upsilon \models_P \xi$ iff $\Theta_P(\Upsilon) \models \Theta(\xi)$.* □

---

[9] Notice that protect statements function in fact as a specific type of constraints.

6 Conclusions and Further Work

We have discussed the qualification problem in the context of scenarios where actions with duration and overlapping actions occur. Two preferential methods have been proposed, CHM and GLM, which amount to chronological (resp. global) maximization of executable actions. We have defined the action language $\mathcal{AL}_1$ to represent these scenarios and have shown the translation from $\mathcal{AL}_1$ into circumscription, sound and complete relative to the semantics of $\mathcal{AL}_1$.

This paper is the starting point for future work. First, properties of the present formalism should be studied. In particular, both preferential methods are to be further investigated in order to assess the range of their correct applicability. For this purpose, the framework proposed in [11] and [12] seems particularly adequate.

The ramification problem has not been considered here. Recent promising results in this topic (e.g. [12]) inspire for further studies.

Continuing investigations on domains with concurrency, we wish to extend our formalism by involving cases where actions communicate with each other. From the standpoint of modeling multi-agents worlds, these problems are certainly of great importance.

References

1. C. Baral, M. Gelfond: Representing Concurrent Actions in Extended Logic Programming, in *Proc IJCAI-93*, Chambery, 1993, pp. 866–871.
2. M. Gelfond, V. Lifschitz: Representing Action and Change by Logic Programs, in *The Journal of Logic Programming*, (17), 1993, pp. 301–322.
3. A. Kakas, R. Miller: A Simple Declarative Language for Describing Narratives with Actions, Imperial College Research Report No DoC95/12, 1995; also in *Journal of Logic Programming* (Special Issue on Reasoning about Action and Change).
4. V. Lifschitz: Pointwise Circumscription, in M. Ginsberg, editor, *Readings in Nonmonotonic Reasoning*, pp. 179–193, Morgan Kaufmann, 1988.
5. F. Lin, R. Reiter: State constraints revisited, in *Journal of Logic and Computation* (Special Issue on Actions and Processes), 4(5), 1994, pp. 655–678.
6. R. Miller, M. Shanahan: Narratives in the Situation Calculus, in *Journal of Logic and Computation*, Special Issue on Actions and Processes, vol. 4, No 5, 1995.
7. A. Radzikowska: Reasoning about Action with Typical and Atypical Effects, in *Proc. of 19th KI-95*, Bielefeld, 1995, pp. 197–209.
8. A. Radzikowska: Formalization of Reasoning about Default Action (Preliminary Report), in *Proc. of FAPR-96*, Bonn, 1996, pp. 540–554.
9. R. Reiter: Natural Actions, Concurrency and Continuous Time in the Situation Calculus, in *Proc. of KR-96*, Boston, MA, 1996.
10. E. Sandewall, R. Röonquist: A representation of action structures, in *Proc. of AAAI-86*, pp. 89–97.
11. E. Sandewall: Features and Fluents: A systematic approach to the representation of knowledge about dynamical systems, Oxford University Press, 1994.
12. E. Sandewall: Comparative Assessment of Update Methods Using Static Domain Constraints, in *Proc. of KR-96*, Boston, MA, 1996.
13. S. E. Bornscheuer, M. Thielscher: Representing Concurrent Actions and Solving Conflicts, in *Proc. of 18th KI-94*, Saarbrücken, 1994, pp. 16–27.

Multiple Subarguments in Logic, Argumentation, Rhetoric and Text Generation*

Chris Reed[1] and Derek Long[2]

[1] Department of Computer Science,
University College London,
Gower St.,
London, WC1E 6BT
C.Reed@cs.ucl.ac.uk
http://www.cs.ucl.ac.uk/staff/C.Reed

[2] Department of Computer Science,
Durham University,
South Road,
Durham
D.P.Long@dur.ac.uk

Abstract. A summary is provided of the problems of representing, determining generating and arranging disjunct multiple subarguments in several fields, including formal systems in uncertain domains, informal logic accounts of argument structure, rhetorical systems for maximising persuasive effect, and the automatic generation of persuasive discourse. Drawing upon the insights, problems, and partial solutions of these fields, a theory of subargument construction and organisation is presented, and is set in a framework for generating natural language argument.

Keywords: aggregating arguments, argumentation theory, rhetoric, defeasible reasoning, natural language generation, planning, rhetoric.

1. Introduction

One key task to be performed during the production of argument for some purpose (interagent negotiation, expert system explanations, decision support systems and other nontrivial human-computer discourse) is that of aggregating subarguments. In particular, it is necessary to make appropriate use of *multiple subarguments*, a term, which for the purposes of this paper, refers not simply to several lines of reasoning contributing to some conclusion, but specifically to lines of reasoning which offer *disjunctive* support. Thus in Fig. 1, only 1b contains multiple subarguments (1a is composed of a single subargument composed of two conjuncts).

(a)	(b)
Fred told me Socrates is human	Fred told me Socrates is mortal
<u>Fred's a reliable source of information</u>	<u>George told me Socrates is mortal</u>
∴Socrates is mortal	∴Socrates is mortal.

Fig. 1. Conjunctive and disjunctive support

*This work has been partly funded by EPSRC grant no. 94313824.

The question of how to deal with such support is one which is relevant not only in text generation, but also in formal approaches to argumentation (including defeasible reasoning and bespoke logics of argument), informal approaches (where the distinction between disjunctive and conjunctive support is particularly problematic) and in rhetoric (where the appropriate use of various forms of support is prescribed). By examining the approaches taken in these areas, it is possible to assail a number of key problems in the generation of natural language argument.

2. Formal approaches

In classical logic, the concept of multiple subarguments is of no use. If it is possible to prove P, say through X and X $\rightarrow$ P, it is unnecessary to then prove X once again with Y and Y $\rightarrow$ P: using both proofs rather than just one would in no way produce a 'better' argument. This accords with superficial intuitions that if a speaker, S, wishes to bring a hearer, H, to believe some proposition, all that is necessary is for S to show that the proposition follows from her beliefs. However, multiple subarguments occur with great frequency, both in natural language (Cohen, 1987), (Reed and Long, 1997) and increasingly in complex argumentation systems such as those of (Fox and Das, 1996). In both cases they are used to create arguments that are in some way 'better' than singly supported alternatives.

There are two (related) key points which necessitate the use of multiple subarguments in these circumstances. In the first place, reasoning about some domain in the real world (such as the medical applications of Fox and Das) can rarely use strict deductive inference; rather, it becomes necessary to use some weaker notion of support - and often then to express the degree of that support (either qualitatively - eg. (Parsons, 1996), or quantitatively - eg. (Sillince and Minors 1992)). If a system no longer relies solely upon strict inference then it can clearly no longer employ classical logic, and as a result may benefit from the use of multiple subarguments. These separate lines of support may then be aggregated under some flattening function, such as those discussed in (Das *et al*, 1996). In the second place, multiple subarguments offer a means of tackling the uncertainty and incompleteness inherent in an environment where agents must communicate. One agent cannot hope to maintain an accurate model of the beliefs of another, and yet it must be able to construct arguments which actually effect changes in other agents' beliefs. By planning several disjunctive supports for a proposition, an agent can increase the likelihood that the conclusion will be accepted. For although ideally an agent would want its model of the hearer's beliefs to be accurate, should there be a discrepancy, the alternative supports may turn out to be of crucial importance in persuading the hearer to concur (rather than those supports playing an ancillary role, as originally planned).

One particularly successful formal approach to dealing with uncertain and incomplete information is defeasible reasoning. Defeasible logics are also inherently able to accommodate multiple subarguments. For example, in Dung's (1995) theory of the acceptability of argumentation, the basic framework, *AF*, comprises two parts, a set of arguments and a set of attacks holding between arguments. Implicit in the former is the possibility that several arguments will contribute to a single conclusion (and it is anticipated that in natural communication, these multiple subarguments would form a *conflict free* set[1]). In other systems, the concept is more explicitly represented: Vreeswijk (1992), for example, introduces the *sub* feature of an argument, which defines the set of all subarguments (the model differs somewhat from that of Dung's, as it aims to capture structural information in addition to the content). Pollock's (1995) theory of defeasible reasoning also makes the concept of subargument explicit and more recently, also distinguishes conjunctive and disjunctive supports (Pollock, 1996).

Clearly, a system of defeasible reasoning such as that of (Pollock, 1995) needs to be extended to cope with the distributed aspects of argumentation occurring between complex, rational, *autonomous* agents (ie. agents that function without human intervention, that can interact with other agents, that can take proactive decisions on the basis of their own selfish goals and intentions, that cannot be guaranteed to have the same information as other agents, cannot be sure of the knowledge they have about other agents beliefs, and cannot be relied upon to be benevolent and cooperative - these issues are discussed in, for example, (Wooldridge and Jennings, 1995)). The notion of agency is touched upon in the work of both Vreeswijk and Dung, *inter alia*. However, it is only relatively recently that argumentation has been seen as a direct means of negotiation (Parsons and Jennings, 1996) and persuasion (Reed *et al*, 1996a) between agents (for after all, these are the primary ends served by argumentative communication between humans - though see (Walton, 1995) for a more detailed coverage).

Parsons and Jennings (1996) extended the defeasible argumentation system of Krause *et al* (1995) such that one agent could propose an argument (for a specific course of action) to another, and the recipient would then evaluate the argument on the basis of its own beliefs, intentions and plans. The evaluation is performed by searching for rebutting and undercutting counter-arguments (a rebuttal occurring when the conclusion of the argument is defeated, undercutting when one of the steps used in reaching the conclusion is defeated). Depending on the evaluation, the second agent may then agree, or communicate a counter-proposal. Thus the defeasibility of the system lies not only within a single agent's reasoning, but also in the subsequent processing in another agent with potentially quite different beliefs and goals. Importantly, the prioritization of the defeasible inferences is localised to each agent: in Parsons' system, this is due to the fact that the ordering relation is based upon the acceptability classes which in turn are dependent upon the beliefs of the agent involved. However, as he points out, it is equally possible to consider a system which performs the ordering based upon some valuation of the inferences (such as that of (Parsons, 1996)), which may offer some advantages in uncertain domains.

---

[1] Though multiple subarguments are unlikely to form an *admissible* set. For this to hold, the set, S, of multiple subarguments would have to attack any argument which itself attacks a component of S (this is Dung's definition of *acceptability*: S is then admissible iff each component of S is acceptable). In practice, this would mean that a speaker would have to create a 'water-tight' argument, anticipating all hearer counter-arguments (and this in turn would reduce the dialectic method to a one-shot monologue).

There are, however, problems with such an operationalization of distributed defeasible reasoning. The system proposed in (Parsons and Jennings, 1996) does not seem to be able to communicate multiple subarguments - a single proposal (albeit a conjunction of separate sub-proposals) is supported by a set of grounds which must *all* be taken together to infer the conclusion. As a result, the arguments which can be generated will be less sophisticated (for although it may be possible for one agent to enumerate all the disjunct subarguments for a particular course of action on successive turns in the dialogue, this is far from an optimal strategy for a resource-bounded agent[2]).

More importantly, defeasible reasoning across rational, autonomous agents seems to miss the intuitions of how the agents are functioning. Consider the scenario in Fig. 2, in which the speaker has two disjunct reasons for inferring[3] p, and the hearer may have a number of reasons for inferring ¬p. Let us assume that S knows that H believes ¬p. Following an account such as that offered by Dung (1995) or Vreeswijk (1992) would lead to S communicating exactly those subarguments which together either defeat, or are undefeated by, all the X_i which S presumes H to believe.

$$
\begin{array}{ll}
\underline{S} & \underline{H} \\
a \rightarrow p & X_i \rightarrow \neg p \\
b \rightarrow p & \neg p \\
p &
\end{array}
$$

Fig. 2. Sample situation between speaker, S and hearer, H

However, in the real world, S may or may not know (or even be aware of the existence of) the X_i by which H believes ¬p. S is certainly unlikely to know the valuation that H places upon the various inferences from the X_i. She is even less likely to be able to anticipate how H will value her own inferences. It is not the case, therefore, that S constructs her argument through anticipating H's possible counter-arguments - rather, she is simply 'building a case' for her conclusion. Clearly, this process is going to involve consideration of what she thinks her hearer believes (in addition to other audience-specific information, such as possible bias and technical competence - see (Reed *et al*, 1996b) for details). But it does not require S to perform 'H-reasoning' to produce the arguments which she must ensure are defeated by her own. Although such reasoning may have a role to play in generating parts of a complex argument, the primary means of generating argument is, of necessity, a process of showing evidence to support a conclusion, resulting in an argument which stands independent of the possible subsequent reasoning by the hearing agent. It is this intuition which escapes a standard defeasible account of why one agent might utter a particular set of subarguments to another.

---

[2] The strategy would also suffer badly if used for communicating with a human: studies of rhetoric such as (Blair, 1838) devote great attention to issues of combining multiple subarguments: an overview is given in §4.

[3] The inference indicated by '→' is intended to be defeasible rather than strict.

3. Informal Approaches

In contrast to the 'bottom up' approach of the formal accounts of argumentation, argumentation theory is fundamentally empirical in its approach, driven by real examples of argumentation expressed in natural language. In naturally occurring argument, disjunct multiple subarguments occur with great frequency, and as a consequence are a feature of almost all theoretical accounts. However, many of these accounts do not explicitly discuss the ways in which subarguments can be combined, and the functional roles that they fulfil as a result. Fogelin (1991) is typical of such informal logic texts, explaining how argument structure can be determined and represented, through analysis of supporting relations holding between premises and conclusions. No distinction is made, however, between arguments supported by a single subargument, and those supported by several disjunctive subarguments - the latter is simply regarded as an extension of the former. Similarly, Fisher (1988) bases his theory of analysis on the *assertibility question*: "What argument or evidence would justify me in asserting the conclusion? (What would I have to know or believe to be justified in accepting it?)". Implicit is the notion that the argument or evidence may be composed of one or more subarguments.

Although there are minor differences in notation and analysis procedures, these and other accounts follow substantially the same approach to determining and describing the structure of argument. In sharp contrast is the theory proposed by Toulmin (1958) which expands upon the conventional premise/conclusion distinction to detail a six-fold division of utterances in argumentation (a *claim* supported by a *datum* to the degree specified by a *qualifier* which may include exceptional conditions which would otherwise cause *rebuttal*; the support between claim and datum is licensed by a *warrant* which may have a *backing*). Toulmin too regards the use of multiple subarguments (presumably multiple D--WB--QR links to the C) as a trivial extension to the theory.

Freeman (1991) discusses at length that the way in which premises can combine to support a conclusion (after discussion of Toulmin's theory, he eschews the six-fold division in favour of the 'standard' premise/conclusion approach). In particular he summarises (p2) the four main types of argument component structure, *viz. divergent* (whereby one premise can support several conclusions[4]), *serial* (whereby a single premise contributes to a single conclusion, which may then act as the single premise to another conclusion, and so on), *convergent* (whereby two or more premises contribute independently to a single conclusion) and *linked* (whereby two or more premises together contribute to a single conclusion). These forms are summarised in Fig. 3, below.

---

[4] He later points out (p93) that under some circumstances, it is more appropriate to view divergent structure as serial, with the premise repeated for each conclusion. This is the approach taken in (Reed *et al*, 1996b).

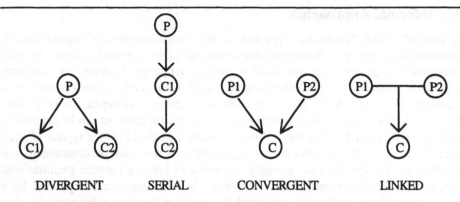

Fig. 3. The four basic argument structures, after (Freeman, 1991, p2).

Clearly, a convergent argument is one in which a conclusion is supported by disjunctive multiple subarguments. However, as Freeman explains, distinguishing between arguments which have a linked structure (such as that in Fig. 1a) from those which have convergent structure (such as that in Fig. 1b) is a particularly difficult task. This is as a direct result of the imprecise terminology used to define the classes[5]. There are two important classes of argumentation which pose a particular problem for classification: inductive generalisation and concluding a conjunction.

Inductive generalisation consists of generalising from a feature known to belong to a sample of a population, to claiming that feature of all members of the population. Thus it can be seen as presenting a number of examples (each member of the sample) from which the conclusion (the population as a whole) is inductively drawn. None of the examples presents particularly convincing evidence that the conclusion is true, but taken together the inference may be judged to be valid. On the one hand, the structure seems to be convergent: each datum is individually contributing to the conclusion, and does not depend on other data for its contribution to be valid. However, on the other hand, this fails to capture the nature of an inductive generalisation, which is that none of the data on their own would license the drawing of the conclusion - only when they are all taken together is the inference valid.

Arguments which involve two premises contributing to the conclusion which is their conjunction pose a similar problem. That is, if $(A \wedge B)$ is to be drawn from the premises A and B, is it the case that A and B have to be taken together (ie. linked structure) or do both A and B independently contribute to the conclusion as multiple subarguments (ie. convergent structure)?

Freeman's solution rests upon Toulmin's notions of argument as a dialectical construction, and potential questions that a challenger may put to the speaker. Consideration of a challenger asking the *relevance* question, *"Why is that reason relevant to the claim?"*, gives rise to additional premises in a linked structure.

---

[5] For example, in the summarised definitions given in parentheses above, it is unclear exactly what the terms *contribute* and *independently* actually mean.

Similarly, consideration of a challenger posing the first *ground adequacy* question, *"Can you give me another reason?"*, gives rise to additional premises in a convergent structure. Classifying an argument becomes a case of identifying which question could have been asked to elicit the additional premises. Crucial to Freeman's account is a distinction between *relevance* and *modality* between a premise and the conclusion it supports (the two aspects giving rise to the relevance and ground adequacy questions). The modality attached to a support arc (ie. the strength of that support, ranging from strict deductive inference to weak inductive inference) can affect intuitions for how a particular example should be classified[6]. In particular, where the strength of support between each premise and the conclusion is high, the temptation is to classify the argument as convergent, since each premise can be seen as providing an independent subargument. Conversely, when strength of support is low (such as from the individual examples of an inductive generalisation), the temptation is to assume a linked structure, since each premise on its own is 'too weak' to contribute to the conclusion by itself.

Using the relevance and ground adequacy questions, Freeman concludes that both inductive generalisation and concluding the conjunction are in fact examples of convergent structure (that is, each premise answers a phantom ground adequacy question). These results are controversial, but if valid may have important ramifications for logical models of natural argumentation.

The fact that concluding a conjunction in natural argument is the product of a convergent structure is particularly surprising, though Freeman's terminology rather hides the fact. By convergent structure, he is referring to premises which provide *disjunctive* support for a conclusion. Thus to claim that a conjunction is supported by premises in a convergent structure, is to claim that individual conjuncts support their conjunction *disjunctively*. It is unclear how such a notion is best represented formally, though in a system such as that of (Das *et al*, 1996), it would presumably mean modifying the aggregation function (- each premise of a conjunction is independently relevant to the conclusion, so it is the modalities which remain to be combined).

4. Rhetorical Approaches

Like argumentation theories, systems of rhetoric are concerned with real examples of argument, and as a consequence, must detail the use of subarguments: their structure, arrangement, presentation and so on. Unlike argumentation theory, however, texts on rhetoric (and in particular, those of the eighteenth and nineteenth centuries) aim to guide the synthesis of good argument - rather than guiding analysis to determine whether or not an argument is good.

The first consideration offered in most texts is the dissection of discourse

---

[6] Indeed, modality formed the basis of a theory of classification espoused by Yanal (1984). There were, however, a number of problems with the approach, as discussed in (Freeman, 1991, pp105-108). In particular, determining the modality of a link involves evaluating an argument: a process which is separate from, and intuitively should follow, the process of argument analysis.

into successive phases. Blair (1838), for example, suggests six stages: introduction, division of subject, narration of facts, argumentative part, pathetic part (essentially constituting an appeal to the emotions of the audience), and conclusion. In addition, he suggests that the argumentative part (seemingly the only part under investigation by either formal or informal logicians[7]) is then further subdivided into invention, arrangement and expression. This subdivision of argumentation follows the classical scheme of the branches of rhetoric, supported by Aristotle, Quintilian and Cicero (Billig, 1996, p81). As Blair notes, however, the task of 'inventing' the arguments is beyond the scope of rhetoric:

> "Art cannot go so far, as to supply a speaker with arguments on every cause, and every subject; though it may be of considerable use in assisting him to arrange and express those, which his knowledge of the subject has discovered. For it is one thing to discover the reasons that are most proper to convince men, and another, to manage these reasons with the most advantage." (Lect. XXXII, p427)

Once the arguments are available, there is the choice of choosing which to use. Blair, again summarises:

> "... one of the first things to be attended to is, among the various arguments which may occur upon a cause, to make a proper selection of such as appear to one's self the most solid; and to employ these as the chief means of persuasion." (Lect. XXXII, p429)

The notion of selecting and employing a number of 'best' subarguments poses several serious problems for computational systems. In a system such as that of (Das *et al*, 1996), it would mean setting some arbitrary threshold; in (Reed *et al*, 1996a), the process of deriving all possible arguments and then selecting the best would be arbitrarily restricted due to resource bounding.

After subarguments are chosen, they must be arranged in such a way as to maximise their persuasive impact. The crucial importance of this stage is noted by many writers (Blair, 1838), (Billig, 1996), but is eloquently explained by Whately:

> "... Arrangement is a more important point than is generally supposed; indeed it is not perhaps of less consequence in Rhetoric than in the Military Art; in which it is well known, that with an equality of forces, in numbers, courage, and every other point, the manner in which they are drawn up, so as either to afford mutual support, or on the other hand, even to impede and annoy each other, may make the difference of victory of defeat."(Whately, 1855, Ch. I, §3, p35)

Various systems of rhetoric provide rules for arranging subarguments, varying as much in specificity as they do in number. Here, a brief summary will be presented of the rules of Blair and Whately, which, between them, are typical of those which are available elsewhere.

Blair suggests four rules. In the first place, he claims, arguments should be grouped together according to their type (where *type* is defined as being one of truth, morality or profitability). Clearly, for such a heuristic to be implemented, it would be necessary to have some means of determining the 'type' of a subargument, something

---

[7] Also, possibly, including the narration: the facts established during the narrative part could be seen as basic grounds for the subsequent argumentation.

which is beyond even a tailored logical argumentation schemes such as LA (Krause *et al*, 1995). The second and third rules concern the strength of individual subarguments: that they should increase in strength (except in particular situations, as discussed in (Reed *et al*, 1996b)) and that any subargument which is particularly strong should be "brought out by itself, placed in its full light, amplified, and rested upon" (Blair, 1838, p431). These heuristics become amenable to implementation as soon as a means of rigorously determining argument strength is devised: in systems such as (Das *et al*, 1996) and (Parsons and Jennings, 1996), strength is indicated explicitly; §5 details how a text generation system might also make use of the distinction, proposed by Freeman (1991), between inferential and persuasive force. Lastly, that subarguments should not be extended too far, nor be too great in number. This is a particularly important fact in human communication, and one that is particularly difficult to deal with formally, where the aim is usually to employ precisely all the subarguments available to lend maximum support to a conclusion[8]. For this last rule builds upon the intuition that, in persuasive communication at least, more can in fact be less: additional subarguments may *weaken* a position rather than bolstering it[9].

Whately's list (1855) is a little longer, and encompasses a slightly different range of variables. Firstly, the ordering is crucially affected by whether the argument is setting out to convince an audience which is already aware of the subject matter, or whether to inform an audience which is not: this is called the *primary aim* in (Reed *et al*, 1996b), and its effects on a computational system are discussed there. Secondly, arguments from cause to effect should precede others (Whately (p34) provides an example: if an honest man were accused of corruption, the evidence is far more likely to be listened to if it is first shown that he may be greedy). The computational implications of this rule are similar to those of Blair's first rule, in that some means of tagging argumentation may be necessary, though in this case, identification of a subargument from cause to effect may be possible on purely structural grounds (if, for example, causality is a distinguished form of support). Thirdly, Whately claims that "Refutations of Objections should generally be placed in the midst of other Arguments, but nearer the beginning than the end" (p38). For, as mentioned above, although an argument is not based entirely on defeating anticipated counter-arguments, the technique should not be eschewed altogether. The same sentiments are echoed by Blair[10] as well as by more recent texts on informal logic[11]. Whately goes on to give two further rules which present a particular problem for a system which is to generate natural language arguments: that well known propositions

---

[8] The system of Parsons and Jennings (1996) system clearly does not suffer from this problem - but only because arguments are exchanged one at a time, at each dialogical turn.

[9] For example, an additional weak subargument may give room for an opponent to build a 'thin-end-of-the-wedge' or 'straw man' counter-argument.

[10] "Every speaker should place himself in the situation of a hearer, and think how he would be affected by those reasons, which he purposes to employ for persuading others" (Blair, 1838, pp429-430)

[11] Eg. Fogelin (1991, p41) discusses the use of *"Discounting*: anticipating criticisms and dismissing them".

should be stated at once, and that a recapitulation of the main points of an argument should occur in reverse order. Both of these techniques entail telling the hearer *something he already knows*, which runs contrary to the intuitions implemented in many natural language generation systems - a point made by Marcu (1996) in his psycholinguistic analysis of what makes a text persuasive. Blair, too, relies upon the fact that repetition is required, when he states (p440) that the conclusion should contain no new subject matter.

This last point makes particularly salient a common criticism of such rhetoric texts - Richards (quoted in (Billing, 1996, p90)) complains that they offer nothing more than

> "prudential Rules about the best sorts of things to say in various argumentative situations ... we get the usual postcard's worth of crude commonsense :- be clear, yet don't be dry: be vivacious, use metaphors when they will be understood not otherwise ..."

However, the careful, exhaustive enumeration of all this commonsense knowledge is a requisite precursor for implementing a system which might be able to produce natural language persuasive discourse.

5. Approaches in Natural Language Generation

In natural language generation (NLG), the problems of multiple subarguments fall naturally into three categories: generation, inclusion and ordering, ie. how subarguments are invented (to borrow the term used by Blair (1838)), how the decisions are made over which to use, and finally, how they are arranged as a persuasive whole.

Ideally, an agent should be able to generate all possible arguments supporting a cause, and then select those which best suit the situation at hand. There are, however, several problems with such an approach. Firstly, intuitions suggest that this is psychologically implausible (despite the implications of Blair's dictum to "make a proper selection of [arguments] as appear to one's self the most solid", as quoted in §4). In addition, the idea of expending significant computational energy on generating arguments which are never used is unappealing. More importantly, for a resource-bounded agent (see, for example, (Bratman *et al*, 1988)), this exhaustive procedure may simply be too costly, and would certainly not make optimal use of the limited resources available.

In practice, though, the reasoning by which arguments are produced is not properly a problem of NLG (in much the same way that Blair claims that the invention of arguments does not come within the purview of rhetoric). Although the boundary is a little unclear, it is important to distinguish the means by which the content of an argument is determined (eg. through the use of *reasoning agents*, (McConachy and Zukerman, 1996)) from the way it is subsequently structured and expressed. Part of the reason that the boundary is unclear is that the processes of invention and expression appear to be interleaved: it is unreasonable (and counter-intuitive) to assume that a full set of complete arguments is available to the NLG system. Rather, for any given belief there are supports which represent final key steps in chains of reasoning, and as the argument progresses, particular chains of support are pursued at greater length, depending upon the strength of the links, and on the hearer's knowledge and sympathy in the areas involved.

Given, then, the existence of argument components in some knowledge base (stored as 'potential', rather than complete, argument units), the first task is to decide which of them to include. As a planning task, this represents a particular problem, due to the nature of the goal to be fulfilled. In standard hierarchical planning, such as that implemented in NOAH (Sacerdoti, 1977), and widely used in discourse planning, (Hovy, 1993), the task is expressed in terms of a number of goals which are then met by applying appropriate operators. In (Reed *et al*, 1996a), for example, the goal BEL(H, P) (that the hearer believe some proposition, P) might be fulfilled by application of the Modus Ponens operator which lists BEL(H, P) on its postcondition list. However, once such an *achievement* goal is fulfilled, it is considered 'finished', such that no further planning is required to support it.

There have been some attempts at provision for a notion of *maintenance* goals. Hovy's (1990) system PAULINE, for example, could express stylistic goals which had continuing effect throughout the planning process. Goals which might give rise to multiple subarguments, however, are neither achievement nor maintenance in nature: they are planned for several times and then considered fulfilled in the classical sense. Several systems have implemented goals of this sort, through the use of some 'for all' function in operator definitions, such as Moore and Paris' (1994) FORALL clause, and Maybury's (1993) explicit use of the $\forall$ symbol: in both these cases, the authors have noted that the clauses have required explicit, unprincipled modification of the plan language.

Even if a rigorous planning foundation were available for such goals, NLG would continue to suffer from a related problem, that with a rich knowledge base available, a NLG system must be able to determine an appropriate level of detail. In argumentation, this becomes even more difficult since, as mentioned above, it is often useful to tell the hearer things he already knows (Marcu, 1996). One partial solution to this problem lies in the notion of coherency (discussed in this regard in more detail in (Reed and Long, 1997)). Including complex or large subarguments is detrimental to the resultant coherency of the argument as a whole. Similarly, every additional subargument employed decreases the overall coherency. The means by which this technique may contribute to a solution is discussed in the next section.

One of the most important factors determining whether or not a subargument is included in an argument is also the primary means of effecting an appropriate ordering between subarguments, namely, the strength of a subargument. Although many formal (and informal) approaches recognise the concept of argument 'strength', few make the distinction between *inferential force* and *persuasive force*, due to Freeman (1991):

> "The persuasive force of an argument is its ability to move an (intended) audience. Inferential strength is a completely normative issue. How well does this inference satisfy the canons of deductive or inductive logic?", p243

Formal systems - (Das *et al*, 1996), (Parsons, 1996), etc. - whether qualitative or quantitative, express only the inferential force. Whereas rhetorical maxims are based almost exclusively on persuasive force. As a consequence, NLG systems such as that of Maybury (1993) have generally been unable to implement persuasion related heuristics. One notable exception was Sycara's (1989) system, PERSUADER, which had a limited notion of persuasive strength, fixed as a rigid, domain-specific hierarchy between her nine argumentation techniques. As a result, PERSUADER was able to perform some degree of choice over the inclusion and ordering of subarguments. However, in order to flexibly plan persuasive arguments, the distinction needs to be made and implemented explicitly.

6. Towards a Solution

The task of generating natural language argument is being approached through the use of the hierarchical framework proposed in (Reed *et al*, 1996b), in which the planning process is conceptually divided into four layers of abstraction. At the highest, *Argument Structure* (AS) level, the structural form of the argument is produced; below this, the *Eloquence Generation* (EG) level performs rhetorical and stylistic refinement; at the next level, the interclausal structure is refined through the use of Rhetorical Structure Theory (Mann and Thompson, 1986); finally, linguistic form is produced at the lowest levels of syntactic and morphological realisation. The problem of the 'generation gap' (Meteer, 1993) is minimised through the underlying use of LOLITA, a large scale, domain-independent, natural language system (Smith *et al*, 1994). The distinction between AS and EG levels mirrors the rhetorical distinction between arrangement and expression (and as mentioned above, both these phases succeed that of invention).

The planning task is carried out by AbNLP (Fox and Long, 1995), which makes use of encapsulation (such that an operator body is completely hidden until the abstract plan is completed) and refinement (which opens up all the operator bodies on completion of the abstract plan). These techniques both correspond closely to the structure of natural argument, and also lead to significant computational savings (Baccus and Yang, 1992). Investigation is under way of an optimal means of implementing the 'forall' problem discussed in the previous section (though it is anticipated that the solution will be similar to that of (Moore and Paris, 1994)).

To deal with the inherent uncertainty of the communicative situation (both in the speaker's own beliefs, and in her beliefs of what her audience believes), a standard approach is taken, such as that proposed by Parsons (1996). However, two important caveats are noted. Firstly, there is no assumption of 'distributed' defeasibility (a concept introduced in §2). That is, agents do not construct their arguments (solely) on the basis of anticipated counter-arguments. This approach to uncertainty is thus similar to that of (Parsons and Jennings, 1996), except that an agent is capable of constructing complex arguments, composed of multiple subarguments. In the second place, it is recognised that the representation is limited to detailing only inferential force, and that the notion of persuasive force needs to be handled explicitly and separately.

Freeman (1991) suggests that persuasive and inferential force are quite distinct and unrelated facets of an argument. However, the claim seems to be rather too strong, certainly for NLG, where the assessment of persuasive force will need to include reference to the inferential force (an utterly illogical argument, for example, is often far less likely to convince an audience). Indeed, in a situation in which the speaker knows nothing about a hearer other than that he is rational, the persuasive force of an argument may very well be the *same* as the inferential force. Clearly, though, persuasive force is normally determined by more than just the inferential strength due only to the structure of an argument. As discussed above, an evaluation of persuasive force is required in order to make various rhetorical decisions over which subarguments to include, and how then to determine an effective ordering between them. Such an evaluation can be conducted by considering both the reception an argument would receive, on the basis of the model of the hearer's beliefs, in addition to various factors which would affect that reception, including hearer bias, scepticism, competence, etc.[12]. Importantly, this process is not equivalent to the counter-intuitive distributed defeasibility which builds arguments on the basis of anticipated hearer counter-arguments. Rather, it gives rise to precisely that functionality suggested by intuition, namely, that arguments can involve refutations of rebuttals, but that those refutations are included (or not) and ordered on the basis of their persuasive force. In many cases, (as expected), such refutations will be an effective means of persuasion, and will play a primary role in an argument as a consequence. But, there will also be instances when other factors (such as high levels of hearer scepticism, or particularly low confidence in aspects of the hearer model) drastically reduce effectiveness of counter-counter-arguments, resulting in their marginalization.

The AS level can thus be seen as mediating between conflicting pressures. On the one hand, coherency constraints (Reed and Long, 1997) aim to minimise complexity, reduce the number of subarguments and avoid deeply nested argumentation. At the same time, structural planning is producing additional complexity; this planning includes persuasion related aspects such as informing the hearer of propositions he already believes and employing fallacies (a technique which is similar to McConachy and Zukerman's (1996) *licentious* mode, and which extends Sycara's (1988) use of fallacies, by granting them the same status as the deductive and inductive operators, thus avoiding the necessity of placing them in a fixed preference hierarchy). This conceptual process of mediation is the primary means of tackling the NLG detail problem, and implementing the associated rhetorical rules.

7. Conclusion

The phenomenon of disjunct multiple subarguments, occurring in many systems of argumentation, present a number of problems: how to reason defeasibly in a multi-agent society without eschewing the intuitions of argument construction; how to distinguish between linked and independent subargument support; how to deal with rhetorical maxims which depend on a notion of persuasive force which is primarily empirical; how to determine an appropriate amount of detail whilst generating an argument in natural language.

This paper has presented a synthesis of these ideas, which has been used as

---

[12] These factors are crucial to the system, at both AS and EG levels, as discussed in (Reed *et al*, 1996b)

the basis for proposing an approach to the generation of natural language argument. As a consequence, it is hoped that the approach benefits from the findings of several disparate fields, whilst avoiding trivialising the problems in any one. The solution put forward is not complete. The conjunction operator used to combine linked subarguments, for example, is currently rather naive and does not fully capture the intuitions discussed by Freeman. The problem of a principled approach to the 'forall' operator required to plan for multiple subarguments remains unresolved. However, a framework has been described which includes a number of important and novel aspects: a clear distinction between disjunctive and conjunctive support; a means of implementing rhetorical heuristics through a definition of 'strength'; a distinction between persuasive and inferential strength; a means of determining an appropriate level of detail. But more importantly, this paper has attempted to clearly identify the problems which have a bearing on the issue of natural language argument generation, and to motivate the continuing research on an interdisciplinary basis.

References

Baccus F. & Yang Q. "The expected value of hierarchical problem-solving", in *Proceedings of the National Conference on AI (AAAI'92)* (1992)

Billig, M., *Arguing and thinking* (Second edition), Cambridge University Press (1996)

Blair, H. *Lectures on Rhetoric and Belles Lettres*, Charles Daly, London (1838)

Bratman, M.E., Israel, D.J., Pollack, M.E., "Plans and resource-bounded practical reasoning", *Computational Intelligence* **4** (1988) 349-355

Cohen, R., "Analyzing the Structure of Argumentative Discourse", *Computational Linguistics* **13** (1) (1987) 11-24

Das, S., Fox, J. & Krause, P., "A Unified Framework for Hypothetical and Practical Reasoning (1): Theoretical Foundations", in Gabbay, D. & Ohlbach, H.J. *Practical Reasoning*, Springer Verlag (1996) 58-72

Dung, P.M., "On the acceptability of arguments and its fundamental role in nonmonotonic reasoning, logic programming and n-person games", *Artificial Intelligence* **77** (1995) 321-357

Fisher, A., *The Logic of Real Arguments*, Cambridge University Press (1988)

Fogelin, R.J. & Sinnott-Armstrong, W., *Understanding Arguments* (Fourth Edition), Harcourt Brace Jovanovich College Publishers (1991)

Fox, J. & Das, S., "A Unified Framework for Hypothetical and Practical Reasoning (2): Lessons from Medical Applications", in Gabbay, D. & Ohlbach, H.J. *Practical Reasoning*, Springer Verlag (1996) 73-92

Fox, M. & Long, D.P. "Hierarchical Planning using Abstraction", *IEE Proceedings on Control Theory and Applications* **142** (3) (1995)

Freeman, J.B., *Dialectics and the Macrostructure of Arguments*, Foris (1991)

Hovy, E.H., "Pragmatics and Natural Language Generation", *Artificial Intelligence* **43** (1991) 153-197

Hovy, E. H. "Automated Discourse Generation Using Discourse Structure Relations", *Artificial Intelligence* **63** (1993) 341-385

Krause, P., Ambler, S., Elvang-Gøransson, M. & Fox, J., "A Logic of Argumentation for Reasoning under Uncertainty", *Computational Intelligence* **11** (1995) 113-131

Mann, W.C. & Thompson, S.A. "Rhetorical structure theory: description and construction of text structures" in Kempen, G., *Natural Language Generation: New Results in AI, Psychology and Linguistics*, Kluwer (1986) 279-300

Marcu, D., "The Conceptual and Linguistic Facets of Persuasive Arguments", in *Working Notes of the ECAI'96 Workshop on Planning and NLG* (1996) 43-46

Maybury, M.T., "Communicative Acts for Generating Natural Language Arguments", in *Proceedings of the 11th National Conference on AI (AAAI'93)* (1993) 357-364

McConachy, R. & Zukerman, I., "Using Argument Graphs to Generate Arguments", in *Proceedings of the 12th European Conference on AI (ECAI'96)* (1996) 592-596

Meteer, M., *Expressibility and the Problem of Efficient Text Planning*, Francis Pinter (1993)

Moore, J.D., Paris, C.L. "Planning Text for Advisory Dialogues: Capturing Intentional and Rhetorical Information", *Computational Linguistics* **19** (4) (1994) 651-694

Parsons, S., "Defining Normative Systems for Qualitative Argumentation", in Gabbay, D. & Ohlbach, H.J. *Practical Reasoning*, Springer Verlag (1996) 449-463

Parsons, S. & Jennings, N.R., "Negotiation Through Argumentation - a Preliminary Report", in *Proceedings of the International Conf on Multi-Agent Systems (ICMAS'96)* (1996)

Pollock, J.L., *Cognitive Carpentry*, MIT Press (1995)

Pollock, J. L., "Implementing Defeasible Reasoning", in *Working Notes of the FAPR'96 Workshop on Computational Dialectics* (1996)

Reed, C.A., Long, D.P., Fox, M. & Garagnani, M., "Persuasion as a Form of Inter-Agent Negotiation", in *Proceedings of the 2nd Australian Workshop on DAI* (1996a, to appear)

Reed, C.A., Long, D.P. & Fox, M. "An Architecture for Argumentative Dialogue Planning", in Gabbay, D. & Ohlbach, H.J. *Practical Reasoning*, Springer Verlag (1996b) 555-566

Reed, C.A. & Long, D.P., "Ordering and Focusing in an Architecture for Persuasive Discourse", under review for *European Workshop on Natural Language Generation (EWNLG'97)* (1997)

Sacerdoti, E.D. *A structure for plans and behaviour*, Elsevier, North Holland (1977)

Sillince, J.A.A & Minors, R.H., "Argumentation, Self-Inconsistency, and Multidimensional Argument Strength", *Communication and Cognition* **25** (4) (1992) 325-338

Smith, M.H., Garigliano, R., Morgan, R.C. "Generation in the LOLITA system: An engineering approach" in *Proceedings of the 7th International Workshop on NLG*, Kennebunkport, Maine (1994)

Sycara, K. "Argumentation: Planning Other Agent's Plans" in *Proceedings of the International Joint Conference on AI (IJCAI'89)* (1989) 517-523

Toulmin, S.E., *The Uses of Argument*, Cambridge University Press (1958)

Vreeswijk, G., "Reasoning with Defeasible Arguments", in Wagner, G. & Pearce, D. (eds) *Proc. of the European Workshop on Logics in AI (JELIA'92)*, Springer Verlag (1992) 189-211

Walton, D.N. & Krabbe, E.C.W., *Commitment in Dialogue*, State University of New York Press (1995)

Whately, R. *Logic*, Richard Griffin, London (1855)

Wooldridge, M. & Jennings, N.R., "Intelligent Agents: Theory and Practice", *The Knowledge Engineering Review* **10** (2) (1995) 115-152

Yanal, R.J., "'Convergent' and 'Linked' Reasons.", *APA Newsletter on Teaching Philosophy*, **4** (5), (1984)1-3

Cactus: A Branching-Time Logic Programming Language*

P. Rondogiannis[1], M. Gergatsoulis[2], T. Panayiotopoulos[3]

[1] Dept. of Computer Science, University of Ioannina,
P.O. BOX 1186, 45110 Ioannina, Greece,
e-mail: prondo@zeus.cs.uoi.gr

[2] Inst. of Informatics & Telecom., N.C.S.R. 'Demokritos',
153 10 A. Paraskevi Attikis, Greece
e-mail: manolis@iit.nrcps.ariadne-t.gr

[3] Dept. of Informatics, University of Piraeus
80 Karaoli & Dimitriou Str., 18534 Piraeus, Greece
e-mail : themisp@unipi.gr

Abstract. Temporal programming languages are recognized as natural and expressive formalisms for describing dynamic systems. However, most such languages are based on linear flow of time, a fact that makes them unsuitable for certain types of applications. In this paper we introduce the new temporal logic programming language **Cactus**, which is based on a branching notion of time. In Cactus, the truth value of a predicate depends on a hidden time parameter which has a tree-like structure. As a result, Cactus appears to be especially appropriate for expressing non-deterministic computations or generally algorithms that involve the manipulation of tree data structures.

Keywords: Logic Programming, Temporal Logic Programming, Branching Time.

1 Introduction

Temporal programming languages [OM94, Org91] are recognized as natural and expressive formalisms for describing *dynamic* systems. For example, consider the following Chronolog [Wad88] program simulating the operation of the traffic lights:

```
first light(green).
next light(amber) ← light(green).
next light(red)   ← light(amber).
next light(green) ← light(red).
```

However, Cronolog as well as most temporal languages [OM94, Hry93, OWD93, Bau93, OW92, Brz91, Brz93, GRP96] are based on linear flow of time, a fact that

---
* This work has been funded by the Greek General Secretariat of Research and Technology under the project "TimeLogic" of $\Pi ENE\Delta'95$, contract no 1134.

makes them unsuitable for certain types of applications. In this paper we present the new temporal logic programming language **Cactus** which is based on a tree-like notion of time; that is, every moment in time may have more than one next moments. The new formalism is appropriate for describing non-deterministic computations or more generally computations that involve the manipulation of trees.

Cactus supports two main operators: the temporal operator **first** refers to the beginning of time (or alternatively to the root of the tree). The temporal operator **next**$_i$ refers to the i-th child of the current moment. Notice that we actually have a family $\{\textbf{next}_i \mid i \in N\}$ of **next** operators, each one of them representing the different next moments that immediately follow the present one.

As an example, consider the following program:

> **first** nat(0).
> **next**$_0$ nat(Y) $\leftarrow$ nat(X),Y is 2*X+1.
> **next**$_1$ nat(Y) $\leftarrow$ nat(X),Y is 2*X+2.

The idea behind the above program is that the set of natural numbers can be mapped on a binary tree of the form shown in figure 1. More specifically, one can think of **nat** as a time-varying predicate. At the beginning of time (at the root of the tree) **nat** is true of the natural number 0. At the left child of the root of the tree, n is true of the value 1, while at the right child it is true of the value 2. In general, if **nat** is true of the value X at some node in the tree, then at the left child of that node **nat** will be true of 2*X+1 while at the right child of the node it will be true of 2*X+2. One can easily verify that the tree created contains all the natural numbers.

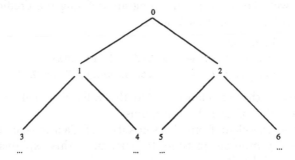

Fig. 1. A mapping of the natural numbers on a binary tree

One could claim that branching time logic programming (or temporal logic programming in general) does not add much to logic programming, because *time* can always be added as an extra parameter to predicates. However, from a theoretical viewpoint this does not appear to be straightforward (see for example [Gab87, GHR94] for a good discussion on this subject). Moreover, temporal languages are very expressive for many problem domains. As it will become apparent in the next sections, one can use the branching time concept in order to represent in a natural way time-dependent data as well as to reason in a lucid manner about these data.

The rest of the paper is organized as follows: in section 2 we present various Cactus programs which demonstrate its potential in expressing tree computations. In section 3, we formally introduce the syntax of the language. Section 4 presents the underlying branching time logic BTL of Cactus. In section 5 we discuss implementation issues, and section 6 gives the concluding remarks.

2 The syntax of Cactus programs

The syntax of Cactus programs is an extension of the syntax of Prolog programs. In the following we assume familiarity with the basic notions of logic programming [Llo87].

A *temporal atom* is an atomic formula with a number (possibly 0) of applications of temporal operators. The sequence of temporal operators applied to an atom is called the *temporal reference* of that atom. A *temporal clause* is a formula of the form:

$$H \leftarrow B_1, ..., B_m$$

where $H, B_1, ..., B_m$ are temporal atoms, $m \geq 0$. If $m = 0$ then the clause is said to be a *unit temporal clause*. A *Cactus program* is a finite set of *temporal clauses*.

A *goal clause* in Cactus is a formula of the form $\leftarrow A_1, ..., A_n$ where A_i, $i = 1, ..., n$ are temporal atoms.

Notice that the syntax of Cactus allows temporal operators to be applied on body atoms as well. For example the program defining the predicate nat in the introduction can be redefined as follows:

```
first nat(0).
next₀ nat(Y) ←  nat(X), Y is 2*X+1.
next₁ nat(Y) ←  next₀ nat(X), Y is X+1.
```

The meaning of the last clause is that the value assigned to the right child of a node is the value of its left sibling plus 1.

As it will become clear from the semantics of Cactus, a clause is assumed to be true at every moment in time. In particular, this explains the difference between a clause of the form

```
first nat(0).
```

and the clause

```
nat(0).
```

The first clause asserts that it is always true that nat(0) is true at the beginning of time while the second clause indicates that it is always true that nat is true of 0 at every moment in time.

3 Cactus Applications

In this section we present various applications showing the expressive power of branching time logic programming.

3.1 Expressing non-deterministic behaviour

Consider the non-deterministic finite automaton shown in figure 2 (taken from [LP81] page 55) which accepts the regular language $L = (01 \cup 010)^*$. We can describe the behaviour of this automaton in Cactus with the following program:

```
first state(q0).
next0 state(q1)  ←  state(q0).
next1 state(q2)  ←  state(q1).
next1 state(q0)  ←  state(q1).
next0 state(q0)  ←  state(q2).
```

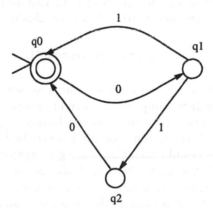

Fig. 2. A non-deterministic finite automaton

Notice that, in this automaton $q0$ is both the initial and the final state. Posing the goal clause:

$$\leftarrow \quad \text{first } next_0 \ next_1 \ next_0 \ \text{state}(q0).$$

will return the answer **yes** which indicates that the string 010 is an acceptable string of the language L.

As we will see in section 5, the proof procedure of Cactus is similar in nature to the well known SLD-resolution of Horn clause logic programming [Llo87].

3.2 Generating sequences

One can write a simple Cactus program for producing the set of all binary sequences. The set of such sequences may be thought of as a tree, which can be described by the following program:

$$\text{first } \text{binseq}([\,]).$$
$$\text{next}_0 \; \text{binseq}([0|X]) \; \leftarrow \; \text{binseq}(X).$$
$$\text{next}_1 \; \text{binseq}([1|X]) \; \leftarrow \; \text{binseq}(X).$$

The goal clause:

$$\leftarrow \quad \text{binseq}(S).$$

will trigger an infinite computation which will generate all possible sequences. More specifically, the underlying proof procedure of Cactus, considers the above goal clause as an infinite set of "temporally ground" goal clauses, each one corresponding to a different point of the time tree.

One can combine the program **binseq** with the program for the nondeterministic automaton given in subsection 3.1. In this way we can produce the language recognized by the automaton. More specifically, the goal clause:

$$\leftarrow \quad \text{state}(q0), \text{binseq}(S).$$

produces the infinite set of all the binary sequences recognized by the automaton. The above goal clause (assuming a left to right computation rule) is not the classical generate-and-test procedure (not all binary sequences are generated but only those for which the automaton reaches the final state q0). This is due to the fact that each succesful evaluation of the goal state(q0) at a specific time point, triggers a corresponding evaluation of **binseq**, at the same time point.

It is worthwhile noting here that in order to generate another language one only needs to change the definition of the automaton and not the definition of **binseq**.

3.3 Representing and manipulating trees

Branching time logic programming is a powerful tool for representing and manipulating trees. A tree can be represented in Cactus as a set of temporal unit clauses. The structure of the tree is expressed through the temporal references of the unit clauses. Moreover, the well known tree manipulation algorithms are easily and naturally expressed through Cactus programs. For example, consider the binary tree of figure 3.

A possible representation of the information included in this tree is given by the following set of Cactus unit clauses:

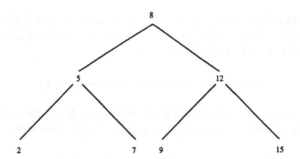

Fig. 3. An (ordered) binary tree containing numeric data

```
first data(8).
first next0 data(5).
first next1 data(12).
first next0 next0 data(2).
first next1 next0 data(9).
first next0 next1 data(7).
first next1 next1 data(15).
```

The following program defines the predicate **descendant(X)**. A temporal atom $< Temporal\ reference >$ **descendant(X)** is true if **data(X)** is true in the time represented by $< Temporal\ reference >$ or in a future moment of this time point.

```
descendant(X) ← data(X).
descendant(X) ← data(Y), next0 descendant(X).
descendant(X) ← data(Y), next1 descendant(X).
```

Notice that the purpose of the existence of the atom **data(Y)** in the bodies of the second and third clause is only to ensure termination of the proof procedure.

A more efficient definition of the predicate **descendant** which takes into account the fact that the binary tree is ordered (binary search) is shown in the following program.

```
descendant(X) ← data(X).
descendant(X) ← data(Y), X < Y, next0 descendant(X).
descendant(X) ← data(Y), X > Y, next1 descendant(X).
```

By posing the goal clause:

$$\leftarrow first\ next_0\ descendant(7).$$

we will get the answer **yes**, because the value 7 is in a node which represents a moment in the future of **first next0**.

Using the definition of the predicate **descendant** we can define the predicate **search** which tests if a specific numeric value is in a node of the data tree. The definition of **search** is given by the clause:

$$\text{search}(X) \leftarrow \text{first descendant}(X).$$

Let us now define a predicate **flattree** which collects the values in the tree nodes into a list. This definition corresponds to the preorder traversal of the tree.

$$
\begin{aligned}
\text{flattree}([\,]) &\leftarrow \text{data(void)}. \\
\text{flattree}([X|L]) &\leftarrow \text{data}(X), \\
&\quad\ \text{next}_0\ \text{flattree}(L1), \\
&\quad\ \text{next}_1\ \text{flattree}(L2), \\
&\quad\ \text{append}(L1, L2, L).
\end{aligned}
$$

Notice that the above program recognizes the tips of the tree when it encounters a **data(void)** unit clause. For this, we have to add the following unit clauses to the program[4]:

$$
\begin{aligned}
&\text{first } \text{next}_0\ \text{next}_0\ \text{next}_0\ \text{data(void)}. \\
&\text{first } \text{next}_0\ \text{next}_0\ \text{next}_1\ \text{data(void)}. \\
&\text{first } \text{next}_0\ \text{next}_1\ \text{next}_0\ \text{data(void)}. \\
&\text{first } \text{next}_0\ \text{next}_1\ \text{next}_1\ \text{data(void)}. \\
&\text{first } \text{next}_1\ \text{next}_0\ \text{next}_0\ \text{data(void)}. \\
&\text{first } \text{next}_1\ \text{next}_0\ \text{next}_1\ \text{data(void)}. \\
&\text{first } \text{next}_1\ \text{next}_1\ \text{next}_0\ \text{data(void)}. \\
&\text{first } \text{next}_1\ \text{next}_1\ \text{next}_1\ \text{data(void)}.
\end{aligned}
$$

4 The branching time logic of Cactus

In this section we describe the branching time logic (BTL) on which Cactus is based. In BTL, time has an initial moment and flows towards the future in a tree-like way. The set of moments in time in BTL, can be modelled by the set $List(N)$ of lists of natural numbers. In this case, each node has a countably infinite number of branches (**next** operators). Similarly, we may choose a finite subset S of N and define the logic $BTL(S)$, which has a finite number of **next**$_i$ operators whose subscript i ranges over the set S. Intuitively, this corresponds to trees in which every node has a finite number of branches. In any case, the empty list [ ] corresponds to the beginning of time and the list $[i|t]$ (that is, the list with head i and tail t) corresponds to the i-th child of the moment identified by the list t.

---

[4] A more compact representation of the above tree (that avoids the use of void nodes) would be to distinguish the (inner) nodes from the leafs of the tree by using two different predicate names e.g. **node(X)** and **tip(X)** instead of the single predicate **data**. In that case we have to change slightly the definition of **flattree**.

BTL uses the temporal operators **first** and **next**$_i$, $i \in N$. The operator **first** is used to express the first moment in time, while **next**$_i$ refers to the i-th child of the current moment in time. The syntax of BTL extends the syntax of first-order logic with two formation rules:

- if A is a formula then so is **first** A, and
- if A is a formula then so is **next**$_i$ A.

BTL is a relatively simple branching time logic. For more on branching time logics one can refer to [BAPM83].

4.1 Semantics of BTL formulas

The semantics of temporal formulas of BTL are given using the notion of *branching temporal interpretation*. Branching temporal interpretations extend the temporal interpretations of the linear time logic of Chronolog [Org91].

Definition 1. A *branching temporal interpretation* or simply a *temporal interpretation* I of the temporal logic BTL comprises a non-empty set D, called the domain of the interpretation, over which the variables range, together with an element of D for each variable; each n-ary function symbol, an element of $[D^n \to D]$; and for each n-ary predicate symbol, an element of $[List(N) \to 2^{D^n}]$.

In the following definition, the satisfaction relation $\models$ is defined in terms of temporal interpretations. $\models_{I,t} A$ denotes that a formula A is true at a moment t in some temporal interpretation I.

Definition 2. The semantics of the elements of the temporal logic BTL are given inductively as follows:

1. If $\mathbf{f}(e_0, \ldots, e_{n-1})$ is a term, then $I(\mathbf{f}(e_0, \ldots, e_{n-1})) = I(\mathbf{f})(I(e_0), \ldots, I(e_{n-1}))$.
2. For any n-ary predicate symbol p and terms $e_0, \ldots, e_{n-1}$,
   $\models_{I,t} \mathbf{p}(e_0, \ldots, e_{n-1})$ *iff* $\langle I(e_0), \ldots, I(e_{n-1}) \rangle \in I(\mathbf{p})(t)$
3. $\models_{I,t} \neg A$ *iff it is not the case that* $\models_{I,t} A$
4. $\models_{I,t} A \wedge B$ *iff* $\models_{I,t} A$ *and* $\models_{I,t} B$
5. $\models_{I,t} A \vee B$ *iff* $\models_{I,t} A$ *or* $\models_{I,t} B$
6. $\models_{I,t} (\forall x)A$ *iff* $\models_{I[d/x],t} A$ *for all* $d \in D$ *where the interpretation* $I[d/x]$ *is the same as* I *except that the variable* x *is assigned the value d.*
7. $\models_{I,t}$ **first** A *iff* $\models_{I,[\,]} A$
8. $\models_{I,t}$ **next**$_i$ A *iff* $\models_{I,[i|t]} A$

If a formula A is true in a temporal interpretation I at all moments in time, it is said to be true in I (we write $\models_I A$) and I is called a *model* of A.

Clearly, Cactus clauses form a subset of BTL formulas. It can be shown that the usual minimal model and fixpoint semantics that apply to logic programs, can be extended to apply to Cactus programs. However, such an investigation is outside the scope of this paper and is reported in a forthcoming paper [RGP97].

4.2 Axioms and Rules of Inference

In this section we present some useful axioms and inference rules that hold for the logic BTL, many of which are similar to those adopted for the case of linear time logics [Org91]. In the following, the symbol ∇ stands for any of **first** and **next**$_i$.

Temporal operator cancellation rules: The intuition behind these rules is that the operator **first** cancels the effect of any other "outer" operator. Formally:
$$\nabla(\textbf{first } A) \leftrightarrow (\textbf{first } A)$$
Notice that this is actually a family of rules, one for each different instantiation of the operator ∇.

Temporal operator distribution rules: These rules express the fact that the branching time operators of BTL distribute over the classical operators $\neg$, $\wedge$ and $\vee$. Formally:
$$\nabla(\neg A) \leftrightarrow \neg(\nabla A)$$
$$\nabla(A \wedge B) \leftrightarrow (\nabla A) \wedge (\nabla B)$$
$$\nabla(A \vee B) \leftrightarrow (\nabla A) \vee (\nabla B)$$
Again, each of the above rules actually represents a family of rules depending on the instantiation of ∇.

From the temporal operator distribution rules we see that if we apply a temporal operator to a whole program clause, the operator can be pushed inside until we reach atomic formulas. This is why we did not consider applications of temporal operators to whole program clauses.

Temporal operator non-commutativity rule: This rule says that the following:

$$\textbf{next}_i \textbf{ next}_j \, A \leftrightarrow \textbf{next}_j \textbf{ next}_i \, A$$

is not a valid axiom of the language when $i \neq j$. The essence of this rule is that in general, two operators **next**$_i$ and **next**$_j$ can not be interchanged when i and j are different.

Rigidness of variables: The following rule states that a temporal operator ∇ can "pass inside" $\forall$:
$$\nabla(\forall X)(A) \leftrightarrow (\forall X)(\nabla A)$$
The above rule holds because variables represent data-values composed of function symbols and constants which are independent of time (i.e. they are *rigid*).

Temporal operator introduction rules: The following rule states that if A is a theorem of BTL then ∇A is also a theorem of BTL.

$$if \ \vdash A \ then \ \vdash \nabla A$$

The validity of the above axioms is easily proved using the semantics of BTL.

5 A proof procedure for branching time logic programs

Cactus programs are executed using a resolution-type proof procedure called *BSLD-resolution* (Branching-time **SLD**-resolution). For practical reasons, we suppose that the underlying logic of Cactus programs is $BTL(S)$, where S is a finite subset of N (i.e. in the time tree every node has a finite number of branches). BSLD-resolution is a refutation procedure which extends SLD-resolution [Llo87], and is similar to TiSLD-resolution [OW93], the proof procedure for Chronolog programs. The following definitions are necessary in order to introduce BSLD-resolution.

Definition 3. A *canonical temporal atom* is a formula $\mathtt{first}\ \mathtt{next}_{i_1} \cdots \mathtt{next}_{i_n}\ A$, where $i_1, \ldots, i_n \in S$ and $n \geq 0$, and A is an atom. A *canonical temporal clause* is a temporal clause whose temporal atoms are canonical temporal atoms.

As in Chronolog [Org91, OWD93], every temporal clause can be transformed into a (possibly infinite) set of canonical temporal clauses. This can be done by applying $\mathtt{first}\ \mathtt{next}_{i_1} \cdots \mathtt{next}_{i_n}$, where $i_1, \ldots, i_n \in S$ and $n \geq 0$, to the clause and then using the axioms of BTL, presented in section 4.2, to distribute the temporal reference so as to be applied to each individual temporal atom of the clause; finally any superfluous operator is eliminated by applying the cancellation rules of BTL.

Intuitively, a canonical temporal clause is an instance in time of the corresponding temporal clause.

Example 1. Consider the following Cactus program:

```
first p(0).
next₀ p(s(X)) ←  p(X).
next₁ p(s(s(X))) ←  p(X).
```

The set of canonical temporal clauses corresponding to the program clauses is as follows:
The clause:

```
first p(0).
```

is the only canonical temporal clause corresponding to the first program clause (because of axiom 1).
The set of clauses:

$$\{\mathtt{first}\ \mathtt{next}_{i_1} \cdots \mathtt{next}_{i_n}\ \mathtt{next}_0\ \mathtt{p}(\mathtt{s}(\mathtt{X})) \leftarrow \mathtt{first}\ \mathtt{next}_{i_1} \cdots \mathtt{next}_{i_n}\ \mathtt{p}(\mathtt{X}) \mid$$
$$n \in N,\ i_1, \ldots, i_n \in S\}$$

corresponds to the second program clause. Finally the set of clauses:

$$\{\mathtt{first}\ \mathtt{next}_{i_1} \cdots \mathtt{next}_{i_n}\ \mathtt{next}_1\ \mathtt{p}(\mathtt{s}(\mathtt{s}(\mathtt{X}))) \leftarrow \mathtt{first}\ \mathtt{next}_{i_1} \cdots \mathtt{next}_{i_n}\ \mathtt{p}(\mathtt{X}) \mid$$
$$n \in N,\ i_1, \ldots, i_n \in S\}$$

corresponds to the third program clause.

The notion of canonical atom/clause is very important since the value of a given formula of a branching time logic $BTL(S)$, for some finite subset S of N, in a temporal interpretation can be expressed in terms of the values of its canonical instances, as the following lemma shows:

Lemma 4. *Let A be a formula and I a temporal interpretation of $BTL(S)$. $\models_I A$ if and only if $\models_I A_t$ for all canonical instances A_t of A.*

BSLD-resolution is applied to canonical instances of program clauses and goal clauses.

Definition 5. Let P be a Cactus program and G be a canonical temporal goal. A *BSLD-derivation* from P with top goal G consists of a (possibly infinite sequence) of canonical temporal goals $G_0 = G, G_1, ..., G_n, ...$ such that for all i the goal G_{i+1} is obtained from the goal:
$$G_i = \leftarrow A_1, ..., A_{m-1}, A_m, A_{m+1}, ..., A_p$$
as follows:

1. A_m is a canonical temporal atom in G_i (called the *selected* atom)
2. $H \leftarrow B_1, ..., B_r$ is a canonical instance of a program clause,
3. there is a substitution θ such that $\theta = mgu(A_m, H)$
4. G_{i+1} is the goal:
$$G_{i+1} = \leftarrow (A_1, ..., A_{m-1}, B_1, ..., B_r, A_{m+1}, ..., A_p)\theta$$

Definition 6. Let P be a Cactus program and G be a canonical temporal goal. A *BSLD-refutation* from P with top goal G is a finite BSLD-derivation of the null clause $\square$ from P with top goal G.

Let us now see an example of the application of BSLD-resolution.

Example 2. Consider the program defining the predicate **nat** presented in the introduction:

$$
\begin{array}{ll}
(1) & \texttt{first nat(0).} \\
(2) & \texttt{next}_0 \texttt{ nat(Y)} \leftarrow \texttt{nat(X),Y is 2*X+1.} \\
(3) & \texttt{next}_1 \texttt{ nat(Y)} \leftarrow \texttt{nat(X),Y is 2*X+2.}
\end{array}
$$

A BSLD-refutation of the canonical temporal goal (in every derivation step the selected temporal atom is the underlined one):

$$\leftarrow \texttt{first next}_0 \texttt{ next}_1 \texttt{ nat(N)}$$

is given below:

$$\leftarrow \texttt{first \underline{next}}_0 \texttt{ next}_1 \texttt{ nat(N)}$$

using clause (3)

$$\leftarrow \texttt{first next}_0 \texttt{ nat(X), first next}_0 \texttt{ (N is 2 * X + 2)}$$

using clause (2)

$\leftarrow$ first nat(X_1), first next$_1$ (X is 2 * X_1 + 1),
        first next$_0$ (N is 2 * X + 2)

(X_1 = 0)   using clause (1)

$\leftarrow$ first next$_1$ (X is 2 * 0 + 1), first next$_0$ (N is 2 * X + 2)

(X = 1)   evaluation of the built-in predicate is[5]

$\leftarrow$ first next$_0$ (N is 2 * 1 + 2)

(N = 4)   evaluation of the built-in predicate is

$\square$

When some of the temporal atoms included in a goal clause are not canonical, we say that we have an *open-ended goal clause* (e.g. the goal clauses in section 3.2). The idea behind open-ended goal clauses was first introduced in the context of Cronolog [OW93]. An open-ended goal clause G represents the infinite set of all canonical queries corresponding to G. Open-ended goal clauses are used to imitate non-terminating computations. An implementation strategy for executing an open-ended goal clause is by enumerating and evaluating (one by one) the set of all canonical instances of the goal clause (e.g. by traversing in a breadth-first way the time-tree).

6 Conclusions

Temporal programming languages, either functional [WA85, DW90, EAAJ91] or logic [OM94, Org91, PG95], have been widely used as a means for describing dynamic systems. However, most temporal languages use a linear notion of time a fact that makes them unsuitable for certain types of applications.

In this paper we introduce the branching time logic programming language **Cactus** which is based on a tree-like notion of time. We demonstrate that Cactus is capable of expressing various problems in a natural way. Moreover, we show that Cactus retains the semantic clarity of logic programming and has a simple procedural interpretation.

The branching time concept has been particularly successful in the functional programming domain [RW97, Ron94, Yag84, Tao94] and we believe that a similar potential exists for the area of logic programming.

References

[BAPM83] M. Ben-Ari, A. Pnueli, and Z. Manna. The Temporal Logic of Branching Time. *Informatica*, pages 207–226, 1983.

---

[5] In order to use Cactus in practical applications it is useful to introduce certain built-in predicates which behave as in classical Prolog. The built-in procedure is is considered to be independent of time (rigid).

[Bau93]   M. Baudinet. A simple proof of the completeness of temporal logic pro-
           gramming. In L. Farinas del Cerro and M. Penttonen, editors, *International
           Logics for Programming*, pages 51–83. Oxford University Press, 1993.

[Brz91]   C. Brzoska. Temporal logic programming and its relation to constraint
           logic programming. In *Proc. of the Logic Programming Symposium*, pages
           661–677. MIT Press, 1991.

[Brz93]   C. Brzoska. Temporal logic programming with bounded universal modality
           goals. In D. S. Warren, editor, *Proc. of the Tenth International Conference
           on Logic Programming*, pages 239–256. MIT Press, 1993.

[DW90]   W. Du and W.W.Wadge. A 3D Spreadsheet Based on Intensional Logic.
           IEEE Software, pages 78–89, July 1990.

[EAAJ91]   A. A. Faustini E. A. Ashcroft and R. Jagannathan. An Intensional Lan-
           guage for Parallel Applications Programming. In B.K.Szymanski, editor,
           Parallel Functional Languages and Compilers, pages 11–49. ACM Press,
           1991.

[Gab87]   Dov Gabbay. Modal and temporal logic programming. In A. Galton, editor,
           Temporal Logics and their applications, pages 197–237. Academic Press,
           London, 1987.

[GHR94]   D. M. Gabbay, I. Hodkinson, and M. Reynolds. *Temporal Logic: Mathe-
           matical Foundations and Computational Aspects*. Clarendon Press-Oxford,
           1994.

[GRP96]   M. Gergatsoulis, P. Rondogiannis, and T. Panayiotopoulos. Disjunctive
           Chronolog. In M. Chacravarty, Y. Guo, and T. Ida, editors, *Proceedings
           of the JICSLP'96 Post-Conference Workshop "Multi-Paradigm Logic Pro-
           gramming"*, pages 129–136, Bonn, 5-6 Sept. 1996.

[Hry93]   T. Hrycej. A temporal extension of Prolog. *The Journal of Logic Program-
           ming*, 15:113–145, 1993.

[Llo87]   J. W. Lloyd. *Foundations of Logic Programming*. Springer-Verlag, 1987.

[LP81]   H. R. Lewis and C. H. Papadimitriou. *Elements of the Theory of Compu-
           tation*. Prentice-Hall, Inc., 1981.

[OM94]   M. A. Orgun and W. Ma. An overview of temporal and modal logic pro-
           gramming. In *Proc. of the First International Conference on Temporal
           Logics (ICTL'94)*, pages 445–479. Springer Verlag, 1994. LNCS No 827.

[Org91]   M. A. Orgun. *Intensional Logic Programming*. PhD thesis, Dept. of Com-
           puter Science, University of Victoria, Canada, December 1991.

[OW92]   M. A. Orgun and W. W. Wadge. Towards a unified theory of intensional
           logic programming. *The Journal of Logic Programming*, 13(4):113–145, Au-
           gust 1992.

[OW93]   M. A. Orgun and W. W. Wadge. Chronolog admits a complete proof pro-
           cedure. In *Proc. of the Sixth International Symposium on Lucid and Inten-
           sional Programming (ISLIP'93)*, pages 120–135, 1993.

[OWD93]   M. A. Orgun, W. W. Wadge, and W. Du. Chronolog($\mathcal{Z}$): Linear-time logic
           programming. In O. Abou-Rabia, C. K. Chang, and W. W. Koczkodaj,
           editors, *Proc. of the fifth International Conference on Computing and In-
           formation*, pages 545–549. IEEE Computer Society Press, 1993.

[PG95]   T. Panayiotopoulos and M. Gergatsoulis. Intelligent information process-
           ing using TRLi. In *6th International Conference and Workshop on Data
           Base and Expert Systems Applications (DEXA' 95), (Workshop Proceed-
           ings) London, UK, 4th-8th September*, pages 494–501, 1995.

[RGP97]   P. Rondogiannis, M. Gergatsoulis, and T. Panayiotopoulos. Theoretical foundations of Branching-Time Logic Programming. 1997. In preparation.

[Ron94]   P. Rondogiannis. *Higher-Order Functional Languages and Intensional Logic*. PhD thesis, Dept. of Computer Science, University of Victoria, Canada, December 1994.

[RW97]   P. Rondogiannis and W. W. Wadge. First-order functional languages and intensional logic. *Journal of Functional Programming*, 1997. (to appear).

[Tao94]   S. Tao. *Indexical Attribute Grammars*. PhD thesis, Dept. of Computer Science, University of Victoria, Canada, 1994.

[WA85]   W. W. Wadge and E. A. Ashcroft. *Lucid, the dataflow Programming Language*. Academic Press, 1985.

[Wad88]   W. W. Wadge. Tense logic programming: A respectable alternative. In *Proc. of the 1988 International Symposium on Lucid and Intensional Programming*, pages 26–32, 1988.

[Yag84]   A. Yaghi. *The Intensional Implementation Technique for Functional Languages*. PhD thesis, Dept. of Computer Science, University of Warwick, Coventry, UK, 1984.

Creating Prototypes for Fast Classification in Dempster-Shafer Clustering

Johan Schubert

Department of Information System Technology,
Division of Command and Control Warfare Technology,
Defence Research Establishment,
SE-172 90 Stockholm, SWEDEN
E-mail: schubert@sto.foa.se

Abstract. We develop a classification method for incoming pieces of evidence in Dempster-Shafer theory. This methodology is based on previous work with clustering and specification of originally nonspecific evidence. This methodology is here put in order for fast classification of future incoming pieces of evidence by comparing them with prototypes representing the clusters, instead of making a full clustering of all evidence. This method has a computational complexity of $O(M \cdot N)$ for each new piece of evidence, where M is the maximum number of subsets and N is the number of prototypes chosen for each subset. That is, a computational complexity independent of the total number of previously arrived pieces of evidence. The parameters M and N are typically fixed and domain dependent in any application.

1 Introduction

In this paper we develop a classification method for incoming pieces of evidence in Dempster-Shafer theory [6]. This methodology is based on earlier work [1] where we investigate a situation where we are reasoning with multiple events that should be handled independently. The approach was, that when we received several pieces of evidence about different and separate events we classified them according to which event they were referring to. That is, we made a partitioning of all sources of evidence into subsets, where all sources of evidence in a given subset concern one of the multiple events, and there is a one-to-one correspondence between the events and the elements of the partitioning. This was done in such a way that the overall conflict across all the subsets was minimal. By a source of evidence we typically mean some kind of sensor generating pieces of evidence, and not an object discovered by the sensor.

The aim of the partitioning is to cluster the sources of evidence in subsets, and select a limited number of prototypical sources of evidence for future classification, in order to speed up computation.

In figure 1 these subsets are denoted by χ_i and the conflict in χ_i is denoted by c_i. Here, thirteen pieces of evidence are partitioned into four subsets. When the number of subsets is uncertain there will also be a "domain conflict" c_0 which is a conflict between the current number of subsets and domain knowledge. The partition is then simply an allocation of all evidence to the different events.

If it is uncertain to which event some evidence is referring we have a problem. It could then be impossible to know directly if two different pieces of evidence are referring to the same event. We do not know if we should put them into the same subset or not.

To solve this problem, we can use the conflict in Dempster's rule when all evidence within a subset are combined, as an indication of whether these pieces of evidence belong together. The higher this conflict is, the less credible that they belong together.

Let us create an additional piece of evidence for each subset where the proposition of this additional evidence states that this is not an "adequate partition". Let the proposition take a value equal to the conflict of the combination within the subset. These new pieces of evidence, one regarding each subset, reason about the partitioning of the original evidence. Just so we do not confuse them with the original evidence, let us call these pieces of evidence "metalevel evidence" and let us say that their combination and the analysis of that combination take place on the "metalevel", figure 1.

In the combination of all metalevel evidence we only receive support stating that this is not an "adequate partition". We may call this support a "metaconflict". The smaller this support is, the more credible the partitioning. Thus, the most credible partitioning is the one that minimizes the metaconflict.

In [2] we further investigated the consequence of transferring different pieces of evidence between the subsets. This was done by observing changes in conflict when we moved a piece of evidence from one subset to another. Such

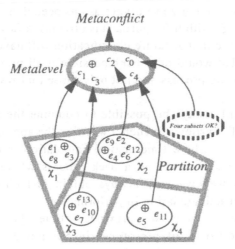

Fig. 1. The conflict in each subset of the partitioning becomes a piece of evidence at the metalevel.

changes in conflict were also interpreted as "metalevel evidence" indicating that the piece of evidence was misplaced.

Based on these items of metalevel evidence we are in this paper able to find potential prototypes as representatives for the subsets.

With the metalevel evidence we can make a partial specification of each original piece of evidence.

With such a partial specification for the potential prototypes we can find a measure of credibility for the correct classification of these prototypes. That credibility is the basis for choosing a fixed number of prototypes among all the potential prototypes as representatives for the subsets.

With the prototypes we will be able to do fast classification of all incoming new items of evidence. The reason why this approach is faster than the previous method in [1] for separating evidence that is mixed up, is that we now make only a simple comparison between an incoming item of evidence and the prototypes of the different subsets. In [1] we did a complete reclustering each time we received some new evidence in order to find a new partitioning, and thus the classification of the newly arrived piece of evidence.

The approach described corresponds in some sense to an idea by T. Denœux [7]. In that article Denœux assumes that a collection of preclassified prototypes are available and compares incoming evidence with the prototypes. He then defines a new item of evidence for each incoming piece of evidence and every prototype. This new item of evidence has a proposition which states that the incoming piece of evidence belongs to the same subset as the prototype. Combining all this new evidence yields the classification.

However, in this paper, we only use prototypes in order to obtain faster classification. These prototypes are not assumed to be preclassified outside of this methodology, instead their classification is derived from a previous clustering process [1]. It is quite possible to cluster and thus classify all incoming evidence without any prototypes as has been done in earlier work [1-5]. While the advantage with this approach is obvious, a fast computation time, there is an disadvantage in that future classification will have a correctness that might by less that what would have been possible if all previous evidence was used in the classification process and not only those that was obtained prior to clustering process.

Of course, it will always be possible to combine the two approaches of this paper and of earlier articles in such a way that the approach in this paper is used in a front process where time-critical calculation is performed and the usual clustering process is made in a back process where we have no time considerations. This way we may always obtain fast classification without suffering any long term degradation in performance.

In Section 2 we review two previous articles [1-2]. They form the foundation for the creation of prototypes in Section 3. A more extensive summary of [1-2] is found in [4]. In Section 4 we develop a method for fast classification based on these prototypes. This method is $O(M \cdot N)$, where M is the maximum number of subsets allowed by an apriori probability distribution regarding the number of subsets, and N is the number of prototypes chosen for each subset. Finally, in Section 5 we draw conclusions.

2 A Summary of Articles [1-2]

In an earlier article [1] we derived a method, within the framework of Dempster-Shafer theory [6], to handle evidence that is weakly specified in the sense that it may not be certain to which of several possible events a proposition is referring. If we receive such evidence about different and separate events and the pieces of evidence are mixed up, we want to classify them according to event.

In this situation it is impossible to directly separate pieces of evidence based only on their proposition. Instead we can use the conflict in Dempster's rule when all pieces of evidence within a subset are combined, as an indication of whether they belong together. The higher this conflict is, the less credible it is that they belong together.

2.1 Separating Nonspecific Evidence

In [1] we established a criterion function of overall conflict called the metaconflict function. With this criterion we can partition the evidence into subsets, each subset representing a separate event. We will use the minimizing of the metaconflict function as the method of partitioning. This method will also handle the situation when the number of events are uncertain.

2.1.1 Metaconflict as a Criterion Function

Let E_i be a domain proposition stating that there are i subsets, $\Theta_E = \{E_0, ..., E_n\}$ a frame of such propositions and $m(E_i)$ the support for proposition E_i.

The metaconflict function can then be defined as:

DEFINITION. *Let the* metaconflict function,

$$Mcf(r, e_1, e_2, ..., e_n) \triangleq 1 - (1 - c_0) \cdot \prod_{i=1}^{r} (1 - c_i),$$

be the conflict against a partitioning of n pieces of evidence into r disjoint subsets χ_i where c_i is the conflict in subset i and c_0 is the conflict between the hypothesis that there are r subsets and our prior belief about the number of subsets.

For a fixed number of subsets a minimum of the metaconflict function can be found by an iterative optimization among partitionings. In each step of the optimization the consequence of transferring a piece of evidence from one subset to another is investigated.

2.2 Specifying Nonspecific Evidence

In [2] we went further by specifying each piece of evidence by observing changes in cluster and domain conflicts if we move a piece of evidence from one subset to another, figure 2.

2.2.1 Evidence About Evidence

A conflict in a subset χ_i is interpreted as a piece of metalevel evidence that there is at least one piece of evidence that does not belong to the subset;

$$m_{\chi_i}(\exists j.e_j \notin \chi_i) = c_i.$$

Note that the proposition always takes a negative form $e_j \notin \chi_i$.

If a piece of evidence e_q in χ_i is taken out from the subset, the conflict c_i in χ_i decreases to c_i^*. This decrease $c_i - c_i^*$ is interpreted as metalevel evidence indicating that e_q does not belong to χ_i, $m_{\Delta\chi_i}(e_q \notin \chi_i)$, and the remaining conflict c_i^* is an other piece of metalevel evidence indicating that there is at least one other piece of evidence e_j, $j \neq q$, that does not belong to $\chi_i - \{e_q\}$,

$$m_{\chi_i - \{e_q\}}\left(\exists j \neq q.e_j \notin \left(\chi_i - \{e_q\}\right)\right) = c_i^*.$$

The unknown bpa, $m_{\Delta\chi_i}(e_q \notin \chi_i)$, is derived by stating that the belief that there is at least one piece of evidence that does not belong to χ_i should be the same irrespective of whether that belief is based on the original metalevel evidence before e_q is taken out from χ_i, or is based on a combination of the other two pieces of metalevel evidence after e_q is taken out from χ_i.

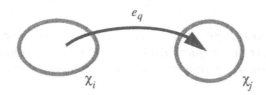

Fig. 2. Moving a piece of evidence changes the conflicts in χ_i and χ_j.

We derive the metalevel evidence that e_q does not belong to χ_i as

$$m_{\Delta\chi_i}(e_q \notin \chi_i) = \frac{c_i - c_i^*}{1 - c_i^*}.$$

We may calculate $m_{\Delta\chi_i}(e_q \notin \chi_i)$ for every χ_i, figure 3.

If e_q after it is taken out from χ_i is brought into another subset χ_k, its conflict will increase from c_k to c_k^*. The increase in conflict is interpreted as if there exists some metalevel evidence indicating that e_q does not belong to $\chi_k + \{e_q\}$, i.e.,

$$\forall k \neq i . m_{\Delta\chi_k}\left(e_q \notin \left(\chi_k + \{e_q\}\right)\right) = \frac{c_k^* - c_k}{1 - c_k}.$$

When we take out a piece of evidence e_q from subset χ_i and move it to some other subset we might see a change in domain conflict. This indicates that the number of subsets is incorrect.

Here, we receive

$$m_{\Delta\chi}(e_q \notin \chi_{n+1}) = \frac{c_0^* - c_0}{1 - c_0}, \; |\chi_i| > 1,$$

$$m_{\Delta\chi}(e_q \notin \chi_i) = \frac{c_0 - c_0^*}{1 - c_0^*}, |\chi_i| = 1, c_0^* < c_0, \text{ and } m_{\Delta\chi}(e_q \in \chi_i) = \frac{c_0}{c_0^*}, |\chi_i| = 1, c_0^* > c_0.$$

2.2.2 Specifying Evidence

We may now make a partial specification of each piece of evidence. We combine all metalevel evidence from different subsets regarding a particular piece of evidence and calculate for each subset the belief and plausibility that this piece of evidence belongs to the subset.

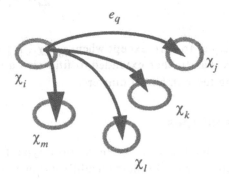

Fig. 3. Test moving e_q to every other subset.

For the case when e_q is in χ_i and $|\chi_i| > 1$ we receive, for example,

$$\forall k \neq n + 1.\text{Bel}(e_q \in \chi_k) = 0$$

and

$$\forall k \neq n + 1.\text{Pls}(e_q \in \chi_k) = \frac{1 - m(e_q \notin \chi_k)}{\displaystyle 1 - \prod_{j=1}^{n+1} m(e_q \notin \chi_j)}.$$

2.3 Finding Usable Evidence

Some pieces of evidence might belong to one of several different subsets. Such an item of evidence is not so useful and should not be allowed to strongly influence a subsequent reasoning process.

If we plan to use a piece of evidence in the reasoning process of some subset, we must find a credibility that it belongs to the subset in question. A piece of evidence that cannot possible belong to a subset has a credibility of zero for that subset, while a piece of evidence which cannot possibly belong to any other subset and is without any support whatsoever against this subset has a credibility of one. That is, the degree to which some piece of evidence can belong to a certain subset and no other, corresponds to the importance it should be allowed to carry in that subset.

The credibility α_j of e_q when e_q is used in χ_j can then be calculated as

$$\alpha_j = [1 - \text{Bel}(e_q \in \chi_i)] \cdot \frac{[\text{Pls}(e_q \in \chi_j)]^2}{\displaystyle\sum_k \text{Pls}(e_q \in \chi_k)}, j \neq i,$$

$$\alpha_i = \text{Bel}(e_q \in \chi_i) + [1 - \text{Bel}(e_q \in \chi_i)] \cdot \frac{[\text{Pls}(e_q \in \chi_i)]^2}{\displaystyle\sum_k \text{Pls}(e_q \in \chi_k)}.$$

Here, $\text{Bel}(e_q \in \chi_i)$ is equal to zero except when $e_q \in \chi_i$, $|\chi_i| = 1$ and $c_0 < c_0^*$.

In [3] this work was further extended to find also a posterior probability distribution regarding the number of clusters.

3 Creating Prototypes

Each piece of evidence is a potential prototype for its most credible subset. When searching for the highest credibility α_j we notice that the term

$$\sum_k \text{Pls}(e_q \in \chi_k)$$

is independent of j. Let us call this term K_q.

When $\text{Bel}(e_q \in \chi_i)$ is equal to zero we may rewrite α_j as

$$\alpha_j = \frac{1}{K_q} \cdot [\text{Pls}(e_q \in \chi_j)]^2 = \frac{1}{K_q} \cdot [1 - m(e_q \notin \chi_j)]^2$$

We see that finding the maximum credibility α_j for e_q is equal to finding the minimum evidence against the proposition that $e_q \in \chi_j$, i.e., maximizing α_j for e_q and all j is equal to minimizing $m(e_q \notin \chi_j)$ for e_q and all j.

If $\text{Bel}(e_q \in \chi_i) \neq 0$ we have found earlier that $\text{Pls}(e_q \in \chi_i) = 1$, [2] Section III, and thus $\alpha_i > \alpha_j$ for $j \neq i$.

We have

$$\alpha_i = \text{Bel}(e_q \in \chi_i) + [1 - \text{Bel}(e_q \in \chi_i)] \cdot \frac{[\text{Pls}(e_q \in \chi_i)]^2}{\sum_k \text{Pls}(e_q \in \chi_k)}$$

$$= \text{Bel}(e_q \in \chi_i) + [1 - \text{Bel}(e_q \in \chi_i)] \cdot \frac{1}{K_q}$$

with $K_q \geq 1$ because $\text{Pls}(e_q \in \chi_i) = 1$.

Since

$$\text{Bel}(e_q \in \chi_i) = m(e_q \in \chi_i) + [1 - m(e_q \in \chi_i)] \cdot \prod_{\chi_j \in \chi^{-i}} m(e_q \notin \chi_j),$$

(see [2] Appendix II.A.3), and

$$\prod_{\chi_j \in \chi^{-i}} m(e_q \notin \chi_j) \leq 1$$

where $\chi^{-i} = \chi - \{\chi_i\}$ and $\chi = \{\chi_1, ..., \chi_n\}$, we find that the maximum value of α_i is obtained when $m(e_q \in \chi_i)$ is maximal. We have $\alpha_i = 1$ when $m(e_q \in \chi_i) = 1$.

A simple decision rule is then:

For every piece of evidence e_q

1. For e_q and all j find the maximum $m(e_q \in \chi_j)$. If the maximum $m(e_q \in \chi_j) \neq 0$ then e_q is a potential prototype for χ_j.

2. However, if the maximum $m(e_q \in \chi_j) = 0$ then for e_q and all j find the minimum $m(e_q \notin \chi_j)$. Now, we have e_q as a potential prototype for χ_j.

However, we must use the credibility itself when determining which pieces of evidence among the potential prototypes for a certain subset will actually by chosen as one of the N prototypes for that subset. While the above approach chooses the best subset χ_j for each piece of evidence by maximizing $m(e_q \in \chi_j)$ or minimizing $m(e_q \notin \chi_j)$, it is still possible that the evidence might be quite useless as a prototype since it could almost have been a potential prototype for some other subset. By ranking the potential prototypes within each subset according to credibility, we are able to find the most appropriate ones.

Note that in [2] we stated that in a subsequent reasoning process within a subset we could use each piece of evidence that had a credibility above zero for that subset, provided that we discounted it to its credibility.

We could then use almost every piece of evidence within each subset and we would get small contributions to the characteristics of the subset from pieces of evidence that most likely does not belong to the subset. A discounted piece of evidence might have something to offer in the reasoning process of a subset even when it is discounted to its credibility, but we do not know for sure that it belongs to the subset.

Thus, a substantially discounted piece of evidence hardly makes an ideal candidate as a prototype for a subset. From the prototypes we want to find the characteristics of the subset. That is something that cannot be guaranteed if we do not know with a high credibility that the piece of evidence belongs to the subset.

Also, the computational complexity would become exponential when making future classification of incoming pieces of evidence, since the number of focal elements in the result of the final combination of all pieces of evidence within the subset, would grow exponentially with the number of pieces of evidence within the subset. That means that we would get an exponential-time algorithm when classifying future incoming pieces of evidence even if all present pieces of evidence within the subset where combined in advance.

Instead we choose the N prototypes with the highest credibility for the subset. They are judged to be the best representatives of the characteristics of the subset. Our future classification will be based on a comparison between them and the new incoming piece of evidence.

A second decision rule can then be formulated as:
For every subset

1. Of the potential prototypes allocated for a subset, choose the N prototypes with highest credibility for that subsets as the actual prototypes.
2. Disregard the other potential prototypes for that subset.

By first applying the first decision rule for all pieces of evidence and then the second decision rule for every subset we are able to find N different prototypes for each subset provided, of course, that there are at least N potential prototypes for each subset.

Finally, we combine all prototypes within each subset into one new basic probability assignment. This way, each subset will now contain only one piece of evidence. While doing that, we also make a note of the conflict c_j received in that combination. We will need it in the fast classification process.

4 Fast Classification

Now, given that we have all the prototypes for each subset, we can make fast classification of future incoming pieces of evidence. We will use the derived items of metalevel evidence $m_{\Delta\chi_i}(e_q \notin \chi_i)$.

If the evidence for e_q against every subset is very high we will not classify e_q as belonging to any of the subsets χ_j, $j \leq n$. We will use a rejection rule if the best subset for e_q is no better than it would be to create an additional subset χ_{n+1}, and take the penalty for that from an increase in domain conflict.

Our rejection rule is:

Reject e_q if the minimum for all j of $m_{\Delta\chi_j}(e_q \notin \chi_j)$ is larger than $m_{\Delta\chi}(e_q \notin \chi_{n+1})$ where

$$m_{\Delta\chi_j}(e_q \notin \chi_j) = \frac{c_j^* - c_j}{1 - c_j} \text{ and } m_{\Delta\chi}(e_q \notin \chi_{n+1}) = \frac{c_0^* - c_0}{1 - c_0}.$$

With

$$c_0 = \sum_{i \neq n} m(E_i) \text{ and } c_0^* = \sum_{i \neq n+1} m(E_i),$$

$m(E_i)$ being the prior support given to the fact there are i subsets, we have

$$m_{\Delta\chi}(e_q \notin \chi_{n+1}) = \frac{c_0^* - c_0}{1 - c_0} = \frac{\sum\limits_{i \neq n+1} m(E_i) - \sum\limits_{i \neq n} m(E_i)}{1 - \sum\limits_{i \neq n} m(E_i)} = \frac{m(E_{n+1}) - m(E_n)}{1 - \sum\limits_{i \neq n} m(E_i)},$$

where $m(E_{n+1})$ is always greater than $m(E_n)$, [1].

If e_q is not rejected by this rule, then e_q is classified as belonging to the subset χ_j for which $m_{\Delta\chi_j}(e_q \notin \chi_j)$ is minimal for all j.

All it takes to find c_j^* is one combination of Dempster's rule for each cluster between the incoming piece of evidence and the already made combination of the prototypes of that cluster. We already have c_j for every subset as well as $m(E_i)$ for all i.

If a fixed maximum number of prototypes N are used for each cluster and an apriori probability distribution regarding the number of clusters has some maximum number of subsets M, i.e., $m(E_i) = 0$ if $i > M$, then the classification can always be done in time $O(M \cdot N)$. That is, independent of the total number of pieces of evidence in the previous clustering process.

5 Conclusions

In this paper we have shown how to create a fixed number of prototypes for each subset. They, and only they, will be the representatives of that subset. If there is also a limit on the maximum number of clusters through the apriori distribution regarding the number of subsets, then we can do a fast classification of all future incoming pieces of evidence. The performance of this methodology will obviously vary with the maximum number of prototypes N. If N is too large the computation time will become too long although it will grow only linearly in N. If on the other hand N it is too small, then the classification process will make too many classification errors. The actual choice of the parameter N will be domain dependent and has to be found for each application separately.

References

1. J. Schubert, On nonspecific evidence, *Int. J. Intell. Syst.* 8(6) (1993) 711-725.
2. J. Schubert, Specifying nonspecific evidence, *Int. J. Intell. Syst.* 11(8) (1996) 525-563.
3. J. Schubert, Finding a posterior domain probability distribution by specifying nonspecific evidence, *Int. J. Uncertainty, Fuzziness and Knowledge-Based Syst.* 3(2) (1995) 163-185.
4. J. Schubert, Cluster-based specification techniques in Dempster-Shafer theory, in *Symbolic and Quantitative Approaches to Reasoning and Uncertainty*, C. Froidevaux and J. Kohlas (Eds.), *Proc. European Conf. Symbolic and Quantitative Approaches to Reasoning and Uncertainty*, Springer-Verlag (LNAI 946), Berlin, 1995, pp. 395-404.
5. J. Schubert, Cluster-based specification techniques in Dempster-Shafer theory for an evidential intelligence analysis of multiple target tracks, Ph.D. thesis, TRITA-NA-9410, Royal Institute of Technology, Stockholm, Sweden, 1994, ISBN 91-7170-801-4.
6. G. Shafer, *A Mathematical Theory of Evidence*. Princeton University Press, Princeton, 1976.
7. T. Denœux, A *k*-nearest neighbor classification rule based on Dempster-Shafer theory, *IEEE Trans. Syst. Man Cyber.* 25(5) (1995) 804-813.

Probabilistic Default Logic Based on Irrelevance and Relevance Assumptions

Gerhard Schurz

Inst. f. Philosophie, Abtlg. Logik und Wissenschaftstheorie, Universität Salzburg*

Abstract. This paper embeds default logic into an extension of Adams' probability logic, called the system $\mathbf{P_{\varepsilon}^{+}DP}$. Default reasoning is furnished with two mechanisms: one generates (ir)relevance assumptions, and the other propagates lower probability bounds. Together both mechanisms make default reasoning probabilistically reliable. There is an exact correspondence between Poole-extensions and $\mathbf{P_{\varepsilon}^{+}DP}$-extensions. The procedure for $\mathbf{P_{\varepsilon}^{+}DP}$-entailment is comparable in complexity with Poole's procedure.

1 Introduction and Motivation

Default reasoning is reasoning from *uncertain laws*, like "Birds *normally* can fly". We formalize these laws as $B \Rightarrow F$, where $\Rightarrow$ stands for uncertain implication and B, F are associated with an invisible variable x (Bx, Fx). Uncertain laws have two characteristics. First, they have *exceptions* (e.g., penguins don't fly), and second, we do not know the conditional probability of flying animals among birds. We just assume that this probability is *high* – otherwise we would not be justified in calling this the *normal* case. We call $p(B/A)$ the conditional probability *associated* with $A \Rightarrow B$.

Default reasoning starts from the *intuitive* principle that one may detach Ga from $F \Rightarrow G$ and Fa as long as $\neg Ga$ is *not* 'implied' by one's knowledge base. Hence the *nonmonotonic* character of default reasoning. But this principle is *ambiguous*. As a result, various *different* systems have been developed (cf. [4]). How are these systems *justified?* In this respect, clear criteria are often missing. The major criterion suggested here is *reliability*. For, in contrast to deductive inferences, default inferences are *uncertain*: they do not preserve truth (from premises to conclusion) for *all* of their individual instances. Still, they should at least preserve truth for *most* of their individual instances. Since uncertain laws are true iff their associated conditional probabilities are high, we can explicate the requirement of reliability as follows:

– *(Reliability:)* If the conditional probabilities associated with the uncertain laws used as premises are high, then the conditional probability of the conclusion given all factual knowledge is high, too.

---

* I am indebted to Ernest Adams for various help.

The probability concept 'p(–)' which underlies this approach is *statistical* (as opposed to subjective) probability: this is essential for the *tie* between reliability and a high predictive success rate. It is important that the premises need not be identical with the uncertain laws contained in the knowledge base, but may be derived from them with help of probabilistic *default assumptions* (irrelevance- and relevance assumptions).

Standard default logics do not satisfy the requirement of reliability. E.g., this is demonstrated by the *Conjunction Problem*: given a set of uncertain laws with the same antecedent $A \Rightarrow B_1, \ldots, A \Rightarrow B_n$ (and without conflicting laws), all standard default logics will reason from A to the conjunction $\bigwedge_{1 \leq i \leq n} B_i$. But the high probabilities of the conjuncts may multiply into a very improbable conjunction. – This approach tries to achieve reliability by superimposing two mechanisms on the inferential reasoning of default logic:

- A mechanism of generating the *minimal* probabilistic default assumptions which must be made in order to derive the conclusion safely – this is called the *assumption generation* mechanism.
- A mechanism of propagating approximately tight lower probability bounds from the premises to the conclusion – this is called the *lower bound propagation* mechanism.

Assumption generation provides *transparency* to the user, and lower bound propagation supplies *risk information*. Both together guarantee reliability. For they ensure that given the premises of the inference are true, then the conclusion's probability (given the facts are true) *must be* greater or equal than a certain value coming along with the conclusion. The decision whether or not this value is high enough to accept the conclusion is left to the user instead of being apriorily decided by some system-inbuilt acceptability threshold.

With help of the assumption generation mechanism we reduce default inferences from a knowledge base $\mathcal{K}$ to probabilistic entailments from $\mathcal{K}$ enriched by the generated default assumptions *Ass*. Probabilistic entailment has been developed by Adams [1, 2]. It is *weaker* than default inference and monotonic w.r.t. the law premises. It preserves high probability for *all* statistical probability distributions. In contrast, standard default inference systems preserve high probability only for certain probability distributions (cf. §3.1). These hidden restrictions become transparent in *Ass*.

So far we have motivated the advantage of probabilistic default reasoning as against standard default reasoning. Its advantage as compared to straightforward probabilistic reasoning in Bayesian networks [12, 13] is that it enables us to reason efficiently in a situation of *partial* and even *minimal probabilistic information* – namely by the mechanism of generating default assumptions. This mechanism is also the main advantage of probabilistic default reasoning as compared to probabilistic 'anytime' deduction systems [7].[2]

---

[2] Also rules and propagation mechanism of these systems are very different; cf. §2.3.

2   Probability Logic: The System $\mathbf{P}_\epsilon^+$

2.1 The Principle of Total Evidence. Our *basic* formal language $BLang$ is a 'first order variant' of the *propositional* language (with $\neg, \wedge, \vee, \rightarrow$ for material implication, $\top$ for Verum and $\bot$ for Falsum). Its formulas are built up from a *finite* set of monadic predicates $F, G \ldots$, a *single* individual constant a and a *single* variable x (by standard formation rules). We omit the variable x in open formulas, hence $F, G \ldots$ abbreviate $Fx, Gx \ldots$; $A, B, \ldots$ (possibly indexed) range over open formulas of $BLang$, and $Aa, Ba, \ldots$ are the associated closed formulas with x replaced by a. If these closed formulas are known, they are called *facts*; if they are derived, they are called singular *conclusions*. Uncertain laws, in short: *laws*, have the form $A \Rightarrow B$.[3] The symbol $\Rightarrow$ does not belong to the basic but to the *extended* formal language $ELang \supset BLang$. $ELang$ includes all propositional combinations of laws, although our reasoning procedure will involve only laws and their negations. $ELang$ *excludes* nested laws (like $A \Rightarrow (B \Rightarrow C)$). A knowledge base $\mathcal{K}$ is a pair $\langle \mathcal{L}, \mathcal{F} \rangle$ with $\mathcal{L}, \mathcal{F}$ *finite* sets of laws and facts, respectively. $\vdash$ denotes classical propositional inference. $\vdash_{\epsilon+}$ stands for (monotonic) probabilistic entailment between laws, and $\vdash\!\sim$ stands for indexed notions of nonmonotonic inference. $r, q, \ldots$ range over real numbers, $\delta, \epsilon, \ldots$ over 'small' ones; ':=' means 'identity by definition'.

We will assume that all laws $L \in \mathcal{L}$ are associated with (reasonably determined) lower probability bounds $b(L)$. We write $A \stackrel{r}{\Rightarrow} B$ to indicate that the lower bound of $p(B/A)$ is r. $\mathcal{K} \vdash\!\sim_{\epsilon+} [Ca, r]$ abbreviates that Ca is ϵ^+-entailed by $\mathcal{K}$ with lower bound r. The ϵ^+-entailment of closed formulas from laws and facts is reduced to the ϵ^+-entailment of laws from laws by the *principle of total evidence* which goes back to Carnap [5, p.211] (cf. [13, ch.10.2.2],[3, p.149]): Let $\mathcal{F}x$ denote the conjunction of all facts with the constant a replaced by x:

$$\textit{Principle of total evidence: } \langle \mathcal{L}, \mathcal{F} \rangle \vdash\!\sim_{\epsilon+} [Ca, r] \text{ iff } \mathcal{L} \vdash_{\epsilon+} \mathcal{F}x \stackrel{r}{\Rightarrow} C.$$

In other words, the subjective probability which $\langle \mathcal{L}, \mathcal{F} \rangle$ conveys to a singular statement Ca is identified with the statistical probability of C conditional to all facts (with a replaced by x) which is ϵ^+-entailed by $\mathcal{L}$. This principle applies in the same way to the system $\mathbf{P}_\epsilon^+\mathbf{DP}$ (§3) where $\mathcal{L}$ gets enriched by default assumptions.

Probability semantics for unrestricted 1st order languages involves various complications [3]. All work on (Adams') probability logic has been carried out in the propositional language. For this reason we work in an essentially propositional framework: our basic language, if restricted to open formulas, is *isomorphic* to a purely propositional language. There is only a difference of *interpretation*. Take the Bird (B) - Canfly (F) - Tweety (a) example. Statistical probabilities apply to *open formulas* (extensionally to classes, like 'being a bird'), while subjective probabilities apply to closed formulas (extensionally to individual states of affairs, like 'Tweety is a bird'). Statistical probability is defined on the par-

---

[3] We do not introduce a separate set of deterministic laws in $\mathcal{K}$. The simplest way to account for them is to treat their ground instances $Aa \rightarrow Ba$ as facts.

tition of *possible states* of a variable individual: $Bx \wedge Fx, Bx \wedge \neg Fx, \neg Bx \wedge Fx, \neg Bx \wedge \neg Fx$. In contrast, subjective probability is defined on the partition of *possible worlds*: $Ba \wedge Fa, Ba \wedge \neg Fa, \neg Ba \wedge Fa, \neg Ba \wedge \neg Fa$. Possible world semantics applies in the same way, except that we have possible states instead of possible worlds, and our probabilities are the frequencies (or frequency limits) of these states in the REAL world.

2.2 Semantics and Calculus of the System $\mathbf{P}_\epsilon^+$.

Some more terminology. $\mathcal{L}$ ranges over finite sets of laws $L \in \mathcal{L}$, and $\mathcal{N}$ ranges over finite sets of *negated laws* $N \in \mathcal{N}$; i.e. N is of the form $\neg(A \Rightarrow B)$. Ω is the finite set of truth assignments $(u, v \ldots)$ to the atomic open formulas of *BLang*; they are also called *possible states*. For p a probability function over Ω and $L := (A \Rightarrow B)$, $p(L) := p(B/A)$ is the (conditional) probability and $u(L) := 1 - p(B/A)$ the (conditional) *uncertainty* associated with $A \Rightarrow B$. We put $p(B/A) = 1$ if $p(A) = 0$. Observe that $p(A) = p(\top \Rightarrow A)$. For reasons of simplicity we identify the probability and the uncertainty of a negated law with that of the law itself: $p(\neg(A \Rightarrow B)) := p(A \Rightarrow B) = p(B/A)$; $u(\neg(A \Rightarrow B)) := u(A \Rightarrow B) = 1 - p(B/A)$. The difference is that the probability of a negated law must be *sufficiently low*. We associate the negated laws N with *upper* bounds $b(N)$ of their probabilities. Thus $\neg(A \overset{r}{\Rightarrow} B)$ means $p(B/A) \leq r$. While the lower probability bounds of laws should be high ($>> 0.5$), the upper bounds of negated laws should be small or at least moderate (≤ 0.5), in order to obtain informative bounds of predicted conclusions.

The basis of our default system are $\mathbf{P}_\epsilon^+$-inferences of the form $\mathcal{L} \cup \mathcal{N} \vdash_{\epsilon^+} L$, having unnegated or negated laws as premises and a(n) (unnegated) law as conclusion. We call them *simple* $\mathbf{P}_\epsilon^+$-inferences. Since the $\mathbf{P}_\epsilon^+$-validity of arbitrary open formulas in *ELang* is reducible to the $\mathbf{P}_\epsilon^+$-validity of simple inferences,[4] there is no essential loss in restricting to simple $\mathbf{P}_\epsilon^+$-inferences. A probability function p over Ω is called *proper* for $\mathcal{L}$ [$\mathcal{N}$] if $p(A) > 0$ holds for all $A \Rightarrow B \in \mathcal{L}$ [or $\neg(A \Rightarrow B) \in \mathcal{N}$ resp.] with logically consistent antecedent A. $\Pi(\mathcal{L}, \mathcal{N})$ denotes the set of *all* probability functions over Ω which are *proper* for $\mathcal{L}$ and $\mathcal{N}$. $\Pi(\mathcal{L}, L)$ abbreviates $\Pi(\mathcal{L} \cup \{L\})$. Based on earlier work (e.g. [1]), Adams [2] has developed a probabilistic semantics and a calculus for *ELang*-inferences. Translated into our framework, it goes as follows.[5]

Definition 1 (ϵ^+-entailment). $\mathcal{L} \cup \mathcal{N}$ ϵ^+-*entails* L iff for every (small) $\delta > 0$ there exists an $\epsilon > 0$ such that for each probability function $p \in \Pi(\mathcal{L}, \mathcal{N}, L)$: if $\forall L \in \mathcal{L}: p(L) \geq 1 - \epsilon$ and $\forall N \in \mathcal{N}: p(N) < 1 - \delta$, then $p(L) \geq 1 - \delta$.

---

[4] By prop. logic, every open *ELang*-formula is equivalent with a conjunction of positive implications of the form $\bigwedge(\mathcal{L} \cup \mathcal{N}) \to L$.

[5] There are two differences of our system as compared to [2]: 1.) Adams [2] states his definitions and theorems for *positive* inferences, which go from sets to disjunctions of (unnegated) laws. By prop. logic, the positive inference $\mathcal{L} \vdash_{\epsilon^+} \bigvee \mathcal{L}^* \vee L$ is covalid with the simple inference $\mathcal{L} \cup \{\neg L \mid L \in \mathcal{L}^*\} \vdash_{\epsilon^+} L$. With help of this transformation, Adams' results are easily translated into our framework. 2.) Adams [2] states his theorems and proof sketches for the slightly weaker system $\mathbf{P}^+$ explained at the end of §2.2.

This definition of *infinitesimal* probabilistic entailment implies that the conclusion probability can be forced to get arbitrarily close to 1 by making the probabilities of the premise laws sufficiently close to 1 while leaving the probabilities of the negated premise laws below some fixed value.[6] But it does not imply any nontrivial lower bound for the conclusion probability if the law probabilities are high but nonextreme (say ≥ 0.95). So, reliability in our sense is not per se guaranteed by infinitesimal probabilistic entailment. But fortunately, infinitesimal ϵ^+-entailment is equivalent with the following inequality for nonextreme probability values: the conclusion's uncertainty is always smaller or equal than the sum of the positive premises' uncertainties divided through the product of the negative premises' uncertainties (see th. 1.2.; of course it is always ≤ 1).

The calculus for ϵ^+-inference ($\vdash_{\epsilon^+}$) is formulated as a natural deduction (sequent) calculus. $OELang$ is the set of open $ELang$-formulas including $\top, \bot$.

The calculus $\mathbf{P}_\epsilon^+$

(Prop)   Propositional axioms and rules for $\vdash_{\epsilon^+}$ among $OELang$-formulas

(CC) $A \Rightarrow B, A \wedge B \Rightarrow C \vdash_{\epsilon^+} A \Rightarrow C$ (Cautious Cut)

(Or) $A \Rightarrow C, B \Rightarrow C \vdash_{\epsilon^+} (A \vee B) \Rightarrow C$ (Disjunction)

(WRM) $\neg(A \Rightarrow \neg B), A \Rightarrow C \vdash_{\epsilon^+} (A \wedge B) \Rightarrow C$ (Weak Rational Monotonicity)

(ϵEFQ) $\not\vdash \neg A, / A \Rightarrow \bot \vdash_{\epsilon^+} \alpha$ (Ex ϵ-Falso Quodlibet)

(SC) $\vdash A \to B / \vdash_{\epsilon^+} A \Rightarrow B$ (Supraclassicality)

'/' indicates a rule. Although $\vdash_{\epsilon^+}$ is monotonic, $\Rightarrow$ is nonmonotonic, whence $\mathrel{\vert\!\sim}_{\epsilon^+}$ is *semi-monotonic*: monotonic w.r.t. the law premises, but nonmonotonic w.r.t. fact-premises. E.g., $\langle \{A \Rightarrow B\}, \{Aa\} \rangle \mathrel{\vert\!\sim}_{\epsilon^+} Ba$, but $\langle \{A \Rightarrow B\}, \{Aa, Ca\} \rangle \mathrel{\not\vert\!\sim}_{\epsilon^+} Ba$, because $A \Rightarrow B \not\vdash_{\epsilon^+} (A \wedge C) \Rightarrow B$. The calculus is restricted to *open* formulas, because we intend a statistical probability interpretation;[7] closed $BLang$-formulas are handled by the total evidence principle. The restriction of $A, B, \ldots$ to (open) $BLang$-formulas prevents the derivability of nested conditionals. The fundamental $\mathbf{P}_\epsilon^+$-theorem is this:[8]

Theorem 1 (System $\mathbf{P}_\epsilon^+$). *For every $\mathcal{L}, \mathcal{N}$ and L, the following conditions are equivalent:*

(1) $\mathcal{L} \cup \mathcal{N}$ ϵ^+-entails L.

(2) For every $p \in \Pi(\mathcal{L}, \mathcal{N}, L): u(L) \leq min\left(\dfrac{\sum_{L' \in \mathcal{L}} u(L')}{\prod_{N \in \mathcal{N}} u(N)}, 1\right).$

(3) $\mathcal{L} \cup \mathcal{N} \vdash_{\epsilon^+} L$ is derivable in the calculus $\mathbf{P}_\epsilon^+$.

In th. 1.2 we assume $min(r, \frac{q}{0}) := r$. To cover empty $\mathcal{L}$'s or $\mathcal{N}$'s we stipulate $\sum_{L \in \emptyset} u(L) := 0$ and $\prod_{N \in \emptyset} u(N) := 1$. Th. 1.2 remains intact if we replace the unknown uncertainty values $u(L')$ and $u(N)$ by the uncertainty bounds

---

[6] The restriction to proper probability functions excludes the possibility of making $p(A \Rightarrow B)$ 1 by making $p(A)$ 0, provided A is not a contradiction.

[7] On the same reason, ϵ^+-entailment is noncompact and thus restricted to *finite* premise sets [1, p.52].

[8] A proof sketch for the system $\mathbf{P}^+$ (cf. fn.s 5,9) is found in [2], a full proof in [18]. The proof for the system $\mathbf{P}_\epsilon^+$ reduces to the proof for $\mathbf{P}^+$ by easy means.

$1 - b(L')[\geq u(L')]$ and $1 - b(N) \leq [u(N)]$; in this form it will figure as the basis of our propagation mechanism. A fourth equivalent representation of ϵ^+-entailment being omitted here is in terms of normal world (ranked model) semantics [2, 10, 6, 9]. So the concept of probabilistic entailment unifies four different perspectives: (1) *infinitesimal probabilistic entailment* (th.1.1), (2) *nonextreme probability propagation* (th.1.2), (3) *logical calculus* (th.1.3) and (4) *normal world semantics*. While Pearl et al. [13, ch.10],[8, 9] concentrate on the first and fourth perspective, this approach focuses on the second and third perspective.

(WRM) is weaker than full rational monotony (RM) of Lehmann and Magidor [10, p.18]. (RM) allows to derive $A \wedge B \Rightarrow C$ from $A \Rightarrow C$ if $A \Rightarrow \neg B$ is *not* derivable from the knowledge base, while (WRM) allows this inference only if $\neg(A \Rightarrow \neg B)$ *is* derivable from the knowledge base. (RM) is nonmonotonic and probabilistically unsafe, while (WRM) is monotonic and probabilistically safe. An important consequence of (WRM) is the cautious monotonicity (CM): $A \Rightarrow B, A \Rightarrow C \vdash_{\epsilon+} A \wedge B \Rightarrow C$; it holds since (Neg): $A \Rightarrow B \vdash_{\epsilon+} \neg(A \Rightarrow \neg B)$ is a $\mathbf{P}_\epsilon^+$-theorem. Further important $\mathbf{P}_\epsilon^+$-theorems are (And): $A \Rightarrow B, A \Rightarrow C \vdash_{\epsilon+} A \Rightarrow B \wedge C$, (LLE, left logical equivalence): $A \dashv\vdash B \,/\, A \Rightarrow C \vdash_\epsilon B \Rightarrow C$, (RW, right weakening): $B \vdash C \,/\, A \Rightarrow B \vdash_\epsilon A \Rightarrow C$ (cf. Adams [1, p.60-65]).

If we restrict $\mathbf{P}_\epsilon^+$ to inferences among unnegated laws and replace (WRM) by (CM), we obtain the system $\mathbf{P}_\epsilon$ (and inference relation $\vdash_\epsilon$) of Adams [1] which also underlies the work of Pearl et al [13, ch.10], [9]. The term 'ϵ-entailment' was introduced in [13, p.481].[9] Important is the notion of ϵ-*consistency*: $\mathcal{L}$ is ϵ-consistent iff for all $\epsilon > 0$ there exists a probability function $p \in \Pi(\mathcal{L})$ such that $p(L) \geq 1 - \epsilon$ for all $L \in \mathcal{L}$. It is well-known that $\mathcal{L}$ ϵ-entails $A \Rightarrow B$ iff $\mathcal{L} \cup \{A \Rightarrow \neg B\}$ is ϵ-inconsistent [1, p.58].

2.3 Tightness of Bounds. The computed lower bounds should not only be *safe*, i.e. be satisfied for all probability distributions. They should also be *tight*, i.e. there should not exist greater lower bounds which are also safe. But there is a problem here. Even if the lower bound (computed by a certain lower bound function f) is tight for the *most general* instances of an inference step (i.e., for instances where different schematic letters are replaced by different atomic formulas), it will *not* be tight for *all* instances of this inference step. For example, no safe lower bound function f can be tight for special (CC)-instances where $A \vdash C$ holds, for in this case the greatest lower bound for the conclusion law $A \Rightarrow C$ is 1. The best what can be achieved is that the lower bound function f is tight for the most general instances of an inference step. In [7, p.109], this

---

[9] Two systems which are slightly weaker than $\mathbf{P}_\epsilon$ and $\mathbf{P}_\epsilon^+$, respectively, are $\mathbf{P}$ and $\mathbf{P}^+$. Semantically, they differ from the corresponding ϵ-systems by admitting *improper* probability functions. Syntactically, $\mathbf{P}_\epsilon$ arises from $\mathbf{P}$ by replacing (ϵEFQ) by (pEFQ): $\top \Rightarrow \bot \vdash_p \alpha$, and $\mathbf{P}^+$ arises from $\mathbf{P}$ by adding (WRM). The system $\mathbf{P}$ is equivalent with the calculus of preferential entailment after [10, p.5f]. The system $\mathbf{P}^+$ is used in Adams [2] and is similar to Delgrande's conditional logic [6] (cf.[18] on the details).

property is called *quasi-tightness*.[10]

In contrast to the rules of [7], not all of our inference axioms are quasi-tight, if furnished with the lower bound function of th.1.2. (WRM) and (SC) [as well as (And), (LLE), (RW)] are indeed quasi-tight, but (CC) and (Or) are not: it can be proved that the quasi-tight bound for (CC) is $1 - \epsilon_1 - \epsilon_2 + \epsilon_1\epsilon_2$ (cf. [7, p.103], rule vii; the ϵ's are the upper uncertainty bounds of the two premises), and the quasi-tight bound for (Or) is $1 - \frac{\epsilon_1 + \epsilon_2 - 2\epsilon_1\epsilon_2}{1 - \epsilon_1\epsilon_2}$. In both cases, the loss is in the *second power* of the ϵ's and thus small.

We argue twofold. *First*, quasi-tightness is *not* always preserved through the concatenation of inference steps. If the lower bounds are computed stepwise along the inference chain, as it is done in [7], then the conclusion's lower bound may fail to be quasi-tight even if all inference steps have been quasi-tight. *Second*, the small loss of the general lower bound inequality of th.1.2 is compensated by a significant advantage. Based on it we may compute the lower bound of the entire inference in *one step*, *independently* from the way in which the conclusion has been proved from the premises.[11] It is easy to prove that for inferences involving only unnegated laws, the one-step calculation produces better (or at least equally good) bounds than the stepwise calculation (roughly speaking, because law premises may be used several times). On the other hand, it is similarly easily to show that for (WRM)-steps the stepwise calculation yields better (or at least equally good) bounds than the one-step calculation.

Our probabilistic default reasoning system exploits these facts. It performs all reasoning steps involving negated laws first. This leads to the so-called *updated laws*, which ϵ-imply the conclusion law $\mathcal{F}x \Rightarrow C$. We calculate the lower bounds for the updated laws by the stepwise method and then apply the one-step calculation. By doing this, we obtain an optimal approximation to tightness, as far as the arrangement of inference steps is concerned. What remains to be achieved is to prove the conclusion from a premise set which is as *small* as possible, and which has lower bounds as *high* as possible. This is described in §3.4.

3 Default Logic Based on $\mathbf{P}_\epsilon^+$

3.1 Irrelevance Assumptions. The calculus $\mathbf{P}_\epsilon^+$ is too weak as a basis for default reasoning. It does not sanction the principle of *default detachment* which is fundamental for nonmonotonic reasoning: to infer Ba from $Aa \in \mathcal{F}$ and $(A \Rightarrow B) \in \mathcal{L}$ as long as $\mathcal{K}$ does not entail $\neg Ba$. For example, $\mathcal{K} = \langle \{\text{Bird} \Rightarrow \text{CanFly}\}, \{\text{Bird(tweety), Female(tweety)}\} \rangle$ *default-implies* CanFly(tweety). But $\mathcal{K}$ does not ϵ-entail CanFly(tweety) because $A \Rightarrow B$ does not ϵ-entail $A \wedge C \Rightarrow B$. In order to give a probabilistic foundation of default reasoning, a probabilistic account of these default detachments is needed.

In the *maximal entropy* approach [13, ch.10.2.3], this is done by an additional constraint on the probability functions: they should have maximal entropy. In

---

[10] The definition in [7, p.109] is equivalent with our formulation, except that the authors state their definition for intervals, while we are only concerned with lower bounds.

[11] On this reason the lower bounds are *not* part of our axioms or rules.

the *system Z* [9] and the equivalent 'rational closure' of [10] it is achieved by an additional constraint on the ranked models: they should minimize the rank (i.e., the abnormality) of worlds. Both approaches give *maximal* justifications of default reasoning in the sense that they assume the world's entropy or normality to be as great as possible.

This approach intends to give a *minimal* justification of default detachment, and it does this by additional syntactic assumptions. In order to derive CanFly(a) from Bird $\Rightarrow$ CanFly and Bird(a) $\wedge$ Female(a) in a probabilistically safe way we only have to assume that Female(a) is irrelevant for Canfly(a), given Bird(a), and need not take care about the entropy or normality of 'the rest of the world'. The irrelevance assumption means probabilistically that p(CanFly/Bird) and p(CanFly/ Bird $\wedge$ Female) are (approximately) equal.

So, the principle which lies behind default detachment is to assume that the remainder facts are *irrelevant* for the conclusion as long as the knowledge base does not tell the contrary.[12] Whenever a default reasoning system makes a default detachment, thereby inferring Ba from $(A \Rightarrow B) \in \mathcal{L}$ and Aa, it has (implicitly) assumed that the (remainder) facts in $\mathcal{F}$ are probabilistically irrelevant for $A \Rightarrow B$. We denote this irrelevance assumption by $Irr(\mathcal{F}x : A \Rightarrow B)$ and say that it has been *generated* by the corresponding default detachment. Since we only deal with lower and not with upper probability bounds, we can 'minimalize' the irrelevance assumptions to *negative* irrelevance assumptions, saying that the remainder facts do not lower the probability of B given A. Hence $Irr(\mathcal{F}x : A \Rightarrow B)$ means $p(B/\mathcal{F}x \wedge A) \geq p(B/A)$.

We include irrelevance assumptions into our extended language *ELang* and add the following axiom to $\mathbf{P}_\epsilon^+$:

(Irr) $A \Rightarrow B, Irr(C : A \Rightarrow B) \vdash_{\epsilon^+} C \wedge A \Rightarrow B$ (Irrelevance updating)

Whenever $(A \overset{r}{\Rightarrow} B) \in \mathcal{L}$ and $Irr(\mathcal{F}x : A \Rightarrow B)$ has been generated, the so-called *updated law* $(\mathcal{F}x \wedge A) \overset{r}{\Rightarrow} B$ is generated. Hence in the rule (Irr), the *same* lower bound is transferred from the law to the updated law. For better readability, conjuncts of $\mathcal{F}x$ which are already contained as conjuncts in A get cancelled in the irrelevance assumption; e.g., $Irr(C \wedge A : A \Rightarrow B)$ is written as $Irr(C : A \Rightarrow B)$, so that C contains only facts which are distinct from A. If $\mathcal{F}x$ coincides with A, the 'reduced' irrelevance assumption $Irr(: A \Rightarrow B)$ becomes *trivial*; throughout the following we assume that all trival irrelevance assumptions get cancelled.

I ranges over irrelevance assumptions and $\mathcal{I}$ over sets of them; D over updated laws and $\mathcal{D}$ over sets of them. $\mathcal{I}$ is *for* $\mathcal{L}$ iff each $I \in \mathcal{I}$ has the form $Irr(C : A \Rightarrow B)$ for some $(A \Rightarrow B) \in \mathcal{L}$. A probability function p is called *proper* for $Irr(C : A \Rightarrow B)$ iff $p(C \wedge A) > 0$. $Irr(C : A \Rightarrow B)$ is said to be *r-satisfied* by p iff $p(B/A) - p(B/C \wedge A) \leq r$. The notion of ϵ-entailment is extended to the entailment from sets of laws $\mathcal{L}$ enriched with irrelevance assumptions $\mathcal{I}$ for $\mathcal{L}$ as follows: $\mathcal{L} \cup \mathcal{I}$ ϵ-entails L iff $\forall \delta > 0 \, \exists \epsilon > 0 \, \forall p \in \Pi(\mathcal{L}, \mathcal{I}, L)$: if $p(L) \geq 1 - \epsilon$

---

[12] The idea is similar to [8], but this approach remains syntactical. Probabilistic irrelevance assumptions have also been suggested by Bacchus [3, ch.5].

for all $L \in \mathcal{L}$ and p ϵ-satisfies all $I \in \mathcal{I}$, then $p(L) \geq 1 - \delta$. Th. 2(i,ii) states that the $\mathbf{P}_\epsilon$-calculus extended by the axiom (Irr) is *correct and complete* for this extended notion of ϵ-entailment, if we restrict it to entailments from sets of laws plus irrelevance assumptions *for* them. Given $\mathcal{L}$ and $\mathcal{I}$ for $\mathcal{L}$, we define $\mathcal{D}(\mathcal{L}, \mathcal{I}) := \{C \wedge A \Rightarrow B \mid Irr(C : A \Rightarrow B) \in \mathcal{I}\}$ as the set of updated laws derivable from $\mathcal{L} \cup \mathcal{I}$ by (Irr). Th. 2(i,iii) shows that the extension of the system $\mathbf{P}_\epsilon$ by irrelevance assumptions is *conservative*: irrelevance assumptions and the corresponding updated default laws ϵ-entail the same laws (in the presence of a given $\mathcal{L}$).

Theorem 2 (Extension of $\mathbf{P}_\epsilon$ by Irr). *For every L, $\mathcal{L}$ and $\mathcal{I}$ for $\mathcal{L}$: (i) $\mathcal{L} \cup \mathcal{I} \vdash_\epsilon L$ iff (ii) $\mathcal{L} \cup \mathcal{I}$ ϵ-entails L iff (iii) $\mathcal{L} \cup \mathcal{D}(\mathcal{L}, \mathcal{I}) \vdash_\epsilon L$.*

Uncertain laws $A \Rightarrow B$ correspond to Reiter's *normal* defaults of the form $(A : MB/B)$ [16, p.95]. In [17] it has been shown that $\mathbf{P}_\epsilon$ extended by irrelevance assumptions supplies probabilistic reliability for Reiter-style default logic. Th. 5 and cor. 2 of [17, p.256] together with theorem 2 imply the following: Ca is in some Reiter-extension of $\langle \mathcal{L}, \mathcal{F} \rangle$ iff $\mathcal{L} \cup \mathcal{I} \vdash_\epsilon \mathcal{F}x \overset{r}{\Rightarrow} C$, where $\mathcal{I}$ is a set of irrelevance assumptions *for $\mathcal{L}$* and $r = 1 - \sum\{1 - b(L) \mid Irr(\mathcal{F}x : L) \in \mathcal{I}\}$. Here is an example.

Example 1 (Irrelevance updating).
Laws ($\mathcal{L}$): Student $\overset{0.9}{\Rightarrow}$ Adult, Adult $\overset{0.95}{\Rightarrow}$ HasCar.
Facts: Student(a), Female(a).
Conclusion: [HasCar(a), 0.85].
Default detachments: Student(a) $\hspace{-0.3em}\sim$ Adult(a), Adult(a) $\hspace{-0.3em}\sim$ HasCar(a).
Irrelevance assumptions generated ($\mathcal{I}$): Irr(Female: Student $\Rightarrow$ Adult),
  Irr(Student $\wedge$ Female: Adult $\Rightarrow$ HasCar).
Updated laws ($\mathcal{D}$): Student $\wedge$ Female $\overset{0.9}{\Rightarrow}$ Adult,
  Adult $\wedge$ Student $\wedge$ Female $\overset{0.95}{\Rightarrow}$ HasCar.
$\mathcal{L} \cup \mathcal{I} \vdash_\epsilon \mathcal{D} \vdash_\epsilon$ Student $\wedge$ Female $\overset{0.85}{\Rightarrow}$ HasCar $[1 - (1 - 0.95) - (1 - 0.9) = 0.85]$.

In the above example the addition of irrelevance assumptions is without problems. This is not so in the case of *conflicting* laws: $\mathcal{K} = \langle \{A \Rightarrow B, C \Rightarrow \neg B\}, \{Aa, Ca\} \rangle$. Here one cannot generate *both* irrelevance assumptions $Irr(C : A \Rightarrow B)$ and $Irr(A : C \Rightarrow \neg B)$ without making $\mathcal{L} \cup \mathcal{I}$ ϵ-*inconsistent* and the set of resulting singular conclusions *logically* inconsistent. The question *which* irrelevance assumptions should be added in the case of conflicting laws is a decisive crossroad for different systems (multiple extensions, sceptical approach, rule priorities, specifity – cf. [4]). Our approach is extremely flexible in this respect. There is a spectrum of possibilities between (1.) irrelevance assumptions completely guided by the *user*, which corresponds formally to a multiple-extension approach, and (2.) irrelevance assumptions completely determined by *system-immanent rationality principles*, which corresponds formally to a specifity-based approach. In this paper we focus on the most liberal default logics which admit multiple extensions, for it is here where the fundamentals of probabilistic default logic have to be laid.

3.2 Poole-Extensions and Maximal Irrelevance-Updates. $\mathcal{M}$ ranges over sets of material implications. $\mathcal{L}^{\rightarrow} := \{A \rightarrow B \mid (A \Rightarrow B) \in \mathcal{L}\}$ denotes the set of material counterparts of laws in $\mathcal{L}$. A Poole-extension of $\langle \mathcal{L}, \mathcal{F} \rangle$ is obtained by adding a maximal set of material counterparts of laws in $\mathcal{L}$ to $\mathcal{F}$ which preserves consistency, and then forming the classical closure.

Definition 2 (Poole-extensions). *1. A materialization of $\langle \mathcal{L}, \mathcal{F} \rangle$ is a set $\mathcal{M} \subseteq \mathcal{L}^{\rightarrow}$ such that $\mathcal{M} \cup \mathcal{F}$ is consistent if $\mathcal{F}$ is so. A materialization $\mathcal{M}$ of $\langle \mathcal{L}, \mathcal{F} \rangle$ is maximal if no proper superset of $\mathcal{M}$ is a materialization of $\langle \mathcal{L}, \mathcal{F} \rangle$.*
2. $\mathcal{E}$ is a Poole-extension of $\langle \mathcal{L}, \mathcal{F} \rangle$ iff $\mathcal{E} = Cn(\mathcal{M} \cup \mathcal{F})$ for some maximal materialization $\mathcal{M}$ of $\langle \mathcal{L}, \mathcal{F} \rangle$. POOLE($\mathcal{K}$) is the set of Poole-extensions of $\mathcal{K}$.

Poole speaks of *scenarios*; they are the unions of consistent $\mathcal{F}$'s with materializations in our sense.[13] – The *maximal $\mathcal{F}$-irrelevance set* of $\mathcal{L}$ is defined as $\mathcal{I}_{\mathcal{F}}(\mathcal{L}) := \{Irr(\mathcal{F}x : A \Rightarrow B) \mid (A \Rightarrow B) \in \mathcal{L}\}$, and the corresponding *maximal $\mathcal{F}$-update* of $\mathcal{L}$ is $\mathcal{D}_{\mathcal{F}}(\mathcal{L}) := \{\mathcal{F}x \wedge A \Rightarrow B \mid (A \Rightarrow B) \in \mathcal{L}\}$. Obviously, $\mathcal{L} \cup \mathcal{I}_{\mathcal{F}}(\mathcal{L}) \vdash_{\epsilon} \mathcal{D}_{\mathcal{F}}(\mathcal{L})$ by the rule (Irr). Theorem 3.1 shows that classical consequence from $\mathcal{L}^{\rightarrow} \cup \mathcal{F}$ implies ϵ-entailment from $\langle \mathcal{D}_{\mathcal{F}}(\mathcal{L}), \mathcal{F} \rangle$, and th. 3.2 supplements that the other direction holds if $\mathcal{D}_{\mathcal{F}}(\mathcal{L})$ is ϵ-consistent.

Theorem 3 (Poole-extensions). *For every $\mathcal{L}$ and $\mathcal{F}$:*
1. If $\mathcal{L}^{\rightarrow} \cup \mathcal{F} \vdash Aa$, then $\mathcal{D}_{\mathcal{F}}(\mathcal{L}) \vdash_{\epsilon} \mathcal{F}x \Rightarrow A$.
2. If $\mathcal{D}_{\mathcal{F}}(\mathcal{L})$ is ϵ-consistent and $\mathcal{D}_{\mathcal{F}}(\mathcal{L}) \vdash_{\epsilon} \mathcal{F}x \Rightarrow A$, then $\mathcal{L}^{\rightarrow} \cup \mathcal{F} \vdash Aa$.

Unfortunately, we cannot make good sense of th. 3 so far. For, the irrelevance assumptions $\mathcal{I}_{\mathcal{F}}(\mathcal{L})$, which together with $\mathcal{L}$ ϵ-entail $\mathcal{D}_{\mathcal{F}}(\mathcal{L})$, will not always be *reasonable*. For example, assume $\mathcal{L} = \{A \Rightarrow B\}$, $\mathcal{F} = \{\neg Ba\}$. Then the (single) Poole-extension of this base contains $\neg Aa$, and the corresponding $\mathcal{F}$-updated law is $A \wedge \neg B \Rightarrow B$. This law indeed ϵ-entails $\neg B \Rightarrow \neg A$, but $\neg B$ lowers the conditional probability of B to 0. So $\neg B$ is certainly *not* irrelevant for $A \Rightarrow B$. This is reflected in the fact that the updated law $A \wedge \neg B \Rightarrow B$ is ϵ-inconsistent. Consider the following rationality principles of *increasing logical strength*:

Rationality principles for irrelevance updating:
P0: $\mathcal{D}_{\mathcal{F}}(\mathcal{L})$ is ϵ-consistent.
P1: For every $(A \Rightarrow B) \in \mathcal{L}$:
  $\mathcal{D}_{\mathcal{F}}(\mathcal{L}) \not\vdash_{\epsilon} \mathcal{F}x \Rightarrow \neg B$ [equivalent: $\mathcal{L}^{\rightarrow} \cup \mathcal{F} \not\vdash \neg Ba$].
P2: For every $(A \Rightarrow B) \in \mathcal{L}$ and elementary disjunct DS of B:
  $\mathcal{D}_{\mathcal{F}}(\mathcal{L}) \not\vdash_{\epsilon} \mathcal{F}x \Rightarrow \neg DS$ [equivalent: $\mathcal{L}^{\rightarrow} \cup \mathcal{F} \not\vdash \neg DSa$].

If $\mathcal{D}_{\mathcal{F}}(\mathcal{L})$ is ϵ-inconsistent, then for any $D := (\mathcal{F}x \wedge A \Rightarrow B) \in \mathcal{D}_{\mathcal{F}}(\mathcal{L})$, $\mathcal{D}_{\mathcal{F}}(\mathcal{L}) \setminus \{D\} \vdash_{\epsilon} \mathcal{F}x \wedge A \Rightarrow \neg B$ will hold (recall §2.2). Hence *P0* is a *minimal* rationality condition: it excludes the case where the remainder updated default laws ϵ-entail that $p(B/\mathcal{F}x \wedge A)$ is low and hence that $\mathcal{F}x$ is *not* irrelevant for $A \Rightarrow B$. *P1* is stronger; it excludes also the case where the updated laws ϵ-entail that $p(B/\mathcal{F}x)$ is low. This does not imply, but gives a *reason* to *believe* that $\mathcal{F}x$ is not irrelevant for $A \Rightarrow B$ (i.e. that $p(B/\mathcal{F}x \wedge A) < p(B/A)$). *P1* implies

---

[13] Poole excludes inconsistent $\mathcal{F}$'s ([14], p.29); we include them for systematic reasons.

P0 because an ϵ-inconsistent $\mathcal{D}_{\mathcal{F}}(\mathcal{L})$ ϵ-implies every law. Therefore th. 3 implies that *P1* has the equivalent but simpler formulation in square brackets (which stands in the same quantifier scope).

P2 is still stronger. It refers to the case where the law consequent is a *disjunction* and the updated laws ϵ-entail that, given $\mathcal{F}x$, the probability of one law disjunct L is low. This gives us a *reason* to *doubt* that $\mathcal{F}x$ is irrelevant for $A \Rightarrow B$. As a simple example, consider the irrelevance assumption $Irr(\neg B : A \Rightarrow B \vee C)$, thus $\mathcal{F} = \{\neg Ba\}$.[14] It is doubtful whether $B \vee C$ (given A) remains highly probable under the additional condition that B is false.

By th. 3, also *P2* has the equivalent but simpler formulation in square brackets. To make *P2* invariant w.r.t. equivalent transformations of laws, we assume from now on that all laws in $\mathcal{L}$ have *standard form*. A law has *standard form* iff its antecedent is a conjunction of disjunctions of literals and its consequent is a disjunction of literals. Every law L can be ϵ-equivalently transformed into a unique set $St(L)$ of laws in standard form (by prop. logic, (LLE), (RW), (And)). For L in standard form, we write $L = (\bigwedge DS(L) \Rightarrow \bigvee LT(L))$, where $DS(L)$ denotes the set of elementary conjuncts of L's antecedent (each such conjunct is itself a disjunction of literals) and $LT(L)$ denotes the set of literals of L's consequent. The empty cases are handled as usual by putting $\bigwedge \emptyset = \top$ and $\bigvee \emptyset = \bot$.

If the materialization $\mathcal{M} := \mathcal{L}^{\rightarrow}$ is not maximal, then not only the satisfaction of principles *P1-2* w.r.t. $\mathcal{M}$ but also w.r.t. $\mathcal{F}$-consistent extensions of $\mathcal{M}$ is important (cf. §3.4). We say that $\mathcal{D}_{\mathcal{F}}(\mathcal{L})$ satisfies principle *P1* w.r.t. $\mathcal{M}' \supseteq \mathcal{L}^{\rightarrow}$ iff $\mathcal{M}' \cup \mathcal{F} \not\vdash \neg Ba$ holds for every $(A \Rightarrow B) \in \mathcal{L}$.

3.3 Relevance Assumptions and Transposition Steps. How can we guarantee that irrelevance updating is reasonable? Our *key idea* is to apply so-called *transposition steps* to the laws in $\mathcal{L}$, in a way such that the resulting set of transposed laws satisfies principles *P0-2*, and satisfies *P1* even w.r.t. all $\mathcal{F}$-consistent $\mathcal{L}^{\rightarrow}$-extensions. For example, $L := (A \Rightarrow B)$ cannot be ϵ-consistently updated with $\mathcal{F}x = \neg B$, but L's contraposition $\neg B \Rightarrow \neg A$ can. The following are the three basic kinds of transposition steps (furnished with examples of lower bound propagation):

$$A \wedge B \overset{0.9}{\Rightarrow} C \vdash_{\epsilon} A \overset{0.9}{\Rightarrow} \neg B \vee C \qquad (1)$$

$$\neg(A \overset{0.5}{\Rightarrow} B), \ A \overset{0.9}{\Rightarrow} (B \vee C) \vdash_{\epsilon+} (A \wedge \neg B) \overset{0.8}{\Rightarrow} C \qquad (2)$$

$$\neg(D \overset{0.4}{\Rightarrow} B), \ (D \wedge A) \overset{0.95}{\Rightarrow} (B \vee C) \vdash_{\epsilon+} (D \wedge \neg B) \overset{0.92}{\Rightarrow} (\neg A \vee C) \qquad (3)$$

Forward transpositions as in (1) are ϵ-valid. *Backward* transpositions, as in (2), require the assumption of an additional negated law for their ϵ^{+}-validity.[15] Finally, a *contraposition* step, as in (3), is obtained by first applying a forward

---

[14] Here $\mathcal{F}x \Rightarrow \neg B$ follows already from $\emptyset$. A more complex example would be, e.g., $Irr(D : A \Rightarrow B \vee C)$ with $\mathcal{L} = \{A \Rightarrow B \vee C, D \Rightarrow \neg B\}$ and $\mathcal{F}x = D$.

[15] (2) is proved by (WRM, And, RW) via $A \wedge \neg B \Rightarrow B \vee C \vdash_{\epsilon} A \wedge \neg B \Rightarrow C$.

and then a backward transposition; it also requires an additional negated law premise. These negated laws are assumed *by default*. Thus, whenever default reasoning has to perform a backward transposition step, it generates the negated law which justifies it. Thereby, the upper probability bounds of the generated negated laws may be either specified in a *user-interactive* fashion, or be identified with a *default value* by the reasoning system. The lower bound calculation is performed via th. 1.2; in (2) the conclusion's upper uncertainty bound is $\frac{1-0.9}{1-0.5}$ $= 0.2$, in (3) it is $\frac{1-0.95}{1-0.4} = 0.08$.

Negated laws are very *natural* default assumptions, in particular in the context of transposition steps: here they directly correspond to probabilistic *relevance* assumptions. For given the law premise $A \overset{r}{\Rightarrow} (B \vee C)$ of (2), $\neg(A \overset{q}{\Rightarrow} B)$ is equivalent with $p(B \vee C/A) - p(B/A) \geq (r - q)$, i.e. with the assertion that B's probability given A is smaller than $B \vee C$'s probability given A for an amount of at least $r - q$. Indeed, when we assert a law with disjunctive consequent $A \Rightarrow B \vee C$ we assume that the disjunctive component C is *relevant* in the sense that $p(B/A)$ is not high, too (and likewise for B). The relevance assumption corresponding to a simple contraposition step ((4) without D) is especially nice: given $A \overset{r}{\Rightarrow} B$, $\neg(\top \overset{q}{\Rightarrow} B)$ is equivalent with $p(B/A) - p(B) \geq (r - q)$, i.e., with the assertion that A increases B's probability for an amount of at least $r - q$. Again, when we assert a law $A \Rightarrow B$ we normally assume that A is positively relevant for B. Because of this correspondence we use the phrases 'generated negated law' and 'generated relevance assumption' synonymously.

The following example illustrates the method of transposition steps, which is systematically developed in the next section:

Example 2 (Relevance-based irrelevance updating).
Laws $\mathcal{L}$: Bird $\overset{0.9}{\Rightarrow}$ Fly.
Facts $\mathcal{F}$: $\neg Fly(a)$, Female(a).
Conclusion Ca: $[\neg Bird(a), 0.83]$.
Relevance assumption generated $\mathcal{N}$: $\neg(\top \overset{0.4}{\Rightarrow} Fly)$.
Transposed laws: $\neg Fly \overset{0.83}{\Rightarrow} \neg Bird$.
Irrelevance assumptions generated $\mathcal{I}$: $Irr(Female: \neg Fly \Rightarrow \neg Bird)$.
Updated transposed laws $\mathcal{D}$: $\neg Fly \wedge Female \overset{0.83}{\Rightarrow} \neg Bird$.
$\mathcal{L} \cup \mathcal{N} \cup \mathcal{I} \vdash_{\epsilon+} \mathcal{D} \vdash_{\epsilon} \mathcal{F}x \overset{0.83}{\Rightarrow} C$, hence $\langle \mathcal{L} \cup \mathcal{N} \cup \mathcal{I}, \mathcal{F} \rangle \hspace{0.1em}\vdash_{\epsilon+} [Ca, 0.83]$.

3.4 The System P_ϵ^+DP. P_ϵ^+DP stands for P_ϵ^+-based Default logic in Poole-style. Given a materialization $\mathcal{M} := \mathcal{L}^{\rightarrow}$ of $\langle \mathcal{L}^\star, \mathcal{F} \rangle$ (not necessarily maximal) and $L \in \mathcal{L}$, we first apply a forward transposition to all antecedent-conjuncts $DS \in DS(L)$ which are *not* logically implied by $\mathcal{M} \cup \mathcal{F}$, and then a backward transposition to all literals $LT \in LT(L)$ the negation of which is logically implied by $\mathcal{M} \cup \mathcal{F}$ (def. 3.1). L^t is the result of this transposition and N_L the negated law which together with L ϵ-entails L^t via transposition step (3). Because the consequent of L^t need not be in standard from, we transform each L^t into $St(L^t)$ before collecting the transposed laws into the set $\mathcal{L}_{\mathcal{F}}^t$ (def. 3.2). The reason for performing forward transpositions is twofold: by doing this, we ensure the ϵ-consistency of $\mathcal{D}_{\mathcal{F}}(\mathcal{L})$ in a computationally simple way and at the same time guarantee that *P1* is satisfied by all $\mathcal{F}$-consistent $\mathcal{M}$-extensions (th. 4.2(ii)). Def.3.3+4 defines $\mathbf{P}_{\epsilon}^{+}\mathbf{DP}$-extensions in terms of maximal Poolean $\mathcal{F}$-updates. For $X \subseteq BLang$, $\neg X := \{\neg A \mid A \in X\}$.

Definition 3 (System $\mathbf{P}_{\epsilon}^{+}\mathbf{DP}$). *Let $\mathcal{L} \subseteq \mathcal{L}^{*}$, $\mathcal{M} := \mathcal{L}^{\rightarrow}$ be a materialization of $\langle \mathcal{L}^{*}, \mathcal{F}\rangle$, and $L \in \mathcal{L}$.*
1. $DS^{+}(L) = \{DS \in DS(L) \mid \mathcal{M} \cup \mathcal{F} \vdash DSa\}$;
$LT^{-}(L) = \{LT \in LT(L) \mid \mathcal{M} \cup \mathcal{F} \vdash \neg LTa\}$;
$L^{t} = \bigwedge[DS^{+}(L) \cup \neg LT^{-}(L)] \Rightarrow \bigvee[(LT \setminus LT^{-}(L)) \cup \neg(DS \setminus DS^{+}(L))]$;
$N_{L} = \neg(\bigwedge DS^{+}(L) \Rightarrow \bigvee LT^{-}(L))$.
2. $\mathcal{L}_{\mathcal{F}}^{t}$ (the $\mathcal{F}$-transposition of $\mathcal{L}$) $= \bigcup_{L \in \mathcal{L}} St(L^{t})$;
$N_{\mathcal{F}}(\mathcal{L})$ (the $\mathcal{F}$-relevance set of $\mathcal{L}$) $= \{N_{L} \mid L \in \mathcal{L}\}$.
3. A Poolean $\mathcal{F}$-update of $\mathcal{L}^{*}$ is any set $\mathcal{D}_{\mathcal{F}}(\mathcal{L}_{\mathcal{F}}^{t})$ for $\mathcal{L} \subseteq \mathcal{L}^{*}$ which is ϵ-consistent. A Poolean $\mathcal{F}$-update $\mathcal{D}_{\mathcal{F}}(\mathcal{L}_{\mathcal{F}}^{t})$ is maximal if there exists no proper extension $\mathcal{L}' \supset \mathcal{L}$ with $\mathcal{L}' \subseteq \mathcal{L}^{*}$ such that $\mathcal{D}_{\mathcal{F}}((\mathcal{L}')_{\mathcal{F}}^{t})$ is ϵ-consistent.
4. $\mathcal{E}$ is a $P_{\epsilon}^{+}DP$-extension of $\langle \mathcal{L}^{*}, \mathcal{F}\rangle$ iff $\mathcal{E} = \mathcal{E}(\mathcal{D})\,[:= \{Ca \mid \mathcal{D} \vdash_{\epsilon} \mathcal{F}x \Rightarrow C\}]$ for some maximal Poolean $\mathcal{F}$-update $\mathcal{D}$ of $\mathcal{L}^{*}$. $P_{\epsilon}^{+}DP(\mathcal{K})$ is the set of all $P_{\epsilon}^{+}DP$-extensions of $\mathcal{K}$.

Theorem 4 states the fundamental facts about the system $\mathbf{P}_{\epsilon}^{+}\mathbf{DP}$.

Theorem 4 (System $\mathbf{P}_{\epsilon}^{+}\mathbf{DP}$). *For given $\langle \mathcal{L}^{*}, \mathcal{F}\rangle$ with consistent $\mathcal{F}$, assume $\mathcal{L} \subseteq \mathcal{L}^{*}$ and $\mathcal{M} = \mathcal{L}^{\rightarrow}$. For claims 2, 4 and 6, assume in addition that $\mathcal{M}$ is a materialization of $\langle \mathcal{L}^{*}, \mathcal{F}\rangle$.*
1. $\mathcal{M}$ is a materialization of $\langle \mathcal{L}^{*}, \mathcal{F}\rangle$ iff $\mathcal{D}_{\mathcal{F}}(\mathcal{L}_{\mathcal{F}}^{t})$ is a Poolean $\mathcal{F}$-update of $\mathcal{L}^{*}$.
2. $\mathcal{D}_{\mathcal{F}}(\mathcal{L}_{\mathcal{F}}^{t})$ satisfies (i) the principles P0-2, and (ii) the principle P1 w.r.t. every materialization $\mathcal{M}^{*} \supseteq \mathcal{M}$ of $\langle \mathcal{L}^{*}\mathcal{F}\rangle$.
3. $\mathcal{L} \cup N_{\mathcal{F}}(\mathcal{L}) \cup \mathcal{I}_{\mathcal{F}}(\mathcal{L}_{\mathcal{F}}^{t}) \vdash_{\epsilon^{+}} \mathcal{D}_{\mathcal{F}}(\mathcal{L}_{\mathcal{F}}^{t})$.
4. For every $Aa \in BLang$: $Aa \in Cn(\mathcal{M} \cup \mathcal{F})$ iff $\mathcal{D}_{\mathcal{F}}(\mathcal{L}_{\mathcal{F}}^{t}) \vdash_{\epsilon} \mathcal{F}x \Rightarrow A$.
5. For every $\mathcal{E} \subseteq BLang$: $\mathcal{E} \in POOLE(\langle \mathcal{L}^{*}, \mathcal{F}\rangle)$ iff $\mathcal{E} \in P_{\epsilon}^{+}DP(\langle \mathcal{L}^{*}, \mathcal{F}\rangle)$.
6. For every $A \in BLang$: (a) $\mathcal{L} \cup \mathcal{L}_{\mathcal{F}}^{t} \cup \mathcal{I}_{\mathcal{F}}(\mathcal{L}_{\mathcal{F}}^{t}) \vdash_{\epsilon} \mathcal{F}x \Rightarrow A$ iff $\mathcal{D}_{\mathcal{F}}(\mathcal{L}_{\mathcal{F}}^{t}) \vdash_{\epsilon} \mathcal{F}x \Rightarrow A$. (b) If $Aa \in Cn(\mathcal{M} \cup \mathcal{F})$, then for all $p \in \Pi(\mathcal{L}, \mathcal{N}, \mathcal{I})$ satisfying $\mathcal{I}_{\mathcal{F}}(\mathcal{L}_{\mathcal{F}}^{t})$: $u(\mathcal{F}x \Rightarrow A) \leq min\left(\sum_{L \in \mathcal{L}} min\left(\frac{u(L)}{u(N_{L})}, 1\right), 1\right)$.

Th.4.1 says, in other words, that the consistency of Poolean scenarios has its exact counterpart in the ϵ-consistency of the corresponding $\mathcal{F}$-updated transformed laws. Concerning th. 4.2, note that *P2* is *not* satisfied for all $\mathcal{F}$-consistent extensions of $\mathcal{M}$. This is an unavoidable consequence of the nonmonotonic character of default reasoning. It follows that the transformed set $\mathcal{L}_{\mathcal{F}}^{t}$ has to be computed newly if the materialization $\mathcal{M}$ found for one query is extended when asking further queries. Th. 4.3 is obvious, and th.4.4-5 tell what is expected, namely that the $\mathcal{F}$-updated transformed laws together with $\mathcal{F}$ ϵ-entail the same singular conclusions as the corresponding Poolean scenario, and that Poolean extensions of a knowledge base are exactly $P_{\epsilon}^{+}DP$-extensions. Th. 4.6a says that from $\mathcal{L} \cup \mathcal{L}_{\mathcal{F}}^{t} \cup \mathcal{I}_{\mathcal{F}}(\mathcal{L}_{\mathcal{F}}^{t})$ not more laws with $\mathcal{F}$-antecedent are ϵ-derivable than from

$\mathcal{D}_{\mathcal{F}}(\mathcal{L}_{\mathcal{F}}^t)$ alone. This immediately entails (via th.4.1) that the enriched premise set $\mathcal{L} \cup \mathcal{L}_{\mathcal{F}}^t \cup \mathcal{I}_{\mathcal{F}}(\mathcal{L}_{\mathcal{F}}^t)$ is always ϵ-consistent (given $\mathcal{M} \cup \mathcal{F}$ is consistent). We cannot prove in a similarly general way that $\mathcal{L} \cup \mathcal{N}_{\mathcal{F}}(\mathcal{L})$ is ϵ-consistent; to guarantee this in a *computationally simple* way is a task for the future.

We finally present a procedure for $\mathbf{P}_{\epsilon}^+\mathbf{DP}$-reasoning. In contrast to def. 3 and th. 4 the procedure takes the *lower bounds* of the laws into account. Its lower bound computation is based on th. 4.6(b) (recall §2.3).

Procedure 1 ($\mathbf{P}_{\epsilon}^+\mathbf{DP}$).
Input: A knowledge base $\langle \mathcal{L}^*, \mathcal{F} \rangle$ with consistent $\mathcal{F}$ and a singular query Q.
Output: If $Q \in \mathcal{E}$ for some $\mathcal{E} \in POOLE(\langle \mathcal{L}^*, \mathcal{F} \rangle)$, then a (good) lower probability bound $b(Q)$ and (small) sets $\mathcal{L}(Q)$, $\mathcal{N}(Q)$ and $\mathcal{I}(Q)$ such that $\mathcal{L}(Q)^{\rightarrow}$ is a materialization of $\langle \mathcal{L}^*, \mathcal{F} \rangle$, and $\langle \mathcal{L}(Q) \cup \mathcal{N}(Q) \cup \mathcal{I}(Q), \mathcal{F} \rangle \vDash_{\epsilon^+} [Q, b(Q)]$. – Else: FAIL.
1. Subprocedure **Poole:** Find a set $\mathcal{L}(Q) \subseteq \mathcal{L}^*$ such that (a) $\mathcal{L}(Q)^{\rightarrow} \cup \mathcal{F} \cup \{\neg Q\}$ resolves to $\perp$ and (b) $\mathcal{L}(Q)^{\rightarrow} \cup \mathcal{F}$ is consistent. If not found: FAIL.
2. Subprocedure **Transpose:** Based on the search tree created in step 1b, determine for each $L \in \mathcal{L}(Q)$, $DS \in DS(L)$ and $LT \in LT(L)$, whether (a) $\mathcal{L}(Q)^{\rightarrow} \cup \mathcal{F} \vdash DS$ and (b) $\mathcal{L}(Q)^{\rightarrow} \cup \mathcal{F} \vdash \neg LT$. Result: $DS^+(L)$, $LT^-(L)$, N_L and L^t (def. 3.1).
3. Subprocedure **Bounds:** For each $L \in \mathcal{L}(Q)$ with $L \neq L^t$, identify $b(N_L)$ (user-interactively or by default) and set $b(L^t) = 1 - min\left(\frac{1-b(L)}{1-b(N_L)}, 1\right)$. For each $L \in \mathcal{L}(Q)$ with $L = L^t$, set $b(L^t) = b(L)$.
4. Subprocedure **Generate:** Based on steps 2. and 3., generate $\mathcal{L}_{\mathcal{F}}^t$, $\mathcal{N}(Q) := \mathcal{N}_{\mathcal{F}}(\mathcal{L}(Q))$ (according to definition 3.2), $\mathcal{I}(Q) := \mathcal{I}_{\mathcal{F}}((\mathcal{L}(Q))_{\mathcal{F}}^t)$ and $b(Q) := 1 - min\left(\sum_{L \in \mathcal{L}(Q)}(1 - b(L^t)), 1\right)$. Output $\mathcal{L}(Q)$, $\mathcal{N}(Q)$, $\mathcal{I}(Q)$ and $b(Q)$.

The procedure is a correct and complete decision procedure, and its complexity is not significantly greater than that of subprocedure **Poole**. For subprocedure **Transpose** does not require any new consistency tests, since the search tree created in step 1b contains every clause provable from $\mathcal{L}(Q)^{\rightarrow} \cup \mathcal{F}$ by resolution. We implement **Transpose** incrementally: for each new node N created in the search tree of step 1b we check whether N covers some $DS(L)$ or is identical with some $LT(L)$ (for some $L \in \mathcal{L}$). After the search tree terminates with failure, the sets $DS^+(L)$ and $LT^-(L)$ will be completely determined.

Recall the considerations on approximating tight bounds at the end of §2.3. Subprocedure **Poole** has been implemented in [15]. It is based on a PROLOG-variant of linear resolution [15, p.10] and thus never encounters premises which are *derivationally irrelevant* for Q (cf. [11, p.139]). We furthermore impose an ordering on the scanning procedure involved in step 1a where the facts come before the laws (cf. [16, p.111]). Usually, this will be sufficient for finding a *minimal* set of law premises $\mathcal{L}(Q)$. By assuming that the laws are ordered with decreasing lower probability bound we take care of the requirement of finding law premises with bounds as high as possible. Implemented in this way, the procedure will generally find good approximations of tight bounds. Of course, a guarantee that an optimal premise set has been found can only be given by backtracking to all alternative subsets of law premises. This is not recommendable for knowledge bases containing many laws.

To implement the procedure in a *user-interactive* fashion has significant advantages. By *refuting* certain (ir)relevance assumptions in $\mathcal{N}(Q)$ and $\mathcal{I}(Q)$ the user can *control* which extensions are *preferred* among mutually incompatible extensions. By asking *bounded queries* $[Q, r_i]$ ('prove Q with lower bound $\geq r_i$') with successively increasing bounds r_i, the user can help to approximate a proof of Q with an optimal bound.

4 Appendix: Proofs of the theorems

Proof of theorem 1: We sketch how the proof of th. 1 for the system $\mathbf{P}_\epsilon^+\mathbf{DP}$ reduces to the proof of th. 1 for the system $\mathbf{P}^+\mathbf{DP}$ (see [2, 18]). For given $\mathcal{L}, \mathcal{N}$ and L, let $Poss$ (for 'possible') be the set of all negated laws $\neg(A \Rightarrow \perp)$ where A is the logically consistent antecedent of some law in $\mathcal{L} \cup \{L\}$ or negated law in $\mathcal{N}$. (1): $\mathcal{L} \cup \mathcal{N} \; \epsilon^+$-entails L iff (2): $Poss \cup \mathcal{L} \cup \mathcal{N} \; p^+$-entails L, because every p over Ω which does not assign 1 to some $N \in Poss$ is proper for $\mathcal{L}, \mathcal{N}, L$. By the proof of th. 1 for the system $\mathbf{P}^+\mathbf{DP}$ it follows that (2) holds iff the uncertainty inequality of th. 1.2 holds for all $p \in \Pi(\mathcal{L}, \mathcal{N}, L)$ iff (3): $Poss \cup \mathcal{L} \cup \mathcal{N} \vdash_{p^+} L$, and it is easy to prove (syntactically) that (3) holds iff $\mathcal{L} \cup \mathcal{N} \vdash_{\epsilon^+} L$. Q.E.D.

Proof of theorem 2:

Let $\mathcal{D} := \mathcal{D}(\mathcal{L}, \mathcal{I})$. We first prove (1) $\mathcal{L} \cup \mathcal{I} \vdash_\epsilon L$ iff (2) $\mathcal{L} \cup \mathcal{D} \vdash_\epsilon L$. The right-to-left-direction holds because $\mathcal{L} \cup \mathcal{I} \vdash_\epsilon \mathcal{D}$ by applications of the inference axiom (Irr). The left-to-right direction holds because (Irr) is the only inference axiom applying to elements of $\mathcal{I}$, and the conclusions of its applications are elements of $\mathcal{D}$. If we replace every instance $L, I \vdash_\epsilon D$ in the proof of (1) by the trivially derivable inference $L, D \vdash_\epsilon D$, we obtain a proof of (2).

Next we prove that (3) $\mathcal{L} \cup \mathcal{I}$ ϵ-entails L iff (4) $\mathcal{L} \cup \mathcal{D}$ ϵ-entails L. For the left-to-right-direction, take some $\delta > 0$. To prove (4) we must show that there exists $\epsilon > 0$ such that for all $p \in \Pi(\mathcal{L}, \mathcal{D}, L)$, if (a) $\forall L \in \mathcal{L}[p(L) \geq 1 - \epsilon]$ and (b) $\forall D \in \mathcal{D}[p(D) \geq 1 - \epsilon]$, then (c) $p(L) \geq 1 - \delta$. By (3), there exists ϵ' such that for all $p \in \Pi(\mathcal{L}, \mathcal{I}, L)$, if (a') $\forall L \in \mathcal{L}[p(L) \geq 1 - \epsilon']$ and (d') $\forall D \in \mathcal{D}[(p(L^D) - p(D)) \leq \epsilon']$, then (c) – where L^D is the unique law $(A \Rightarrow B) \in \mathcal{L}$ such that $Irr(C : A \Rightarrow B) \in \mathcal{I}$ and $D = (C \wedge A \Rightarrow B)$. $\Pi(\mathcal{L}, \mathcal{I}, L) = \Pi(\mathcal{L}, \mathcal{D}, L)$, since (by def.) p is proper for $Irr(C : A \Rightarrow B)$ iff p is proper for $C \wedge A \Rightarrow B$. We put $\epsilon = \epsilon'$. Then the assumption that (a)+(b) hold for some given p implies that (a') and (d') hold for p. Hence (c) follows by (3); which proves (4). – For the right-to-left-direction, take again some $\delta > 0$. To prove (3) we must show that there is $\epsilon > 0$ such that for all $p \in \Pi(\mathcal{L}, \mathcal{I}, L)$, if (a) holds and (d) $\forall D \in \mathcal{D}[(p(L^D) - p(D)) \leq \epsilon]$, then (c). By (4) there is $\epsilon' > 0$ such that for all $p \in \Pi(\mathcal{L}, \mathcal{D}, L)$, if (a') holds and (b') $\forall D \in \mathcal{D}[p(D) \geq 1 - \epsilon']$, then (c). Again $\Pi(\mathcal{L}, \mathcal{I}, L) = \Pi(\mathcal{L}, \mathcal{D}, L)$. (a) and (d) imply that $p(D) \geq 1 - 2\epsilon$ for all $D \in \mathcal{D}$. We put $\epsilon = \frac{\epsilon'}{2}$. Then the assumption that (a) and (d) hold for given p implies that (a') and (b') hold for p. Hence (c) follows by (4), which proves (3).

By th.1.1+3, (2) is equivalent with (4), so (1) is equivalent with (3), Q.E.D.

For the next theorems we need two well-known lemmata about ϵ-entailment and ϵ-consistency and one lemma about material counterparts. Some terminology: A truth valuation u verifies [falsifies] $A \Rightarrow B$ iff $u(A \wedge B) = 1$ $[u(A \wedge \neg B) = 1$, respectively]; u falsifies $\mathcal{L}$ iff u falsifies some $L \in \mathcal{L}$; u confirms $\mathcal{L}$ iff u verifies some $L \in \mathcal{L}$ and does not falsify $\mathcal{L}$; $\mathcal{L}$ is confirmable iff there exists u which confirms $\mathcal{L}$; L is nontrivial iff L's antecedent is consistent; finally $\mathcal{L}$ is nontrivial iff $\mathcal{L}$ is nonempty and every $L \in \mathcal{L}$ is nontrivial. For lemma 1 see [1, p.61]; for lemma 2 [1, p.52], [13, p.488].

Lemma 1 $\mathcal{L}$ ϵ-entails $A \Rightarrow B$ iff either (a) $\mathcal{L}$ is ϵ-inconsistent or (b) some subset $\mathcal{L}^* := \{A_i \Rightarrow B_i \mid 1 \leq i \leq n\} \subseteq \mathcal{L}$ yields $A \Rightarrow B$, which means that the following holds: (i) each truth valuation confirming $\mathcal{L}^*$ verifies $A \Rightarrow B$, and (ii) each truth valuation falsifying $A \Rightarrow B$ falsifies $\mathcal{L}^*$.

Lemma 2 $\mathcal{L}$ is ϵ-inconsistent iff some nontrivial subset of $\mathcal{L}$ is not confirmable iff for some consistent $A \in BLang$, $\mathcal{L} \vdash_\epsilon A \Rightarrow \perp$.

Lemma 3 If $\mathcal{L}$ is ϵ-consistent, then $\mathcal{L} \vdash_\epsilon L$ implies $\mathcal{L}^\rightarrow \vdash L^\rightarrow$.

Proof of lemma 3: If $\mathcal{L}$ is ϵ-consistent, then every L derivable from $\mathcal{L}$ by the rules of $\mathbf{P}_\epsilon$ is derivable without use of the rule (ϵEFQ), because this rule applies only to laws $A \Rightarrow \perp$ with consistent antecedent, and whenever $\mathcal{L}$ implies such a law it is ϵ-inconsistent by lemma 2. All rules of $\mathbf{P}_\epsilon$ distinct from (ϵEFQ) are propositionally valid for the material counterparts of the uncertain laws. Hence $L^\rightarrow$ is derivable from $\mathcal{L}^\rightarrow$ by rules of propositional logic. Q.E.D.

Proof of theorem 3: For th.3.1: Let $\mathcal{L} = \{A_i \Rightarrow B_i \mid 1 \leq i \leq n\}$. Given $\mathcal{L}^\rightarrow \cup \mathcal{F} \vdash Aa$, then (a): $\mathcal{F}x \wedge \bigwedge_{1 \leq i \leq n}(A_i \rightarrow B_i) \vdash \mathcal{F}x \wedge A$ by prop. logic and universal generalization. This implies that (b): $\mathcal{F}x \wedge \bigwedge_{1 \leq i \leq n}(\mathcal{F}x \wedge A_i \rightarrow B_i) \vdash \mathcal{F}x \wedge A$, because by prop. logic, (b)'s antecedent implies (a)'s antecedent. Moreover, (b) implies (c): $\bigwedge_{1 \leq i \leq n}(\mathcal{F}x \wedge A_i \rightarrow B_i) \vdash \mathcal{F}x \rightarrow A$ as well as (d): $(\mathcal{F}x \vee \bigvee_{1 \leq i \leq n} A_i) \wedge \bigwedge_{1 \leq i \leq n}(\mathcal{F}x \wedge A_i \rightarrow B_i) \vdash \mathcal{F}x \wedge A$ by prop. logic. But (c)+(d) imply that $\mathcal{D}_\mathcal{F}(\mathcal{L})$ yields $\mathcal{F} \Rightarrow A$ in the sense of lemma 1(b) (by prop. logic). This implies by lemma 1 that $\mathcal{D}_\mathcal{F}(\mathcal{L}) \vdash_\epsilon \mathcal{F}x \Rightarrow A$.

To prove th. 3.2, we note that by lemma 3, $\mathcal{D}_\mathcal{F}(\mathcal{L}) \vdash_\epsilon \mathcal{F}x \Rightarrow A$ implies (e): $(\mathcal{D}_\mathcal{F}(\mathcal{L}))^\rightarrow \vdash \mathcal{F}x \rightarrow A$, since $\mathcal{D}_\mathcal{F}(\mathcal{L})$ is ϵ-consistent by assumption. From (e) it follows that (f): $\mathcal{L}^\rightarrow \vdash \mathcal{F}x \rightarrow A$ (since $\mathcal{L}^\rightarrow \vdash (\mathcal{D}_\mathcal{F}(\mathcal{L}))^\rightarrow$), and (f) implies $\mathcal{L}^\rightarrow \cup \mathcal{F} \vdash Aa$ by universal instantiation. Q.E.D.

Proof of theorem 4:

For 1: We must show that $\mathcal{M} \cup \mathcal{F}$ is consistent iff $\mathcal{D}_\mathcal{F}(\mathcal{L}_\mathcal{F}^t)$ is ϵ-consistent. *Left-to-right:* Assume $\mathcal{D}_\mathcal{F}(\mathcal{L}_\mathcal{F}^t)$ is ϵ-inconsistent. Then by lemma 2 and prop. logic, there is a nontrivial subset $\{\mathcal{F}x \wedge A_i^t \Rightarrow B_i^t \mid 1 \leq i \leq n\}$ of $\mathcal{D}_\mathcal{F}(\mathcal{L}_\mathcal{F}^t)$ such that (a) $\{\mathcal{F}x \wedge A_i^t \rightarrow B_i^t \mid 1 \leq i \leq n\} \vdash \bigwedge_{1 \leq i \leq n} \neg(\mathcal{F}x \wedge A_i^t)$. Each $A_i^t \Rightarrow B_i^t$ is in $\mathcal{L}_\mathcal{F}^t$, and following from def.3.1-2 and prop. logic, $\mathcal{L}^\rightarrow \dashv\vdash (\mathcal{L}_\mathcal{F}^t)^\rightarrow$. Hence

$\{A_i^t \rightarrow B_i^t \mid 1 \leq i \leq n\}$ and thus also (b) $\{\mathcal{F}x \wedge A_i^t \rightarrow B_i^t \mid 1 \leq i \leq n\}$ is implied by $\mathcal{M}$. (a)+(b) imply that $\mathcal{M} \cup \mathcal{F} \vdash \bigwedge_{1 \leq i \leq n} \neg A_i^t a$, by prop. logic. By def. 6.1, $\mathcal{M} \cup \mathcal{F} \vdash A_i^t a$ holds for every $1 \leq i \leq n$. Thus, $\mathcal{M} \cup \mathcal{F}$ is inconsistent. – *Right-to-left:* Assume $\mathcal{M} \cup \mathcal{F}$ is inconsistent. Since $(\mathcal{L}_{\mathcal{F}}^t)^{\rightarrow} \dashv\vdash \mathcal{M}$ (by prop. logic), it follows that $(\mathcal{L}_{\mathcal{F}}^t)^{\rightarrow} \cup \mathcal{F}$ is inconsistent. Hence $\mathcal{D}_{\mathcal{F}}(\mathcal{L}_{\mathcal{F}}^t) \vdash_{\epsilon} \mathcal{F}x \Rightarrow \bot$ by th. 3.1. Since $\mathcal{F}$ is consistent, lemma 2 implies that $\mathcal{D}_{\mathcal{F}}(\mathcal{L}_{\mathcal{F}}^t)$ is ϵ-inconsistent.

For 2(i): $\mathcal{D}_{\mathcal{F}}(\mathcal{L}_{\mathcal{F}}^t)$ is ϵ-consistent by th. 4.1 and assumption and thus satisfies *P0*. Next we prove that $\mathcal{D}_{\mathcal{F}}(\mathcal{L}_{\mathcal{F}}^t)$ satisfies *P2* (and thus, *P1*). For reductio, assume $(A \Rightarrow B) \in \mathcal{L}_{\mathcal{F}}^t$, and for some literal $LT \in LT(B)$, $\mathcal{M} \cup \mathcal{F} \vdash \neg LTa$ holds (i.e. *P2* is violated). Now $(A \Rightarrow B) \in St(L^t)$ for some tranformed law L^t. Thus L^t's consequent either contains LT as an elementary disjunct, or it contains an elementary disjunct of the form $\neg(-LT \wedge C)$, where $-F := \neg F$ and $--\neg F := F$. In both cases $\mathcal{M} \cup \mathcal{F} \vdash \neg DS$ must hold for some elementary disjunct DS of L^t's consequent. But this is excluded by def. 3.1; contradiction. Finally, we prove that every $\mathcal{F}$-consistent $\mathcal{M}$-extension $\mathcal{M}'$ satisfies *P1*. Again, assume for reduction that (a) $\mathcal{M}' \cup \mathcal{F} \vdash \neg Ba$ for some $(A \Rightarrow B) \in \mathcal{L}_{\mathcal{F}}^t$. Since $\mathcal{M} \cup \mathcal{F} \vdash Aa$ (by def. 3.1), also (b) $\mathcal{M}' \cup \mathcal{F} \vdash Aa$ must hold. But (a)+(b) imply that $\mathcal{M}'$ is $\mathcal{F}$-inconsistent; a contradiction.

For 3: By applications of inference axiom (Irr) and transposition step (3).

For 4: $\mathcal{M} \dashv\vdash (\mathcal{L}_{\mathcal{F}}^t)^{\rightarrow}$ by prop. logic, and $\mathcal{D}_{\mathcal{F}}(\mathcal{L}_{\mathcal{F}}^t)$ is ϵ-consistent by th. 4.1; so th.4.3 follows by th.3.

For 5: $\mathcal{E} \in POOLE(\langle \mathcal{L}^*, \mathcal{F} \rangle)$ iff $\mathcal{E} = Cn(\mathcal{L}^{\rightarrow} \cup \mathcal{F})$ for some maximal materialization $\mathcal{L}^{\rightarrow}$ of $\langle \mathcal{L}^*, \mathcal{F} \rangle$ iff (i): $\mathcal{E} = \mathcal{E}[\mathcal{D}_{\mathcal{F}}(\mathcal{L}_{\mathcal{F}}^t)]$ by th. 3, since $(\mathcal{L}_{\mathcal{F}}^t)^{\rightarrow} \dashv\vdash \mathcal{L}^{\rightarrow}$ and $\mathcal{D}_{\mathcal{F}}(\mathcal{L}_{\mathcal{F}}^t)$ is ϵ-consistent (i.e. is a Poolean update) by th. 4.1. It remains to show that $\mathcal{D}_{\mathcal{F}}(\mathcal{L}_{\mathcal{F}}^t)$ is maximal. Assume for reductio that for some $L \in (\mathcal{L}^* \setminus \mathcal{L})$, $\mathcal{D}_{\mathcal{F}}((\mathcal{L} \cup \{L\})_{\mathcal{F}}^t)$ would be ϵ-consistent. Then $(\mathcal{L} \cup \{L\})^{\rightarrow} \cup \mathcal{F}$ must be consistent by th. 4.1. Because $(\mathcal{L} \cup \{L\})^{\rightarrow}$ properly extends $\mathcal{M}$, $\mathcal{M}$ would then not be a maximal materialization; a contradiction. – So $\mathcal{D}_{\mathcal{F}}(\mathcal{L}_{\mathcal{F}}^t)$ is a maximal Poolean update, whence (i) holds iff $\mathcal{E} \in P_{\epsilon}^+ DP(\langle \mathcal{L}^*, \mathcal{F} \rangle)$ (def. 3.3-4).

For 6(a): By th. 2, $\mathcal{L} \cup \mathcal{L}_{\mathcal{F}}^t \cup \mathcal{I}_{\mathcal{F}}(\mathcal{L}_{\mathcal{F}}^t)$ ϵ^+-implies the same laws as $\mathcal{L} \cup \mathcal{L}_{\mathcal{F}}^t \cup \mathcal{D}_{\mathcal{F}}(\mathcal{L}_{\mathcal{F}}^t)$. To prove 6(a) we first have to prove:

(A): $\quad \mathcal{L} \cup \mathcal{L}_{\mathcal{F}}^t \cup \mathcal{D}_{\mathcal{F}}(\mathcal{L}_{\mathcal{F}}^t)$ is ϵ-consistent.

For reductio, assume that the negation of (A) is true. Then by lemma 2, there is a *nonempty* subset $\mathcal{X} \subset \mathcal{L} \cup \mathcal{L}_{\mathcal{F}}^t \cup \mathcal{D}_{\mathcal{F}}(\mathcal{L}_{\mathcal{F}}^t)$ such that (B): $\mathcal{X}^{\rightarrow} \vdash \bigwedge \mathcal{Y}$, where $\mathcal{Y} = \{\neg P \mid (P \Rightarrow Q) \in \mathcal{X}\}$. Now note that (C): $\mathcal{L}^{\rightarrow} \vdash [\mathcal{L} \cup \mathcal{L}_{\mathcal{F}}^t \cup \mathcal{D}_{\mathcal{F}}(\mathcal{L}_{\mathcal{F}}^t)]^{\rightarrow}$ must hold, because $(\mathcal{L}_{\mathcal{F}}^t)^{\rightarrow} \vdash (\mathcal{D}_{\mathcal{F}}(\mathcal{L}_{\mathcal{F}}^t))^{\rightarrow}$ and $(\mathcal{L}_{\mathcal{F}}^t)^{\rightarrow} \dashv\vdash \mathcal{L}^{\rightarrow}$ (by prop. logic). (C) and (B) imply (D): $\mathcal{L}^{\rightarrow} \vdash \bigwedge \mathcal{Y}$. For each element of $Y \in \mathcal{Y}$, either (i): $Y = \neg A$ for $(A \Rightarrow B) \in \mathcal{L}$, or (ii): $Y = \neg A$ or $Y = \neg(\mathcal{F}x \wedge A)$ for $(A \Rightarrow B) \in \mathcal{L}_{\mathcal{F}}^t$. Case (ii) implies that $\mathcal{M} \cup \mathcal{F} \vdash \neg Aa$, but this is impossible because also $\mathcal{M} \cup \mathcal{F} \vdash Aa$ holds by def. 3.1 and $\mathcal{M} \cup \mathcal{F}$ is consistent by assumption. So only case (i) can be true. This implies that $\mathcal{X} \subseteq \mathcal{L}$ and hence that $\mathcal{L}$ would be ϵ-inconsistent (by lemma 2). But this contradicts the assumption that $\mathcal{M} \cup \mathcal{F}$ is consistent and th.4.1.

Right-to-left of th.4.6(a) is easy, since $\mathcal{L} \cup \mathcal{N}_{\mathcal{F}}(\mathcal{L}) \cup \mathcal{I}_{\mathcal{F}}(\mathcal{L}_{\mathcal{F}}^t) \vdash_{\epsilon} \mathcal{D}_{\mathcal{F}}(\mathcal{L}_{\mathcal{F}}^t)$ by applications of (Irr) and transposition step (3). *Concerning left-to-right:* Assume

(1): $\mathcal{L} \cup \mathcal{L}_{\mathcal{F}}^{t} \cup \mathcal{D}_{\mathcal{F}}(\mathcal{L}_{\mathcal{F}}^{t}) \vdash_{\epsilon} \mathcal{F}x \Rightarrow A$. The premises of (1) are ϵ-consistent by (A) above. Hence by lemma 3, (2): $\mathcal{L}^{\rightarrow} \cup (\mathcal{L}_{\mathcal{F}}^{t})^{\rightarrow} \cup (\mathcal{D}_{\mathcal{F}}(\mathcal{L}_{\mathcal{F}}^{t}))^{\rightarrow} \vdash \mathcal{F}x \rightarrow A$. (2) and (C) above imply that $\mathcal{L}^{\rightarrow} \vdash \mathcal{F}x \rightarrow A$ and hence (3): $\mathcal{M} \cup \mathcal{F} \vdash Aa$. (3) implies by th.3.1 that $\mathcal{D}_{\mathcal{F}}(\mathcal{L}_{\mathcal{F}}^{t}) \vdash_{\epsilon} \mathcal{F}x \Rightarrow A$.

For 6(b): The antecedent of 6(b) implies $\mathcal{D}_{\mathcal{F}}(\mathcal{L}_{\mathcal{F}}^{t}) \vdash_{\epsilon} \mathcal{F}x \Rightarrow A$ by th.4.6(a). Every $D \in \mathcal{D}_{\mathcal{F}}(\mathcal{L}_{\mathcal{F}}^{t})$ results from a unique $L \in \mathcal{L}$ (by transposing and updating) such that for all $p \in \Pi(\mathcal{L}, \mathcal{N}, \mathcal{I})$ satisfying all irrelevance assumptions in $\mathcal{I}_{\mathcal{F}}(\mathcal{L})$, $u(D) \leq min\left(\frac{u(L)}{u(N_L)}, 1\right)$ (by th. 1.2). By th. 1.2 again, this implies our claim. Q.E.D.

References

1. E.W. Adams. *The Logic of Conditionals*. Reidel, Dordrecht, 1975.
2. E.W. Adams. On the logic of high probability. *Journal of Philosophical Logic*, 15:255–279, 1986.
3. F. Bacchus. *Representing and Reasoning with Probabilistic Knowledge*. MIT Press, Cambridge, Mass., 1990.
4. G. Brewka. *Nonmonotonic Reasoning. The Logic of Commonsense*. Cambridge University Press, 1991.
5. R. Carnap. *Logical Foundations of Probability*. University of Chicago Press, second edition, 1962.
6. J.P. Delgrande. An approach to default reaoning based on a first-order conditional logic: revised report. *Artificial Intelligence*, 36:63–90, 1988.
7. A.M. Frisch and P. Haddawy. Anytime deduction for probabilistic logic. *Artificial Intelligence*, 69:93–122, 1994.
8. H. Geffner and J. Pearl. A framework for reasoning with defaults. In H.E. Kyburg et al., editors, *Knowledge Representation and Defeasible Reasoning*, pages 69–87. Kluwer, The Netherlands, 1990.
9. M. Goldszmidt and J. Pearl. Qualitative probabilities for default reasoning, belief revision and causal modeling. *Artificial Intelligence*, 84:57–112, 1996.
10. D. Lehmann and M. Magidor. What does a conditional knowledge base entail? *Artificial Intelligence*, 55:1–60, 1992.
11. A. Y. Levy and Y. Sagiv. Exploiting irrelevance reasoning to guide problem solving. In *Proceedings IJCAI-93*, pages 138–144, Santa Mateo, 1993.
12. J. Pearl. Fusion, propagation, and structuring in belief networks. *Artificial Intelligence*, 29:241–288, 1986.
13. J. Pearl. *Probabilistic Reasoning in Intelligent Systems: Networks of Plausible Inference*. Morgan Kaufmann, San Mateo, CA, 1988.
14. D. Poole. A logical framework for default reasoning. *Artificial Intelligence*, 36:27–47, 1988.
15. D. Poole. Compiling a default reasoning system into prolog. *New Generation Computing*, 9:3–38, 1991.
16. R. Reiter. A logic for default reasoning. *Artificial Intelligence*, 13:81–132, 1980.
17. G. Schurz. Probabilistic justification of default reasoning. In B. Nebel and L. Dreschler-Fischer, editors, *KI-94: Advances in Artificial Intelligence*, pages 248–259. Springer, Berlin, 1994.
18. G. Schurz. Research note: an examination of Delgrande's conditional default logic. IPS preprints 1/97, University of Salzburg, 1997.

Logic for Two: The Semantics of Distributive Substructural Logics

John Slaney and Robert Meyer

Australian National University, Canberra 0200, Australia
{John.Slaney,Robert.Meyer}@anu.edu.au

Abstract. This is an account of the semantics of a family of logics whose paradigm member is the relevant logic R of Anderson and Belnap. The formal semantic theory is well worn, having been discussed in the literature of such logics for over a quarter of a century. What is new here is the explication of that formal machinery in a way intended to make sense of it for those who have claimed it to be esoteric, 'merely formal' or downright impenetrable. Our further goal is to put these logics in the service of practical reasoning systems, since the basic concept of our treatment is that of an agent *a* reasoning to conclusions using as assumptions the theory of agent *b*. This concept requires true multi-agent reasoning, as opposed to what is merely reasoning by multiple agents.

This paper is a companion-piece to [14] which contains an account of a general mechanism for formulating a range of non-classical logics somewhat similar to what are now often called Labelled Deductive Systems. Since [14] was intended for a philosophy journal and since its major message was that many supposedly esoteric logics are really nothing of the kind, it made no attempt to include everything. In fact, it outlined a handful of systems only, treating these as indicative of a much larger class. It also made few claims to originality, being concerned to gather together long-known facts about deduction and present them in an accessible place and form.

The most glaring omission from [14] is a semantic treatment of the logics considered. The natural deductive calculi are given a reading in terms of sequents and motivated by ruminations on "bodies of information", but the formal theory is resolutely syntactic. The present aim is to perform a similar service for the semantics of distributive substructural logic as [14] did for the syntax: to explain the constructions involved and thus to make the case that, far from being contrived and incomprehensible, these logics sustain a very intuitive interpretation.

As before, it must be emphasised that little of the formal material to be found here is new. The main source on which it draws is our work with Routley [10, 11] which in turn was part of a body of research involving Dunn, Fine, Urquhart, Belnap and further back Curry and others. See [1] for a detailed bibliography. We have chosen not to attempt a survey of recent work on semantics for substructural systems. Nonetheless, the relevance to contemporary logical research should be evident. Note that we are neither objecting to other nonclassical systems nor claiming that only the ones treated here make sense; our present aim is only to make sense *at least* of these.

1 Multi-agent Inference

Agents have perspectives not only on what is the case but also on what inferences are warranted. That being so, the implicit beliefs, or *theory*, of agent a according to agent b may be different from those according to agent c or from those of b according to a. Here we do not mean that b and c may make different *errors* about a's beliefs, but that even given perfect knowledge as to what a believes they may differ as to the *upshot*. It is therefore important to study not only a's theory, but more generally a's theory according to b, which we might take to be what follows by taking a's theory as premises and b's as the determinant of implication. Evidently, the logic of multi-agent systems needs a more elaborate notion of closure than the usual one. Where classically we may write $X \vdash A$ to indicate that A is a logical consequence of set X, we need in the more general setting some notation like $X \vdash_Y A$ meaning that A is a consequence of X licensed by Y. We are thinking of X and Y as theories held by agents, but they could equally be generated by possible worlds, situations, pieces of information, programs or other objects; for the present, we are exploring the idea of a double starting point for inference without tying it ontologically to a particular realisation.

Note first what the closure of X under $\vdash_Y$ is *not*. It is not the closure of $X \cup Y$ under $\vdash$. Plausibly, we do not even expect the weaker condition

$$X \vdash_Y A \iff Y \vdash_X A$$

Therefore we do not expect to be able to represent the notion of (binary) closure by means of a unary modal connective and we do not expect the usual apparatus of possible worlds semantics to be adequate to model it.

Nonetheless, it is equally clear that there is a lot of good logic to be extracted from the idea of a generalised closure. Just as some readings of the modal box make it plausible that two boxes take you no further than one—that the accessibility relation is transitive—while others do not, so for instance there are readings of the putting together of agents or their theories which make it reasonable to impose a commutativity condition and readings which do not. And just as modal logic is the right tool for investigating what difference it makes whether or not a transitivity postulate is imposed, so we want formal logic to deliver the machinery necessary to elucidate the difference between commutative and non-commutative combination of sources. Below we shall see more examples of postulates corresponding to systematic properties of logical systems, but before coming to those we need to develop the semantic ideas a little further.

Let us therefore consider agents to have available to them certain information which they may use in various ways. Most significantly for logic:

- They may use it to tell them how the world stands. That is, it warrants them in making assertions.
- They may use it to tell them how the world does not stand. That is, it warrants them in making denials.
- They may use it to tell them what follows from what. That is, it warrants them in making inferences.

Applying a as inference
to b as assertion source
(consequences of $a.b$)

A more complicated example:
consequences of $(a.b).(a.c)$

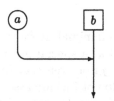
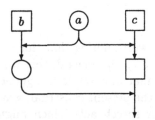

Fig. 1. Multi-agent inferences

The inferences warranted by a piece of information need not be logically valid of course. For instance, the inference from "Timmy is a tiger" to "Timmy eats meat" is a legitimate one for many agents although the warrant for it is fallible biological theory rather than Pure Reason.

Figure 1 shows diagrams of simple examples. The circled pieces of information are being used as sources of inference tickets and the boxed ones are providing premises of those inferences. The diagram is intended as an aid only: this graphical representation is not supposed to have the generality and rigour of a calculus.

We now adopt some more notation. Where i and j are bodies of information, we write

$$i \sqsubseteq j$$

to mean that j is at least as strong as i, so that everything warranted by i is also warranted by j. By extension, we write

$$i \sqsubseteq_k j$$

to mean that j is at least as strong as the result of applying k to i. That is, j warrants everything obtainable from i by means of inferences warranted by k (rather than just those warranted by logic). We shall identify some bodies of information with the agents whose information states they are, but allow that there may be others. In particular, it is useful to consider the information made available by applying one body of information k to another one i and we shall write $k.i$ for this. So evidently,

$$i \sqsubseteq_k j \iff k.i \sqsubseteq j$$

The last of our primitives is a special body of information $\mathcal{L}$ (Logic). It warrants assertion of exactly the logical truths and inference of exactly the logical consequences of that to which it is applied. On this reading, the formal properties of $\mathcal{L}$ are obvious. Inference according to $\mathcal{L}$ is logical inference, so

$$i \sqsubseteq_{\mathcal{L}} j \iff i \sqsubseteq j$$

Put otherwise, what is warranted by applying Logic to i is just what is (really) warranted by i, so we may assume

$$\mathcal{L}.i = i$$

It is worth noting that this is *not* to assume that $\mathcal{L}$ is minimal. If we were to require $\mathcal{L} \sqsubseteq i$ in general, this would have to be imposed as a special postulate. In a similar way, we do not generally suppose that $i.j = j.i$ or even that $i.i = i$.

On the present construal, warrants are strict: they do not include assertions or other speech acts which might be made available to agents under default assumptions which are revised with increased information. Consequently the accumulation of warrants is monotonic. Hence, if $i \sqsubseteq i'$, $j \sqsubseteq j'$, and $k' \sqsubseteq k$ then

$$i' \sqsubseteq_{j'} k' \implies i \sqsubseteq_j k$$

Moreover, the unsubscripted (binary) relation $\sqsubseteq$ is reflexive and transitive.

A set equipped with a ternary relation $\sqsubseteq$ satisfying these conditions and with a distinguished member $\mathcal{L}$ as above shall be called an *agent frame*.

2 Assertion and Logic

Among the acts which may be warranted by a state of information, assertion has a certain priority from the standpoint of logic. Logic deals with theories, and theories are assertoric. In the previous section we were careful to allow that inference is a means of passing between bodies of information where these are construed as something more general than theories. However, in order to be more precise about logic and to understand what sort of reasoning our semantic structures allow we should now focus the account on the assertability of propositions.

By way of notation, we write

$$i \models A$$

to indicate that the formula A is assertible given the information i. This notation is fairly standard. If information i is included in information j then everything warranted by i is warranted by j, because that is what 'included' means (and because we are considering only monotonic reasoning at this stage). So if $i \sqsubseteq j$ then we impose the obvious modelling constraint that for any (atomic) p if $i \models p$ then $j \models p$. An *interpretation* of a propositional language in an agent frame is a determination of which atoms are warranted by which information states in accordance with this heredity constraint.

Some particles exist in languages for the specific purpose of reducing other speech acts to assertions. These particles are of interest to logical theory because it is through them that structural properties of logic affect the content of theories. The simplest of these particles is negation. The negation of a proposition is what is asserted in order to deny that proposition, so the purpose of negation in a language is to reduce denial to assertion. In the same way, the implication connective exists to give us something to assert where an inference is warranted.

558

This last observation is the key to the semantics of implication as well as to its behaviour in proofs. Proof theoretically, a theory contains an implication $A \to B$ if and only if assuming A in the context of that theory suffices for the derivation of B, so the rule for introducing the implication in a proof is exactly that there be a subproof in which B was derived from A. That is, the logic of implication is caught in the deduction theorem:

$$\Gamma \vdash A \to B \iff \Gamma, A \vdash B$$

As we have been at pains to point out many times, in [10] and [14] for instance, this familiar equivalence, while in a sense the whole story about the implication connective, does not suffice to determine which inferences involving that connective are valid. The reason is that the structural rules of the logic are essentially involved. The recent fashion for substructural logics has raised general awareness that there are systems in which the operation which joins Γ and A above lacks some of the familiar properties of set union and hence some of the familiar principles of inference. For example, the idempotence of set union gives rise to the structural rule of contraction: that if a conclusion follows from two assumptions of the same thing then it follows from just one. Logics in which premise combination is not idempotent may invalidate contraction, with the result that such classically valid pure implication formulae as

$$(p \to (p \to q)) \to (p \to q)$$

are not theorems. Again, if the function symbolised by the comma is not commutative then the logic will not permit the structural rule of exchange which allows the order of assumptions to be shuffled. Without exchange, the inference from p to $(p \to q) \to q$ is blocked. Below we shall consider yet more exotic systems in which that operation is not even associative.[1]

At any rate, in terms of the account of agent-based reasoning, the semantics of implication are obvious. When is an agent in possession of a warrant for asserting an implication? Exactly when that agent has a warrant for making the inference from its antecedent to its consequent. That is to say, exactly when applying the agent's information to a warrant for the antecedent yields a warrant for the consequent. In symbols,

$$k \models A \to B \iff \forall i \, (i \models A \implies k.i \models B)$$

which is to say

$$k \models A \to B \iff \forall i \, (i \models A \implies \forall j \, (i \sqsubseteq_k j \implies j \models B))$$

So far we have not mentioned truth, which seems something of an omission in what purports to be an account of semantics. We are coming to that; but first note that we can already give some account of logical validity, for we have "Logic"

---

[1] This possibility is in fact not particularly exotic, since where the arrow is the strict implication of a normal modal logic such as S4 the corresponding fusion operation is not associative, and S4 is just about the most mundane and classical system we know. However, we still find that the idea of a non-associative way in which assumptions may be put together for the sake of argument tends to shock the susceptible.

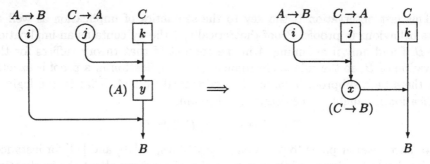

Fig. 2. How associativity validates prefixing

as one of the information states, and presumably this warrants the assertion of exactly the logical theorems. Let us, then, fix this as a definition: A is *verified* by an interpretation in a frame iff $\mathcal{L} \models A$ on that interpretation, and *valid* on a set of agent frames iff it is verified by all interpretations in all of them. This already gives some logical laws concerning the implication connective. The only generally valid formulae so far are the instances of the identity schema

$$A \to A$$

but we do have a number of derivable rules clustered around the affixing principles

$$\models A \to B \implies \models (C \to A) \to (C \to B)$$
$$\models A \to B \implies \models (B \to C) \to (A \to C)$$

Note that the affixing principles are derivable in general only in the form of rules. The corresponding implicational formulae

$$(A \to B) \to ((C \to A) \to (C \to B))$$
$$(A \to B) \to ((B \to C) \to (A \to C))$$

are not valid unless the structures satisfy some additional conditions. For the former, prefixing, law the additional condition is part of associativity for the operation of combining information:

$$i.(j.k) \sqsubseteq (i.j).k$$

which is to say

$$k \sqsubseteq_{i.j} m \implies j.k \sqsubseteq_i m$$

Fairly clearly, we can write this without using dots to indicate combination explicitly:

$$\exists x(j \sqsubseteq_i x \land k \sqsubseteq_x m) \implies \exists y(k \sqsubseteq_j y \land y \sqsubseteq_i m)$$

Figure 2 shows the situation diagrammatically. As before, the circled information states are applied as inference warrants and the boxed ones as assertion warrants.

Whatever four agents derive like this...   ...they could derive like this

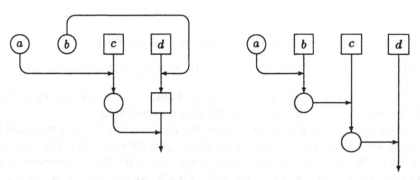

Fig. 3. Postulate for our favourite theorem

Re-associating takes us from the trivial reasoning on the left to the nontrivial prefixing theorem on the right.

There is a well known correspondence between implicational formulae and combinators in type theory. The idea of formulas as types has become so familiar that we shall not rehearse it again in any detail here. What it means for us is that if our implicational logic is to validate a certain set of principles then the corresponding combinatory moves must be the ones allowed in virtue of the structural rules of our deductive system. On the semantic side, structural rules go over into postulates on the relation $\sqsubseteq_i$ strongly reminiscent of those imposed on the binary accessibility relation in modal logic.

At least for logics contained in intuitionist logic, each possible theorem in the pure implication vocabulary corresponds exactly to a combinator (definable in terms of **S** and **K**) and hence to a postulate in the algebra of the application function, and hence also to a condition on the subscripted inclusion relation. To illustrate, suppose we consider validating our favourite theorem

$$(p \to (q \to r)) \to ((s \to q) \to (p \to (s \to r)))$$

This has a simple natural deduction proof:

$$\begin{array}{cc}
x_1 : p \to (q \to r) \qquad x_3 : p & \qquad x_2 : s \to q \qquad x_4 : s \\
\hline
x_1 x_3 : q \to r & \qquad x_2 x_4 : q
\end{array}$$

$$\frac{(x_1 x_3)(x_2 x_4) : r}{\lambda x_4.(x_1 x_3)(x_2 x_4) : s \to r}$$

$$\lambda x_3 x_4.(x_1 x_3)(x_2 x_4) : p \to (s \to r)$$

$$\lambda x_2 x_3 x_4.(x_1 x_3)(x_2 x_4) : (s \to q) \to (p \to (s \to r))$$

$$\lambda x_1 x_2 x_3 x_4.(x_1 x_3)(x_2 x_4) : (p \to (q \to r)) \to (s \to q) \to (p \to (s \to r))$$

Evidently, the combinator **X** corresponding to this proof must satisfy

$$\mathbf{X} x_1 x_2 x_3 x_4 > (x_1 x_3)(x_2 x_4)$$

and the corresponding algebraic postulate is

$$(a.c).(b.d) \sqsubseteq ((a.b).c).d$$

which in the purely relational vocabulary, without the dots, is

$$\exists x \exists y \, (b \sqsubseteq_a x \wedge c \sqsubseteq_x y \wedge d \sqsubseteq_y e) \implies \exists x' \exists y' \, (c \sqsubseteq_a x' \wedge d \sqsubseteq_b y' \wedge y' \sqsubseteq_{x'} e)$$

The diagram of this postulate is given in Figure 3. Clearly, an agent frame validates our favourite theorem iff it satisfies this postulate.

We have often enough[2] listed commonly used axioms for propositional logics along with their corresponding semantic postulates in algebraic and relational form. We shall refrain from repeating such a list here. However, it may be worthwhile to offer a few comments. Firstly, as everyone now knows, the usual structural rules of weakening, contraction and exchange correspond to the combinators **K**, **W** and **C** respectively and hence to the relational postulates:

Weakening: $\quad a \sqsubseteq_b c \implies b \sqsubseteq c$

Contraction: $\quad a \sqsubseteq_b c \implies \exists x (a \sqsubseteq_b x \wedge a \sqsubseteq_x c)$

Exchange: $\quad \exists x (a \sqsubseteq_c x \wedge b \sqsubseteq_x d) \implies \exists x (b \sqsubseteq_c x \wedge a \sqsubseteq_x d)$

The corresponding formulae are of course:

Weakening: $\quad A \to (B \to A)$

Contraction: $\quad (A \to (A \to B)) \to (A \to B)$

Exchange: $\quad (A \to (B \to C)) \to (B \to (A \to C))$

Next, each of these implications can if desired be weakened to the corresponding rule of inference (for example, contraction to the rule that if $A \to (A \to B)$ is a theorem then $A \to B$ is a theorem). In each case, the corresponding combinator is obtained by applying the given one to **I**, thus giving the simple postulates:

KI: $\quad a \sqsubseteq_b c \implies a \sqsubseteq c$

WI: $\quad a \sqsubseteq_a a$

CI: $\quad a \sqsubseteq_b c \implies b \sqsubseteq_a c$

Note that **KI** may be replaced by the even simpler postulate $\mathcal{L} \sqsubseteq a$. **KI** and **CI** correspond via the Curry-Howard isomorphism to the formulae

KI: $\quad A \to (B \to B)$

CI: $\quad A \to ((A \to B) \to B)$

However, there is no normal form for **WI** using just '$\to$': to find a formula to express the rule form of contraction we need conjunction, which will be introduced in the next section.[3]

---

[2] For instance, in [10] and in [12].

[3] In terms of type theory, this amounts to requiring the *intersection types* introduced by Barendregt, Coppo and Dezani [2, 3]. The appropriate formula for **WI** turns out to be $((A \to B) \wedge A) \to B$.

3   Worlds

So far, although there are multiple agents in a frame, these may all be considered as inhabiting a single world. We wish to hang onto that feature for a while, but in order to recast the semantics in a more familiar and more convenient way we shall now consider structures with multiple worlds. The agents may as well all be in one of the worlds (the real world) but of course their *theories* may take us to other possible, or even impossible, worlds. By associating with each agent the set of worlds in which its theory holds, we may abstract from the agents to formulate the semantics purely in terms of worlds. We shall call the resulting structures *world frames*.

What is most significant about a world, from the present standpoint, is the information it contains. That is, the worlds may conveniently be identified with certain of the information states. What distinguishes them from arbitrary information states is a certain completeness. This does not emerge clearly in the pure implication fragment of logic, but in richer languages it does. For instance, information may be disjunctive (we may know that a coin will show either heads or tails, without knowing which) but there cannot be a disjunctive world. The theory of a world need not decide every question, but it must decide those on which it pronounces that an answer exists. So if it supports a disjunction $A \vee B$ then it does so because it supports one of the disjuncts, and it never warrants the assertion of an existential claim $\exists x A$ unless there is an individual in that world of which A is assertible. That is, warranted assertion at a world is truth-like; and there is no such concept as barely disjunctive truth.[4]

We make minimal assumptions here about truth-likeness. It is such as to treat connectives such as disjunction constructively, but we do not suppose that it requires any particular claims to be assertible. Indeed, we allow even the null information state, that warrants nothing at all (the belief state of the empty database, for instance) to count as a world, since it does no violence to the logical constants. However, the logical truths are (necessarily) true, so if a world is a *possible* one then it will at least verify those.

One of the advantages of the move to worlds is precisely that they yield very natural semantics for extensional connectives. The disadvantage is that carrying over the treatment of such particles as implication from agent frames is not trivial. For one thing, there has to be an account in terms of worlds of information states which are not themselves worlds, such as the special state $\mathcal{L}$ which is needed to define validity.[5] A more difficult problem is that while for information states i and j there exists the information state $i.j$ got by applying the one to the other, so that we can think of $j \sqsubseteq_i k$ as meaning that $i.j$ is contained in k, the same is not true of worlds: even if i and j are worlds, $i.j$ generally is

---

[4] We are aware that this opposition to nondeterministic truth is one of the distinctive features of the present account; so such logics as these are good for *that kind* of reasoning. We do not see this as a vitiating limitation.

[5] In many of the logics, even with negation present, the set of theorems *can* be the theory of a world. See [13, 6] for this result. In general, however, respect for Logic should not be so great that we take the logical truths to be the *only* truths.

not and we must prove that we can get by without it. The solution, roughly speaking, is to show that the semantic purposes served by any information state i may be served by the set of worlds w such that $i \sqsubseteq w$. In particular, the set $\mathcal{N}$ (for 'normal') of worlds which extend $\mathcal{L}$ is picked out and made into one of the defining features of a world frame. Notice that the purely relational versions of postulates, written without using dots, do make sense for worlds, which is why we introduced that notation.

For the purposes of this section, we restrict attention to propositional logic in the connectives $\rightarrow$, $\wedge$ and $\vee$ only and thus to *positive* world frames. Formally, a positive world frame is a 4-tuple $\langle G, \mathcal{N}, K, \sqsubseteq \rangle$ where $G \in \mathcal{N} \subseteq K$ (the real world is normal and normal worlds are worlds) and $\sqsubseteq$ is a ternary relation on K. We define inclusion and postulate reflexivity and monotonicity:

Df.1 $a \sqsubseteq b \equiv \exists x (x \in \mathcal{N} \wedge a \sqsubseteq_x b)$

P1. $a \sqsubseteq a$

P2.   If $a \sqsubseteq a'$, $b \sqsubseteq b'$ and $c' \sqsubseteq c$ then if $a' \sqsubseteq_{b'} c'$ then $a \sqsubseteq_b c$

As for agent frames, valuation in a world frame is a function v assigning to each propositional variable a subset of K closed under (binary) $\sqsubseteq$. Each valuation v induces a relation $\models_v$ between worlds and formulae:

Ip $w \models_v p$ iff $w \in v(p)$

I$\wedge$ $w \models_v A \wedge B$ iff $w \models_v A$ and $w \models_v B$

I$\vee$ $w \models_v A \vee B$ iff $w \models_v A$ or $w \models_v B$

I$\rightarrow$ $w \models_v A \rightarrow B$ iff $\forall x \forall y$ if $x \sqsubseteq_w y$ and $x \models_v A$ then $y \models_v B$

Formula A is true on v iff $G \models_v A$ and valid in a set of frames iff true on every valuation in every frame in the set.

The logic of the extensional connectives $\wedge$ and $\vee$ according to world frames is very simple. The following formulae are valid in all frames:

$\wedge$E1 $(A \wedge B) \rightarrow A$

$\wedge$E2 $(A \wedge B) \rightarrow B$

$\wedge$I $((A \rightarrow B) \wedge (A \rightarrow C)) \rightarrow (A \rightarrow (B \wedge C))$

$\vee$I1 $A \rightarrow (A \vee B)$

$\vee$I2 $B \rightarrow (A \vee B)$

$\wedge$I $((A \rightarrow C) \wedge (B \rightarrow C)) \rightarrow ((A \vee B) \rightarrow C)$

$\wedge$V $(A \wedge (B \vee C)) \rightarrow ((A \wedge B) \vee (A \wedge C))$

Moreover, if A and B are both valid in a frame then so is $A \wedge B$. To obtain a Hilbert formulation of one of the logics characterised as in this paper, it suffices to take an axiomatisation of its implicational fragment with detachment as a rule of inference and conservatively extend it by adding the above axioms and the rule of adjunction. Thus the theory of extensional reasoning is very classical and remains stable over a wide range of logics.

To get sets of world frames modelling particular logics it normally suffices to lift the postulates on $\sqsubseteq$ (in the dot-free notation, of course) directly from the corresponding agent frames. Reduction to dot-free notation is achieved by rewriting according to the definitions:

Df.2 $\alpha \sqsubseteq \beta.\gamma \equiv \forall x(\gamma \sqsubseteq_\beta x \implies \alpha \sqsubseteq x)$

Df.3 $\alpha.\beta \sqsubseteq_\gamma y \equiv \exists x(\beta \sqsubseteq_\alpha x \wedge x \sqsubseteq_\gamma y)$

Df.4 $\alpha \sqsubseteq_{\beta.\gamma} y \equiv \exists x(\gamma \sqsubseteq_\beta x \wedge \alpha \sqsubseteq_x y)$

The questions of which conditions can be transferred in this way and of what sorts of logic they determine have not been systematically addressed, though there is a body of work on related questions in modal logic.

In order to establish that no generality is lost in the move to world frames a completeness theorem is needed. We shall not prove it here, since variations on it have been published in several places, most comprehensively in [12] and it would be inappropriate for this paper. In outline, it proceeds in the standard way by concocting frames from the language. Where S is any set of formulae, an S theory may be defined as a set T of formulae closed under adjunction and such that $S \circ T \subseteq T$. Here $S \circ T$ is $\{B : \exists A \in T\,(A \to B \in S)\}$. If in addition $S \subseteq T$ then T is said to be a *regular* S theory. The notion is most important when S is itself a regular L theory for some logic L. A theory is *prime* iff whenever it contains a disjunction $A \vee B$ it contains one of the disjuncts A or B. The key idea for completeness is to define a frame whose worlds are the prime S theories for an appropriately chosen S, letting $T \sqsubseteq_V U$ mean $V \circ T \subseteq U$ and letting $\mathcal{N}$ be the regular prime S theories. A canonical valuation $\mathcal{V}$ is then chosen to make $T \models_\mathcal{V} A$ hold iff $A \in T$. The two crucial lemmas, related to Lindenbaum's lemma and proved in a similar way, are the Priming and Squeeze lemmas. Where L is the logic whose completeness is being proved and S is a regular L theory:

Priming: Every S theory is the intersection of its prime S supertheories.

Squeeze: Let T and U be S theories and let P be a prime S theory such that $T \circ U \subseteq P$. Then there exist prime S theories X and Y such that $T \subseteq X$, $U \subseteq Y$ and $X \circ Y \subseteq P$.

The upshot of these lemmas is that prime theories can make all the distinctions made by any theories, which is to say worlds can perform in place of arbitrary information states. With a little work, for which see [12], the completeness results flow from this.

4 Denial

Our last topic for the present account is negation. We started by indicating three speech acts—assertion, denial and inference—which are of importance to logic. Of these, assertion and inference are treated in the theory of positive world frames and the agent-based reasoning schemes which they incorporate, but so far we have said little about denial. The account of the standard propositional logics will not be adequate until we fill this gap.

Our view of denial is that it is an operation on propositions which reverses their sense. Re-reversing the reversed sense restores it to the original state. Consequently, while we do not impose on agents any sort of completeness requirement, that they pronounce on every issue, nor any sort of rationality constraint that they deny only what they fail to assert, we do suppose that re-denying a denial undoes it so that the denial operation is of period 2.

The semantic constructions of the present paper are in terms of information states, which are positively oriented towards logical closure and the like, but what is denied by an agent is naturally negative and so is not conveniently interpreted as another information state. One solution would be to associate two entities with each agent: the theory representing what is accepted and the anti-theory representing what is rejected. A neater solution, however, is to work instead with two positive objects, one corresponding to what is actively asserted and the other to what is "passively" asserted by failing to be denied.

With each information state i, therefore, we associate the dual information state i^* which warrants the assertion of A exactly when i does not warrant the denial of A, and conversely warrants the denial of exactly what is not assertible given i. If an information state j is stronger than i then not only are the assertions warranted by i a subset of those warranted by j, but so are the denials. Hence in that case, what *fails* to be deniable given j also fails to be deniable given i. The interaction of assertion and denial is therefore summed up in the two principles

$$i^{**} = i$$
$$i \sqsubseteq j \iff j^* \sqsubseteq i^*$$

Moreover, since denial is reversal of sense, we expect that it will interact not only with assertion but also with inference. Whatever information warrants the assertion of A given B equally warrants the denial of B given that of A. That is, we may generalise the second postulate above:

$$i \sqsubseteq_k j \iff j^* \sqsubseteq_k i^*$$

As already noted, the negation of A is what is asserted in order to deny A, so the semantic postulate for negation is obvious: i warrants the assertion of $\neg A$ if and only if i^* does not warrant the assertion of A. However, not everything in the semantic picture is as pretty. The problem is that in introducing the star as a new primitive we have dualised everything, so that the assertibility conditions for conjunctions are now as awkward as those for disjunctions. It will not do to suppose that $A \wedge B$ is assertible just according to information that suffices for A and B individually, for the information state in question may be a^*, warranting just what agent a fails to reject. And of course nothing prevents an agent from rejecting a conjunction even if neither conjunct on its own is rejectable. As might be expected, this situation is exactly dual to the usual semantic difficulty raised by disjunction, and the solution is the same. While the image under $\star$ of a theory is not in general a theory, the image of a *prime* theory is always a prime theory.

That is, while it is inconvenient to work with ⋆ as an operation on arbitrary information states, it makes excellent sense to work with it as an operation on worlds.

As we showed long ago [8] the addition of negation in this manner conservatively extends the distributive substructural logics. An essential feature for this to hold is the existence of worlds which are not fixed points for ⋆ and hence which are incomplete or inconsistent with respect to negation. However, we might expect that the real world, where truth is determined, is better behaved than most. Of course, unless the logic in question validates the law of the excluded middle $A \lor \neg A$ the real world cannot generally be assumed to be complete, but we may reasonably demand that it be consistent. It is a nontrivial metatheorem (nearly thirty years old but still nontrivial) that for a wide range of these substructural logics such a consistency assumption may be imposed without loss of generality.[6]

5 Glimpses beyond

In this paper, we have explicated the standard semantics of distributive substructural propositional logics as it has been known, though not well enough known, since the early 1970s. The originality has been in our less formal remarks, where we have striven to emphasise the naturalness of the account by relating it to agent-based systems. The idea that inference comes from the interaction of *two* inputs, the facts and the laws, seems so obvious to us that we have long been surprised that it should not seem so to everybody. We hope to have shown that the frame semantics of these logics, are much simpler and more intuitive than a superficial glance at all the stars and ternary relations might suggest.

In closing, we wish to point more to the future than the past, by gesturing at two new investigations which arise naturally from the foregoing discussion and which must be pursued if logics like these are really to find their place in practical reason.

Firstly, we have been at pains to point out that the logics treated here have monotonicity deeply built in, though many of them sustain "resource-sensitive" readings on which (multiplicative) extensions of the premises of an inference may not preserve its validity. What might be done with the present semantics to model nonmonotonic reasoning such as occurs in belief revision or in update is an open question. For some preliminaries, see [15] where it is shown that there are good prospects of coping with nonmonotonicity in a paraconsistent context. A fuller account, making use of the monotonic implication operator to express lawlike connection, would give rise to more structure than we found in [15], where we simply assumed that some system of spheres would be imposed in a rather classical fashion on a set of possibly inconsistent worlds.

---

[6] See [4]. What makes it nontrivial is that while a theory is the intersection of its prime supertheories, a consistent theory need not be the intersection of its *consistent* prime supertheories. Peano arithmetic based on reasonable distributive substructural logics provides an important counter-example [5].

Secondly, the effect of taking worlds as semantically primitive is to remove from the story the agents who were supposed to be at the base of it all. A very natural development is to put them back in the manner of the usual epistemic logics. As well as giving direct expression to notions such as agents' beliefs in various worlds, this may be combined with the more elaborate apparatus of the above semantics to express, for instance, the consequences according to agent *b* of what *b* *thinks* is in the theory of agent *a*. This may, after all, have direct applications in multi-agent systems. Put another way, the logics considered in this paper allow for reasoning that is genuinely *multi-agent* in that it essentially requires information states to interact. It may be useful to combine this with more familiar *multiple agent* reasoning, in which the theories of several agents are treated with unary modal operators. Even single-agent epistemic logic with a substructural base is little investigated and likely to be more difficult than might be imagined. See [7, 9] for an account of how even a simple modal logic like S4 raises hard problems for this project. The effects obtainable by extending the logics of this paper with multiple epistemic operators are yet unknown.

References

1. A.R. Anderson, N.D. Belnap and J.M. Dunn, *Entailment, vol. II.* Princeton, 1992.
2. H.P. Barendregt, M. Coppo and M. Dezani-Ciancaglini, A Filter Lambda Model and the Completeness of Type Assignment. *Journal of Symbolic Logic* 48 (1983) 931–940.
3. M. Dezani-Ciancaglini and J.R. Hindley, Intersection Types for Combinatory Logic. *Theoretical Computer Science* 100 (1992) 303–324.
4. J.M. Dunn, Relevance Logic and Entailment. Gabbay & Günthner (ed) *Handbook of Philosophical Logic, vol. 3.* Dordrecht, 1986, 117–229.
5. H. Friedman and R.K. Meyer, Whither Relevant Arithmetic? *Journal of Symbolic Logic* 57 (1992) 824-831.
6. S. Giambrone, Real Reduced Models for Relevant Logics without WI. *Notre Dame Journal of Formal Logic* 33 (1992) 442-449
7. E.D. Mares and R.K. Meyer, The Admissibility of Gamma in R4. *Notre Dame Journal of Formal Logic* 33 (1992) 197-206.
8. R.K. Meyer, Intuitionism, Entailment, Negation. Leblanc (ed) *Truth, Syntax and Modality.* Amsterdam, 1973, 168–198.
9. R.K. Meyer and E.D. Mares, The Semantics of Entailment 0, Kosta Došen & Peter Schroeder-Heister (Eds.), *Substructural Logics,* Oxford, OUP, 1993 239-258.
10. R.K. Meyer and F.R. Routley, Algebraic Analysis of Entailment 1, *Logique et Analyse* 15 (1972) 407–428.
11. F.R. Routley and R.K. Meyer, Semantics of Entailment. Leblanc (ed) *Truth, Syntax and Modality.* Amsterdam, 1973, 199–243.
12. F.R. Routley, R.K. Meyer, R.T. Brady and V. Plumwood, *Relevant Logics and their Rivals,* Atascadero, CA, 1983.
13. J.K. Slaney, Reduced Models for Relevant Logics Without WI, *Notre Dame Journal of Formal Logic* 28 (1987) 395–407.
14. J.K. Slaney, A General Logic, *Australasian Journal of Philosophy* 68 (1990) 74–88.
15. J.K. Slaney and G.A. Restall, Realistic Belief Revision. *Proc. Second World Conference on the Fundamentals of AI.* Paris, 1995, 367–378.

Multivalued Extension of Conditional Belief Functions *

Anna Slobodová

Institute of Control Theory and Robotics, Slovak Academy of Sciences,
Dúbravská cesta 9, 842 37 Bratislava, Slovakia
E-mail: utrraslo@nic.savba.sk

Abstract. Uncertainty is present in most tasks that require intelligent behaviour, such as reasoning, decision making, etc. Therefore, several methods were proposed for handling uncertainty. One of them is the Dempster-Shafer theory. Its central problem is conditioning. We want to study here a multivalued extension of conditional belief function and examine its properties.

1 Introduction

The advantage of the Dempster-Shafer theory is its ability to model the narrowing of the hypothesis set with accumulation of evidence.

This theory uses a number in the range [0, 1] to indicate belief in a hypothesis given a piece of evidence. This number is the degree to which the evidence supports the hypothesis. The impact of each distinct piece of evidence on the subsets of Ω is represented by a function called *a basic probability assignment* (bpa).

A bpa is a generalization of the traditional probability density function: the latter assigns a number in the range [0, 1] to every singleton of Ω such that the numbers sum to 1. Using 2^Ω, the enlarged domain of all subsets of Ω, a bpa denoted m assigns a number in [0 , 1] to every subset of Ω such that the numbers sum to 1. The quantity m(A) is a measure of that portion of the total belief committed exactly to A, where A is an element of 2^Ω and the total belief is 1. This portion of belief cannot be further subdivided among the subsets of A and does not include portions of belief committed to subsets of A. Since belief in a subset certainly entails belief in subsets containing that subset, it is useful to define a function that computes a total amount of belief in A. This quantity will include not only belief committed exactly to A, but belief committed to all subsets of A. Such a function is called a *belief function*.

How should one update one's belief given new evidence? If beliefs are expressed in terms of probability, then the standard approach is to use conditioning.

---
* This work has been supported by the grant of Scientific Grant Agency of Ministry of Slovak Republic and Slovak Academy of Sciences No. 2/1189/96

It was shown by Spies in [8] that a full generalization of Bayesian methods used in probabilistic expert systems to belief functions is possible, where an updating rule with milder conditions is proposed. Spies considered to express uncertainty via discrete random sets, defined conditional belief functions and presented Jeffrey's rule for belief functions.

First, we shall introduce several necessary definitions and notions.([8])

2 Conditional Events

Conditional events were considered, following Boole's ideas, by several researchers. (see, for instance, the monograph [3], [2]).

Assume a finite probability space $(\Omega, \mathcal{A}, P)$ and a multivalued mapping T: $\Omega \to \mathcal{P}(\Omega')$, where the finite range Ω' carries σ-algebra $\mathcal{A}'$. Such a multivalued mapping is called a random set. It gives rise to probability intervals defined by a lower bound, called lower probability in Dempster [1], degree of belief in Shafer [6]. Let us say, for $A' \in \mathcal{A}'$,

$$Bel(A') = P(T \subseteq A') = P(\{\omega | T(\omega) \subseteq A'\}) . \tag{1}$$

Similarly, an upper bound of the probabilities of events under random set T can be defined. It is called upper probability in Dempster [1], degree of plausibility in Shafer [6]. Let us define

$$Pl(A') = P(T \cap A' \neq \emptyset) = P(\{\omega | T(\omega) \cap A' \neq \emptyset\}) . \tag{2}$$

Let us take a look at the events for which a conditional probability measure P_A assumes a common value. There is a set of events for which this measure vanishes, the so-called P_A-null sets: $\mathcal{N}(P_A) = \{B | B \in \mathcal{A}, P_A(B) = 0\}$.

Call the maximal null set N_A. Taking the power set operator $\mathcal{P}$, we may write $\mathcal{N}(P_A) = \mathcal{P}(N_A)$. We also can write the P_A-null sets as $\mathcal{P}(N_A) = \{X | \exists Y \in \mathcal{A} : X = Y \cap N_A\}$. In the theory of Boolean rings, such a set of subsets is called a principal ideal. A principal ideal in a Boolean ring has the important property to give rise an equivalence relation on the underlying universe, each equivalence class being obtained by taking, for a fixed set Z, the symmetric differences to all members of $\mathcal{P}(N_A)$; shorthand notation: $Z \triangle \mathcal{P}(N_A)$. The equivalence classes thus obtained are also called residue classes. The partition obtained this way is called quotient ring of the universe modulo the ideal. It is peculiar to Boolean rings that any two sets are in the same equivalence class if their symmetric difference is in the principal ideal.

The important observation linking principal ideals and null sets under conditioning is that by forming the set of symmetric differences of any fixed element of $\mathcal{A}$ with the members of $\mathcal{N}(P_A)$ we obtain an equivalence class of sets with a single value of P_A.

Thus, it is reasonable to view the equivalence classes generated by the principal ideal as conditional events, because the sets of equal probability under P_A form a single such class or a union of them.

We will assume from now on that A, the antecedent, contains only atoms with nonzero probability. As a consequence of this assumption $N_A = \overline{A}$ and $\mathcal{N}(P_A) = \mathcal{P}(\overline{A})$. Now we define a conditional event.

Definition 1. [3]: A *conditional event* $[B|A]$ with $A, B \subseteq \Omega$ is a set of events with the same conditional probability, given A, as $P_A(B)$. Assuming $N_A = \overline{A}$, it is given by the residue class

$$[B|A] = \{C | \exists Z \in \mathcal{N}(P_A) : C = B \bigtriangleup Z\} := B \bigtriangleup \mathcal{N}(P_A) \ . \tag{3}$$

This equivalence class belongs to a partition obtained with the principal ideal generated by $\mathcal{N}(P_A) = \mathcal{P}(\overline{A})$.

Besides this view of conditional events there is another view, which is related to the logical meaning of conditioning.

Definition 2. [3]: A *conditional event* $[B|A]$ is an interval of sets ranging from $B \cap A$ to $\overline{A} \cup B$

$$[B|A] = \{C | A \cap B \subseteq C \subseteq \overline{A} \cup B\} \ . \tag{4}$$

We can mention a consequence of the fact that conditional events are residue classes. Such equivalence classes enjoy the property of preserving the algebraic structure given by the operations $\cup$ and $\cap$. Thus, equivalent events remain equivalent if the same event is joined to or intersected with each of them. Such equivalence relations are called congruence relations.([7])

3 Conditional Belief Function

Definition 3. [8]: Let T: $\Omega \to \mathcal{P}(\Omega')$ be a multivalued mapping. A *conditional multivalued mapping* T_B for a fixed $B \subseteq \Omega'$ is such a mapping restricted to have its images in the set of conditional events $[W|B]$, where $W \subseteq \Omega'$. We have $T_B : \omega \mapsto [W|B]$, where $W = T(\omega)$. In short, $T_B(\omega) = [T(\omega)|B]$.

Using the definitions and observations on conditional events, it is plausible to say that conditional belief refers to an uncertainty with respect not to a single set but to a whole interval of sets, namely those in the conditional object. This motivates the following simple definition.

Definition 4. [8]: A *conditional belief function w.r.t. the antecedent* (conditioning subset) $B \subseteq \Omega'$ is a belief function whose focal elements are in the set of conditional events $\Omega'_B = \{X | \exists Y \subseteq \Omega' s.t. X = [Y|B]\}$. Such a belief function is given by

$$Bel_{T_B}(C) = P(T_B = [C|B]) \ , \tag{5}$$

where T_B is a conditional multivalued mapping and the righthand side is short for $P(\{\omega | T_B(\omega) = [C|B]\})$. Notation: $Bel([.|B])$.

The important consequence of this definition is that a conditional belief function, given some event $B \subseteq \Omega'$ is not a belief function on a subalgebra on $\{Y|Y = C \cap B, C \subseteq \Omega'\}$. Thus, using conditional events, conditioning is expressed by simplifying an existing algebra using equivalence classes, while, in the conventional description, conditioning is expressed by restricting the universe of events.

4 Properties of Conditional Belief Functions

Sometimes we have the situation: we obtain conditional degrees of belief without access to a marginal belief function to which Bayesian conditioning could be applied. In order to accomplish marginalization and combination in this case, we need to see what happens if we apply Dempster's rule of combination to conditional belief functions.

It is possible to show ([8]), conditional belief functions are closed under Dempster's rule of combination.

Now we will examine whether conditional belief functions obey the following requirements:

First, the following property will allow repeated updating:

$$Bel_B([A|C]) = Bel([A|B \cap C]) \tag{6}$$

where $Bel_B([.]) = Bel([.|B])$.

Second, if two diametrically opposed assumptions impart two different degrees of belief onto a proposition A, then the unconditional degree of belief merited by A should be somewhere between the two. In other words, if every possible outcome of an experiment would lead you to choose the same action, then you ought to choose that action without running the experiment. So $Bel(A)$ should be sandwiched between $Bel([A|B])$ and $Bel([A|\overline{B}])$, i.e. $Bel([.|.])$ should satisfy the sandwich principle:

$$\min(Bel([A|B]), Bel([A|\overline{B}])) \le Bel(A) \le \max(Bel([A|B]), Bel([A|\overline{B}])) \tag{7}$$

for any $A, B \in \mathcal{A}'$ with $Bel(B) \ne 0$ and $Bel(\overline{B}) \ne 0$, where $\overline{B}$ is the complement of B.

Now, we will deal with the first requirement. It is possible to find examples that conditional belief functions, in general *do not* satisfy the first requirement.

Example 1. Let us take a probability space $(\Omega, \mathcal{A}, P)$ with five atoms $\Omega = \{a, b, c, d, e\}$. For simplicity, assume $P(x) = 0, 2; x \in \Omega$. A multivalued mapping can be introduced from our sample space Ω to a new space $\Omega' = \{x_0, x_1, x_2, x_3\}$. We take a random set T with the following definition:

$$
\begin{aligned}
T(a) &= \{x_0, x_1, x_2, x_3\} \\
T(b) &= \{x_0\} \\
T(c) &= \{x_1\} \\
T(d) &= \{x_2\} \\
T(e) &= \{x_3\} \ .
\end{aligned}
\tag{8}
$$

Suppose we distinguish only between pairs of even in Ω' and define:

$$A = \{x_0, x_2\} \tag{9}$$
$$B = \{x_1, x_3\} \tag{10}$$
$$C = \{x_0, x_3\} . \tag{11}$$

Evidently, T induces the following degrees of belief on Ω'

$$Bel(A) = P(b) + P(d) \tag{12}$$
$$Bel(B) = P(c) + P(e) \tag{13}$$
$$Bel(C) = P(b) + P(e) . \tag{14}$$

According to Bayesian conditioning this leads to the following conditional beliefs:

$$Bel_B([A|C]) = \frac{1}{3}, \tag{15}$$
$$Bel([A|B \cap C]) = 0 . \tag{16}$$

So, we can see the first requirement is not fulfilled.

Further, it is possible to find examples that conditional belief functions, in general, *do not* satisfy the second requirement.

Example 2. Let us take a probability space $(\Omega, \mathcal{A}, P)$ with five atoms $\Omega = \{a, b, c, d, e\}$. For simplicity, assume $P(x) = 0, 2$; $x \in \Omega$. A multivalued mapping can be introduced from our sample space Ω to a new space $\Omega' = \{x_0, x_1, x_2, x_3\}$. We take a random set T with the following definition

$$
\begin{aligned}
T(a) &= \{x_0, x_1, x_2, x_3\} \\
T(b) &= \{x_0\} \\
T(c) &= \{x_1\} \\
T(d) &= \{x_2\} \\
T(e) &= \{x_3\} .
\end{aligned}
\tag{17}
$$

Suppose we distinguish only between pairs of even in Ω' and define

$$A = \{x_0, x_2\} \tag{18}$$
$$B = \{x_0, x_1\} \tag{19}$$
$$C = \{x_0, x_3\} . \tag{20}$$

Evidently, T induces the following degrees of belief on Ω'

$$Bel(A) = P(b) + P(d) \tag{21}$$
$$Bel(B) = P(b) + P(c) \tag{22}$$
$$Bel(C) = P(b) + P(e) . \tag{23}$$

Suppose now that we wish to condition our beliefs on subset B of Ω'. Then we have to change our frame Ω' to a partition induced by conditional events

$[.|B]$. According to Bayesian conditioning this leads to the following conditional beliefs:

$$Bel([A|B]) = \frac{P(b)}{1 - P(d) - P(e)},\tag{24}$$

$$Bel([A|\overline{B}]) = \frac{P(d)}{1 - P(b) - P(c)},\tag{25}$$

$$Bel([\Omega'|A]) = 1 \ .\tag{26}$$

Now we have

$$\min(Bel([A|B]), Bel([A|\overline{B}])) = \frac{1}{3},\tag{27}$$

$$\max(Bel([A|B]), Bel([A|\overline{B}])) = \frac{1}{3},\tag{28}$$

$$Bel(A) = 0,4 \ .\tag{29}$$

Thus, the sandwich principle is not satisfied.

References

1. Dempster, A. P.: Upper and Lower Probabilities Induced by a Multivalued Mapping. Ann. Math. Stat. **38** (1967) 325–339.
2. Goodman, I., Nguyen, H.: Conditional Objects and the Modelling of Uncertainties. In: Fuzzy Computing. Elsevier Science: New York (1988) 119–138.
3. Goodman, I., Nguyen, H. and Walker, E.: Conditional Inference and Logic for Intelligent Systems. Elsevier: Amsterdam. (1991).
4. Kyburg, H.: Bayesian and Non-Bayesian Evidential Updating. Artificial Intelligence. **31** (1987) 271–293.
5. Pearl, J.: Probabilistic Reasoning in Intelligent Systems. San Mateo: Morgan Kaufmann. (1988).
6. Shafer, G.: A Mathematical Theory of Evidence. Princeton University Press. Princeton. (1976).
7. Spies, M.: Combination of Evidence with Conditional Objects and its Application to Cognitive Modeling. In: Conditional Logic in Expert Systems. North-Holland. (1991) 181–210.
8. Spies, M.: Conditional Events, Conditioning, and Random Sets. IEEE Transactions on Systems, Man, and Cybernetics. **24** (1994) 1755–1763.

Combining Evidence under Partial Ignorance

Frans Voorbraak*

ILLC, Department of Mathematics and Computer Science
University of Amsterdam, Plantage Muidergracht 24
1018 TV Amsterdam, The Netherlands
fransv@wins.uva.nl

Abstract. In this paper, we discuss the problem of combining several pieces of uncertain evidence, such as provided by symptoms, expert opinions, or sensor readings. Several of the proposed methods for combining evidence are reviewed and criticized. We argue for the position that (1) in general these proposed methods are inadequate, (2) strictly speaking, the only justifiable solution is to carefully model the situation, (3) a careful modelling of the situation requires a distinction between ignorance and uncertainty, and (4) drawing useful conclusions in the presence of ignorance may require additional assumptions which are not derivable from the available evidence.

1 Introduction

Combining and weighing evidence is an important aspect of reasoning, but it is hard to formalize, at least when the evidence is uncertain. In AI and robotics, the problem of combining pieces of uncertain evidence is an important problem, which appears in many guises. (Sensor fusion for autonomous vehicles, coordination of several agents in an uncertain environment, diagnosis based on several inconclusive symptoms, et cetera.) Although defeasible reasoning may include some resolution of conflicting uncertain evidence, we feel that numeric approaches to uncertainty are best suited to handle subtleties of uncertain reasoning such as the possible mutual reinforcement of weak pieces of evidence.

Ideally, the available evidence leads to degrees of belief which can be represented by a probability function. To cover also less ideal situations, we propose a representation of the belief states of ideally rational agents with probabilistic information, without assuming that the available information determines a probability function. In other words, we allow agents to be partially ignorant about the uncertainty of events.

The assumption that a probability function is the unique correct representation of uncertainty has also been attacked by the proponents of Dempster-Shafer theory (DS theory). The belief functions of DS theory also allow the representation of ignorance next to uncertainty. However, the (partial) probabilistic belief

* The investigations were carried out as part of the PIONIER-project Reasoning with Uncertainty, subsidized by the Netherlands Organization of Scientific Research (NWO), under grant pgs-22-262.

states proposed in this paper are more general than the belief functions of DS theory.

We will frequently compare our approach with DS theory, since this is the most popular theory for combining evidence under partial ignorance. The main argument of the paper can probably be understood without any prior knowledge of DS theory. Readers not familiar with the theory can find the definitions of the essential notions in [4, 9, 10, 15] and in many other publications on DS theory or uncertainty in AI.

In the remainder of this paper, we first review and criticize several schemes for combining evidence in probability theory. We then argue for partial probability theory (PPT) in which in general only constraints on probabilities are represented. We further argue that the bounds on probability values sanctioned by the information may very well be too wide to be of practical use, but that the tighter bounds obtained by the proposed combination schemes (including Dempster's rule of DS theory) rely on (implicit) assumptions concerning the interaction of the pieces of evidence which cannot be justified in general, and are often highly counterintuitive.

We claim that in concrete situations, it is often possible to come up with intuitively acceptable assumptions which allow useful conclusions to be obtained by probabilistic reasoning only (without Dempster's rule). To illustrate our position we treat an example of sensor fusion in robotics. We further discuss the notion of conditioning in the context of partial probability theory.

2 Combining Probability Distributions

We briefly review some combination schemes for probability distributions that appear in the literature. A more thorough treatment can be found in [3]. Throughout this paper, $P_1, P_2, \ldots, P_n$ denote probability functions over a finite sample space Θ, and T denotes a combination scheme. In other words, $T(P_1, P_2, \ldots, P_n)$ is a probability function over Θ which is proposed as a summary or combination of $P_1, P_2, \ldots, P_n$.

The first scheme for combining probability distributions we consider is the linear opinion pool, which takes a "weighted average" of the probability functions to be combined.

Definition 1 (linear opinion pool). The linear opinion pool T_{lin} is given by

$$T_{lin}(P_1, \ldots, P_n) = \sum_{i=1}^{n} w_i P_i, \text{where } w_i \geq 0 \text{ and } \sum_{i=1}^{n} w_i = 1. \qquad (1)$$

It is easy to check that $T_{lin}(P_1, \ldots, P_n)$ is a probability function over Θ. McConway and Wagner independently proved that the linear opinion pool is unavoidable if one wants the combined probability to be a function of the individual probabilities. In other words, $T = T_{lin}$, whenever there exists a function $f : [0,1]^n \longrightarrow [0,1]$ such that

$$\text{for every } A \subseteq \Theta, T(P_1, \ldots, P_n)(A) = f(P_1(A), \ldots, P_n(A)). \qquad (2)$$

Another popular combination scheme is the independent opinion pool defined below.

Definition 2 (independent opinion pool.) The independent opinion pool T_{ind} assigns to $P_1, \ldots, P_n$ the probability function given by

$$\text{for every } \theta \in \Theta, T_{ind}(P_1, \ldots, P_n)(\{\theta\}) = \frac{\prod_{i=1}^{n} P_i(\{\theta\})}{\sum_{\theta \in \Theta} \prod_{i=1}^{n} P_i(\{\theta\})}. \tag{3}$$

Notice that (3) indeed determines a probability function over Θ, since by the choice of the denominator the values of the function $T_{ind}(P_1, \ldots, P_n)$ on elementary events sum to 1. The independent opinion pool can be generalized by allowing the different opinions or pieces of evidence to have different weights. This is achieved in the following combination scheme.

Definition 3 (logarithmic opinion pool.) The logarithmic opinion pool T_{log} assigns to $P_1, \ldots, P_n$ the probability function given by

$$\text{for every } \theta \in \Theta, T_{ind}(P_1, \ldots, P_n)(\{\theta\}) = k \cdot \prod_{i=1}^{n} (P_i(\{\theta\}))^{w_i}, \tag{4}$$

where the w_i are positive and k is the appropriate normalizing constant.

The above defined combination schemes have serious weaknesses. For example, the linear opinion pool does not generalize the combination of certain evidence by means of the logical conjunction.

Example 1. Let $\Theta = \{a, b, c\}$, and let P_1 and P_2 be the probability functions over Θ given by the following table.

	a	b	c
P_1	0.9	0.1	0
P_2	0	0.1	0.9
$T_{lin}(P_1, P_2)$	0.45	0.1	0.45

It follows from P_1 that a or b is the case, and c is not the case. It follows from P_2 that b or c is the case, and a is not the case. By logical reasoning one can conclude that b must be the case. However, there are no weights such that the linear opinion pool admits this conclusion. For example, the above table describes $T_{lin}(P_1, P_2)$ in case P_1 and P_2 are given equal weight.

The linear opinion pool does not allow reinforcement of opinion, since the value $T_{lin}(P_1, \ldots, P_n)$ assigned to any particular event A is at most the maximum of all values $P_i(A)$. Intuitively, it should be possible for several weak pieces of evidence to 'add up'. This is possible in the case of the independent or logarithmic opinion pool. The problem in this case is that the reinforcement of opinion sometimes occurs when it is not appropriate.

Example 2. Let $\Theta = \{a, b, c, d\}$, and let P_1 and P_2 be two identical probability functions over Θ given by the following table.

	a	b	c	d
$P_1 = P_2$	0.4	0.2	0.2	0.2
$T_{ind}(P_1, P_2)$	$\frac{4}{7}$	$\frac{1}{7}$	$\frac{1}{7}$	$\frac{1}{7}$

Suppose P_1 and P_2 reflect the opinions of two medical experts. Further suppose that a patient first learns about the opinion of one expert and then goes to the other for a second opinion. Intuitively, the agreement between the experts should merely confirm the original opinion without reinforcing or changing any original conclusion.

However, $T_{ind}(P_1, P_2)$ puts more confidence in a than the original probability functions. Moreover, $T_{ind}(P_1, P_2)(\{a\}) > T_{ind}(P_1, P_2)(\{b, c, d\})$, although both $P_1(\{a\}) < P_1(\{b, c, d\})$ and $P_2(\{a\}) < P_2(\{b, c, d\})$.

The linear opinion pool is perhaps best viewed as a way of estimating the 'correct' probability distribution based on several unreliable estimates or measurements represented by $P_1, \ldots, P_n$. In the independent opinion pool each P_i seems to be interpreted as representing (only) a part of the total evidence which is independent from the parts represented by the other probability functions. This latter interpretation best fits most of the applications of combining uncertain evidence in AI and robotics, although the pieces of evidence to be combined are practically never (completely) independent.

Dempster's rule of combination used in DS theory is a generalization of the independent opinion pool in the sense that both combination rules coincide when applied to probability functions. Of course, Dempster's rule is also defined for non-Bayesian belief functions, i.e., belief functions that are not (equivalent to) probability functions. Since DS theory can be viewed as a special case of partial probability theory, one can evaluate the rule in a probabilistic setting. We will argue that the application of Dempster's rule can only be justified probabilistically under strong unintuitive assumptions in addition to the already mentioned strong independence assumptions.

It should be mentioned that in [9] Shafer stressed the fact that the belief functions of his evidence theory should not be interpreted probabilistically. A similar position is defended by Smets [11] who has developed his transferable belief model in an attempt to provide belief functions with a coherent non-probabilistic interpretation. On the other hand, in [2], Dempster introduces belief functions as lower envelopes of partially determined probability measures, and this is in our view still the best understood interpretation.

In any case, as soon as belief functions are seen as generalizations of probability functions, Dempster's rule implies a general way of pooling probability functions, and in the following section we will discuss some arguments against the possibility of such a general combination scheme.

3 Impossibility Results

Property (2), characterizing the linear opinion pool, is related to the property of independence of irrelevant alternatives in social choice theory. This latter property has been proposed as a minimal condition on any reasonable function aggregating individual preferences into a joint preference ordering, and a famous theorem of Arrow shows no aggregation function can simultaneously satisfy this and a some other minimal conditions. Similar impossibility results can be derived in the case of combining probability distributions.

Definition 4. A combination scheme T is said to *preserve independence* iff it holds that $T(P_1, \ldots, P_n)(A, B) = T(P_1, \ldots, P_n)(A) \cdot T(P_1, \ldots, P_n)(B)$, whenever for every $i \in \{1, \ldots, n\}$, $P_i(A, B) = P_i(A) \cdot P_i(B)$.

Proposition 5 (Laddaga). *If T_{lin} preserves independence, then T_{lin} is dictatorial, i.e., one of the weights $w_i = 1$. In other words, there is no non-dictatorial combination scheme which preserves independence and satisfies (2).*

Proposition 6 (Dalkey). *Suppose there exists a function $f : [0,1]^n \longrightarrow [0,1]$ such that the following generalization of property (2) holds.*

$$\text{For every } A, B \subseteq \Theta, T(P_1, \ldots, P_n)(A|B) = f(P_1(A|B), \ldots, P_n(A|B)). \tag{5}$$

Then T is a dictatorial linear opinion pool.

A much more general impossibility result can be given. If we think of P_i as the conditional probability function $P(\cdot|E_i)$, then a combination scheme T can be viewed as a way of computing $P(\cdot|E_1, \ldots, E_n)$ from $P(\cdot|E_i)$. The following result from [7] shows that in general this is not possible.

Proposition 7 (Neapolitan). *Let Σ be the algebra generated by the events A, B, and C. There is in general no function F which computes for any probability function P on Σ the value $P(A|B,C)$ from the values of P on the algebras generated by at most two elements from $\{A, B, C\}$. In particular, there is no general rule for computing $P(A|B,C)$ from the values of $P(A|B)$, $P(A|C)$, $P(A)$, $P(B)$, $P(C)$, $P(A,B)$, et cetera.*

This result still holds if the events A, B, and C are pairwise independent and if the function F is only required to approximate the value of $P(A|B,C)$. One can conclude from the result that the value of $P(A|B,C)$ is only guaranteed to be derivable if some information is available (or assumed) concerning the probability of some event involving all three elements from $\{A, B, C\}$. Such information can for example be coded in strong conditional independence properties, such as "the pieces of evidence (B, C) are independent given the hypothesis (A), and given its negation $(\overline{A})$".

However, as is well-known from the study of the assumptions underlying some of the methods for updating probabilities in rule-based systems, such conditional independence properties may easily result in constraints which can only be satisfied in trivial situations. See, for example, [5]. Moreover, for most applications such assumptions are not even approximately true. (Different experts usually share a common body of knowledge, sensors are often influenced or disturbed by the same environmental factors, et cetera.)

4 Partial Probability Theory

Authors recognizing the ad hoc nature of the discussed combination schemes sometimes propose, as an alternative to their use, a careful probabilistic modelling of the situation, combined with ordinary probabilistic reasoning. See, for example, [1]. To a large degree, we agree with this proposal, but in our opinion a careful modelling of the situation does not necessarily lead to a probability function.

In many situations, the available evidence does not determine a unique probability function. Choosing a so-called "least informative" probability function satisfying the evidence may be the best general solution, in case one insists on using probability functions to represent belief states. However, in our opinion it is better to represent the available information, and no more than the available information, by means of probabilistic constraints. Absence of information, or ignorance, should lead to undetermined probabilities.

Example 3. Suppose an echo is received at time t_1 from a single firing at t_0 of a sonar mounted on an autonomous vehicle. This provides some evidence that there is an obstacle somewhere on an arc at range R, where R depends on $t_1 - t_0$ and the arc depends on R and the angle of the sonar beam. There is also evidence that the area covered by the beam is free from obstacles up to range R. But the received echo does not provide any evidence concerning the presence of obstacles beyond range R or outside the sonar beam.

A careful modelling of such a sensor reading should represent this absence of information, and not assign some "least informative" probability of occupancy outside the area for which evidence is available. (Recently, several authors have argued that using DS theory instead of insisting on using probability functions may result in improved performance of systems fusing such sensor data. See for example [8].)

We claim that in the case of combining evidence it is especially unreasonable to assume that there is sufficient information available about the interactions between the different pieces of evidence to determine a joint probability distribution. A partially determined probabilistic belief state can be formalized as follows.

Definition 8. A *(partial) probabilistic belief state (pbs)* is a triple $\langle \Theta, \mathcal{B}, C \rangle$, where Θ is a sample space, $\mathcal{B}$ is a set of constraints on probability measures over Θ, and $C \subseteq \Theta$.

Throughout the paper, the sample space Θ is assumed to be finite. The constraints in $\mathcal{B}$ represent general information about the relevant probabilities, whereas the subset C of the sample space Θ represents specific information concerning the case at hand. The members of $\mathcal{B}$ constrain the possible probability measures, and the specific information C is incorporated by essentially conditioning the probability measures on C. For example, in the context of example 3, information about the reliability of the sonar is of a general nature, and a particular reading of the sonar provides specific information.

The conditioning information C may seem to be redundant, since at any time the pbs $\langle \Theta, \mathcal{B}, C \rangle$ can be replaced with $\langle C, \mathcal{B}', C \rangle$, where $\mathcal{B}'$ is a set of constraints on probability measures over C. However, the generality of definition 8 will turn out to be convenient, in particular when belief changes are studied. We return to this issue in section 6.

If $\mathcal{B} = \emptyset$ and $C = \Theta$, then the pbs $\langle \Theta, \mathcal{B}, C \rangle$ represents the ignorant belief state with respect to Θ. Of course, if $\mathcal{B}$ determines a single probability measure over Θ, say P, and if $P(C) > 0$, then one obtains the classical Bayesian belief state represented by the probability measure P_C, defined by $P_C(A) = P(A|C)$.

In general, the set of constraints $\mathcal{B}$ naturally determines a set of probability measures over the sample space satisfying the constraints. Let us write $\mathcal{B}(\Theta)$ for this set of probability measures. The following definition extends Bayesian conditioning to sets of probability measures.

Definition 9 (extended Bayesian conditioning). Let Π be a set of probability measures over Θ, and let $C \subseteq \Theta$, such that for some $P \in \Pi$, $P(C) > 0$. Then Π_C is defined to be the set $\{P_C : P \in \Pi, P(C) > 0\}$.

Given this definition, the pbs $\langle \Theta, \mathcal{B}, C \rangle$ gives rise to the set $\mathcal{B}(\Theta)_C$ of probability measures over Θ. We write $(\mathcal{B}(\Theta)_C)_{low}$ for the lower envelope of $\mathcal{B}(\Theta)_C$, that is, the function defined by $(\mathcal{B}(\Theta)_C)_{low}(A) = \inf\{P(A) : P \in \mathcal{B}(\Theta)_C\}$. Any belief function of DS theory can be obtained as a lower envelope induced by a pbs, but not every such lower envelope is a belief function. (See, for example, [6].)

The notion of pbs generalizes the formalization of a belief state in (propositional) logic, (Bayesian) probability theory, and DS theory. However, we want to generalize the formalization one step further. In our opinion, the available probabilistic information is not only often insufficient to determine a single probability measure, it may also be insufficient to justify conclusions that are of practical use. In that case, one might try to strengthen one's conclusions by making some reasonable assumptions. This suggests the following.

Definition 10. A *(partial) probabilistic belief state with assumptions (pbsa)* is a quadruple $\langle \Theta, \mathcal{B}, \mathcal{A}, C \rangle$, where Θ is a sample space, $\mathcal{B}$ and $\mathcal{A}$ are sets of constraints on probability measures over Θ, and $C \subseteq \Theta$. The members of $\mathcal{B}$ are called belief or probability constraints, and $\mathcal{A}$ is called the set of assumptions.

If $\langle \Theta, \mathcal{B}, \mathcal{A}, C \rangle$ is a pbsa, then $\langle \Theta, \mathcal{B}, C \rangle$ is a pbs. We write $\mathcal{B}^{\mathcal{A}}(\Theta)$ for the set of all probability measures over Θ which satisfy all the constraints in $\mathcal{A} \cup \mathcal{B}$. By definition, $\mathcal{B}^{\mathcal{A}}(\Theta) \subseteq \mathcal{B}(\Theta)$. Therefore, the conclusions warranted by $\mathcal{B}^{\mathcal{A}}(\Theta)$ are (weakly) stronger than those warranted by $\mathcal{B}(\Theta)$. However, it is intended to be understood that the conclusions warranted by $\mathcal{B}^{\mathcal{A}}(\Theta)$ depend on the assumptions represented in $\mathcal{A}$.

By partial probability theory (PPT) we mean the approach to (probabilistic) reasoning in which belief states are represented by pbsa's. Notice that PPT generalizes both probability theory and DS theory. Therefore, it is to be expected that there is no simple rule for combining evidence in PPT.

The general procedure for combining evidence in PPT is to merge the pbsa's representing the two pieces of evidence into one pbsa with a sample space which is a common refinement of the two original sample spaces, with the induced belief constraints and assumptions of both the original pbsa's, and with as much additional belief constraints as possible, and as much assumptions concerning the interaction of the evidence as necessary for drawing useful conclusions. Let us first give the formal definition for a special case.

Definition 11. The combination of the probabilistic belief states $\langle \Theta, \mathcal{B}_1, C_1 \rangle$ and $\langle \Theta, \mathcal{B}_2, C_2 \rangle$ is defined to be the pbs $\langle \Theta, \mathcal{B}_1 \cup \mathcal{B}_2, C_1 \cap C_2 \rangle$.

The above can be extended to the case of different sample spaces by first refining both belief states to a common refinement of the original sample spaces, and the applying definitiion 11. It may be the case that there is evidence concerning the more refined sample space that is not represented in the original belief states. Such a complication can of course be avoided by assuming that the original sample spaces are sufficiently fine. From now on, let us assume for simplicity that the belief states to be copmbined have the same sample space.

Definition 12. The minimal combination of $\langle \Theta, \mathcal{B}_1, \mathcal{A}_1, C_1 \rangle$ and $\langle \Theta, \mathcal{B}_2, \mathcal{A}_2, C_2 \rangle$ is defined to be the pbsa $\langle \Theta, \mathcal{B}_1 \cup \mathcal{B}_2, \emptyset, C_1 \cap C_2 \rangle$.

Proposition 13. *The result of the minimal combination of* $\langle \Theta, \mathcal{B}_1, \mathcal{A}_1, C_1 \rangle$ *and* $\langle \Theta, \mathcal{B}_2, \mathcal{A}_2, C_2 \rangle$ *is equivalent to the combination of* $\langle \Theta, \mathcal{B}_1, C_1 \rangle$ *and* $\langle \Theta, \mathcal{B}_2, C_2 \rangle$.

It follows that the minimal combination abstracts from the assumptions present in the belief states to be combined, and the resulting combined belief state should be completed with as much assumptions as are necessary for drawing useful conclusions. Notice that in particular we do not include the assumptions of the original belief states in the minimal combination. The reason for this is that these assumptions may be unreasonable and even lead to inconsistency after combining evidence. The exact need for assumptions is (at least for a large part) determined by the kind of decisions the (probabilistic) belief are to be used for.

In [16] we discuss how Bayesian decision analysis can be extended to the case where the relevant probabilities can only be partially determined. We argue that in some cases deciding upon a reasonable action may be impossible without making assumptions supplementing the constraints based on the available evidence. For example, if a robot has no prior knowledge about the probability that a particular door is open, then it has at least to eliminate (by assumption) the extreme probability values (0 and 1) to make reasonable decisions about whether to use its (not completely reliable) sensors to acquire (specific) information about the state of the door.

Of course, assuming that the uncertainty is represented by a "least informative" probability measure satisfying the evidence can be seen as an extreme special case of our proposal. However, often much weaker assumptions will suffice to make a reasonable decision, and by explicitly representing the underlying assumptions, one can distinguish decisions based on evidence from decisions based on (additional) assumptions.

In the following section, the general procedure for combining evidence in PPT is illustrated and compared to using Dempster's combination rule. For this purpose, we use a concrete, admittedly greatly simplified, example of sensor fusion.

5 Combining Unreliable Sensors

The following example is discussed more extensively in [13]. Consider an autonomous vehicle which has to perform some subtle manoeuvres for which it needs to know its distance to its nearest obstacle with an accuracy of 0.01 metre. To measure this distance, it can use three sensors, let us call them S_1, S_2, and S_3. They each provide the vehicle's CPU with an integer between 0 and 999, representing the measurement in centimetres. They cannot detect objects which are removed 10 metre or more.

Each sensor is reliable 50% of the time. That is, 50% of the time, the sensor is working properly and the returned number is the correct distance. The remaining 50% of the time, the sensor is in some way disturbed and the returned number is not properly related to the true distance, although the number may of course happen to be correct by shear luck.

Suppose that on a particular occasion, all three sensors return the number 454. Intuitively, this provides strong support that 454 is the actual distance to the nearest obstacle, even though the sensors are rather unreliable when taken in isolation. Of course, such sensor information never gives complete certainty about the actual distance, but let us say that a 5% failure rate would still be acceptable. So if there is at least a 95% chance that a reading is correct, the vehicle should use this reading to guide its manoeuvres.

The question is whether the required confidence can be obtained in the case of the three agreeing sensors (or perhaps already when at least two sensors agree). That is, can one use the number returned by three agreeing sensors (unanimous decision rule) or even the number returned by at least two agreeing sensors (the majority rule)? We first give the answer provided by DS theory, which includes a rule for combining evidence (Dempster's combination rule) generalizing the independent opinion pool.

Suppose all three sensors S_1, S_2 and S_3 return the same number, say 454. Then we obtain three identical mass functions m_1, m_2 and m_3. Each of them assigns 0.5 mass to $\{454\}$ and 0.5 mass to $\Theta = \{0, 1, 2, \ldots, 999\}$. Combining these three mass functions by means of Dempster's combination rule leads to the mass function $m_1 \oplus m_2 \oplus m_3$, given by $m_1 \oplus m_2 \oplus m_3(\{454\}) = 0.875$, and $m_1 \oplus m_2 \oplus m_3(\Theta) = 0.125$. This does not give the required 95% guarantee that 454 is correct, even if all sensors return this number.

In order to apply Dempster's rule one has to assume that the pieces of evidence are (DS) independent, in the sense that the sources behave independently. That is, the (un)reliability of one sensor is independent form the (un)reliability of the other sensors. More precisely, if we write R_i for the proposition that sensor S_i is reliable, and r_i for the propositional variable denoting either R_i or $\overline{R_i}$. Then one has to assume that $P(r_i, r_j) = P(r_i)P(r_j)$, whenever $1 \le i < j \le 3$.

This assumption of DS independence is quite strong, and practically never satisfied by sensors, since they usually have common causes for their unreliable behaviour (the movement of the vehicle, the size and shape of the obstacles, et cetera). Moreover, as is shown in [12] this assumption of DS independence is not sufficient to justify Dempster's rule. In addition, one has to assume that each instantiation of r_1, r_2, r_3 is equally confirmed by the evidence. This additional assumption is explained below.

Let x_1 (y_2, z_3) denote the fact that the reading x (y, z) is obtained from sensor S_1 (S_2, S_3). In the case of receiving evidence $454_1, 454_2, 454_3$, the assumption of equal confirmation is represented by the equation $P(r_1, r_2, r_3|454_1, 454_2, 454_3) = P(r_1, r_2, r_3)$.

The reasoning giving a confidence of 0.875 in 454 goes as follows. The chance of 454 being the actual distance is at least the probability that at least one of the sensors is reliable. By DS independence the prior probability of this event is 0.875, and by the assumption of equal confirmation, this probability is not changed after receiving the evidence.

The assumption of equal confirmation is not plausible since agreeing sensors are more likely to be reliable (in which case they *have* to agree) than unreliable (in which case there is only a very small chance of agreeing). In fact, if a similar assumption is made simultaneously for every possible reading x_1, x_2, x_3 ($0 \le x \le 999$), then absurd conclusions can be derived. (See [13].)

In partial probability theory, the evidence provided by the reading 454 of sensor S_1 can be captured by the pbs $\langle \Theta_1, \mathcal{B}_1, C_1 \rangle$, where Θ_1 is the set of triples $\langle w_0, x_1, r_1 \rangle$ such that w denotes the real distance to the nearest obstacle, and x_1 and r_1 are as defined above. It follows that $0 \le w \le 999$, $0 \le x \le 999$, $r_1 \in \{R_1, \overline{R_1}\}$, and if $r_1 = R_1$, then $w = x$. Further, $\mathcal{B}_1 = \{P(R_1) = 0.5\}$, and $C_1 = \{\langle w_0, x_1, r_1 \rangle \in \Theta | x = 454\}$.

Combining the three pieces of evidence gives the pbs $\langle \Theta, \mathcal{B}, C \rangle$, where Θ is the set of tuples $\langle w_0, x_1, y_2, z_3, r_1, r_2, r_3 \rangle$, where $w, x, y, z \in \{0, \ldots, 999\}$, $r_i \in \{R_i, \overline{R_i}\}$, and if $r_i = R_i$, then $w = $ the reading of sensor S_i. Further, $\mathcal{B} = \{P(R_1) = P(R_2) = P(R_3) = 0.8\}$, and C is the subset of Θ characterized by $x = y = z = 454$.

In PPT one cannot conclude much based on the evidence alone, since the available information about the sensors simply does not allow strong conclusions about the value of $P(r_1, r_2, r_3|454_1, 454_2, 454_3)$. However, relatively weak additional assumptions are sufficient to obtain at least as strong conclusions as DS theory. For example, one could assume that $P(\overline{R_1}, \overline{R_2}, \overline{R_3}|454_1, 454_2, 454_3) \leq P(\overline{R_1}, \overline{R_2}, \overline{R_3})$, which is much weaker and far more plausible than the corresponding equality. What's more, one can argue for assumptions which justify both the unanimous decision rule and the majority rule.

It is reasonable to assume that the reading of an *unreliable* sensor does not depend on whether the other sensors are reliable or on what their reading is. We can then conclude, for example,

$$P(x_1|\overline{R_1}) = P(x_1|\overline{R_1}, r_2) = P(x_1|\overline{R_1}, r_2, y_2) = P(x_1|\overline{R_1}, r_2, y_2, r_3, z_3) \qquad (6)$$

If we further assume that an unreliable sensor is modelled by a uniform probability distribution over the possible readings, then a simple calculation shows that a 95% confidence in the unanimous reading 454 can already be obtained when the prior probability of 454 being the actual distance is assumed to be $1.9 \cdot 10^{-8}$, or higher. If this prior probability is assumed to be around 0.001, then $P(454_0|454_1, 454_2, 454_3) \approx 0.999999$. (See [13].)

Let us now turn to the majority decision rule, and let the evidence be $454_1, 454_2$, and 673_3. Then the following five possibilities remain.
(1) $R_1, R_2, \overline{R_3}$, (2) $R_1, \overline{R_2}, R_3$, (3) $\overline{R_1}, R_2, \overline{R_3}$, (4) $\overline{R_1}, \overline{R_2}, R_3$, and (5) $\overline{R_1}, \overline{R_2}, \overline{R_3}$.
The first three possibilities support 454, the fourth possibility supports 673, and the fifth possibility supports $\Theta = \{0, 1, 2, \ldots, 999\}$. If, in addition to the previously mentioned assumptions, one assumes that the prior probabilities $P(454_0)$ and $P(673_0)$ are both approximately equal to 0.001, then it can be shown that $P(454_0|454_1, 454_2, 673_3) \approx 0.998$.

Thus also in the case of *two* agreeing sensors one can have a high confidence in their reading. Of course, the confidence level is not as high as in the case *three* agreeing sensors, and one needs to assume that the prior probability of the distance indicated by the disagreeing sensor is not much larger than the prior probability of the distance indicated by the two agreeing sensors.

It has long been recognized that the application of Dempster's rule (interpreted as rule for combining lower probabilities) results in bounds on probabilities that are tighter than those justified by probability theory. (See, for example, [6].) We claim that the sensor fusion example shows that often the bounds justified by probabilistic *evidence* are too wide to be of practical use, and making assumptions may be unavoidable. However, the assumptions underlying Dempster's rule are strong, hard to justify, and may lead to unintuitive conclusions.

In PPT, the assumptions supplementing the evidence are represented explicitly. In principle, the assumptions underlying Dempster's rule can be added, and the conclusions of DS theory can be obtained. However, in situations like the sensor fusion example, more reasonable assumptions can be made, leading to intuitively more acceptable conclusions.

Whenever practical applications of DS theory seem successful, as for example is the case in the experiments reported on in [8], one can also obtain this success by means of PPT, with the additional bonus of having to make the assumptions explicit, which provides information concerning other contexts where similar success can be expected.

6 Conditioning

In PPT, conditioning can be viewed as a special case of evidence combination.

Definition 14. Conditioning the pbs $\langle \Theta, \mathcal{B}_1, \mathcal{A}_1, C_1 \rangle$ on C_2 results in the pbs $\langle \Theta, \mathcal{B}_1, \mathcal{A}_1, C_1 \cap C_2 \rangle$. The minimal conditioning of the pbsa $\langle \Theta, \mathcal{B}_1, \mathcal{A}_1, C_1 \rangle$ on C_2 is the pbsa $\langle \Theta, \mathcal{B}_1, \emptyset, C_1 \cap C_2 \rangle$.

As in the case of combination, we cannot simply keep the assumptions of the original belief state, since they may be unreasonable in the light of the new evidence.

Proposition 15. *The result of conditioning the pbs $\langle \Theta, \mathcal{B}_1, C_1 \rangle$ on C_2 is equivalent to the combination of $\langle \Theta, \mathcal{B}_1, C_1 \rangle$ and pbs $\langle \Theta, \emptyset, C_2 \rangle$. The result of minimal conditioning the pbsa $\langle \Theta, \mathcal{B}_1, \mathcal{A}_1, C_1 \rangle$ on C_2 is equivalent to the minimal combination of $\langle \Theta, \mathcal{B}_1, \mathcal{A}_1, C_1 \rangle$ and $\langle \Theta, \emptyset, \emptyset, C_2 \rangle$.*

In Dempster-Shafer theory, conditioning can also be defined as a special case of the rule for combination. If the mass function m_2 assigns the total mass to some $A \subseteq \Theta$, then the combination $m_1 \oplus m_2$ can be viewed as the result of conditioning m_1 on A. However, the following example, deriving from Smets [10] and also discussed in [4], shows that this notion of conditioning is different from the one in PPT.

Example 4 (The three assassins). Mr. Jones has been murdered by one of the assassins Peter, Paul, and Mary under orders of Big Boss, who has chosen between these three possible killers as follows. He decided between a male and a female killer by means of tossing a fair coin. A male killer was chosen in case the coin landed heads. Otherwise, a female killer was chosen. No information is available on how he decided between the two male assassins in case the coin landed heads.

Based on the information above, it seems reasonable to say that the possibility of the killer being male and that of the killer being female are equally likely. Now suppose that you learn that at the time of the murder, Peter was at the police station, where he was questioned about some other crime. So you can rule out Peter as the killer. How should this new evidence be modelled? In particular, is it still equally likely for the killer to be male or female?

To formalize this example, let $\Theta = \{a, b, c\}$, where a : Peter is the killer, b : Paul is the killer, and c : Mary is the killer. The information that a fair coin toss decided the choice between a male and female killer leads to the set of constraints $\mathcal{B} = \{P(\{a, b\}) = 0.5, P(\{c\}) = 0.5\}$. This agrees with the interpretation in Dempster-Shafer theory, where the information is encoded in the mass function m given by $m(\{a, b\}) = 0.5, m(\{c\}) = 0.5$, which induces a belief function Bel

such that $Bel = \mathcal{B}(\Theta)_{low}$. Strict Bayesians will opt for the probability function P given by $P(\{a\}) = 0.25$, $P(\{b\}) = 0.25$, and $P(\{c\}) = 0.5$, which is the 'least informative' member of $\mathcal{B}(\Theta)$.

How to take account of the information that Peter is not the killer? Strict Bayesians use Bayesian conditioning and arrive at $P_{\{b,c\}}$ given by $P_{\{b,c\}}(\{b\}) = \frac{1}{3}$, and $P_{\{b,c\}}(\{c\}) = \frac{2}{3}$, which implies that it is twice as likely for the killer to be female than to be male. However, as argued by Smets [10] and Halpern and Fagin [4], the 'least informative' prior P on which this answer is based makes some (unjustified) assumptions about how the choice between Peter and Paul is made.

Starting from the partial probabilistic belief state, the information that Peter is not the killer can be added by conditioning on $\{b, c\}$, or by adding the constraint $P(\{b, c\}) = 1$. This latter possibility results in the single probability measure P, determined by $P(\{b\}) = 0.5$, and $P(\{c\}) = 0.5\}$. This result, which implies that the possibility of the killer being male and that of the killer being female are still equally likely, completely agrees with the answer given be Dempster's rule in Dempster-Shafer theory, and is defended by Smets [10].

Conditioning the pbs on $\{b, c\}$ results in the set $\mathcal{B}(\Theta)_{\{b,c\}}$ of probability measures P over Θ satisfying $0 \le P(\{b\}) \le 0.5, 0.5 \le P(\{c\}) \le 1$, and $P(\{b, c\}) = 1$. This implies that the possibility of the killer being female is at least as likely as that of the killer being male. This answer, which is defended by Halpern and Fagin [4], agrees with our intuition that finding out that Peter has an alibi makes it less likely that the coin landed heads to a degree which equals one's degree of belief that Peter would have been chosen in case of heads. The ignorance concerning $P(\{a\}|\{a, b\})$ makes it impossible to justify a specific answer, but, if necessary, one can add assumptions to arrive at stronger conclusions.

Notice however, that in this case it need not be necessary to add assumptions to draw useful conclusions, since based on the evidence as represented in PPT one can argue that it is reasonable for the police to first investigate Paul. However, as soon one has to decide upon a specific distribution of resources (such as the number of detectives) assigned to investigating Paul, respectively Mary, then one needs information or assumptions about $P(\{a\}|\{a, b\})$.

In [14], we further study these two possible ways of incorporating new information into a pbs, and we argue that the operation of constraining, i.e., adding a constraint, is the proper probabilistic version of the expansion operation studied in the AGM theory of belief revision. Conditioning in DS theory by means of Dempster's rule is more strongly related to constraining than to the notion of conditioning in PPT.

In our opinion, the specific information in example 4 that Peter is not the killer should be incorporated by conditioning (in PPT). We do not claim that one should never use constraining. For example, constraining should be used if one does not learn that Peter is not the killer from Peter's alibi, but from a report of an undercover agent saying that Big Boss decided to choose Paul in the event that a male killer had to be chosen. Constraining is applicable in case of (general) information referring to the prior probabilities, and (specific) information about the case at hand should be incorporated using (extended) conditioning. More on this issue can be found in [14].

7 Conclusion and Further Work

The example of the unreliable sensors illustrates that without any additional assumptions the result of combining two pieces of evidence in PPT is typically highly indecisive. One simply needs some information or assumptions concerning the interaction of the pieces of evidence. In the case of the unreliable sensors one can argue for such assumptions which are more reasonable than those underlying Dempster's rule.

We propose to use a pbsa as a general formalization of a belief state in which belief constraints and assumptions are clearly separated, in order to keep track of the assumptions used in evidence combination. This gives a theoretical framework generalizing both probability theory and DS theory. Perhaps the main challenge for PPT in the context of AI and robotics is to automate the process of choosing the additional assumptions. However, this challenge does not necessarily impose insuperable difficulties.

It can be argued that the assumption leading to (6) should already be part of a careful modelling of the unreliable sensors. The other assumptions used in the PPT treatment of the sensor fusion example are assumptions of (approximately) uniform probability distribution. The main differences with strict Bayesians choosing least informative probability distributions is that in PPT (1) the assumptions are represented explicitly, and (2) the uniformity assumption is not applied to the global joint probability distribution, but is applied locally, only as far as is necessary for arriving at useful conclusions.

Some indication of where and when to apply these assumptions is given by the decision problem under consideration. For example, if it is crucial to know the exact distance to the nearest obstacle, then the relevant sensor readings should be obtained and combined. To do the latter one needs information, or assumptions, about the interaction between the relevant sensors. Of course, the use of additional assumptions comes at a cost (the chosen decisions are not unconditionally justified), but it is not yet clear how this cost should affect the decision problem. In [16] we essentially propose to choose, if possible, any satisfactory (i.e., close to optimal) action and only make assumptions if without them no such action can be found. But this is certainly not the final word about this issue.

Acknowledgments

Thanks to Michiel van Lambalgen for requesting a simple concrete argument against DS theory and to Philippe Smets for many valuable discussions on belief functions. A preliminary version of this paper has been presented at the Eight Dutch Conference on Artificial Intelligence (NAIC'96).

References

1. J.O. Berger, *Statistical Decision Theory and Bayesian Analysis* (Springer, Berlin, 1985).
2. A.P. Dempster, Upper and lower probabilities induced by a multivalued mapping, *Annals of Mathematical Statistics* 38 (1967) 325-339.
3. C. Genest and J. Zidek, Combining probability distributions: a critique and an annotated bibliography, *Statistical Science* 1 (1986) 114-148.
4. J.Y. Halpern and R. Fagin, Two views of belief: belief as generalized probability and belief as evidence, *Artificial Intelligence* 54 (1992) 275-317.
5. R. Johnson, Independence and Bayesian updating methods, *Artificial Intelligence* 29 (1986) 217-222.
6. H.E. Kyburg, Bayesian and non-Bayesian evidential updating, *Artificial Intelligence* 31 (1987) 271-293. (Addendum: *Artificial Intelligence* 36 (1988) 265-266.)
7. R.E. Neapolitan, A note of caution on combining certainties, *International Journal of Pattern Recognition and Artificial Intelligence* 1 (1987) 427-433.
8. D. Pagac, E.M. Nebot, and H. Durrant-Whyte, An evidential approach to probabilistic map-building, in: L. Dorst, M. van Lambalgen, and F. Voorbraak eds., *Reasoning with Uncertainty in Robotics* (Springer, Berlin, 1996) 164-170.
9. G. Shafer, *A Mathematical Theory of Evidence* (Princeton U.P., Princeton, 1976).
10. P. Smets, Belief functions versus probability functions, in: B. Bouchon, L. Aitt, R.R. Yager, eds., *Uncertainty and Intelligent Systems*, LNCS 313 (Springer, Berlin, 1988) 17-24.
11. P. Smets and R. Kennes, The transferable belief model, *Artificial Intelligence* 66 (1994) 191-234.
12. F. Voorbraak, On the justification of Dempster's rule of combination, *Artificial Intelligence* 48 (1991) 499-515.
13. F. Voorbraak, Combining unreliable pieces of evidence, *research report* CT-95-07 (ILLC, University of Amsterdam, 1995).
14. F. Voorbraak, Probabilistic belief expansion and conditioning, *research report* LP-96-07 (ILLC, University of Amsterdam, 1996).
15. F. Voorbraak, Reasoning with uncertainty in AI, in: L. Dorst, M. van Lambalgen, and F. Voorbraak eds., *Reasoning with Uncertainty in Robotics* (Springer, Berlin, 1996) 164-170.
16. F. Voorbraak, Decision analysis using partial probability theory, *Proceedings AAAI 1997 Spring Symposium Qualitative Preferences in Deliberation and Practical Reasoning* LP-96-07 (Stanford University, 1997).

Rational Default Quantifier Logic

A canonical framework for monotonic reasoning about first-order default knowledge

- Extended abstract -

Emil Weydert

Max-Planck-Institute for Computer Science
Im Stadtwald, D-66123 Saarbrücken, Germany
emil@mpi-sb.mpg.de

Abstract. We introduce a powerful new framework for monotonic reasoning about general first-order default knowledge. It is based on an extension of standard predicate logic with a new generalized quantifier, called the rational default quantifier, whose meaning is grasped by quasi-probabilistic κπ-ranking measure constraints over product domains. It subsumes and refines the original propositional notion of a rational default conditional, admits a sound and complete axiomatization, **RDQ**, and overcomes some basic problems of other first-order conditional approaches.

1. INTRODUCTION

In recent times, default conditionals interpreted as constraints over order-of-magnitude or ranking measure distributions have become an increasingly popular tool for encoding normal or plausible relationships. Conceptual adequacy, the correct and natural handling of specificity, semantic transparency, the availability of probabilistic justifications, the existence of proof-theoretic characterizations, and the appearance of promising non-monotonic inference relations [Weydert 96a, 97] are strong arguments in favour of this approach. However, like most formalisms for defeasible reasoning, they have been primarily designed for propositional or quantifier-free contexts. But this may not be enough to meet the representational needs of areas like natural language processing, user modeling or nonmonotonic reasoning about action and time in open domains. In fact, if we are dealing with incomplete and uncertain world descriptions, it seems methodologically appropriate – modulo practical considerations – to represent and use every available bit of knowledge.

Unfortunately, the traditional first-order approaches to default reasoning – at least in their usual formulations – do not only need an artificial ad hoc machinery to satisfy even modest desiderata for defeasible inference (e.g. specificity), they are also haunted by oversimplification – like circumscription, which seeks to identify abnormality with nonexistence – or syntax-dependence – like Reiter's open default logic, which is affected by Skolemization artefacts [Baader and Schlechta 93]. However, first-order conditional logics based on a propositional default connective with [Delgrande 88, Asher and Morreau 91b, Morreau 92, Alechina 95] or without [Lehmann and Magidor 90] quantifying-in, have their own problems. First of all, as we shall see, their usefulness is diminished by serious representational limitations. Secondly, they may have to confront the complexities of quantification into modal contexts, inducing arbitrary and cumbersome formal commitments, e.g. about the possible world domains.

A careful representation of first-order default knowledge, especially in the presence of explicit or implicit exceptions, therefore requires more fine-grained and powerful tools.

In the following, we are going to introduce, discuss, interpret and axiomatize an appropriate alternative, notably a very expressive intrinsic default quantifier which may be seen as a variable-binding or proper first-order version of a default conditional. A default quantifier is a rather flexible concept, which may also be used to encode propositional default implication but doesn't require a modal context. Generalized quantifiers have been well-known in natural language semantics [Westerstahl 89] and mathematical logic [Barwise and Feferman 85], but those proposals have a different punch line and do not directly meet our needs. Compared to the interest for default conditionals, there hasn't been much work on default quantifiers. The pioneers in this area have been Lorenz [89], Schlechta [90, 92, 95], Asher and Morreau [91a], Weydert [93], followed in a second wave by Brafman [96], Friedman, Halpern and Koller [96], as well as Weydert [96b]. But, whereas there is a large consensus about the correct principles for propositional default conditionals, as implemented in rational conditional logic, a canonical system for default quantification has yet to emerge. The present paper tries to reach this goal.

To start, we explain the need for strong, multidimensional – i.e. binding several variables at once – default quantifiers and introduce a correspondingly extended first-order language. Similarly to what we have done for propositional default conditionals [Weydert 91, 94], we present a default quantification semantics based on a quasi-probabilistic ranking measure concept, generalizing – just as far as necessary – Spohn and Pearl's κ as well as Dubois and Prade's possibilistic notions. We then propose a sound and complete axiomatization for our default quantifier logic, which may be seen as a qualitative counterpart of probabilistic logic [Bacchus 90] as well as a canonical extension of rational conditional logic [KLM 90, Boutilier 94] to the first-order level. To conclude, we are going to compare our proposal with several other accounts from the literature.

2. DEFAULT QUANTIFICATION

Let L be a standard language of first-order predicate logic with identity. We extend L by a new multi-dimensional dyadic quantifier $\Rightarrow$, called the *rational default quantifier*. The language $L(\Rightarrow)$ is defined to be the smallest set of formulas extending L, closed under the usual connectives and quantifiers (T, F, $\neg$, &, v, $\rightarrow$, $\leftrightarrow$, $\forall$, $\exists$), and including for each pair of formulas φ, ψ and each finite sequence of object variables $x = x_1, ...,$ x_n, the expression $\varphi \Rightarrow_x \psi$. We may read $\varphi \Rightarrow_x \psi$ as *"the x verifying φ normally satisfy ψ"* (objective reading) or *"the x verifying φ can be expected to satisfy ψ"* (epistemic reading). This language allows us to represent complex, i.e. nested and multivariate, genuinely first-order default knowledge.

A major advantage of default quantifiers is their ability to adequately handle (necessary) exceptions. Suppose, we want to state that birds can normally fly, but also that those birds living on Blurb Island necessarily can't fly – gravitation may be too strong, and that Fanny is necessarily one of them – this name may have been reserved for Blurb birds. In $L(\Rightarrow)$, this can be expressed by

- $\text{Bird}(x) \Rightarrow_x \text{Canfly}(x)$, $\text{Blurb}(x)\&\text{Bird}(x)\&\text{Canfly}(x) \Rightarrow_x F$,

¬(Bird(*Fanny*)&Blurb(*Fanny*)) $\Rightarrow_X$ **F** (equivalent to Bird(*Fanny*)&Blurb(*Fanny*)).

According to our intuitions, these three statements should be fully compatible and consistent with each other. And in fact, our semantics will support this view. However, if we restrict ourselves to propositional default conditionals, we obtain

- $\forall x(\text{Bird}(x) \Rightarrow \text{Canfly}(x))$, $\forall x(\text{Blurb}(x)\&\text{Bird}(x)\&\text{Canfly}(x) \Rightarrow \textbf{F})$,

  ¬(Bird(*Fanny*)&Blurb(*Fanny*)) $\Rightarrow$ **F**.

Exploiting only the laws of preferential conditional logic, we may then derive

- $\forall x(\textbf{T} \Rightarrow \neg(\text{Bird}(x)\&\text{Blurb}(x)))$.

Because universal quantification without instantiation doesn't make much sense, we get **T** $\Rightarrow$ ¬(Bird(*Fanny*)&Blurb(*Fanny*)). Together with ¬(Bird(*Fanny*)&Blurb(*Fanny*)) $\Rightarrow$ **F**, this gives us **T** $\Rightarrow$ **F**, i.e. triviality, which is definitely inadmissible.

In [Weydert 93], we have tried to address the problems caused by propositional default conditionals with a single-variable default quantifier. The initial assumption was that unary quantifiers and pairing functions might be enough to express multidimensional default quantification. However, it has turned out that this way to proceed is not only impractical and cumbersome, but also inappropriate for more fine-grained modeling tasks. Consider for instance the following example, concerned with human-animal relationships at home, which illustrates the inadequacy of simple propositional approaches and unary default quantification in multidimensional contexts.

- $\Phi = \{\text{Cat}(\textit{Pussy})\&\text{Feed}(\textit{Al, Pussy}), \text{Cat}(x)\&\text{Feed}(y, x) \Rightarrow_{x,y} \text{Like}(x, y),$

  $\text{Cat}(x)\&\text{Feed}(\textit{Al}, x) \Rightarrow_x \neg\text{Like}(x, \textit{Al})\}.$

Restricting ourselves to one-dimensional quantifiers, natural translations of Cat(x)& Feed(y, x) $\Rightarrow_{x,y}$ Like(x, y) would be **T** $\Rightarrow_y$ (Cat(x)&Feed(y, x) $\Rightarrow_x$ Likes(x, y)) or Cat(x) $\Rightarrow_x$ (Feed(y, x) $\Rightarrow_y$ Likes(x, y)). But these three formulas may take a totally different meaning. Within our general ranking measure semantics, for instance, it is easily possible to disconnect these expressions, i.e. to prevent that one of them is entailed by another one.

However, with universally quantified conditionals, things would become even worse. The corresponding knowledge base there would be

- $\Phi' = \{\text{Cat}(\textit{Pussy})\&\text{Feed}(\textit{Al, Pussy}), \forall x \forall y(\text{Cat}(x)\&\text{Feed}(y, x) \Rightarrow \text{Like}(x, y)),$

  $\forall x(\text{Cat}(x)\&\text{Feed}(\textit{Al}, x) \Rightarrow \neg\text{Like}(x, \textit{Al}))\}.$

Instantiation then gives us Cat(*Pussy*)&Feed(*Al, Pussy*) $\Rightarrow$ Likes(*Pussy, Al*), but also Cat(*Pussy*)&Feed(*Al, Pussy*) $\Rightarrow$ ¬Likes(*Pussy, Al*), which entails Cat(*Pussy*)&Feed(*Al, Pussy*) $\Rightarrow$ **F**, i.e. the impossibility of Pussy to be a cat fed by Al, in contradiction with our basic assumptions. Again, this seems to be completely inacceptable. That is, multi-variable default quantifiers turn out to be a necessary ingredient for ensuring flexibility and expressiveness.

When we try to encode default information like "*birds can fly*" within this fairly expressive framework, we have to think about what we really mean by such an assertion.

Is the flying ability ascribed to almost all existing birds once and for all, i.e. Bird(x) $\Rightarrow_x$ Canfly(x), or do we refer to the flying potential of birds at different time-points or in different actual or possible contexts ? Furthermore, as soon as variables for birds and time-points are involved, we must decide which variables to bind by universal and which ones by default quantification. For instance, we may have to choose between

- $\forall t$(Timepoint(t) $\rightarrow$ (Bird(t, x) $\Rightarrow_x$ Canfly(t, x))),

- Time-point(t) $\Rightarrow_t$ (Bird(t, x) $\Rightarrow_x$ Canfly(t, x)),

- Time-point(t) $\Rightarrow_t$ $\forall x$(Bird(t, x) $\rightarrow$ Canfly(t, x)).

A comprehensive first-order formalization of normal relationships not only permits but also forces us to be much more precise than we were used to be when dealing with de facto propositional accounts. That's quite important because qualitative reasoning with uncertainty tends to have a rather shaky character, so that we should try to exploit every knowledge item available to us. This task is considerably simplified by default quantification. These observations indicate that, despite the additional inferential complexity, knowledge engineering may well profit from default quantifier logic.

As our formulations suggest, default statements bear some resemblance to well-known generic constructions in natural language. The work of Asher and Morreau [91a,b], for instance, was primarily motivated by this application area. In the present paper, however, we want to abstract from linguistic considerations and restrict ourselves to the default reasoning part. If all we know about a is $\varphi(a)$, then $\varphi \Rightarrow_x \psi$ just tells us that it may be plausible to assume $\psi(a)$. In fact, sometimes, direct translations may fail – e.g., "*birds lay eggs*" implicitly only refers to the female instances and should be correctly formalized by Bird(x)&Female(x) $\Rightarrow_x$ Egg-layer(x). Nevertheless, default quantifiers supplemented by pragmatic accommodation mechanisms could still be an interesting description tool for particular forms of genericity. Conversely, the linguistic analysis of generic sentences could provide important clues for our more abstract notion. Of course, when we are going to evaluate benchmark problems, we have to be aware of the differences.

Propositional defaults like "*if Tweety is a bird, then by default Tweety can fly*"

- Bird(*Tweety*) $\Rightarrow$ Canfly(*Tweety*),

are best interpreted modally. Fortunately, our approach is flexible enough to implement the modal viewpoint. This is achieved by the introduction of world parameters and default quantification over worlds. The above conditional, for instance, would become

- World(ω)&Bird(ω, *Tweety*) $\Rightarrow_\omega$ Canfly(ω, *Tweety*),

i.e. "*worlds where Tweety is a bird are normally worlds where Tweety can fly*". In particular, where required, we may emulate the more traditional accounts and model universally quantified default conditionals. Note that because x is absent from the component formulas, Bird(*Tweety*) $\Rightarrow_x$ Canfly(*Tweety*) itself is equivalent to the material implication Bird(*Tweety*) $\rightarrow$ Canfly(*Tweety*). So, whatever our basic commitments, a first-order logic with a suitable multi-variable default quantifier seems to be a good starting point.

3. RATIONAL DEFAULT QUANTIFIER SEMANTICS

The most general framework for interpreting flat rational default conditionals in a boolean propositional context is the ranking measure semantics [Weydert 91, 94]. Ranking measures are implausibility valuations associating with each proposition a value in a suitable linearly ordered scale structure $(V, o, \infty, \min, *, <)$, whose role is analoguous to that of $([0, 1], 1, 0, +, x, >)$ in probability theory. We may interpret the values in V as degrees of implausibility, surprise, exceptionality or disbelief. Spohn's natural conditional functions [Spohn 88], i.e. Pearl's κ-rankings [Pearl 89], and Dubois and Prade's possibility distributions [Dubois and Prade 88] are special instances of ranking measures.

Following Spohn and Pearl, we are going to interpret ranking measures as subjective — in the sense of non-frequentist — semi-qualitative order-of-magnitude probability distributions. This approach allows us to exploit important results and tools from classical probability theory. More precisely, let R be the standard and R^* be any nonstandard model of the reals (i.e. with ε s.t. $0 < \varepsilon < ... < 1/n < ... < 1/2 < 1$, called infinitesimals). Let ε be a strictly positive infinitesimal in R^* and $stan : R^* \to R \cup \{-\infty, \infty\}$ be the function which maps each x to its standard part $stan(x)$, i.e. the closest standard real, resp. ∞ or $-\infty$, if x is infinite. Let $B \subseteq 2^S$ be a boolean set field on S and $P : B \to [0, 1]^*$ be a (finitely additive) nonstandard probability measure over B. Then we can define a ranking measure R_P by stipulating that $R_P(A)$ is the standard part of $P(A)$'s order of magnitude w.r.t. ε. That is, for all $A \in B$, we define $R_P : B \to R^+ \cup \{\infty\}$ with

- $R_P(A) = stan(\log_\varepsilon P(A))$, by convention $\log_\varepsilon 0 = \infty$ and $stan(\log_\varepsilon 0) = \infty$.

For instance, if $P(A) = \varepsilon^{r+\varepsilon} + \varepsilon^{r+1}$ $(0 < r \in R)$, we get $R_P(A) = r$. Note that R_P may fail to be a κ-measure, because r doesn't have to be an integer. In fact, there are well-motivated nonmonotonic inference notions implementing entropy maximization within the ranking measure framework which necessitate the inclusion of all the positive rational numbers. This suggests the following extension of the κ-measure concept.

Definition 3.1 Let $B \subseteq 2^S$ be a boolean set field on S. A map $R : B \to V$ is called a $\kappa\pi$–*ranking measure* over B iff, for all $A, B \in B$,

- $V = R^+ \cup \{\infty\}$ or $V = R^{*+} \cup \{\infty\}$, where R^* is a nonstandard model of the reals,

- $R(A \cup B) = \min\{R(A), R(B)\}$,

- $R(\emptyset) = \infty$, $R(S) = 0$,

- if $X \subseteq B$, for all $A \in X$, $R(A) = \infty$, and $\cup X \in B$, then $R(\cup X) = \infty$.
The *conditional* $\kappa\pi$–*ranking measure* $R(|) : B x B \to V$ is defined by

- $R(B \mid A) = R(A \cap B) - R(A)$, for $R(A) \neq \infty$, and $R(B \mid A) = \infty$, for $R(A) = \infty$.

R and $R(|)$ are called *standard* if $\text{Im}(R) \subseteq R^+ \cup \{\infty\}$, *nonstandard* otherwise. R_O is the uniform $\kappa\pi$-ranking measure over B, i.e. $R_O(A) = 0$ for $A \neq \emptyset$.

For finite B, standard $\kappa\pi$–measures are structurally equivalent to real-valued possibility measures (assuming the multiplicative independence notion). We prefer the $\kappa\pi$-notation because our approach stands in the tradition of the κ–calculus and we want to avoid possibilistic-fuzzy connotations or a possible confusion with probability values. Note that because nonstandard reals are not closed under infima, we cannot have

- $R(\cup X) = \inf\{R(A) \mid A \in X\}$ for $X \subseteq \mathbf{B}$ and $\cup X \in \mathbf{B}$ (continuity).

Nonstandard $\kappa\pi$–measures are necessary to obtain a sound and complete axiomatization of default quantifier logic. This even holds if we restrict ourselves to finite theories. On the other hand, in propositional default conditional logic, nonstandard valuations are only required to handle infinite theories. Furthermore, it is important to see that $\kappa\pi$–ranking measures are universal in the sense that any ranking algebra $\mathbf{V}$, according to the definitions in [Weydert 94], can be embedded into some $\mathbf{R}^{*+} \cup \{\infty\}$.

Propositional default conditionals may be interpreted as inequality constraints over the ranking measure values of model sets. We are willing to accept $\varphi \Rightarrow \psi$ iff the implausibility degree of $\varphi \& \neg\psi$ is (sufficiently) higher than that of $\varphi \& \psi$, or if φ is simply impossible (maximal implausibility). The most liberal interpretation is given by

- R satisfies $\varphi \Rightarrow \psi$ iff $R(\text{Mod}(\neg\psi) \mid \text{Mod}(\varphi)) > 0$

  iff $R(\text{Mod}(\varphi \& \psi)) < R(\text{Mod}(\varphi \& \neg\psi))$ or $R(\text{Mod}(\varphi \& \neg\psi)) = \infty$.

Let $\text{lM} = (M, ..., h)$ be a classical first-order interpretation over L, where $(M, ...)$ is an L-structure and h an interpretation of the L-variables by elements of M. The interpretation of multi-variable default quantifiers $\varphi \Rightarrow_x \psi$ over lM requires a coherent hierarchy of ranking measures over the powersets of all the domain cross-products M^n (or at least over their algebras of definable sets). This is similar to what has been done in probability logic [Bacchus 90].

Definition 3.2 (lM, lR) is called a $\kappa\pi$-*interpretation* over L iff

1. $\text{lM} = (M, ..., h)$ is a first-order interpretation over L.

2. $\text{lR} = (R_n \mid n \in \text{Nat})$ is a family of $\kappa\pi$-ranking measures $R_n : 2^{Mn} \to V$ s.t. for all n, m,

- if $A \in 2^{Mn}$, $B \in 2^{Mm}$, then $R_{n+m}(A \times B) = R_n(A) + R_m(B)$,
- if $A \in 2^{Mn}$, π is a permutation on $\{1, ..., n\}$, $A^\pi = \{(a_{\pi(1)}, ..., a_{\pi(n)}) \mid (a_1, ... a_n) \in A\}$, then $R_n(A) = R_n(A^\pi)$.

(lM, lR) is called standard iff all the R_n are standard.

Our default quantifier and default conditional semantics are based on the same ideas. A first-order default statement $\varphi \Rightarrow_x \psi$ should express – for a given interpretation lM – that the set of x-tuples satisfying $\varphi \& \neg\psi$ is negligible w.r.t. or has a lower qualitative magnitude than the set of x-tuples verifying $\varphi \& \psi$, i.e. it is less plausible to get x from the first than from the second set. This condition easily translates into ranking measure constraints over sets of object tuples. To account for nested $\Rightarrow_x$, satisfaction must be defined inductively.

Definition 3.3 Let $\|=$ be the unique satisfaction relation linking $\kappa\pi$-interpretations (lM, lR) over L and L($\Rightarrow$)-formulas which handles standard first-order concepts as usual and default quantification as follows. Let $n = \lg(x)$ (length(x)).

- $(\text{lM}, \text{lR}) \|= \varphi \Rightarrow_x \psi$ iff

$$R_n(\{a \in M^n \mid (lM[x/a], lR) \Vert= \neg\psi\} \mid \{a \in M^n \mid (lM[x/a], lR) \Vert= \phi\}) > 0.$$

The corresponding monotonic entailment relation $\Vert\!\vdash$, called *rational default quantifier entailment*, is defined in the usual Tarskian way.

- $\Sigma \Vert\!\vdash \phi$ iff for every $\kappa\pi$-interpretation $(lM, lR) \Vert= \Sigma$, we have $(lM, lR) \Vert= \phi$.

The main question is now whether $\Vert\!\vdash$ has a natural axiomatization, reminiscent of Lehmann's rationality postulates [KLM 90].

4. RATIONAL DEFAULT QUANTIFIER LOGIC

For practical purposes, we need a proof theory for our $\kappa\pi$-entailment concept $\Vert\!\vdash$. If possible, it should reflect our intuitions about the rational default quantifier as a first-order extension of the rational conditional.

Definition 4.1 Let L be a language of first-order predicate logic with identity. Then the *rational default quantifier logic* over L, also called **RDQ**-logic, is the smallest derivability notion $\vdash$ on $L(\Rightarrow)$ verifying all the axioms and rules of predicate logic with identity, which we denote by **RDQ 0**, as well as the universally quantified schemes **RDQ 1 - 8**. Let $\phi, \psi, \xi, \eta \in L(\Rightarrow)$ and x, y be finite sequences of object variables.

RDQ 1 $\phi \Rightarrow_x \psi \rightarrow \phi[x/y] \Rightarrow_y \psi[x/y]$ y not in ϕ, ψ and $lg(x) = lg(y)$ *(Renaming)*

RDQ 2 $\forall x(\phi \leftrightarrow \eta) \& \forall x(\psi \leftrightarrow \xi) \rightarrow (\phi \Rightarrow_x \psi \leftrightarrow \eta \Rightarrow_x \xi)$ *(Equivalence)*

RDQ 3 $\forall x(\phi \rightarrow \psi) \rightarrow \phi \Rightarrow_x \psi$ *(Suprainclusivity)*

RDQ 4 $\phi \& \neg\psi \Rightarrow_x F \rightarrow \forall x(\phi \rightarrow \psi)$ *(Necessity)*

RDQ 5 $\phi \Rightarrow_x \psi \& \phi \Rightarrow_x \xi \rightarrow \phi \Rightarrow_x \psi \& \xi$ *(And)*

RDQ 6 $\phi \Rightarrow_x \xi \& \psi \Rightarrow_x \xi \rightarrow \phi \lor \psi \Rightarrow_x \xi$ *(Or)*

RDQ 7 $\phi \Rightarrow_x \xi \& \neg(\phi \Rightarrow_x \neg\psi) \rightarrow \phi \& \psi \Rightarrow_x \xi$ *(Rational monotony)*

RDQ 8 $\exists y \psi \rightarrow (\phi \Rightarrow_x \xi \leftrightarrow \phi \& \psi \Rightarrow_{x,y} \xi)$ y not in ϕ, ξ, x not in ψ *(Irrelevance monotony)*

The expressive power of first-order logic allows some simplifications w.r.t. the rational conditional axioms. In particular, we can derive,

- $\phi \Rightarrow_x \xi \& \phi \Rightarrow_x \psi \rightarrow \phi \& \psi \Rightarrow_x \xi$ *(Cautious monotony)*.

RDQ 1, 2 are common standard equivalence or substitution principles. **RDQ 3, 4** establish the links between strict and defeasible inclusion. **RDQ 5** guarantees coherence on the right-hand-side and witnesses our qualitative, logical commitments, as opposed to classical probabilistic threshold reasoning. **RDQ 6** implements reasoning by cases and reflects a formally realist philosophy. That is, the antecedent is not considered to be in the scope of an implicit belief modality, as it is in most epistemic or rule-oriented appoaches. **RDQ 7** enforces inheritance to non-exceptional subclasses. It is one of the strongest forms of antecedent strengthening still compatible with the default reading and implementable at the propositional level. To these axioms, we add a further restricted monotony principle, **RDQ 8**, which may be described as inheritance to explicitly irrelevant subclasses, i.e. whose irrelevance is supported by the variable configuration and the corresponding independency assumptions. This postulate is of central importance for multi-dimensional default quantification and is the key the following representation theorem.

Theorem 4.1 ⊢ *is sound and complete w.r.t.* ‖− *on* L(⇒).

To prove the completeness part, the main difficulty is to construct a suitable valuation algebra **V** to be be embedded into some $\mathbf{R}^{+*} \cup \{\infty\}$. Here, we make heavy use of **RDQ** 8. Because of this completeness result and the universality of κπ-ranking measures, we consider **RDQ** to be the minimal reasonable default quantifier logic. Given that stronger systems, like Brafman's logic **NS 1-17**, are not general enough to model everything we may want to model (cf. below), **RDQ** may even be seen as the maximal reasonable default quantifier logic. Additional assumptions about the application domain would then have to be added to the contingent knowledge base.

Of course, that's not the whole story. What is still missing is a nonmonotonic plausible inference relation ‖≈, allowing us to exploit these powerful default statements in a rational way, i.e. to draw reasonable (defeasible) conclusions not sanctioned by ‖−. However, because first-order logic enables much more sophisticated interactions, this is a very difficult task, at least if we do not impose severe restrictions on the knowledge input. In particular, universal quantifying-in may force us to deal with infinite default sets, and we may have to handle models with different domains. Under these conditions, even the realization of standard patterns like defeasible transitivity becomes nontrivial. We are currently investigating whether and how system **J** and its more sophisticated relatives, some very promising propositional nonmonotonic entailment notions [Weydert 96a, 97] based on iterated Jeffrey-conditionalization and close to qualitative entropy maximization, can be generalized to the first-order level. But this issue will be addressed elsewhere.

5. COMPARISONS

In first-order default conditional logic, we may distinguish two kinds of theories, those based on a modal default implication, i.e. a propositional connective, and those exploiting a genuine default quantifier, i.e. a variable-binding connective. The work of Delgrande [88], Lehmannn and Magidor [90], Asher and Morreau [91b], and Alechina [95], belongs to the first category. With the exception of Lehmann and Magidor, where default conditionals may not appear within the scope of a quantifier, all these accounts interpret first-order defaults as universally quantified default implications. As we have explained before, this approach is inappropriate because it doesn't allow an adequate, intuitive encoding of normal relationships.

Schlechta was the first researcher to exploit generalized quantifiers for modeling default knowledge. His normal case semantics for default quantifier is based on a weak filter concept and hierarchies of corresponding big set systems [Schlechta 90, 92, 95, Lorenz 89]. He has also provided a sound and complete axiomatization. Although his approach stands in a different tradition, it may be interpreted as a rough counterpart to our fine-grained ranking measure semantics. His framework is very general but, unfortunately, rather cumbersome and it doesn't pay much attention to the multi-dimensionality issue.

The work of Brafman [Brafman 91, 96] extends the classical ranked model semantics for default conditionals [KLM 90] to the context of multiple-variable default quantifiers. He presents several variants and offers for each of them an axiomatization. Unfortunately, his approach inherits the well-known problems of the original minimal model semantics, in particular the need for an impractical and unintuitive smoothness condition. It certainly would be preferable to use the limit evaluation scheme for conditionals, which goes back to Lewis and has been pushed in the default reasoning community by Boutilier [94]. However, this step might complicate the axiomatization task. Furthermore, it doesn't solve another major problem of any total pre-order semantics, notably the inability to model the quite natural situation where all the singletons are assumed to be equally exceptional, e.g. a fair infinitary lottery. Its $L(\Rightarrow)$-description is

- $\forall y(T \Rightarrow_x \neg\, x = y) \,\&\, \forall x \forall y \neg (w = x \vee w = y \Rightarrow_w \neg\, w = x).$

But this sentence is inconsistent in Brafman's logics. Contrarily, it is easily satisfiable in our $\kappa\pi$-ranking measure framework — set $R(\{a\}) = 1$ for each individual a — or in the parametrized probability semantics [GMP 90]. In fact, by violating **NS 12**, this example seems to falsify Brafman's Theorem 5 as well as his Conjecture 1 [Brafman 96]. On the other hand, Brafman's bottom-up perspective allows us to evaluate the plausibility of a set by looking only at the plausibility of its singleton subsets. This doesn't hold for our general $\kappa\pi$-semantics. **RDQ** is incomparable with Brafman's weakest axiom set, **NS 1-12**, but it is subsumed by his strongest one, **NS 1-17**. More precisely, $\kappa\pi$-ranking measures may violate the Universal Interchange (**NS 12**), the first-order generalization of right conjunction,

- $\forall y(\varphi \Rightarrow_x \psi) \;\rightarrow\; (\varphi \Rightarrow_x \forall y \psi)$ if y not in φ, x.

the Distribution (**NS 16**) and the Projection (**NS 17**) principle. Another pattern not supported by **RDQ** but entailed by NS 1-17 is commutativity for absolute normality,

- $T \Rightarrow_x (T \Rightarrow_y \varphi) \;\rightarrow\; T \Rightarrow_y (T \Rightarrow_x \varphi).$

Friedman, Halpern and Koller have developed a general framework for quasi-probabilistic valuations called plausibility measures, which includes probability and $\kappa\pi$-ranking measures as well as much weaker notions. However, their classification is unable to grasp the full power of our concept, because they are looking at valuation scales without considering the additive structure necessary for adequately modeling independency. In [FHK 96], they discuss a plausibility semantics for one-dimensional default quantification and compare the behaviour of different semantic approaches, e.g. w.r.t. infinitary lottery paradox. Similarly to what we have done, they have shown the necessity to use general top-down variants of plausibility measures, i.e. not determined by the values for singletons. However, their paper does neither offer an axiomatization of default quantifiers, which they call statistical conditionals, nor does it address the multi-dimensionality issue in sufficient detail.

REFERENCES

[Alechina 95] N. Alechina. For all typical. In *Proceedings of ECSQARU 95*. Springer.

[Asher and Morreau 91a] N. Asher, M. Morreau. Commonsense entailment: a modal theory of nonmonotonic reasoning. In *Logics in AI*, ed. J. van Eijck. Springer.

[Asher and Morreau 91b] N. Asher, M. Morreau. Commonsense entailment: a modal theory of nonmonotonic reasoning. In *Proceedings of the 12th. IJCAI*. Morgan Kaufmann.

[Baader and Schlechta 93] F. Baader, K. Schlechta. A semantics for open normal defaults via a modified preferential approach. In *Proceedings of ECSQARU 93.*. Springer.

[Bacchus 90] F. Bacchus. *Representing and Reasoning with Probabilistic Knowledge*. MIT-Press, Cambridge, Massachusetts.

[Barwise and Feferman 85] J. Barwise, S. Feferman (eds.). *Model-theoretic Logics.*. Omega-series, Springer.

[Boutilier 94] C. Boutilier. Conditional logics of normality : a modal approach. *Artificial intelligence*, 68:87-154, 1994.

[Brafman 91] R. Brafman. *A logic of normality : Predicate calculus incorporating assertions*. Master's thesis, Hebrew University of Jerusalem, 1991.

[Brafman 96] R. Brafman. "Statistical" First Order conditionals. In *Proceedings of the Fifth Conference on the Principles of Knowledge Representation and reasoning KR 96*. Morgan Kaufmann.

[Delgrande 88] J. Delgrande. An approach to default reasoning based on a first-order conditional logic: a revised report. *Artificial Intelligence*, 36:63-90, 1988.

[GMP 90] M. Goldszmidt, P. Morris, J. Pearl. A maximum entropy approach to non-monotonic reasoning. In *Proceedings of AAAI 90*. Morgan Kaufmann.

[FHK 96] N. Friedman, J.Y. Halpern, D. Koller. First-order conditional logic revisited. In *Proceedings of AAAI 96*, Morgan Kaufmann.

[KLM 90] S. Kraus, D. Lehmann, M. Magidor. Nonmonotonic reasoning, preferential models and cumulative logics. *Artificial Intelligence*, 44: 167-207, 1990.

[Lehmann and Magidor 90] D. Lehmann, M. Magidor. Preferential logics: the predicate logical case. In *Proceedings of the 3rd Conference on Theoretical Aspects of Reasoning about Knowledge TARK 1990*, ed. R. Parikh. Morgan Kaufmann.

[Lorenz 89] S. Lorenz. Skeptical reasoning with order-sorted defaults. In *Proceedings of the Workshop on Nonmonotonic Reasoning*, eds. G. Brewka, H. Freitag, TR 443, GMD, Sankt Augustin, Germany, 1989.

[Morreau 92] M. Morreau. *Conditionals in Philosophy and Artificial Intelligence*. Working papers of the SFB 340, Report 26, Stuttgart, 1992.

[Pearl 89] J. Pearl. Probabilistic semantics for nonmonotonic reasoning: a survey. In *KR 89*, Morgan Kaufmann.

[Schlechta 90] K. Schlechta. Semantics for defeasible inheritance. In *Proceedings of ECAI 90*, ed. L.G. Aiello, Pitman, London.

[Schlechta 92] K. Schlechta. *Results on Nonmonotonic Logics*. IWBS-Report 204.

[Schlechta 95] K. Schlechta. Defaults as generalized quantifiers. *Journal of Logic and Computation*, 5 (4), 473-494, 1995.

[Spohn 88] W. Spohn. Ordinal conditional functions : A dynamic theory of epistemic states. In *Causation in Decision, Belief Change and Statistics*, W.L. Harper, B. Skyrms (eds.). Kluwer Academic Publishers, Dordrecht, 1988.

[Westerstahl 89] D. Westerstahl. Quantifiers in formal and natural languages. In *Handbook of Philosophical Logic*, Vol. IV, eds. D. Gabbay and F. Guenthner. Kluwer, 1989.

[Weydert 91] E. Weydert. Qualitative magnitude reasoning. Towards a new syntax and semantics for default reasoning. In *Nonmonotonic and Inductive Logics*, eds. Dix, Jantke, Schmitt. Springer, 1991.

[Weydert 93] E. Weydert. Default quantifiers. About plausible reasoning in first-order contexts. *Proceedings of the Dutch-German Workshop on Nonmonotonic Reasoning Techniques and Applications*, Aachen, 1993.

[Weydert 94] E. Weydert. General belief measures. In *Tenth Conference on Uncertainty in Artificial Intelligence*. Morgan Kaufmann, 1994.

[Weydert 96a] E. Weydert. System J - Revision entailment. Default reasoning through ranking measure updates. In *Proceedings of the First Conference on Formal and Applied Practical Reasoning*, Bonn 1996, Springer-Verlag.

[Weydert 96b] E. Weydert. Default Quantifier Logic. In *Proceedings of the ECAI 96 Workshop on Integrating Nonmonotonicity into Automated reasoning systems*.

[Weydert 97] E. Weydert. Qualitative Entropy Maximization. *Proceedings of the Third Dutch-German Workshop on Nonmonotonic Reasoning Techniques and their Applications*, eds. G. Brewka et al., MPI for Computer Science, Saarbrücken, 1997.

Disjunctive Update, Minimal Change, and Default Reasoning

Yan Zhang

Department of Computing
University of Western Sydney, Nepean
Kingswood, NSW 2747, Australia
E-mail: yan@st.nepean.uws.edu.au

Abstract. It is well known that the minimal change principle was widely used in knowledge base updates. However, recent research has shown that conventional minimal change methods, eg. the PMA [9], are generally problematic for updating knowledge bases with disjunctive information under some circumstances. In this paper, we propose a new approach, which is called the *minimal change with exceptions* (the MCE), to deal with this problem in propositional knowledge base updates. We show that the MCE generalizes the PMA and still satisfies the standard Katsuno and Mendelzon's update postulates. Furthermore, we also investigate the relationship between update and default reasoning. In particular, we represent a translation from the MCE to extended disjunctive default theories and prove the soundness and completeness of such translation relative to the semantics of the MCE.

1 Introduction

Recent research has shown that traditional minimal change approaches, eg. the PMA [9], are generally problematic for updating knowledge bases with disjunctive information [2, 5, 10]. In this paper, we propose an alternative approach called *Minimal Change with Exceptions* (MCE) to handle this problem. The MCE remains the minimal change criterion for updates generally but with exceptions to the disjunctive effect of updates. We show that the MCE still satisfies the standard Katsuno-Mendelzon's update postulates (U1) - (U8) [6]. We then investigate the connection between update and default reasoning. We represent a translation from the MCE to extended disjunctive default theories and prove the soundness and completeness of the translation relative to the semantics of the MCE.

The paper is organised as follows. The next section first reviews the PMA and discusses the problem with the PMA in updates with disjunctive information. Section 3 proposes an alternative approach called the MCE to handle this problem. Section 4 investigates the property of the MCE and shows that the MCE still satisfies standard Katsuno and Mendelzon's update postulates. Based on Gelfond *et al*'s disjunctive default theories [4], section 5 presents extended disjunctive default theories which can derive every possible interpretation for a disjunction during the default reasoning. Section 6 describes a translation from the MCE to extended disjunctive default theories and shows the soundness and completeness of this translation relative to the semantics of the MCE. Finally, section 7 concludes the paper with some concluding remarks.

2 The Minimal Change Approach: A Review

2.1 The Language

Consider a finitary propositional language $\mathcal{L}$. We represent a *knolwedge base* by a propositional formula ψ. A propositional formula ϕ is *complete* if ϕ is consistent and for any propositional formula μ, $\phi \models \mu$ or $\phi \models \neg\mu$. $Models(\psi)$ denotes the set of all models of ψ, i.e. all interpretations of $\mathcal{L}$ in which ψ is true. We also consider *state constraints* about the world. Let C be a satisfiable propositional formula that represents all state constraints about the world[1]. Thus, for any knowledge base ψ, we require $\psi \models C$. Let I be an interpretation of $\mathcal{L}$. We say that I is a *state* of the world if $I \models C$. A knowledge base ψ can be treated as a *description* of the world, where $Models(\psi)$ is the set of all *possible states* of the world with respect to ψ.

Let ψ be the current knowledge base and μ a propositional formula which is regarded as a new knowledge (information) about the world. Then, informally, the general question of updating ψ with μ is how to specify the new knowledge base after combining the new knowledge (information) μ into the current knowledge base ψ. Consider a student enrollment domain where a constraint states that a second year student must enroll at least two CS courses with level 200 or 300. This constraint may be expressed by a formula C:

$$(Enrolled(CS201) \vee Enrolled(CS202) \vee Enrolled(CS301)) \wedge$$
$$(Enrolled(CS201) \vee Enrolled(CS202) \vee Enrolled(CS303)) \wedge$$
$$(Enrolled(CS201) \vee Enrolled(CS301) \vee Enrolled(CS303)) \wedge$$
$$(Enrolled(CS202) \vee Enrolled(CS301) \vee Enrolled(CS303)).$$

An initial knowledge base of the student's enrollment information may be expressed by formula $\psi \equiv Enrolled(CS201) \wedge Enrolled(CS301) \wedge C$. Now suppose we want to *update* ψ with a new knowledge $\mu \equiv \neg Enrolled(CS301)$. The resulting knowledge base after updating ψ with μ is denoted as $\psi \diamond \mu$.

2.2 Winslett's PMA

Now we review the PMA (*possible models approach*) – a classical minimal change approach for update proposed by Winslett[2] [9]. In the PMA, the knowledge base update is achieved by updating *every possible state* of the world with respect to

---

[1] Usually, we use a set of formulas to represent state constraints. In this case, C can be viewed as a conjunction of all such formulas.

[2] Note that the PMA was originally based on a first order language. Here we restrict the PMA to the propositional case.

ψ with μ, and such state update is constructed based on the *principle of minimal change* on models.

Formally, let I_1 and I_2 be two interpretations of $\mathcal{L}$. We say that I_1 and I_2 differs on a propositional letter l if l appears in exactly one of I_1 and I_2. $Diff(I_1, I_2)$ denotes the set of all different propositional letters between I_1 and I_2. Let I be an interpretation and $\mathcal{I}$ a set of interpretations. We define the set of all *minimally different interpretations* of $\mathcal{I}$ with respect to I as follows:

$$Min(I, \mathcal{I}) = \{I' \mid I' \in \mathcal{I}, \text{ and there does not exist other } I'' \in \mathcal{I} \text{ such}$$
$$\text{that } Diff(I, I'') \subset Diff(I, I')\}.$$

Then we can present the formal definition of the state update in the PMA as follows.

Definition 1. Let C be the state constraint, S a state of the world, i.e. $S \models C$, and μ a propositional formula. Then the *set of all possible states* of the world resulting from updating S with μ by the PMA, denoted as $Res(S, \mu)$, is defined as follows:

$$Res(S, \mu) = Min(S, Models(C \wedge \mu)). \qquad (1)$$

Based on the definition of state update, we can then define the PMA update operator $\diamond_{pma}$ for knowledge bases.

Definition 2. Let ψ be a knowledge base and μ a propositional formula. $\psi \diamond_{pma} \mu$ denotes the update of ψ with μ by the PMA[3], where

1. If ψ is inconsistent, then $\psi \diamond_{pma} \mu \equiv \psi$, otherwise
2. $Models(\psi \diamond_{pma} \mu) = \bigcup_{S \in Models(\psi)} Res(S, \mu)$.

In the above definition, condition 1 says that , if ψ is inconsistent, then any update can not change it into a consistent knowledge base [6]. Condition 2 says that if ψ is consistent and does not entail μ, then ψ should be changed, and this change is made by updating every model of ψ with μ as defined in Definition 1. It has been shown that under many circumstances, the PMA is powerful and effective for representing knowledge updates and reasoning about action [9, 6]. However, as will be shown next, the PMA is problematic sometimes for update with disjunctive information.

Example 1. Suppose a round table is painted with three equal parts of red color, white color and black color respectively[4]. Intuitively, a box on the table implies that it may be entirely within one of these three regions, *or* touching any two of these three regions, *or* touching all of these three regions. Also, a constraint to formalize this domain is specified as:

$$Ontable(Box) \supset Inred(Box) \vee Inwhite(Box) \vee Inblack(Box). \qquad (2)$$

Now suppose the current knowledge base is

---

[3] Here we only consider the *well-defined* update, that is, μ is consistent with the state constraint C.

[4] This is an extension of the example originally raised by Reiter [5].

$$\psi \equiv \neg Ontable(Box) \wedge \neg Inred(Box) \wedge \neg Inwhite(Box) \wedge$$
$$\neg Inblack(Box) \wedge (2),$$

which corresponds to a unique state:

$$S = \{\neg Ontable(Box), \neg Inred(Box), \neg Inwhite(Box), \neg Inblack(Box)\}.$$

Consider updating state S with $\mu \equiv Ontable(Box)$ (i.e. the box is dropped on the table). Using the PMA, it is not difficult to see that from Definitions 1 and 2, the resulting knowledge base is:

$$\psi \diamond_{pma} \mu \equiv Ontable(Box) \wedge$$
$$(Inred(Box) \wedge \neg Inwhite(Box) \wedge Inblack(Box) \vee$$
$$\neg Inred(Box) \wedge Inwhite(Box) \wedge Inblack(Box) \vee$$
$$\neg Inred(Box) \wedge \neg Inwhite(Box) \wedge Inblack(Box)) \wedge (2),$$

which means that the box will only be in one of these three regions. Obviously, this solution is not reasonable from our intuition.

3 MCE: Minimal Change with Exceptions

To overcome the problem with the PMA, in this section we propose an approach for update based on the principle of *Minimal Change with Exceptions*, which we abbreviate as the MCE. In fact, our approach is based on the PMA but with some modifications. The idea is described as follows. Consider the state update[5]. Generally, during the update, the truth value of any literal in the state changes minimally by default. But if the truth value of a literal is *logically indefinite* with respect to the update, then this literal is treated as an *exception* to the minimal change principle. In this case, the change of this literal's truth value will not obey the rule of minimal change.

Informally, we say that the truth value of a literal is *logically indefinite* with respect to an update, if this literal occurs in a disjunction which is entailed by the constraint and the update effect and not satisfied in the initial knowledge base (or the state of initial knowledge base). Consider the example presented in section 2 where the constraint is (2) and the update effect is $Ontable(Box)$. As $Inred(Box)$, $Inwhite(Box)$ and $Inblack(Box)$ are not true in the initial knowledge base but the disjunction $Inred(Box) \vee Inwhite(Box) \vee Inblack(Box)$ is entailed by (2) and $Ontable(Box)$, we know that $Inred(Box) \vee Inwhite(Box) \vee Inblack(Box)$ should be true in the resulting knowledge base but we can not determine the truth values of $Inred(Box)$, $Inwhite(Box)$ and $Inblack(Box)$ exactly. In this case, we say literals $Inred(Box)$, $Inwhite(Box)$ and $Inblack(Box)$ are *logically indefinite* with respect to the update. According to our idea described above, $Inred(Box)$, $Inwhite(Box)$ and $Inblack(Box)$ should be regarded as exceptions to the minimal change principle. Thus, $Inred(Box)$, $Inwhite(Box)$

---

[5] Similar to the PMA, in our approach, updating a knowledge base is achieved by updating every possible model of the knowledge base.

and $Inblack(Box)$ are *not* forced to change minimally during the update, from which we get the desired solution including the case that the box can be on the region connecting two colors or can be on the region connecting three colors (the central part of the round table) after updating the knowledge base with $Ontable(Box)$.

Formally, let EXC be a set of propositional letters that we represent to be exceptional to the minimal change, I_1 and I_2 two interpretations. $Diff(I_1, I_2)^{EXC}$ denotes the set of all different propositional letters, which are *not* in EXC, between I_1 and I_2. That is, $l \in Diff(I_1, I_2)^{EXC}$ iff $[\neg]l \notin I_1 \cap I_2$ and $l \notin EXC$, where notation $[\neg]$ means that the negation sign $\neg$ may or may not occur. For example, let $I_1 = \{a, b, \neg c, \neg d\}$, $I_2 = \{\neg a, b, c, \neg d\}$ and $EXC = \{a, b\}$. Then $Diff(I_1, I_2)^{EXC} = \{c\}$. Let I be an interpretation and $\mathcal{I}$ a set of interpretations. We define the set of all minimal different interpretations of $\mathcal{I}$ with respect to I *with the exception EXC* as follows:

$$Min(I, \mathcal{I})^{EXC} = \{I' \mid I' \in \mathcal{I}, \text{ and there does not exist other } I'' \in \mathcal{I} \\ \text{ such that } Diff(I, I'')^{EXC} \subset Diff(I, I')^{EXC}\}.$$

Let C be a propositional formula used to represent the state constraint and μ a propositional formula. We say a disjunction $\bigvee_{i=1}^{n}[\neg]l_i$[6] $(1 < n)$ satisfying $C \wedge \mu \models \bigvee_{i=1}^{n}[\neg]l_i$, where l_i is a propositional letter $(1 \le i \le n)$, is a *non-trivial disjunction* entailed by $C \wedge \mu$ if for any $M \subset \{1, \cdots, n\}$, $C \wedge \mu \not\models \bigvee_{j \in M}[\neg]l_j$. We denote the set of all non-trivial disjunctions entailed by $C \wedge \mu$ as $D(\mu)$. If $d \equiv \bigvee_{i=1}^{n}[\neg]l_i$ in $D(\mu)$, then we denote $|d| = \{l_1, \cdots, l_n\}$.

In the example presented in section 2, as we have $(2) \wedge Ontable(Box) \models Inred(Box) \vee Inwhite(Box) \vee Inblack(Box)$, $(2) \wedge Ontable(Box) \not\models Inred(Box)$, $(2) \wedge Ontable(Box) \not\models Inwhite(Box)$, and $(2) \wedge Ontable(Box) \not\models Inblack(Box)$, we then get $D(Ontable(Box)) = \{d\} = \{Inred(Box) \vee Inwhite(Box) \vee Inblack(Box)\}$, and $|d| = \{Inred(Box), Inwhite(Box), Inblack(Box)\}$. Now we give the definition of state update in the MCE as follows.

Definition 3. Let C be the state constraint, S a state of the world, i.e. $S \models C$, μ a propositional formula, and $D(\mu)$ the set of non-trivial disjunctions entailed by $C \wedge \mu$. We define the *exceptional letters* with respect to S and μ as follows:

$$EXC(S, \mu) = \bigcup_{d \in D(\mu), S \not\models d} |d|. \tag{3}$$

Then the set of possible states resulting from updating S with μ by the MCE, denoted as $Res(S, \mu)^{EXC(S, \mu)}$, is defined as

$$Res(S, \mu)^{EXC(S, \mu)} = Min(S, Models(C \wedge \mu))^{EXC(S, \mu)}. \tag{4}$$

Let us examine Definition 3 in detail. Firstly, (3) defines a set of propositional letters that should be viewed as exceptions to the minimal change principle

---

[6] $[\neg]$ means that the negation sign $\neg$ may or may not occur.

during the state update. If $d \in D(\mu)$ is already satisfied in S, then any letters which or whose negations occur in d will not be specified in $EXC(S, \mu)$, otherwise the letters should be in $EXC(S, \mu)$. For instance, suppose $S = \{\neg a, \neg b, c, \neg d\}$ and $D(\mu) = \{a \vee b, b \vee c\}$, then $EXC(S, \mu) = \{a, b\}$ while c is not in $EXC(S, \mu)$ as $S \models b \vee c$. Secondly, (4) defines the set of possible resulting states after updating S with μ. Note that any literals in S whose corresponding letters are in $EXC(S, \mu)$ will not obey the minimal change principle during the update. In the above example, if $C \equiv (d \supset a \vee b) \wedge (d \supset b \vee c)$ and $\mu \equiv d^7$, then we get

$Res(S, \mu)^{EXC(S, \mu)} = \{S_1, S_2, S_3\}$, where
$S_1 = \{a, \neg b, c, d\}$,
$S_2 = \{\neg a, b, c, d\}$, and
$S_3 = \{a, b, c, d\}$.

Based on Definition 3, we define the knowledge base update in the MCE as follows.

Definition 4. Let ψ be a knowledge base, μ a propositional formula. $\psi \diamond_{mce} \mu$ denotes the update of ψ with μ by the MCE, where

1. If ψ is inconsistent, then $\psi \diamond_{mce} \mu \equiv \psi$, otherwise
2. $Models(\psi \diamond_{mce} \mu) = \cup_{S \in Models(\psi)} Res(S, \mu)^{EXC(S, \mu)}$.

Comparing with Definition 1 and 2, it is easy to see that the MCE is defined based on the PMA but with exception $EXC(S, \mu)$. Clearly, if $EXC(S, \mu) = \emptyset$ for any $S \in Models(\psi)$, then the MCE reduces to the PMA. Consider the dropping-box example once again. As the knowledge base ψ corresponds to a unique state $S = \{\neg Ontable(Box), \neg Inred(Box), \neg Inwhite(Box), \neg Inblack(Box)\}$, we have $EXC(S, Ontable(Box)) = \{Inred(Box), Inwhite(Box), Inblack(Box)\}$, then from Definition 3 and 4, we get the desired result:

$\psi \diamond_{mce} \mu \equiv Ontable(Box) \wedge (Inred(Box) \vee Inwhite(Box) \vee Inblack(Box)) \wedge$
(2).

From the above discussion, we can see that the MCE overcomes the problem with the PMA of updating knowledge bases with disjunctive information.

4 Properties of the MCE

In this section, we first illustrate one more example to show the application of the MCE in propositional knowledge base update with disjunctive information, and then discuss how our update approach relates to Katsuno and Mendelzon's update theory [6, 7].

---

[7] This implies that $D(\mu) = \{a \vee b, b \vee c\}$.

Example 2. We still consider the student's enrollment scenario described in section 2.1, but with a bit more complex situation. The set of constraints C includes the following formulas:

$$Enrolled(CS301) \supset Prestudy(CS401), \tag{5}$$

$$Enrolled(CS302) \supset Prestudy(CS401) \vee Prestudy(CS403), \tag{6}$$

$$Enrolled(CS303) \supset Prestudy(CS403) \vee Prestudy(CS405), \tag{7}$$

where (5) means that if some one enrolls CS301, it implies that he/she is preparing to study course CS401 in the future, similarly, (6) and (7) state that if someone enrolls CS302 (or CS303 respectively), then he/she is preparing to study course CS401 or CS403 (or CS403 or CS405 respectively).

We assume that currently the student enrolled CS301 but did not enrolled CS302 and CS303. Further, the student did not prepare to study courses CS403 and CS405. So, the knowledge base of this student's enrollment is

$$\psi \equiv Enrolled(CS301) \wedge \neg Enrolled(CS302) \wedge \neg Enrolled(CS303) \wedge \\ \neg Prestudy(CS403) \wedge \neg Prestudy(CS405) \wedge C.$$

Clearly, $\psi \models Enrolled(CS401)$. Now suppose that after the semester began two weeks, the student decided to enroll courses CS302 and CS303 in order to prepare to study CS403 and CS405. Hence, we need to update ψ with $\mu \equiv Enrolled(CS302) \wedge Enrolled(CS303)$.

Let us first use the MCE to derive the resulting knowledge base. Obviously, ψ has a unique model[8]:

$$S = \{Enrolled(CS301), \neg Enrolled(CS302), \neg Enrolled(CS303), \\ Prestudy(CS401), \neg Prestudy(CS403), \neg Prestudy(CS405)\}.$$

Clearly, we have

$$D(\mu) = \{Prestudy(CS401) \vee Prestudy(CS403), \\ Prestudy(CS403) \vee Prestudy(CS405)\}.$$

Since $S \models Prestudy(CS401) \vee Prestudy(CS403)$ and $S \not\models Prestudy(CS403) \vee Prestudy(CS405)$, the set of exceptional letters with respect to S and μ is

$$EXC(S, \mu) = \{Prestudy(CS403), Prestudy(CS405)\}.$$

Therefore, we have

$Res(S, \mu)^{EXC(S,\mu)} = \{S_1, S_2, S_3\}$, where
$S_1 = \{Enrolled(CS301), Enrolled(CS302), Enrolled(CS303), \\ \phantom{S_1 = \{}Prestudy(CS401), Prestudy(CS403), \neg Prestudy(CS405)\},$
$S_2 = \{Enrolled(CS301), Enrolled(CS302), Enrolled(CS303), \\ \phantom{S_2 = \{}Prestudy(CS401), \neg Prestudy(CS403), Prestudy(CS405)\},$
$S_3 = \{Enrolled(CS301), Enrolled(CS302), Enrolled(CS303), \\ \phantom{S_3 = \{}Prestudy(CS401), Prestudy(CS403), Prestudy(CS405)\}.$

---

[8] we assume that there is no more propositional letters in our language

Finally, the resulting knowledge is

$$\psi \diamond_{mce} \mu \equiv Enrolled(CS301) \wedge Enrolled(CS302) \wedge Enrolled(CS303) \wedge$$
$$(Prestudy(CS403) \vee Prestudy(CS405)) \wedge C.$$

Ignoring the detail, if we use the PMA to this example, the final result would be:

$$\psi \diamond_{pma} \mu \equiv Enrolled(CS301) \wedge Enrolled(CS302) \wedge Enrolled(CS303) \wedge$$
$$(\neg Prestudy(CS403) \wedge Prestudy(CS405) \vee$$
$$(Prestudy(CS403) \wedge \neg Prestudy(CS405)) \wedge C,$$

which seems implausible in the sense that enrolling CS302 and CS303 implies that the student is just preparing to study one of CS403 and CS405.

The following theorem first shows the relation between the PMA and MCE.

Theorem 5. *Let ψ be a propositional knowledge base and μ a propositional formula. Then for any propositional formula ϕ, $\psi \diamond_{mce} \mu \models \phi$ implies $\psi \diamond_{pma} \mu \models \phi$.*

Now let us consider the relationship between the MCE and Katsuno and Mendelzon's update postulates. The motivation of Katsuno and Mendelzon's proposal for update is an observation on the difference between revision and update. In particular, let ψ be a knowledge base and μ a formula. Gardenfors *el al.* proposed the following postulates which, as they argued, should be satisfied by any revision operator $\circ$ [1, 6].

(R1) $\phi \circ \mu$ implies μ.
(R2) If $\phi \wedge \mu$ is satisfiable then $\psi \circ \mu \equiv \psi \wedge \mu$.
(R3) If μ is satisfiable then $\psi \circ \mu$ is also satisfiable.
(R4) If $\psi_1 \equiv \psi_2$ and $\mu_1 \equiv \mu_2$ then $\psi_1 \circ \mu_1 \equiv \psi_2 \circ \mu_2$.
(R5) $(\psi \circ \mu) \wedge \phi$ implies $\psi \circ (\mu \wedge \phi)$.
(R6) If $(\psi \circ \mu) \wedge \phi$ is satisfiable then $\psi \circ (\mu \wedge \phi)$ implies $(\psi \circ \mu) \wedge \phi$.

However, Katsuno and Mendelzon revealed that updating a knowledge base is quite different from revising it, and ignoring such difference may lead to unreasonable solutions for knowledge base updates [6]. Based on this observation, they proposed the following alternative postulates for any update operator $\diamond$.

(U1) $\psi \diamond \mu$ implies μ.
(U2) If ψ implies μ then $\psi \diamond \mu \equiv \psi$.
(U3) If both ψ and μ are satisfiable then $\psi \diamond \mu$ is also satisfiable.
(U4) If $\psi_1 \equiv \psi_2$ and $\mu_1 \equiv \mu_2$ then $\psi_1 \diamond \mu_1 \equiv \psi_2 \diamond \mu_2$.
(U5) $(\psi \diamond \mu) \wedge \phi$ implies $\psi \diamond (\mu \wedge \phi)$.
(U6) If $\psi \diamond \mu_1$ implies μ_2 and $\psi \diamond \mu_2$ implies μ_1 then $\psi \diamond \mu_1 \equiv \psi \diamond \mu_2$.
(U7) If ψ is complete then $(\psi \diamond \mu_1) \wedge (\psi \diamond \mu_2)$ implies $\psi \diamond (\mu_1 \vee \mu_2)$.
(U8) $(\psi_1 \vee \psi_2) \diamond \mu \equiv (\psi_1 \diamond \mu) \vee (\psi_2 \diamond \mu)$.

In fact, the Katsuno-Mendelzon's update postulates characterize the update semantics for a class of update operators that are based on the principle of minimal change. For instance, the PMA update operator $\diamond_{pma}$ satisfies all postulates (U1) – (U8).

Theorem 6. *The MCE update operator $\diamond_{mcd}$ satisfies Katsuno-Mendelzon's postulates (U1) – (U8).*

Theorem 6 reveals an important fact that Katsuno-Mendelzon's postulates (U1) – (U8) still correctly capture the semantics of updating knowledge base with disjunctive information. In other words, Katsuno-Mendelzon's postulates are independent from the update with disjunctive information, which is quite different from the arguments of some previous work (eg. [11]). The following result shows the relationship between the MCD and AGM revision postulates.

Theorem 7. *The MCE update operator $\diamond_{mcd}$ satisfies AGM postulates (R1), (R4) and (R5), but violates (R2), (R3) and (R6).*

5   Extended Disjunctive Default Theories

In previous sections, it has been shown that the principle of minimal change with exceptions on disjunctive information provides an alternative way to handle the problem of disjunctions in propositional knowledge base updates. It is also known that the classical minimal change approaches, eg. the PMA, usually can be interpreted by different nonmonotonic logics appeared in the literature, eg. the default logic or circumscriptions. Now our question is: Can our alternative minimal change approach proposed earlier be also interpreted by some nonmonotonic logic? This is the issue we will investigate in the rest of the paper.

In this section, we first describe extended disjunctive default theories based Gelfond *et al.*'s disjunctive default theories (*ddt*) [4], and then discuss the difference between our extended disjunctive default theories and Gelfond *et al.*'s disjunctive default theories[9].

An *extended disjunctive default*[10] is an expression of the form

$$\frac{\alpha : \beta_1, \cdots, \beta_m}{\gamma_1 \uparrow \cdots \uparrow \gamma_n},\tag{8}$$

where $\alpha, \beta_1, \cdots \beta_m, \gamma_1, \cdots, \gamma_n$ $(m, n \geq o)$ are propositional formulas[11]. Formula α is called the *prerequisite* of the default, $\beta_1, \cdots, \beta_m$ are its *justifications*, and $\gamma_1 \uparrow \cdots \uparrow \gamma_n$ is its *consequent*. As special cases, if $\alpha \equiv True$, we often write

---

[9] Due to the space limitation, the detail about Gelfond *et al.*'s disjunctive default theories is referred to [4].

[10] We will alternatively use phrase *extended disjunctive default* and *default* if there is no confusion in the context.

[11] As we only consider propositional knowledge base updates, we restrict the default theory to propositional case.

(8) as $\frac{:\beta_1,\cdots,\beta_m}{\gamma_1\uparrow\cdots\uparrow\gamma_n}$, and if $m = 0$, (i.e. the set of justifications is empty), we often write $\frac{\alpha}{\gamma_1\uparrow\cdots\uparrow\gamma_n}$. If $\alpha \equiv True$ and $m = 0$, we simply identify default (8) with $\gamma_1 \uparrow \cdots \uparrow \gamma_n$. An *extended disjunctive default theory* (an *eddt*, for short) is a set of extended disjunctive defaults.

Definition 8. Let D be an extended disjunctive default theory and E a set of formulas. E is an *extension* for D if it is one of the smallest deductively closed sets of formulas E' satisfying the condition: For any extended disjunctive default (8) from D, if $\alpha \in E'$ and $\neg\beta_1, \cdots, \neg\beta_m \notin E$ then, for some $\eta \in \Pi$ ($\eta \neq \emptyset$), where $\Pi = 2^{\{\gamma_1,\cdots,\gamma_n\}}$, $\eta \subseteq E'$. Formula ϕ is called a *theorem* of D, denoted as $D \models_{df} \phi$, if ϕ belongs to all extensions of D.

It is important to notice the difference between our *eddt* and Gelfond *et al.*'s *ddt*. In Gelfond *et al.*'s *ddt*, a *disjunctive default* is an expression of the form

$$\frac{\alpha : \beta_1, \cdots, \beta_m}{\gamma_1 | \cdots | \gamma_n}, \tag{9}$$

and a *disjunctive default theory (ddt)* is a set of disjunctive defaults. An extension of a ddt is defined exactly the same as Definition 5 except $\Pi = \{\{\gamma_1\}, \cdots, \{\gamma_n\}\}$, and the theorem of a ddt is defined the same as Definition 5.

Semantically, the difference between (8) and (9) is that the latter requires an extension to contain one of $\gamma_1, \cdots, \gamma_n$, while (8) requires an extension to contain one of non-empty subsets of $\{\gamma_1, \cdots, \gamma_n\}$. If we view extensions as interpretations for disjunction $\gamma_1 \vee \cdots \vee \gamma_n$, then our eddt derives *maximal* possible interpretations of this disjunction, while ddt derives *minimal* possible interpretations for it. The following example further shows such difference.

Example 3. The disjunctive default theory

$$\{\frac{a : b}{b}, \frac{c : d}{d}, a|c\}$$

has two extensions: the deductive closure of $\{a, b\}$ and the deductive closure of $\{c, d\}$. On the contrary, the extended disjunctive default theory

$$\{\frac{a : b}{b}, \frac{c : d}{d}, a \uparrow c\}$$

has three extensions: the deductive closure of $\{a, b\}$, the deductive closure of $\{c, d\}$, and deductive closure of $\{a, b, c, d\}$.

Theorem 9. *Let D be an extended disjunctive default theory. We define a disjunctive default theory D' by replacing each extended disjunctive default of the form (8) from D with disjunctive default of the form (9). Then $D \models_{df} \phi$ implies $D' \models_{df} \phi$.*

6 Translating MCE to Extended Disjunctive Default Theories

Given a domain constraint C, a knowledge base ψ and a propositional formula μ, we define a class of extended disjunctive default theories which, as will be shown later, can capture the semantics of updating ψ with μ by using the MCE. We first need to define a new language $\mathcal{L}_{old-new}$ based on the propositional language $\mathcal{L}$ of the MCE. In particular, $\mathcal{L}_{old-new}$ is obtained from $\mathcal{L}$ by replacing each propositional letter l in $\mathcal{L}$ with two different propositional letters $Old\text{-}l$ and $New\text{-}l$. Our eddts will be constructed from $\mathcal{L}_{old-new}$. The translation is described by the following steps.

Step 1. For each $m \in Models(\psi)$, specify a set of literals, denoted as $Old(m)$, as follows:

$$Old(m) = \{[\neg]Old\text{-}l \mid [\neg]l \in m\},$$

where l is a propositional letter of $\mathcal{L}$, and notation $[\neg]l \in m$ means that l or $\neg l$ is in m. Therefore, if $l \in m$, then $Old\text{-}l \in Old(m)$, otherwise $\neg Old\text{-}l \in Old(m)$. Intuitively, $Old(m)$ is viewed as a possible model for knowledge base ψ in language $\mathcal{L}_{old-new}$.

Step 2. Replace every propositional letter l occurring in C by $New\text{-}l$. Then a new formula $New\text{-}C$ is obtained from C.

Step 3. Replace each propositional letter l occurring in μ by $New\text{-}l$. Therefore, a new formula $New\text{-}\mu$ is obtained from μ.

Step 4. For each *non-trivial* disjunction $\bigvee_{i=1}^{k}[\neg]l_i$ $(1 < k)$ entailed by $\psi \wedge \mu$, that is, $\psi \wedge \mu \models \bigvee_{i=1}^{k}[\neg]l_i$ and for any $M \subset \{1, \cdots, k\}$ $\psi \wedge \mu \not\models \bigvee_{j \in M}[\neg]l_j$, if $m \not\models \bigvee_{i=1}^{k}[\neg]l_i$, specify an extended disjunctive default

$$[\neg]New\text{-}l_1 \uparrow \cdots \uparrow [\neg]New\text{-}l_k.$$

Denote the set of all such extended disjunctive defaults as $New\text{-}dis(\psi \wedge \mu, m)$.

Step 5. For each propositional letter l in $\mathcal{L}$, specify two defaults:

$$\frac{Old\text{-}l : New\text{-}l}{New\text{-}l}, \qquad \frac{\neg Old\text{-}l : \neg New\text{-}l}{\neg New\text{-}l}. \tag{10}$$

Denote the set of all such defaults as IR. Obviously, these defaults form *inertia rules* which are used to derive persistent facts during the update.

Step 6. Define an extended disjunctive default theory $D(m)$ as

$$D(m) = Old(m) \cup \{New\text{-}C, New\text{-}\mu\} \cup New\text{-}dis(\psi \wedge \mu, m) \cup IR. \tag{11}$$

Step 7. Finally, we define a class of extended disjunctive default theories with respect to knowledge base ψ and formula μ as follows:

$$\mathcal{D}(\psi, \mu) = \bigcup_{m \in M_{oldes}(\psi)} D(m). \tag{12}$$

For a given $D(m) \in \mathcal{D}(\psi, \mu)$, an extension of $D(m)$ represents a possible solution of updating state m with μ under the semantics of the MCE, in which *Old* literals (i.e. $[\neg]Old\text{-}l$) represent the facts that are true in m before the update and *New* literals (i.e. $[\neg]New\text{-}l$) represent the facts that are true after the update. It is also observed that during the default reasoning, *New* literals are generally derived from their corresponding *Old* literals by applying inertia rules (10) except those *New* literals occurring in some $[\neg]New\text{-}l_1 \uparrow \cdots \uparrow [\neg]New\text{-}l_k$.

The following example shows how our extended disjunctive default theories represent the update under the semantics of the MCE.

Example 4. Example 1 continued. In language $\mathcal{L}$ the initial knowledge base is

$$\psi \equiv \neg Ontable(Box) \wedge \neg Inred(Box) \wedge$$
$$\neg Inwhite(Box) \wedge \neg Inblack(Box) \wedge (2),$$

and $\mu \equiv Ontable(Box)$. Now we translate our knowledge base update under the MCE to an extended default theory following the steps described previously[12]. Ignoring the detail, we have the following extended disjunctive defaults:

$$Old(m) = \{\neg Old\text{-}Ontable(Box), \neg Old\text{-}Inred(Box),$$
$$\neg Old\text{-}Inwhite(Box), \neg Old\text{-}Inblack(Box)\},$$

$$New\text{-}C \equiv New\text{-}Ontable(Table) \supset$$
$$New\text{-}Inred(Box) \vee New\text{-}Inwhite(Box) \vee New\text{-}Inblack(Box),$$

$$New\text{-}\mu \equiv New\text{-}Ontable(Table),$$

$$New\text{-}dis(\psi \wedge \mu, m) = \{New\text{-}Inred(Box) \uparrow New\text{-}Inwhite(Box) \uparrow$$
$$New\text{-}Inblack(Box)\}.$$

Then, we define an extended disjunctive default theory $D(m)$ as follows:

$$D(m) = Old(m) \cup \{New\text{-}C, New\text{-}\mu\} \cup New\text{-}dis(\psi \wedge \mu, m) \cup IR,$$

where IR is a set of defaults with the form (10) for propositional letters $Ontable(Box)$, $Inred(Box)$, $Inwhite(Box)$, and $Inblack(Box)$ respectively.

It is not difficult to see that $D(m)$ has seven extensions including every possible interpretation for disjunction $New\text{-}Inred(Box) \vee New\text{-}Inwhite(Box) \vee New\text{-}Inblack(Box)$ respectively[13], and the only effect of the update we can derive is

---

[12] As ψ only has one model, obviously there will be only one eddt.

[13] Recall that we view extensions as interpretations for the disjunction.

$D(m) \models_{df} New\text{-}Ontable(Box) \land$
$(New\text{-}Inred(Box) \lor New\text{-}Inwhite(Box) \lor New\text{-}Inblack(Box)),$

which represents the desired solution of Box's position with respect to updating ψ with μ under the MCE's semantics.

The following theorem shows that the translation from the MCE to extended disjunctive default theories under the semantics of the MCE is sound and complete.

Theorem 10. *Let C be the domain constraint, ψ a knowledge base, μ a propositional formula in propositional language $\mathcal{L}$. Suppose $\mathcal{D}(\psi, \mu)$ is a class of extended disjunctive default theories in language $\mathcal{L}_{old-new}$ defined through Step 1 to Step 7 described above. For any propositional formula ϕ in $\mathcal{L}$, let $New\text{-}\phi$ be a propositional formula in $\mathcal{L}_{old-new}$ obtained by replacing each propositional letter l occurring in ϕ with $New\text{-}l$. Then for each $D(m)$ in $\mathcal{D}(\psi, \mu)$, $D(m) \models_{df} New\text{-}\phi$ iff $\psi \diamond_{mcd} \mu \models \phi^{14}$.*

7 Conclusion

In this paper, we proposed an alternative approach called the MCE to deal with the problem of update with disjunctive information, which overcomes the difficulty of the PMA.

There some desired features of our approach proposed in this paper. First, our approach can deal with both direct and indirect disjunctive effects of updates. For instance, in Example 1, the update effect $\mu \equiv Ontable(Box)$ is definite. Together with the constraint (2), however, it implies an indirect disjunctive effect $Inred(Box) \lor Inwhite(Box) \lor Inblack(Box)$. In this case, our approaches produce the desired result. Second, as the MCE satisfy Katsuno and Mendelzon's postulates (U1) - (U8), we can see that in general, the principle of minimal change[15] still captures the semantics of update with disjunctive information. Finally, by extending Gelfond *et al*'s disjunctive default theories, we then revealed the important connection between the disjunctive update and default reasoning - the knowledge base (belief) update with disjunctive information may be interpreted in terms of the underlying nonmonotonic reasoning.

We should mention that our original motivation to this problem came from our early work on reasoning about indefinite actions [10]. We notice that the issue of update with disjunctive information has also been addressed by other researchers recently. For instance, T. Eiter, G. Gottlob and Y. Gurevich proposed a circumscriptive approach to deal with disjunctions in nonmonotonic reasoning [3]. In fact, a semantic approach we proposed in our AAAI paper [13] is similar to theirs though the two representations are quite different. However, one of the

---

[14] Note that here we require $New\text{-}\phi$ be a theorem for *each* extended disjunctive default theory in $\mathcal{D}(\psi, \mu)$.

[15] Obviously, we should define an appropriate criterion for minimal change.

most significant results we obtained, which was omitted by other researchers, is that the principle of minimal change is general enough to capture the semantics of disjunctive information in reasoning about change.

Acknowledgements

This research is supported in part by a grant from the Australian Research Council and a grant from University of Western Sydney, Nepean.

References

1. P. Gardenfors, *Knowledge in Flux*. MIT Press, 1988.
2. G. Brewka and J. Hertzberg, How do things with worlds: On formalizing actions and plans. *Journal of Logic and Computation*, **3(5)** (1993) 517–532.
3. T. Eiter, G. Gottlob and Y. Gurevich, Curb your theory! A circumscriptive approach for inclusive interpretation of disjunctive information. In *Proceedings of 13th International Joint Conference on Artificial Intelligence (IJCAI'93)*, Morgan Kaufmann, Inc. (1993) 634–639.
4. M. Gelfond, V. Lifschitz, H. Przymusinska and M. Truszczynski, Disjunctive defaults. In Proceedings of the 2nd International Conference on Principles of Knowledge Representation and Reasoning (KR'91), Morgan Kaufmann, Inc. (1991) 230–237.
5. G.N. Kartha and V. Lifschitz, Actions with indirect effects. in *Proceedings of the Fourth International Conference on Principles of Knowledge Representation and Reasoning (KR'94)*. Morgan Kaufmann, Inc. (1994) 341–350.
6. H. Katsuno and A.O. Mendelzon, On the difference between updating a knowledge database and revising it. In *Proceedings of KR'91*. Morgan Kaufmann, Inc. (1991) 387-394.
7. H. Katsuno and A.O. Mendelzon, Propositional knowledge base revision and minimal change. In *Artificial Intelligence* **52** (1991) 263-294.
8. R. Reiter, A logic for default reasoning. *Artificial Intelligence* **13** (1980) 81–132.
9. M. Winslett, Reasoning about action using a possible models approach. In *Proceedings of the Seventh National Conference on Artificial Intelligence (AAAI'88)*. Morgan Kaufmann Publisher, Inc. (1988) 89–93.
10. Y. Zhang and N.Y. Foo, Reasoning about persistence: A theory of actions. In *Proceedings of 13th International Joint Conference on Artificial Intelligence (IJCAI'93)*, Morgan Kaufmann, Inc. (1993) 718–723.
11. Y. Zhang and N.Y. Foo, Applying the persistent set approach in temporal reasoning. *Annals of Mathematics and Artificial Intelligence* 14 pp 75–98, 1995.
12. Y. Zhang, Semantical considerations for knowledge base updates. In *Proceedings of the Ninth International Symposium on Methodologies for Intelligent Systems (ISMIS'96)*. Also in *Lecture Notes in Artificial Intelligence*, pp. 88-97, Vol. 1079, Springer-Verlag, 1996.
13. Y. Zhang and N.Y. Foo, Updating knowledge bases with disjunctive information. In *Proceedings of the 14th National Conference on Artificial Intelligence (AAAI'96)*, pp 562-568. AAAI/MIT Press, 1996.

Toward a Uniform Logical Representation of Different Kinds of Integrity Constraints

Robert Demolombe[1] and Andrew J.I. Jones[2], Jose Carmo[3]

[1] ONERA/CERT, Toulouse,France
[2] Department of Philosophy and Norwegian Research Centre for Computers and Law, University of Oslo, Oslo, Norway
[3] Department of Mathematics, Instituto Superior Tecnico, Lisboa, Portugal

1 Introduction

There is no agreement in the literature about the definition of integrity constraints in the context of data and/or knowledge bases. What kinds of properties are integrity constraints supposed to be able to guarantee? The consistency of database content, or the completeness of database content [7]? The validity or completeness of database content with respect to the world [3]? Or some properties which should be satisfied in the world for a given application domain [1]? There is no single set of answers to these questions; it depends on the intended use of the database. However, even though we accept that integrity constraints concern the preservation of a number of different kinds of properties, we think it is interesting both to ask whether integrity constraints might nevertheless share some common features, and to examine the implicit assumptions on which each kind are based.

The objective of this paper is to offer a first proposal in this direction. Since the point here is just to try to gain a better understanding of the nature of integrity constraints, we do not consider specific techniques for efficiently checking whether constraints are violated [5], or for repairing violations [6, 8].

In order to provide a uniform representation of the different kinds of constraints, we adopt a logical framework. In this framework, database content is represented by a set db of formulas in a language of classical first order logic. The world which the database is supposed to represent is itself represented by a set w of formulas of the same language. In what follows, we call the "world" a correct representation of the world in the language used for a given application domain.

To be able to make the distinction between a sentence p that represents part of database content, and the same sentence p that represents part of the world, we use a doxastic logic; the fact that p is part of the database content is represented by the sentence Bp, which is to be read: "the database believes that p". The representation of database content in this doxastic logic is a set of sentences dbb, defined on the basis of db in the following way:

$$\text{dbb} = \{\text{Bp} : \vdash \text{db} \rightarrow \text{p}\} \cup \{\neg\text{Bp} : \nvdash \text{db} \rightarrow \text{p}\}$$

We adopt for the modality B the modal system (KD) [2].

Since, in our view, constraints may be expressed by normative sentences, we also need to be able to make the distinction between the fact that p is the case, and the fact that p should be the case. For this purpose we include a deontic logic, and read sentences of the form Op as: "it should be the case that p" or "it ought to be the case that p". Many different proposals are to be found for the formal treatment of the deontic modality (see, e.g.[1, 4, 9]); in this paper we do not elaborate on the choice of the appropriate deontic logic, and - to simplify matters - we employ standard deontic logic (SDL), (see [2], ch. 6).

For each different kind of constraint, we shall consider the definition of property that it is intended to preserve, the kinds of associated sentences that may be assumed true of the world, and the characterisation both of constraint violation and of the secondary obligations which indicate what is to be done to repair, or recover from, a state of violation.

We shall not here consider constraints about the consistency of the set of database beliefs, because it is generally agreed that database content should be consistent.

2 Constraints About the World

To illustrate this kind of constraint let us consider the context of library management in a research laboratory. In this context managers want to control situations happening in the world. For instance, it may happen that members of the department leave the department without returning books they have borrowed. To prevent such situations library managers may explicitly characterize ideal situations by the constraint:

(1) Books can be borrowed only by members of the department.

which can be formally represented by:

(1') $\forall x \forall y O(\text{borrowed}(x, y) \rightarrow \text{dept}(x))$

A database could here be used as a means for detecting possible violations of this constraint. That is possible only if it is guaranteed that database content is a correct representation of the world, in the sense that if the database believes that some member a of the department has borrowed a book b then this fact is true in the world (inft. $B(\text{borrowed}(a, b)) \rightarrow \text{borrowed}(a, b)$), and if it is true in the world that a is member of the department then this fact is represented in the database (inft. $\text{dept}(a) \rightarrow B(\text{dept}(a))$) [4].

The property $B(\text{borrowed}(a, b)) \rightarrow \text{borrowed}(a, b)$ is called "validity" of database content in regard to the fact borrowed(a,b), and, in general, we call validity of database content in regard to p the property: $Bp \rightarrow p$. In a similar way, $\text{dept}(a) \rightarrow B(\text{dept}(a))$ is called "completeness" of database content in regard to the fact dept(a), and, in general, we call completeness of database content in regard to q the property: $q \rightarrow Bq$.

Now, on the basis of the assumption of validity of the database for the fact borrowed(a,b) and of completeness for the fact dept(a), if the database is in a

---

[4] Along the paper we use "inft." as an abbreviation for "in formal terms".

state where it believes that a has borrowed book b and it does not believe that a is member of the department (inft. B(borrowed(a, b)) $\land \neg$B(dept(a))), we can infer that it is the case in the world that a has borrowed b and a is not member of the department (inft. borrowed(a, b) $\land \neg$dept(a)), which is a situation that violates constraint (1).

In general, if we have Bp $\land \neg$Bq, from assumptions Bp $\rightarrow$ p and q $\rightarrow$ Bq, we can infer p $\land \neg$q, which characterizes a violation of a constraint of the form O(p $\rightarrow$ q). This can be informally represented by Figure 1 (thin arrows represent implications that should hold, while thick arrows represent implications that are assumed to hold).

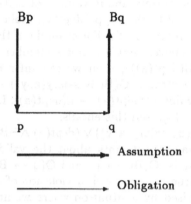

Fig. 1. Constraints about the the world.

To repair a violation of constraint (1) secondary obligations are imposed. They may be, for instance, to send a letter to a, asking him to return borrowed books. At the moment books have been returned the fact borrowed(a,b) is no longer true, and since database content is supposed to be valid in regard to this fact, it has to be removed from the database. However, in the mean time, it is accepted to have B(borrowed(a, b)) $\land \neg$B(dept(a)).

3 Constraints About the Links Between Database Content and the World

A constraint requiring validity of the database content must surely hold generally, for all facts, since it would be nonsense to consider acceptable a database whose content was false of the world - excluding here, of course, the possibility that a database might be designed to serve the function of deceiving its users.

Constraints about completeness, on the other hand, will ordinarily be optional, depending on the needs of the user. Thus a company might require that its database must be complete with respect to the age of each employee, but readily accept incompleteness with respect to a number of other facts about its employees.

To check whether these kinds of constraints regarding validity and completeness have been violated, we need to assume that some database beliefs are true beliefs - that is, that they are true of the world. In many cases this assumption concerns generalizations of which it may reasonably be supposed that they have no counter-instances. For instance, it may be reasonable to assume that the following rule is always true, and in that sense cannot be violated:

(2) Only members of the department have a salary from the department which can be formally represented by:

(2') $\forall x((\exists y\ salary(x, y)) \rightarrow dept(x))$

Then, if we have constraints about validity of facts of type $\exists y\ salary(x, y)$, and about completeness of facts of type $dept(x)$, violations of these constraints may be detected by reference to (2'). For example, if the database believes that a has a salary, and it believes that a is not a member of the department (inft. $B(\exists y\ salary(a, y)) \wedge \neg B(dept(a)))$, then we can infer that there is a violation either of the validity constraint: $O(B(\exists y\ salary(a, y)) \rightarrow (\exists y\ salary(a, y))$, or of the completeness constraint: $O(dept(a) \rightarrow B(dept(a))$. Indeed, from (2') we have: $\neg(\exists y\ salary(a, y)) \vee dept(a)$, and this entails:

$(\neg(\exists y\ salary(a, y)) \wedge B(\exists y\ salary(a, y)) \vee (dept(a) \wedge \neg B(dept(a)))$.

In general, if we have constraints about the validity of p and about the completeness of q, that is: $O(Bp \rightarrow p)$ and $O(q \rightarrow Bq)$, and if it is assumed that $p \rightarrow q$ is always true in the world, a violation of at least one of these two constraints is characterised by a situation where we have: $Bp \wedge \neg Bq$. This can be informally represented by Figure 2.

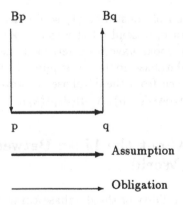

Fig. 2. Constraints about the links between data base content and the world.

In that case, to repair a violation of $O(Bp \rightarrow p)$ or $O(q \rightarrow Bq)$, secondary obligations will require (respectively) either deletion of p from the database or insertion of q into the database. Here, in contrast to the case of constraints about the world, the situation where we have $Bp \land \neg Bq$ cannot be accepted, and secondary obligations require not change to the world, but change to the database.

4   Constraints About Database Content

A particular species of completeness constraint (cf. [7]) is worthy of special attention. Supposing, again, the library scenario, it may be that, in fact, the following generalization is true: every departmental member is either a professor or a student. And it may then well be the case that the library manager wants his classification of the departmental members to be complete with respect to the categories of student and professor. (Perhaps lending entitlements are dependent on this classification, for instance.)

Thus the library database will be subject to the constraint:

(3) If the database believes that some x is a member of department, then the database should believe that x is a student or it should believe that x is a professor

which can be represented in formal terms by:

(3') $\forall x O(B(dept(x)) \rightarrow B(student(x)) \lor B(professor(x)))$

Another example of this type of constraint would be:

(4) If the database believes that some x is a member of department, then the database should be aware of the address of x

which can be represented in formal terms by:

(4') $\forall x O(B(dept(x) \rightarrow \exists y B(address(x, y)))$

The general forms of the constraints exhibited by (3') and (4') [7] are: $O(Bp \rightarrow Bq \lor Br)$ and $O(Bp \rightarrow \exists x Bq(x))$.

The process of checking whether a constraint of these general kinds is violated does not itself involve an assumption to the effect that some beliefs are true beliefs; it is sufficient to consider just the set of database beliefs. For instance, considering the first of these two general forms, if the database content is such that we have: $Bp \land \neg Bq \land \neg Br$, then it is clear that the constraint $O(Bp \rightarrow Bq \lor Br)$ has been violated. Obviously, to repair the state of violation, the appropriate secondary obligation would require either the deletion of the sentence p, or the retention of p and the addition of either the sentence q or the sentence r. And which choice is made here will, of course, depend on what is true in the world.

In general, acceptance of a constraint of the form $O(Bp \rightarrow Bq \lor Br)$ implicitly assumes that the sentence $p \rightarrow q \lor r$ is true in the actual world. This can be informally represented by Figure 3.

5   Conclusion

We have outlined a logical framework for the representation of some species of database integrity constraints. In each case, the constraints were represented as

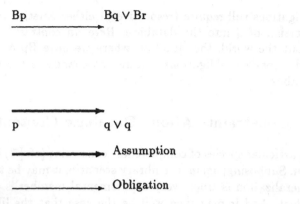

Fig. 3. Constraints database containt.

obligation sentences. For examples of type (1'), the obligation sentence pertains to how things ought to be in the actual world, whereas in examples of what we have called validity constraints and completeness constraints (including those constraints exhibited by (3') and (4')) the obligation sentences pertain to how things ought to be in the database.

Since all of these integrity constraints are represented as obligation sentences, they are all violable; but there is an important distinction between the kind of response taken to violation of constraints of type (1'), and the kind of response taken to violation of the other types. For when a constraint of type (1') is violated, we tolerate the fact that this state of violation is represented in the database, but recovery procedures will be designed to change the state of the world so that the violation is repaired (and then the database must also be changed accordingly: the borrowed books, say, are finally returned, and so we delete from the database the sentence saying that books are on loan to a non-member). But when a violation of the other kinds of integrity constraints occurs, it is the resulting database state which is not tolerable, and must be changed, either because the information registered in the database is not true, or because it is incomplete.

The discussion in [1] of what they (following Sergot) called "soft" integrity constraints, was primarily focussed on examples of type (1'). What they called "hard" constraints are not deontic constraints of the kind considered in this abstract, but rather necessary constraints which no database state will ever be allowed to violate. For instance, a constraint to the effect that no person is both male and female may be a "hard" constraint for a given database in the sense that the database will never be allowed to register some person x as both male and female.

In future work we plan to develop further this attempted classification of types of database constraints, reflecting (as we have done here) some differences between types of violable constraints, but incorporating in addition those constraints which - as far as the database is concerned - are deemed inviolable, necessary truths.

References

1. J. Carmo and A.J.I. Jones. Deontic Database Constraints and the Characterisation of Recovery. In A.J.I.Jones and M.Sergot, editors, *2nd Int. Workshop on Deontic Logic in Computer Science*, pages 56–85. Tano A.S., 1994.
2. B. F. Chellas. *Modal Logic: An introduction*. Cambridge University Press, 1988.
3. R. Demolombe and A.J.I. Jones. Integrity Constraints Revisited. *Journal of the Interest Group in Pure and Applied Logics*, 4(3), 1996.
4. A.J.I. Jones and M. Sergot. On the role of Deontic Logic in the characterization of normative sytems. In J-J. Meyer and R.J. Wieringa, editors, *Proc. First Int. Workshop on Deontic Logic in Computer Science*, 1991.
5. J-M. Nicolas and K. Yazdanian. Integrity checking in Deductive Databases. In H. Gallaire and J. Minker, editors, *Logic and Databases*. Plenum, 1982.
6. A. Olivé. Integrity checking in Deductive Databases. In *17th Int. Conf. on Very Large Data Bases*, 1991.
7. R. Reiter. What Should a Database Know? *Journal of Logic Programming*, 14(2,3), 1992.
8. E. Teniente and A. Olivé. The Events Method for View Updating in Deductive Databases. In *Int. Conf. on Extending Data Base Technology*, 1992.
9. R.J. Wieringa and J-J. Meyer. Applications of Deontic Logic in Computer Science: a concise overview. In J-J. Meyer and R.J. Wieringa, editors, *Proc. First Int. Workshop on Deontic Logic in Computer Science*, 1991.

Authors Index

Springer
and the
environment

At Springer we firmly believe that an international science publisher has a special obligation to the environment, and our corporate policies consistently reflect this conviction.

We also expect our business partners – paper mills, printers, packaging manufacturers, etc. – to commit themselves to using materials and production processes that do not harm the environment. The paper in this book is made from low- or no-chlorine pulp and is acid free, in conformance with international standards for paper permanency.

Lecture Notes in Artificial Intelligence (LNAI)

Lecture Notes in Computer Science